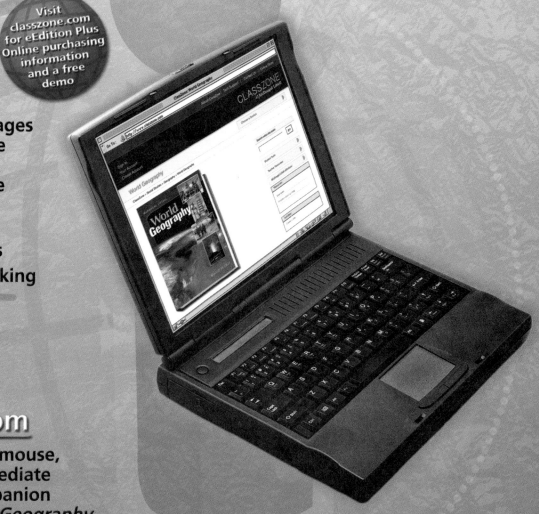

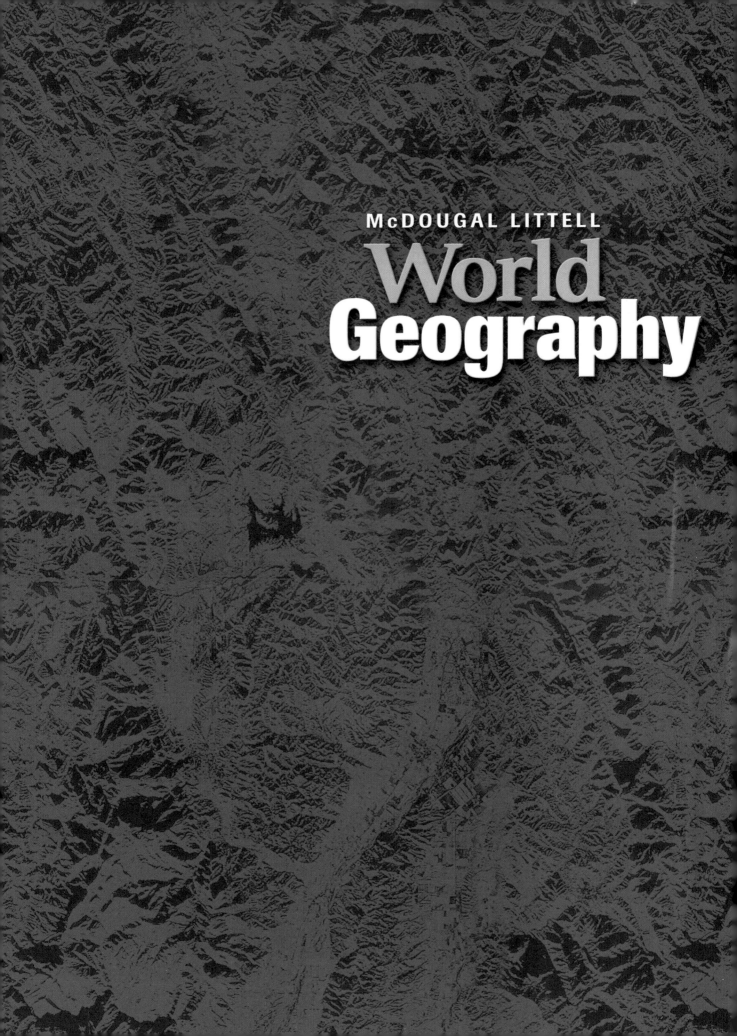

McDOUGAL LITTELL

World
Geography

Namche Bazaar, Nepal

Leaning tower, Pisa, Italy

Masai warriors, Kenya

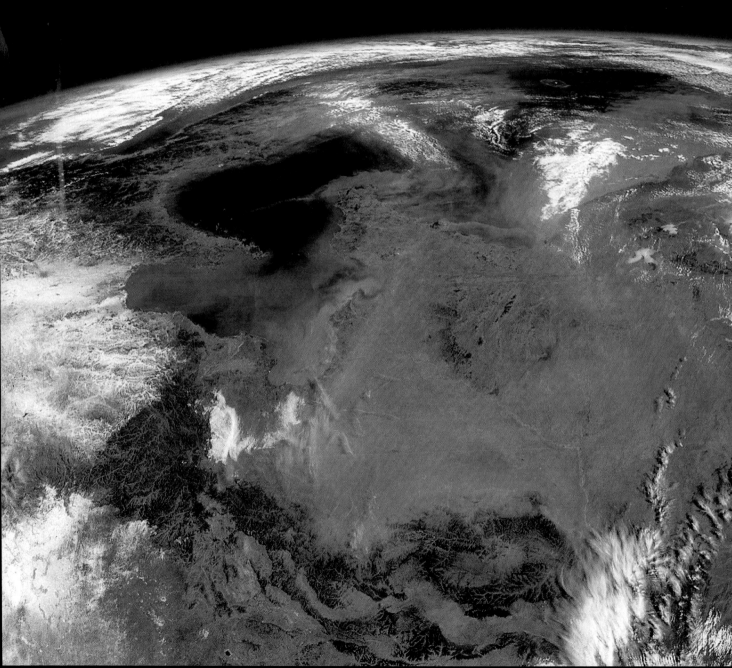

olcano, Honolulu, Hawaii

World
Geography

Daniel D. Arreola

Marci Smith Deal

James F. Petersen

Rickie Sanders

McDougal Littell

A DIVISION OF HOUGHTON MIFFLIN COMPANY

Maps on pages A1–A37, 64, 102–115, 131, 179, 190–199, 206, 248, 262–271, 281, 322, 336–343, 357, 391, 402–413, 419, 464, 478–485, 494, 528, 542–549, 559, 596, 610–617, 624, 664, 678–687, 693, 733 © Rand McNally & Company. All rights reserved.

Acknowledgments begin on page R66.

ISBN 0-618-37763-8

Printed in the United States of America

4 5 6 7 8 9 – VEI – 09 08 07 06 05

Senior Consultants

Daniel D. Arreola is Professor of Geography and an affiliate faculty member of the Center for Latin American Studies at Arizona State University. He has taught world regional geography for more than a decade at universities in Arizona and Texas. Dr. Arreola has published extensively on topics relating to the cultural geography of the Mexican-American borderlands. He is co-author of *The Mexican Border Cities: Landscape Anatomy and Place Personality* and author of *Tejano South Texas: A Mexican American Cultural Province.*

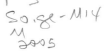

Marci Smith Deal is the K-12 Social Studies Curriculum Coordinator for Hurst-Euless-Bedford Independent School District in Texas. She received the 2000 Distinguished Geographer Award for the State of Texas, and was one of the honorees of the 2001 National Council for Geographic Education Distinguished Teacher Award. She has served as president for the Texas Council for Social Studies Supervisors and as vice-president for the Texas Council for Social Studies. She currently serves as a teacher consultant for National Geographic Society.

James F. Petersen is Professor of Geography at Southwest Texas State University. He served as president of the National Council for Geographic Education in 2000. As a charter member of the National Geographic Society's Alliance, Dr. Petersen has directed summer institutes and national conferences for teachers and educational organizations. He is the author of many articles on geographic education, as well as media/book reviews, textbooks, and curricular materials.

Rickie Sanders is Professor and Chair of Geography/Urban Studies at Temple University. She recently served on the team that directed the National Science Foundation/National Council for Geographic Education's "Finding A Way" project, which produced learning modules for integrating gender into geography classrooms. Dr. Sanders has received numerous awards for teaching, including the NCGE Distinguished Teaching Award and the Temple University Distinguished Teaching Award. She has numerous publications and is co-author of *Growing Up in America: An Atlas of Youth in the U.S.A.*

Consultants and Reviewers

Content Consultants

C. Cindy Fan
Department of Geography
UCLA
Los Angeles, California

Howard Johnson
Department of Physical and Earth
 Sciences
Jacksonville State University
Jacksonville, Alabama

Cheryl Johnson-Odim
Liberal Education Division
Columbia College
Chicago, Illinois

Charles Kovacik
Department of Geography
University of South Carolina
Columbia, South Carolina

Barbara McDade
Department of Geography
University of Florida
Gainesville, Florida

Inés Miyares
Department of Geography
Hunter College
New York City, New York

Joseph Stoltman
Department of Geography
Western Michigan University
Kalamazoo, Michigan

Donald Zeigler
Department of Political Science and
 Geography
Old Dominion University
Norfolk, Virginia

Multicultural Advisory Board

Betty Dean
Social Studies Consultant
Pearland, Texas

C. Cindy Fan
Department of Geography
UCLA
Los Angeles, California

Cheryl Johnson-Odim
Liberal Education Division
Columbia College
Chicago, Illinois

Barbara McDade
Department of Geography
University of Florida
Gainesville, Florida

Inés Miyares
Department of Geography
Hunter College
New York City, New York

Pat Payne
Office of Multicultural Education
Indianapolis Public Schools
Indianapolis, Indiana

Betto Ramirez
Region One Education Service
 Center
Edinburg, Texas

Jon Reyhner
Department of Education
Northern Arizona University
Flagstaff, Arizona

Teacher Consultants
The following educators reviewed manuscript or wrote classroom activities.

Deborah Althouse
Thomas J. Anderson High School
Southgate, Michigan

Jamie Berlin
South High School
Sheboygan, Wisconsin

Heather Berry
Hazelwood East High School
St. Louis, Missouri

Jewel Berryman
Kashmere High School
Houston, Texas

Deborah Bittner
Sandra Day O'Connor High School
Helotes, Texas

Dora Bradley
Lakewood Middle School
North Little Rock, Arkansas

Denise Butler
Hillcrest High School
Dallas, Texas

Deborah Canales
Austin High School
Houston, Texas

Fred Cibik
Brown Deer High School
Brown Deer, Wisconsin

Jim Curtis
Antioch High School
Antioch, Illinois

Sam Eigel
Cody High School
Detroit, Michigan

Jan Ellersieck
Ft. Zummalt South High School
St. Peters, Missouri

The Basics of Geography

**Volcano in
Costa Rica** (p. 2)

**Nanjing Road,
Shanghai, China**
(p. 81)

For more information on the basics of geography . . .

CLASSZONE.COM

The United States and Canada

Mesa Verde
National Park,
Colorado (p. 135)

Parliament
guards, Ottawa,
Ontario (p. 158)

For more information on the United States and Canada . . .

CLASSZONE.COM

ix

Unit 3 Latin America

Chacobo Indians on the Amazon River, Bolivia (p. 186)

São Paulo, Brazil (p. 251)

For more information on Latin America . . .

CLASSZONE.COM

Europe

Unit 4

The Eiffel Tower, Paris, France (p. 259)

The Wetterhorn, Switzerland (p. 310)

For more information on Europe . . .
CLASSZONE.COM

Unit 5

Russia and the Republics

Frozen Lake
Baikal, Russia
(p. 350)

St. Basil's
Cathedral,
Moscow,
Russia
(p. 362)

For more information on Russia and the Republics . . .

CLASSZONE.COM

Africa

Mount Kilimanjaro, Tanzania (p. 399)

Masai Girl, Kenya (p. 434)

For more information on Africa . . .

CLASSZONE.COM

Sahara Desert, North Africa (p. 420)

Unit 7 Southwest Asia

Oasis on caravan route from Yemen to Palestine (p. 475)

Kurdish family, Turkey (p. 476)

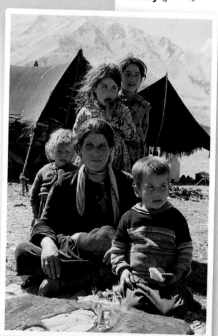

For more information on Southwest Asia . . .

CLASSZONE.COM

South Asia

**Taj Mahal,
Agra, India**
(p. 538)

**Tea Plantation,
Sri Lanka**
(p. 586)

For more information on South Asia . . .

CLASSZONE.COM

Unit 9 East Asia

Mount Fuji, Japan (p. 607)

Potala Palace, Tibet (p. 619)

Crowded urban street, Hong Kong (p. 668)

For more information on East Asia . . .

 CLASSZONE.COM

Southeast Asia, Oceania, and Antarctica

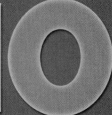

Unit 10

Ice Cliffs, Antarctica (p. 674)

Fishing on Cook Island (p. 714)

Uluru, or Ayers Rock, Australia (p. 729)

For more information on Southeast Asia, Oceania, and Antarctica . . .

CLASSZONE.COM

Features

Comparing Cultures

Disasters!

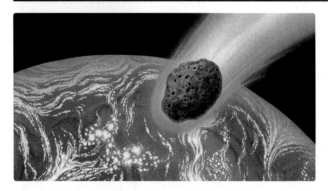

RAND McNALLY Map and Graph Skills

growing up in

5 THEMES

5 THEMES

HUMAN-ENVIRONMENT INTERACTION

The Gaucho

Geography TODAY

Maps

Unit 4

Unit 5

Unit 6

Climates of Africa

Maps

Graphs, Charts, and Tables

GRAPHS

CHARTS AND TABLES

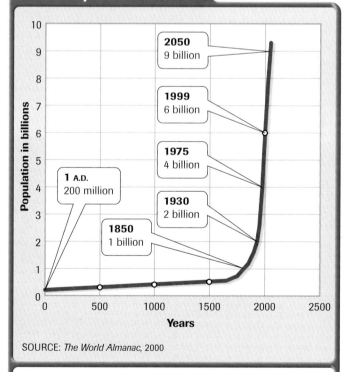

World Population Growth

Population in billions

- **1 A.D.** 200 million
- **1850** 1 billion
- **1930** 2 billion
- **1975** 4 billion
- **1999** 6 billion
- **2050** 9 billion

Years

SOURCE: *The World Almanac*, 2000

SKILLBUILDER: Interpreting Graphs

1. **ANALYZING DATA** How long did it take for the population to reach one billion?

2. **MAKING GENERALIZATIONS** How have the intervals between increases changed?

Infographics and Time Lines

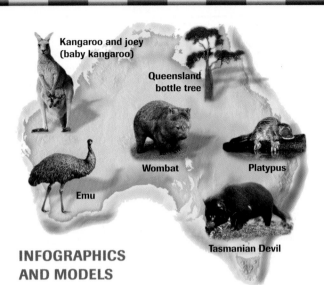

Kangaroo and joey (baby kangaroo)

Queensland bottle tree

Wombat

Platypus

Emu

Tasmanian Devil

INFOGRAPHICS AND MODELS

TIME LINES

Primary Sources

"My goodness, if I'd known how badly you wanted democracy I'd have given it to you ages ago."

Primary Sources/Videos

The Voyageur Experience in World Geography Videos

Strategies for Taking Standardized Tests

This section of the textbook helps you develop and practice the skills you need to study geography and to take standardized tests. Part 1, **Strategies for Studying Geography**, takes you through the features of the textbook and offers suggestions on how to use these features to improve your reading and study skills.

Part 2, **Test-Taking Strategies and Practice,** offers specific strategies for tackling many of the items you'll find on a standardized test. It gives tips for answering multiple-choice, constructed-response, extended-response, and document-based questions. In addition, it offers guidelines for analyzing primary and secondary sources, political cartoons, maps, charts, graphs, including population pyramids, and time lines. Each strategy is followed by a set of questions you can use for practice.

CONTENTS

Part 1: Strategies for Studying Geography

Reading is the central skill in the effective study of geography or any other subject. You can improve your reading skills by using helpful techniques and by practicing. The better your reading skills are, the more you will remember what you read. Below you will find several strategies that involve built-in features of *World Geography*. Careful use of these strategies will help you learn and understand geography more effectively.

Preview Chapters Before You Read

Each chapter begins with a one-page introduction. Study this introductory material to help you get ready to read.

1 Read the chapter and section titles. These provide a brief outline of what will be covered in the chapter.

2 Study the chapter-opening photograph. It often illustrates a major theme of the chapter. Regional human geography chapters open with a map rather than a photograph. Examine the map to get an idea of the location and size of the region discussed in the chapter.

3 Read the **GeoFocus** question and activity. These items will help focus your reading of the chapter.

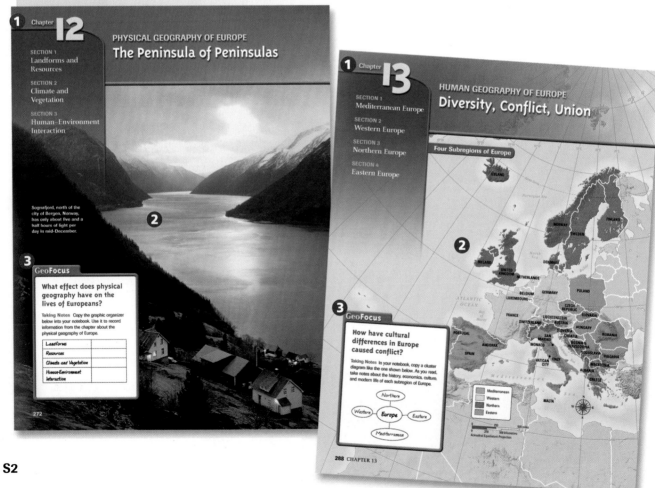

Preview Sections Before You Read

Each chapter consists of three, four, or five sections. Depending on the chapter, these sections focus on particular aspects of physical or human geography. Use the section openers to help you prepare to read.

1 Study the points under the **Main Ideas** heading. These identify the major topics discussed in the section.

2 Preview the **Places & Terms** list. It will give you an idea of the locations and concepts you will read about in the section.

3 Read the **Connect to the Issues** feature. This feature connects the section content to one of the major geographic issues covered in each unit of *World Geography.*

4 Skim through the section and look at the maps and illustrations. They will give you a quick visual overview of the section content.

5 Notice the structure of the section. **Blue** heads label the major topics covered in the section; **green** subheads signal smaller topics within these major topics. Together, these heads provide you with a quick outline of the section.

Landforms and Resources

A HUMAN PERSPECTIVE Elephants in Europe? In 218 B.C., Hannibal, a general from Carthage in North Africa, attacked the Roman Empire, which was at war with Carthage. He moved 38 war elephants and an estimated 60,000 troops across the Mediterranean Sea to Spain. To reach Italy, his armies had to cross the Pyrenees Mountains, the Rhone River, and the Alps. Hannibal used rafts to float the elephants across the Rhone. In the Alps, steep paths and slick ice caused men and animals to fall to their deaths. Despite this, Hannibal arrived in Italy with 26,000 men and a few elephants, and he defeated Rome in many battles. His crossing of the Alps was a triumph over geographic barriers.

Peninsulas and Islands

On a map you will see that Europe is a large peninsula stretching to the west of Asia. Europe itself has many smaller peninsulas, so it is sometimes called a "peninsula of peninsulas." Because of these peninsulas, most locations in Europe are no more than 300 miles from an ocean or sea. As you can imagine, the European way of life involves using these bodies of water for both business and pleasure.

NORTHERN PENINSULAS In northern Europe is the Scandinavian Peninsula. Occupied by the nations of Norway and Sweden, it is bounded by the Norwegian Sea, the North Sea, and the Baltic Sea. More than almost any other place in Europe, this peninsula shows the results of the movement of glaciers during the Ice Age. The glaciers scoured away the rich topsoil and left only thin, rocky soil that is hard to farm.

In Norway, glaciers also carved out **fjords** (fyawrdz), which are steep U-shaped valleys that connect to the sea and that filled with seawater after the glaciers melted. Fjords provide excellent harbors for fishing boats. The fjords are often separated by narrow peninsulas.

The Jutland Peninsula is directly across the North Sea from Scandinavia. Jutland forms the largest part of Denmark and a small part of Germany. This peninsula is an extension of a broad

Main Ideas **1**
- Europe is composed of many peninsulas and islands.
- Europe's landforms also include large plains and mountain ranges.

Places & Terms **2**
fjord   *Massif Central*
uplands   peat
Meseta

CONNECT TO THE ISSUES **3**
UNIFICATION Resources helped Western Europe develop industry before other regions. The European Union began in Western Europe.

Major European Peninsulas

4

SKILLBUILDER: Interpreting Maps
1 LOCATION Where are Europe's major peninsulas located in relation to each other?
2 REGION Why might each peninsula be considered a region?

Landforms and Resources **273**

plain that reaches across northern Europe. Its gently rolling hills and swampy low-lying areas are very different from the rocky land of the Scandinavian Peninsula.

SOUTHERN PENINSULAS The southern part of Europe contains three major peninsulas:

- The Iberian Peninsula is home to Spain and Portugal. The Pyrenees Mountains block off this peninsula from the rest of Europe.
- The Italian Peninsula is home to Italy. It is shaped like a boot, extends into the Mediterranean Sea, and has 4,700 miles of coastline.
- The Balkan Peninsula is bordered by the Adriatic, Mediterranean, and Aegean Seas. It is mountainous, so transportation is difficult.

ISLANDS Another striking feature of Europe is its islands. The larger islands are Great Britain, Ireland, Iceland, and Greenland. Although far from mainland Europe, Iceland and Greenland were settled by Scandinavians and have maintained cultural ties with the mainland. Over the centuries, many different groups have occupied the smaller Mediterranean Sea islands of Corsica, Sardinia, Sicily, and Crete. All of Europe's islands have depended upon trade.

Geographic Thinking
Seeing Patterns
What geographic advantages do islands have that help to promote trade?

Mountains and Uplands **5**

The mountains and uplands of Europe may be viewed as walls because they separate groups of people. They make it difficult for people, goods, and ideas to move easily from one place to another. These landforms also affect climate. For example, the chilly north winds rarely blow over the Alps into Italy, which has a mild climate as a result.

MOUNTAIN CHAINS The most famous mountain chain in Europe is the Alps. On a map you can see that the Alps arc across France, Italy, Germany, Switzerland, Austria, and the northern Balkan Peninsula. They cut Italy off from the rest of Europe. Similarly, the Pyrenees restrict movement from France to Spain and Portugal. Both ranges provide opportunities for skiing, hiking, and other outdoor activities.

Running like a spine down Italy, the Apennine Mountains divide the Italian Peninsula between east and west. The Balkan Mountains block

HUMAN-ENVIRONMENT INTERACTION
The Wetterhorn in the Swiss Alps stands 12,142 feet above the city in the valley below. How do the mountains affect the lives of the people in the valley?

4

274

Use Active Reading Strategies as You Read

Now you are ready to read the chapter. Read one section at a time, from beginning to end.

❶ Ask and answer questions as you read. Look for the **Geographic Thinking** and **Connect to the Issues** questions in the margin. Answering these questions will show whether you understand what you have just read.

❷ Try to visualize the places, events, activities, and people you read about. Studying the pictures and any illustrated features will help you do this.

❸ Read to build your vocabulary. Look for the **boldfaced, underlined** terms in the text and note their meaning.

❹ Study the **Background** notes in the margin for additional information on the section content.

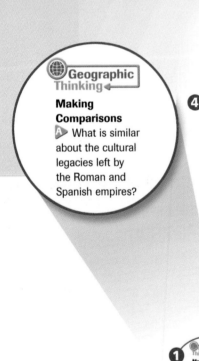

Geographic Thinking

Making Comparisons

A What is similar about the cultural legacies left by the Roman and Spanish empires?

(also called the Byzantine Empire) for nearly 1,000 years. Beginning in the 1300s, Italy saw the birth of the Renaissance, and in the 1400s, Portugal and Spain launched the Age of Exploration.

ITALIAN CITY-STATES The invaders who overran the Italian Peninsula had no tradition of strong central government. Italy eventually became divided into many small states and remained so for centuries.

❸ In 1096, European Christians launched the <u>**Crusades**</u>, a series of wars to take Palestine from the Muslims. Italians earned large profits by supplying the ships that carried Crusaders to the Middle East. Italian cities such as Florence and Venice became rich from banking and foreign trade. This wealth helped them grow into powerful city-states.

❹ BACKGROUND The Renaissance shaped modern life by stressing classical culture, material comfort, and the value of individuals.

The <u>**Renaissance**</u>, which began in the Italian city-states, was a time of renewed interest in learning and the arts that lasted from the 14th through 16th centuries. It was inspired by classical art and writings. Renaissance ideas spread north to the rest of Europe.

But the wealth of Italy did not protect it from disease. In 1347, the bubonic plague reached Italy from Asia and in time killed millions of Europeans. (See pages 294–295.)

SPAIN'S EMPIRE In the 700s, Muslims from North Africa conquered the Iberian Peninsula. Muslims controlled parts of the Iberian Peninsula for more than 700 years. Spain's Catholic rulers, Ferdinand and Isabella, retook Spain from the Muslims in 1492.

Also in 1492, Queen Isabella paid for Christopher Columbus's first voyage. Portugal had already sent out many voyages of exploration. Both Spain and Portugal established colonies in the Americas and elsewhere. Their empires spread Catholicism and the Spanish and Portuguese languages throughout the world.

REGION Italian Renaissance paintings often show the Virgin Mary and baby Jesus. Muslim art, like the Spanish wall design below (*bottom*), often uses calligraphy to praise God.

The Virgin and Child Surrounded by Five Angels, Sandro Botticelli

A Rich Cultural Legacy

Mediterranean Europe's history shaped its culture by determining where languages are spoken and where religions are practiced today. And the people of the region take pride in the artistic legacy of the past.

ROME'S CULTURAL LEGACY Unlike many areas of Europe that Rome conquered, Greece retained its own language. Greek was, in fact, the official language of the Byzantine Empire. In contrast, Portuguese, Spanish, and Italian are Romance languages that evolved from Latin, the language of Rome.

The two halves of the Roman Empire also developed different forms of Christianity. The majority religion in Greece today is Eastern Orthodox Christianity. Roman Catholicism is strong in Italy, Spain, and Portugal.

CENTURIES OF ART This region shows many signs of its past civilizations. Greece and Italy have ancient ruins, such as the Parthenon, that reveal what classical

Alhambra Palace, Granada, Spain

Geographic Thinking

Making Comparisons
A What is similar about the cultural legacies left by the Roman and Spanish empires?

Mediterranean Europe **291**

Review and Summarize What You Have Read

When you finish reading a section, review and summarize what you have read. If necessary, go back and reread information that was not clear the first time through.

1 Reread the blue heads and green subheads for a quick summary of the major points covered in the section.

2 Study any charts, graphs, or maps. These visual materials usually provide a condensed version of information in the section.

3 Review the visuals—photographs, charts, graphs, maps, and time lines—and any illustrated features, and note how they relate to the section content.

4 Complete all the questions in the **Section Assessment.** They will help you think critically about what you have just read.

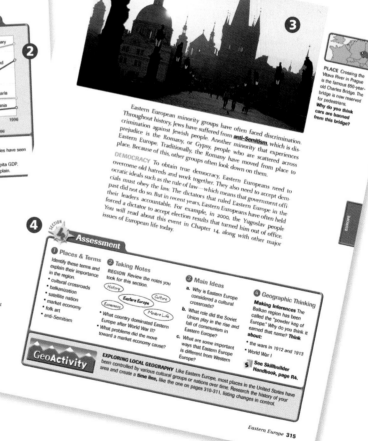

returned to ethnic loyalties. That was especially true in Yugoslavia, a nation consisting of six republics. In the early 1990s, four of the six Yugoslav republics voted to become separate states. Serbia objected, leading to civil war. (See Chapter 14 for details.) In contrast, Czechoslovakia peacefully split into the Czech Republic and Slovakia.

1 Developing the Economy

Because of its fertile plains, Eastern Europe has traditionally been a farming region. After 1948, the Soviet Union promoted industry there.

1 INDUSTRY Under communism, the government owned all factories and told them what to produce. This system was inefficient because industries had little motive to please customers or to cut costs. Often, there were shortages of goods. Eastern European nations traded with the Soviet Union and each other, so they didn't keep up with the technology of other nations. As a result, they had difficulty selling goods to nations outside Eastern Europe. And their outdated factories created heavy pollution.

After 1989, most of Eastern Europe began to move toward a **market economy**, in which industries make the goods consumers want to buy. Many factories in Eastern Europe became privately owned instead of state owned. The changes caused problems, such as inflation, the closing of factories, and unemployment. Since then, however, many factories have cut their costs and improved production. As a result, the Czech Republic, Hungary, and Poland have all grown economically.

CONNECT TO THE ISSUES
UNIFICATION
▷ Do you think the nations of Eastern Europe will want to join the European Union? Why or why not?

LINGERING PROBLEMS Some Eastern European nations have had trouble making economic progress—for many different reasons.

- Albania's economic growth is slowed by old equipment, a lack of raw materials, and a shortage of educated workers.
- Few of Romania's citizens have money to invest in business. In addition, the Romanian government still owns some industries. Foreigners don't want to invest their money in those industries.
- The civil wars of the 1990s damaged Yugoslavia and its former republics of Bosnia and Herzegovina and Croatia. Equipment and buildings were destroyed; workers were killed or left the country.

In general, it will take years for Eastern Europe to overcome the damage caused, in part, by decades of Communist control.

East

Per Capita GDP in Eastern Europe

SOURCE: *United Nations Statistical Yearbook, 1996*

SKILLBUILDER: Interpreting Graphs
1 SEEING PATTERNS Which of these four countries have seen economic improvement since 1990? Explain.
2 DRAWING CONCLUSIONS In terms of per capita GDP, which country has the best standard of living? Explain.

Eastern European minority groups have often faced discrimination. Throughout history, Jews have suffered from **anti-Semitism**, which is discrimination against Jewish people. Another minority that experiences prejudice is the Romany, or Gypsy, people who are scattered across Eastern Europe. Traditionally, the Romany have moved from place to place. Because of this, other groups often look down on them.

DEMOCRACY To obtain true democracy, Eastern Europeans need to overcome old hatreds and work together. They also need to adopt democratic ideals such as the rule of law—which means that government officials must obey the law. The dictators that ruled Eastern Europe in the past did not do so. But in recent years, Eastern Europeans have often held their leaders accountable. For example, in 2000, the Yugoslav people forced a dictator to accept election results that turned him out of office. You will read about this event in Chapter 14, along with other major issues of European life today.

PLACE Crossing the Vltava River in Prague is the famous 650-year-old Charles Bridge. The bridge is now reserved for pedestrians. **Why do you think cars are banned from this bridge?**

4 SECTION Assessment

1 Places & Terms
Identify these terms and explain their importance in the region.
- cultural crossroads
- balkanization
- satellite nation
- market economy
- folk art
- anti-Semitism

2 Taking Notes
REGION Review the notes you took for this section.

History — Eastern Europe — Culture
Economics — Modern Life

- What country dominated Eastern Europe after World War II?
- What problems did the move toward a market economy cause?

3 Main Ideas
a. Why is Eastern Europe considered a cultural crossroads?
b. What role did the Soviet Union play in the rise and fall of communism in Eastern Europe?
c. What are some important ways that Eastern Europe is different from Western Europe?

4 Geographic Thinking
Making Inferences The Balkan region has been called the "powder keg of Europe." Why do you think it earned that name? **Think about:**
- the wars in 1912 and 1913
- World War I

$ See Skillbuilder Handbook, page R4.

GeoActivity
EXPLORING LOCAL GEOGRAPHY Like Eastern Europe, most places in the United States have been controlled by various cultural groups or nations over time. Research the history of your area and create a **time line**, like the one on pages 310-311, listing changes in control.

Eastern Europe 315

Part 2: Test-Taking Strategies and Practice

Improve your test-taking skills by practicing the strategies discussed in this section. Read the tips on the left-hand page. Then apply them to the practice items on the right-hand page.

Multiple Choice

A multiple-choice question consists of a stem and a set of choices. The stem is usually in the form of a question or an incomplete sentence. One of the choices correctly answers the question or completes the sentence.

1 Read the stem carefully and try to answer the question or complete the sentence without looking at the choices.

2 Pay close attention to key words in the stem. They may direct you toward the correct answer.

3 Read each choice with the stem. Don't jump to conclusions about the correct answer until you've read all the choices.

4 Think carefully about questions that include *All of the above* among the choices.

5 After reading all of the choices, eliminate any that you know are incorrect.

6 Use modifiers to help narrow your choices further.

7 Look for the best answer among the remaining choices.

answers: 1 (C), 2 (D), 3 (C)

stem

1 1 Which of the following statements *best* characterizes the economies of the Arabian Peninsula nations?

2 *Best* is a key word here. It means you should look not just for a true statement but for the most important true statement.

choices **3**

A Their economies depend on subsistence agriculture.

B A lack of water has kept their economies from growing.

C Income from oil exports dominates their economies.

D Almost all goods are sold in traditional marketplaces called *souks*.

2 Which of the following is a cause of the continuing conflict in the Balkans?

A The desire of different ethnic groups to control the same land

B The attempt by Serbia to dominate Yugoslavia

C The opposition of many Serbs to the breakup of Yugoslavia

D All of the above

4 If you select this answer, be sure that all the choices are correct.

3 Japan is an example of a global economy because

A it became an international economic powerhouse in the 1820s.

B *all* of its people work in international business.

C it imports natural resources from other countries and sells manufactured goods around the globe.

D it rapidly industrialized after World War II.

6 Absolute words, such as *all, never, always, every,* and *only,* often signal an incorrect choice.

7 Both **C** and **D** describe facts. Only **C**, however, fits the definition of a global economy. Therefore, **C** is the best answer.

5 You can eliminate **A** if you remember that Japan remained relatively isolated from the West until 1853 when U.S. Commodore Perry arrived.

Directions: Read the following questions and choose the *best* answer from the four choices.

1   Which of the following was a result of the migration of Europeans to North America?

   A Native Americans were displaced.

   B The United States became a "nation of immigrants."

   C Plants, animals, and diseases moved between the Eastern and Western hemispheres.

   D All of the above

2   Which of the following is *not* an effect of the rapid destruction of rain forests in Latin America and other parts of the world?

   A The temperature of the atmosphere is rising.

   B Plants and animals are becoming extinct.

   C More oxygen is building up in the atmosphere.

   D The earth's biodiversity is being reduced.

3   After the Soviet Union collapsed in 1991 and Russia changed from a command economy to a market economy, economic control over the production of goods and services began to shift from

   A the central government to private businesses.

   B local workers to workers from abroad.

   C the legislature to the banks.

   D regional governments to the national government.

4   The Bantu migrations in Africa resulted in

   A the spread of Bantu languages and culture.

   B the end of the international slave trade.

   C the colonization of Africa by numerous European nations.

   D an AIDS epidemic in Africa.

Primary Sources

Primary sources are materials produced by people who traveled to the places they describe or who took part in or witnessed the events they portray. Letters, diaries, speeches, newspaper and magazine articles, travelogues, and autobiographies are all primary sources. So, too, are legal documents, such as wills, deeds, and financial records.

❶ Look at the source line and identify the author. Consider what qualifies the author to write about the places or events discussed in the passage.

❷ Skim the document to form an idea of what it is about.

❸ Note special punctuation. Ellipses indicate that words or sentences have been removed from the original passage.

❹ Carefully read the passage and distinguish between facts and the author's opinions. (Note that the author's use of a metaphor, *like layers in a slice of cake*, conveys a clear image of the land to the reader who cannot see Hadar in person.)

❺ Consider for whom the author was writing. The intended audience may influence what and how an author writes.

❻ Before rereading the passage, skim the questions to identify the information you need to find.

In 1974, the oldest and most well-preserved skeleton of an erect-walking human ancestor was found in Ethiopia. Paleoanthropologist Donald Johanson, who named the skeleton Lucy, describes the geography of the remote Afar desert region where Lucy was found. The region is rich with geological and paleontological information. Eventually, Johanson and his colleagues discovered the bones of at least 13 ancient individuals, now known as the First Family, in this desert area.

❷ At Hadar, which is a wasteland of bare rock, gravel and sand, the fossils that one finds are almost all exposed on the surface of the

❸ ground. Hadar is ⟨...⟩ an ancient lake bed now dry and filled with sediments that record the history of past geological events. You can trace volcanic-ash falls there, deposits of mud and silt washed down from distant mountains, episodes of volcanic dust, more mud, and so on. Those events reveal themselves like layers in a slice of **❹** cake in the gullies of new young rivers that recently have cut through the lake bed here and there. It seldom rains at Hadar, but when it does it comes in an overpowering gush—six months' worth overnight. The soil, which is bare of vegetation, cannot hold all that water. It roars down the gullies, cutting back their sides and bringing more fossils into view.

❶ —Donald Johanson, *Lucy: The Beginnings of Humankind*

❶ Johanson's description is very detailed because he took several field expeditions to Hadar looking for fossils.

❺ Although the author is a scientist, he wrote this book for a general audience to explain the work and the excitement of finding fossils.

❻

1 Now a wasteland of bare rock, gravel, and sand, Hadar was once a

   A volcano.

   B mountain chain.

   C lake bed.

   D river.

2 The author most likely describes Hadar and its geological history for which of the following reasons?

   A Because knowing about the area's geological past might help to locate and identify fossils

   B To illustrate how the area's current climate and geography reveal its past geological events

   C To explain why fossils are found on the surface of the ground at Hadar

   D All of the above

answers: 1 (C), 2 (D)

Directions: Read the following excerpt from a letter written by the Spanish conquistador Hernán Cortés in which he describes the Aztec capital city. Use the passage and your knowledge of world geography to answer the questions.

The Aztec Capital: The Great City of Tenochtitlán in Mexico

The great city of Tenochtitlán is built in the midst of this salt lake, and it is two leagues from the heart of the city to any point on the mainland. Four causeways lead to it, all made by hand and some twelve feet wide. The city itself is as large as Seville or Córdova. The principal streets are very broad and straight, the majority of them being of beaten earth, but a few and at least half the smaller thoroughfares are waterways along which they pass in their canoes. Moreover, even the principal streets have openings at regular distances so that the water can freely pass from one to another, and these openings which are very broad are spanned by great bridges of huge beams, very stoutly put together, so firm indeed that over many of them ten horsemen can ride at once.

—Hernán Cortés, in a letter to the King of Spain

Excerpt from "The Second Letter of Hernán Cortés," from *Five Letters of Hernán Cortés,* 1519–1526, translated by J. Bayard Morris (New York: W. W. Norton and Company, Inc.). Norton Paperback Edition published in 1969, reissued in 1991. Reprinted by permission of W. W. Norton and Company, Inc.

1 Which of the following statements *best* describes the location of Tenochtitlán?

A It was built on a peninsula, and all of its roads were waterways.

B It was built next to a lake, which the people crossed over by ferry boats.

C It was built on an island connected to the mainland by four hand-built causeways.

D It was built on the mainland with several bridges connecting it to an island in the nearby salt lake.

2 The letter contains the information that the Aztec citizens and the Spanish conquistadors traveled around the city by

A canoe and horse.

B canoe only.

C foot only.

D wagons and foot.

3 Which of the following statements reveals that Cortés admires the city of Tenochtitlán and its builders?

A "The city itself is as large as Seville or Córdova."

B "These openings . . . are spanned by great bridges of huge beams, very stoutly put together."

C "The principal streets are very broad and straight."

D All of the above

4 Eventually, Cortés and the Spanish destroyed most of Tenochtitlán. On its ruins, they built what became the present-day city of

A Seville.

B Baja.

C Mexico City.

D Tijuana.

Secondary Sources

Secondary sources are descriptions of places, people, cultures, and events. Usually, secondary sources are made by people who are not directly involved in the event or living in the place being described or discussed. The most common types of written secondary sources are textbooks, reference books, some magazine and newspaper articles, and biographies. A secondary source often combines information from several primary sources.

1 Read the title to preview the content of the passage.

2 Look at the source line to learn more about the document and its origin. (The spelling of the word *organized* indicates that the magazine is probably from Great Britain.)

3 Look for topic sentences. Ask yourself what the main idea is.

4 As you read, use context clues to guess at the meaning of difficult or unfamiliar words. (You can use the description of crime in the rest of the passage to understand that the word *pervasiveness* most likely means "being everywhere" or "existing throughout.")

5 Read actively by asking and answering questions about the passage.

6 Before rereading the passage, skim the questions to identify the information you need to find.

You might ask: What makes organized crime in Russia different from organized crime in other countries? Are crime and corruption in all levels of society new to Russian culture? **5**

1 Organized Crime in Russia

3 This highlights the key feature of Russian criminality: its **4** pervasiveness. "Organised crime usually deals with [minor] economic issues . . . [but] in Russia it's the mainstream," notes Toby Latta of Control Risks, a London security [firm]. Russian criminality reaches the highest levels of government—is, indeed, often indistinguishable from it. And it affects the humblest activity. Buy a jar of coffee? More likely than not, you are feeding organised crime: according to a grumbling Nestlé, most coffee sold in Russia has evaded full import duties. Give money to a beggar? He will have paid the local mafia for his spot on the street. Build a factory? You will pay one lot of bureaucrats to get it going, another to keep it running. In Russia, organised crime and corruption are everywhere. **3**

The last sentence restates the main idea.

2 Excerpt from "Russian Organised Crime," from *The Economist,* August 28, 1999. Copyright © 1998 The Economist. Reprinted by permission.

1 What is the main idea of this passage?

A The Russian economy is in a depression.

B The Russian government is ineffective.

C Organized crime operates in all areas of the Russian economy.

D Russia is on the verge of collapse.

6

2 Which of the following conclusions can you draw from this passage?

A Anyone who wants to start a business in Russia may have to pay the mafia first.

B The Russian government loses money because some import taxes are not paid.

C The Russian mafia operates within the government.

D All of the above

answers: 1 (C), 2 (D)

For more test practice online . . .

TEST PRACTICE
CLASSZONE.COM

Directions: Use the passage about Mohandas K. Gandhi's work for social reform in India and your knowledge of world geography to answer the questions below.

Gandhi's Work in the 1920s

Gandhi's understanding of economic relations was shot through with emphasis originating in Hindu tradition, such as the duty of the wealthy to extend charity. . . . But in the 1920s he was forced to confront very precisely some of the aspects of India's social order which were rooted in Hindu tradition. . . .

His primary social concern at this time was the problem of untouchability, the rejection of a whole group of the poorest and most menial in society as a result of Hindu ideas of hierarchy. . . . Now, as he travelled widely, he saw in harsh practice the power of this social division, and the poverty and degradation it caused. . . .

Personal example was one of Gandhi's strategies to end untouchability. He mixed freely with [the "untouchables"], as everybody knew; he ate with them. . . . But Gandhi did not expect everyone to go this far. For most caste Hindus the obligation was to treat the untouchables as a caste *within* Hindu society, affording them citizens' rights. They should be allowed to use wells, roads and public transport, attend schools and enter temples, though conventions prohibiting marriage or meals with them would remain.

—Judith M. Brown, *Gandhi: Prisoner of Hope*

1 In Hindu tradition, there are four main classes in the social hierarchy known as the caste system. You can tell from the passage that the "untouchables" are

  A the highest social group.

  B the lowest social group.

  C priests and scholars.

  D merchants, traders, and farmers.

2 As the passage explains, Gandhi broke with Hindu tradition by

  A trying to convert the poorest people to Islam.

  B extending charity to the poorest people.

  C spending time with the poorest people.

  D rejecting the poorest people.

3 According to the author, which of the following ideas did Gandhi promote?

  A Citizens' rights for members of the lowest caste

  B Intermarriage among members of low and high castes

  C Abolishing the caste system altogether

  D Scholarships for members of the lowest caste

4 You can infer from the last paragraph that low caste Indians in the 1920s were *not* usually allowed to

  A use public wells.

  B ride on public buses.

  C attend schools.

  D All of the above

Political Cartoons

Political cartoons are drawings made to express a point of view on political issues of the day. Cartoonists use words, symbols, and such artistic styles as caricature—exaggerating a person's physical features—to get their message across.

1 Identify the subject of the cartoon. Titles, captions, and labels are often clues to the subject matter. (The subject here is Chechnya's fight for independence from Russia.)

2 Identify the main characters in the cartoon. (The main character is Russian President Vladimir Putin.)

3 Note the symbols—ideas and images that stand for something else—used in the cartoon. (The bear is an often-used symbol of Russia.)

4 Study labels and other written information in the cartoon.

5 Analyze the point of view. How cartoonists use caricature often shows how they feel.

6 Interpret the cartoonist's message.

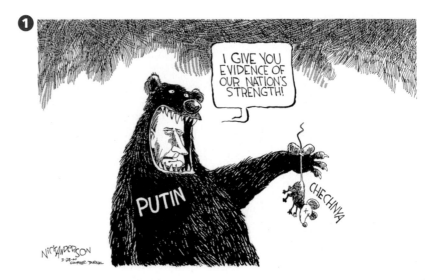

1

© 2000 Washington Post Writers Group. Reprinted with permission.

5 Putin represents the Russian government and army. The exaggeration of Putin's nose makes him appear ridiculous.

3 Drawing Chechnya as a mouse exaggerates the difference in size of Chechnya compared to Russia.

1 Chechnya is portrayed as a mouse because

A it is so much smaller and less powerful than Russia.

B the region has no natural resources.

C its rebel leaders lack courage and the will to fight.

D the region produces so much cheese.

2 Which of the following statements best represents the cartoonist's point of view?

A Russia should maintain firm control of Chechnya.

B Chechnya is not important to Russia.

C Russia is a military powerhouse and should be feared by other countries.

D Russia used more military might than necessary in fighting the rebellion in Chechnya.

answers: 1 (A), 2 (D)

Directions: Use the political cartoon and your knowledge of geography to answer the questions below.

Copyright © 1989 Rob Rogers/The Pittsburgh
Press/United Feature Syndicate

ROB ROGERS
Courtesy Pittsburgh Press

1 The cartoonist has drawn the "Berlin Mall," to refer to the

A main shopping district in the center of Berlin.

B seat of city government in East Germany.

C Berlin Wall, which divided the city of Berlin into democratic and communist sections.

D World War II division of Germany.

2 What does the "Berlin Mall" most likely stand for in the cartoon?

A the European Union

B Western capitalism

C Eastern philosophy

D Soviet communism

3 The cartoonist is implying that

A free-market countries and corporations looked for new markets in Berlin.

B the people of former communist countries in Europe were eager to buy products not previously available to them.

C the fall of the Berlin Wall changed economics and politics in Europe.

D All of the above

4 The father's statement to his son implies that this Berlin site

A was recently built on the site of an old market.

B is very different from what it used to be.

C is the only shopping area located in Berlin.

D All of the above.

Charts

Charts present information in a visual form. Geography textbooks use several types of charts, including tables, flow charts, Venn diagrams, and infographics. The type of chart most commonly found in standardized tests is the table, which organizes information in columns and rows for easy viewing.

❶ Read the title to identify the broad subject of the chart.

❷ Read the column and row headings and any other labels. The headings and labels will provide more details on the subject of the chart.

❸ Compare and contrast the information from column to column and row to row.

❹ Try to draw conclusions from the information in the chart. Ask yourself: What trends or patterns does the chart show?

❺ Read the questions and then study the chart again.

❶ Adult Literacy Rates in South Asia by Gender, 2003

❷

Country	Male	Female	Total
Bangladesh	54%	32%	43%
Bhutan*	56%	28%	42%
India	70%	48%	59%
Maldives	97%	97%	97%
Nepal	63%	28%	46%
Pakistan	60%	31%	46%
Sri Lanka	95%	90%	92%

❹ Based on the data in this chart, you might conclude that males in most of these countries receive more education than females.

*1995 estimate   **Source**: CIA, *The World Fact Book 2003*

❸ Compare and contrast the literacy rates of males and females in each country.

❺ **1** What is the general pattern in the literacy rates for males and females of this region?

A The rates for males and females are similar.

B The rates for males are generally much higher than those for females.

C The rates for females are generally much higher than those for males.

D The rates for both sexes are extremely low in all the countries.

2 One observation that you can make about the literacy rate in these countries is that the

A higher the female literacy rate is, the higher the total literacy rate is.

B higher the literacy rate, the less interest females have in reading and writing.

C literacy rate in mountainous countries is higher than the rate in island countries.

D lower the total literacy rate is, the higher the female literacy rate is.

answers: 1 (B), 2 (A)

Directions: Use the chart and your knowledge of world geography to answer the questions below.

Comparison of European, American, and Japanese Workers' Hours

Country	Scheduled Weekly Hours	Number of Annual Days Off/Holidays	Annual Hours Worked
Germany	39	42	1,708
Netherlands	40	43.5	1,740
Austria	39.3	38	1,751
France	39	34	1,771
Italy	40	39	1,776
United Kingdom	39	33	1,778
Sweden	40	37	1,792
United States	40	22	1,912
Portugal	45	36	2,025
Japan	44	23.5	2,116

Source: "Comparison of European, American, and Japanese Workers' Hours," from *Hammond New Century World Atlas.* Copyright © 2000 by Hammond World Atlas Corporation. All rights reserved. Reprinted by permission.

1 People are scheduled to work the most hours annually in

A the United States.

B Portugal.

C Germany.

D Japan.

2 If Germany has a five-day work week, the Germans' time off equals how many work weeks?

A More than 2 work weeks

B More than 4 work weeks

C More than 6 work weeks

D More than 8 work weeks

3 People have the least number of holidays and days off work in

A the United States.

B Portugal.

C the United Kingdom.

D Japan.

4 Compared to the Americans and Japanese, Europeans work

A fewer hours per week.

B fewer days per year.

C more hours per week.

D more days per year.

Line and Bar Graphs

Graphs show statistics in a visual form. Line graphs are particularly useful for showing changes over time. Bar graphs make it easy to compare numbers or sets of numbers.

1 Read the title to identify the broad subject of the graph.

2 Study the labels on the vertical and horizontal axes to see the kinds of information presented in the graph. Note the intervals between amounts and between dates.

3 Study any keys or legends.

4 Look at the source line and evaluate the reliability of the information in the graph. Federal and state government statistics, as well as those from universities, tend to be reliable.

5 Study the information in the graph and note any trends.

6 Draw conclusions and make generalizations based on these trends.

7 Read the questions carefully and then study the graph again.

1 Three Fastest Growing States in the United States, 1990–2000

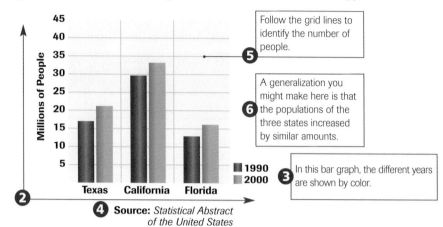

Follow the grid lines to identify the number of people. **5**

A generalization you might make here is that **6** the populations of the three states increased by similar amounts.

1990 **2000** In this bar graph, the different years **3** are shown by color.

4 **Source:** *Statistical Abstract of the United States*

7 1 The population of Texas increased between 1990 and 2000 by about —

A 4 million

B 8 million

C 10 million

D 100,000

1 Projected Population Growth in China, 1950 to 2050

Source: U.S. Census Bureau, International Data Base **4**

7 1 What is expected to happen to China's population after the year 2040?

A It will decline sharply.

B It will begin to decline slowly.

C It will continue to increase slowly.

D It will increase very sharply.

answers: 1 (A), 2 (B)

Directions: Use the graphs and your knowledge of world geography to answer the questions below.

World's Major Energy Consumers, 2001

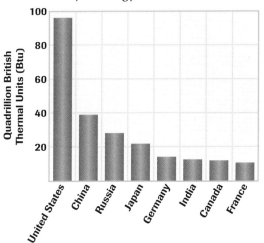

Source: Energy Information Administration

Global Average Temperatures, 1880–2000

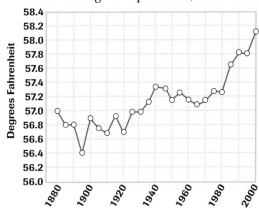

Source: Goddard Institute for Space Studies

1 Which of the following statements is true according to the graph?

   A All of the countries consume about the same amount of energy.

   B The country with the largest area consumes the most energy.

   C The United States consumes the most energy.

   D The country with the largest population consumes the most energy.

2 Which of the following statements is *not* accurate?

   A The United States consumes more energy than China, Russia, and Japan combined.

   B The five top energy consumers are all large countries.

   C India consumes more energy than France.

   D Japan consumes more energy than India.

3 What has been the general trend in the global average temperature since 1960?

   A It has been rising every five years.

   B It has been decreasing.

   C It has risen and fallen, but overall it has risen.

   D It has stayed fairly steady.

4 Which of the following statements accurately reflects information in the graph?

   A Global average temperatures go up and down over time.

   B The temperature has not changed by more than one degree in any 20-year period.

   C It is difficult to make long-term climate predictions from this graph alone.

   D All of the above

Population Pyramids

A population pyramid is a type of graph that shows the gender and age distribution of a population. It is useful in showing patterns in these and other categories, such as ethnicity. The size of one age group compared to another may have important economic, social, and political consequences. For example, if the number of working-age adults in a country is small, the labor pool might be small.

1 Read the title to identify the population that the graph represents.

2 Study the age groups labeled along the vertical axis in the center of the pyramid. Each horizontal bar represents the size of an age-and-gender group. Note that the intervals between the numbers along the base of the pyramid identify the size of each age-gender group.

3 Compare the sizes of the gender groups and note any patterns. Then compare the sizes of the age groups and note any patterns.

4 Draw conclusions and make generalizations based on the patterns you see.

5 Read the questions carefully and then refer to the graph again to answer them.

1 Population Pyramid for Bolivia, 2000

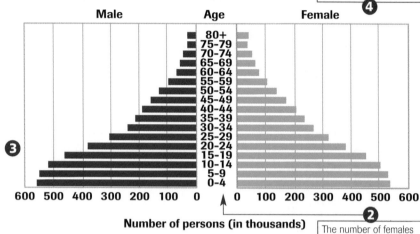

A generalization you might make here is that the population is not evenly distributed. The very young age groups greatly outnumber the older age groups.
4

The number of females aged 5–9, for example, is about 530,000.
2

Source: U.S. Census Bureau, International Data Base

1 Most Bolivians are

A between the ages of 35 and 39.

B below the age of 40.

C between the ages of 45 and 49.

D older than 59.

2 Which statement *best* characterizes the gender distribution of Bolivia's population?

A Males greatly outnumber females.

B Females greatly outnumber males.

C The population has about an equal number of males and females.

D Females outnumber males in the youngest age groups.

answers: 1 (B), 2 (C)

Directions: Use the graph and your knowledge of world geography to answer the questions below.

Population Pyramid for France, 2000

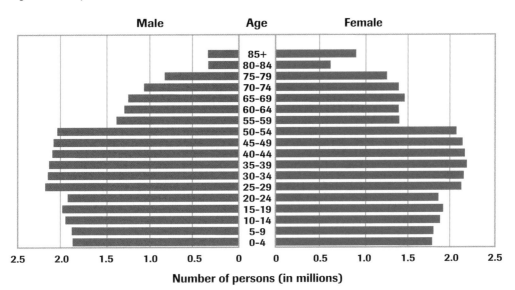

Source: U.S. Census Bureau, International Data Base

1 The largest age group in France is composed of people

  A from 10 to 24 years of age.

  B under 10 years of age.

  C from 25 to 54 years of age.

  D 55 years of age and over.

2 Which statement *best* characterizes the population distribution between the genders in France?

  A Males outnumber females in all age groups.

  B Females outnumber males in all age groups.

  C The genders are roughly equal except in the youngest age group.

  D As the population ages, it changes from slightly more males to more females.

3 Which statement accurately reflects the information in this graph?

  A French women live longer than French men.

  B French men live longer than French women.

  C Very few French people live past the age of 54.

  D There are fewer French teenagers than any other age group.

4 Which of the following conclusions can you draw from this graph?

  A Large families are common in France.

  B France has a high infant mortality rate.

  C There was a "baby boom" in France after 1945.

  D France has a labor shortage.

Pie Graphs

A pie, or circle, graph shows relationships among the parts of a whole. These parts look like slices of a pie. The size of each slice is proportional to the percentage of the whole that it represents.

1 Read the title and identify the broad subject of the pie graph.

2 Look at the legend to see what each of the slices of the pie represents.

3 Read the source line and note the origin of the data shown in the pie graph.

4 Compare the slices of the pie, and try to make generalizations and draw conclusions from your comparisons.

5 Read the questions carefully and review difficult terms.

6 Think carefully about questions that have *not* in the stem.

7 Eliminate choices that you know are wrong.

1 Typical Growing Season Work Day for 10-Year-Old Girl in Rural Nepal

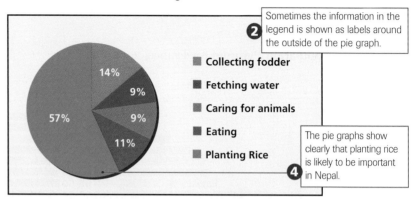

Sometimes the information in the legend is shown as labels around the outside of the pie graph. **2**

■ Collecting fodder
■ Fetching water
■ Caring for animals
■ Eating
■ Planting Rice

The pie graphs show clearly that planting rice is likely to be important in Nepal. **4**

3 **Source:** Adapted from "A working day in the life of a 10-year old girl in Nepal," from *Listening to Smaller Voices* by Victoria Johnson, Joanna Hill, and Edda Ivan-Smith. Copyright © 1995 by ActionAid Nepal. Reprinted by permission.

1 A typical 10-year-old girl in rural Nepal spends the greatest percentage of her time

A planting rice.

B eating.

C collecting (fodder. **5**

The word *fodder* refers to feed for livestock. It is usually coarsely chopped straw or hay.

D fetching water.

6

2 Which of the following is (*not*) a conclusion you can draw from the information in this pie graph?

A Young girls spend no time raising animals in rural Nepal.

B During the growing season, children in rural Nepal do farm chores most of the day.

C Rice is an important part of the diet in Nepal.

D Children in Nepal do not attend school during the growing season.

7 You can eliminate **B** because the pie graph shows they do spend most of their day doing farm chores.

Directions: Use the pie graphs and your knowledge of world geography to answer the questions below.

Trends in World Urbanization, 1900 and 2015

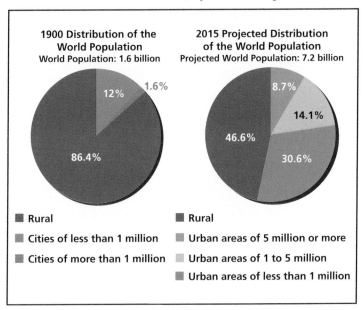

1900 Distribution of the World Population
World Population: 1.6 billion

1.6%
12%
86.4%

2015 Projected Distribution of the World Population
Projected World Population: 7.2 billion

8.7%
14.1%
46.6%
30.6%

■ Rural
■ Cities of less than 1 million
■ Cities of more than 1 million

■ Rural
■ Urban areas of 5 million or more
■ Urban areas of 1 to 5 million
■ Urban areas of less than 1 million

Source: "Trends in World Urbanization," from *Introduction to Geography*, Sixth Edition by Arthur Getis, Judith Getis, and Jerome D. Fellman. Copyright © 1998 by McGraw-Hill Companies, Inc. Reprinted by permission.

1 In 1900, most people of the world lived in

A cities of more than one million people.

B cities of less than one million people.

C suburban areas.

D rural areas.

2 Which of the following statements *best* describes the projected change in the distribution of people in 2015?

A The same number of people will live in urban as live in rural areas.

B The largest percentage of people will live in urban areas of over one million people.

C More people will live in urban than in rural areas.

D Forty percent of people will live in urban areas of all sizes.

3 The percentage of people living in rural areas in 2015, as compared to the percentage in 1900, is projected to decline by approximately

A 10 percentage points.

B 20 percentage points.

C 40 percentage points.

D 60 percentage points.

4 The current rise in the number of cities and the lifestyle changes that result from it are called

A land-use patterns.

B urbanization.

C industrialization.

D suburbanization.

Political Maps

Political maps show features on the earth's surface that are created by humans. Included on a political map may be the location of cities, states, provinces, territories, or countries. There also may be some physical features, such as rivers, seas, oceans, and lakes. You can use these features to show an area's shape and size and where it is located on the earth's surface. You can also look at its location in relation to other areas, and how all of these physical facts might affect a place in ways such as its economy or population.

1 Read the title to determine the subject and purpose of the map.

2 Review the map labels, which reveal specific features that further illustrate the subject and purpose of the map.

3 Study the legend to find the meaning of the symbols used on the map.

4 Look at the lines of latitude and longitude. This grid makes locating places easier.

5 Use the compass rose to determine directions on the map.

6 Use the scale to measure the actual distances between places shown on the map.

7 Read the questions and then carefully study the map to determine the answers.

1 Kenya: Political

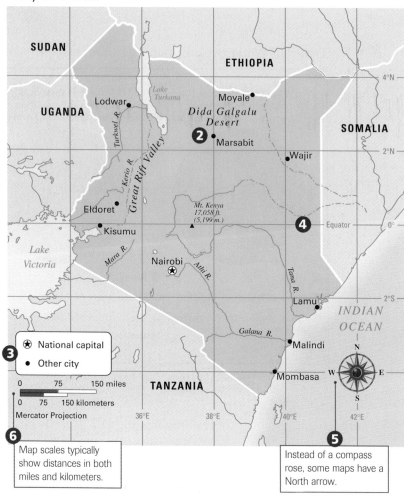

Map scales typically show distances in both miles and kilometers.

Instead of a compass rose, some maps have a North arrow.

1 About how far is Mombasa from the capital of Kenya?

  A About 100 miles

  B About 200 miles

  C About 300 miles

  D About 400 miles

2 The country that borders Kenya on the south is

  A Somalia.

  B Tanzania.

  C Ethiopia.

  D Uganda.

answers: 1 (C), 2 (B)

S22

Directions: Use the map and your knowledge of world geography to answer the questions below.

Mexico: Political

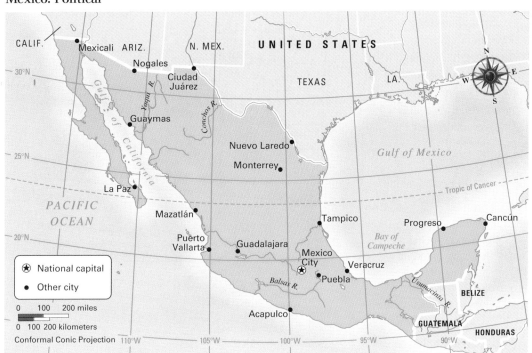

1   Which statement *best* describes the location of the capital of Mexico?

  A It is located on the Gulf of Mexico.

  B It is located on the Pacific Ocean.

  C It is located near the U.S.-Mexico border.

  D It is centrally located within the country.

2   Which of the following countries does *not* share a border with Mexico?

  A Honduras

  B Belize

  C Guatemala

  D The United States

3   Mexico is bordered on the north by

  A Louisiana and Texas.

  B California, Arizona, New Mexico, and Texas.

  C Arizona, New Mexico, and Texas.

  D only Texas.

4   The popular resort cities of Acapulco, Puerto Vallarta, and Cancún all have in common their location

  A on the Pacific Ocean.

  B on a coast.

  C north of the Tropic of Cancer.

  D in the interior of the country.

Physical Maps

Physical maps show the landforms and bodies of water in a specific area. They use color, shading, or contour lines to indicate elevation or altitude, which is also called relief. Many maps combine features of both physical and political maps—that is, they show physical characteristics as well as political boundaries.

① Read the title to determine the area shown on the map.

② Study the legend to find the meaning of the colors used on the map. Typically, different colors are used to indicate levels of elevation. Match the legend colors to places on the map.

③ Review the labels on the map to see what physical features are shown.

④ Look at the lines of latitude and longitude. You can use this grid to identify the location of physical features.

⑤ Use the compass rose to determine directions on the map.

⑥ Use the scale to measure the actual distances between places shown on the map.

⑦ Read the questions and then carefully study the map to determine the answers.

answers: 1 (C), 2 (C)

① Australia: Physical

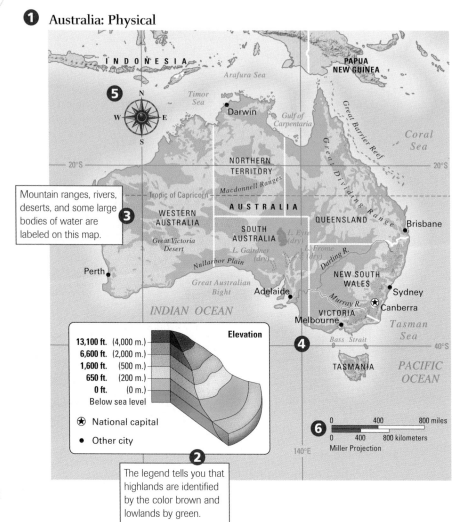

Mountain ranges, rivers, deserts, and some large bodies of water are labeled on this map.

13,100 ft. (4,000 m.)
6,600 ft. (2,000 m.)
1,600 ft. (500 m.)
650 ft. (200 m.)
0 ft. (0 m.)
Below sea level

Elevation

⊛ National capital
• Other city

The legend tells you that highlands are identified by the color brown and lowlands by green.

1 South Australia, Victoria, and New South Wales contain mostly

A mountains.

B plateaus.

C lowlands.

D deserts.

2 Where is the Great Barrier Reef located?

A Along the Nullarbor Plain

B In the Great Australian Bight

C In the Coral Sea

D Near the Great Victoria Desert

Directions: Use the map and your knowledge of world geography to answer the questions below.

Egypt: Physical

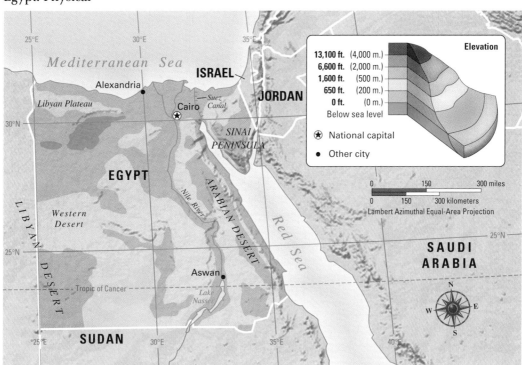

1   The location of Egypt's capital is approximately

   A 30°N 31°E.

   B 30°S 31°W.

   C 25°N 33°E.

   D 25°S 33°W.

2   The physical feature that dominates Egypt's landscape is

   A mountains.

   B deserts.

   C mesas.

   D lakes.

3   Which of the following statements *best* characterizes the Nile River?

   A It is the longest river in Egypt.

   B It extends the full length of the country.

   C It is one of the few rivers in Egypt.

   D All of the above

4   Which of the following conclusions can you draw from this map?

   A Egypt has a well-distributed water supply.

   B Agriculture is important in Egypt's southwest area.

   C Much of Egypt has a dry climate.

   D All of Egypt is sparsely populated.

Thematic Maps

A thematic map, or special-purpose map, focuses on a particular topic. The location of state parks, a country's natural resources, the vegetation of a region, voting patterns, migration routes, and economic activities are all topics you might see illustrated on a thematic map.

1 Read the title to determine the subject and purpose of the map.

2 Examine the labels on the map to find more detailed information on the map's subject and purpose.

3 Study the legend to find the meaning of the symbols and colors used on the map.

4 Look at the symbols and colors on the map, and try to identify patterns.

5 Read the questions and then carefully study the map to determine the answers.

1 Ethnic Diversity in the Former Yugoslavia

3 The map and legend show you nine distinct ethnic groups that reside in the former Yugoslavia, which is now eight republics or provinces.

4 This map is a visual tool that might help you understand the ethnic and religious conflicts in Bosnia.

1 According to the map, which of the following ethnic groups live in Bosnia and Herzegovina?

A Croats, Macedonians, and Slovenes

B Serbs, Albanians, and Hungarians

C Bulgarians, Italians, and Albanians

D Croats, Serbs, and Muslims

2 The former Yugoslavia did *not* include

A Kosovo.

B Slovenia.

C Romania.

D Croatia.

answers: 1 (D), 2 (C)

Directions: Use the satellite image below and your knowledge of world geography to answer the following questions. Your answers need not be complete sentences.

Satellite Image: Glacier in Patagonia, Chile, May 2, 2000

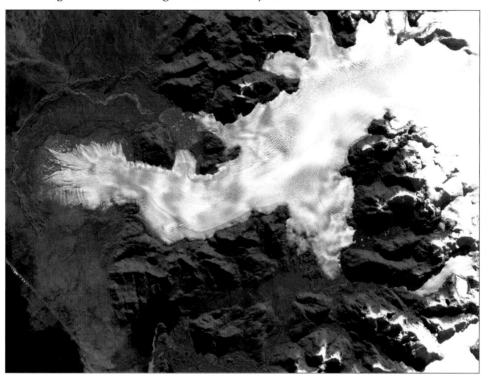

Satellite image: NASA/GSFC/MITI/ERSDAC/JAROS, and U.S./Japan ASTER Science Team

Image acquired 440 miles above Earth's surface by NASA *Terra* spacecraft over the North Patagonia Ice Sheet near latitude 47°S, longitude 73°W, covering an area of 36 by 30 kilometers.

The satellite image is relayed to scientists who are studying most of the world's 16,000 glaciers. By comparing these new, more detailed images with older ones, scientists have found that the glaciers in Patagonia are melting. Some have receded as much as one mile in the past 13 years. Here, vegetation is shown in red and the glacier in white. The semicircular gray area around the spoon-like end of the glacier is a terminal moraine, which shows that the glacier was once larger than it is now.

1 What does the satellite image taken over Patagonia, Chile, show clearly?

2 What is the significance of the semicircular terminal moraine shown on the image?

3 How might satellite images help geographers and climatologists study global changes in climate?

Extended Response

Extended-response questions, like constructed-response questions, usually focus on one type of document. However, they are more complex and require more time to complete than typical short-answer constructed-response questions. Some extended-response questions ask you to present information from the document in a different form. Others ask you to write an essay or report or some other extended piece of writing. Sometimes you are required to apply your knowledge of geography to information contained in the document.

1 Read the title of the document to get an idea of the subject.

2 Study and analyze the document.

3 Carefully read each extended-response question.

4 If the question calls for a drawing, such as a diagram, graph, or chart, make a rough sketch on scrap paper first. Then make a final copy of your drawing on the answer sheet.

5 If the question requires an essay, jot down your ideas in outline form. Use this outline to write your answer.

1 Causes of Death in Developed and Developing Countries, 1993

2

Cause	Developed countries (percentage of deaths)	Developing countries (percentage of deaths)
Infections and parasites	1.2	41.5
Respiratory diseases	7.8	5.0
Cancers	21.6	8.9
Circulatory diseases	46.7	10.7
Childbirth	0	1.3
Infant mortality	0.7	7.9
Injury	7.5	7.9
Other causes	14.5	16.8

Source: "Causes of Death, 1993," from *Oxford Atlas of World History,* edited by Patrick K. O'Brien. Copyright © 1999 by Oxford University Press. All rights reserved. Reprinted by permission of Oxford University Press.

3 **1** Use the information in the chart to create a bar graph showing the causes of death in developed and developing countries.

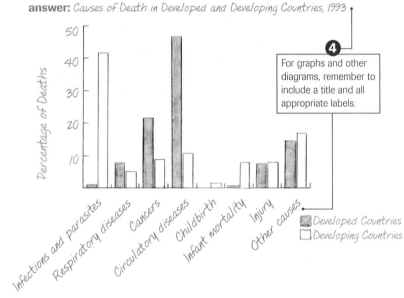

answer: Causes of Death in Developed and Developing Countries, 1993

4 For graphs and other diagrams, remember to include a title and all appropriate labels.

Developed Countries
Developing Countries

3 **2** Write a short essay summarizing what the chart and graph show about the major causes of death in developed and developing countries. Give a possible explanation for the data.

5 **Essay Rubric** The best essays will point out that infections and parasites claim the most lives in developing countries, whereas circulatory diseases and cancers are the main killers in developed countries. Poor sanitation and a lack of access to health care might account for the high death rate from infections and parasites in developing countries. In developed countries, a longer life expectancy as well as a fatty diet, smoking, and a lack of exercise might be factors in the high incidence of circulatory diseases and cancers.

For more test practice online . . .
TEST PRACTICE
CLASSZONE.COM

Directions: Use the following passage and your knowledge of world geography to answer the questions below.

Subregions of Canada

Canada and the United States share a similar history and culture. Canada's location in the northern latitudes, however, has affected its population distribution and its economic growth in ways that make the country different from the United States.

1 The chart below lists the four subregions of Canada. Complete the chart by briefly describing the population and economic activities of each subregion. (Note that some of the answers have been written for you.)

Subregion	Population	Economic activities
Atlantic Provinces	small population due to rugged terrain and severe weather; most people living in coastal cities	logging, fishing, mining
Core Provinces–Quebec and Ontario		agriculture, mining, manufacturing
Prairie Provinces	populated by diverse immigrant groups	
Pacific Province and the Territories		logging, mining, and hydroelectric production in British Columbia; mining, fishing, and logging in territories

2 In a short essay, compare and contrast population distribution in Canada with population distribution in the United States. Note any similarities between subregions of Canada and subregions of the United States, and describe the outstanding differences.

Document-Based Questions

A document-based question (DBQ) requires you to analyze and interpret a variety of documents. These documents often are accompanied by short-answer questions. You use these answers and information from the documents to write an essay on a specified subject.

1 Read the "Context" section to get a sense of the issue addressed in the question.

2 Read the "Task" section and note the action words. This will help you understand exactly what the essay question requires.

3 Study and analyze each document. Consider what connection the documents have to the essay question. Take notes on your ideas.

4 Read and answer the document-specific questions. Think about how these questions connect to the essay topic.

Introduction

1 **Context:** In recent years, population densities in the urban areas of the United States have been falling. This is due, in large part, to urban sprawl—widespread low-density urban development, such as strip malls, large office buildings, and housing subdivisions, in areas well beyond city boundaries.

2 **Task:** (Define) the term *urban sprawl* and (explain) why this recent development has become an issue of concern, particularly in the Sunbelt states.

Part 1: Short Answer

Study each document carefully and answer the questions that follow.

3 Document 1: Urban Sprawl in Nashville, Tennessee

Copyright © Gary Layda/Metropolitan Planning Department of Nashville—Davidson County

4 How does this photograph illustrate urban sprawl?

The photograph shows low-density urban development well beyond city boundaries—a huge single-family housing subdivision butting up against the mountains.

Document 2: Developed Land in the United States

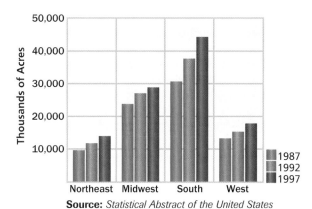

Source: *Statistical Abstract of the United States*

What trends does the bar graph show?

The area of developed land is increasing in all regions of the United States, but particularly in the South.

Document 3: The Impact of Sprawl in the United States

Sprawl increases traffic.	Sprawl lengthens trips and forces people to drive more.
Sprawl increases pollution.	As sprawl increases, people rely more and more on cars and driving. Cars are a major source of air pollution.
Sprawl increases the risk of flooding.	Developments sometimes are built on floodplains and in wetland areas.
Sprawl consumes parks, farms, and open space.	Over one million acres of parks, farms, and open space are developed each year to accommodate sprawl.
Sprawl drains the tax coffers.	Sprawl requires millions of tax dollars for new infrastructure. These tax dollars could be spent on improving existing communities.
Sprawl overcrowds schools	Sprawl puts more children in suburban schools, causing overcrowding.

Source: The Sierra Club

What is the environmental impact of the sprawl-related increase in traffic?

The increase in traffic causes an increase in air pollution.

Part 2: Essay

5 Using information from the documents, your answers to the questions in Part 1, and your knowledge of world geography, write an essay that defines the term *urban sprawl* and explains why this recent development has become an issue of concern, particularly in the Sunbelt states.

5 Carefully read the essay question. Then write an outline for your essay.

6 Write your essay. Be sure that it has an introductory paragraph that introduces your argument, main body paragraphs that explain it, and a concluding paragraph that restates your position. In your essay, include quotations or details from specific documents to support your ideas. Add other supporting facts or details that you know from your study of world geography.

6
Essay Rubric The best essays will point out that urban sprawl involves widespread low-density urban development (such as strip malls, large office buildings, and housing subdivisions) in areas well beyond city boundaries (Document 1). They will go on to mention that this largely unplanned and uncontrolled development is cause for concern for several reasons. These include increased traffic and related air pollution, increased risk of flooding, increased costs to government, and school overcrowding (Document 3). They will conclude by pointing out that the rapid increase in the amount of developed land in the South indicates that urban sprawl is of particular concern in the Sunbelt states (Document 2).

Introduction

Context: For more than 2,500 years, the city of Istanbul has been a center of civilization and a place of passage, where languages, crafts, goods, and necessities have exchanged hands and enriched the cultures of the world. It began as Byzantium and was later known as Constantinople. Istanbul today is the commercial center of Turkey. It has a population of over 8 million people.

Task: Discuss how the Istanbul of today is like and unlike the Constantinople of the 1300s. Discuss its role as a crossroads that connects vastly different cultures. Explain how the unique location of Istanbul is important to its development as a world port.

Part 1: Short Answer

Study each document carefully and answer the questions that follow.

Document 1: Constantinople, Center of Trade and Travel in the 1300s

[Constantinople] is enormous in size, and in two parts separated by a great river. . . . The part of the city on the eastern bank of the river is called Istanbul. . . . Its bazaars and streets are spacious and paved with flagstones; each bazaar has gates which are closed upon it at night, and the majority of the artisans and sellers in them are women. The city lies at the foot of a hill which projects about nine miles into the sea. . . . Round this hill runs the city-wall, which is very strong and cannot be taken by assault from the sea front. Within its circuit there [are] about thirteen inhabited villages. The principal church is in the midst of this part of the city. The second part, on the western bank of the river . . . is reserved to the Frankish Christians who dwell there. They are of different kinds, including Genoese, Venetians, Romans and people of France. . . . They are all men of commerce and their harbour is one of the largest in the world; I saw there about a hundred galleys and other large ships.

—Excerpt from *The Adventures of Ibn Battuta: A Muslim Traveler of the 14th Century,* translated and edited by Ross W. Dunn (Berkeley: University of California Press, 1989), page 3. Reprinted by permission of the University of California Press.

What are three of the characteristics of Constantinople described by Ibn Battúta that might explain its long history as a major commercial, cultural, religious, and political center to the world?

Document 2: Istanbul, Turkey, June 16, 2000, Satellite Image

The urban areas appear blue-green; vegetation appears red; water, blue. Istanbul is divided by the Bosporus Strait, which is a deep, twisting waterway, about 19 miles long, and about 800 yards wide at places. The city is a major port for Europe and Asia.

NASA/GSFC/MITI/ERSDAC/JAROS, and U.S./Japan ASTER Science Team

What geographic factors explain the growth of Istanbul into a large city?

Document 3: Number of Ships Traveling the Bosporus, 1995–2000

Years	Tankers	Total Passages	Monthly Average	Daily Average
1995	unknown	46,954	3,912	128
1996	4,248	49,952	4,162	137
1997	4,303	50,942	4,245	142
1998	5,142	49,304	4,109	137
1999	4,452	47,906	3,992	133
2000	4,937	48,079	4,007	134

Source: "Number of Ships Traveling the Bosporus," from the Turkish Maritime Pilots' Association Web site. Reprinted by permission of the Turkish Maritime Pilots Association.

How does the chart show that Istanbul is a major port?

Part 2: Essay

Using information from the documents, your answers to the questions in Part 1, and your knowledge of world geography, write an essay that discusses how the Istanbul of today is like and unlike the Constantinople of the 1300s. Discuss its role as a crossroads that connects vastly different cultures. Explain how the unique location of Istanbul is important to its development as a world port.

RAND McNALLY
World Atlas

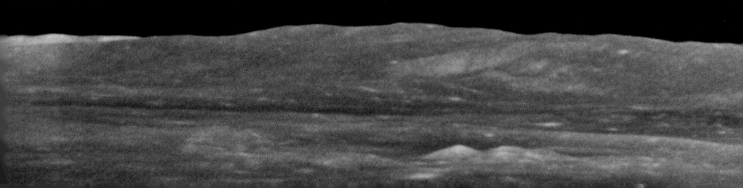

Contents

Complete Legend for Physical and Political Maps

Symbols

 Lake

 Salt Lake

 Seasonal Lake

 River

\ Waterfall

—— Canal

△ Mountain Peak

▲ Highest Mountain Peak

Cities

■ Los Angeles   City over 1,000,000 population

▣ Calgary   City of 250,000 to 1,000,000 population

• Haifa   City under 250,000 population

✸ Paris   National Capital

★ Vancouver   Secondary Capital (State, Province, or Territory)

Type Styles Used to Name Features

CHINA   Country

ONTARIO   State, Province, or Territory

PUERTO RICO (U.S.)   Possession

ATLANTIC OCEAN   Ocean or Sea

A l p s   Physical Feature

Borneo   Island

Boundaries

 International Boundary

 Secondary Boundary

Land Elevation and Water Depths

Land Elevation

Meters		Feet
3,000 and over --		-- 9,840 and over
2,000 - 3,000 --		-- 6,560 - 9,840
500 - 2,000 --		-- 1,640 - 6,560
200 - 500 --		-- 656 - 1,640
0 - 200 --		-- 0 - 656

Water Depth

Less than 200 --		-- Less than 656
200 - 2,000 --		-- 656 - 6,560
Over 2,000 --		-- Over 6,560

A1

ARCTIC OCEAN

Greenland
Jan Mayen
Baffin Island
Baffin Bay
Arctic Circle
Iceland
Faroe Is.
Hudson Bay
British Isles
London
Newfoundland
Azores
Iberian Peninsula
Atlas Mts.

Mt. McKinley △ 20,320 Ft. 6,194m
Yukon
Mackenzie
Canadian Shield
Aleutian Islands
N O R T H
Vancouver
Rocky Mountains
Great Plains
St. Lawrence
Appalachian Mts.
A M E R I C A
Washington D.C.
Cape Hatteras
Los Angeles
Colorado
Mississippi
ATLANTIC
Canary Islands
Midway Is.
Baja California
Gulf of Mexico
Yucatan Peninsula
Cuba
Hispaniola
Puerto Rico
Cape Verde Islands
Tropic of Cancer
Hawaiian Islands
Jamaica
Caribbean Sea
Cape Verde
Niger
PACIFIC
Trinidad
O C E A N
Orinoco
Palmyra
Amazon
Amazon
Equator
Galapagos Islands
Basin
S O U T H
Kiribati
Andes
A M E R I C A
Marquesas Is.
Samoa Islands
Mato Grosso Plateau
St. Helena
Tonga Is.
Cook Islands
Tahiti
Rio de Janeiro
Tropic of Capricorn
Andes
Paraná
Easter Island
△ Mt. Aconcagua 22,831 Ft. 6,959m
Chatham Is.
Archipiélago Juan Fernández
Buenos Aires
N
Patagonia
Falkland Is.
South Georgia
South Sandwich Is.
Tierra del Fuego
South Orkney Is.
Cape Horn
South Shetland Is.
Antarctic Circle
Antarctic Peninsula
Weddell Sea
Ross Sea
Marie Byrd Land
△ Vinson Massif 16,066 Ft. 4,897m

0   1000   2000 Miles
0   1000   2000   3000 Kilometers
Copyright by Rand McNally & Co.
Robinson Projection

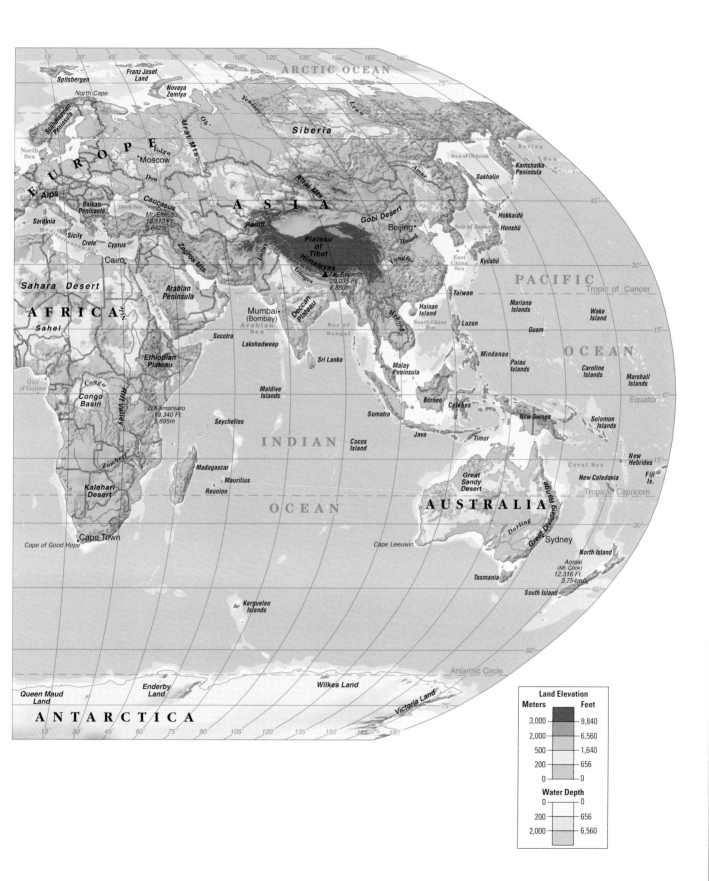

ATLAS

ARCTIC OCEAN

Spitsbergen

Franz Josef
Land

North Cape

Novaya
Zemlya

75°

60°

Scandinavian Peninsula

North
Sea

Yenisei

Lena

Bering
Sea

EUROPE

Ural Mts.

Ob'

Siberia

Sea of Okhotsk

Volga

Moscow

Kamchatka
Peninsula

45°

Alps

Don

Balkan
Peninsula

Black Sea

Caucasus

ASIA

Gobi Desert

Altai Mts.

Sakhalin

Hokkaidō

Sardinia

Mt. Elbrus
18,510 Ft.
5,642m

Pamir

Plateau
of
Tibet

Beijing

Honshū

Sicily

Crete

Cyprus

Mediterranean Sea

Zagros Mts.

Indus

Himalayas

Huang

Yangtze

Sea of Japan

East
China
Sea

Kyūshū

30°

Cairo

Red Sea

Sahara Desert

Arabian
Peninsula

Deccan
Plateau

Ganges

Mt. Everest
29,035 Ft.
8,850m

Mekong

PACIFIC

Tropic of Cancer

AFRICA

Sahel

Nile

Mumbai
(Bombay)

Arabian
Sea

Socotra

Lakshadweep

Bay of
Bengal

Hainan
Island

Taiwan

South China
Sea

Mariana
Islands

Wake
Island

15°

Sri Lanka

Luzon

Guam

OCEAN

Gulf
of Guinea

Ethiopian
Plateau

Malay
Peninsula

Mindanao

Palau
Islands

Caroline
Islands

Marshall
Islands

Congo

Rift Valley

Congo
Basin

Kilimanjaro
19,340 Ft.
5,895m

Maldive
Islands

Seychelles

Sumatra

Borneo

Celebes

Java

New Guinea

Solomon
Islands

0°

Equator

INDIAN

Cocos
Island

Timor

Zambezi

Madagascar

Mauritius

Reunion

OCEAN

New
Hebrides

15°

Coral Sea

New Caledonia

Fiji
Is.

Kalahari
Desert

Great
Sandy
Desert

AUSTRALIA

Tropic of Capricorn

Cape Town

Cape of Good Hope

Cape Leeuwin

Darling

Great Dividing Range

Sydney

30°

North Island

Aoraki
(Mt. Cook)
12,316 Ft.
3,754m

Tasmania

South Island

45°

Kerguelen
Islands

60°

Antarctic Circle

Queen Maud
Land

Enderby
Land

Wilkes Land

Victoria Land

75°

ANTARCTICA

15°   30°   45°   60°   75°   90°   105°   120°   135°   150°   165°   180°

Land Elevation		
Meters		Feet
3,000		9,840
2,000		6,560
500		1,640
200		656
0		0

Water Depth		
0		0
200		656
2,000		6,560

RAND McNALLY

A3

ARCTIC OCEAN

GREENLAND
(Den.)

Baffin
Bay

Arctic Circle

RUSSIA
ALASKA
Yukon (U.S.)
Anchorage•

ICELAND
FAROE IS.
(Den.)

UNITED
KINGDOM
IRELAND
London

CANADA

Hudson
Bay

Newfoundland

Vancouver
Missouri
Montréal•
Ottawa•

FRANCE

Madrid
SPAIN

Chicago•
UNITED STATES

New York
Washington D.C.

PORTUGAL
Azores
(Port.)

Los Angeles•
Colorado
Mississippi

Casablanca•

Houston•

MIDWAY IS.
(U.S.)

ATLANTIC

Canary
Islands
(Sp.)

MEXICO

Gulf of Mexico

BAHAMAS

MAURITANIA
MALI

Tropic of Cancer

Hawaiian
Islands
(U.S)

Mexico City

CUBA
HAITI
DOM. REP.
PUERTO RICO (U.S.)

CAPE
VERDE

SENEGAL

BELIZE
GUAT. HOND.
JAMAICA

GAMBIA
GUINEA-BISSAU
GUINEA

BURK.
FASO

PACIFIC
EL. SAL.
NIC.
Caribbean
Sea

Caracas•
TRINIDAD AND TOBAGO

SIERRA LEONE

COTE
D'IVOIRE

LIBERIA

COSTA
RICA
PANAMA
VENEZUELA
GUYANA
SURINAME
FRENCH GUIANA

COLOMBIA

Equator
KIRIBATI

Galapagos Islands
(Ecuador)
ECUADOR

Amazon

BRAZIL

OCEAN

SAMOA

AMERICAN
SAMOA
COOK
ISLANDS (N.Z.)
TONGA

OCEAN
PERU
Lima•

BOLIVIA

ST. HELENA
(U.K.)

Tropic of Capricorn

FRENCH POLYNESIA

PARAGUAY
Rio de Janeiro

Easter Island
(Chile)

ARGENTINA
URUGUAY

Santiago•
Buenos
Aires

N

0   1000   2000 Miles
0   1000   2000   3000 Kilometers
Copyright by Rand McNally & Co.
Robinson Projection

FALKLAND IS.
(U.K.)

South
Georgia
(U.K.)

South
Orkney Is.
(U.K.)

Antarctic Circle

South
Shetland Is.
(U.K.)

Weddell
Sea

15° 30° 45° 60° 75° 90° 105° 120° 135° 150° 165° 180°

ARCTIC OCEAN

Spitsbergen
(Nor.)

Franz Josef
Land

Novaya
Zemlya

75°

Yenisey

Lena

NORWAY   FINLAND

R U S S I A

Ob'

Volga

SWEDEN
EST.
LAT.
LITH.
DEN.
Novosibirsk

60°

North
Sea
NETH.
GERMANY
POLAND
BELARUS
Moscow

Bering   Sea
Sea of Okhotsk

SWITZ.
CZ.
SLVK.
AUS. HUNG.
UKRAINE
MOLD.

KAZAKHSTAN

MONGOLIA

45°

ITALY
CRO. YUG. ROM.
BOS. BUL.
ALB. MA.
GREECE

Black Sea
GEO.
ARM. AZER.

UZBEKISTAN
KYRG.

NORTH
KOREA

Sea of Japan

Rome

Mediterranean Sea
Crete
TURKEY
CYPRUS LEB.
ISRAEL
JORDAN
SYRIA
IRAQ

TURKMENISTAN
TAJIK.

Beijing

SOUTH
KOREA
JAPAN

Tunisia

IRAN
AFGHANISTAN

C H I N A

Chang Jiang
Yangtze

Tokyo

30°

ALGERIA
LIBYA
EGYPT
Cairo

SAUDI
ARABIA
QATAR
U.A.E.
KUWAIT

PAKISTAN
NEPAL
BHU.
Ganges
Kolkata
(Calcutta)
BNGL.

Shanghai

TAIWAN

P A C I F I C

Tropic of Cancer

NIGER
CHAD
SUDAN

OMAN
YEMEN
Nile
Red Sea
ERITREA

INDIA

Mumbai
(Bombay)
Arabian
Sea

MYANMAR
LAOS
THAILAND

Guangzhou

NORTHERN
MARIANA ISLANDS
(U.S.)

WAKE ISLAND
(U.S.)

15°

BENIN
NIGERIA
Lagos

CENTRAL
AFRICAN
REPUBLIC
Addis
Ababa
DJIBOUTI
ETHIOPIA

Bay of
Bengal

Bangkok
CAMBODIA
VIETNAM

South China
Sea

PHILIPPINES

GUAM (U.S.)

O C E A N

EQUATORIAL
GUINEA
CAMEROON

SRI LANKA

BRUNEI

PALAU

FED. STATES OF
MICRONESIA

MARSHALL
ISLANDS

GABON
REP. OF
CONGO
Congo
RWANDA
DEM. REP.
OF CONGO
BURUNDI
UGANDA
KENYA
SOMALIA

MALDIVES

MALAYSIA
SINGAPORE
Borneo
New Guinea

Equator   0°

TANZANIA

SEYCHELLES

Sumatra
Java
Jakarta
INDONESIA
EAST TIMOR

PAPUA
NEW GUINEA

SOLOMON
ISLANDS

ANGOLA
ZAMBIA
COMOROS

I N D I A N

Darwin

Coral Sea

15°

NAMIBIA
ZIMBABWE
MOZAMBIQUE
BOTSWANA
MALAWI
MADAGASCAR
MAURITIUS

REUNION
(Fr.)

VANUATU
NEW CALEDONIA
(Fr.)
FIJI

Tropic of Capricorn

SWAZILAND

O C E A N

A U S T R A L I A

SOUTH
AFRICA
LESOTHO

Perth
Darling
Sydney

30°

Cape Town

Kerguelen
Islands
(Fr.)

Melbourne

Tasmania

NEW ZEALAND
Wellington

45°

60°

Antarctic Circle

A N T A R C T I C A

75°

15° 30° 45° 60° 75° 90° 105° 120° 135° 150° 165° 180°

✳ National Capital

• Major Cities

TROPICAL
Hot with Rain All Year
Hot with Seasonal Rain

DRY
Desert
Some Rain

MODERATE (RAINY WINTER)
Hot, Dry Summer
Hot, Humid Summer
Mild, Rainy Summer

CONTINENTAL (SNOWY WINTER)
Long, Warm, Humid Summer
Short, Cool, Humid Summer
Very Short, Cool, Humid Summer

POLAR
Tundra – Very Cold and Dry
Ice Cap

HIGHLANDS
Varies with Altitude

0   1000   2000 Miles
0   1000   2000   3000 Kilometers
Copyright by Rand McNally & Co.
Robinson Projection

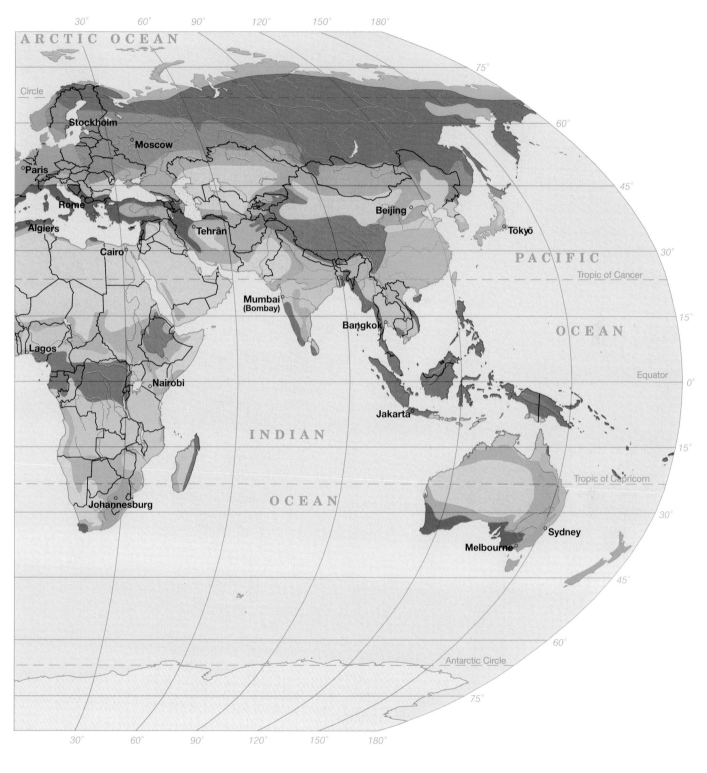

180° 150° 120° 90° 60° 30°

75°

60°

45°

30°

Tropic of Cancer

15°

Equator 0°

15°

Tropic of Capricorn

30°

Arctic

London

Madrid

Vancouver
Montréal
Chicago
New York
Los Angeles
Houston
ATLANTIC
Mexico City
PACIFIC
Caracas
OCEAN
OCEAN
Rio de Janeiro
Buenos Aires
OCEAN

Antarctic Circle

75°

180° 150° 120° 90° 60° 30°

N

Legend

- Forest
- Swamp
- Crop and Woodland
- Cropland
- Crop and Grazing Land
- Grassland
- Desert
- Tundra
- Barren
- Urban

0   1000   2000 Miles

0   1000   2000   3000 Kilometers

Copyright by Rand McNally & Co.
Robinson Projection

ARCTIC OCEAN

Circle

Stockholm

Moscow

Paris

Rome

Algiers

Cairo

Tehrān

Beijing

Tōkyō

PACIFIC

75°

60°

45°

30°

Tropic of Cancer

OCEAN

15°

Mumbai
(Bombay)

Bangkok

Lagos

Nairobi

INDIAN

Jakarta

Equator

0°

15°

OCEAN

Tropic of Capricorn

Johannesburg

Sydney

30°

Melbourne

45°

60°

Antarctic Circle

75°

30°   60°   90°   120°   150°   180°

RAND MCNALLY

Arctic

75°

60°

45°

30°

Vancouver

Montréal

Chicago

New York

Los Angeles

Houston

London

Madrid

Tropic of Cancer

15°

Mexico City

ATLANTIC

Caracas

PACIFIC

OCEAN

Equator

OCEAN

15°

Tropic of Capricorn

30°

Rio de
Janeiro

Buenos Aires

N

Per square mile
(per square kilometer)

Under 2 (Under 1)

2-60 (1-25)

60-125 (25-50)

125-250 (50-100)

Over 250 (Over 100)

Antarctic Circle

75°

0   1000   2000 Miles

0   1000   2000   3000 Kilometers

Copyright by Rand M?Nally & Co.
Robinson Projection

180°   150°   120°   90°   60°   30°

180°   150°   120°   90°   60°   30°

ARCTIC OCEAN

Circle

Stockholm

Moscow

Yekaterinburg

Paris

Rome

Algiers

Cairo

Tehrān

Beijing

Tōkyō

PACIFIC

Tropic of Cancer

Mumbai
(Bombay)

Bangkok

OCEAN

Lagos

Nairobi

INDIAN

Jakarta

Equator

OCEAN

Johannesburg

Tropic of Capricorn

Sydney

Melbourne

Antarctic Circle

75°
60°
45°
30°
15°
0°
15°
30°
45°
60°
75°

30°   60°   90°   120°   150°   180°

Arctic

180° 150° 120° 90° 60° 30°

75°

60°

45°

30°

Tropic of Cancer

15°

Equator 0°

15°

Tropic of Capricorn

30°

Antarctic Circle

75°

PACIFIC OCEAN

OCEAN

ATLANTIC OCEAN

Vancouver
Montréal
Chicago
New York
Los Angeles
Houston
Mexico City
Caracas
Rio de Janeiro
Buenos Aires

London
Madrid

N

Little or no activity
Nomadic Herding
Hunting, Forestry, Subsistence Farming
Forestry
Agriculture
Stock Raising
Manufacturing, Commerce
Fishing

0 1000 2000 Miles
0 1000 2000 3000 Kilometers
Copyright by Rand McNally & Co.
Robinson Projection

ATLAS

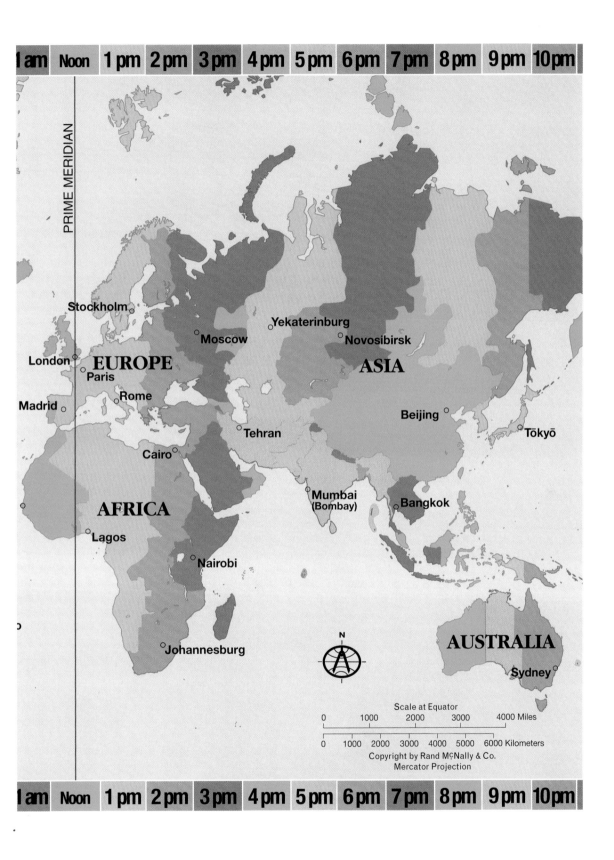

11 am | Noon | 1 pm | 2 pm | 3 pm | 4 pm | 5 pm | 6 pm | 7 pm | 8 pm | 9 pm | 10 pm

PRIME MERIDIAN

Stockholm

Yekaterinburg

Moscow   Novosibirsk

London   EUROPE   ASIA
Paris

Madrid   Rome   Beijing   Tōkyō

Tehran

Cairo

Mumbai   Bangkok
(Bombay)

AFRICA

Lagos

Nairobi   AUSTRALIA

Johannesburg   Sydney

N

Scale at Equator

0   1000   2000   3000   4000 Miles

0   1000 2000 3000 4000 5000 6000 Kilometers
Copyright by Rand McNally & Co.
Mercator Projection

11 am | Noon | 1 pm | 2 pm | 3 pm | 4 pm | 5 pm | 6 pm | 7 pm | 8 pm | 9 pm | 10 pm

RAND M^CNALLY

A17

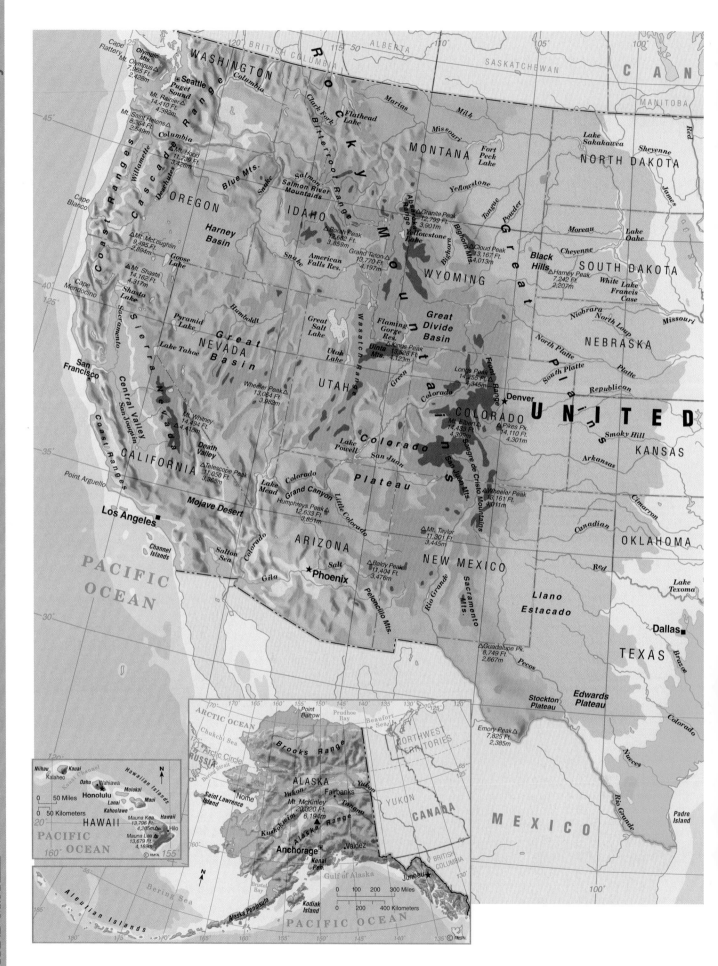

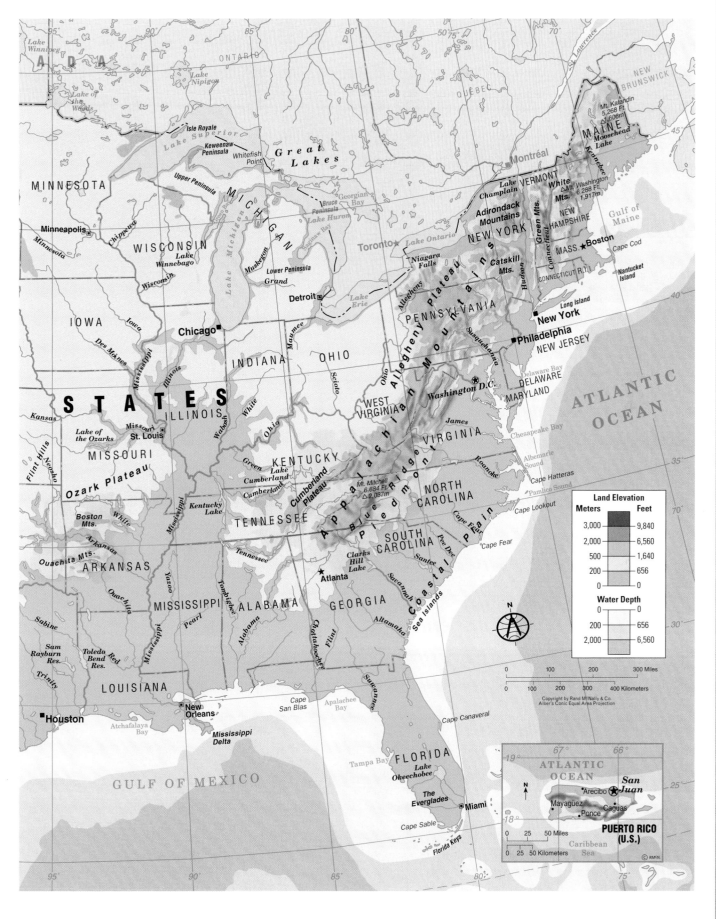

Land Elevation

Meters		Feet
3,000		9,840
2,000		6,560
500		1,640
200		656
0		0

Water Depth

0		0
200		656
2,000		6,560

0   100   200   300 Miles
0   100   200   300   400 Kilometers

Copyright by Rand McNally & Co.
Alber's Conic Equal Area Projection

ATLANTIC
OCEAN

San
Juan
• Arecibo
Mayagüez •
• Ponce   • Caguas
PUERTO RICO
(U.S.)

0   25   50 Miles
0   25   50 Kilometers

Caribbean
Sea

A19

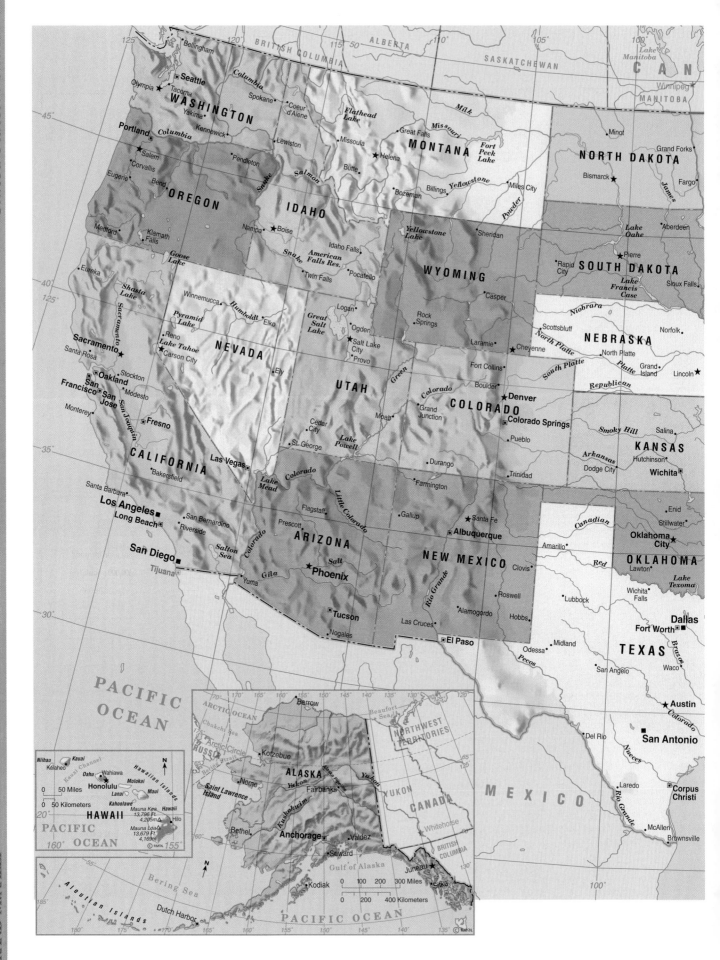

A20

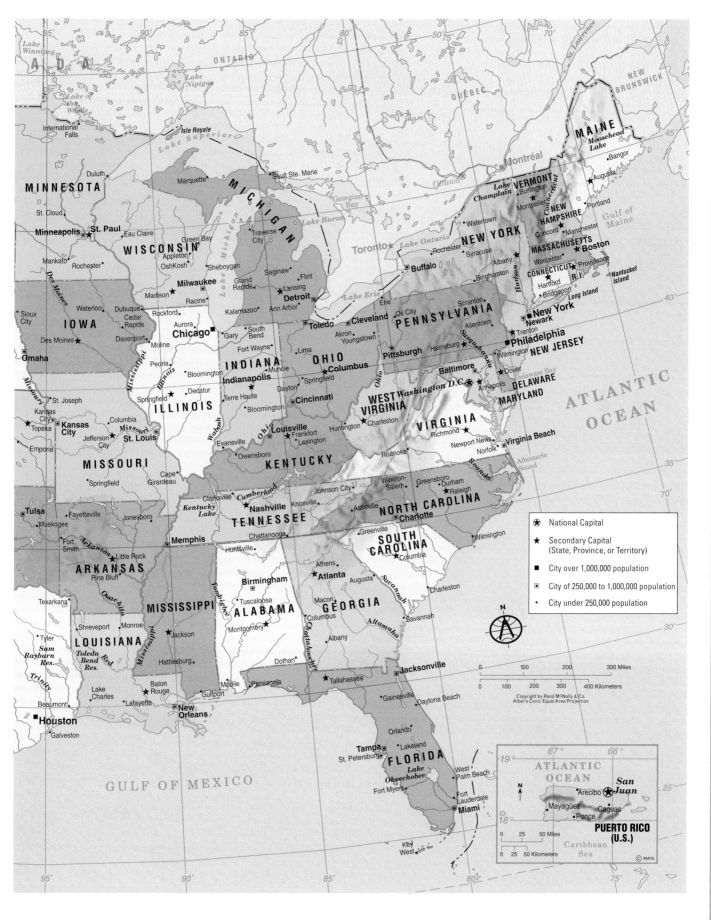

National Capital

Secondary Capital
(State, Province, or Territory)

City over 1,000,000 population

City of 250,000 to 1,000,000 population

City under 250,000 population

100   200   300 Miles

100   200   300   400 Kilometers

Copyright by Rand McNally & Co.
Alber's Conic Equal Area Projection

PUERTO RICO
(U.S.)

0   25   50 Miles

0   25   50 Kilometers

RAND McNALLY

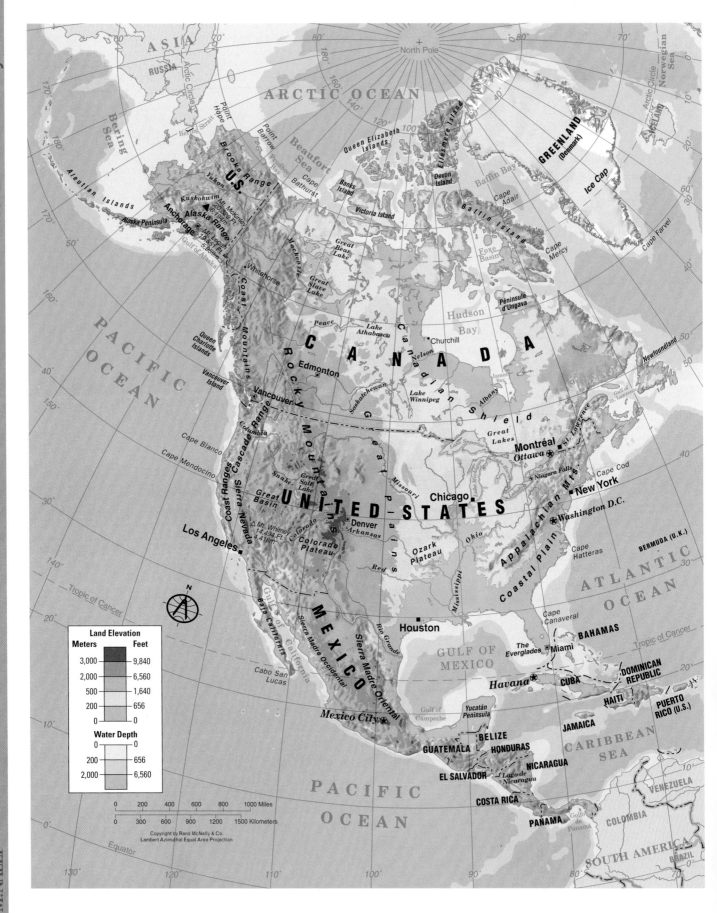

ASIA

RUSSIA

Arctic Circle

ARCTIC OCEAN

North Pole

Bering Strait

Point Hope

Point Barrow

Prudhoe Bay

Beaufort Sea

Cape Bathurst

Queen Elizabeth Islands

Banks Island

Devon Island

Ellesmere Island

Baffin Bay

Cape Adair

GREENLAND (Denmark)

Ice Cap

Arctic Circle

ICELAND

Norwegian Sea

Brooks Range

U.S.

Yukon

Kuskokwim

Mt. McKinley 20,320 m.

Anchorage Mt. Logan 19,551 ft. 5,959 m.

Alaska Range

Gulf of Alaska

Aleutian Islands

Alaska Peninsula

Whitehorse

Mackenzie

Great Bear Lake

Great Slave Lake

Victoria Island

Baffin Island

Cape Mercy

Cape Farvel

Foxe Basin

Péninsule d'Ungava

PACIFIC OCEAN

Queen Charlotte Islands

Coast Mountains

Vancouver Island

Vancouver

Rocky Mountains

Edmonton

Peace

Lake Athabasca

Saskatchewan

Lake Winnipeg

C A N A D A

Nelson

Canadian Shield

Hudson Bay

Churchill

Albany

James Bay

Great Lakes

Newfoundland

Gulf of St. Lawrence

Cape Blanco

Cape Mendocino

Columbia

Cascade Range

Coast Ranges

Snake

Sierra Nevada

Great Basin

Great Salt Lake

Colorado

Mt. Whitney 14,494 Ft. 4,418m.

Los Angeles

Colorado Plateau

Denver

Arkansas

Missouri

Red

U N I T E D - S T A T E S

Great Plains

Chicago

Mississippi

Ozark Plateau

Ohio

Lake Michigan

Lake Superior

St. Lawrence

Montréal

Ottawa

Niagara Falls

Appalachian Mts

New York

Washington D.C.

Cape Cod

Coastal Plain

Cape Hatteras

ATLANTIC OCEAN

BERMUDA (U.K.)

Tropic of Cancer

Cabo San Lucas

Gulf of California

Baja California

M E X I C O

Sierra Madre Occidental

Sierra Madre Oriental

Rio Grande

Houston

GULF OF MEXICO

Gulf of Campeche

Yucatán Peninsula

Mexico City

Cape Canaveral

The Everglades

Miami

BAHAMAS

Tropic of Cancer

Havana

CUBA

DOMINICAN REPUBLIC

HAITI

PUERTO RICO (U.S.)

JAMAICA

CARIBBEAN SEA

BELIZE

GUATEMALA

HONDURAS

EL SALVADOR

NICARAGUA

Lago de Nicaragua

COSTA RICA

PANAMA

Golfo de Panamá

VENEZUELA

COLOMBIA

SOUTH AMERICA

BRAZIL

PACIFIC OCEAN

Equator

Land Elevation

Meters	Feet
3,000	9,840
2,000	6,560
500	1,640
200	656
0	0

Water Depth

0	0
200	656
2,000	6,560

N

0   200   400   600   800   1000 Miles
0   300   600   900   1200   1500 Kilometers

Copyright by Rand McNally & Co.
Lambert Azimuthal Equal Area Projection

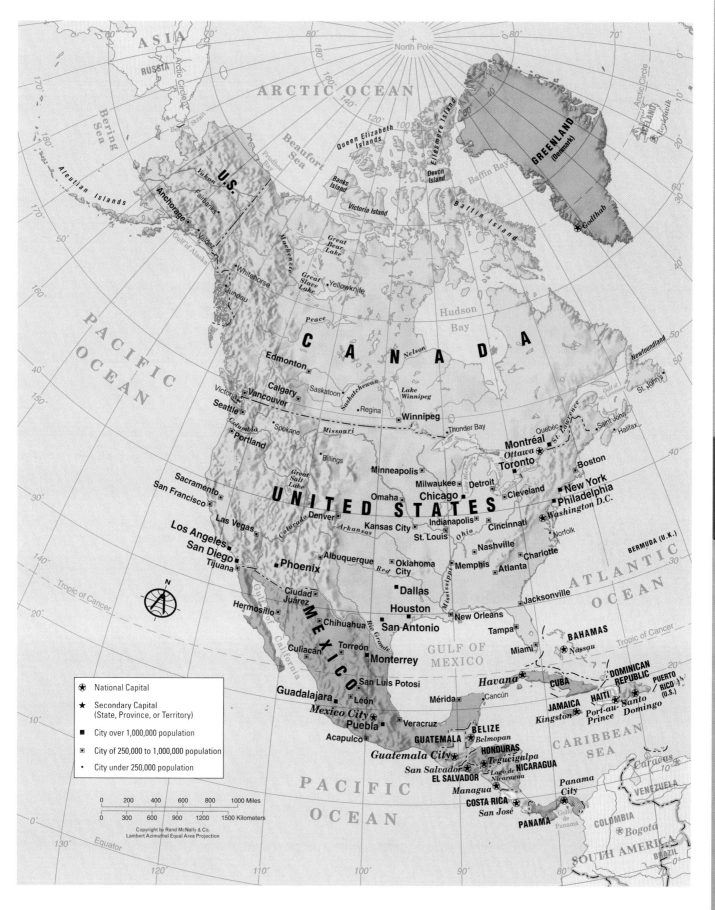

ASIA
RUSSIA
Arctic Circle
Bering Strait
Bering Sea
Aleutian Islands

ARCTIC OCEAN
North Pole
Arctic Circle
ICELAND
Reykjavik

Beaufort Sea
Prudhoe Bay
Queen Elizabeth Islands
Ellesmere Island
Devon Island
Baffin Bay
GREENLAND (Denmark)

U.S.
Anchorage
Fairbanks
Valdez
Gulf of Alaska
Yukon
Whitehorse
Juneau
Banks Island
Victoria Island
Great Bear Lake
Mackenzie
Great Slave Lake
Yellowknife
Peace

Baffin Island
Godthåb

PACIFIC OCEAN

Hudson Bay

CANADA
Edmonton
Calgary
Saskatoon
Saskatchewan
Regina
Lake Winnipeg
Winnipeg
Nelson
Thunder Bay
Newfoundland
St. John's
Gulf of St. Lawrence
Saint John
Halifax

Victoria
Vancouver
Seattle
Columbia
Spokane
Portland
Missouri
Billings
Minneapolis
Milwaukee
Lake Superior
Lake Michigan
Detroit
Cleveland
Quebec
St. Lawrence
Montréal
Ottawa
Toronto
Boston
New York
Philadelphia
Washington D.C.

Sacramento
San Francisco
UNITED STATES
Omaha
Chicago
Indianapolis
Cincinnati
Ohio
Norfolk
BERMUDA (U.K.)

Denver
Kansas City
St. Louis
Nashville
Charlotte
ATLANTIC OCEAN

Las Vegas
Colorado
Arkansas
Red
Memphis
Atlanta

Los Angeles
San Diego
Tijuana
Phoenix
Albuquerque
Oklahoma City
Dallas
Mississippi
Jacksonville
Tropic of Cancer

Ciudad Juárez
Hermosillo
Gulf of California
MEXICO
Chihuahua
Rio Grande
San Antonio
Houston
New Orleans
Tampa
GULF OF MEXICO
Miami
BAHAMAS
Nassau
Tropic of Cancer

Culiacán
Torreón
Monterrey
San Luis Potosí
Havana
CUBA
DOMINICAN REPUBLIC
PUERTO RICO (U.S.)

Guadalajara
León
Mérida
Cancún
JAMAICA
Kingston
HAITI
Port-au-Prince
Santo Domingo

Mexico City
Puebla
Veracruz
BELIZE
Belmopan
GUATEMALA
CARIBBEAN SEA

Acapulco
Guatemala City
HONDURAS
Tegucigalpa
NICARAGUA
Lago de Nicaragua
Caracas
VENEZUELA

San Salvador
EL SALVADOR
Managua
Panama City

PACIFIC OCEAN
COSTA RICA
San José
PANAMA
Golfo de Panamá
COLOMBIA
Bogotá

SOUTH AMERICA
BRAZIL
Equator

Legend:
- ⊛ National Capital
- ★ Secondary Capital (State, Province, or Territory)
- ■ City over 1,000,000 population
- ▣ City of 250,000 to 1,000,000 population
- • City under 250,000 population

0 200 400 600 800 1000 Miles
0 300 600 900 1200 1500 Kilometers

Copyright by Rand McNally & Co.
Lambert Azimuthal Equal Area Projection

RAND MCNALLY

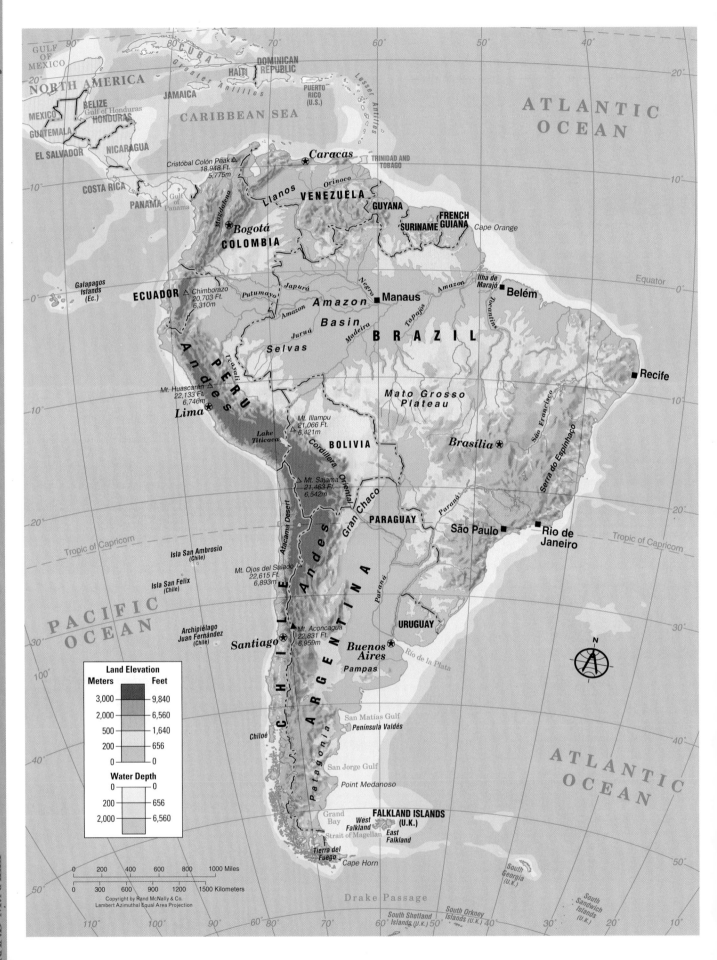

GULF OF MEXICO

CUBA

NORTH AMERICA

MEXICO
BELIZE
GUATEMALA
HONDURAS
Gulf of Honduras
EL SALVADOR
NICARAGUA
COSTA RICA
PANAMA
Gulf of Panama

JAMAICA

HAITI
DOMINICAN REPUBLIC
Greater Antilles
PUERTO RICO (U.S.)
Lesser Antilles

CARIBBEAN SEA

ATLANTIC OCEAN

Cristóbal Colón Peak △ 18,948 Ft. 5,775m

✪ Caracas
TRINIDAD AND TOBAGO

Llanos
Orinoco
VENEZUELA
GUYANA
SURINAME
FRENCH GUIANA
Cape Orange

Magdalena
✪ Bogotá
COLOMBIA

Galapagos Islands (Ec.)

ECUADOR
△ Chimborazo 20,703 Ft. 6,310m
Putumayo
Japurá
Amazon
Negro
Amazon
Manaus
Ilha de Marajó
Belém

Equator

Amazon Basin
Juruá
Madeira
Tapajós
Tocantins

Selvas

BRAZIL

Ucayali

ANDES

Mt. Huascarán △ 22,133 Ft. 6,746m

Mato Grosso Plateau

Recife

PERU

Lima ✪

Mt. Illampu △ 21,066 Ft. 6,421m

São Francisco

Serra do Espinhaço

Lake Titicaca

BOLIVIA

Brasília ✪

Cordillera Oriental

△ Mt. Sajama 21,463 Ft. 6,542m

Gran Chaco

Atacama Desert

PARAGUAY

São Paulo
Rio de Janeiro

Tropic of Capricorn

Isla San Ambrosio (Chile)

Mt. Ojos del Salado △ 22,615 Ft. 6,893m

Paraná

Isla San Felix (Chile)

ANDES

Paraná

PACIFIC OCEAN

Archipiélago Juan Fernández (Chile)

Mt. Aconcagua △ 22,831 Ft. 6,959m

URUGUAY

CHILE

ARGENTINA

Santiago ✪

Buenos Aires ✪
Río de la Plata

Pampas

San Matías Gulf
Península Valdés

N

Chiloé

Patagonia

San Jorge Gulf

Point Medanoso

ATLANTIC OCEAN

Grand Bay
West Falkland
East Falkland
FALKLAND ISLANDS (U.K.)

Strait of Magellan

Tierra del Fuego
Cape Horn

South Georgia (U.K.)

Drake Passage

South Shetland Islands (U.K.)
South Orkney Islands (U.K.)

South Sandwich Islands (U.K.)

Land Elevation

Meters		Feet
3,000		9,840
2,000		6,560
500		1,640
200		656
0		0

Water Depth

0		0
200		656
2,000		6,560

0   200   400   600   800   1000 Miles
0   300   600   900   1200   1500 Kilometers

Copyright by Rand McNally & Co.
Lambert Azimuthal Equal Area Projection

RAND McNALLY

ATLAS

GULF OF MEXICO

NORTH AMERICA

MEXICO
BELIZE
GUATEMALA
HONDURAS
EL SALVADOR
NICARAGUA
COSTA RICA
PANAMA

Havana
CUBA
JAMAICA
HAITI
DOMINICAN REPUBLIC
PUERTO RICO (U.S.)

CARIBBEAN SEA

Lesser Antilles

TRINIDAD AND TOBAGO

ATLANTIC OCEAN

Barranquilla
Cartagena
Cúcuta
Medellín
Maracaibo
Barquisimeto
Valencia
Caracas
Ciudad Guayana
Bucaramanga
VENEZUELA
Georgetown
GUYANA
Paramaribo
SURINAME
Cayenne
FRENCH GUIANA
Bogotá
COLOMBIA
Cali

Orinoco

Equator

Galapagos Islands (Ec.)

Quito
ECUADOR
Guayaquil

Putumayo
Japurá
Iquitos
Amazon
Juruá

Negro
Manaus
Santarém

Amazon

Macapá

Belém
São Luís
Fortaleza
Teresina
Imperatriz
Natal

Chiclayo
Trujillo

PERU

Ucayali

Madeira
Pôrto Velho

B R A Z I L

Tapajós
Tocantins

Recife
Maceió
Aracaju
Feira de Santana
Salvador

Lima

Cusco
Lake Titicaca
Arequipa

BOLIVIA
La Paz
Cochabamba
Santa Cruz
Sucre

Cuiabá

Goiânia

Brasília
Montes Claros

Uberlândia

Belo Horizonte
Vitória

Campo Grande

PARAGUAY

Paraná

Campinas
Rio de Janeiro
São Paulo
Curitiba
Caxias do Sul
Pôrto Alegre

Antofagasta

Salta
San Miguel de Tucumán

Asunción

Tropic of Capricorn

Isla San Ambrosio (Chile)

Isla San Felix (Chile)

Paraná

Archipiélago Juan Fernández (Chile)

Valparaíso
Santiago
Mendoza

Córdoba
Rosario
Santa Fe
Buenos Aires
La Plata

URUGUAY
Montevideo
Río de la Plata

PACIFIC OCEAN

Concepción

Bahía Blanca

Mar del Plata

C H I L E

A R G E N T I N A

Chiloé

Archipiélago de los Chonos

Comodoro Rivadavia

ATLANTIC OCEAN

⊛ National Capital

★ Secondary Capital
(State, Province, or Territory)

■ City over 1,000,000 population

⊡ City of 250,000 to 1,000,000 population

• City under 250,000 population

FALKLAND ISLANDS (U.K.)
West Falkland
East Falkland

Strait of Magellan

Punta Arenas
Tierra del Fuego

South Georgia (U.K.)

N

| 0 | 200 | 400 | 600 | 800 | 1000 Miles |
| 0 | 300 | 600 | 900 | 1200 | 1500 Kilometers |

Copyright by Rand McNally & Co.
Lambert Azimuthal Equal Area Projection

Drake Passage

South Shetland Islands (U.K.)
South Orkney Islands (U.K.)
South Sandwich Islands (U.K.)

⊛ RAND MCNALLY

A25

ICELAND

Horn

Fontur

Surtsey

ATLANTIC

OCEAN

NORWEGIAN SEA

Arctic Circle

Lofoten Islands

Torneälven

Kebnekaise
6,926 Ft.
2,111m

Lap

Scandinavian

Peninsula

Umeälven

Lule älven

FAROE ISLANDS
(Den.)

NORWAY

SWEDEN

Galdhøpiggen
8,100 Ft.
2,469m

Glåma

Klarälven

Dalälven

Stockholm

Gulf of Bothnia

Hebrides

Orkney
Islands

Grampian
Mts.

UNITED

NORTH

SEA

Skagerrak

DENMARK

Vänern

Vättern

Öland

BALTIC SEA

Cheviot
Hills

KINGDOM

IRELAND

Irish
Sea

Great
Britain

Bornholm
(Den.)

RUSSIA

St. George's Channel

Thames

London

NETHERLANDS

Elbe

Oder

Berlin

Northern

Europ

Land Elevation

Meters		Feet
3,000		9,840
2,000		6,560
500		1,640
200		656
0		0

Water Depth

0		0
200		656
2,000		6,560

N

English Channel

Strait of Dover

BELGIUM

Rhine

GERMANY

POLAND

Wisła

0	100	200	300	400 Miles
0	200	400		600 Kilometers

Copyright by Rand McNally & Co.
Lambert Conformal Conic Projection

Paris

LUX.

CZECH
REPUBLIC

Paris
Basin

Seine

SLOVAKIA

Loire

Black
Forest

Bohemian
Forest

Danube

Bay of Biscay

FRANCE

Saône

Jura

SWITZERLAND

LIECH.

AUSTRIA

HUNGARY

Great Hungarian
Plain

Drava

Cantabrian Mts.

Rhône

Massif
Central

Mt. Blanc
15,771 Ft.
4,808m

A l p s

SLOVENIA

Dordogne

Pyrenees

Rhine

Po

CROATIA

Duero

Iberian Mts.

Ebro

ANDORRA

MONACO

SAN
MARINO

Apennines

BOSNIA AND
HERZEGOVINA

Dinaric Alps

Balkan

Deiro

Corsica
(Fr.)

ADRIATIC SEA

YUGOSLAVIA

Lisbon

PORTUGAL

Iberian

Peninsula

SPAIN

Tejo

Rome

ITALY

ALBANIA

MACE-
DONIA

Sierra Morena

Balearic Islands

Minorca

Sardinia
(It.)

Vesuvius
4,190 Ft.
1,277m

Pindus Mts.

Ibiza

Majorca

Strait of Gibraltar

GIBRALTAR
(U.K.)

TYRRHENIAN
SEA

IONIAN
SEA

Algiers

Mt. Etna
10,902 Ft.
3,323m

Sicily

MEDITERRANE

MOROCCO

AFRICA

ALGERIA

TUNISIA

MALTA

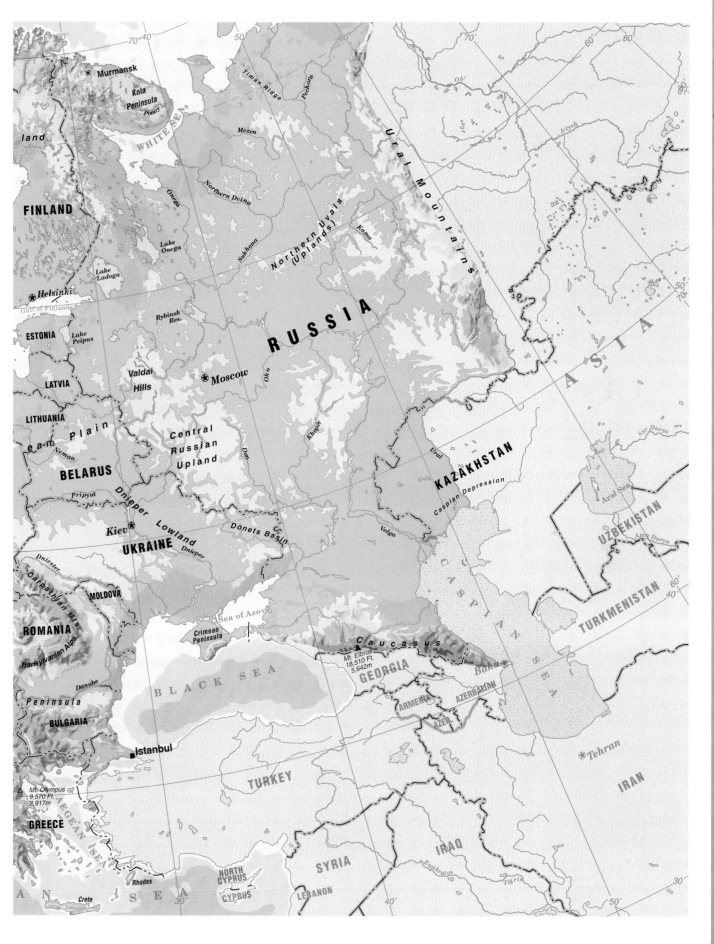

Murmansk

Kola
Peninsula
Ponoy

WHITE SEA

Timan Ridge

Pechora

Ob'

Irtysh

land

FINLAND

Mezen

Northern Dvina

Ural Mountains

ASIA

Onega

Lake
Onega

Northern Uvals
(Uplands)

Kama

Lake
Ladoga

Helsinki

Gulf of Finland

RUSSIA

Rybinsk
Res.

ESTONIA

Lake
Peipus

Sukhona

Valdai
Hills

Moscow

Oka

LATVIA

LITHUANIA

Central
Russian
Upland

Khopёr

Ural

KAZAKHSTAN

Syr Darya

e a n Plain

Neman

Don

Caspian Depression

Aral Sea

BELARUS

Pripyat

Dnieper Lowland

UZBEKISTAN

Dniester

Kiev

Donets Basin

Volga

Amu Darya

UKRAINE

Dnieper

MOLDOVA

TURKMENISTAN

Carpathian Mts.

Sea of Azov

CASPIAN

ROMANIA

Crimean
Peninsula

Caucasus

SEA

Transylvanian Alps

Mt. Elbrus
18,510 Ft.
5,642m

GEORGIA

Baku

AZERBAIJAN

Danube

BLACK SEA

ARMENIA

AZER.

Peninsula

Tehran

BULGARIA

Istanbul

IRAN

Mt. Olympus
9,570 Ft.
2,917m

TURKEY

GREECE

AEGEAN SEA

SYRIA

IRAQ

Euphrates

Rhodes

NORTH
CYPRUS

LEBANON

Tigris

Crete

SEA

CYPRUS

ATLANTIC OCEAN

NORWEGIAN SEA

ICELAND
Reykjavík

Arctic Circle

FAROE ISLANDS
(Den.)

Hammerfest

Trondheim

Umeå

NORWAY **SWEDEN**

Gulf of Bothnia

Bergen

Tampere

Oslo

Vänern   Vättern

Stockholm

Aberdeen

Göteborg

SCOTLAND

Glasgow   ★Edinburgh

BALTIC SEA

UNITED

NORTH

DENMARK

Skagerrak

LITHUANIA

NORTHERN
IRELAND

★Belfast

KINGDOM

SEA

Copenhagen

Kaliningrad ⊛RUSSIA

Dublin⊛

Irish
Sea

Gdańsk

Liverpool ◻ Manchester

Hamburg

Szczecin

POLAND

IRELAND

Cork

WALES

Birmingham

ENGLAND

Amsterdam
NETHERLANDS
The Hague

Berlin⊛

Warsaw⊛

St. George's Channel

Cardiff

Thames

London⊛

Elbe

Oder

Wisła

Plymouth

GERMANY

Dresden ◻ Wrocław

Brussels⊛

Cologne ◻

English Channel

Strait of Dover

BELGIUM Bonn

Frankfurt ◻

Prague⊛

Kraków

Le Havre

LUX.
Luxembourg⊛

CZECH
REPUBLIC

France Paris⊛

Strasbourg

Stuttgart ◻

Nantes

Loire

Danube

SLOVAKIA

Munich ◻

Vienna⊛ ★Bratislava

Seine

Zürich

AUSTRIA

Bay of Biscay

A Coruña

Lyon

Bern⊛ **LIECH.**

SWITZERLAND

Geneva

SLOVENIA Ljubljana⊛

Budapest⊛

HUNGARY

Gijón

Bordeaux

Rhône

⊛Zagreb

Porto

Bilbao

Toulouse

Turin ◻ ◻ Milan

Venice

CROATIA

Belgrade⊛

ANDORRA

Genoa

Po

Valladolid

★Zaragoza

Nice

SAN
MARINO

BOSNIA AND
HERZEGOVINA

⊛Sarajevo

Lisbon

PORTUGAL

Tagus

Madrid⊛

SPAIN

◻Barcelona

Marseille

MONACO

Florence ◻

YUGOSLAVIA

Ebro

Valencia

Corsica
(Fr.)

ITALY

Skopje⊛

Córdoba

Palma

Rome⊛

VATICAN CITY

ALBANIA

MACE-
DONIA

Seville

Sardinia
(It.)

Naples ◻

Bari ◻

Tiranë⊛

Strait of Gibraltar

Málaga

GIBRALTAR
(U.K.)

Cagliari

TYRRHENIAN
SEA

IONIAN
SEA

ADRIATIC SEA

Rabat

Algiers

Palermo ◻

AFRICA

Sicily

Catania

MOROCCO

ALGERIA

Tunis

TUNISIA

MEDITERRANEAN

Valletta⊛ **MALTA**

Legend

⊛ National Capital

★ Secondary Capital
(State, Province, or Territory)

◻ City over 1,000,000 population

⊡ City of 250,000 to 1,000,000 population

· City under 250,000 population

| 0 | 100 | 200 | 300 | 400 Miles |

| 0 | 200 | 400 | 600 Kilometers |

Copyright by Rand McNally & Co.
Lambert Conformal Conic Projection

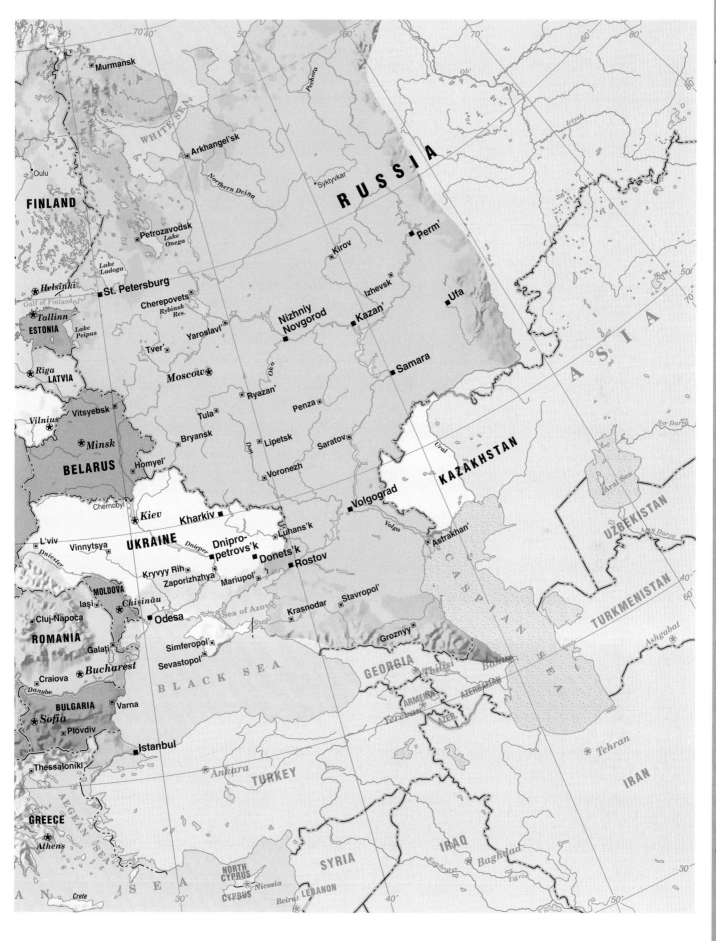

Murmansk

Oulu

FINLAND

WHITE SEA

Arkhangel'sk

Northern Dvina

Pechora

RUSSIA

Syktyvkar

Ob'

Irtysh

Petrozavodsk
Lake
Onega

Kirov

Perm'

Izhevsk

Ufa

A S I A

Helsinki

Lake
Ladoga

St. Petersburg

Cherepovets
Rybinsk
Res.

Nizhniy
Novgorod

Kazan'

Gulf of Finland

Tallinn

ESTONIA

Lake
Peipus

Yaroslavl'

Oka

Tver'

Samara

50°

Rīga

LATVIA

Moscow

Ryazan'

Vitsyebsk

Tula

Penza

Ural

Vilnius

Bryansk

Lipetsk

Saratov

KAZAKHSTAN

Aral Sea

Minsk

Homyel'

Don

Voronezh

BELARUS

Syr Darya

Chernobyl

Kiev

Kharkiv

Volgograd

UKRAINE

Dnieper

Dnipro-
petrovs'k

Luhans'k

Volga

Astrakhan'

UZBEKISTAN

Amu Darya

L'viv

Vinnytsya

Donets'k

Rostov

Dniester

Kryvyy Rih

Zaporizhzhya

Mariupol'

C A S P I A N

TURKMENISTAN

40°

MOLDOVA

Iaşi

Chişinău

Sea of Azov

Krasnodar

Stavropol'

S E A

Ashgabat

Cluj-Napoca

Odesa

ROMANIA

Galaţi

Simferopol'

Groznyy

Bucharest

Sevastopol'

GEORGIA

Tbilisi

Baku

Craiova

BLACK SEA

Danube

ARMENIA

AZERBAIJAN

BULGARIA

Varna

Yerevan

AZER.

Sofia

Plovdiv

Tehran

Istanbul

Thessaloniki

Ankara

TURKEY

IRAN

GREECE

AEGEAN SEA

Athens

SYRIA

IRAQ

Baghdad

NORTH
CYPRUS

Nicosia

Crete

CYPRUS

Beirut

LEBANON

Euphrates

Tigris

30°

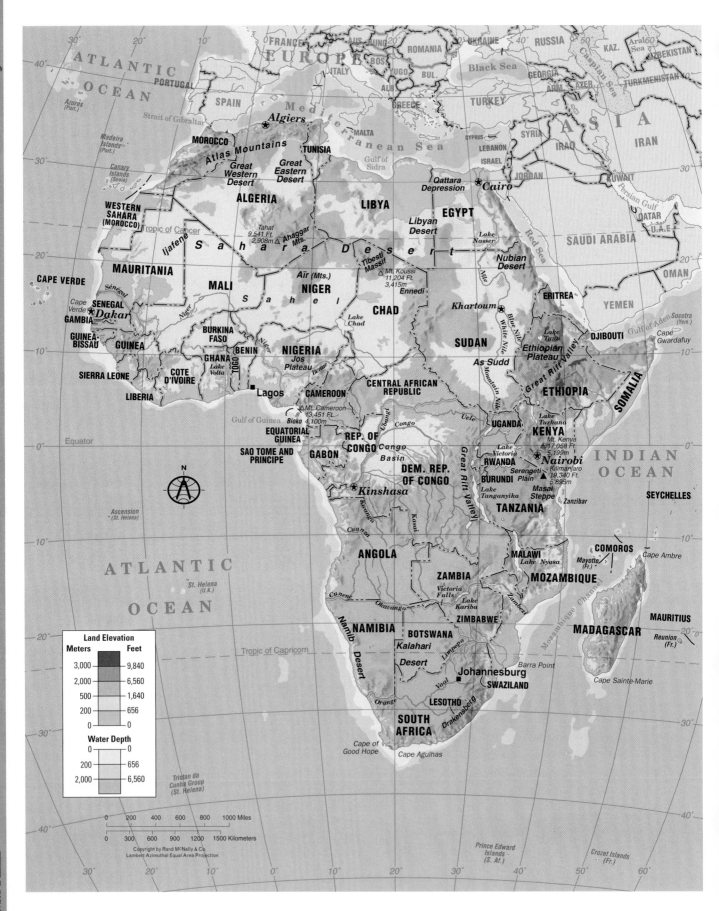

Land Elevation

Meters		Feet
3,000		9,840
2,000		6,560
500		1,640
200		656
0		0

Water Depth

0		0
200		656
2,000		6,560

0 200 400 600 800 1000 Miles

0 300 600 900 1200 1500 Kilometers

Copyright by Rand McNally & Co.
Lambert Azimuthal Equal Area Projection

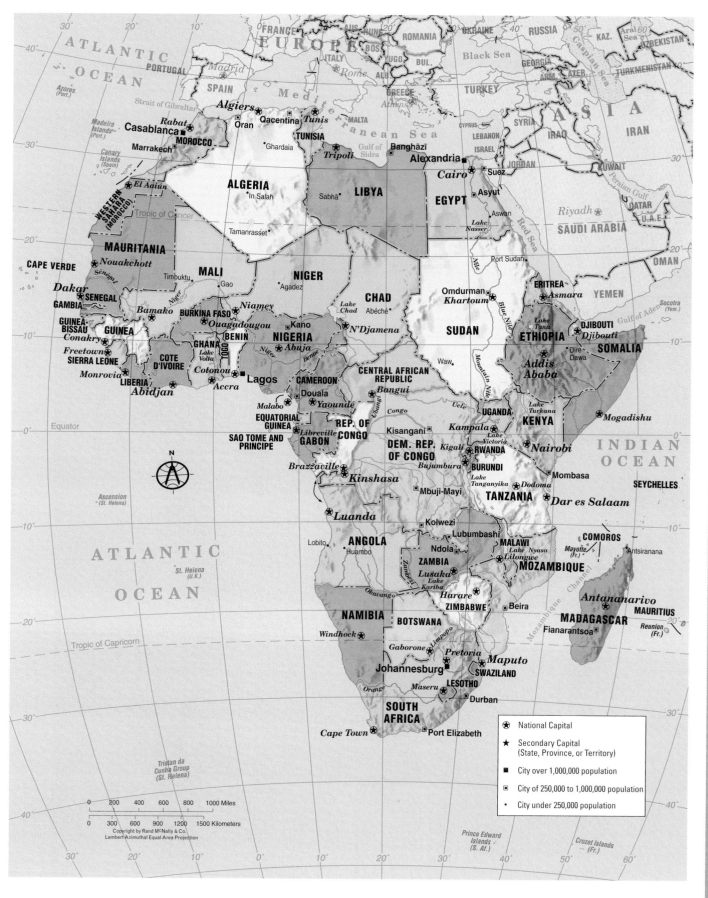

ATLAS

National Capital

★ Secondary Capital
(State, Province, or Territory)

■ City over 1,000,000 population

▫ City of 250,000 to 1,000,000 population

• City under 250,000 population

Copyright by Rand McNally & Co.
Lambert Azimuthal Equal Area Projection

| 0 | 200 | 400 | 600 | 800 | 1000 Miles |

| 0 | 300 | 600 | 900 | 1200 | 1500 Kilometers |

RAND McNALLY

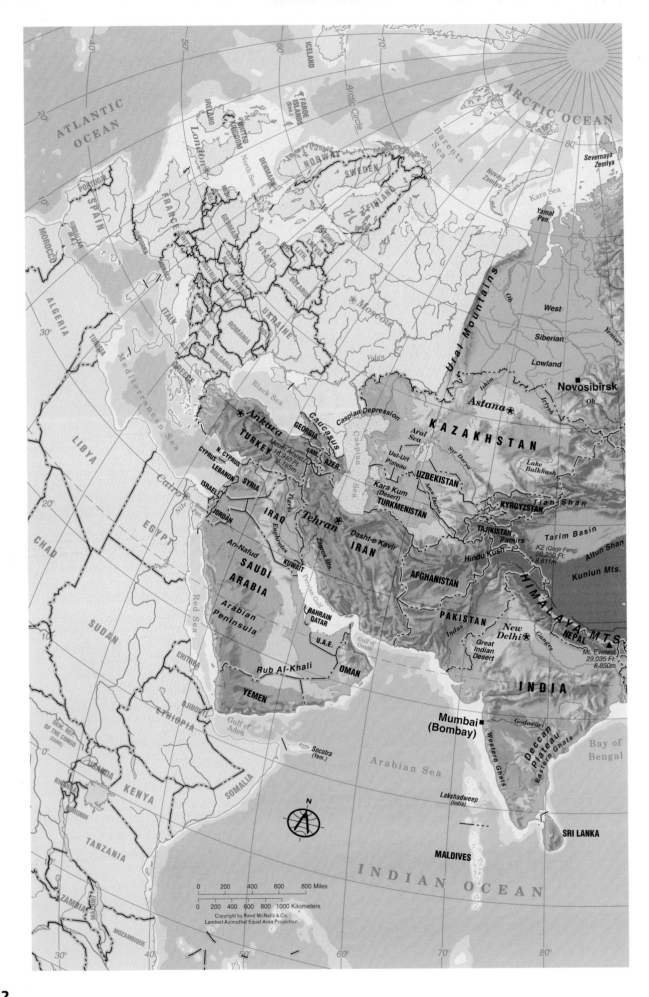

RAND McNALLY

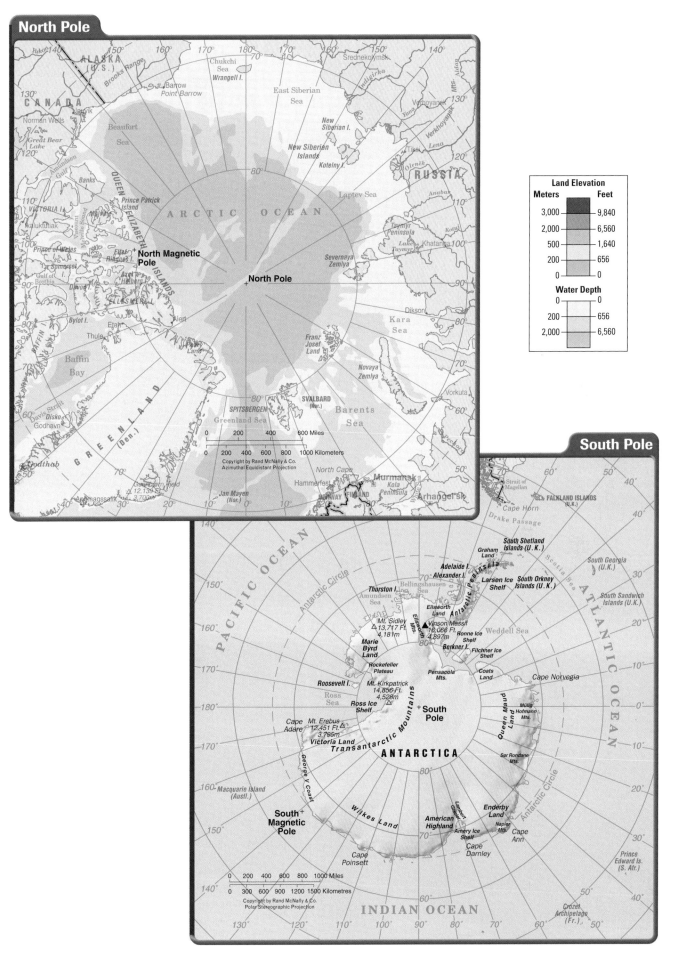

North Pole

ALASKA (U.S.)
CANADA
Chukchi Sea
Wrangell I.
Pt. Barrow
Point Barrow
Brooks Range
Yukon
Norman Wells
Great Bear Lake
Banks
Amundsen Gulf
VICTORIA I.
Kalukutiak
Prince of Wales
M'Clure
Beaufort Sea
QUEEN ELIZABETH ISLANDS
Prince Patrick Island
Melville I.
Axel Heiberg
Ellef Ringnes I.
North Magnetic Pole
North Pole
Viscount
Somerset I.
Gulf of Boothia
Devon I.
ELLESMERE I.
Alert
Bylot I.
BAFFIN
Etah
Thule
Peary Land
Kane Basin
Baffin Bay
Davis Strait
Disko
Godhavn
Godthab
Angmagssalik
GREENLAND (Den.)
Gunnbjorn Field 12,139 ft. 3,700 m.
ARCTIC OCEAN
New Siberian I.
New Siberian Islands
Kotelny I.
East Siberian Sea
Srednekolymsk
Indigirka
Verkhoyansk
Yana
Kolyma
Verkhoyansk Mts.
Aldan
RUSSIA
Lena
Tiksi
Olenek
Laptev Sea
Anabar
Taymyr Peninsula
Lake Taymyr
Khatanga
Kotui
Severnaya Zemlya
Franz Josef Land
Dikson
Kara Sea
Novaya Zemlya
Vorkuta
Pechora
SVALBARD (Nor.)
SPITSBERGEN
Greenland Sea
Jan Mayen (Nor.)
North Cape
Hammerfest
Murmansk
Kola Peninsula
Arhangel'sk
NORWAY
FINLAND
Barents Sea

0   200   400   600 Miles
0   200   400   600   800   1000 Kilometers
Copyright by Rand McNally & Co.
Azimuthal Equidistant Projection

Land Elevation

Meters		Feet
3,000		9,840
2,000		6,560
500		1,640
200		656
0		0

Water Depth

0		0
200		656
2,000		6,560

South Pole

Strait of Magellan
Cape Horn
Drake Passage
FALKLAND ISLANDS (U.K.)
South Georgia (U.K.)
South Shetland Islands (U.K.)
Graham Land
Adelaide I.
Alexander I.
Antarctic Peninsula
Larsen Ice Shelf
South Orkney Islands (U.K.)
South Sandwich Islands (U.K.)
Scotia Sea
PACIFIC OCEAN
ATLANTIC OCEAN
Antarctic Circle
Thurston I.
Bellingshausen Sea
Amundsen Sea
Ellsworth Land
Ellsworth Mts.
Vinson Massif 16,066 Ft. 4,897m
Ronne Ice Shelf
Weddell Sea
Mt. Sidley 13,717 Ft. 4,181m
Marie Byrd Land
Berkner I.
Filchner Ice Shelf
Rockefeller Plateau
Pensacola Mts.
Coats Land
Roosevelt I.
Mt. Kirkpatrick 14,856 Ft. 4,528m
Ross Sea
Ross Ice Shelf
Transantarctic Mountains
South Pole
ANTARCTICA
Queen Maud Land
Cape Norvegia
Mühlig Hofmann Mts.
Ser Rondane Mts.
Cape Adare
Mt. Erebus 12,451 Ft. 3,795m
Victoria Land
George V Coast
Macquarie Island (Austl.)
South Magnetic Pole
Wilkes Land
American Highland
Lambert Glacier
Amery Ice Shelf
Enderby Land
Napier Mts.
Cape Ann
Cape Darnley
Cape Poinsett
Prince Edward Is. (S. Afr.)
Crozet Archipelago (Fr.)
INDIAN OCEAN

0   200   400   600   800   1000 Miles
0   300   600   900   1200   1500 Kilometres
Copyright by Rand McNally & Co.
Polar Stereographic Projection

RAND McNALLY

Unit 1

The Basics of Geography

The earth is a unique planet capable of supporting a wide variety of life forms. Human beings adapt and alter the environments on earth.

PHYSICAL GEOGRAPHY Internal and external forces constantly change the earth's surface. Here the volcano Arenal, located in Costa Rica, spews molten rock that will cool and alter the land.

2

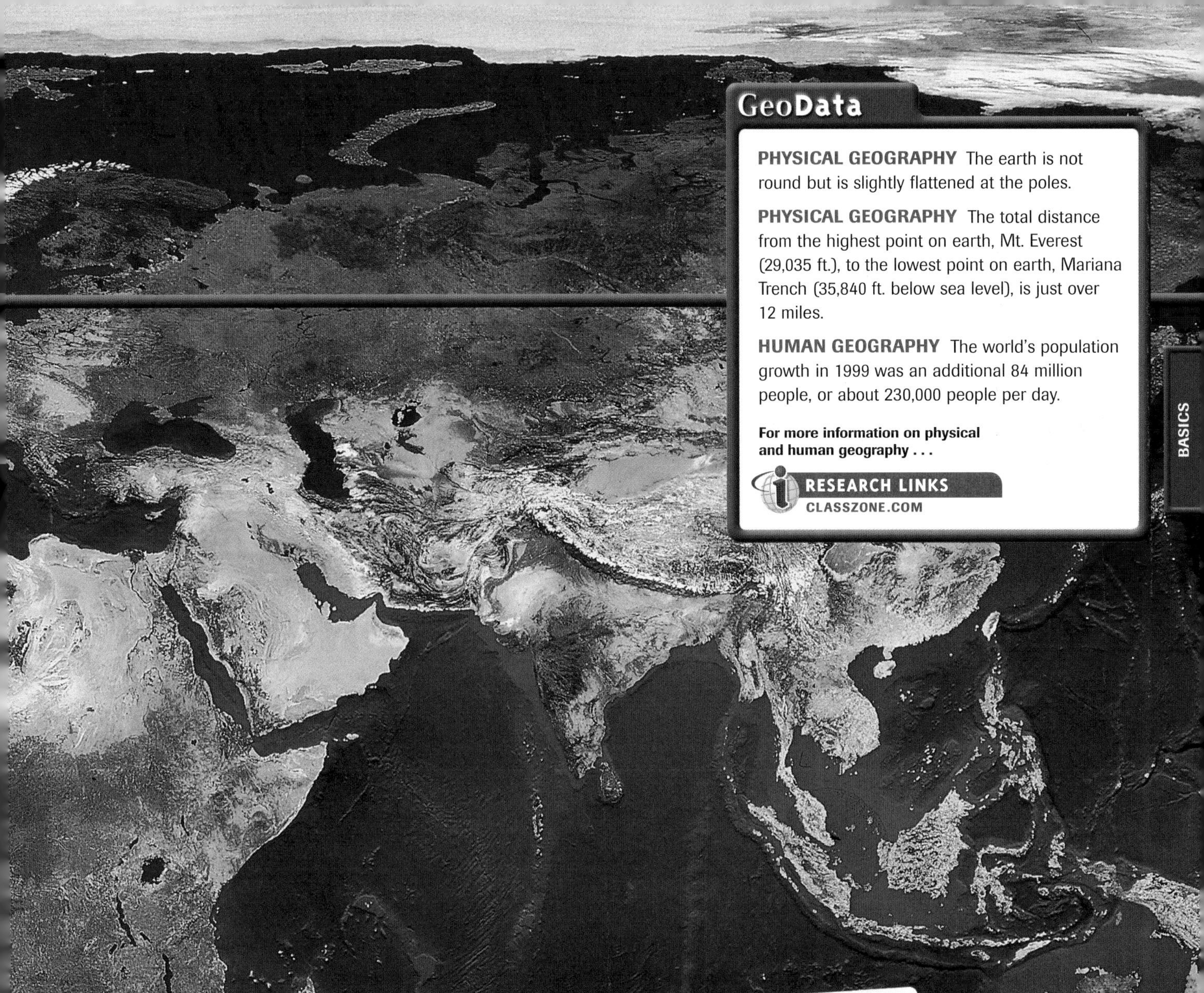

PHYSICAL GEOGRAPHY The earth is not round but is slightly flattened at the poles.

PHYSICAL GEOGRAPHY The total distance from the highest point on earth, Mt. Everest (29,035 ft.), to the lowest point on earth, Mariana Trench (35,840 ft. below sea level), is just over 12 miles.

HUMAN GEOGRAPHY The world's population growth in 1999 was an additional 84 million people, or about 230,000 people per day.

For more information on physical and human geography . . .

RESEARCH LINKS CLASSZONE.COM

BASICS

HUMAN GEOGRAPHY Pictured here is Adnan Nevic, the child officially identified by the United Nations as the six billionth person on earth. He was born on October 12, 1999, in Sarajevo, Bosnia.

The Basics of Geography **3**

PHYSICAL GEOGRAPHY
Looking at the Earth

SECTION 1
The Five Themes of Geography

SECTION 2
The Geographer's Tools

GEOGRAPHY SKILLS HANDBOOK

Seneca Falls, New York, is represented in a road map, a heat sensing (thermal) scan, and a satellite image.

Heat Sensing Scan

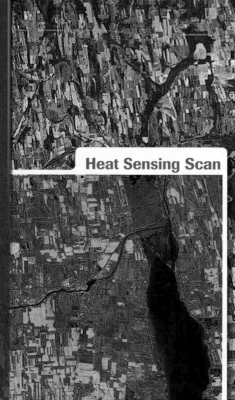

Road Map

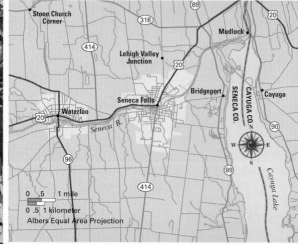

Satellite Image

GeoFocus

How do geographers view the world?

Taking Notes Copy the graphic organizer below into your notebook. Use it to record information about the work of geographers and the themes of geography.

5 Themes:

Tools:

TYPES OF MAPS The three types of maps are general reference maps, thematic maps, and navigational maps. One kind of general reference map is called a **topographic map,** which is a representation of natural and man-made features on the earth. Thematic maps emphasize specific kinds of information, such as climate or population density. Sailors and pilots use the third type of map—navigation maps. You can learn more about using different maps in the Geography Skills Handbook, pages 20–23.

The Science of Mapmaking

A cartographer decides what type of map to create by considering how the map will be used. Keeping that purpose in mind, he or she then determines how much detail to show and what size the map should be.

SURVEYING The first step in making a map is to complete a field survey. Surveyors observe, measure, and record what they see in a specific area. Today, most mapping is done by remote sensing, the gathering of geographic information from a distance by an instrument that is not physically in contact with the mapping site. These data are gathered primarily by aerial photography or by satellites.

The data gathered includes information such as elevation, differences in land cover, and variations in temperature. This information is recorded and converted to a gray image. Cartographers then use these data and computer software to construct maps. See the illustration below to learn more about satellite surveying.

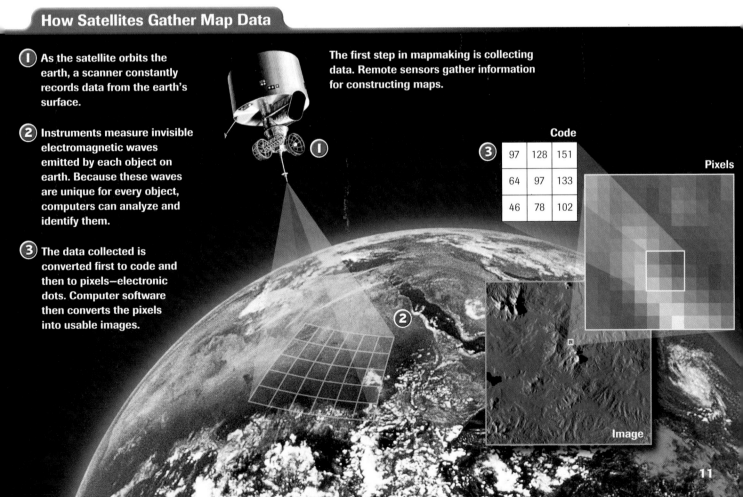

How Satellites Gather Map Data

1. As the satellite orbits the earth, a scanner constantly records data from the earth's surface.

2. Instruments measure invisible electromagnetic waves emitted by each object on earth. Because these waves are unique for every object, computers can analyze and identify them.

3. The data collected is converted first to code and then to pixels—electronic dots. Computer software then converts the pixels into usable images.

The first step in mapmaking is collecting data. Remote sensors gather information for constructing maps.

Code

97	128	151
64	97	133
46	78	102

Pixels

Image

11

SATELLITES Today, geographers rely heavily on satellites to provide geographic data. Two of the best-known satellites are Landsat and GOES. **Landsat** is actually a series of satellites that orbit more than 100 miles above Earth. Each time a satellite makes an orbit, it picks up data in an area 115 miles wide. Landsat can scan the entire Earth in 16 days.

Geostationary Operational Environment Satellite (GOES) is a weather satellite. This satellite flies in orbit in sync with Earth's rotation. By doing so, it always views the same area. It gathers images of atmospheric conditions that are useful in forecasting the weather.

Geographic Information Systems

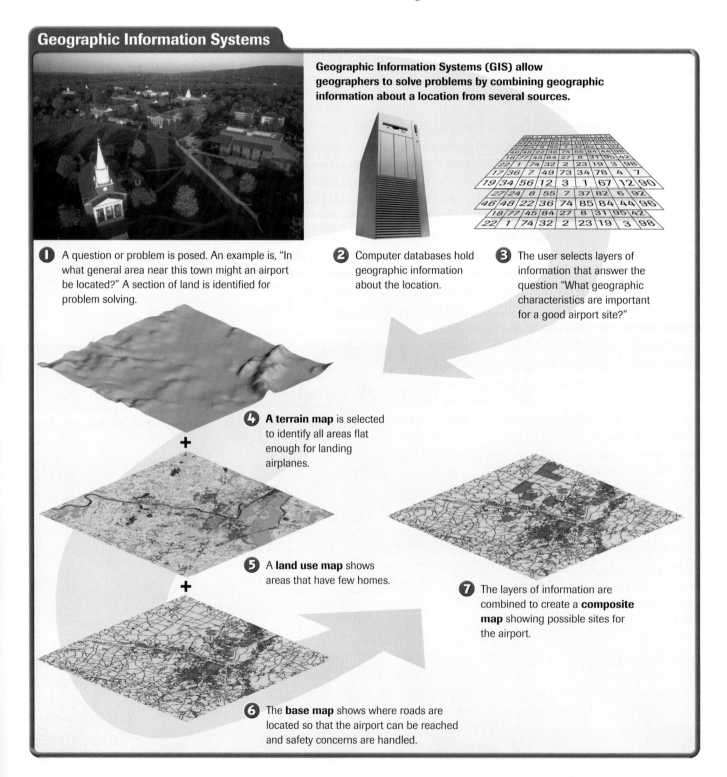

Geographic Information Systems (GIS) allow geographers to solve problems by combining geographic information about a location from several sources.

1 A question or problem is posed. An example is, "In what general area near this town might an airport be located?" A section of land is identified for problem solving.

2 Computer databases hold geographic information about the location.

3 The user selects layers of information that answer the question "What geographic characteristics are important for a good airport site?"

4 **A terrain map** is selected to identify all areas flat enough for landing airplanes.

5 A **land use map** shows areas that have few homes.

6 The **base map** shows where roads are located so that the airport can be reached and safety concerns are handled.

7 The layers of information are combined to create a **composite map** showing possible sites for the airport.

GEOGRAPHIC INFORMATION SYSTEMS The newest tool in the geographer's toolbox is **Geographic Information Systems (GIS).** GIS stores information about the world in a digital database. GIS has the ability to combine information from a variety of sources and display it in ways that allow the user to visualize the use of space in different ways.

When using the system, geographers must look at a problem and decide what types of geographic information would help them solve the problem. The information could include maps, aerial photographs, satellite images, or other data. Next, they select the appropriate layers of information. Then, GIS creates a composite map combining the information. Study the diagram on page 12 to learn more about the way GIS works.

GLOBAL POSITIONING SYSTEM (GPS) A familiar tool of geographers is GPS or Global Positioning System. It was originally developed to help military forces know exactly where they were on the earth's surface. The system uses a series of 24 satellites called Navstars, which beam information to the earth. The exact position—latitude, longitude, altitude, and time—is displayed on a hand–held receiver. Hikers, explorers, sailors, and drivers use GPS devices to determine location. They are also used to track animals.

Geographers use a variety of other tools including photographs, cross sections, models, cartograms, and population pyramids. These tools help geographers to visualize and display information for analysis. They are looking for patterns and connections in the data they find. You will learn how to use these tools in the Geography Skills Handbook, which follows, and in the Map and Graph Skills pages in this book.

MOVEMENT Scientists use a GPS device to track this bear in Minnesota.
What other uses could be found for a GPS device?

🌐 **Geographic Thinking**

Making Comparisons
A How might the military use both GOES and GPS?

SECTION 2 ▶ Assessment

1 Places & Terms

Explain the meaning of each of the following terms.

- globe
- map
- cartographer
- map projection
- topographic map
- GIS

2 Taking Notes

REGION Review the notes you took for this section.

Tools:

- How would a globe show a region differently than a map?
- How does GIS aid in understanding a region?

3 Main Ideas

a. What are the three basic types of maps?

b. What are some geographers' tools in addition to maps and globes?

c. How does a cartographer decide which type of map is needed?

4 Geographic Thinking

Making Generalizations
How does modern technology help geographers? **Think about:**

- digital information
- satellite images

S See Skillbuilder Handbook, page R6.

MAKING COMPARISONS Choose a place on the earth and in an atlas, and find three maps that show the place in three different ways. Create a **chart** that lists the similarities and differences in the way the place is shown on the three maps.

This handbook covers the basic map skills and information that geographers rely on as they investigate the world—and the skills you will need as you study geography.

Finding Location

Mapmaking depends on surveying the earth's surface. Until recently, that activity could only happen on land or sea. Today, aerial photography and satellite imaging are the most popular ways to gather data.

A personal **GPS** device provides the absolute location to the user.

Magnetic compasses introduced by the Chinese around the 1100s helped to accurately determine direction.

Nigerian surveyors use a **theodolite,** a type of surveying instrument. It precisely measures angles and distances on the earth.

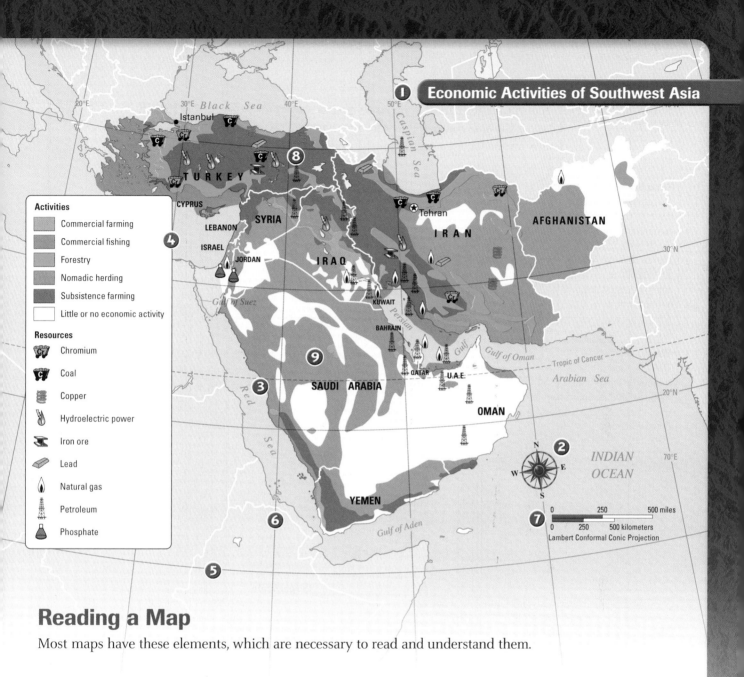

Activities

Commercial farming

Commercial fishing

Forestry

Nomadic herding

Subsistence farming

Little or no economic activity

Resources

Chromium

Coal

Copper

Hydroelectric power

Iron ore

Lead

Natural gas

Petroleum

Phosphate

Reading a Map

Most maps have these elements, which are necessary to read and understand them.

① TITLE The title explains the subject of the map and gives you an idea of what information the map conveys.

② COMPASS ROSE The compass rose shows you the north (N), south (S), east (E), and west (W) directions on the map. Sometimes only north is indicated.

③ LABELS Labels are words or phrases that explain features on the map.

④ LEGEND A legend or key lists and explains the symbols and use of color on the map.

⑤ LINES OF LATITUDE These are imaginary lines that measure distance north or south of the equator.

⑥ LINES OF LONGITUDE These are imaginary lines that measure distance east or west of the prime meridian.

⑦ SCALE A scale shows the ratio between a unit of length on the map and a unit of distance on the earth.

⑧ SYMBOLS Symbols represent such items as capital cities, economic activities, or natural resources. Check the map legend for more details.

⑨ COLORS Colors represent a variety of information on a map. The map legend indicates what the colors mean.

Scale

A geographer decides what scale to use by determining how much detail to show. If many details are needed, a large scale is used. If fewer details are needed, a small scale is used.

Ratio Scale

This shows the ratio of distance on the map compared to real earth measurement. Here, 1 inch on the map equals 30,000,000 inches (500 miles) in actual distance on the earth.

Bar Scale

This bar shows the ratio of distance on the map to distance on the earth. Here, 1 inch equals 500 miles.

EASTERN UNITED STATES
Scale: 1:30,000,000
1"= 500 miles

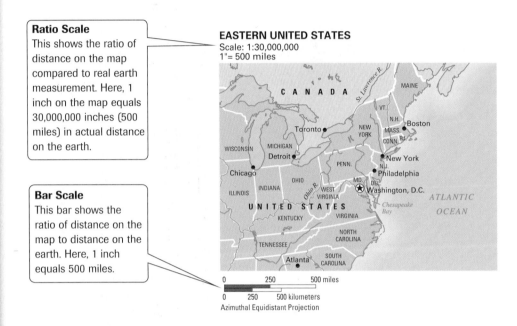

Azimuthal Equidistant Projection

Small Scale

A small scale map shows a large area but without much detail. A small scale is used to see relative location in a region or between regions.

WASHINGTON, D.C., METRO AREA
Scale: 1:3,000,000
1"= 50 miles

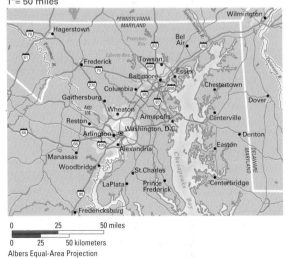

Albers Equal-Area Projection

Large Scale

A large scale map shows a small area with much more detail. A large scale is used to see relative location within a region.

WASHINGTON, D.C.
Scale: 1:62,500
1"= 1 mile

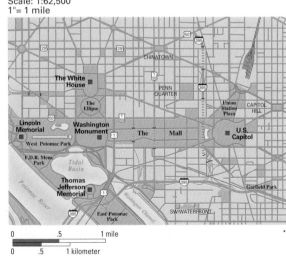

Albers Equal-Area Projection

Using the Geographic Grid

As you learned in Chapter 1, geographers use a grid system to identify absolute location. The grid system uses two kinds of imaginary lines:

- latitude lines, also called parallels because they run parallel to the equator
- longitude lines, also called meridians because, like the prime meridian, they run from pole to pole

Latitude
There are 90° in North latitude and 90° in South latitude.

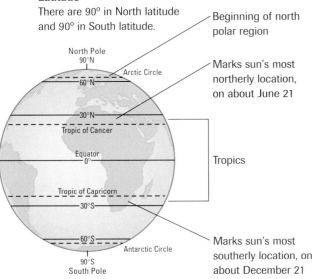

Beginning of north polar region

Marks sun's most northerly location, on about June 21

Tropics

Marks sun's most southerly location, on about December 21

Longitude
There are 180° in West longitude and 180° in East longitude. Lines also mark the hours of the day as the earth rotates. Every 15° east or west is equal to one hour.

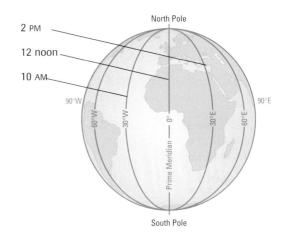

Global Grid
Absolute location can be determined by noting where latitude and longitude lines cross. For more precision, each degree is divided into 60 minutes.

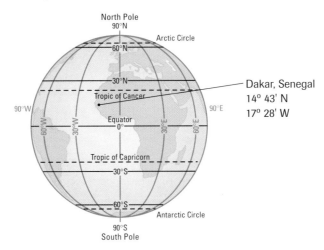

Dakar, Senegal
14° 43' N
17° 28' W

Projections

A projection is a way of showing the curved surface of the earth on a flat map. Because the earth is a sphere, a flat map will distort some aspect of the earth's surface. Distance, shape, direction, or area may be distorted by a projection. Be sure to check the projection of a map so you are aware of how the areas are distorted.

PLANAR PROJECTIONS

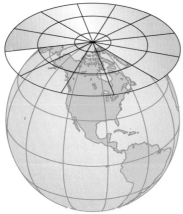

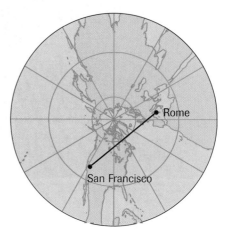

A planar projection is a projection on a flat surface. This projection is also called an azimuthal projection. It distorts size and shape. To the right is a type of planar projection.

The **azimuthal** projection shows the earth so that a line from the central point to any other point on the map gives the shortest distance between the two points. Size and shape are distorted.

CONICAL PROJECTIONS

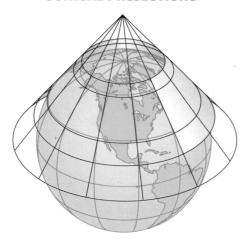

A conical projection is a projection onto a cone. This projection shows shape fairly accurately, but it distorts landmasses at the edges of the map.

Conical projections are often used to show landmasses that extend over large areas going east and west.

COMPROMISE PROJECTIONS

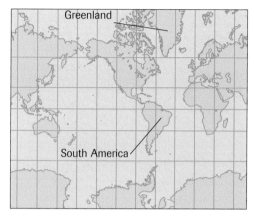

A compromise projection is a projection onto a cylinder. This projection shows the entire earth on one map. Included here are three types of compromise projections.

In the compromise projection called **Mercator,** the shapes of the continents are distorted at the poles and somewhat compressed near the equator. For example, the island of Greenland is actually one-eighth the size of South America.

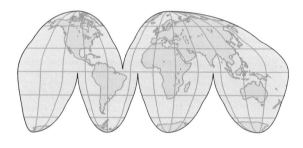

The compromise projection called **homolosine** is sometimes called an "interrupted map," because the oceans are divided. This projection shows the accurate shapes and sizes of the landmasses, but distances on the map are not correct.

A **Robinson** projection is a type of compromise projection, commonly used in textbooks. It shows the entire earth with nearly the true sizes and shapes of the continents and oceans. However, the shapes of the landforms near the poles appear flat.

 Map Practice

Use pages 14–19 to help you answer these questions. Look at the map on page 15 to answer questions 1–3.

1. How are colors used on this map?

2. Is the map a large-scale or a small-scale map? How do you know?

3. What is the approximate longitude of Buenos Aires?

4. What are the names of three lines of latitude besides the equator?

5. Which projections show shape of landmasses most accurately?

MAKING COMPARISONS Look at the maps in the atlas in this book. Create a **database** that shows the projection and scale of each map. Write a summary of your findings.

Using Different Types of Maps

PHYSICAL MAPS Physical maps help you see the types of landforms and bodies of water found in a specific area. By studying the map, you can begin to understand the relative location and characteristics of a place or region.

On a physical map, color, shading, or contour lines are used to indicate elevation or altitude, also called relief.

Ask these questions about the physical features shown on a map:

- Where on the earth's surface is this area located?
- What is its relative location?
- What is the shape of the region?
- In which direction do the rivers flow? How might the direction of flow affect travel and transportation in the region?
- Are there mountains or deserts? How do they affect the people living in the area?

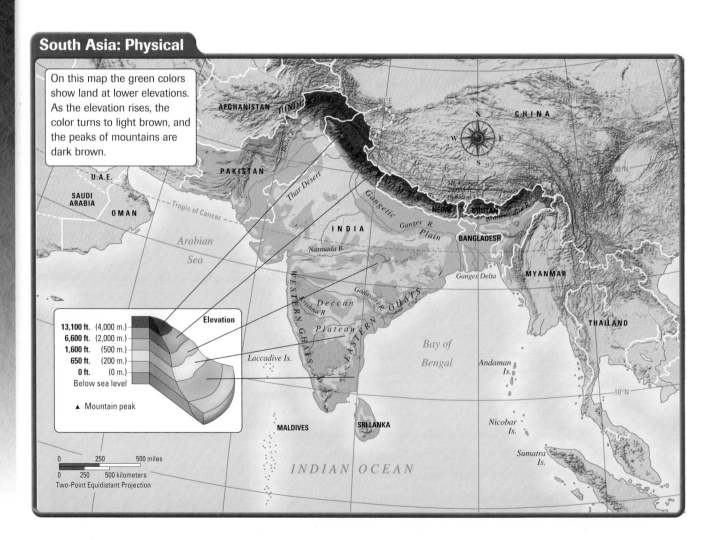

South Asia: Physical

On this map the green colors show land at lower elevations. As the elevation rises, the color turns to light brown, and the peaks of mountains are dark brown.

Elevation

13,100 ft. (4,000 m.)
6,600 ft. (2,000 m.)
1,600 ft. (500 m.)
650 ft. (200 m.)
0 ft. (0 m.)
Below sea level

▲ Mountain peak

0   250   500 miles
0   250   500 kilometers
Two-Point Equidistant Projection

POLITICAL MAPS Political maps show features on the earth's surface that humans created. Included on a political map may be cities, states, provinces, territories, or countries.

Ask these questions about the political features shown on a map:

- Where on the earth's surface is this area located?
- What is its relative location? How might the location affect the economy or foreign policy of a place?
- What is the shape and size of the country? How might shape or size affect the people living in the country?
- Who are the neighbors in the region, country, state, or city?
- How populated does the area seem to be? How might that affect activities there?

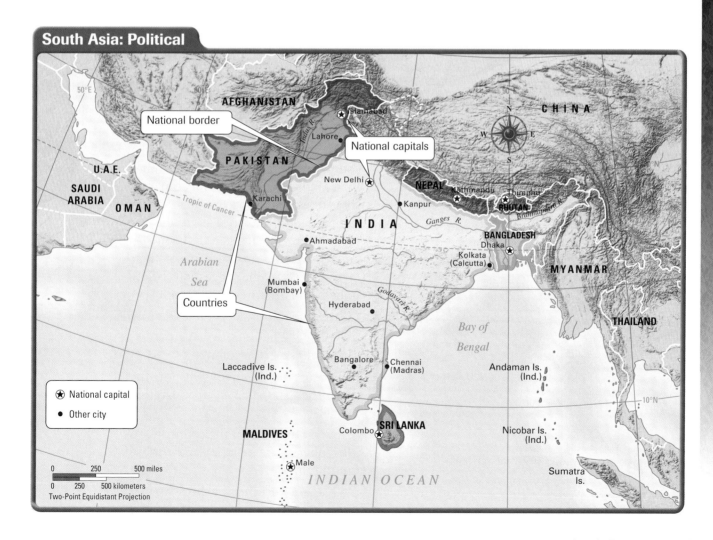

South Asia: Political

Thematic Maps

Geographers also rely on thematic maps, which focus on specific types of information. For example, in this textbook you will see thematic maps that show climate, vegetation, natural resources, population density, and economic activities. Some thematic maps illustrate historical trends, and others may focus on the movement of people or ideas. These maps may be presented in a variety of ways.

Cultural Legacy of the Roman Empire

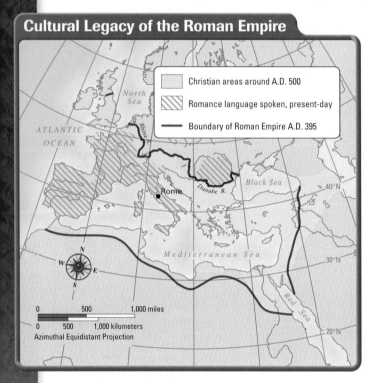

Christian areas around A.D. 500

Romance language spoken, present-day

Boundary of Roman Empire A.D. 395

QUALITATIVE MAPS Qualitative maps use colors, symbols, dots, or lines to help you see patterns related to a specific idea. The map shown to the left shows the influence of the Roman Empire on Europe, North Africa, and Southwest Asia. Use the suggestions below to help you interpret a map.

- Check the title to identify the theme and data being presented.
- Study the legend to understand the theme and the information presented.
- Look at physical or political features of the area. How might the theme of the map affect them?
- What are the relationships among the data?

CARTOGRAMS In a cartogram, geographers present information about a country based on a set of data other than land area. The size of each country is drawn in proportion to that data rather than to its land size. On the cartogram shown to the left, the countries are represented on the basis of their oil reserves. Use the suggestions below to help you interpret a cartogram.

- Check the title and legend to identify the data being presented.
- What do sizes represent?
- Look at the relative sizes of the countries shown. Which is largest? smallest?
- How do the sizes of the countries on the physical map differ from those in the cartogram?
- What are the relationships among the data?

Estimated World Oil Reserves

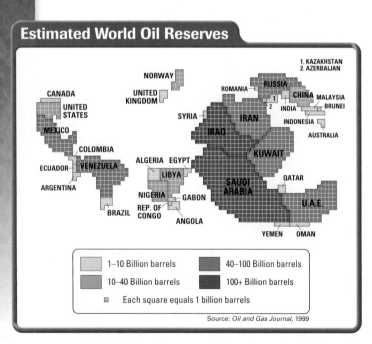

1–10 Billion barrels
10–40 Billion barrels
40–100 Billion barrels
100+ Billion barrels

Each square equals 1 billion barrels

Source: *Oil and Gas Journal*, 1999

FLOW-LINE MAPS Flow-line maps illustrate movement of people, goods, ideas, animals, or even glaciers. The information is usually shown in a series of arrows. Location, direction, and scope of movement can be seen. The width of the arrow may show how extensive the flow is. Often the information is given over a period of time. The map shown to the right portrays the movement of the Bantu peoples in Africa. Use the suggestions below to help you interpret a flow-line map.

- Check the title and legend to identify the data being presented.
- Over what period of time did the movement occur?
- In what direction did the movement occur?
- How extensive was the movement?

Remember that the purpose of a map is to show a location and provide additional information. Be sure to look at the type of map, scale, and projection. Knowing how maps present the information will help you interpret the map and the ideas it presents.

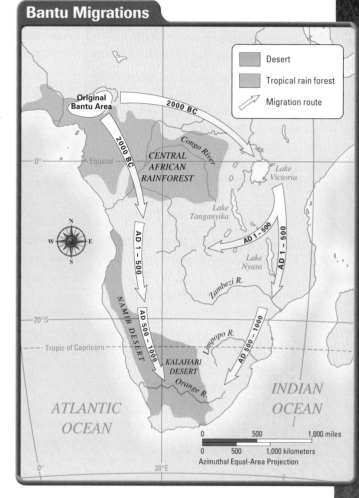

Bantu Migrations

Desert
Tropical rain forest
Migration route

Original Bantu Area
2000 BC
Congo River
CENTRAL AFRICAN RAINFOREST
Equator
Lake Victoria
Lake Tanganyika
AD 1–500
AD 1–500
Lake Nyasa
Zambezi R.
NAMIB DESERT
AD 500–1000
Tropic of Capricorn
Limpopo R.
AD 500–1000
KALAHARI DESERT
Orange R.
INDIAN OCEAN
ATLANTIC OCEAN

0   500   1,000 miles
0   500   1,000 kilometers
Azimuthal Equal-Area Projection

20°E

Map Practice

Use pages 20–23 to help you answer these questions. Use the maps on pages 20–21 to answer questions 1–3.

1. In what direction does the Ganges River flow?

2. China is the northern neighbor of which countries?

3. Which city is closer to the Thar Desert—Lahore, Pakistan or New Delhi, India?

4. Why are so few nations shown on the cartogram?

5. Which of the thematic maps would best show the location of climate zones?

GeoActivity

EXPLORING LOCAL GEOGRAPHY Obtain a physical–political map of your state. Use the data on it to create two separate **maps.** One should show physical features only, and one should show political features only.

VISUAL SUMMARY
LOOKING AT THE EARTH

The Five Themes of Geography

Location
- **Absolute Location** uses latitude and longitude.
- **Relative Location** uses relationships to other places.

Place This explains the characteristics of an area.

Region This looks at a larger area with similar characteristics.

Movement People, plants, animals, and ideas move through time and across space.

Human–Environment Interaction Humans interact with the environment to adjust to it or to alter it.

The Geographer's Tools

Globe A three-dimensional representation of the earth

Map A two-dimensional representation of the earth

Mapmaking
- Area is surveyed.
- High-tech tools, including satellites, are used to gather data and create maps.

Geography Skills Handbook

Map Elements Maps have elements such as a legend to aid in interpreting them.

Scale This determines how much detail is shown on a map.

Grid Gridlines help to determine absolute location.

Projection This shows the earth's surface in two dimensions but distorts either size, shape, direction, or area.

Types of Maps These include physical, political, and thematic, such as qualitative, cartographic, or flow-line.

Reviewing Places & Terms

A. Briefly explain the importance of each of the following.

1. geography
2. hemisphere
3. equator
4. prime meridian
5. latitude
6. longitude
7. globe
8. map
9. cartographer
10. map projection

B. Answer the questions about vocabulary in complete sentences.

11. Which of the above terms indicate imaginary parallel lines that circle the earth?
12. Which term marks the beginning of longitude?
13. Which of the above terms has 180° in each hemisphere?
14. How may hemispheres be divided?
15. What imaginary line separates the Northern Hemisphere from the Southern Hemisphere?
16. Which term is also known as a meridian line?
17. Would a cartographer work on a map or a globe?
18. Why are map projections needed?
19. Which of the above terms are associated with the geographic grid?
20. Which term characterizes the study of the use of land space?

Main Ideas

The Five Themes of Geography (pp. 5–9)

1. How is absolute location different from relative location?
2. What are some examples of information that would be included in a place description?
3. How is place different from region?
4. Why do geographers study movement?

The Geographer's Tools (pp. 10–13)

5. What is the purpose of a map?
6. How do satellites aid in mapmaking?
7. Why is GIS a valuable tool for examining the geography of a place?

Geography Skills Handbook (pp. 14–23)

8. How is the use of small-scale maps different from the use of large-scale maps?
9. In what ways may relief be shown on a map?
10. What are three types of thematic maps?

Critical Thinking

1. Using Your Notes

Use your completed chart to answer these questions.

5 Themes:
Tools:

a. How are relative location and place related?

b. How do thematic maps help geographers understand the five themes?

2. Geographic Themes

a. **REGION** Write a sentence describing a region that your community is a part of. Be sure to identify the region and give reasons for your answer.

b. **MOVEMENT** How are linear and time distances related to the theme of movement?

3. Identifying Themes

Into which two hemispheres would an island at 50°S, 60°W be placed? Which of the five themes are reflected in your answer?

4. Drawing Conclusions

Why was it necessary for geographers to develop a grid system?

5. Seeing Patterns

Into which formal region, functional region, and perceptual region might your community be placed?

Additional Test Practice, pp. S1–S37

TEST PRACTICE
CLASSZONE.COM

Geographic Skills: Interpreting Maps

Continents of the World

Use the map to answer the following questions.

1. **LOCATION** What is the absolute location of the continent of Australia?

2. **LOCATION** What is the relative location of South America?

3. **PLACE** What body of water is located at 45° N, 45° W?

GeoActivity

With a partner, choose and record the latitude and longitude of five locations on the map at left. Then trade your list with another set of partners. Have them search for the coordinates on your list, and do the same with their list. Then check the accuracy of the findings.

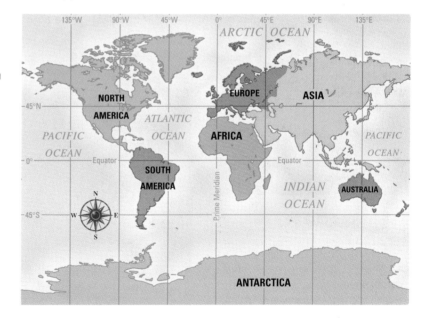

INTERNET ACTIVITY

Use the links at **classzone.com** to do research about GIS. Take notes on the ways GIS can be used to provide geographic information for mapmaking, site selection, and simulating environmental effects.

Creating a Multimedia Presentation Using the information you gathered about GIS, create a multimedia presentation explaining the various aspects of GIS and how it helps geographers and others solve problems.

PHYSICAL GEOGRAPHY
A Living Planet

Third planet from the
Sun: Earth appears as
a blue and white ball in
the darkness of space.

GeoFocus

What forces shape the earth?

Taking Notes Copy the graphic organizer
below into your notebook. Use it to record
information from the chapter about the
structure of the earth.

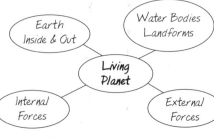

The Earth Inside and Out

Main Ideas
- The earth is the only habitable planet in the sun's solar system.
- The drifting of the continents shaped the world we live in today.

Places & Terms

continent	atmosphere
solar system	lithosphere
core	hydrosphere
mantle	biosphere
magma	continental drift
crust	

BASICS

A HUMAN PERSPECTIVE A quick look at a world map will convince you that the **continents,** landmasses above water on Earth, fit together like a huge jigsaw puzzle. South America and Africa are good examples. With imagination, you can see how other continents might fit together as well. The first person to suggest that the seven continents were once all one supercontinent was Englishman Francis Bacon in 1620. Bacon's idea received support in the early 1900s, when scientists found rocks in Africa that matched rocks in South America. Other evidence also supported the idea of a supercontinent millions of years ago.

The Solar System

The "home address" of the earth is the third planet in the solar system of the sun, which is a medium-sized star on the edge of the Milky Way galaxy. Its distance from the sun is 93 million miles. The **solar system** consists of the sun and nine known planets, as well as other celestial bodies that orbit the sun. The solar system also contains comets, spheres covered with ice and dust that leave trails of vapor as they race through space. Asteroids—large chunks of rocky material—are found in space as well. As you can see in the diagram, our solar system has an asteroid belt between the orbits of Jupiter and Mars.

LOCATION This not-to-scale illustration shows the nine planets and other objects in our solar system.
What is the earth's relative location in the solar system?

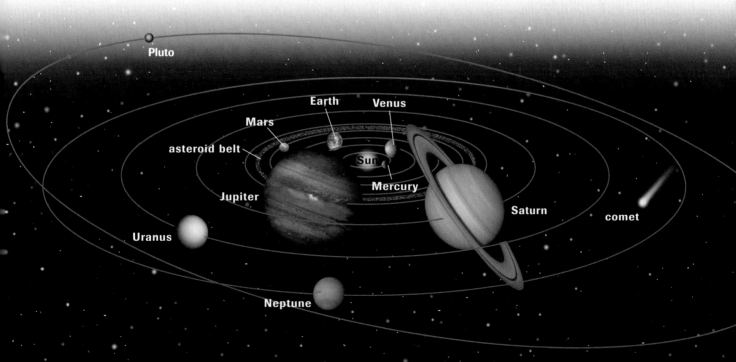

The Structure of the Earth

The earth is about 24,900 miles in circumference and about 7,900 miles in diameter. Although the earth seems like a solid ball, it is really more like a series of shells that surround one another.

INSIDE THE EARTH The **core** is the center of the earth and is made up of iron and nickel. The outer core is liquid, but the inner core is solid. Surrounding the core is the **mantle,** which has several layers. The mantle contains most of the earth's mass. **Magma,** which is molten rock, can form in the mantle and rise through the **crust,** the thin layer of rock at the earth's surface. Study the diagram below to learn more about the earth's interior.

ON AND ABOVE THE EARTH Surrounding the earth is a layer of gases called the **atmosphere.** It contains the oxygen we breathe, protects the earth from radiation and space debris, and provides the medium for weather and climate. The solid rock portion of the earth's surface is the **lithosphere,** which includes the crust and uppermost mantle. Under the ocean, the lithosphere forms the seafloor. The huge landmasses above water are called continents. There are seven continents: North America, South America, Europe, Asia, Africa, Australia, and Antarctica. The **hydrosphere** is made up of the water elements on the earth, which include oceans, seas, rivers, lakes, and water in the atmosphere. Together, the atmosphere, the lithosphere, and the hydrosphere form the **biosphere,** the part of the earth where plants and animals live.

BACKGROUND
Part of the upper portion of the mantle is known as the asthenosphere. It is the hot, but still mostly solid, rock below the cold, brittle rock of the lithosphere.

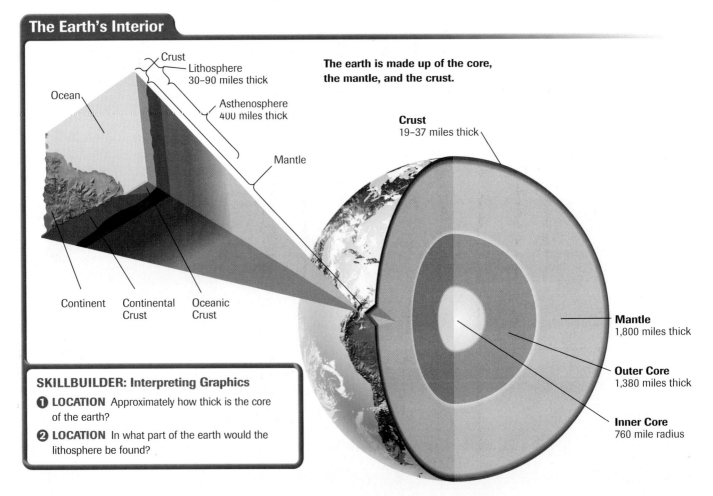

The Earth's Interior

Crust
Lithosphere
30–90 miles thick

Asthenosphere
400 miles thick

Ocean

Mantle

Continent Continental Crust Oceanic Crust

The earth is made up of the core, the mantle, and the crust.

Crust
19–37 miles thick

Mantle
1,800 miles thick

Outer Core
1,380 miles thick

Inner Core
760 mile radius

SKILLBUILDER: Interpreting Graphics

❶ **LOCATION** Approximately how thick is the core of the earth?

❷ **LOCATION** In what part of the earth would the lithosphere be found?

CONTINENTAL DRIFT In 1912, Alfred Wegener of Germany presented a new idea about continents—the **continental drift** hypothesis. It maintained that the earth was once a supercontinent that divided and slowly drifted apart over millions of years. Wegener called the supercontinent Pangaea (from a Greek word meaning "all earth"). An ocean called Panthalassa surrounded it. The supercontinent split into many plates that drifted, crashed into each other, and split apart several times before they came to their current positions. This process occurred over millions of years.

In the 1960s, scientists studying the sea floor discovered that the youngest rocks were in the middle of the ocean, at long cracks in the crust. This suggested that the new sea floor was being added, pushing the continents apart. Later in this chapter, you will learn how the rocks of Earth's surface are broken into giant plates that move and continue to shape the earth.

Continental Drift Hypothesis

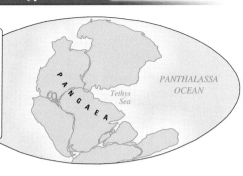

200 million years ago The super-continent was named *Pangaea.* An ocean called Panthalassa surrounded it.

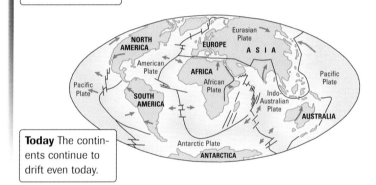

65 million years ago The super-continent split apart and began moving in different directions. Notice that India broke away from Antarctica and Australia and drifted toward Asia.

Today The continents continue to drift even today.

Assessment

❶ Places & Terms

Identify and explain where on the earth these terms would be found.

• continent
• mantle
• magma
• crust
• biosphere

❷ Taking Notes

PLACE Review the notes you took for this section.

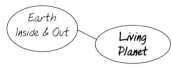

• What are the three basic parts of the earth's interior?
• What are four basic spheres found on or above the earth?

❸ Main Ideas

a. What makes up the interior of the earth?

b. What makes up the biosphere?

c. How can the presence of seven continents on the earth's surface be explained?

❹ Geographic Thinking

Making Inferences How do the earth's spheres influence one another? **Think about:**

• the function of the atmosphere
• the makeup of the biosphere

See Skillbuilder Handbook, page R4.

MAKING COMPARISONS Study the diagrams of continental drift on this page. Write a **description** of the location of the continents in the past in comparison with their current location.

Disasters!

Asteroid Hit!

For years, scientists speculated that the extinction of dinosaurs was due to one very large "environmental event." Today we know that event was most likely the impact of an asteroid about six miles wide. Sixty-five million years ago it slammed into the earth traveling a thousand times faster than a rifle bullet. Fallout from the asteriod impact changed the environment so drastically that 50 to 70 percent of all living species on earth were wiped out.

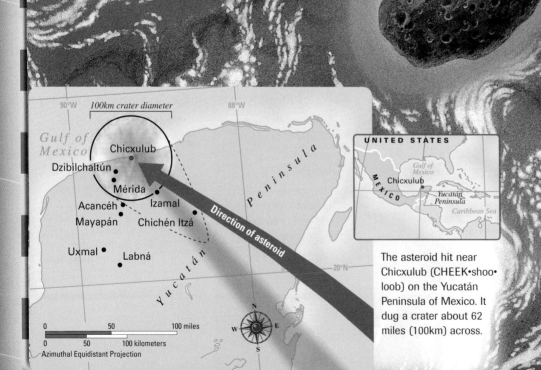

90°W | 100km crater diameter | 88°W

Gulf of Mexico

Chicxulub

Dzibilchaltún

Mérida

Acancéh · Izamal

Mayapán · Chichén Itzá

Uxmal · Labná

Y u c a t á n   P e n i n s u l a

Direction of asteroid

20°N

0   50   100 miles
0   50   100 kilometers
Azimuthal Equidistant Projection

UNITED STATES

Gulf of Mexico

MEXICO

Chicxulub

Yucatán Peninsula

Caribbean Sea

The asteroid hit near Chicxulub (CHEEK•shoo•loob) on the Yucatán Peninsula of Mexico. It dug a crater about 62 miles (100km) across.

The asteroid plows into the earth at 150,000 mph, vaporizing limestone and seawater. It creates an immense fireball that causes fires thousands of miles away.

The Earth's skies are darkened for several months by 25 trillion tons of rock, dust, and smoke from the impact. Acid rain created by vaporized minerals poisons lakes and rivers. Food chains collapse, and plants and animals die.

A thick layer of carbon dioxide is trapped in the atmosphere, creating a "greenhouse effect" for perhaps a thousand years or more. Ferns, burrowing mammals, and some freshwater animals survive. Some even thrive in the new climate.

GeoActivities

CREATING A FRONT PAGE
With a small group, use the Internet to research the Chicxulub event. Then create the front page of a **newspaper** describing the event.

- Create a map showing the impact area.
- Add an article describing the destruction caused by the asteroid.
- Write an interview with a scientist who predicts event results.

RESEARCH LINKS
CLASSZONE.COM

GeoData

ASTEROIDS
- Asteroids are small planetary bodies that orbit the sun.
- There are an estimated 50,000 asteroids in our solar system.
- Asteroids range in size from 20 feet to 580 miles in diameter.
- Fragments of asteroids that reach the earth are called meteorites.

TUNGUSKA EVENT
On June 30, 1908, at about 7:30 A.M., an explosion occurred over the Tunguska region of Siberia. This event might have been an asteroid hit.

- The force of the explosion was estimated at between 10 and 20 megatons of TNT.
- The fireball and explosion were seen and felt 500 miles away.
- Five hundred thousand acres of forest were flattened and burned.
- More than 600 grazing reindeer were roasted instantly.
- No crater could be found.

Bodies of Water and Landforms

Main Ideas
- Water covers about three-fourths of the earth's surface.
- The earth's surface displays a variety of landforms.

Places & Terms

hydrologic cycle

drainage basin

ground water

water table

landform

continental shelf

relief

topography

A HUMAN PERSPECTIVE In July 1971, astronaut James Irwin was lifted into space on the Apollo 15 mission. As he circled the earth, he was deeply moved by the beauty of our planet. Later he wrote this:

> Anyone passing through our solar system would be attracted to the blue planet. They would know that the blue color indicated water on Earth. They would know that where there is water there is probably life. They might try to meet us. We, the blue planet, stand out as a beacon to all.

The earth is unlike any other observable planet in our solar system. It is a living planet.

Bodies of Water

Without both freshwater and saltwater, life on this planet would be impossible. Water not only supports plants and animals, it helps distribute heat on the earth.

OCEANS AND SEAS The ocean is an interconnected body of salt water that covers about 71 percent of our planet. It covers a little more than 60 percent of the Northern Hemisphere and about 81 percent of the Southern Hemisphere. Even though it is one ocean, geographers divide it into four main parts: the Atlantic Ocean, the Pacific Ocean, the Indian Ocean, and the Arctic Ocean, which is sometimes considered part of the Atlantic. The largest of the oceans is the Pacific. The waters near Antarctica are sometimes called the Southern Ocean.

OCEAN MOTION The salty water of the ocean circulates through three basic motions: currents, waves, and tides. Currents act like rivers flowing through the ocean. Waves are swells or ridges produced by winds. Tides are the regular rises and falls of the ocean created by the gravitational pull of the moon or the sun. The motion of the ocean helps distribute heat on the planet. Winds blow over the ocean and are either heated or cooled by the water. When the winds eventually blow over the land, they moderate the temperature of the air over the land.

HYDROLOGIC CYCLE The **hydrologic cycle** is the continous circulation of water between the atmosphere, the oceans, and the earth. As you can see in

PLACE Iguaçu Falls at the Argentina-Brazil border has 275 separate waterfalls varying between 200 and 269 feet high. It is nearly three times wider than Niagara Falls in North America.

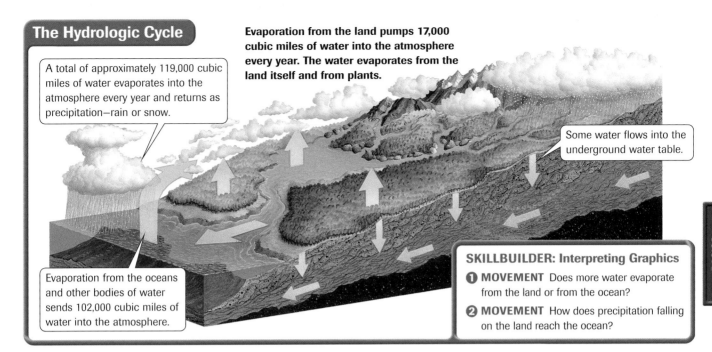

the diagram above, water evaporates into the atmosphere from the surface of the oceans, other bodies of water, and from plants. The water exists in the atmosphere as vapor. Eventually, the vapor cools, condenses, and falls to earth as precipitation—rain or snow. The water soaks into the ground, evaporates to the atmosphere, or flows into rivers to be recycled.

LAKES, RIVERS, AND STREAMS Lakes hold more than 95 percent of all the earth's fresh water supply. The largest freshwater lake is Lake Baikal in Russia. Its volume of water equals 18 percent of all freshwater on earth. Freshwater lakes like the Great Lakes of North America are the result of glacial action thousands of years ago. Saltwater lakes result from changes in the earth's surface that cut off outlets to the sea. Saltwater lakes are created when creeks and rivers carry salts into a lake, and there is no outlet to carry the salt away. The Great Salt Lake in Utah is the remnant of a large freshwater lake—Lake Bonneville. Its water outflows were cut off, causing the remaining water to become more salty as the water evaporated. The largest saltwater lake is the Caspian Sea in Western Asia.

Rivers and streams flow through channels and move water to or from larger bodies of water. Rivers and streams connect into drainage systems that work like the branches of a tree, with smaller branches, called tributaries, feeding into larger and larger ones. Geographers call an area drained by a major river and its tributaries a **drainage basin.**

GROUND WATER Some water on the surface of the earth is held by the soil, and some flows into the pores of the rock below the soil. The water held in the pores of rock is called **ground water.** The level at which the rock is saturated marks the rim of the **water table.** The water table can rise or fall depending on the amount of precipitation in the region and on the amount of water pumped out of the ground.

BACKGROUND
Rock layers that store water are called aquifers. The largest U.S. aquifer is the Ogallala Aquifer, which runs from South Dakota south to Texas.

Landforms

Landforms are naturally formed features on the surface of the earth. The diagram on pages 34–35 shows the different kinds of landforms.

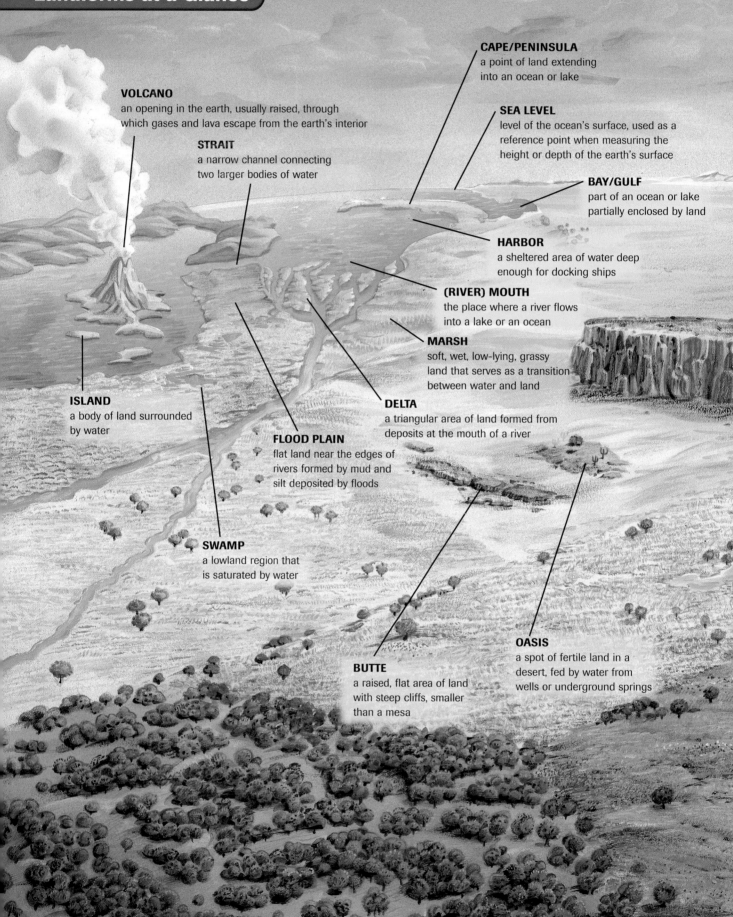

Landforms at a Glance

VOLCANO
an opening in the earth, usually raised, through which gases and lava escape from the earth's interior

STRAIT
a narrow channel connecting two larger bodies of water

CAPE/PENINSULA
a point of land extending into an ocean or lake

SEA LEVEL
level of the ocean's surface, used as a reference point when measuring the height or depth of the earth's surface

BAY/GULF
part of an ocean or lake partially enclosed by land

HARBOR
a sheltered area of water deep enough for docking ships

(RIVER) MOUTH
the place where a river flows into a lake or an ocean

MARSH
soft, wet, low-lying, grassy land that serves as a transition between water and land

ISLAND
a body of land surrounded by water

DELTA
a triangular area of land formed from deposits at the mouth of a river

FLOOD PLAIN
flat land near the edges of rivers formed by mud and silt deposited by floods

SWAMP
a lowland region that is saturated by water

BUTTE
a raised, flat area of land with steep cliffs, smaller than a mesa

OASIS
a spot of fertile land in a desert, fed by water from wells or underground springs

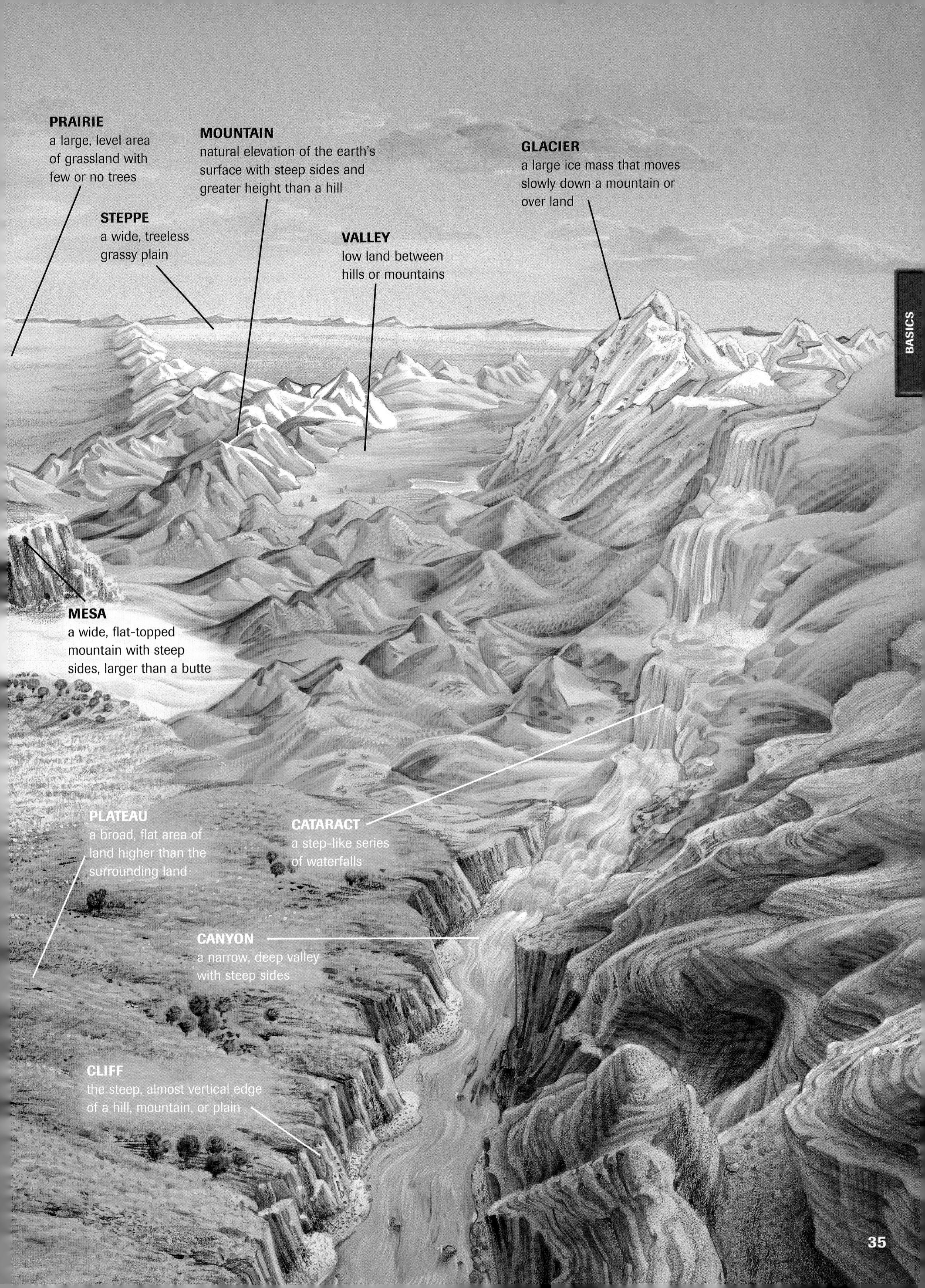

PRAIRIE
a large, level area of grassland with few or no trees

STEPPE
a wide, treeless grassy plain

MOUNTAIN
natural elevation of the earth's surface with steep sides and greater height than a hill

VALLEY
low land between hills or mountains

GLACIER
a large ice mass that moves slowly down a mountain or over land

MESA
a wide, flat-topped mountain with steep sides, larger than a butte

PLATEAU
a broad, flat area of land higher than the surrounding land

CATARACT
a step-like series of waterfalls

CANYON
a narrow, deep valley with steep sides

CLIFF
the steep, almost vertical edge of a hill, mountain, or plain

The Everglades

Native Americans called it Pa-May-Okee ("Grassy Water"). Today, we call it the Everglades. It is a region of wetlands, a marshy landform, that covers 4,000 square miles of Florida. Live oak, palms, and pines cover the region. Saw grass with tiny sharp teeth stands over 10 feet high among cypress. Swampy waters hide snakes, alligators, and turtles. Huge numbers of birds call the Everglades home.

Drainage projects have met with protest because they threaten the plants and animals that flourish there. Plans are now under way to restore the Everglades to its natural habitat.

Everglades National Park

OCEANIC LANDFORMS The sea floor has landforms similar to those above water. The earth's surface from the edge of a continent to the deep part of the ocean is called the **continental shelf.** The floor of the ocean has ridges, valleys, canyons, and plains. Ridges mark places where new crust is being formed on the edges of the tectonic plates. Mountain chains similar to those on the continents themselves cover parts of the ocean floor. The longest continuous range is the Mid-Atlantic Ridge, which extends for thousands of miles north to south through the middle of the Atlantic Ocean. Islands dot the ocean surface. Islands can be formed by volcanic action, deposits of sand, or deposits of coral skeletons.

CONTINENTAL LANDFORMS To understand the types of landforms, study the illustration on pages 34–35. The major geographic feature that separates one type of landform from another is relief. **Relief** is the difference in elevation of a landform from its lowest point to its highest point. There are four categories of relief: mountains, hills, plains, and plateaus. A mountain, for instance, has great relief compared with a plain, which displays very little difference between its high and low points. Ⓐ

Topography is the combination of the surface shape and composition of the landforms and their distribution in a region. A topographic map shows the landforms with their vertical dimensions and their relationship to other landforms.

In the next section, you will learn how internal forces of the earth help to build and change the landforms on the earth—and how those forces affect humans.

🌐 Geographic Thinking

Using the Atlas
Ⓐ Use the map on page A10 to determine the relief of your state.

---

2 Assessment

❶ Places & Terms

Explain the meaning of each of the following terms.

- hydrologic cycle
- ground water
- continental shelf
- relief
- topography

❷ Taking Notes

MOVEMENT Review the notes you took for this section.

- How does the hydrologic cycle circulate water?
- How does ocean water circulate?

❸ Main Ideas

a. How do the winds and the ocean distribute heat on the earth's surface?

b. How are relief and topography related?

c. How are islands formed?

❹ Geographic Thinking

Making Comparisons How is the floor of the ocean similar to land above sea level? **Think about:**

- mountain chains
- other landforms

S See Skillbuilder Handbook, page R3.

GeoActivity

SEEING PATTERNS Study the Landforms at a Glance diagram on pages 34–35. Choose a part of it to reproduce in a three-dimensional **relief map.** Be sure to label the landforms clearly.

Internal Forces Shaping the Earth

Main Ideas

• Internal forces reshape the earth's surface.

• Internal forces shaping the earth often radically alter the lives of people as well.

Places & Terms

tectonic plate	Richter scale
fault	tsunami
earthquake	volcano
seismograph	lava
epicenter	Ring of Fire

BASICS

A HUMAN PERSPECTIVE Sally Ride, America's first female astronaut, wrote the following after one of her trips into space:

> I also became an instant believer in plate tectonics; India really is crashing into Asia, and Saudi Arabia and Egypt really are pulling apart, making the Red Sea even wider. Even though their respective motion is really no more than mere inches a year, the view from overhead makes the theory come alive.

From space, Ride was seeing evidence of the internal forces that have shaped the earth's surface.

Plate Tectonics

The internal forces that shape the earth's surface begin beneath the lithosphere. Rock in the asthenosphere is hot enough to flow slowly. Heated rock rises, moves up toward the lithosphere, cools, and circulates downward. Riding above this circulation system are the **tectonic plates,** enormous moving pieces of the earth's lithosphere. You can see the position of the tectonic plates in the map below.

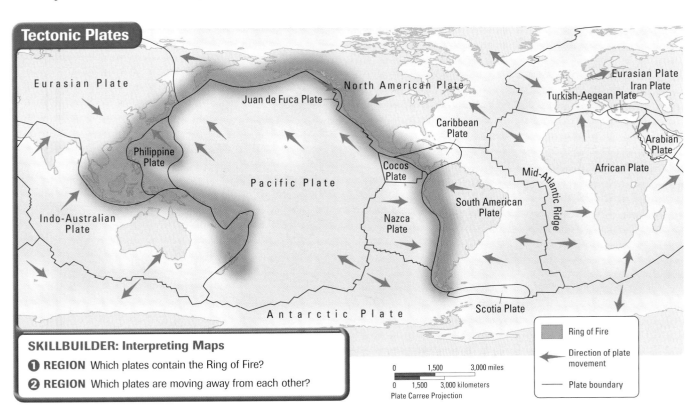

Tectonic Plates

Eurasian Plate
North American Plate
Juan de Fuca Plate
Eurasian Plate
Iran Plate
Turkish-Aegean Plate
Caribbean Plate
Philippine Plate
Arabian Plate
Cocos Plate
African Plate
Pacific Plate
Mid-Atlantic Ridge
Indo-Australian Plate
Nazca Plate
South American Plate
Antarctic Plate
Scotia Plate

Ring of Fire
Direction of plate movement
Plate boundary

0   1,500   3,000 miles
0   1,500   3,000 kilometers
Plate Carree Projection

SKILLBUILDER: Interpreting Maps

❶ **REGION** Which plates contain the Ring of Fire?

❷ **REGION** Which plates are moving away from each other?

Geographers study the movement of the plates and the changes they cause in order to understand how the earth is continually being reshaped—and how earthquakes and volcanoes occur.

PLATE MOVEMENT Tectonic plates move in one of four ways: 1) spreading, or moving apart; 2) subduction, or diving under another plate; 3) collision, or crashing into one another; 4) sliding past each other in a shearing motion. The diagrams below show details about plate movement.

When tectonic plates come into contact, changes on the earth's surface occur. Three types of boundaries mark plate movements:

- **Divergent boundary**—Plates move apart, spreading horizontally.
- **Convergent boundary**—Plates collide, causing either one plate to dive under the other or the edges of both plates to crumple.
- **Transform boundary**—Plates slide past one another.

An example of a divergent boundary is the one between Saudi Arabia and Egypt. The two plates on which those countries sit are spreading apart, making the Red Sea even wider. The Red Sea is actually a part of the Great Rift Valley in Africa. If you look at the map of Africa on page A18, you will see a string of lakes along the eastern side of Africa, including Lake Tanganyika and Lake Nyasa. These lakes, along with the Red Sea, were formed in the spreading boundary.

An example of a convergent boundary can be found in South Asia. The plate where India is located is crashing into the Asian continent and building up the Himalayas. One of the most famous examples of a transform boundary is in North America—the San Andreas Fault in

Geographic Thinking

Making Comparisons
Ⓐ Which of the plate boundaries involves a collision of plates?

Plate Movement and Boundaries

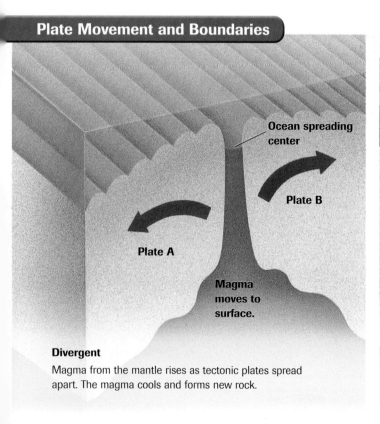

Ocean spreading center

Plate B

Plate A

Magma moves to surface.

Divergent
Magma from the mantle rises as tectonic plates spread apart. The magma cools and forms new rock.

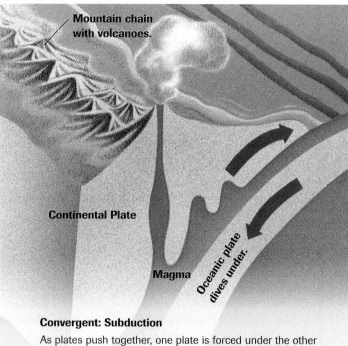

Mountain chain with volcanoes.

Continental Plate

Magma

Oceanic plate dives under.

Convergent: Subduction
As plates push together, one plate is forced under the other in a process called subduction. As the bottom plate starts to melt, magma rises and forms volcanoes at the surface.

California. Study the diagrams below to understand the movement of the plates and their effect on the surface of the earth.

FOLDS AND FAULTS When two plates meet each other, they can cause folding and cracking of the rock. The transformation of the crust by folding or cracking occurs very slowly, often only a few centimeters or inches in a year. Because the movement is slow, the rocks, which are under great pressure, become more flexible and bend or fold, creating changes in the crust. However, sometimes the rock is not flexible and will crack under the pressures exerted by the plate movement. This fracture in the earth's crust is called a **fault.** It is at the fault line that the plates move past each other.

Earthquakes

As the plates grind or slip past each other at a fault, the earth shakes or trembles. This sometimes violent movement of the earth is an **earthquake.** Thousands of earthquakes occur every year, but most are so slight that people cannot feel them. Only a special device called a **seismograph** (SYZ•muh•GRAF) can detect them. A seismograph measures the size of the waves created by an earthquake.

EARTHQUAKE LOCATIONS The location in the earth where an earthquake begins is called the focus. The point directly above the focus on the earth's surface is the **epicenter.** The map on page 37 outlines the major plate boundaries. Nearly 95 percent of all recorded earthquakes occur around those boundaries. Plate movement along the Pacific Rim

BACKGROUND
Seismographs measure earthquakes, but no accurate device for predicting quakes has been developed.

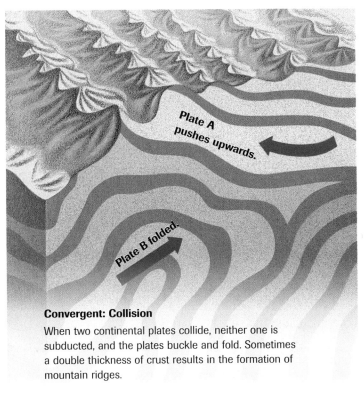

Convergent: Collision
When two continental plates collide, neither one is subducted, and the plates buckle and fold. Sometimes a double thickness of crust results in the formation of mountain ridges.

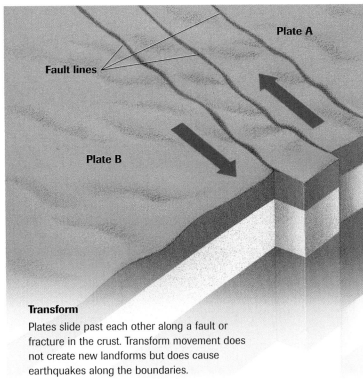

Transform
Plates slide past each other along a fault or fracture in the crust. Transform movement does not create new landforms but does cause earthquakes along the boundaries.

and from southern Asia westward to southern Europe makes this region especially vulnerable to quakes.

EARTHQUAKE DAMAGE Earthquakes result in squeezing, stretching, and shearing motions of the earth's crust that damage land and structures.

The changes are most noticeable in places where people live. Landslides, displacement of land, fires (from broken gas lines), and collapsed buildings are major outcomes of the ground motion. Aftershocks, or smaller-magnitude quakes, may occur after an initial shock and can sometimes continue for days afterward.

An earthquake is the sudden release of energy in the form of motion. C.F. Richter developed a scale to measure the amount of energy released. The **Richter Scale** uses information collected by seismographs to determine the relative strength of an earthquake. The scale has no absolute upper limit. Most people would not notice a quake that measured 2 on the scale. A 4.5 quake will probably be reported in the news. A major quake has a measurement of 7 or more. The largest quake ever measured was 8.9 in the Kermadec Islands of the South Pacific in 1986.

TSUNAMI Sometimes an earthquake causes a **tsunami** (tsu·NAH·mee), a giant wave in the ocean. A tsunami can travel from the epicenter of a quake at speeds of up to 450 miles per hour, producing waves of 50 to 100 feet or higher. The world record for a tsunami was set in 1971 off the Ryukyu Islands near Japan, where the wall of water reached 238 feet—more than 20 stories high. Tsunamis may travel across wide stretches of the ocean and do damage on distant shores. For example, in 1960 a quake near Chile created a tsunami that caused damage in Japan, almost half a world away. A tsunami from a quake near Alaska killed 159 people in Hilo, Hawaii, in 1946.

LOCATION Victims of the 1995 earthquake in Kobe, Japan, wait out aftershocks. More than 5,000 people died in this quake.
Why does the location of Japan make it vulnerable to earthquakes?

Geographic Thinking

Using the Atlas
B Using the map on pages A2–3, calculate the distance these two tsunamis traveled.

Volcanoes

Volcanoes are among the most spectacular of natural events. Magma, gases, and water from the lower part of the crust or the mantle collect in underground chambers. Eventually the materials pour out of a crack in the earth's surface called a **volcano.** Most volcanoes are found along the tectonic plate boundaries.

VOLCANIC ACTION When the magma flows out onto the land slowly, it may spread across an area and cool. Magma that has reached the earth's surface is called **lava.** The most dramatic volcanic action is an eruption, in which hot lava, gases, ash, dust, and rocks explode out of vents in the earth's crust. Often a hill or a mountain is created by lava. The landform may also be called a volcano.

Volcanoes do not erupt on a predictable schedule; they may be active over many years and then stop. Sometimes they remain inactive for

Building Soil

Weathering and erosion are a part of the process of forming soil. Soil is the loose mixture of weathered rock, organic matter, air, and water that supports plant growth. Organic matter in the soil helps to support the growth of plants by providing needed plant food. Water and air share tiny pore-like spaces in the soil. When it rains, the pores are filled with water. As the water evaporates, drains away, or is used by the plants, the pores are filled with air. The texture of the soil, the amount of organic material called **humus,** and the amount of air and water in the soil all contribute to the soil's fertility—its ability to nurture plants.

SOIL FACTORS When geographers study soil, they look at five factors:

- **Parent material** The chemical composition of the original rock, or parent rock, before it decomposes affects its fertility.

- **Relief** Steeper slopes, such as mountainsides, are eroded easily and do not produce soil quickly.

- **Organisms** Organisms include plants, small animals like worms, ants, and bacteria that decompose material. They help to loosen soil and supply nutrients for plants.

- **Climate** Hot climates produce a soil different from that produced by cold climates. Wet climates and dry climates produce soils that are different from each other as well.

- **Time** The amount of time to produce soil varies, but a very rough average is about 2.5 cubic centimeters per century.

The variety of soils—and the climates in which they are found—determine the types of vegetation that can grow in a location. Agricultural activities, such as farming, ranching, and herding, depend on this complex relationship. In the next chapter, you will learn about the climate and vegetation on the earth and how it affects human life.

BACKGROUND
In some soils, as many as a million or more bacteria inhabit each cubic centimeter of soil.

SECTION 4 Assessment

1 Places & Terms

Explain the meaning of each of the following terms.

- weathering
- sediment
- erosion
- delta
- glaciation
- humus

2 Taking Notes

REGION Review the notes you took for this section.

Living Planet

External Forces

- How does weathering vary according to climate?
- What are five factors affecting soil composition?

3 Main Ideas

a. What is the difference between mechanical weathering and chemical weathering?

b. What are three types of eroding action by water?

c. What factors contribute to soil fertility?

4 Geographic Thinking

Making Inferences In what ways does erosion affect the lives of humans? **Think about:**

- water, wind, and glacial action
- results of erosion

S See Skillbuilder Handbook, page R4.

EXPLORING LOCAL GEOGRAPHY Choose a type of erosion that occurs in your community. Do some research to find examples of that type of erosion. Make **sketches** or take **photographs** of the effects of the erosion. Write captions for the pictures describing the type of erosion and where it was found.

VISUAL SUMMARY
A LIVING PLANET

The Earth Inside and Out

- The earth's interior is made up of a series of layers that float on one another.
- The exterior of the earth is the crust.
- The presence of air and water make life on earth possible.

Bodies of Water and Landforms

- Almost three-fourths of the earth is covered with water.
- The hydrologic cycle circulates water.
- Landforms on the land and under the ocean are similar.

Internal Forces Shaping the Earth

- Huge plates on the earth's crust move because of the circulation of magma.
- Earthquakes and volcanoes are the results of plate movement.

External Forces Shaping the Earth

- Weathering and erosion cause changes in the earth's surface and build soil.
- Actions of wind, water, ice, and gravity shape the earth's surface.

Reviewing Places & Terms

A. Briefly explain the importance of each of the following.

1. continent
2. magma
3. hydrologic cycle
4. landform
5. relief
6. tectonic plate
7. earthquake
8. volcano
9. weathering
10. erosion

B. Answer the questions about vocabulary in complete sentences.

11. How are continents and tectonic plates related?
12. Where is magma found?
13. Lava is a form of which term listed above?
14. What is an example of a landform?
15. What does relief tell you about a landform?
16. What is the purpose of the hydrologic cycle?
17. What causes earthquakes?
18. How are magma and volcanoes related?
19. What are the two types of weathering?
20. What must be present for erosion to occur?

Main Ideas

The Earth Inside and Out (pp. 27–31)

1. What layers are found in the earth's interior?
2. What is the continental drift theory?

Bodies of Water and Landforms (pp. 32–36)

3. How does water reach a drainage basin?
4. What is topography?

Internal Forces Shaping the Earth (pp. 37–41)

5. What are three types of plate boundaries?
6. How are the Richter scale and a seismograph used?
7. What is the Ring of Fire?

External Forces Shaping the Earth (pp. 42–45)

8. What is the difference between weathering and erosion?
9. What are three transporting agents of erosion?
10. Why are there many different types of soil?

Critical Thinking

1. Using Your Notes

Use your completed chart to answer these questions.

a. Why is water a critical element on the earth?

b. How do internal and external forces shape the earth?

2. Geographic Themes

a. **MOVEMENT** How does the movement of wind, water, or ice reshape the earth's surface?

b. **HUMAN–ENVIRONMENT INTERACTION** How do volcanoes and earthquakes affect human life?

3. Identifying Themes

What might be the hazards of living near the Ring of Fire? Which of the five themes apply to this situation?

4. Determining Cause and Effect

What is the relationship between tectonic plates, earthquakes, and volcanoes?

5. Making Comparisons

How is a valley created by water different from a valley created by a glacier?

Additional Test Practice, pp. S1–S37    **TEST PRACTICE** CLASSZONE.COM

Geographic Skills: Interpreting Charts

Ten Most Deadly Earthquakes in the 20th Century

Use the information in the chart to answer the following questions.

1. **LOCATION** Which location suffered two deadly earthquakes in the 20th century?

2. **MOVEMENT** How is the magnitude of a quake related to loss of life?

3. **PLACE** What reasons might there be for so great a loss of life in Tangshan, China?

Date	Location	Deaths	Magnitude*
1976, July 27	Tangshan, China	255,000	8.0
1920, Dec. 16	Gansu, China	200,000	8.6
1927, May 22	Nan-Shan, China	200,000	8.3
1923, Sept. 1	Yokohama, Japan	143,000	8.3
1908, Dec. 28	Messina, Italy	83,000	7.5
1932, Dec. 25	Gansu, China	70,000	7.6
1970, May 31	Northern Peru	66,000	7.8
1935, May 30	Quetta, India	50,000	7.5
1990, June 20	Western Iran	40,000	7.7
1988, Dec. 7	Armenia	25,000	7.0

*Magnitude of earthquakes measured on the Richter scale developed in 1935.
SOURCES: Global Volcanism Network, Smithsonian Institution, U.S. Geological Survey, *World Almanac*

GeoActivity

Using a base map of the world and an atlas, plot the locations of the ten most deadly earthquakes. Write a sentence describing the pattern you see in the locations.

 INTERNET ACTIVITY

Use the links at **classzone.com** to do research about volcanic action. Focus on a variety of volcanic activities, including eruptions, geysers, hot springs, and island formation.

Creating a Multimedia Presentation Put together a presentation about the variety of volcanic activity. Include diagrams of several different types of activity and give examples of locations where the activity takes place.

PHYSICAL GEOGRAPHY
Climate and Vegetation

A tornado roars through the countryside. Tornado winds may reach speeds up to 300 miles per hour.

GeoFocus

How do climate and vegetation affect life on earth?

Taking Notes Copy the graphic organizer below into your notebook. Use it to record information about weather, climate, and vegetation.

Seasons & Weather	
Climate	
World Climates	
Soils & Vegetation	

Seasons and Weather

Main Ideas

• Seasons and weather occur because of the changing position of the earth in relation to the sun.

• Weather extremes are related to location on earth.

Places & Terms

solstice   hurricane

equinox   typhoon

weather   tornado

climate   blizzard

precipitation   drought

rain shadow

BASICS

A HUMAN PERSPECTIVE The smell of thousands of decaying corpses hung in the air in what was once the thriving seaport of Galveston, Texas. The day before, winds estimated at 130 miles per hour roared through the city. A storm surge of seawater more than 15 feet high pushed a wall of debris across the island of Galveston. Through this turmoil, Isaac Cline's family huddled in their home. A trolley trestle rammed the house until at last it collapsed, and the waves poured in. Cline survived, but some of his family did not. With a toll of 8,000 human lives, the "Great Galveston Hurricane" would be the deadliest hurricane to hit the United States. The storm date was September 8, 1900.

Seasons

Hurricanes occur frequently in the southern and eastern United States during summer and fall. During these seasons, storm systems with strong winds form over warm ocean water.

EARTH'S TILT Seasons have an enormous impact on us, affecting the conditions in the atmosphere and on the earth that create our weather. As the earth revolves around the sun, it is tilted at a 23.5° angle in relation to the sun. Because of the earth's revolution and its tilt, different parts of the earth receive the direct rays of the sun for more hours of the day at certain times in the year. This causes the changing seasons on the earth. Notice in the diagram to the right that the northern half of the earth tilts toward the sun in summer and away from the sun in winter.

Two lines of latitude—the tropic of Cancer and the tropic of Capricorn—mark the points farthest north and south that the sun's rays shine directly overhead at noon. The day on which this occurs is called a **solstice.** In the Northern Hemisphere, the summer solstice, or the beginning of summer, is the longest day of the year. Winter solstice, the beginning of winter, is the shortest.

Another signal of seasonal change are the equinoxes. Twice a year on the **equinox,** the days and nights all over the world are equal in length. The equinoxes mark the beginning of spring and autumn.

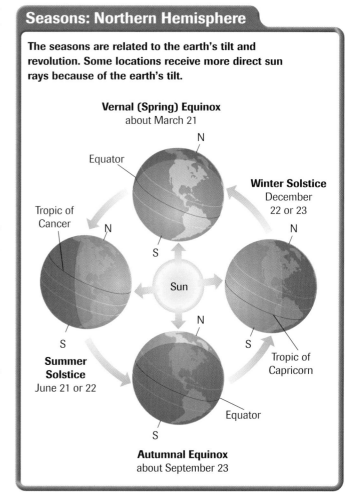

Seasons: Northern Hemisphere

The seasons are related to the earth's tilt and revolution. Some locations receive more direct sun rays because of the earth's tilt.

Vernal (Spring) Equinox
about March 21

Winter Solstice
December 22 or 23

Equator

Tropic of Cancer

Sun

Summer Solstice
June 21 or 22

Tropic of Capricorn

Autumnal Equinox
about September 23

Equator

Weather

Weather and climate are often confused. **Weather** is the condition of the atmosphere at a particular location and time. **Climate** is the term for weather conditions at a particular location over a long period of time. Northern Russia, for example, has a cold climate. ▷

WHAT CAUSES THE WEATHER? Daily weather is the complex result of several conditions. For example, the amount of solar energy received by a location varies according to the earth's position in relation to the sun. Large masses of air absorb and distribute this solar energy, which in turn affects the weather. Other factors include:

- **water vapor** This determines whether there will be **precipitation**—falling water droplets in the form of rain, sleet, snow, or hail.
- **cloud cover** Clouds may hold water vapor.
- **landforms and bodies of water** Water heats slowly but also loses heat slowly. Land heats rapidly but loses heat quickly as well.
- **elevation** As elevation above sea level increases, the air becomes thinner and loses its ability to hold moisture.
- **air movement** Winds move the air and the solar energy and moisture that it holds. As a result, weather can change very rapidly.

PRECIPITATION Precipitation depends on the amount of water vapor in the air and the movement of that air. As warm air rises, it cools and loses its ability to hold water vapor. The water vapor condenses, and the water droplets form into clouds. When the amount of water in a cloud is too heavy for the air to hold, rain or snow falls from the cloud. Geographers classify precipitation as convectional, orographic, or frontal, as illustrated in the diagram below.

Geographic Thinking

Making Comparisons
◁ Why might geographers be more interested in the climate of a place than its weather?

Types of Precipitation

Convectional Typical of hot climates, convection occurs after morning sunshine heats warm moist air. Clouds form in the afternoon and rain falls.

Orographic Associated with mountain areas, orographic storms drop more rain on the windward side of a mountain and create a rain shadow on the leeward side.

Frontal Mid-latitude frontal storms feature cold dense air masses that push lighter warm air masses upward, causing precipitation to form.

Global Ocean Currents

NORTH AMERICA · **EUROPE** · **ASIA** · **AFRICA** · **SOUTH AMERICA** · **AUSTRALIA** · **ANTARCTICA**

Alaska · North Pacific · California · Labrador · North Atlantic Drift · Oyashio · Japan · North Equatorial · Equatorial Counter Current · South Equatorial · Gulf Stream · Canary · Guinea · North Equatorial · South Equatorial · Peru · Brazil · Benguela · West Wind Drift · West Australian · East Australian · West Wind Drift · West Wind Drift

PACIFIC OCEAN · ATLANTIC OCEAN · INDIAN OCEAN · PACIFIC OCEAN

Arctic Circle · Tropic of Cancer · Equator · Tropic of Capricorn · Antarctic Circle

60°N · 30°N · 0° · 30°S · 60°S

150°W · 120°W · 90°W · 60°W · 30°W · 0° · 30°E · 60°E · 90°E · 120°E · 150°E

Robinson Projection

➤ Warm current
➤ Cool current

BASICS

SKILLBUILDER: Interpreting Maps

❶ **MOVEMENT** What happens to the Peru Current as it reaches the equator?

❷ **LOCATION** Where does the West Wind Drift flow?

OCEAN CURRENTS Ocean currents are like rivers flowing in the ocean. Moving in large circular systems, warm waters flow away from the equator toward the poles, and cold water flows back toward the equator. Winds blowing over the ocean currents affect the climate of the lands that the winds cross. For example, the warmth of the Gulf Stream and the North Atlantic Drift help keep the temperature of Europe moderate. Even though much of Europe is as far north as Canada, it enjoys a much milder climate than Canada.

Ocean currents affect not only the temperature of an area, but also the amount of precipitation received. Cold ocean currents flowing along a coastal region chill the air and sometimes prevent warm air and the moisture it holds from falling to earth. The Atacama Desert in South America and the Namib Desert in Africa, for example, were formed partly because of cold ocean currents nearby. ◀

ZONES OF LATITUDE Geographers divide the earth into three general zones of latitude: low or tropical, middle or temperate, and high or polar. Tropical zones are found on either side of the equator. They extend to the tropic of Cancer in the Northern Hemisphere and the tropic of Capricorn in the Southern Hemisphere. Lands in tropical zones are hot all year long. In some areas, a shift in wind patterns causes variations in the seasons. For example, Tanzania experiences both a rainy season and a dry season as Indian Ocean winds blow in or away from the land.

Geographic Thinking

Making Comparisons
▶ How are wind and ocean currents similar in their effect on climate?

Climate **55**

Latitude and elevation influence climate. Notice that as you move along the latitude line, the climates at the lower altitude change. However, the greater the altitude, the fewer the climate zones no matter what latitude a location may be.

SKILLBUILDER: Interpreting Graphics

❶ **LOCATION** At about what latitude and altitude would you find a desert climate?

❷ **REGION** How do the climate zones change as latitude gets higher?

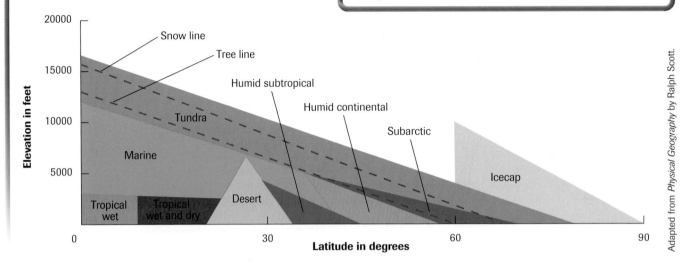

Adapted from *Physical Geography* by Ralph Scott.

The high-latitude polar zones, which encircle the North Pole and South Pole, are cold all year. Summer temperatures in the polar regions may reach a high of only 50°F.

The earth's two temperate zones lie at the middle latitudes, between the tropics and the polar regions. Within the temperate zones, climates can vary greatly, ranging from relatively hot to relatively cold. These variations occur because solar heating is greater in the summer than in the winter. So summers are much warmer.

ELEVATION Another factor in determining the climate of a region is elevation, or distance above sea level. You would think that the closer you get to the sun, the hotter it would become. But as altitude increases, the air temperature drops about 3.5°F for every 1,000 feet. Therefore, the climate gets colder as you climb a mountain or other elevated location. Climates above 12,000 feet become like those in Arctic areas—with snow and ice. For example, Mt. Kilimanjaro in east Africa is capped by snow all year long. The diagram above will help you see how latitude and elevation are related.

TOPOGRAPHY Landforms also affect the climate. This is especially true of mountain areas. Remember that moisture-laden winds cool as they move up the side of a mountain, eventually releasing rain or snow. By the time the winds reach the other side of the mountain, they are dry and become warmer as they flow down the mountain.

Changes in Climate

Climates change over time. Scientists studying ice-core samples from thousands of years ago have noted a variety of changes in temperature and precipitation. Some of the changes in climate appear to be natural while others are the result of human activities.

EL NIÑO The warming of the waters off the west coast of South America—known as **El Niño**—is a natural change in the climate. About every two to seven years, prevailing easterly winds that blow over the central Pacific Ocean slow or reverse direction, changing the ocean temperature and affecting the weather worldwide. Normally, these easterlies bring seasonal rains and push warm ocean water toward Asia and Australia. In El Niño years, however, the winds push warm water and heavy rains toward the Americas. This can cause floods and mudslides there, while Australia and Asia experience drought conditions.

When the reverse occurs—that is, when the winds blow the warmer water to the lands on the western Pacific rim—the event is called La Niña. La Niña causes increases in precipitation in places such as India and increased dryness along the Pacific coasts of the Americas.

El Niño and La Niña

El Niño and La Niña act to transfer the heat on the earth's surface and in the atmosphere to other parts of the globe.

SKILLBUILDER: Interpreting Graphics

❶ **MOVEMENT** In which direction do winds and water move in El Niño?

❷ **LOCATION** Where does flooding occur during La Niña?

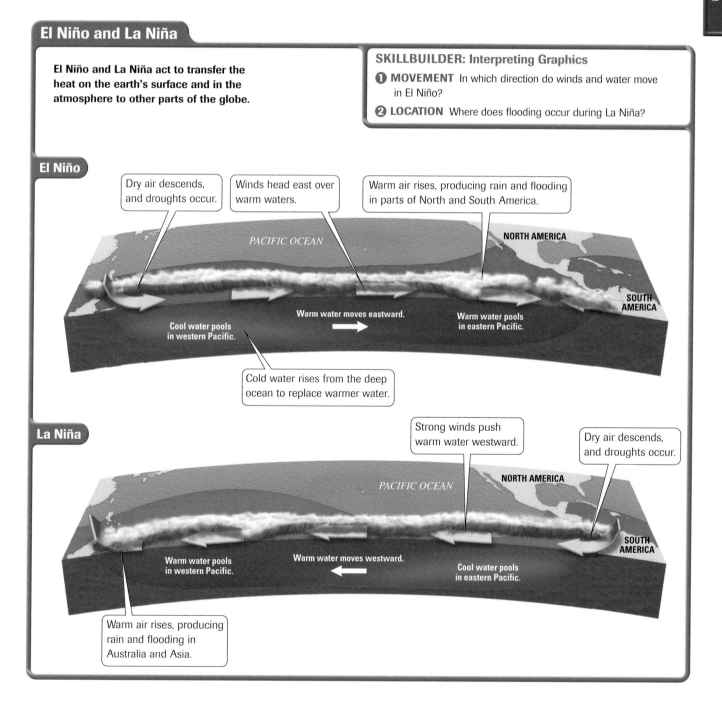

El Niño

Dry air descends, and droughts occur.

Winds head east over warm waters.

Warm air rises, producing rain and flooding in parts of North and South America.

PACIFIC OCEAN

NORTH AMERICA

SOUTH AMERICA

Warm water moves eastward.

Cool water pools in western Pacific.

Warm water pools in eastern Pacific.

Cold water rises from the deep ocean to replace warmer water.

La Niña

Strong winds push warm water westward.

Dry air descends, and droughts occur.

NORTH AMERICA

PACIFIC OCEAN

SOUTH AMERICA

Warm water pools in western Pacific.

Warm water moves westward.

Cool water pools in eastern Pacific.

Warm air rises, producing rain and flooding in Australia and Asia.

Greenhouse Effect

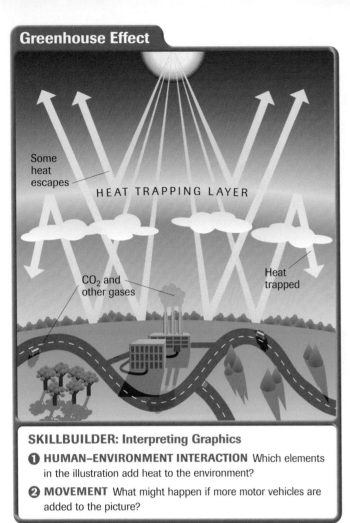

Some heat escapes

HEAT TRAPPING LAYER

CO_2 and other gases

Heat trapped

SKILLBUILDER: Interpreting Graphics

❶ **HUMAN–ENVIRONMENT INTERACTION** Which elements in the illustration add heat to the environment?

❷ **MOVEMENT** What might happen if more motor vehicles are added to the picture?

GLOBAL WARMING Although controversy exists over the causes of global warming, scientists agree that air temperatures are increasing. Since the late 1800s, the temperature of the earth has increased by one degree. However, estimates for the next century suggest that the increase will be almost 3.5 degrees.

Some scientists believe that this warming is part of the earth's natural warming and cooling cycles. For example, 18,000 to 20,000 years ago, the earth was in the last of several ice ages, when vast glaciers advanced over huge portions of the land mass.

Other scientists argue that global temperature increases are caused by the **greenhouse effect.** The layer of gases released by the burning of coal and petroleum traps some solar energy, causing higher temperatures in the same way that a greenhouse traps solar energy.

As more and more nations become industrialized, the amount of greenhouse gases will also increase. Scientists predict that, if global warming continues, ice caps will melt, flooding some coastal areas, covering islands, and changing the global climate. In the next section, you will learn about world climate regions.

BACKGROUND
The air temperature in the period between about 1500 and 1850 was so much cooler than today that it is known as a "Little Ice Age."

SECTION 2 Assessment

❶ **Places & Terms**

Explain the meaning of each of the following terms.

- convection
- El Niño
- greenhouse effect

❷ **Taking Notes**

MOVEMENT Review the notes you took for this section.

Climate

- What are four factors that affect climate?
- What are examples of forces that produce climate changes?

❸ **Main Ideas**

a. What role do wind and ocean currents play in climate?

b. How do latitude and altitude affect climate?

c. How do El Niño and La Niña affect climate?

❹ **Geographic Thinking**

Drawing Conclusions
Which of the factors affecting climate has the greatest impact on the climate in your region? **Think about:**

- the four factors affecting climate
- the climate where you live

RESEARCH LINKS CLASSZONE.COM

GeoActivity

SEEING PATTERNS Review the information and diagram about El Niño and La Niña on page 57. Use the Internet to find more information on these events. Create a **multimedia presentation** explaining one of the events and how it affects the world-wide weather conditions.

World Climate Regions

Main Ideas
- Temperature and precipitation define climate regions.
- Broad climate definitions help to identify variations in weather at a location over the course of a year.

Places & Terms
tundra
permafrost

BASICS

A HUMAN PERSPECTIVE Songs have been written celebrating April in Paris. Springtime there is mild, with temperatures in the 50°F range. But no songs have been written about April in Winnipeg, Canada. Temperatures in April there are only slightly above freezing. If you look at the two locations on a map, you will find the cities are almost the same distance north of the equator. To understand why two cities at the same latitude are so different, you need to understand climate regions. When studying climate, one of the key words is location.

Defining a Climate Region

Climate regions act like a code that tells geographers much about an area without giving many local details. To define a climate region, geographers must make generalizations about what the typical weather conditions are like over many years in a location.

The two most significant factors in defining different climates are temperature and precipitation. A place's location on a continent, its topography, and its elevation may also have an impact on the climate.

Geographers use a variety of methods to describe climate patterns. The most common method uses latitude to help define the climate. There are five general climate regions: tropical (low latitude), dry, mid-latitude, high latitude, and highland. Dry and highland climates occur at several different latitudes. Within the five regions, there are variations that geographers divide into smaller zones. You can see the varied climate regions on the map on pages 60–61.

Although the map shows a distinct line between each of the climate regions, in reality there are transition zones between the regions. As you read about climate regions, refer to the climate map. You should see the latitude-related patterns that emerge in world climate regions.

PLACE This highland climate zone in Patagonia, South America, has several different climate regions, including tundra and subartic.

Types of Climates

World climates are generally divided into five large regions: tropical, dry, mid-latitude, high latitude, and highland. The regions are divided into smaller subregions that are described below.

TROPICAL WET This subregion has little variation in temperature over the year—it is always hot, with an average temperature of 80°F. The days begin sunny but by afternoon have clouded up, and rain falls almost daily. The average amount of rain in a year is more than 80 inches. Tropical wet climates are found in Central and South America as well as Africa and Southwest Asia.

Climate Regions

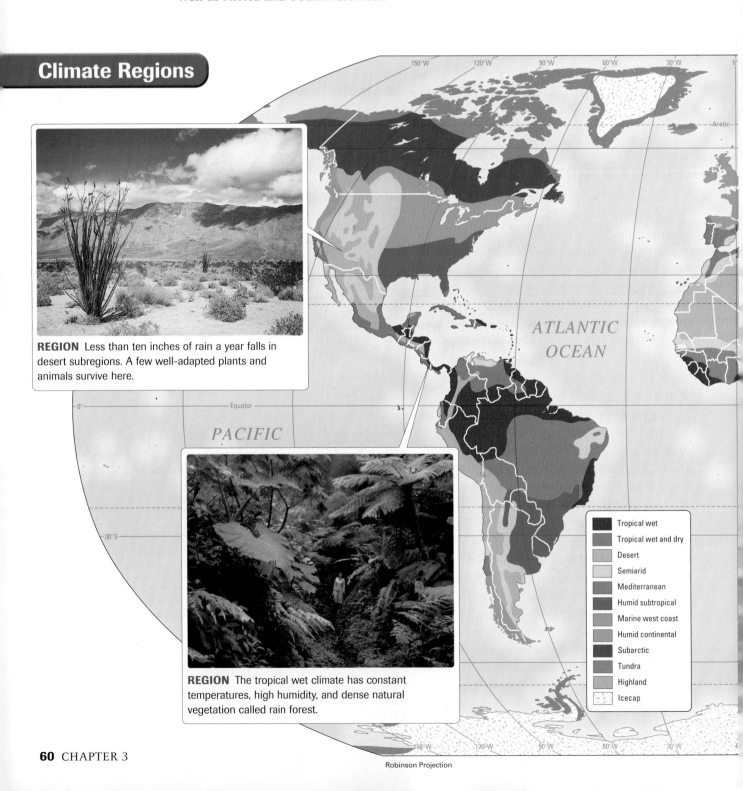

REGION Less than ten inches of rain a year falls in desert subregions. A few well-adapted plants and animals survive here.

REGION The tropical wet climate has constant temperatures, high humidity, and dense natural vegetation called rain forest.

- Tropical wet
- Tropical wet and dry
- Desert
- Semiarid
- Mediterranean
- Humid subtropical
- Marine west coast
- Humid continental
- Subarctic
- Tundra
- Highland
- Icecap

ATLANTIC OCEAN

PACIFIC

Robinson Projection

TROPICAL WET AND DRY This climate is called "tropical wet and dry" because the subregion has a rainy season in summer and a dry season in winter. Temperatures are cooler in the dry season and warmer in the wet season. Rainfall is less than in the tropical wet climate subregion and occurs mostly in the wet season. Tropical wet and dry climates are found next to tropical wet climates in Africa, South and Central America, and parts of Asia.

SEMIARID This climate subregion does receive precipitation, just not very much: about 16 inches per year. Summers are hot. Winters are mild to cold, and some semiarid locations can produce snow. The climate is found in the interior of continents, or in a zone around deserts. The region contains some of the most productive agricultural lands in the world.

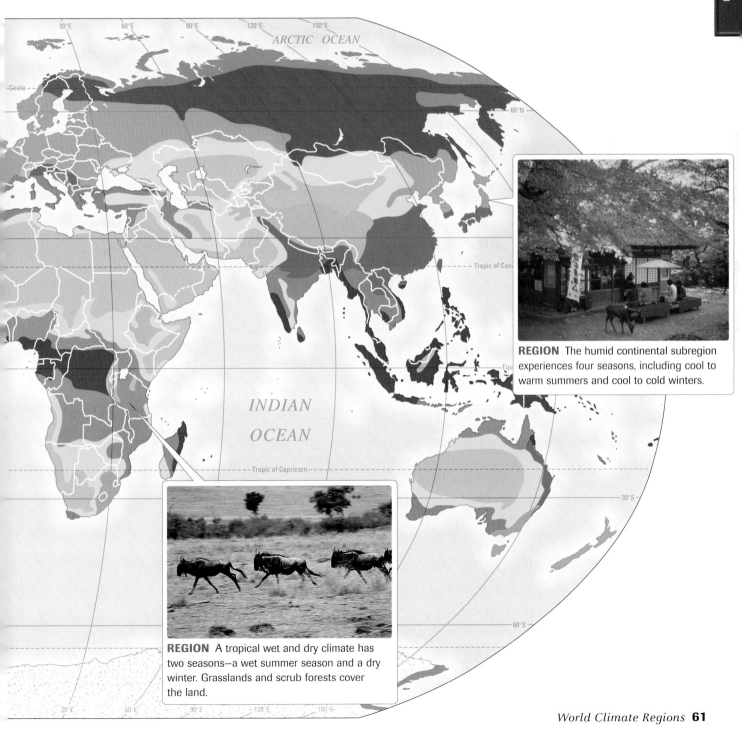

REGION The humid continental subregion experiences four seasons, including cool to warm summers and cool to cold winters.

REGION A tropical wet and dry climate has two seasons—a wet summer season and a dry winter. Grasslands and scrub forests cover the land.

DESERT Some people think a desert is nothing but sand dunes. However, deserts are categorized according to the amount of rainfall, rather than by landforms, and can be hot or cool/cold. Deserts receive less than ten inches of rain per year. Hot deserts, like the Sahara and the Arabian Desert, regularly have low humidity and high temperatures during the day. At night, temperatures drop because the dry air cannot hold heat well.

Cool/cold deserts are found in the mid-latitudes mostly in the Northern Hemisphere, often in the rain shadow of nearby mountain ranges. Summer temperatures are warm to hot, and winter temperatures range from quite cool to below freezing.

MEDITERRANEAN This climate subregion is named for the land around the Mediterranean Sea where it is located. It also exists elsewhere, such as the west coast of the United States and parts of Australia. Its summers are dry and hot, and its winters cool and rainy. This climate region supports a dense population and rich agricultural activity.

MARINE WEST COAST This climate subregion, which is located close to the ocean, is frequently cloudy, foggy, and damp. The winds over the warm ocean moderate the temperatures and keep them relatively constant. Parts of the west coast of the United States and Canada and most of Western Europe experience this climate. Precipitation in marine west coast climate regions is evenly distributed throughout the year. Industrial regions with marine west coast climate may have smog (a mixture of smoke and fog). Ⓐ

PLACE These Italian vineyards thrive in the hot dry summers and cool rainy winters of the Mediterranean climate. The climate also supports the cultivation of citrus fruit, olives, and vegetables.

Geographic Thinking

Making Comparisons
Ⓐ How are Mediterranean and marine west coast climates different?

HUMID SUBTROPICAL Long periods of summer heat and humidity characterize the humid subtropics. These areas are found on the east coast of continents and are often subject to hurricanes in late summer and autumn. The southeastern part of the United States and large areas of China are examples. Winters are mild to cool, depending on latitude. The climate is very suitable for raising crops, especially rice.

HUMID CONTINENTAL A great variety in temperature and precipitation characterizes this climate, which is found in the mid-latitude interiors of Northern Hemisphere continents. For example, Winnipeg, Manitoba, in Canada is located deep in the North American continent. It has a humid continental climate. Air masses chilled by Arctic ice and snow flow south over these areas and frequently collide with tropical air masses, causing changing weather conditions. These areas experience four seasons. However, the length of each season is determined by the region's latitude.

SUBARCTIC Evergreen forests called taiga cover the lands in the subarctic subregion, especially in Canada and Russia. Huge temperature variations occur in this subregion between summer and winter. Although the summers are short and cool, the winters are always very cold.

Temperatures at freezing or below freezing last five to eight months of the year.

TUNDRA The flat, treeless lands forming a ring around the Arctic Ocean are called **tundra.** The climate subregion is also called tundra. It is almost exclusively located in the Northern Hemisphere. Very little precipitation falls here, usually less than 15 inches per year. The land has **permafrost**—that is, the subsoil is constantly frozen. In the summer, which lasts for only a few weeks, the temperature may reach slightly above 40°F. B

ICE CAP Snow, ice, and permanently freezing temperatures characterize the region, which is so cold that it rarely snows. These subregions are sometimes called polar deserts since they receive less than ten inches of precipitation a year. The coldest temperature ever recorded, 128.6°F below zero, was on the ice cap at Vostok, Antarctica.

HIGHLANDS The highlands climate varies with latitude, elevation, other topography, and continental location. In rugged mountain areas such as the Andes of South America, climates can vary based on such factors as whether a slope faces north or south and whether it is exposed to winds carrying moisture.

Understanding climate helps you understand about the general weather conditions in an area. In the next section, you will learn about the variety of soils and vegetation on the earth.

Geographic Thinking

Making Comparisons
B How are precipitation amounts in a tundra climate similar to those of a desert climate?

REGION Life is hard during the long, cold, and dark winter in the subarctic. The only places where the temperatures are colder are the icecaps of Greenland and the Antarctic.

Assessment

1 Places & Terms

Identify and explain where in the region these would be found.

- tundra
- permafrost

2 Taking Notes

REGION Review the notes you took for this section.

World Climates

- What are the five basic climate regions?
- What are the factors that determine climate?

3 Main Ideas

a. How do tropical climates differ from each other?

b. How do desert regions differ from each other?

c. How are Humid subtropical and Mediterranean climates different from each other?

4 Geographic Thinking

Making Generalizations
How are the climates of the Northern Hemisphere different from the climates of the Southern Hemisphere?
Think about:

- sizes and locations of the continents

S **See Skillbuilder Handbook, page R6.**

MAKING COMPARISONS Study the descriptions of climates in this chapter. Then either draw pictures or find pictures that illustrate the climates. Using a hanger and string, create a **mobile** displaying world climate regions.

Interpreting Climographs

How many seasons are in a year where you live? In some parts of the world the climate is the same all year long. Other places have only two seasons—wet and dry. Still others experience changes in temperature and precipitation almost every month. A climograph allows you to quickly determine what the climate is like in a place. If you have two climographs you may compare two different places.

THE LANGUAGE OF GRAPHS A **climograph** shows the average daily temperature and precipitation for each month of the year for a specific location. This information shows what the climate is like over a year. Use the green line on the graph to find the average temperature and the blue bars to find average rainfall.

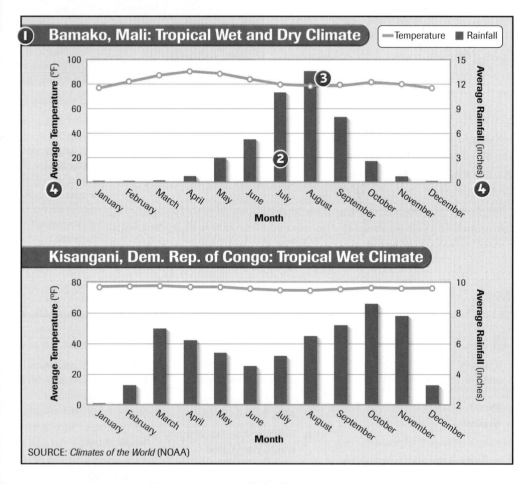

❶ Bamako, Mali: Tropical Wet and Dry Climate — Temperature ■ Rainfall

Average Temperature (°F) / Average Rainfall (inches)
Month

Kisangani, Dem. Rep. of Congo: Tropical Wet Climate

Average Temperature (°F) / Average Rainfall (inches)
Month

SOURCE: *Climates of the World* (NOAA)

❶ The title indicates the place, sometimes its absolute location, and the type of climate.

❷ Each blue bar shows average rainfall for one month of the year. For example, more than 13 inches of rain falls in Bamako in August.

❸ The green line shows the average temperature. For example, the July temperature in Bamako is 80°F.

❹ Precipitation can be shown in inches (in.) or centimeters (cm.). Temperature can be shown in Fahrenheit (F°) or Celsius (C°) degrees.

Map and Graph Skills Assessment

1. Analyzing Data
What information is shown on each side of the vertical axis?

2. Analyzing Data
What are the rainy months in Bamako? How much rain falls in the rainiest month?

3. Drawing Conclusions
How is the tropical wet and dry climate of Bamako different from the tropical wet climate of Kisangani?

Critical Thinking

1. Using Your Notes

Use your completed chart to answer these questions.

Seasons & Weather	
Climate	
World Climates	
Soils & Vegetation	

a. How are seasons, weather, and climate connected to each other?

b. How would knowing about the climate of a region help you determine the vegetation of the region?

2. Geographic Themes

a. **REGION** Why are there few subarctic climate zones in the Southern Hemisphere?

b. **LOCATION** How does location affect climate?

3. Identifying Themes

How might the climate of an area be affected by global warming? Which of the five themes apply to this situation?

4. Drawing Conclusions

What is incorrect about defining a desert by landforms such as sand dunes?

5. Making Inferences

Why is a hurricane such a deadly storm?

Additional Test Practice, pp. S1–S37

TEST PRACTICE
CLASSZONE.COM

Additional Test Practice, pp. S1–S37

Geographic Skills: Interpreting Graphs

Temperature Variations

Use the graph to answer the following questions.

1. **MOVEMENT** Which decade (10-year span) had the highest temperatures?

2. **MOVEMENT** In approximately which year did temperatures begin to consistently rise above the average?

3. **HUMAN–ENVIRONMENT INTERACTION** What impact might the greenhouse effect have on the temperature changes?

GeoActivity

Using straws, devise a three-dimensional model to show the information on the graph. Be sure to provide time frames and temperature information on your model.

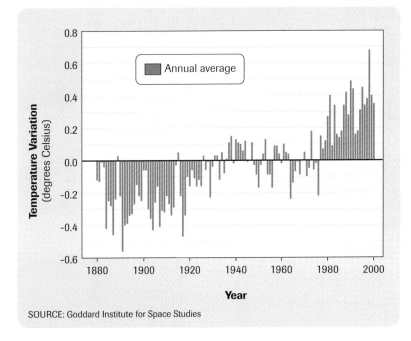

SOURCE: Goddard Institute for Space Studies

INTERNET ACTIVITY

Use the links at **classzone.com** to do research about global warming. Choose one of the nine regions in this textbook. Focus on determining the effects of global warming on the region, especially on coastal areas.

Creating a Multimedia Presentation Combine charts, maps, or other visual images in an electronic presentation showing how the earth will be affected by global warming.

HUMAN GEOGRAPHY
People and Places

Petroglyphs like this one from the Fremont culture (found in Dinosaur National Monument) offer evidence of human life in the desert of the Colorado Plateau.

GeoFocus

What is human geography about?

Taking Notes Copy the graphic organizer below into your notebook. Use it to record information about human geography.

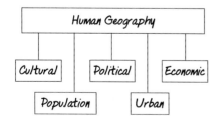

The Elements of Culture

Main Ideas

• Human beings are members of social groups with shared and unique sets of behaviors and attitudes.

• Language and religion are two very important aspects of culture.

Places & Terms

culture   cultural hearth

society   acculturation

ethnic group   dialect

innovation   religion

diffusion

BASICS

A HUMAN PERSPECTIVE In an article titled "The 100% American," anthropologist Ralph Linton described how a typical American, in eating breakfast, had borrowed from other cultures.

> He has coffee, an Abyssinian plant, with cream and sugar. Both the domestication of cows and the idea of milking them originated in the Near East, while sugar was first made in India. . . . As a side dish he may have the egg of a species of bird domesticated in Indo-China, or thin strips of the flesh of an animal domesticated in Eastern Asia.

Borrowing from other cultures is common around the world, even if we are not aware of it.

Defining Culture

What makes us similar to some people in the world but different from most others? The answer is culture. **Culture** is the total of knowledge, attitudes, and behaviors shared by and passed on by the members of a specific group. It includes all products of human work and thought. Culture acts as a blueprint for how a group of people should behave if they want to fit in with the group. It ties us to one group and separates us from other groups—and helps us to solve the problems that all humans face. Culture involves the following factors:

- food and shelter
- religion
- relationships to family and others
- language
- education
- security/protection
- political and social organization
- creative expression

A group that shares a geographic region, a sense of identity, and a culture is called a **society.** Sometimes the term **ethnic group** is used to refer to a group that shares a language, customs, and a common heritage. An ethnic group has an identity as a separate group of people within the region where they live. For example, the San peoples—known as the Bushmen of the Kalahari Desert in Africa—live in a specific territory, speak their own language, and have a social organization distinct from other groups in the region.

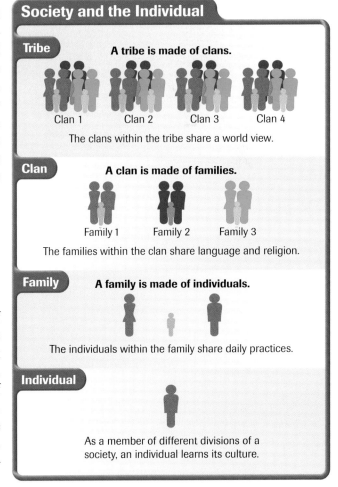

Society and the Individual

Tribe   **A tribe is made of clans.**

Clan 1   Clan 2   Clan 3   Clan 4

The clans within the tribe share a world view.

Clan   **A clan is made of families.**

Family 1   Family 2   Family 3

The families within the clan share language and religion.

Family   **A family is made of individuals.**

The individuals within the family share daily practices.

Individual

As a member of different divisions of a society, an individual learns its culture.

Culture Change and Exchange

Cultures and societies are always in the process of changing. Change comes very slowly to some societies and rapidly to others. It can come about through innovation or the spread of ideas or behaviors from one culture to another.

INNOVATION Taking existing technology and resources and creating something new to meet a need is called **innovation.** For example, to solve the need for storage of goods, some societies invented baskets woven from reeds because reeds were abundant. Other cultures developed clay pots to solve the same problem.

Innovation and invention may happen on purpose or by accident. History is filled with examples of "accidents" that changed the life of a society. For example, the first cooked meat may have happened by accident, but it led to the practice of cooking most food rather than eating it raw.

MOVEMENT A satellite dish brings the outside world to a Mongolian family living in this traditional house called a yurt. **How does this picture show acculturation?**

DIFFUSION Good ideas or inventions are hard to keep secret—they spread when people from different societies, or their ideas and inventions, come into contact with one another. This spread of ideas, inventions, or patterns of behavior is called **diffusion.** In an age of electronic technology, diffusion can happen very quickly. Television and the Internet speed ideas and facilitate the sale of goods around the globe. Almost no group of people can avoid some kind of contact with other societies. Ⓐ

A **cultural hearth** is a site of innovation from which basic ideas, materials, and technology diffuse to many cultures. River civilizations such as those along the Indus River in South Asia, Huang He in East Asia, the Nile River in Africa, and the Tigris and Euphrates in Southwest Asia are the best known cultural hearths.

ACCULTURATION Exposure to an innovation does not guarantee that a society will accept that innovation. Individuals in the society must decide whether the innovation is useful and consistent with its basic principles. **Acculturation** occurs when a society changes because it accepts or adopts an innovation. An example of acculturation might be wearing jeans instead of traditional garments.

Sometimes individuals or a group adopt innovations that radically change the society. The resulting changes may have a positive or a negative effect on the society, depending on how the change came about. If change is forced on a group, it may have negative consequences. On the other hand, if the individuals or a group accept the change, it may lead to a better life for everyone. For example, the lives of thousands of people in Somalia were saved when they were persuaded to be vaccinated for smallpox in the 1970s.

Geographic Thinking

Seeing Patterns
Ⓐ In which locations would diffusion happen less frequently?

Language

Language is one of the most important aspects of culture because it allows the people within a culture to communicate with each other. Language reflects all aspects of culture, including the physical area occupied by the society. For example, a society that lives in the subarctic or tundra region may have many different words to describe various forms of snow. However, those words would be useless for a culture in a place with no snow.

LANGUAGE AND IDENTITY Language helps establish a cultural identity. It builds a group identity and a sense of unity among those who speak the language. If a language is spoken throughout a political region, a spirit of unity and sometimes nationalism (a strong feeling of pride in one's nation) grows. Language can also divide people. If more than one language is spoken in an area, but one language seems to be favored, then conflict sometimes results. In Canada, for example, where both English and French are spoken, French Canadians pressured the government to recognize both French and English as official languages.

LANGUAGE FAMILIES Geographers estimate that between 3,000 and 6,500 languages are spoken across the world today. The languages are categorized by placing them with other similar languages in language families. (See page 74.) Today's languages evolved from earlier languages. One of the earlier languages, called Nostratic, developed in the area known today as Turkey. Nostratic is believed to be the basis of the Indo-European languages that you see on the chart on page 74. Languages as different as English, Russian, Hindi, and Greek all developed from the Indo-European family.

Versions of a language are called dialects. A **dialect** reflects changes in speech patterns related to class, region, or other cultural changes. For example, in the United States, dialects might include a Southern drawl, a Boston accent, or even street slang.

LANGUAGE DIFFUSION Like other aspects of culture, language can be diffused in many ways. It may follow trade routes or even be invented. For example, Swahili developed as a trade language between Arabic traders and Bantu-speaking tribes on Africa's east coast. Sometimes a blended language develops to aid communication among groups speaking several languages. In Louisiana, the presence of French, African, and North American peoples resulted in a blended language called Louisiana Creole.

A second way diffusion occurs is through migration. As people settle in new locations, the language they carry with them sometimes takes hold in the region. For example, colonists from Europe brought the English, Spanish, French, and Dutch languages to North and South America, Africa, Australia, and parts of Asia.

BACKGROUND
The language spoken by the largest number of native speakers is Mandarin Chinese, with an estimated 885 million speakers.

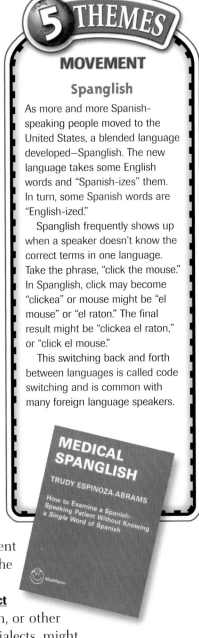

5 THEMES

MOVEMENT

Spanglish

As more and more Spanish-speaking people moved to the United States, a blended language developed—Spanglish. The new language takes some English words and "Spanish-izes" them. In turn, some Spanish words are "English-ized."

Spanglish frequently shows up when a speaker doesn't know the correct terms in one language. Take the phrase, "click the mouse." In Spanglish, click may become "clickea" or mouse might be "el mouse" or "el raton." The final result might be "clickea el raton," or "click el mouse."

This switching back and forth between languages is called code switching and is common with many foreign language speakers.

MEDICAL SPANGLISH
TRUDY ESPINOZA-ABRAMS
How to Examine a Spanish-Speaking Patient Without Knowing a Single Word of Spanish

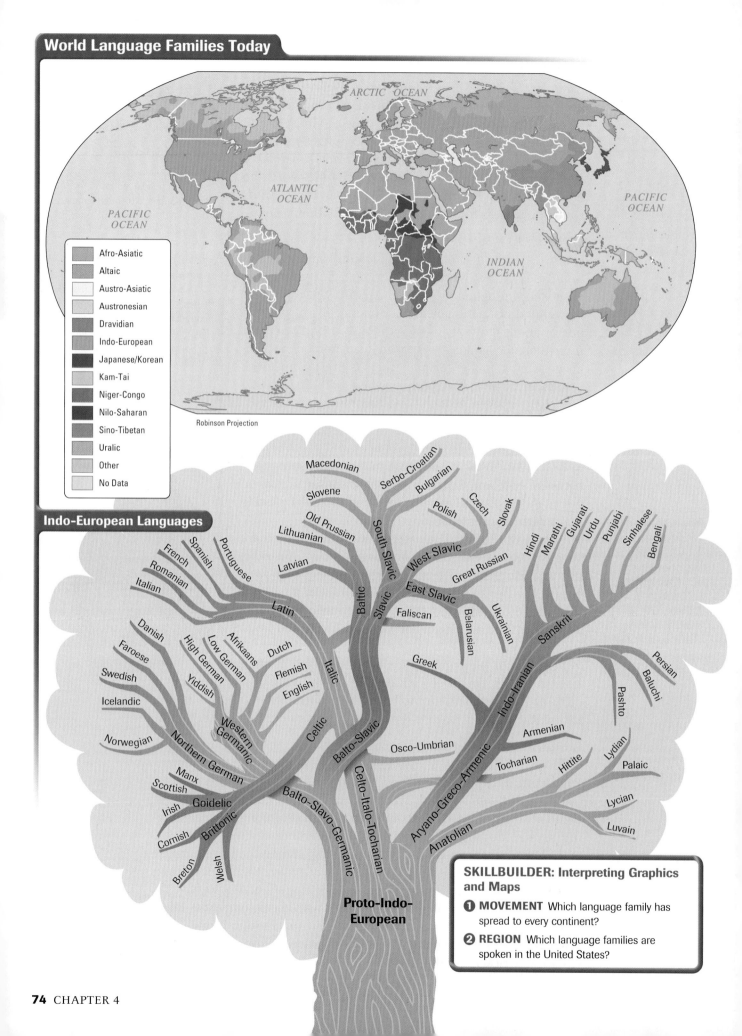

World Language Families Today

ARCTIC OCEAN

ATLANTIC OCEAN

PACIFIC OCEAN

PACIFIC OCEAN

INDIAN OCEAN

Robinson Projection

- Afro-Asiatic
- Altaic
- Austro-Asiatic
- Austronesian
- Dravidian
- Indo-European
- Japanese/Korean
- Kam-Tai
- Niger-Congo
- Nilo-Saharan
- Sino-Tibetan
- Uralic
- Other
- No Data

Indo-European Languages

Macedonian
Serbo-Croatian
Bulgarian
Slovene
Czech
Slovak
Old Prussian
Polish
Lithuanian
South Slavic
West Slavic
Hindi
Marathi
Gujarati
Urdu
Punjabi
Sinhalese
Bengali
Latvian
Baltic
Slavic
Great Russian
Spanish
French
Portuguese
Romanian
Italian
Latin
East Slavic
Faliscan
Belarusian
Ukrainian
Sanskrit
Persian
Danish
Afrikaans
Dutch
Greek
Baluchi
Faroese
High German
Low German
Flemish
Italic
Indo-Iranian
Pashto
Swedish
Yiddish
English
Armenian
Lydian
Icelandic
Western Germanic
Celtic
Balto-Slavic
Osco-Umbrian
Tocharian
Hittite
Palaic
Norwegian
Northern German
Aryano-Greco-Armenic
Lycian
Manx
Scottish
Goidelic
Balto-Slavo-Germanic
Celto-Italo-Tocharian
Anatolian
Luvain
Irish
Brittonic
Cornish
Breton
Welsh

Proto-Indo-European

SKILLBUILDER: Interpreting Graphics and Maps

① MOVEMENT Which language family has spread to every continent?

② REGION Which language families are spoken in the United States?

Religion

An aspect of culture that has a great deal of influence on people's lives is religion. **Religion** consists of a belief in a supernatural power or powers that are regarded as the creators and maintainers of the universe. Religions establish beliefs and values that define how people worship the divine being or divine forces and how they behave toward each other. Traditionally, religions have been categorized as one of three types:

- **monotheistic,** with a belief in one god
- **polytheistic,** with a belief in many gods
- **animistic** or traditional, often with a belief in divine forces in nature

SPREAD OF RELIGION Religions spread across the world through diffusion and through converts, people who give up their former beliefs for a new religion. Some religions, such as Christianity, Islam, and Buddhism, actively seek to convert people to their beliefs. Other religions, such as Judaism and Hinduism, do not. Finally, isolated pockets of religions, mostly animist, are found in Japan, Central Africa, Oceania, and among Native Americans of both North and South America. ◄ⓑ

Geographic Thinking◄
Seeing Patterns
ⓑ How does location contribute to the isolation of animist practice?

Major Religions

Three major religions of the world began in Southwest Asia and two in South Asia. The religions of Southwest Asia—Judaism, Christianity, and Islam—are monotheistic and share similar basic beliefs, and some prophets and teachers. Of the South Asian religions, Buddhism represents an adaptation of Hinduism.

JUDAISM The oldest of the Southwest Asian religions, Judaism is concentrated in Israel. Followers, called Jews, live in Israel, the United States, Canada, South America, and many European cities. Established more than 3,200 years ago, Judaism is the oldest monotheistic religion. It is considered an ethnic religion with a long tradition of faith and culture tied tightly together. The basic laws and teachings come from a holy book called the Torah. The religious center of Judaism is the city of Jerusalem in Israel.

CHRISTIANITY Christianity evolved about 2,000 years ago from the teachings of Judaism. It, too, is monotheistic. Christianity is based on the teachings of Jesus Christ, whom Christians believe was the Son of God. The teachings of Jesus are recorded in the New Testament of the Bible. The religion spread from Jerusalem, first through the work of the Apostle Paul and, later, by many missionaries. It is the largest of all the religions with 2 billion followers. Christians live on every continent. Christianity has three major groups: Roman Catholic, Protestant, and Eastern Orthodox.

ISLAM The third religion that originated in Southwest Asia is Islam. It is based on the teachings of the Prophet Muhammad, who began teaching around 613 A.D. Its followers are known as Muslims. Islam is a monotheistic religion in which followers worship God, who is called Allah in Arabic. The religion has close ties to the prophets and teachers of Judaism and Christianity. The holy book of the Muslims is the

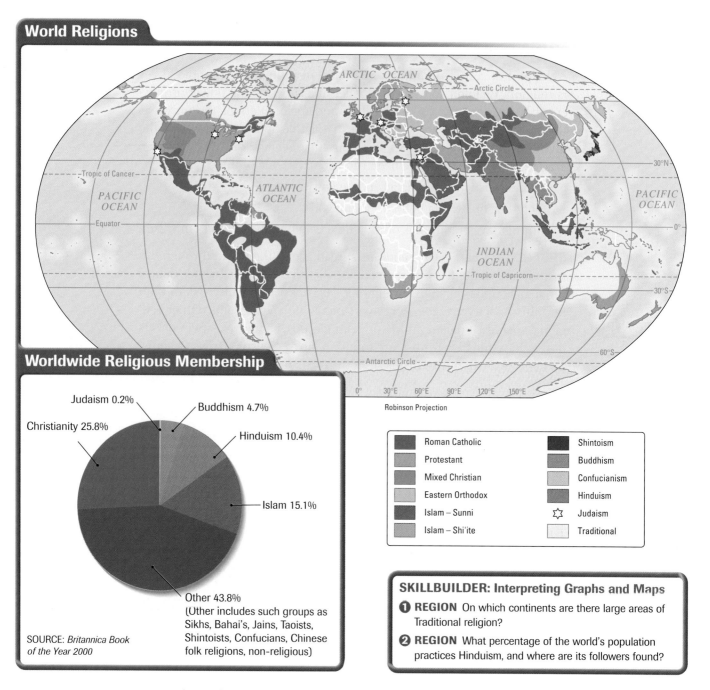

Worldwide Religious Membership

Judaism 0.2%

Buddhism 4.7%

Christianity 25.8%

Hinduism 10.4%

Islam 15.1%

Other 43.8%
(Other includes such groups as Sikhs, Bahai's, Jains, Taoists, Shintoists, Confucians, Chinese folk religions, non-religious)

SOURCE: *Britannica Book of the Year 2000*

Robinson Projection

Roman Catholic	Shintoism
Protestant	Buddhism
Mixed Christian	Confucianism
Eastern Orthodox	Hinduism
Islam – Sunni	☆ Judaism
Islam – Shi'ite	Traditional

SKILLBUILDER: Interpreting Graphs and Maps

1 REGION On which continents are there large areas of Traditional religion?

2 REGION What percentage of the world's population practices Hinduism, and where are its followers found?

Qur'an. Islam spread from Southwest Asia to Africa, Central, South, and Southeast Asia, and parts of the Balkans in Europe. The two major divisions of Islam are Sunni and Shiíte.

HINDUISM One of the world's oldest religions, Hinduism dates back about 5,000 years. It is an ethnic religion concentrated in India, but has followers elsewhere. Hinduism is usually considered polytheistic because a Hindu may believe in one god or many gods, each of whom represents an aspect of the divine spirit, Brahman. The religious requirements of a caste system—levels of fixed social classes with specific rites and duties—shape many aspects of Hindus' lives and culture.

BUDDHISM An offshoot of Hinduism, Buddhism developed about 563 B.C. in India, near the Nepal border. Its founder, Siddhartha Gautama (also called the Buddha or Enlightened One), rejected the Hindu idea of caste. Buddha's teachings promote the correct way of living in order to reach an enlightened spiritual state called nirvana. Missionaries spread the Buddha's teaching from India to Southeast Asia, China, Japan, and Korea. Buddhism has several branches, the largest of which are Theravada, Mahayana, Lamaism, and Zen.

OTHER ASIAN PRACTICES In parts of East Asia, three belief systems are widely practiced. They are Confucianism, Taoism, and Shinto. Sometimes those belief systems are thought of as religions and sometimes as philosophies of life. All of them have specific ways of life and behaviors associated with them.

Creative Cultural Expressions

All cultures have ways of expressing themselves creatively. The environment and culture in which an artist lives is reflected in the artistic product. Cultures produce performing arts, visual arts, and literature.

Performing arts developed by a culture often include music, dance, theater, and film. Music is a cultural aspect found in all societies. The instruments on which the music is played and the style of music are unique to each group.

Seeing Patterns
⚪ How might climate affect the visual arts of a region?

Visual arts include architecture, painting, sculpture, and textiles. The style of the visual arts will reflect materials available in the region and cultural themes. ⚪

Oral and written literature, such as poems, folk tales, and stories, often illustrate aspects of the culture such as attitudes and behaviors. They can also be a reflection of the environment in which they are produced.

Throughout this book, you will find discussions of creative cultural expressions. As you study them, remind yourself that each culture is unique—as are the artistic expressions that the people from that culture produce.

HUMAN–ENVIRONMENT INTERACTION This Peruvian bone flute dates back to sometime before 700 A.D. Bone flutes are among the oldest of all musical instruments.
In what way does this instrument show human-environment interaction?

SECTION ① Assessment

① Places & Terms

Explain the meaning of each of the following terms.
- culture
- society
- ethnic group
- diffusion
- acculturation
- dialect

② Taking Notes

MOVEMENT Review the notes you took for this section.

> Human Geography
> |
> Cultural

- In what ways is culture diffused?
- Which religions have spread from the place where they were founded?

③ Main Ideas

a. What factors make up culture?

b. In what ways is language spread?

c. What are the major religions of the world?

④ Geographic Thinking

Determining Cause and Effect What role do innovation and diffusion play in changing a culture? **Think about:**
- contact with other groups
- acculturation

RESEARCH LINKS CLASSZONE.COM

MAKING COMPARISONS Choose one of the factors of culture listed on page 71. Then select three countries. Use the Internet to find information on how each culture solves the problems associated with the factor you selected. Create a **database** showing the results of your research.

Population Geography

Main Ideas
• People are not distributed equally on the earth's surface.
• The world's population continues to grow, but at different rates in different regions.

Places & Terms

birthrate

fertility rate

mortality rate

infant mortality rate

rate of natural increase

population pyramid

push-pull factors

population density

carrying capacity

A HUMAN PERSPECTIVE In 1999, the world's population reached 6 billion people. To get an idea of how many people that is, consider this:

If you had a *million* dollars in thousand dollar bills, the stack would be 6.3 inches high. If you had a *billion* dollars in thousand dollar bills, the stack would be 357 feet high, or about the length of a football field including the end zones. Now multiply by 6. Six billion dollars would be almost 6 football fields high.

At the world's natural growth rate in 1999, that 6 billion population figure was reached by the births of 230,000 people each day.

Worldwide Population Growth

The earth's population hit the one billion mark in the early 1800s. As the world industrialized, people grew more and better food and improved sanitation methods, and the population of the world began to soar. As more and more women reached childbearing age, the number of children added to the population also increased. As you can see in the diagram at the right, by 1930 two billion people lived on the earth. Notice that the number of years between each billion mark gets smaller.

BIRTH AND DEATH RATES A population geographer studies aspects of population such as birth and death rates, distribution, and density. To understand population growth, geographers calculate several different statistics. One is the **birthrate,** which is the number of live births per thousand population. In 2000, the highest birthrate in the world was more than 54 per thousand in Niger, and the lowest rate was about 8 per thousand in Latvia. The world average birthrate is 22 per thousand.

Another way to study population is to look at the fertility rate. The **fertility rate** shows the average number of children a woman of childbearing years would have in her lifetime, if she had children at the current rate for her country. A fertility rate of 2.1 is necessary just to replace current population. Today, the worldwide average fertility rate is about 3.0.

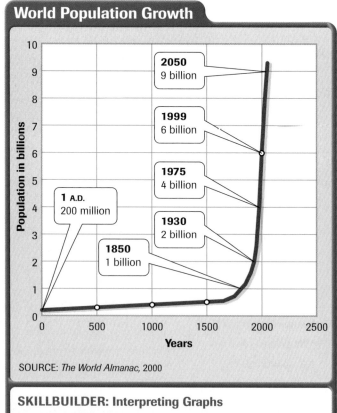

World Population Growth

- **2050** 9 billion
- **1999** 6 billion
- **1975** 4 billion
- **1 A.D.** 200 million
- **1930** 2 billion
- **1850** 1 billion

SOURCE: *The World Almanac*, 2000

SKILLBUILDER: Interpreting Graphs

❶ **ANALYZING DATA** How long did it take for the population to reach one billion?

❷ **MAKING GENERALIZATIONS** How have the intervals between increases changed?

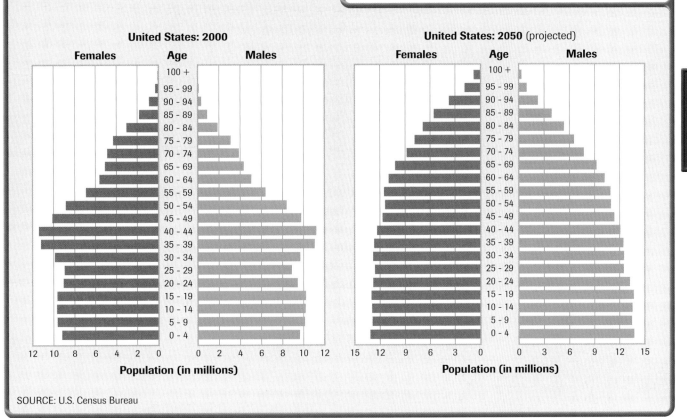

U.S. Population Pyramids, 2000 and 2050

A population pyramid presents a quick picture of a country's population distribution by age and sex. The effects of events in society can also be seen. Notice that in the year 2000 pyramid there is a bulge between ages 35 to 49. This reflects the "baby boom" generation born after World War II.

SKILLBUILDER: Interpreting Graphs

❶ **ANALYZING DATA** How old are the people in the "baby boom" generation in the 2000 pyramid?

❷ **DRAWING CONCLUSIONS** Why will the numbers for the very elderly (85+) increase so much by the year 2050?

SOURCE: U.S. Census Bureau

BASICS

The **mortality rate**—also called the death rate—is the number of deaths per thousand people. In general, a society is considered healthy if it has a low mortality rate. However, some healthy nations have higher mortality rates because they have large numbers of elderly people.

For this reason, geographers also look at infant mortality rates in measuring how healthy the people of a nation are. The **infant mortality rate** shows the number of deaths among infants under age one per thousand live births. In the 1800s, the worldwide infant mortality rate was about 200 to 300 deaths per thousand live births. At the beginning of the 21st century, improved health care and nutrition led to a much lower rate worldwide. However, some parts of the world still record as many as 110 infant deaths per thousand. To find the rate at which population is growing, subtract the mortality rate from the birthrate. The difference is the **rate of natural increase,** or population growth rate. ◀

POPULATION PYRAMID Another way to analyze populations is to use a **population pyramid,** a graphic device that shows sex and age distribution of a population. A population pyramid allows geographers to examine how events in society, such as wars, famine, or epidemics, affect the population of a country or region. Study the population pyramids shown above to learn how to interpret these graphics.

Geographic Thinking

Seeing Patterns
A⟩ What will the rate of natural increase be like if the birthrate is high and the mortality rate is low?

Population Distribution

The billions of people in the world are not distributed equally across the earth. Some lands are not suitable for human habitation. In fact, almost 90 percent of the world's population lives in the Northern Hemisphere. One in four people in the world lives in East Asia, and one of every two people lives in either East Asia or South Asia. Several factors, including climate, altitude, and access to water, influence where people live.

HABITABLE LANDS Almost two-thirds of the world's population lives in the zone between 20° N and 60°N latitude. Some of the lands in this zone have suitable climate and vegetation for dense human habitation. They are warm enough and wet enough to make agriculture possible. In addition, populations are concentrated along coastal regions and river valleys. The lightly populated areas are in polar regions, heavily mountainous regions, and desert regions. **B**

Geographic Thinking

Seeing Patterns
B Why are populations concentrated along coastal regions and river valleys?

URBAN–RURAL MIX Currently, more than half of the world's population lives in rural areas, but that number is changing rapidly. More people are moving into cities—particularly cities with populations of more than one million people. Twenty-six giant cities, called megacities, are home to a total of more than 250 million people. The largest of these is Tokyo, with more than 28 million inhabitants. These huge cities struggle with overcrowded conditions and immense demand for water and sanitation. You'll learn more about cities and their populations in the Urban Geography section of this chapter.

World Population Density

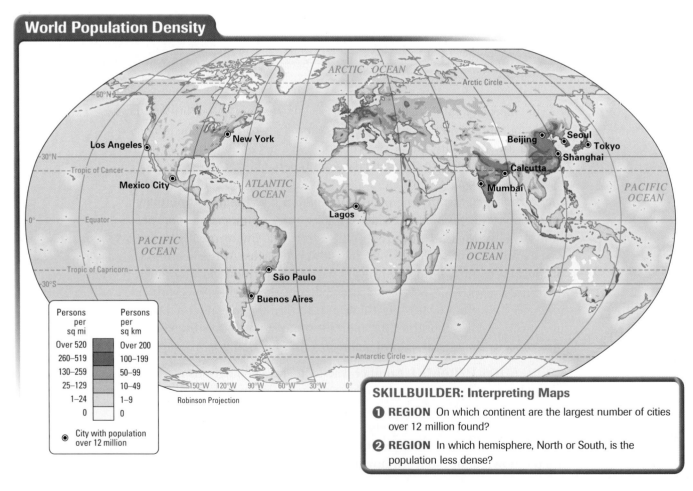

Persons per sq mi | Persons per sq km
Over 520 | Over 200
260–519 | 100–199
130–259 | 50–99
25–129 | 10–49
1–24 | 1–9
0 | 0

● City with population over 12 million

Robinson Projection

SKILLBUILDER: Interpreting Maps

❶ **REGION** On which continent are the largest number of cities over 12 million found?

❷ **REGION** In which hemisphere, North or South, is the population less dense?

4 Urban Geography

A HUMAN PERSPECTIVE Around 4500 B.C. in Sumer, an ancient country in what today is Iraq, the city of Ur was settled. Eventually it grew to be home to as many as 34,000 people. Archaeologists believe that it was one of the first cities in the world. Within the city walls, a broad avenue led up to an immense temple with a roof that loomed 80 feet above the ground. Surrounding the temple were private homes and large open markets with shops on streets resembling those in cities of Southwest Asia today. Some people lived in two-story houses with balconies and even had clay-lined drains for waste disposal. A canal ran through the city from the river to a harbor built on its northern edge. This was not an overgrown village, but a real city.

In the centuries since, cities have grown so important that geographers have developed the field of **urban geography**—the study of how people use space in cities.

Growth of Urban Areas

Today, much of the population of the world lives in cities. **Cities** are not just areas with large populations—they are also centers of business and culture. Cities are often the birthplace of innovation and change in a society. Urban lifestyles are different from those of towns, villages, or rural areas. When geographers study urban areas, they consider location, land use, and functions of the city.

URBAN AREAS An urban area develops around a main city called the central city. The built-up area around the central city may include **suburbs,** which are political units touching the borders of the central city or touching other suburbs that touch the city. These suburbs are within commuting distance of the city. Some suburbs are mostly residential, while others have a whole range of urban activities.

Smaller cities or towns with open land between them and the central city are called exurbs. The city, its suburbs, and exurbs link together economically to form a functional area called a **metropolitan area.** A megalopolis is formed when several metropolitan areas grow together. An example of a megalopolis is the corridor in the northeastern United States including Boston, New York, Philadelphia, Baltimore, and Washington, D.C.

PLACE Both the old city and the new parts of Cairo, Egypt, can be seen in this view.

Why do you think the old parts of the city were not torn down and replaced with new buildings?

Main Ideas
- Nearly half the world's population lives in urban areas.
- Cities fulfill economic, residential, and cultural functions in different ways.

Places & Terms
urban geography

city

suburb

metropolitan area

urbanization

central business district (CBD)

BASICS

87

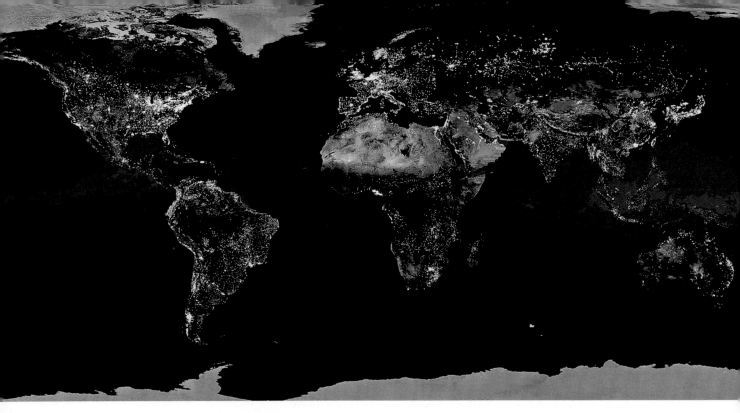

PLACE Urban areas are clearly visible in this satellite view of earth at night. The light blue areas are "reflective" areas with either snow pack or sand.
Which regions of the earth have few urban areas?

URBANIZATION The dramatic rise in the number of cities and the changes in lifestyle that result is called **urbanization.** The trend to live in cities increased rapidly over the last two centuries. As more and more people moved into cities to find work, the cities and their surrounding areas grew. Today, some cities are enormous in physical area and have populations exceeding 10 million residents. As you can see above, cities are found on all continents except Antarctica.

City Locations

Around the world, cities have certain geographic characteristics in common. Many cities are found in places that allow good transportation, such as on a river, lake, or coast. Others are found in places with easy access to natural resources. Sacramento, California, for instance, grew rapidly after gold was discovered in 1848 in north-central California. Because of their geographic advantages, cities serve as economic bases, attracting businesses and people to work in those businesses.

Cities are often places where goods are shifted from one form of transportation to another. For example, the city of Chicago, Illinois, is a transportation hub for goods produced in the upper Great Lakes states. Goods are sent by air, truck, or train to Chicago on Lake Michigan, then to the U.S. east coast and the rest of the world.

Cities may specialize in certain economic activities because of their location. For example, the city of Pittsburgh, Pennsylvania, which is located close to iron ore and coal sources, became a steel-producing center. The same is true for the city of Sheffield in England. Some urban areas may grow or expand because of economic activities located in the city. Brasília, the capital of Brazil, has grown to 1.8 million people since 1960 because of all the government agencies and activities there. Cultural, educational, or military activities may also attract people to a specific location.

Geographic Thinking

Using the Atlas
Ⓐ Use the map of North America on page A10. What waterway leads from the Great Lakes to the Atlantic Ocean?

Land Use Patterns

Urban geographers also study land use, the activities that take place in cities. Basic land use patterns found in all cities are:

- **residential,** including single-family housing and apartment buildings
- **industrial,** areas reserved for manufacturing of goods
- **commercial,** used for private business and the buying and selling of retail products

The core of a city is almost always based on commercial activity. This area of the city is called the <u>**central business district (CBD).**</u> Business offices and stores are found in this part of the city. In some cities, very expensive housing may also be found there. Predictably, the value of the land in the CBD is very high. In fact, the land is so expensive that skyscrapers are often built to get the most value from the land.

As you move away from the CBD, other functions become more important. For example, residential housing begins to dominate land use. Generally, the farther you get from the CBD, the lower the value of the land. Lower land values may lead to less expensive housing. Tucked into these less expensive areas are industrial activities and retail areas, such as shopping centers, markets, or bazaars. However, the patterns for urban activities vary by culture and geography. Study the models below to learn more about urban land use patterns.

Geographic Thinking

Seeing Patterns
B Why do industrial activities take place where land is less expensive?

Urban Area Models

Geographers may use a model to illustrate patterns they find in the use of space. The models below are patterns of land use in urban areas.

Concentric Zone Model	Sector Model	Multiple Nuclei Model

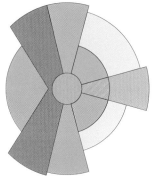

Multiple Nuclei Model

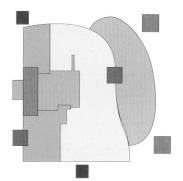

An early model showed the CBD as the "bull's-eye" of the urban area. It is surrounded by other activities.
by E. W. Burgess

Activities are concentrated in wedges or sectors, which may follow transportation lines or natural features such as a river.
by H. Hoyt

Districts, called nuclei, specialize in one urban activity, and are found throughout the urban area.
by C. D. Harris and E. L. Ullman

- Central business district
- Wholesale and light manufacturing
- Low-income housing
- Middle-income housing
- High-income housing
- Low-income and high-income housing
- Middle-income and high-income housing
- Heavy manufacturing
- Outlying business district
- Outer suburban housing
- Outer suburban industry
- High-income commuter zone

SKILLBUILDER: Interpreting Graphics

❶ **MAKING GENERALIZATIONS** Where is low-income housing found in each of the models?

❷ **MAKING COMPARISONS** What has happened to business and industry activities in the multiple nuclei model as compared to the other two models?

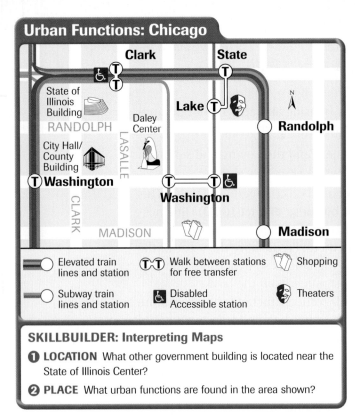

Urban Functions: Chicago

Clark

State

State of Illinois Building

RANDOLPH

Daley Center

Lake

Randolph

City Hall/ County Building

LA SALLE

T Washington

CLARK

Washington

Madison

MADISON

Key:
- Elevated train lines and station
- Subway train lines and station
- T~T Walk between stations for free transfer
- Disabled Accessible station
- Shopping
- Theaters

SKILLBUILDER: Interpreting Maps

❶ LOCATION What other government building is located near the State of Illinois Center?

❷ PLACE What urban functions are found in the area shown?

The Functions of Cities

The city is the center of a variety of functions. The map at the left shows a portion of the CBD of Chicago, Illinois. Notice that shopping, entertainment, and government services are located there. Large office buildings occupy much of the rest of the area shown.

Many cities also have educational and cultural activities such as libraries or museums located in the CBD. The Manhattan section of New York City, for example, is home to about 70 museums. Other functions of the city—such as manufacturing, wholesaling, residential, recreation, and a variety of religious and social services—may be located in other parts of the city.

Cities need a great deal of space to accomplish these functions, which makes good transportation absolutely essential. Major cities may have several forms of mass transit, such as bus systems, subways, or commuter trains, to move thousands of people to and from the areas of the city where the various functions take place. In some areas, freeway systems link people in the suburbs to the activities in the city. Geographers often study a city's transportation system to understand how well the city is fulfilling its functions.

In the next section, you'll learn more about economic geography that takes place across the globe.

Geographic Thinking

Making Comparisons
ⓒ How are city transportation systems different from those of towns or villages?

SECTION 4 Assessment

❶ Places & Terms

Explain the meaning of each of the following terms.

- city
- suburb
- metropolitan area
- urbanization
- central business district (CBD)

❷ Taking Notes

LOCATION Review the notes you took for this section.

Human Geography

Urban

- What functions or activities are located away from the CBD?
- In what types of relative locations are many cities found?

❸ Main Ideas

a. What components make up a metropolitan area?

b. What are some basic land use patterns in cities?

c. What are some functions of an urban area?

❹ Geographic Thinking

Making Inferences How does land value influence the activities that take place on a piece of urban land? **Think about:**

- land use patterns
- the CBD

S **See Skillbuilder Handbook, page R4.**

EXPLORING LOCAL GEOGRAPHY Survey the CBD of the city you live in or one close to you. Make notes of the urban functions you see there. Create a **sketch map** of your CBD. Be sure to label the areas or buildings, and the urban functions they fill.

Economic Geography

Main Ideas
- Economic activities depend on the resources of the land and how people use them.
- The level of economic development can be measured in different ways.

Places & Terms
economy
economic system
command economy
market economy
natural resources
infrastructure
per capita income
GNP
GDP

BASICS

A HUMAN PERSPECTIVE One of the most valuable of natural resources—petroleum—wasn't always used as a source of energy. Until the world began to run on gasoline-powered machinery, oil was used for a variety of purposes. Native Americans, for instance, used "rock oil" for medicinal purposes. Egyptians used oil as a dressing for wounds. Ancient Persians wrapped oil-soaked fibers around arrows, lit them, and fired them into the city of Athens in 480 B.C.

Sometimes a resource only becomes valuable after the technology to use it is developed. In today's world, petroleum is vital to providing power for industry, commerce, and transportation. Petroleum plays a powerful role in the economies of nations that supply it and consume it.

Economic Systems

An **economy** consists of the production and exchange of goods and services among a group of people. Economies operate on a local, regional, national, or international level. Geographers study economic geography by looking at how people in a region support themselves and how economic activities are linked across regions.

TYPES OF ECONOMIC SYSTEMS The way people produce and exchange goods and services is called an **economic system.** In the world today, there are four basic types of economic systems:

- **Traditional Economy** Goods and services are traded without exchanging money. Also called "barter."
- **Command Economy** Production of goods and services is determined by a central government, which usually owns the means of production. Production does not necessarily reflect the consumer demand. Also called a planned economy.
- **Market Economy** Production of goods and services is determined by the demand from consumers. Also called a demand economy or capitalism.
- **Mixed Economy** A combination of command and market economies provides goods and services so that all people will benefit.

Economic behaviors and activities to meet human needs take place within these economic systems.

PLACE A woman sells goods on a Moscow street. Russia is changing from a command economy to a market economy.
Is the activity in this photograph an example of a command or market economy?

Economic Activities

People may choose from a variety of methods to meet their basic needs. Some groups simply raise enough food or animals to meet their need to eat, but have little left over to sell to others. This is called subsistence agriculture. In other areas, market-oriented agriculture produces crops or animals that farmers sell to markets.

In some places, industries dominate economic activities. Small industries often involve a family of craftspersons who produce goods to be sold in a local area. Since they often take place in the home, these businesses are referred to as cottage industries. Finally, commercial industries meet the needs of people within a very large area. Economic behaviors are related to the economic activities described below.

LEVELS OF ECONOMIC ACTIVITY No matter how small or large a business is, it operates at one of four economic levels. The four levels of economic activity describe how materials are gathered and processed into goods or how services are delivered to consumers.

Primary Activities involve gathering raw materials such as timber for immediate use or to use in the making of a final product.

Secondary Activities involve adding value to materials by changing their form. Manufacturing automobiles is an example.

Tertiary Activities involve providing business or professional services. Salespeople, teachers, or doctors are examples.

Quaternary Activities provide information, management, and research services by highly-trained persons.

The more developed an economy is, the greater the number and variety of activities you will find.

Geographic Thinking

Making Comparisons
Ⓐ Into which level of activity would insurance sales fit?

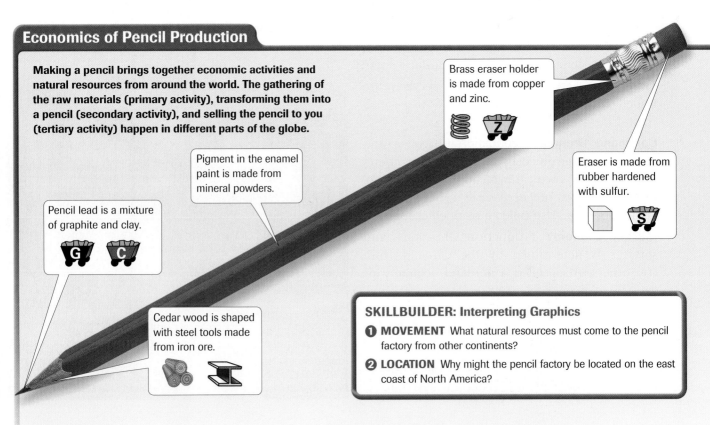

Economics of Pencil Production

Making a pencil brings together economic activities and natural resources from around the world. The gathering of the raw materials (primary activity), transforming them into a pencil (secondary activity), and selling the pencil to you (tertiary activity) happen in different parts of the globe.

Brass eraser holder is made from copper and zinc.

Pigment in the enamel paint is made from mineral powders.

Eraser is made from rubber hardened with sulfur.

Pencil lead is a mixture of graphite and clay.

Cedar wood is shaped with steel tools made from iron ore.

SKILLBUILDER: Interpreting Graphics

❶ **MOVEMENT** What natural resources must come to the pencil factory from other continents?

❷ **LOCATION** Why might the pencil factory be located on the east coast of North America?

The Economics of Natural Resources

An important part of economic geography is understanding which resources a nation possesses. **Natural resources** are materials on or in the earth—such as trees, fish, or coal—that have economic value. Materials from the earth become resources only when the society has the technology and ability to transform those resources into goods. So, iron ore is useless until people have the technology to produce steel from it.

Natural resources are abundant but are not distributed equally around the world. As a result, when geographers study the economy of a country, they look closely at the location, quality, and quantity of its natural resources. They also divide natural resources into three basic types:

- **Renewable**—These resources can be replaced through natural processes. Examples include trees and seafood.

- **Non-renewable**—These resources cannot be replaced once they have been removed from the ground. Examples include metals, such as gold, silver, and iron, and non-metals, such as gemstones, limestone, or sulfur. Also included are fossil fuels, petroleum, natural gas, and coal. They are the basis of energy production.

- **Inexhaustible energy sources**—These resources, which are used for producing power, are the result of solar or planetary processes and are unlimited in quantity. They include sunlight, geothermal heat, winds, and tides. ◄ B

Natural resources are a major part of world trade. This is especially true of the fossil fuels, since industry relies on them for both power and raw materials in manufacturing. The value of a natural resource depends on the qualities that make it useful. For example, trees can provide lumber for building or pulp for paper. Countries trade for raw materials that they need for energy and to manufacture products.

Making Comparisons
B What advantage do inexhaustible energy sources have over fossil fuels?

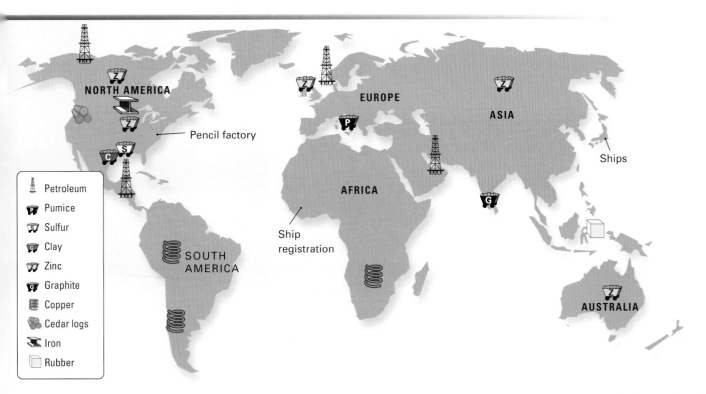

Pencil factory

Ships

Ship registration

- Petroleum
- Pumice
- Sulfur
- Clay
- Zinc
- Graphite
- Copper
- Cedar logs
- Iron
- Rubber

NORTH AMERICA

EUROPE

ASIA

AFRICA

SOUTH AMERICA

AUSTRALIA

BASICS

Levels of economic development are measured in the numbers of goods and services available in a country. This graphic compares the availability of televisions and passenger cars in three countries at different stages of development.

SOURCE: *2004 World Development Indicators* (World Bank)

DEVELOPING NATION
Ethiopia

6 televisions per 1,000 people

1 passenger car per 1,000 people

Economic Support Systems

Producing and distributing goods and services requires a series of support systems. The most important of these services is infrastructure.

INFRASTRUCTURE A nation's **infrastructure** consists of the basic support systems needed to keep an economy going, including power, communications, transportation, water, sanitation, and education systems. The more sophisticated the infrastructure, the more developed the country.

One of the most important systems in the infrastructure is transportation. Geographers look at the patterns of roads and highways, ports, and airports to get an idea of how transportation affects economic growth. For example, the country of Honduras has only one major north-south highway. The highway leads to port cities where a major export, bananas, is shipped out of the country. Areas not accessible to the major highway remain undeveloped.

Communications systems give geographers an idea of how a country is linked internally as well as with the outside world. Countries with a strong economy are linked internally and externally by high-speed Internet and satellite communications.

The level of available technology and access to it is also an indicator of the development of a country. A country may have valuable natural resources but be unable to profit from them because its people lack the skills to make use of them. Technology may be available, but a country may lack educated workers to run and maintain sophisticated equipment.

Measuring Economic Development

Geographers use a variety of standards to make comparisons among economies. One is **per capita income,** the average amount of money earned by each person in a political unit. Another way of comparing economies examines levels of development based on economic activities such as industry and commerce. Still others use a standard of living that reflects a society's purchasing power, health, and level of education.

GNP AND GDP A commonly-used statistic to measure the economy of a country is the **gross national product (GNP).** The **GNP** is the total value of all goods and services produced by a country over a year or some other specified period of time.

Because economies have become so interconnected, the GNP may reflect the value of goods or services produced in one country by a com-

251 televisions per 1,000 people

 785 televisions per 1,000 people

 16 passenger cars per 1,000 people

 283 passenger cars per 1,000 people

pany based in another country. For example, the value of sport shoes produced in Thailand by an American company is counted as U.S. production, even though the shoes were not produced in the United States. To adjust for situations like this, a second statistic is used—**GDP, or gross domestic product**—which is the total value of all goods and services produced *within* a country in a given period of time.

DEVELOPMENT LEVELS Countries of the world have different levels of economic development. Developing nations are nations that have a low GDP and limited development on all levels of economic activities. These countries lack an industrial base and struggle to provide their residents with items to meet their basic needs.

BACKGROUND
Developing countries that have greatly improved their GDP are called countries in transition.

Developed nations, on the other hand, are countries with a high per capita income and varied economy, especially with quaternary activities such as computer software development. Western European nations, Japan, Canada, and the United States have highly developed economies.

In this chapter, you've learned that human geography is a complex mix of human activities and the earth's resources. As you study the regions of the world, remember that a geographer views those regions by looking at the space and the interactions that take place there.

SECTION 5 Assessment

1 Places & Terms

Explain the meaning of each of the following terms.

- economy
- natural resources
- infrastructure
- per capita income
- GDP

2 Taking Notes

PLACE Review the notes you took for this section.

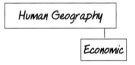

- What are the four basic economic systems?
- What are the three types of resources?

3 Main Ideas

a. What are the basic activities in each of the four economic activity levels?

b. What role do natural resources play in the economy of a country?

c. What systems are a part of a country's infrastructure?

4 Geographic Thinking

Drawing Conclusions Fossil fuels are non-renewable resources. What does this suggest about worldwide supplies of this energy?

Think about:

- industrial need for power
- alternative sources of power

MAKING COMPARISONS Study the types of economic systems on page 91. Create a series of **illustrations** showing the differences among the systems. Be sure your illustrations show the role of the consumer and the government in determining what goods or services are produced in each type of economy.

VISUAL SUMMARY
PEOPLE AND PLACES

The Elements of Culture

- All human groups have a culture.
- Language and religion are a part of culture.

Population Geography

- The world's population is expanding rapidly.
- Most of the world's population lives in the Northern Hemisphere.

Political Geography

- Size, shape, and location influence political geography.
- States of the world have a variety of political systems.

Urban Geography

- Urban areas have expanded rapidly and now are home to about one half of the world's population.
- Functions of cities are similar.
- Land use patterns are unique to a place.

Economic Geography

- Resources, available technology, and economic systems shape the economy of a state.
- Economic activities are based on how goods or services are produced and traded.

Reviewing Places & Terms

A. Briefly explain the importance of each of the following.

1. culture
2. diffusion
3. rate of natural increase
4. population density
5. state
6. nation
7. urbanization
8. economy
9. infrastructure
10. GDP

B. Answer the questions about vocabulary in complete sentences.

11. What is the growth in the number of cities called?
12. Which term above refers to the blueprint for the behaviors of a group?
13. How is the birthrate different from the rate of natural increase?
14. How is population density determined?
15. How is a nation different from a state?
16. Which term refers to the spread of ideas, innovations and inventions, and patterns of behavior?
17. How are the economy and the infrastructure related to each other?
18. What does the GDP number tell you about a country's economy?
19. Which terms above are associated with population geography?
20. What are some examples of infrastructure?

Main Ideas

The Elements of Culture (pp. 71–77)

1. What is the purpose of culture?
2. Why is language so important to a culture?

Population Geography (pp. 78–82)

3. What geographic factors influence population distribution?
4. How is population density different from population distribution?

Political Geography (pp. 83–86)

5. What are the geographic characteristics of a state?
6. What is the difference between a country with a democracy and one with a dictatorship?

Urban Geography (pp. 87–90)

7. What are some characteristics of city locations?
8. What are the basic land use patterns in cities?

Economic Geography (pp. 91–95)

9. Why does a country need an infrastructure?
10. How are natural resources related to a country's economy?

Critical Thinking

1. Using Your Notes

Use your completed chart to answer these questions.

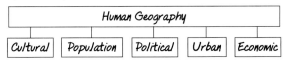

a. Which type of human geography focuses on how goods and services are produced and distributed by a country?

b. What do population geographers study?

2. Geographic Themes

a. **MOVEMENT** How might migration affect both population distribution and density?

b. **PLACE** What are some characteristics of an urban area?

3. Identifying Themes

How do landform and climate affect the distribution of population? Which of the five themes apply to this situation?

4. Making Inferences

Why might two groups of people living in the same area develop different cultures?

5. Identifying and Solving Problems

What reasons might countries have to form a regional political unit?

Additional Test Practice, pp. S1–S37

TEST PRACTICE
CLASSZONE.COM

Geographic Skills: Interpreting Maps

Dominant World Cities*

Use the map to answer the following questions.

1. **REGION** Which continent has the most dominant world cities shown?

2. **REGION** Which continents do not have dominant world cities?

3. **MOVEMENT** Into which continent does the most activity appear to flow? Give a reason for your answer.

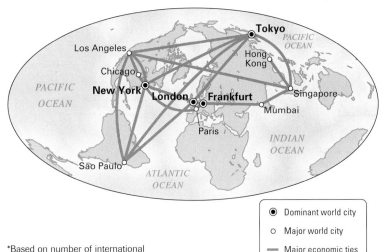

*Based on number of international banks and transactions

- ● Dominant world city
- ○ Major world city
- — Major economic ties

Molleweide Projection

Using a blank map of the world, mark in the cities shown on this map. Then go to page 80. Add the cities with more than 12 million shown on that map. On the back of your map, write two observations about the cities on your map.

Use the links at **classzone.com** to do research about population growth. Focus on the projected growth by 2050. Identify ten places where predicted growth will be the greatest and ten with little predicted growth.

Creating a Database Create a **database** showing your findings about worldwide growth. Create separate databases for the fastest growth and for the slowest growth. Be sure to label your databases.

Unit 2

The United States and Canada

The United States and Canada are two of the world's largest countries with vast lands and abundant resources. They occupy four-fifths of the continent of North America.

MOVEMENT Thousands of cars enter and leave Chicago, Illinois, daily. They use the vast expressway system that links the city to its surrounding suburbs and to interstate highways.

PLACE Cowhands and tourists from around the world gather in the western Canadian city of Calgary, Alberta, each July for the Calgary Stampede—the world's largest rodeo.

LOCATION The majestic falls of the Niagara River are shared by both the United States and Canada. The American Falls, to the left, are in New York state; Horseshoe Falls, to the right, are in Ontario, Canada.

Today's Issues in the United States and Canada

Today, the United States and Canada face the issues previewed here. As you read Chapters 5, 6, and 7, you will learn helpful background information. You will study the issues themselves in Chapter 8.

In a small group, answer the questions below. Then have a class discussion of your answers.

Exploring the Issues

1. **TERRORISM** Consider news stories that you have heard about terrorist groups in other countries. Make a list of the countries and the type of terrorist activity in each.

2. **URBAN SPRAWL** Why is the ever-expanding spread, or sprawl, of cities and suburbs a problem? What can be done to improve the quality of life in these areas?

3. **DIVERSE SOCIETIES** Search the Internet for information about diversity in the United States or Canada. What strategies or actions are being taken to help these many cultures unify?

For more on these issues in the United States and Canada . . .

CURRENT EVENTS
CLASSZONE.COM

TERRORISM

How can a country protect itself from terrorism?

A surprise attack, such as the one on the World Trade Center in New York City, is just one way terrorists attempt to intimidate governments and civilian populations to further their objectives.

URBAN SPRAWL

How can urban sprawl be controlled?

Urban communities, such as Las Vegas shown here, are trying to solve problems caused by urban areas spreading farther and farther out.

CASESTUDY

DIVERSE SOCIETIES

How can many cultures form a unified nation?

The diverse population of the United States is reflected in this group of California students. How to bring many cultures together as one nation is a continuing challenge for the United States, and for Canada, as well.

Patterns of Physical Geography

Use the Unit Atlas to add to your knowledge of the United States and Canada. As you look at the maps and charts, notice geographic patterns and specific details about the region.

After studying the illustrations, graphs, and physical map on these two pages, jot down answers to the following questions in your notebook.

Making Comparisons

1. Compare the world's longest river, the Nile, to the Mississippi. How much difference is there in the lengths of the two rivers?

2. Compare the landmass and population of the United States to those of Canada. What statement can be made about the two countries?

3. Compare the mountain peaks of the United States to those of Canada. What statement can be made about the height of these mountains?

For updated statistics on the United States and Canada . . .

DATA UPDATE
CLASSZONE.COM

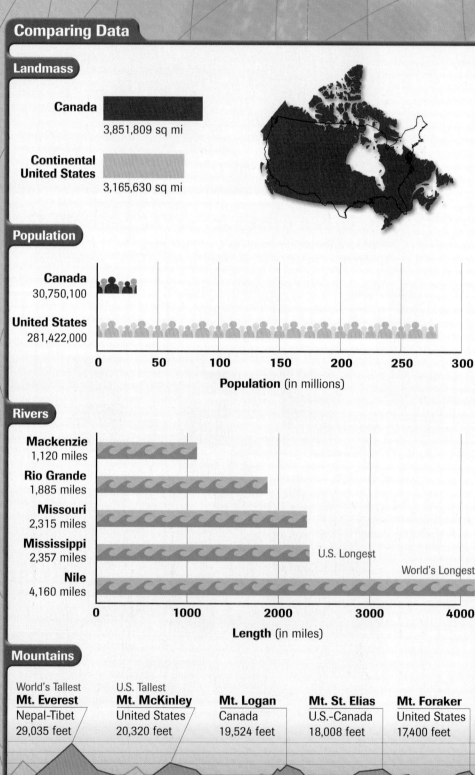

Comparing Data

Landmass

Canada 3,851,809 sq mi

Continental United States 3,165,630 sq mi

Population

Canada 30,750,100

United States 281,422,000

| 0 | 50 | 100 | 150 | 200 | 250 | 300 |

Population (in millions)

Rivers

Mackenzie 1,120 miles

Rio Grande 1,885 miles

Missouri 2,315 miles

Mississippi 2,357 miles — U.S. Longest

Nile 4,160 miles — World's Longest

| 0 | 1000 | 2000 | 3000 | 4000 |

Length (in miles)

Mountains

World's Tallest **Mt. Everest**	U.S. Tallest **Mt. McKinley**	**Mt. Logan**	**Mt. St. Elias**	**Mt. Foraker**
Nepal-Tibet 29,035 feet	United States 20,320 feet	Canada 19,524 feet	U.S.-Canada 18,008 feet	United States 17,400 feet

RUSSIA

ARCTIC OCEAN

Greenland Sea

ICELAND

Chukchi Sea

Beaufort Sea

Ellesmere Island

Greenland

Queen Elizabeth Islands

Baffin Bay

Bering Sea

Arctic Circle

Denmark Strait

BROOKS RANGE

ALASKA RANGE
Mt. McKinley
20,320 ft.
(6,194 m)

Yukon R.

Victoria Island

Baffin Island

Davis Strait

Kodiak Is.

Mt. Logan
19,551 ft.
(5,959 m)

MACKENZIE MTS.

Mackenzie R.

Great Bear Lake

Labrador Sea

Aleutian Islands

Gulf of Alaska

COAST MOUNTAINS

Great Slave Lake

Hudson Bay

Queen Charlotte Is.

Lake Athabasca

CANADA

CANADIAN SHIELD

Newfoundland

Vancouver I.

Lake Winnipeg

James Bay

Gulf of St. Lawrence

ROCKY MOUNTAINS

COAST RANGES

CASCADES

Columbia R.

GREAT PLAINS

Superior

St. Lawrence R.

APPALACHIAN MOUNTAINS

PACIFIC OCEAN

SIERRA NEVADA

GREAT BASIN

Missouri R.

Mississippi R.

Michigan

Huron

Ontario

Mt. Whitney
14,494 ft.
(4,421 m)

Death Valley
282 ft.
(86 m)

Colorado R.

UNITED STATES

Ohio R.

Bermuda Is.

Channel Is.

ATLANTIC OCEAN

Mississippi R.

COASTAL PLAIN

Gulf of California

Rio Grande

Gulf of Mexico

BAHAMAS

Tropic of Cancer

West Indies

MEXICO

CUBA

Hispaniola

Caribbean Sea

BELIZE

GUATEMALA

HONDURAS

NICARAGUA

VENEZUELA

EL SALVADOR

COSTA RICA

PANAMA

COLOMBIA

Hawaiian Islands

Nihau Kauai
Oahu
HAWAII Molokai
Lanai
Kahoolawe Maui

Hawaii

PACIFIC OCEAN

| 0 | 75 | 150 miles |
| 0 | 75 | 150 kilometers |

Scale

| 0 | 250 | 500 miles |
| 0 | 250 | 500 kilometers |

Azimuthal Equal–Area Projection

Elevation

13,100 ft.	(4,000 m.)
6,600 ft.	(2,000 m.)
1,600 ft.	(500 m.)
650 ft.	(200 m.)
0 ft.	(0 m.)
Below sea level	

▲ Mountain peak

Glacier

N S E W

US & CANADA

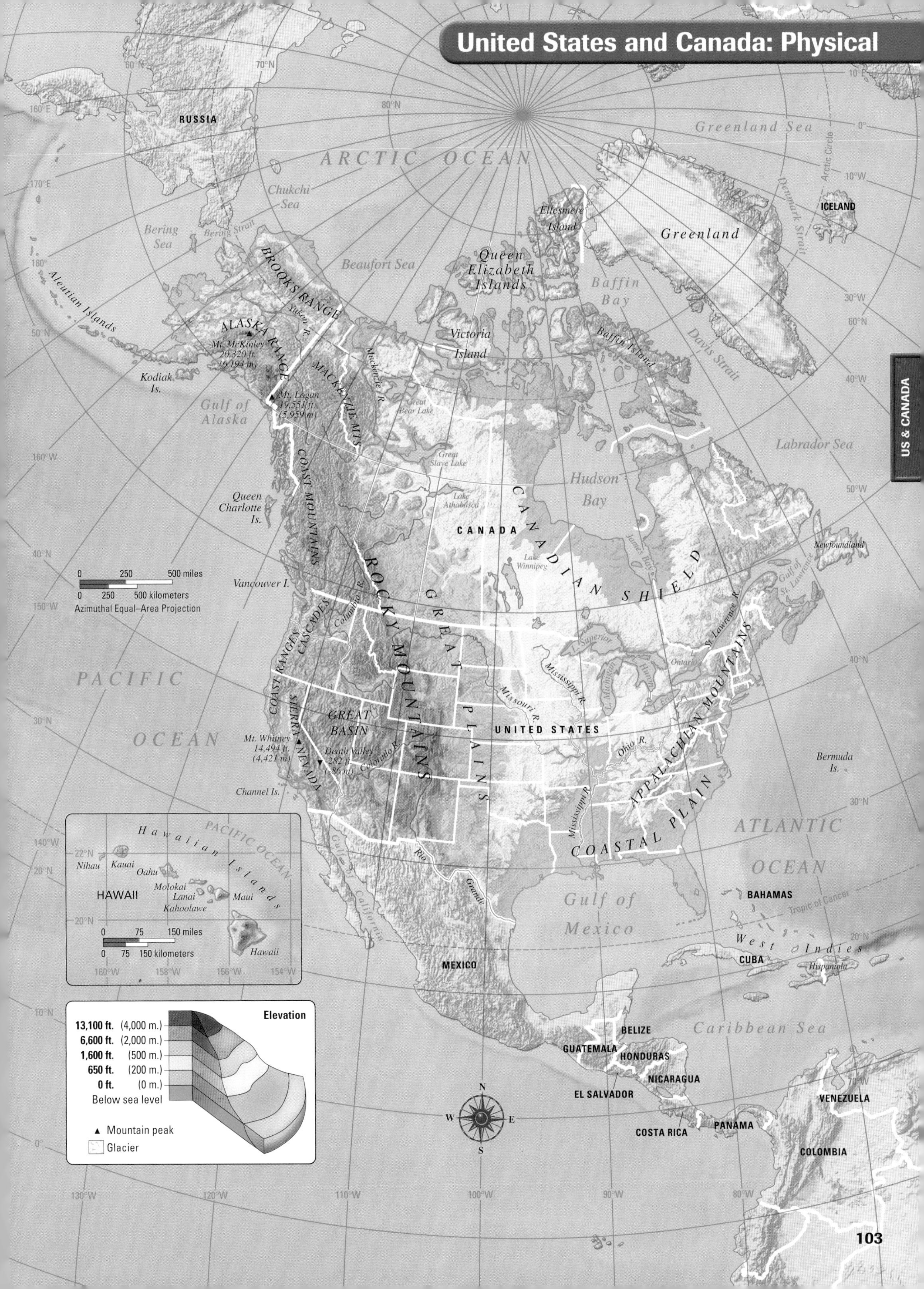

Unit ATLAS

Patterns of Human Geography

After the coming of European settlers in the 17th century, the political map of North America changed quickly and significantly. Study the historical and political maps of the United States and Canada on these two pages. In your notebook, answer these questions.

Making Comparisons

1. What differences do you notice when you compare the map of 1600 with the map of the United States and Canada today?

2. Which names of native peoples are found as geographic names on the map on page 105?

3. Which country was more sparsely settled by native peoples in 1600?

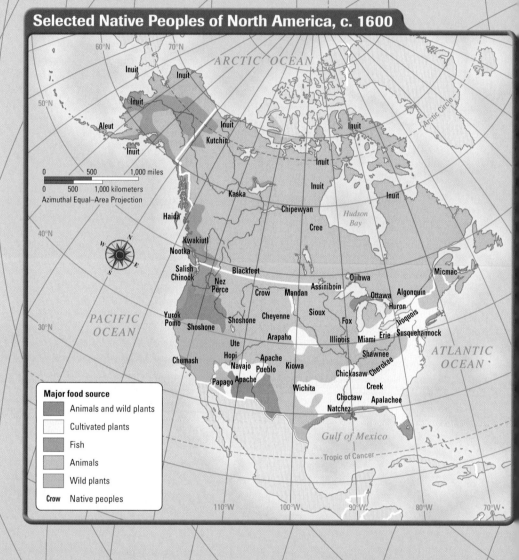

Selected Native Peoples of North America, c. 1600

Major food source
- Animals and wild plants
- Cultivated plants
- Fish
- Animals
- Wild plants
- **Crow** Native peoples

RUSSIA

ARCTIC OCEAN

Chukchi Sea

Bering Strait

Bering Sea

Beaufort Sea

Greenland Sea

Denmark Strait

ICELAND

GREENLAND (Den.)

Baffin Bay

Davis Strait

ALASKA (U.S.)

Anchorage

Yukon R.

Gulf of Alaska

YUKON TERRITORY

NORTHWEST TERRITORIES

Great Bear Lake

Mackenzie R.

NUNAVUT

Labrador Sea

Hudson Bay

James Bay

BRITISH COLUMBIA

Great Slave Lake

Lake Athabasca

C A N A D A

ALBERTA

SASKATCHEWAN

MANITOBA

Lake Winnipeg

ONTARIO

QUEBEC

NEWFOUNDLAND

Gulf of St. Lawrence

PRINCE EDWARD ISLAND

NEW BRUNSWICK

NOVA SCOTIA

Edmonton

Calgary

Winnipeg

Vancouver

Seattle

WASHINGTON

Columbia R.

OREGON

MONTANA

IDAHO

WYOMING

N. DAKOTA

S. DAKOTA

MINN.

WIS.

L. Superior

Mississippi R.

Missouri R.

Minneapolis

L. Michigan

L. Huron

St. Lawrence R.

Ottawa ★

Toronto

Montreal

VT.

MAINE

N.H.

MASS.

Boston

R.I.

CONN.

N.Y.

New York

L. Ontario

L. Erie

MICH.

Detroit

PENN.

Philadelphia

N.J.

DEL.

MD.

Washington, D.C. ★

Chicago

IOWA

NEBRASKA

IND.

OHIO

W.VA.

VIRGINIA

Salt Lake City

NEVADA

UTAH

U N I T E D   S T A T E S

COLORADO

Colorado R.

Kansas City

KANSAS

St. Louis

MO.

ILL.

Ohio R.

KENTUCKY

N.C.

San Francisco

PACIFIC OCEAN

CALIFORNIA

Los Angeles

San Diego

ARIZONA

Phoenix

NEW MEXICO

OKLAHOMA

ARK.

TENNESSEE

Mississippi R.

Atlanta

S.C.

GEORGIA

MISS.

ALA.

Dallas

TEXAS

LA.

Rio Grande

Houston

New Orleans

FLORIDA

Miami

Gulf of California

Gulf of Mexico

Bermuda (U.K.)

ATLANTIC OCEAN

BAHAMAS

Tropic of Cancer

CUBA

HAITI

DOMINICAN REPUBLIC

MEXICO

Caribbean Sea

BELIZE

HONDURAS

GUATEMALA

EL SALVADOR

NICARAGUA

COSTA RICA

PANAMA

COLOMBIA

Arctic Circle

US & CANADA

★ National capital
• Other city

0   250   500 miles
0   250   500 kilometers
Azimuthal Equal-Area Projection

Hawaiian Islands

PACIFIC OCEAN

Nihau

Kauai

Oahu

Honolulu

Molokai

Lanai

Kahoolawe

Maui

HAWAII

Hawaii

0   75   150 miles
0   75   150 kilometers

N   W   E   S

Regional Patterns

These pages contain three thematic maps and an infographic. The infographic illustrates economic connections between the United States and Canada. The maps show economic activities, population density, and areas affected by natural hazards.

Study these two pages and then answer the questions below in your notebook.

Making Comparisons

1. Where are the areas of greatest population density found in each country? Do settlement patterns have any relationship to the threat of natural hazards?

2. Where are manufacturing and trade concentrated in the United States and Canada? Why might this be so?

Canada-U.S. Connections

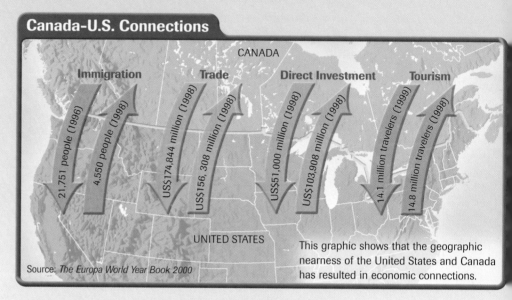

CANADA

Immigration
21,751 people (1996)
4,550 people (1998)

Trade
US$174,844 million (1998)
US$156,308 million (1998)

Direct Investment
US$51,000 million (1998)
US$103,908 million (1998)

Tourism
14.1 million travelers (1999)
14.8 million travelers (1998)

UNITED STATES

Source: *The Europa World Year Book 2000*

This graphic shows that the geographic nearness of the United States and Canada has resulted in economic connections.

Economic Activities of the U.S. and Canada

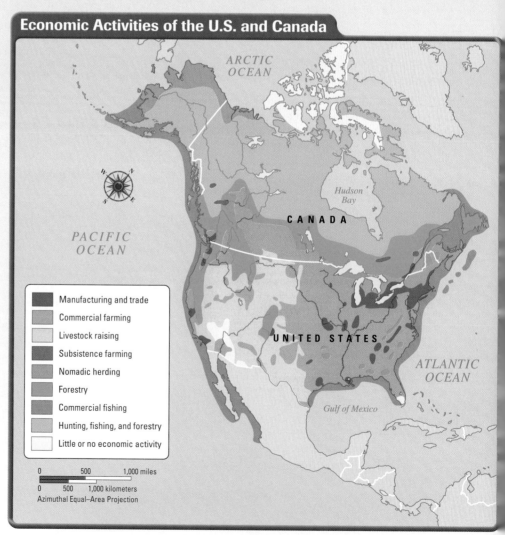

ARCTIC OCEAN

Hudson Bay

CANADA

PACIFIC OCEAN

UNITED STATES

ATLANTIC OCEAN

Gulf of Mexico

- Manufacturing and trade
- Commercial farming
- Livestock raising
- Subsistence farming
- Nomadic herding
- Forestry
- Commercial fishing
- Hunting, fishing, and forestry
- Little or no economic activity

0   500   1,000 miles
0 500   1,000 kilometers
Azimuthal Equal–Area Projection

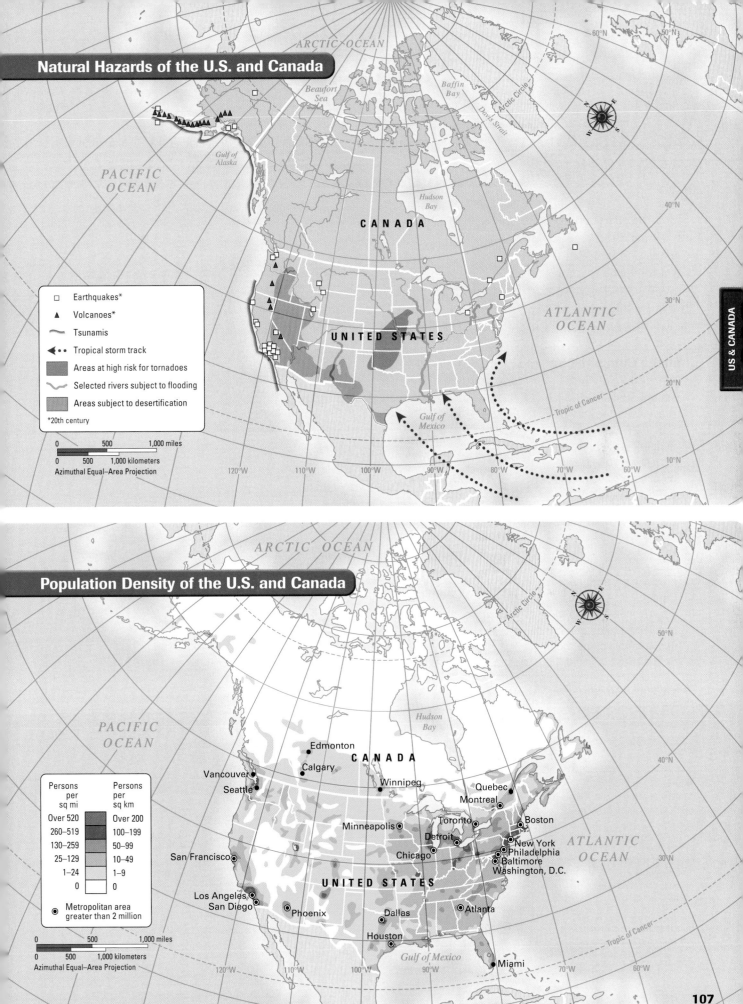

Natural Hazards of the U.S. and Canada

ARCTIC OCEAN

Beaufort Sea

Baffin Bay

David Strait

Arctic Circle

60°N
50°N
40°N
30°N
20°N
10°N

Gulf of Alaska

PACIFIC OCEAN

Hudson Bay

CANADA

ATLANTIC OCEAN

UNITED STATES

Gulf of Mexico

Tropic of Cancer

120°W 110°W 100°W 90°W 80°W 70°W 60°W

□ Earthquakes*
▲ Volcanoes*
〜 Tsunamis
◄··· Tropical storm track
▨ Areas at high risk for tornadoes
〜 Selected rivers subject to flooding
▨ Areas subject to desertification
*20th century

0   500   1,000 miles
0   500   1,000 kilometers
Azimuthal Equal–Area Projection

US & CANADA

Population Density of the U.S. and Canada

ARCTIC OCEAN

Arctic Circle

50°N
40°N
30°N

PACIFIC OCEAN

Hudson Bay

Edmonton
Calgary
CANADA
Vancouver
Seattle
Winnipeg
Quebec
Montreal
Minneapolis
Toronto
Boston
Detroit
New York
Chicago
Philadelphia
San Francisco
Baltimore
Washington, D.C.
ATLANTIC OCEAN
UNITED STATES
Los Angeles
San Diego
Phoenix
Dallas
Atlanta
Houston
Gulf of Mexico
Miami

Tropic of Cancer

120°W 110°W 100°W 90°W 70°W 60°W

Persons per sq mi	Persons per sq km
Over 520	Over 200
260–519	100–199
130–259	50–99
25–129	10–49
1–24	1–9
0	0

◉ Metropolitan area greater than 2 million

0   500   1,000 miles
0   500   1,000 kilometers
Azimuthal Equal–Area Projection

Regional Data File

Study the charts on the United States and Canada and their political subdivisions—states, provinces, and territories. In your notebook, answer these questions.

Making Comparisons

1. Which state of the United States and which province or territory of Canada have the most people? Is each also the largest in total area in its country? Locate them on the map. What is significant about their locations?

2. Which state of the United States and which province or territory of Canada have the least people? Is each also the smallest in total area in its country? Locate them on the map.

(continued on page 110)

Notes:

* The federal district of Washington, D.C., is the capital city of the United States.

ᵃ In constant 1996 dollars.

ᵇ Percentage of the population, 25 years old or older, with high school diploma or higher.

ᶜ Includes land and water, when figures are available.

For updated statistics on the United States and Canada . . .

DATA UPDATE
CLASSZONE.COM

Flag	State or Territory/ Capital	Population (2000)	Population Rank (2000)	Infant Mortality (per 1,000 live births) (1998)
	Alabama Montgomery	4,447,100	23	10.2
	Alaska Juneau	626,900	48	5.9
	Arizona Phoenix	5,130,600	20	7.5
	Arkansas Little Rock	2,673,400	33	8.9
	California Sacramento	33,871,600	1	5.8
	Colorado Denver	4,301,300	24	6.7
	Connecticut Hartford	3,405,600	29	7.0
	Delaware Dover	783,600	45	9.6
	District of Columbia*	572,100	–	12.5
	Florida Tallahassee	15,982,400	4	7.2
	Georgia Atlanta	8,186,500	10	8.5
	Hawaii Honolulu	1,211,500	42	6.9
	Idaho Boise	1,294,000	39	7.2
	Illinois Springfield	12,419,300	5	8.4
	Indiana Indianapolis	6,080,500	14	7.6
	Iowa Des Moines	2,926,300	30	6.6
	Kansas Topeka	2,688,400	32	7.0
	Kentucky Frankfort	4,041,800	25	7.5
	Louisiana Baton Rouge	4,469,000	22	9.1

Doctors (per 100,000 pop.) (1998–1999)	Population Density (per square mile)	Urban/Rural Population (%) (1990)	Per Capita Income[a] ($US) (1999)	High School Graduates[b] (%) (1998)	Area Rank (2000)	Total Area[c] (square miles)	
291	266.6	69 / 31	27,420	84.1	33	46,058	
338	851.6	86 / 14	24,418	80.7	50	1,231	
207	128.6	55 /45	22,467	78.6	40	31,189	
184	9.8	50 / 50	24,007	86.3	17	77,121	
246	135.0	61 / 39	24,461	76.9	36	42,146	
203	78.0	80 / 20	25,363	78.3	2	267,277	
200	26.3	87 / 13	22,333	89.3	13	84,904	
305	63.3	32 / 68	24,758	86.7	43	9,615	
241	167.2	69 / 31	28,193	82.6	35	42,326	
235	83.4	76 / 24	28,968	92.0	19	70,637	
215	74.6	36 / 64	19,973	76.4	41	24,231	
227	82.0	66 / 34	26,212	88.0	22	65,499	
171	5.0	65 / 35	24,864	90.0	9	97,818	
0.3 (1996)	727.2	33 / 67	3,270 (1995)	61.3 (1995)	—	90	
0.9 (1995)	712.5	38 / 62	19,000 (1996)	73.1 (1995)	—	217	
1.8	1,116.2	72 / 28	9,800 (1995)	49.7 (1989)	—	3,508	
1.1 (1989)	707.1	37 / 63	10,942 (1995)	58.6 (1995)	—	171	

Making Comparisons
(continued)

7. Which state and which province or territory is the most densely populated? Which state and which territory is the least densely populated? Are the most densely populated the smallest in area and the least populated the largest in area?

Flag	Province or Territory/ Capital	Population (2000)	Population Rank (2000)	Infant Mortality (per 1,000 live births) (1997)
	Alberta Edmonton	2,997,200	4	4.8
	British Columbia Victoria	4,063,800	3	4.7
	Manitoba Winnipeg	1,147,900	5	7.5
	New Brunswick Fredericton	756,600	8	5.7
	Newfoundland St. John's	538,800	9	5.2
	Northwest Territories Yellowknife	42,100	11	10.9
	Nova Scotia Halifax	941,000	7	4.4
	Nunavut Iqaluit	27,700	13	N/A
	Ontario Toronto	11,669,300	1	5.5
	Prince Edward Island Charlottetown	138,900	10	4.4
	Quebec Quebec City	7,372,400	2	5.6
	Saskatchewan Regina	1,023,600	6	8.9
	Yukon Territory Whitehorse	30,700	12	8.4
	Canada Ottawa, Ontario	30,750,100	–	5.5
	United States Washington, D.C.	281,422,000	–	7.0

Sources:
Bureau of Economic Analysis, U.S. Dept. of Commerce
Canadian Institute for Health Information, online
Census 2000, U.S. Census Bureau, online
Digest of Educational Statistics 2000, online
Europa World Year Book 2000
Merriam-Webster's Geographical Dictionary, 1997
Northwest Territories Bureau of Statistics, online
Pan-American Health Organization, online
Statistical Abstract of the United States, 1999 and 2000
Statistics Canada, online
World Factbook 2000, CIA online
N/A = not available

Notes:
ª In constant 1996 dollars.
ᵇ Percentage of the population, 25 years old or older, with high school diploma or higher.
ᶜ Includes land and water, when figures are available.

Doctors (per 100,000 pop.) (1998)	Population Density (per square mile)	Urban/Rural Population (%) (1996)	Per Capita Income[a] ($US) (1996)	High School Graduates[b] (%) (1998)	Area Rank (2000)	Total Area[c] (square miles)	
162	11.7	80 / 20	30,038	86	6	255,285	
193	11.1	82 / 18	31,592	87	5	366,255	
177	4.6	72 / 28	26,829	79	8	250,934	
153	26.7	49 / 51	26,607	78	11	28,345	
171	12.4	57 / 43	27,692	71	10	43,359	
92	0.08	42 / 58	33,738 (1994)	64 (1996)	3	503,951	
196	44.0	55 / 45	25,712	78	12	21,425	
N/A	0.03	N/A	27,421 (1994)	N/A	1	818,959	
178	28.3	83 / 17	32,537	84	4	412,582	
128	49.4	44 / 56	25,534	74	13	2,814	
211	12.4	78 / 22	28,826	78	2	594,860	
149	4.1	63 / 34	26,463	82	7	251,700	
149	0.2	60 / 40	36,130	67 (1996)	9	186,661	
185	8.0	78 / 22	23,000 (1999)	82	–	3,851,809	
251	74.3	76 / 24	33,900 (1999)	83	–	3,787,319	

Chapter 5

PHYSICAL GEOGRAPHY OF THE UNITED STATES and CANADA

A Land of Contrasts

SECTION 1
Landforms and Resources

SECTION 2
Climate and Vegetation

SECTION 3
Human–Environment Interaction

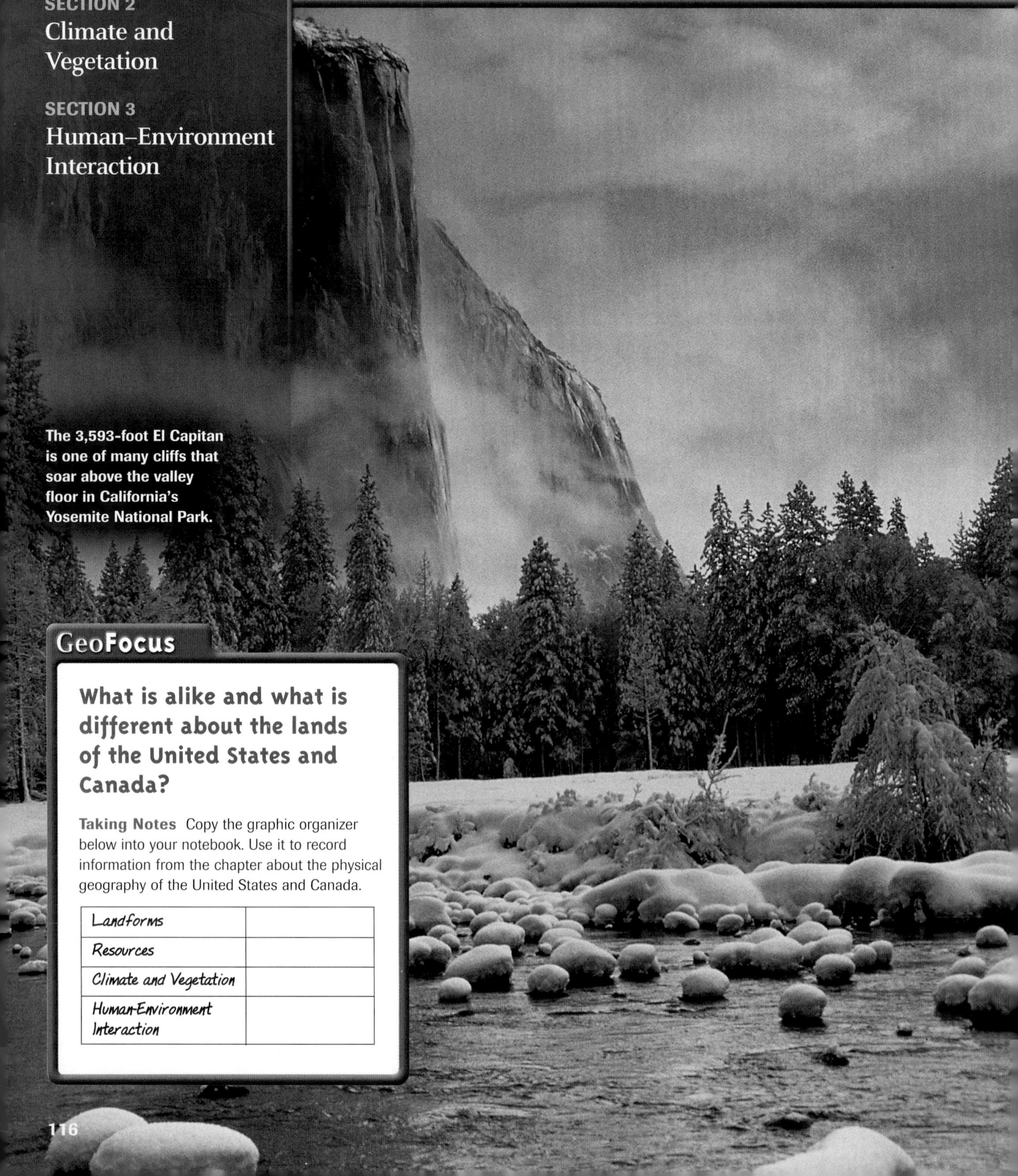

The 3,593-foot El Capitan is one of many cliffs that soar above the valley floor in California's Yosemite National Park.

GeoFocus

What is alike and what is different about the lands of the United States and Canada?

Taking Notes Copy the graphic organizer below into your notebook. Use it to record information from the chapter about the physical geography of the United States and Canada.

Landforms	
Resources	
Climate and Vegetation	
Human-Environment Interaction	

Landforms and Resources

Main Ideas

• The United States and Canada have vast lands and abundant resources.

• These two countries share many of the same landforms.

Places & Terms

Appalachian Mountains

Great Plains

Canadian Shield

Rocky Mountains

Great Lakes

Mackenzie River

US & CANADA

CONNECT TO THE ISSUES

URBAN SPRAWL Urban development in the United States is generally determined by the location of landforms and the abundance of natural resources.

A HUMAN PERSPECTIVE The beauty and abundance of the land was a source of wonder to early explorers of North America. One who traveled the Atlantic coast referred to the "amazing extent of uncultivated land, covered with forests, and intermixed with vast lakes and marshes." A 17th–century French expedition described "a beautiful river, large, broad, and deep" (the Mississippi). Still others found "an unbounded prairie" (the Great Plains), "shining mountains" (the Rocky Mountains), and "an infinite number of fish" (along the Pacific coast). To the continent's first settlers, the land was "strong and it was beautiful all around," according to an old Native American song.

Landscape Influenced Development

The United States and Canada occupy the central and northern four-fifths of the continent of North America. Culturally, the region is known as Anglo America because both countries were colonies of Great Britain at one time and because most of the people speak English. (The southern one-fifth of the continent—Mexico—is part of Latin America.) The two countries are bound together not only by physical geography and cultural heritage, but also by strong economic and political ties.

VAST LANDS The United States and Canada extend across North America from the Atlantic Ocean on the east to the Pacific on the west, and from the Arctic Ocean on the north to the Gulf of Mexico on the south (only the United States). In total area, each ranks among the largest countries of the world. Canada ranks second, behind Russia, and the United States is third. Together, they fill one-eighth of the land surface of the earth.

ABUNDANT RESOURCES In addition to their huge landmass, the United States and Canada are rich in natural resources. They have fertile soils, ample supplies of water, vast forests, and large deposits of a variety of minerals. This geographic richness has for centuries attracted immigrants from around the world and has enabled both countries to develop into global economic powers.

LOCATION Pittsburgh, Pennsylvania, is located where the Allegheny and Monongahela rivers meet to form the Ohio River.

ARCTIC OCEAN

170°E

180°

160°W

50°N

40°N

160°W

PACIFIC
OCEAN

30°N

ARCTIC
COASTAL PLAIN

INTERMOUNTAIN BASINS AND PLATEAUS

R O C K Y

M O U N T A I N S

PACIFIC MOUNTAINS AND VALLEYS

INTERMOUNTAIN BASINS AND PLATEAUS

Mackenzie R.

Columbia R.

Colorado R.

Rio
Grande

C A N A D I A N   S H I E L D

I N T E R I O R

G R E A T   P L A I N S

Hudson
Bay

ARCTIC COASTAL PLAIN

L. Superior

Mississippi R.

Missouri R.

L. Michigan

Huron

L. Erie

Ontario

St. Lawrence R.

APPALACHIAN HIGHLANDS

PIEDMONT

Ohio R.

INTERIOR
HIGHLANDS

Mississippi R.

GULF-ATLANTIC COASTAL PLAIN

Gulf of Mexico

Tropic of Cancer

ATLANTIC
OCEAN

60°N   40°W

50°N   50°W

40°N   40°W

30°N

1 East Quoddy Head in New Brunswick

2 Great Regina Plain in Saskatchewan

3 Cypress Gardens in Florida

N
W   E
S

0   250   500 miles
0   250   500 kilometers
Azimuthal Equal-Area Projection

0°

130°W   120°W   110°W   100°W   90°W

Many and Varied Landforms

All major types of landforms are found in the United States and Canada. If you look at the map on the opposite page, you will see that both countries share many of these landforms. The most prominent are eastern and western mountain chains and enormous interior plains.

THE EASTERN LOWLANDS A flat, coastal plain runs along the Atlantic Ocean and the Gulf of Mexico. One section, called the Atlantic Coastal Plain, begins as narrow lowland in the northeastern United States and widens as it extends southward into Florida. This area features many excellent harbors. A broader section of the plain—the Gulf Coastal Plain—stretches along the Gulf of Mexico from Florida into Texas. The Mississippi River empties into the Gulf from this region.

Between these plains and the nearby Appalachian (A·puh·LAY·chun) Highlands is a low plateau called the Piedmont (PEED·MAHNT). This area of rolling hills contains many fast-flowing rivers and streams.

BACKGROUND
The word *piedmont* comes from *pied*, meaning "foot," and *mont*, for "mountain." A piedmont is found at the foot of a mountain chain.

THE APPALACHIAN HIGHLANDS West of the coastal plain are the Appalachian highlands. The gently sloping **Appalachian Mountains** are in this region. They are one of the two major mountain chains in the United States and Canada. Both chains run north to south. The Appalachian Mountains extend some 1,600 miles from Newfoundland in Canada to Alabama. There are several mountain ranges in the Appalachian system. Among them are the Green and the Catskill mountains in the north and the Blue Ridge and the Great Smoky mountains in the south.

Because the Appalachians are very old—more than 400 million years old—they have been eroded by the elements. Many peaks are only between 1,200 and 2,400 feet high. The Appalachian Trail, a scenic hiking path 2,160 miles long, spans almost the entire length of the chain.

THE INTERIOR LOWLANDS A huge expanse of mainly level land covers the interior of North America. It was flattened by huge glaciers thousands of years ago. The terrain includes lowlands, rolling hills, thousands of lakes and rivers, and some of the world's most fertile soils.

The interior lowlands are divided into three subregions: the Interior Plains, the Great Plains, and the Canadian Shield. The Interior Plains spread out from the Appalachians to about 300 miles west of the Mississippi River. They gradually rise from a few hundred feet above sea level to about 2,000 feet. To the west are the **Great Plains,** a largely treeless area that continues the ascent to about 4,000 feet. The **Canadian Shield** lies farther north. This rocky, mainly flat area covers nearly 2 million square miles around Hudson Bay. It averages 1,500 feet above sea level but reaches over 5,000 feet in Labrador. ◀A

THE WESTERN MOUNTAINS, PLATEAUS, AND BASINS West of the plains are the massive, rugged **Rocky Mountains,** the other major mountain system of the

Geographic Thinking

Making Comparisons
A▶ Which of the interior lowlands has the highest elevation?

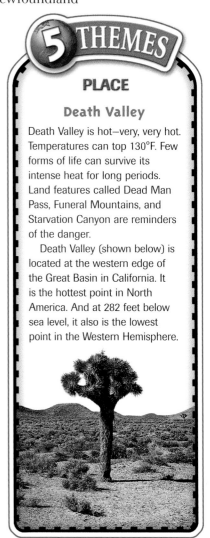

5 THEMES

PLACE

Death Valley

Death Valley is hot—very, very hot. Temperatures can top 130°F. Few forms of life can survive its intense heat for long periods. Land features called Dead Man Pass, Funeral Mountains, and Starvation Canyon are reminders of the danger.

Death Valley (shown below) is located at the western edge of the Great Basin in California. It is the hottest point in North America. And at 282 feet below sea level, it also is the lowest point in the Western Hemisphere.

United States and Canada. The Rockies are a series of ranges that extend about 3,000 miles from Alaska south to New Mexico. Because they are relatively young—about 80 million years old—the Rockies have not been eroded like the Appalachians. Many of their jagged, snow-covered peaks are more than 12,000 feet high. The **Continental Divide** is the line of highest points in the Rockies that marks the separation between rivers flowing eastward and westward.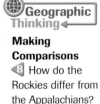

Between the Rockies and the Pacific Ocean is an area of mixed landforms. A series of ranges, including the Sierra Nevada and the Cascade Range, run parallel to the Pacific coastline from California to Alaska. North America's highest peak—Mt. McKinley (also called by its Native American name, Denali)—is in Alaska, towering 20,320 feet above sea level. Major earthquakes occur near the Pacific ranges. Between these

Geographic Thinking

Making Comparisons
B How do the Rockies differ from the Appalachians?

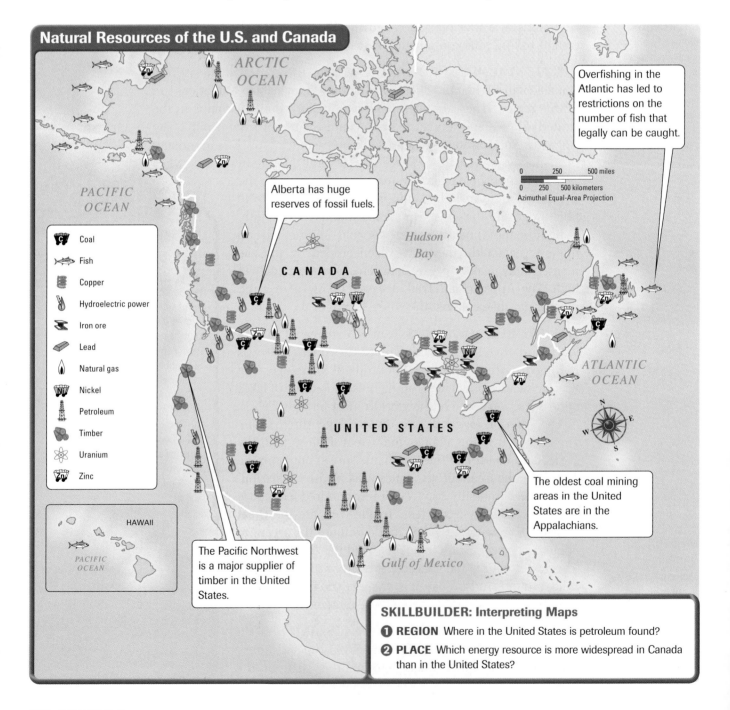

Natural Resources of the U.S. and Canada

Overfishing in the Atlantic has led to restrictions on the number of fish that legally can be caught.

Alberta has huge reserves of fossil fuels.

0 250 500 miles
0 250 500 kilometers
Azimuthal Equal-Area Projection

Coal
Fish
Copper
Hydroelectric power
Iron ore
Lead
Natural gas
Nickel
Petroleum
Timber
Uranium
Zinc

ARCTIC OCEAN
PACIFIC OCEAN
CANADA
Hudson Bay
UNITED STATES
ATLANTIC OCEAN
Gulf of Mexico
HAWAII
PACIFIC OCEAN

The oldest coal mining areas in the United States are in the Appalachians.

The Pacific Northwest is a major supplier of timber in the United States.

SKILLBUILDER: Interpreting Maps
❶ **REGION** Where in the United States is petroleum found?
❷ **PLACE** Which energy resource is more widespread in Canada than in the United States?

ranges and the Rockies are steep cliffs, deep canyons, and lowland desert areas called basins.

THE ISLANDS Canada's northernmost lands are islands riding the icy seas near the Arctic Circle. Three of the islands—Ellesmere, Victoria, and Baffin—are huge. In North America, only Greenland is larger.

Two island chains created by volcanic activity are part of the westernmost United States. The rugged, treeless Aleutian Islands extend in an arc off the coast of Alaska. The lush, tropical Hawaiian Islands, though politically part of the United States, are not geographically part of North America. They lie in the central Pacific, about 2,400 miles to the southwest.

Resources Shape Ways of Life

The landforms of the United States and Canada hold a rich variety and abundance of natural resources. Both countries are leading agricultural and industrial nations because of this wealth of resources.

OCEANS AND WATERWAYS The United States and Canada possess ample water resources. They are bounded by three oceans—Atlantic, Pacific, and Arctic. The United States is also bounded by the Gulf of Mexico. As a result, both countries have important shipping and fishing industries.

Inland, large rivers and lakes serve as sources of transportation, hydroelectric power, irrigation, fresh water, and fisheries. Eight of the world's 15 largest lakes are found in this region. Among these are the **Great Lakes**— Huron, Ontario, Michigan, Erie, and Superior. As you will see on page 129, these lakes and the St. Lawrence River form one of the world's major shipping routes.

The continent's longest and busiest river system is the Mississippi-Missouri-Ohio. The Mississippi River runs almost the north-south length of the United States, from Minnesota to the Gulf of Mexico. (See map at right.) The Mississippi's main tributaries, the Ohio and Missouri rivers, are major rivers in their own right. Canada's longest river is the **Mackenzie River,** which is part of a river system that flows across the Northwest Territories to the Arctic Ocean. ◄

LAND AND FORESTS One of the richest natural resources of the United States and Canada is the land itself. Both countries are large and contain some of the most fertile soils in the world. In fact, the land is so productive that North America is the world's leading food exporter. Much of this agricultural land is found in the plains regions and in river valleys.

Geographic Thinking

Using the Atlas
▷ Use the map on page 103. Find the Mackenzie River. Into which body of water does it empty?

The Mississippi River

The Mississippi begins its 2,357-mile journey southward in Lake Itasca, Minnesota.

About 500 million tons of freight, including grain, are carried on the river each year.

Ocean-going ships can only navigate the Mississippi to Baton Rouge—80 miles north of New Orleans—because of low water.

0   100   200 miles
0   100   200 kilometers
Albers Equal-Area Projection

SKILLBUILDER: Interpreting Maps

❶ **LOCATION** What states have the Mississippi River for at least part of their border?

❷ **MOVEMENT** What rivers empty into the Mississippi?

REGION This West Virginia coal mine is in one of the world's most important coal-producing regions—the Appalachian highlands. **What other region in North America is an important coal producer?**

The United States and Canada also have huge forests. About one-half of Canada is covered by woodlands, as is one-third of the United States. Canada's forests cover more land than those of the United States, but the United States has more kinds of trees because of its more varied climate. Both countries are major producers of lumber and forest products.

MINERALS AND FOSSIL FUELS As you saw on the map on page 120, the United States and Canada have large quantities and varieties of minerals and fossil fuels. These resources gave both countries the means to industrialize rapidly.

Valuable deposits of iron ore, nickel, copper, gold, and uranium are found in the Canadian Shield. Scattered among the western mountains are gold, silver, copper, and uranium. Both countries also have substantial deposits of coal, natural gas, and oil, and well-developed networks for distributing these energy-producing fossil fuels. Important coal-producing areas are the Appalachian highlands and the northern Great Plains. Significant deposits of oil and natural gas are found in the Great Plains, Alaska, and along the Gulf of Mexico. Ⓓ

The United States is the world's biggest consumer of energy resources. Its need for these fuels is so great that it is a major importer. In fact, most of Canada's energy exports go to its neighbor to the south.

In the next section, you will read how some landforms of the United States and Canada have affected climate and vegetation patterns.

Geographic Thinking

Seeing Patterns
Ⓓ Why are oil and natural gas important to highly-industrialized nations?

SECTION

⬥ Assessment

❶ Places & Terms

Identify and explain where in the region these would be found.

- Appalachian Mountains
- Great Plains
- Canadian Shield
- Rocky Mountains
- Great Lakes

❷ Taking Notes

LOCATION Review the notes you took for this section.

Landforms	
Resources	

- What is the relative location of the Great Lakes?
- What is the relative location of most of Canada's islands?

❸ Main Ideas

a. What landforms are shared by the United States and Canada?

b. Why are the Great Lakes important to both the United States and Canada?

c. Why do most of Canada's energy exports go to the United States?

❹ Geographic Thinking

Making Generalizations
What makes the United States and Canada leading industrial nations? **Think about:**

- available resources
- oceans and waterways

Ⓢ **See Skillbuilder Handbook, page R6.**

EXPLORING LOCAL GEOGRAPHY Using the maps on pages 103 and 118, identify the landforms located in your state. Then draw a **sketch map** of your state showing the major landforms and water bodies.

Climate and Vegetation

Main Ideas
- Almost every type of climate is found in the 50 United States because they extend over such a large area north to south.
- Canada's cold climate is related to its location in the far northern latitudes.

Places & Terms

permafrost

prevailing westerlies

Everglades

CONNECT TO THE ISSUES
URBAN SPRAWL The rapid spread of urban sprawl has led to the loss of much vegetation in both the United States and Canada.

US & CANADA

A HUMAN PERSPECTIVE A little gold and bitter cold—that is what thousands of prospectors found in Alaska and the Yukon Territory during the Klondike gold rushes of the 1890s. Most of these fortune hunters were unprepared for the harsh climate and inhospitable land of the far north. Winters were long and cold, the ground frozen. Ice fogs, blizzards, and avalanches were regular occurrences. You could lose fingers and toes—even your life—in the cold. But hardy souls stuck it out. Legend has it that one miner, Bishop Stringer, kept himself alive by boiling his sealskin and walrus-sole boots and then drinking the broth.

Shared Climates and Vegetation

The United States and Canada have more in common than just frigid winter temperatures where Alaska meets northwestern Canada. Other shared climate and vegetation zones are found along their joint border at the southern end of Canada and the northern end of the United States.

If you look at the map on page 125, you will see that the United States has more climate zones than Canada. This variety, ranging from tundra to tropical, occurs because the country extends over such a large area north to south. Most of the United States is located in the mid-latitudes, where the climates are moderate. Canada is colder because so much of it lies far north in the higher latitudes.

COLDER CLIMATES The Arctic coast of Alaska and Canada have tundra climate and vegetation. Winters are long and bitterly cold, while summers are brief and chilly. Even in July, temperatures are only around 40°F. The land is a huge, treeless plain. Much of the rest of Canada and Alaska have a subarctic climate, with very cold winters and short, mild summers. A vast forest of needle-leafed evergreens covers the area. In some places, there is **permafrost,** or permanently frozen ground.

The Rocky Mountains and the Pacific ranges have highland climate and vegetation. Temperature and vegetation vary with elevation and latitude. Generally, the temperature is colder and the vegetation is more sparse in the higher, more northerly mountains. The mountains also influence the temperature and precipitation of surrounding lower areas. For example, the

MOVEMENT The snowmobile has replaced the dogsled as transportation in many parts of the Northwest Territories. Here, a mother picks up her children from school.

coastal ranges protect the coast from cold Arctic air from the interior. In the United States, the western mountains trap Pacific moisture. This makes lands west of the mountains rainy and those east very dry.

MODERATE CLIMATES The north central and northeastern United States and southern Canada near the U.S. border have a humid continental climate. Winters are cold and summers warm. Climate and soil make this one of the world's most productive agricultural areas, yielding an abundance of dairy products, grain, and livestock. In the northern part of this climate zone, summers are short. There are mixed forests of deciduous and needle-leafed evergreen trees. Most of the population of Canada is concentrated here. In the southern part of this zone, which is in the United States, summers are longer. For the most part, deciduous forests are found east of the Mississippi River and temperate grasslands are found to the west.

The Pacific coast from northern California to southern Alaska, which includes British Columbia, has a climate described as marine west coast. This climate is affected by Pacific Ocean currents, the coastal mountains, and the **prevailing westerlies**—winds that blow from west to east in the middle of the latitudes. The summers are moderately warm. The winters are long and mild, but rainy and foggy. Vegetation is mixed, including dense forests of broad-leafed deciduous trees, needle-leafed evergreens, and giant California redwoods. The Washington coast even has a cool, wet rain forest.

Geographic Thinking

Seeing Patterns
A Why is most of Canada's population clustered in the humid continental region?

Differences in Climate and Vegetation

The milder, dry, and tropical climates of North America are found south of 40°N latitude. Much of the United States is located in these climate zones; little of Canada is.

MILDER CLIMATES Most southern states have a humid subtropical climate. This means that summers are hot and muggy, with temperatures ranging from about 75°F to 90°F. Winters are usually mild and cool. Moist air from the Gulf of Mexico brings rain during the winter. The combination of mild temperatures and adequate rainfall provides a long growing season for a variety of crops—from citrus fruits in Florida to peanuts in Georgia. Broad-leafed evergreen trees and needle-leafed evergreen trees are found in this region. The central and southern coasts of California have a Mediterranean climate. Summers are dry, sunny, and warm. Winters are mild and somewhat rainy. Temperatures range from 50°F to 80°F year-round. A long growing season and irrigation make this a rich farming area for fruits and vegetables.

Geographic Thinking

Making Comparisons
B Why don't central and southern California have a marine west coast climate?

DRY CLIMATES The Great Plains and dry northern parts of the Great Basin have a semiarid climate. This means dry weather—only about 15 inches of rain annually—and vegetation that is mainly short grasses and shrubs. The southwestern states have a desert climate. In these states, the weather is usually hot and dry. Less than 10 inches of rain falls each year. Some cactus plants thrive, but much of the area is barren rock or sand. Large desert areas are the Mojave and the Sonoran.

TROPICAL CLIMATES In the United States, only Hawaii and southern Florida have tropical climates. The islands of Hawaii have a tropical wet climate that supports lush rain forests. Temperatures vary only

Climate and Vegetation of the U.S. and Canada

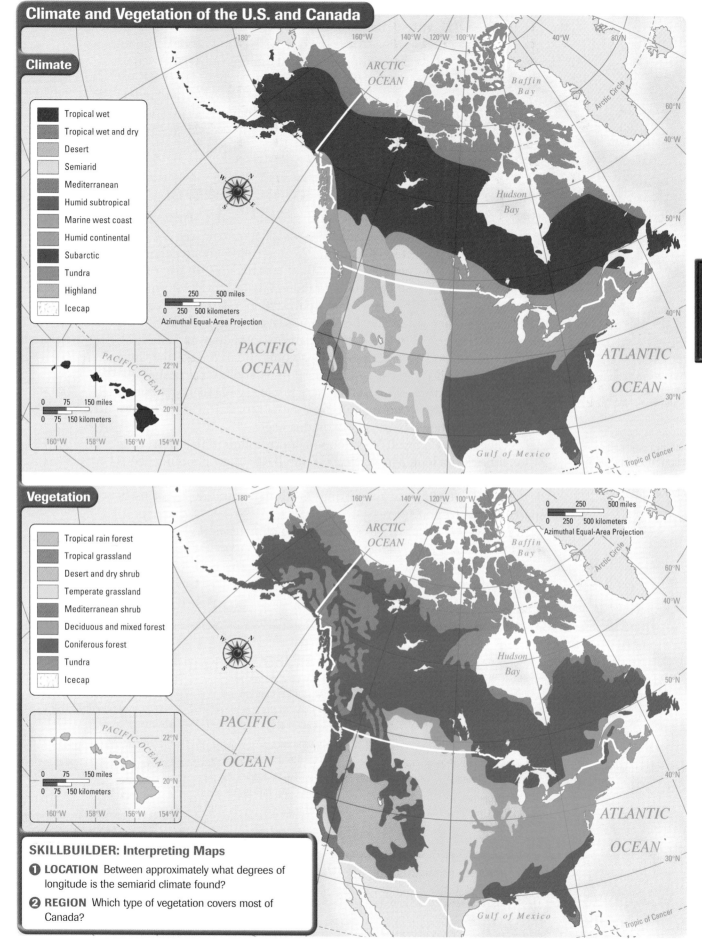

Climate

- Tropical wet
- Tropical wet and dry
- Desert
- Semiarid
- Mediterranean
- Humid subtropical
- Marine west coast
- Humid continental
- Subarctic
- Tundra
- Highland
- Icecap

0   250   500 miles
0   250   500 kilometers
Azimuthal Equal-Area Projection

0   75   150 miles
0   75   150 kilometers

Vegetation

- Tropical rain forest
- Tropical grassland
- Desert and dry shrub
- Temperate grassland
- Mediterranean shrub
- Deciduous and mixed forest
- Coniferous forest
- Tundra
- Icecap

0   250   500 miles
0   250   500 kilometers
Azimuthal Equal-Area Projection

0   75   150 miles
0   75   150 kilometers

SKILLBUILDER: Interpreting Maps

❶ **LOCATION** Between approximately what degrees of longitude is the semiarid climate found?

❷ **REGION** Which type of vegetation covers most of Canada?

US & CANADA

REGION Deadly ice storms like this one in Watertown, New York, create chaos each winter, especially in heavily populated areas. **What are some of the hazards of this form of extreme weather?**

a few degrees in the 70s°F. Mount Waialeale (wy•AH•lay•AH•lay) on Kauai island receives about 460 inches of rain annually, and is one of the wettest spots on earth. Southern Florida has a tropical wet and dry climate. It is nearly always warm, but has wet and dry seasons. Vegetation is mainly tall grasses and scattered trees, like those in the **Everglades,** a huge swampland that covers some 4,000 square miles. ▶

Effects of Extreme Weather

Weather in the United States and Canada can be harsh and sometimes deadly. You can see the areas affected by extreme weather and climate conditions by looking at the natural hazards map on page 107.

In both cold and mild climates, severe storms can trigger widespread devastation. Warm air from the Gulf of Mexico and cold Canadian air masses sometimes clash over the plains region to produce violent thunderstorms, tornadoes, and blizzards. As you read in Unit 1, tornadoes strike so often in an area of the Great Plains that it is called "Tornado Alley." In summer and fall, hurricanes that sweep along the Atlantic and Gulf coasts can cause great damage. Winter snowstorms may bring normal life to a temporary halt in many cities, such as the one shown in the photo on this page.

Disasters can also result from too much precipitation in a short time or too little over a long period. Heavy rainfall can cause flooding. Lands along major rivers, such as the Mississippi, are especially at risk. Too little rain or too much heat may bring on droughts and dust storms or spark destructive forest fires.

In this section, you read about the varied climates and vegetation of the United States and Canada. In the next section, you will learn how physical geography has shaped life in these countries.

Geographic Thinking

Making Comparisons
◀ How do climate and vegetation differ between Mediterranean and tropical climates?

SECTION 2 Assessment

① Places & Terms

Identify and explain where in the region these would be found.

- permafrost
- prevailing westerlies
- Everglades

② Taking Notes

REGION Review the notes you took for this section.

Climate and Vegetation

- What climate regions do the United States and Canada share?
- What climate regions are found in the United States but not in Canada?

③ Main Ideas

a. How do the prevailing westerlies change the climate of parts of the United States and Canada?

b. In which region would you find the dry climates?

c. In which climate type would you find the Everglades?

④ Geographic Thinking

Seeing Patterns Why doesn't all of Alaska have cold, snowy winters? **Think about:**

- location
- prevailing westerlies

RESEARCH LINKS
CLASSZONE.COM

MAKING COMPARISONS Make a list of five Canadian cities and five U.S. cities. Then use the Internet to find out the average monthly temperature and monthly rainfall for each city. Create a **database** with the information. Then summarize your findings.

Human–Environment Interaction

Main Ideas
- Humans have dramatically changed the face of North America.
- European settlements in the United States and Canada expanded from east to west.

Places & Terms
nomad
Beringia
St. Lawrence Seaway
lock

US & CANADA

CONNECT TO THE ISSUES
URBAN SPRAWL The spreading of cities and suburbs over wider areas—urban sprawl—is causing problems.

A HUMAN PERSPECTIVE The sun-baked American Southwest was a harsh environment for its early inhabitants, the ancestors of today's Pueblo peoples. But these early settlers made good use of available resources. From the land, they took clay and stone building materials. They built multi-room, apartment-like dwellings in cliffs. This gave protection against daytime heat, nighttime cold, and human and animal enemies. From plants and animals, the early settlers got food and clothing. They survived because they adapted to their environment.

Settlement and Agriculture Alter the Land

Before humans came, North American landforms were changed only by natural forces, such as weathering and erosion. That changed when the first settlers—the ancestors of the native peoples of North America—arrived thousands of years ago.

SETTLEMENT The first inhabitants of the area of North America now known as the United States and Canada were **nomads,** people who move from place to place. Most archaeologists believe that they probably migrated from Asia over **Beringia,** a land bridge that once connected Siberia and Alaska. These migrants moved about the land. They hunted game, fished, and gathered edible wild plants. Since water was necessary for survival, these first Americans made temporary settlements along coastlines and near rivers and streams. They adjusted to extremes of temperature and climate. They also adapted to the region's many natural environments, including mountains, forests, plains, and deserts.

AGRICULTURE Many early settlements became permanent after agriculture replaced hunting and gathering as the primary method of food production about 3,000 years ago. When people began to cultivate crops, they changed the landscape to meet their needs. In wooded areas, early farmers cut down trees for lumber to build houses and to burn as fuel. To plant crops, they plowed the rich soil of river valleys and flood plains using hoes of wood, stone, and bone. They dug ditches for irrigation. Vegetables they first cultivated—corn, beans, and squash—are now staples around the world.

Agriculture remains an important economic activity in the United States and Canada. In fact, both countries are leading exporters of agricultural products.

REGION Irrigation has opened land in dry areas to farming. Tracts such as these in New Mexico are watered by a method called center-pivot, which taps underground water.
What are some other ways water can be brought to dry land?

127

Building Cities

Where a city is built and how it grows depends a great deal on physical setting. As you read, living near water was crucial to early settlers, as it would be to those who followed. Other factors that can affect the suitability of a site are landscape, climate, weather, and the availability of natural resources. Some of these factors played a role in the development of two major cities of the region.

MONTREAL—ADAPTING TO THE WEATHER Montreal, Quebec, is Canada's second largest city and a major port—even though its temperature is below freezing more than 100 days each year. Montreal's location on a large island where the St. Lawrence and Ottawa rivers meet made it an appealing site to early French explorers. The French built a permanent settlement there in 1642. The community was founded at the base of Mount Royal and grew by spreading around the mountain. To make the city's severe winters more endurable, people went inside and underground. In fact, large areas of Montreal have been developed underground, including a network of shops and restaurants.

LOS ANGELES—CREATING URBAN SPRAWL Unlike Montreal, Los Angeles, California, has a mild climate year-round. It also has a desirable location on the Pacific coast. Hundreds of thousands of people were pouring into this once small Spanish settlement by the early 1900s. As a result, the city expanded farther and farther into nearby valleys and desert-like foothills. During the 1980s, Los Angeles became the second most populous city in the United States. However, rapid population expansion brought problems. These included air pollution, inadequate water supplies, and construction on earthquake-threatened land. But such problems did not stop the city's growth. Los Angeles itself now covers about 469 square miles. Its metropolitan area spreads over 4,060 square miles. Ⓐ▶

Building cities was just one way humans interacted with their environment. Another was in the construction of transportation systems to make movement from place to place less difficult.

Overcoming Distances

The native peoples and the Europeans who followed encountered many obstacles when they moved across the land. They faced huge distances,

HUMAN–ENVIRONMENT INTERACTION Los Angeles sprawls out almost as far as the eye can see in this photo. **What changes were made to the environment as the city grew?**

Geographic Thinking◀

Making Comparisons
Ⓐ How has climate influenced the development of Los Angeles and Montreal?

large bodies of water, formidable landforms, and harsh climates. But they spanned the continent and changed the natural environment forever.

TRAILS AND INLAND WATERWAYS Some of the early peoples who came across the land bridge from Siberia blazed trails eastward. Others followed the Pacific coast south toward warmer climates. Still others remained in the northwest, in what are now Alaska and northern Canada.

When Europeans from England and France crossed the Atlantic to North America, they set up colonies along the coast. Then, they moved inland. As they did, they carved overland trails, including the National and Wilderness roads and the Oregon and Santa Fe trails. They also used inland waterways, such as the Mississippi and Ohio rivers. To connect bodies of water, they built a network of canals. The Erie Canal across upstate New York opened in 1825 and made the first navigable water link between the Atlantic and the Great Lakes. ◀**B**

North America's most important deepwater ship route—the **St. Lawrence Seaway**—was completed in the 1950s as a joint project of the United States and Canada. As you can see from the map on this page, the seaway connects the Great Lakes to the Atlantic Ocean by way of the St. Lawrence River. Ships are raised and lowered some 600 feet by a series of **locks,** sections of a waterway with closed gates where water levels are raised or lowered. The seaway enables huge, oceangoing vessels to sail into the industrial and agricultural heartland of North America.

Geographic Thinking

Seeing Patterns
B ▷ Why was it important to link waterways?

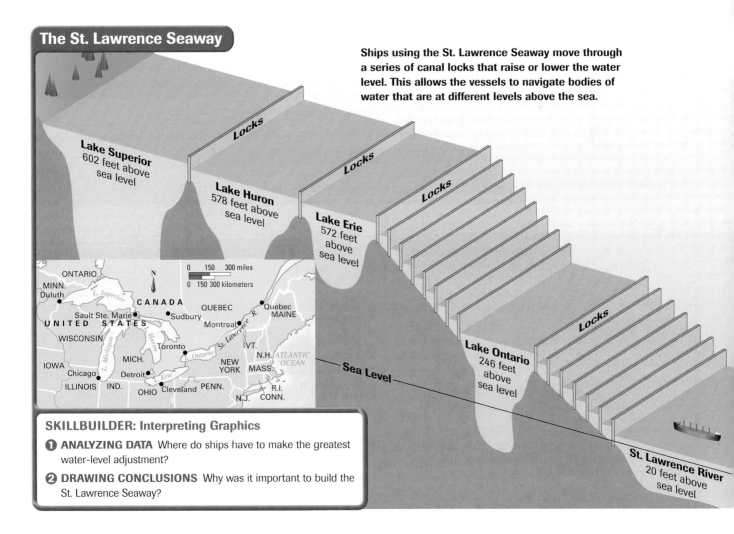

The St. Lawrence Seaway

Ships using the St. Lawrence Seaway move through a series of canal locks that raise or lower the water level. This allows the vessels to navigate bodies of water that are at different levels above the sea.

Lake Superior 602 feet above sea level

Locks

Lake Huron 578 feet above sea level

Lake Erie 572 feet above sea level

Locks

Locks

Locks

Sea Level

Lake Ontario 246 feet above sea level

St. Lawrence River 20 feet above sea level

SKILLBUILDER: Interpreting Graphics

❶ ANALYZING DATA Where do ships have to make the greatest water-level adjustment?

❷ DRAWING CONCLUSIONS Why was it important to build the St. Lawrence Seaway?

Human–Environment Interaction **129**

TRANSCONTINENTAL RAILROADS The marriage of the steam locomotive and the railroads made crossing the continent from the Atlantic to the Pacific quicker and easier. Railroad building began in North America in the early 19th century. But many of the physical features shown on the map on page 103 presented natural barriers. To make way, railroad workers had to cut down forests, build bridges over streams, and blast tunnels through mountains.

The first transcontinental railroad was completed across the United States in 1869. A trans-Canada railroad, from Montreal to British Columbia, was completed in 1885. These railroads carried goods and passengers cross-country, promoting economic development and national unity as they went. Today, the United States has the world's largest railway system, and Canada the third largest.

NATIONAL HIGHWAY SYSTEMS Before the railroads came, there were roads that connected towns and cities and provided pathways to the interior. But it was the development of the automobile in the early 20th century that spurred roadbuilding. Today, both the United States and Canada have extensive roadway systems. The United States has about 4 million miles of roads, while Canada has about 560,000 miles.

As you read earlier, much of Canada's population is concentrated in the south. So, Canadians built their major highways east to west in the southern part of the country, connecting principal cities. The Trans-Canada Highway, Canada's primary roadway, stretches about 4,860 miles from St. John's, Newfoundland, to Victoria, British Columbia. In the United States, the interstate highway system is a network of more than 46,000 miles of highways that crisscross the country. Begun in the 1950s, it connects the United States with Canada on the north and Mexico on the south, and also runs east-west across the country.

In this chapter, you read about the physical geography of the United States and Canada. In the next chapter, you will learn about the human geography of one of these countries—the United States.

Geographic Thinking

Making Comparisons
How is the Trans-Canada Highway similar to and different from the U.S. interstate highway system?

SECTION 3 Assessment

① Places & Terms

Identify and explain where in the region these would be found.

- nomad
- Beringia
- lock
- St. Lawrence Seaway

② Taking Notes

MOVEMENT Review the notes you took for this section.

Human-Environment Interaction

- Why are railroads important to a nation's development?
- In what ways did settlers in Canada and the United States move across the continent?

③ Main Ideas

a. What factors affect the choice of location of a city?

b. Why is the St. Lawrence Seaway important?

c. How did methods of moving people and goods across the continent change over time?

④ Geographic Thinking

Making Inferences In what ways have transportation systems crossing the continent altered the environment? **Think about:**

- construction of canals and railroads
- building cities

See Skillbuilder Handbook, page R4.

GeoActivity

ASKING GEOGRAPHIC QUESTIONS Obtain and study a highway map of your state. Then come up with a geographic question about the map, perhaps one considering geographic features that caused the location of a highway. Answer the question and make a **class presentation** using visuals.

Reading a Highway Map

San Antonio, Texas, is a part of a metropolitan area of more than one million people, located in south central Texas. It has been a crossroads for much of its history—for its earliest Native American settlers, the Spanish who came later, and finally, the Texans who won independence from Mexico not long after the battle of the Alamo. Looking at the map below, you can see that the city remains a meeting point, crisscrossed by interstate, U.S., state, and county highways.

THE LANGUAGE OF MAPS The primary purpose of a **highway map** is to show the location of roadways in an area and the distance between places. But highway maps usually include much other information. For example, they may identify important sites, such as airports, parks, and universities.

San Antonio and Vicinity, 2001 ❶

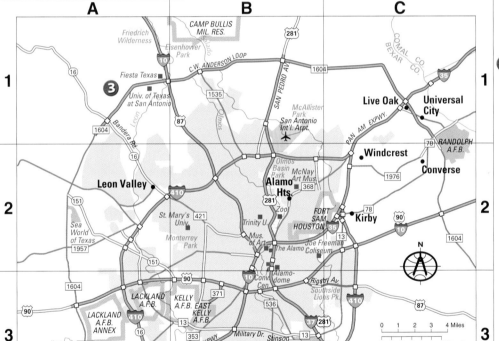

Copyright by Rand McNally & Co.

❶ The title identifies the area covered by the map.

❷ The key shows the symbols used on the map and explains what they mean. For example, the ✈ symbol shows where airports are located.

❸ Points of interest, such as the Alamo (B–2) or Sea World (A–2), are marked by small red squares or by pink ribbons, depending on their size.

Map and Graph Skills Assessment

1. Seeing Patterns
Which interstate highways pass through the center of San Antonio?

2. Making Decisions
Which interstate highway and U.S. highway would you take to the Alamo when coming from the southeast?

3. Analyzing Data
By the most direct route, how far is Live Oak from Leon Valley by highway?

VISUAL SUMMARY
PHYSICAL GEOGRAPHY OF THE UNITED STATES AND CANADA

Landforms

Major Mountain Ranges:
Rocky Mountains, Appalachian Mountains

Major Waterways:
Mississippi-Missouri-Ohio river system, Great Lakes, Mackenzie River, Columbia River, Rio Grande River, Colorado River

Interior Lowlands:
Great Plains, Canadian Shield, Interior Plains

Resources

• Both the United States and Canada have huge mineral and fossil fuel resources.

• Forest lands cover about one-third of the United States and one-half of Canada.

Climate and Vegetation

• Canada's climates and vegetation are related to its far northern location.

• The United States includes regions that are in almost every climate and vegetation zone.

Human-Environment Interaction

• Movement westward altered the land in both the United States and Canada.

• Transportation networks helped develop the land and economy of the region.

Reviewing Places & Terms

A. Briefly explain the importance of each of the following.

1. Appalachian Mountains
2. Rocky Mountains
3. Great Plains
4. Canadian Shield
5. Great Lakes
6. Mackenzie River
7. prevailing westerlies
8. Everglades
9. lock
10. St. Lawrence Seaway

B. Answer the questions about vocabulary in complete sentences.

11. Which of the places listed above are found both in the United States and Canada?
12. Which of the mountain chains form a boundary with the Canadian Shield?
13. The Great Plains are bounded on one side by which landform listed above?
14. The Hudson Bay is found in which place listed above?
15. Which two waterways are linked?
16. Which place above is a huge swampland?
17. Which of the places are subregions of the Interior Lowlands?
18. What climate region in North America is influenced by the prevailing westerlies?
19. Why are the Great Lakes and the St. Lawrence Seaway important?
20. Why are locks needed on the St. Lawrence Seaway?

Main Ideas

Landforms and Resources (pp. 117–122)

1. How do the Eastern Lowlands differ from the Interior Lowlands?
2. What is the Continental Divide?
3. Why are the United States and Canada leading food producers?
4. What are the most abundant natural resources in the United States and Canada?

Climate and Vegetation (pp. 123–126)

5. In what type of climate would you expect to find permafrost?
6. Which climates are found in the United States and not in Canada?
7. What type of vegetation covers most of Canada?

Human-Environment Interaction (pp. 127–131)

8. How did the earliest inhabitants of the United States and Canada, those who arrived before the Europeans, alter the land?
9. What problems arose in Los Angeles with rapid expansion?
10. How did the settlers of the United States and Canada overcome the distances across the continent?

Critical Thinking

1. Using Your Notes

Use your completed chart to answer these questions.

Landforms	
Resources	

   a. How is the location of cities related to landforms and to climate?

   b. How is Canada's economy affected by its climate and vegetation?

2. Geographic Themes

   a. **MOVEMENT** Write a sentence describing the movement of people and goods across the United States and Canada over the last 200 years.

   b. **PLACE** How have the Great Lakes contributed to the development of both the United States and Canada?

3. Identifying Themes

In developing their city, how did the people of Montreal solve the problems of a severe climate? Which of the five themes apply to this situation?

4. Making Inferences

What aspects of physical geography have contributed to the economic success of the United States and Canada?

5. Seeing Patterns

How did the presence of north-to-south flowing rivers in the United States affect its development?

Additional Test Practice, pp. S1–S37

TEST PRACTICE
CLASSZONE.COM

Geographic Skills: Interpreting Maps

Physical Profile of the United States

Use the map below to answer the following questions.

1. **REGION** What might be said about the land between the Appalachians and the Mississippi?

2. **PLACE** What is the difference in altitude between the Coastal ranges and the Sierra Nevada?

3. **REGION** What happens to the land as you move west of the Mississippi?

GeoActivity

Create a three-dimensional model of the cross section on this page. Use colors to indicate elevations and label the physical features you show. Create a legend for your model.

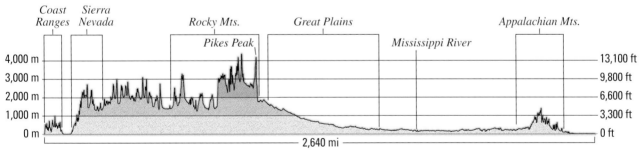

INTERNET ACTIVITY

Use the links at **classzone.com** to conduct research on the landforms of the United States and Canada. Focus on finding pictures of major and well-known landforms and waterways.

Creating a Multimedia Presentation From your research, select a series of pictures to include in a presentation on the theme "A Land of Contrasts." List the Web sites you used in preparing your report.

HUMAN GEOGRAPHY OF THE UNITED STATES
Shaping an Abundant Land

Four Subregions of the United States

Northeast
South
Midwest
West

GeoFocus

What factors shaped the development of the United States?

Taking Notes In your notebook, copy a cluster diagram like the one shown. As you read, take notes about the history, economy, culture, and modern life of the United States and its subregions.

History
and Government

Economy
and Culture

The
United States

Subregions

History and Government of the United States

Main Ideas
- The United States is a "nation of immigrants," settled by people from all over the world.
- The United States is the most diverse and highly industrialized and urbanized nation in the world.

Places & Terms
migration
Columbian Exchange
Louisiana Purchase
frontier
suburb
representative democracy

US & CANADA

CONNECT TO THE ISSUES
TERRORISM Beginning in the late 20th century, the United States has been subjected to terrorist attacks by individuals and groups opposed to its policies.

A HUMAN PERSPECTIVE Women were North America's first farmers. In all early cultures except the hunter-gatherer culture of the Southwest, women cultivated the land. They discovered which wild plants could be used as food for the family. They planted the seeds, tended the garden, harvested the crops, and prepared food for meals. Corn, beans, and squash were the first of these foods. Women also learned which leaves, bark, roots, stems, and berries could be used for medicines. Their efforts helped to ensure the survival of human settlement in North America—and the part of the land that became the United States.

Creating a Nation

The United States occupies nearly two-fifths of North America. It is the world's third largest country in both land area and population. It is rich in natural resources and is also fortunate to have a moderate climate, fertile soil, and plentiful water supplies. For thousands of years, this bounty has attracted waves of immigrants who came to find a better life. This continuing immigration is a recurring theme in the country's history; so is the constant **migration,** or movement, of peoples within the United States.

MANY PEOPLES SETTLE THE LAND As you read in Chapter 5, the first inhabitants of North America were believed to be nomads who came from Asia at least 13,000 or more years ago. These people settled the continent, spreading south along the Pacific coast and east to the Atlantic. Over the centuries, they developed separate cultures, as the map on page 104 shows. These native peoples occupied the land undisturbed until the 15th century, when Europeans began to explore what they called the "New World." The Spanish arrived first. They searched the present-day Southeast and Southwest for gold and other treasure. In 1565, they founded St. Augustine, Florida, the oldest permanent European settlement in the United States.

The French and English came later. France was interested in fisheries and the fur trade. In the early 1600s, the French settled along the northern Atlantic Coast and the St. Lawrence River in what is now Canada. The English arrived at about the same time. During the 1600s and 1700s,

HUMAN-ENVIRONMENT INTERACTION Early Native American settlers in the Southwest often built their dwellings into canyon walls. The dwellings shown are in Mesa Verde National Park in Colorado. **Why did the earliest settlers choose such locations for their dwellings?**

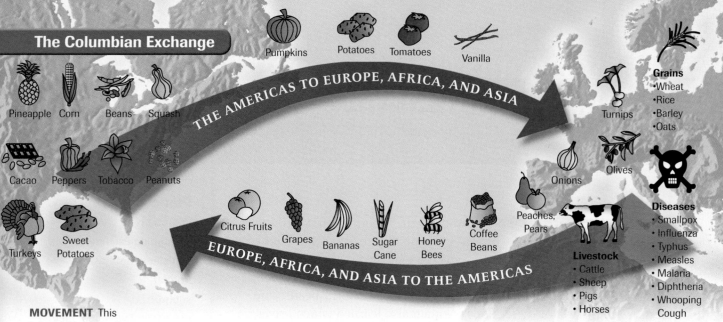

The Columbian Exchange

THE AMERICAS TO EUROPE, AFRICA, AND ASIA

Pineapple Corn Beans Squash

Cacao Peppers Tobacco Peanuts

Turkeys Sweet Potatoes

Pumpkins Potatoes Tomatoes Vanilla

EUROPE, AFRICA, AND ASIA TO THE AMERICAS

Citrus Fruits Grapes Bananas Sugar Cane Honey Bees Coffee Beans Peaches, Pears

Turnips Onions Olives

Grains
•Wheat
•Rice
•Barley
•Oats

Diseases
• Smallpox
• Influenza
• Typhus
• Measles
• Malaria
• Diphtheria
• Whooping Cough

Livestock
• Cattle
• Sheep
• Pigs
• Horses

MOVEMENT This infographic shows how plants, animals, and diseases were transferred between the Eastern and Western hemispheres as trade followed the voyages of Christopher Columbus to the Americas.

they settled to the south—on rivers and bays along the Atlantic coast from present-day Maine to Georgia. The English made their first permanent settlement in Jamestown, Virginia, in 1607.

European colonies often displaced Native Americans. In 1617, the Europeans brought Africans to America to work as slave laborers on cotton and tobacco plantations in the South. The coming of the Europeans also began what historians call the **Columbian Exchange.** The infographic above shows how the arrival of Europeans in the Western Hemisphere affected the lives of both Europeans and the native peoples.

ESTABLISHING AND MAINTAINING THE UNION The French and the English eventually fought in North America over trade and territory. In 1763, Great Britain gained control of all of North America east of the Mississippi River. But its control was short-lived. Britain's 13 American colonies soon began to resent the policies forced on them by a government thousands of miles away across the Atlantic. Their protests led to the American Revolution (1775–1783) and the founding of the United States of America. The new nation grew rapidly, and settlers pushed westward to the Mississippi. In 1803, the United States nearly doubled in size when the government purchased the vast plains region between the Mississippi and the Rocky Mountains from France. This territory became known as the **Louisiana Purchase.**

In the early 1800s, immigrants from Western Europe arrived in great numbers. They settled in cities in the Northeast, where industrialization was beginning. One such city was Lowell, Massachusetts, which had become a booming textile center by the 1840s. The newcomers also moved to rich farmlands in what is now the Midwest.

Meanwhile, sectionalism was growing. People were placing loyalty to their region, or section, above loyalty to the nation. The result was rising political and economic tensions between an agricultural South dependent on slave labor and the more industrialized North. These tensions led to the Civil War (1861–1865). It took four years of bloody fighting and many more years of political conflict to reunite the country.

BACKGROUND
About 600,000 Africans were brought to the United States to work as slave laborers from 1617 until the importation of slaves was banned in 1808.

Agriculture and Industry of the United States

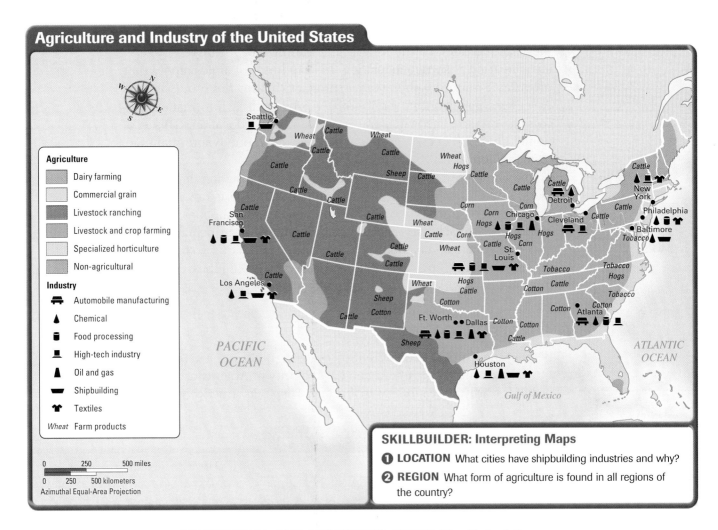

Agriculture

- Dairy farming
- Commercial grain
- Livestock ranching
- Livestock and crop farming
- Specialized horticulture
- Non-agricultural

Industry

- Automobile manufacturing
- ▲ Chemical
- Food processing
- High-tech industry
- Oil and gas
- Shipbuilding
- Textiles
- *Wheat* Farm products

0   250   500 miles
0   250   500 kilometers
Azimuthal Equal-Area Projection

SKILLBUILDER: Interpreting Maps

❶ LOCATION What cities have shipbuilding industries and why?

❷ REGION What form of agriculture is found in all regions of the country?

AN AGRICULTURAL AND INDUSTRIAL GIANT The United States not only feeds itself but also helps to feed the world. American farms and ranches supply about 40 percent of the world's production of corn, 20 percent of its cotton, and about 10 percent of its wheat, cattle, and hogs. Fertile soil, a favorable climate, and the early mechanization of the country's farms are mainly responsible for this bounty. Different areas of the country produce different products, as you can see from the map on this page. The Midwest and South, for example, specialize in crop farming, while livestock ranching is concentrated in the West.

The industrial output of the United States is larger than that of any other country. Advances in technology, especially in electronics and computers, revolutionized industry and led to the creation of new products and methods of production. Leading industries are petroleum, steel, transportation equipment, chemicals, electronics, food processing, telecommunications, consumer goods, lumber, and mining. ◁

Major industrial centers have long been located along the Atlantic Coast and around the Great Lakes. In recent decades, a variety of industries have also started up in urban areas in the South and along the Pacific coast. Over time, some areas have become associated with certain products, such as Detroit (automobiles), Seattle (aircraft), and northern California, in an area called Silicon Valley (computers).

A POSTINDUSTRIAL ECONOMY The graphs on page 140 show the rich farming and manufacturing traditions of the United States. But

Geographic Thinking

Seeing Patterns

Ⓐ Why might industrial centers be located near bodies of water?

they also indicate that the American economy today is driven by service industries. A **service industry** is any kind of economic activity that produces a service rather than a product. Nearly three out of four Americans now work in service-related jobs, such as information processing, finance, medicine, transportation, and education. This economic phase is called a **postindustrial economy,** one where manufacturing no longer plays a dominant role.

The United States is the world's major trading nation, leading the world in the value of its exports and imports. It exports raw materials, agricultural products, and manufactured goods. Automobiles, electronic equipment, machinery, and apparel are some of its principal imports. Its North American neighbors, Canada and Mexico, are two of its most important trading partners. Many American corporations engage in business worldwide and are called **multinationals.** Ⓑ

A Diverse Society

Because the United States is a nation of immigrants, it is a nation of different races and ethnic traditions. The majority of Americans, about 70 percent, trace their ancestry to Europe. Hispanic Americans, mainly

Geographic Thinking ◄

Seeing Patterns
Ⓑ Where do some of the natural resources of the United States go?

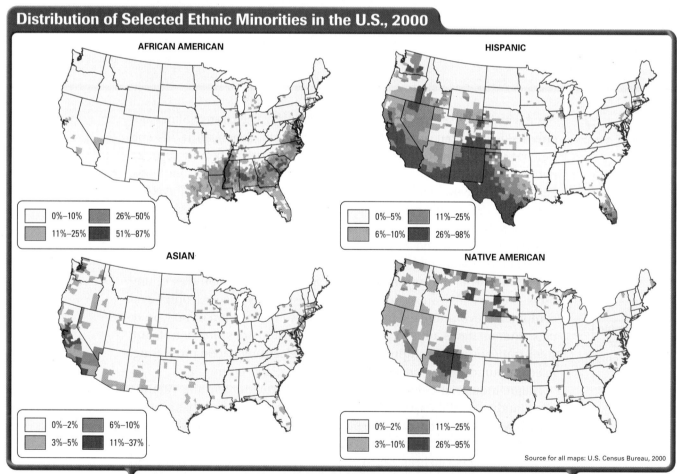

Distribution of Selected Ethnic Minorities in the U.S., 2000

AFRICAN AMERICAN

0%–10% 26%–50%
11%–25% 51%–87%

HISPANIC

0%–5% 11%–25%
6%–10% 26%–98%

ASIAN

0%–2% 6%–10%
3%–5% 11%–37%

NATIVE AMERICAN

0%–2% 11%–25%
3%–10% 26%–95%

Source for all maps: U.S. Census Bureau, 2000

SKILLBUILDER: Interpreting Maps

❶ **REGION** What subregion has significant numbers of most of the ethnic groups shown?

❷ **MOVEMENT** Compare this map with the map on page 104. What has changed about the distribution of Native Americans since 1600?

from Central and South America, make up about 13 percent of the population; African Americans, about 12 percent; Asian Americans, 4 percent; and Native Americans, 1 percent. The largest ethnic groups are English, German, Irish, African, French, Italian, Scottish, Polish, and Mexican. The maps on page 142 show the distribution of some groups.

BACKGROUND
English is the dominant language in countries, including Canada, the United States, and Australia, that cover one-fifth of the earth's land surface.

LANGUAGES AND RELIGION English has been the dominant language of the United States since its founding. Spanish is the second most commonly spoken language. Typically, immigrants have spoken their native language until they learned English.

Religious freedom has been a cornerstone of American society. Today, more than 1,000 different religious groups practice their faiths in the United States. By far, the majority of the American people—85 percent—are Christians. About 56 percent are Protestants and 28 percent Roman Catholics. Jews and Muslims each account for about 2 percent of the religious population.

THE ARTS AND POPULAR CULTURE The United States has a rich artistic heritage, the product of its diverse population. Its first artists were Native Americans, who made pottery, weavings, and carvings. Early European settlers brought with them the artistic traditions of their homelands. Truly American styles developed in painting, music, literature, and architecture in the 19th century. Artists depicted the country's expansive landscape and scenes of American life both on the western frontier and in the cities. One 19th-century American creation, the skyscraper, changed urban architecture all over the world.

Today, motion pictures and popular music are two influential American art forms. Hollywood, California, is the center of the movie industry in the United States. American films provide entertainment for the world. Many ethnic groups contributed to the musical heritage of the United States. For example, jazz, blues, gospel, and rock 'n' roll have African-American origins. Country and bluegrass music developed among Southern whites whose ancestors came from the British Isles.

5 THEMES

MOVEMENT

Moving the Blues

Blues music developed among African Americans in the rural South around the beginning of the 20th century. This expressive folk music, usually played on a guitar or harmonica, had its roots in Africa. The blues spread throughout the United States, as African Americans migrated to urban areas to find jobs.

The form of the blues born in the delta region of Mississippi was taken north by rural migrants to cities like Memphis, St. Louis, and Chicago (where the blues guitarist Muddy Waters settled). Blues from the Carolinas reached New York City, while the Texas blues went west to Los Angeles and Oakland.

Muddy Waters

US & CANADA

American Life Today

More than 280 million people live in the United States. The majority enjoy a high standard of living. Despite coming from many ethnic and racial groups, they generally live and work together. They are pursuing what attracted their ancestors to the New World and came to be called "the American dream," a better life for themselves and their children.

WHERE AMERICANS LIVE About 80 percent of Americans live in cities or surrounding suburbs. Americans moved first from rural areas to cities and then from cities to suburbs. The shift to the suburbs was made possible by the widespread ownership of automobiles. There is one auto

HUMAN–ENVIRONMENT INTERACTION Lake Michigan and its shoreline provide Chicago residents with many opportunities for recreation.
What might some of these recreational opportunities be?

for every 1.3 Americans. A highly developed transportation network that includes highways, expressways, railroads, and airlines aids mobility.

HOW AMERICANS LIVE, WORK, AND PLAY Nearly 50 percent of American adults of working age are employed. Almost half of them are women. As you read earlier, about seven out of ten Americans in the workforce hold service industry jobs. Many are highly skilled positions, which require advanced education. Americans have always valued education, seeing it as a means to provide equality and opportunity. As a result, all children from the ages of 6 or 7 to age 16 are required to attend school. Nine out of ten students are in the public school system, where education is free through secondary school. The United States also has more than 2,300 four-year public and private colleges and universities.

Americans have a wide range of choices for leisure-time activities. As either spectators or players, they take part in sports such as baseball, basketball, football, golf, soccer, tennis, and skiing. Most major cities have professional sports teams. Americans of all ages also use their free time to engage in hobbies, visit museums and libraries, and watch television and movies. Another favorite activity is spending time on the computer, surfing the Internet or playing video games.

Unfortunately, not all Americans live well. More than one in ten lives in poverty. It is a continuing challenge for government and society to try to bring these people into the mainstream of American life. In the next section, you will learn about life in the country's subregions.

SECTION 2 Assessment

❶ Places & Terms

Explain the meaning of each of the following terms.

- export
- free enterprise
- service industry
- postindustrial economy
- multinational

❷ Taking Notes

REGION Review the notes you took for this section.

- Where are the industrial centers in the United States?
- Where do the majority of Americans live?

❸ Main Ideas

a. What three factors have contributed to the success of the American economy?

b. What are the geographic origins of some American musical styles?

c. What invention made life in the suburbs possible?

❹ Geographic Thinking

Making Comparisons How is the economy of the United States today different from its economy 50 years ago?
Think about:

- postindustrial economy
- multinational trade

S See Skillbuilder Handbook, page R3.

EXPLORING LOCAL GEOGRAPHY Study the maps on page 142. Find your state. Create a **sketch map** of your state and show the location of major ethnic groups that live in your state.

Subregions of the United States

Main Ideas
• The United States is divided into four major economic and cultural subregions.
• There are both similarities and differences among the subregions of the United States.

Places & Terms
New England   metropolitan
megalopolis   area
the Midwest   the West
the South

CONNECT TO THE ISSUES
DIVERSE SOCIETIES While diversity can be a strength, it has also been the cause of tension and conflict among regions.

A HUMAN PERSPECTIVE America's back roads were the beat of reporter and author Charles Kuralt for more than 20 years. Beginning in the 1960s, he traveled by van through every region of the country. In his "On the Road" series for television, he reported on the uniqueness of the lives of ordinary Americans. He said that he wanted to make these trips off the beaten path because most people traveled across the country on interstate highways without seeing the "real" America. Whether he visited Minnesota's lake country or a small New England town, Kuralt spotlighted America's regional diversity. In fact, one of the key strengths of the United States is the variety of life in its subregions—the Northeast, the Midwest, the South, and the West.

The Northeast

As you can see on the map on page 134, the Northeast covers only 5 percent of the nation's land area. But about 20 percent of the population lives there. The six northern states of the subregion—Maine, Vermont, New Hampshire, Massachusetts, Rhode Island, and Connecticut—are called **New England.** The other three—Pennsylvania, New York, and New Jersey—are sometimes referred to as Middle Atlantic states. (Maryland and Delaware, which are included in the South in this book, are sometimes included in the Middle Atlantic states.)

AMERICA'S GATEWAY Because of its location along the Atlantic coast, the Northeast contains many of the areas first settled by Europeans. The region served as the "gateway" to America for millions of immigrants from all over the world. Many people still engage in fishing and farming,

LOCATION BosWash is the name given to the highly urbanized northeastern seaboard of the United States.

Urbanization in the Northeast

Washington, D.C.
pop. 523,800

Philadelphia
pop. 1,452,300

New York
pop. 7,405,400

Boston
pop. 559,100

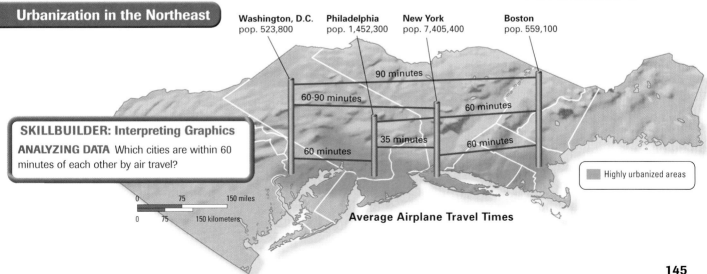

90 minutes

60-90 minutes

60 minutes

35 minutes

60 minutes

60 minutes

60 minutes

SKILLBUILDER: Interpreting Graphics
ANALYZING DATA Which cities are within 60 minutes of each other by air travel?

| 0 | 75 | 150 miles |
| 0 | 75 | 150 kilometers |

Average Airplane Travel Times

Highly urbanized areas

as the Northeast's early settlers did. But the region's coastal and inland waters turned it into the heart of trade, commerce, and industry for the nation. In fact, the Northeast is one of the most heavily industrialized and urbanized areas in the world. The Atlantic seaboard cities of Philadelphia, Boston, and New York City serve as international trade centers.

Coal, iron ore, and oil—found mainly in Pennsylvania—fueled the industrialization of the region. Traditional industries, such as iron and steel, petroleum, and lumber, still play a role in the region's economy. But most Northeasterners are now employed in such manufacturing and service industries as electronics, communications, chemicals, medical research, finance, and tourism. Pennsylvania, New York, and New Jersey have rich farmlands, but much of New England is too hilly or rocky to grow crops easily. Ⓐ

Parts of the Middle Atlantic states are often referred to as the "rust belt" because of their declining and abandoned traditional industries. They share this term with some of the states of the Midwest. In recent times, many "rust belt" industries have moved to the warmer climates of the "sunbelt" in the South and West.

GROWTH OF THE MEGALOPOLIS The nation's first megalopolis developed in the Northeast. A **megalopolis** is a region in which several large cities and surrounding areas grow together. You can see the extent of the "BosWash" megalopolis, as it is called, in the illustration on page 145.

Geographic Thinking

Using the Atlas
Ⓐ Refer to the map on page 106. What economic activities are shown for the Northeast?

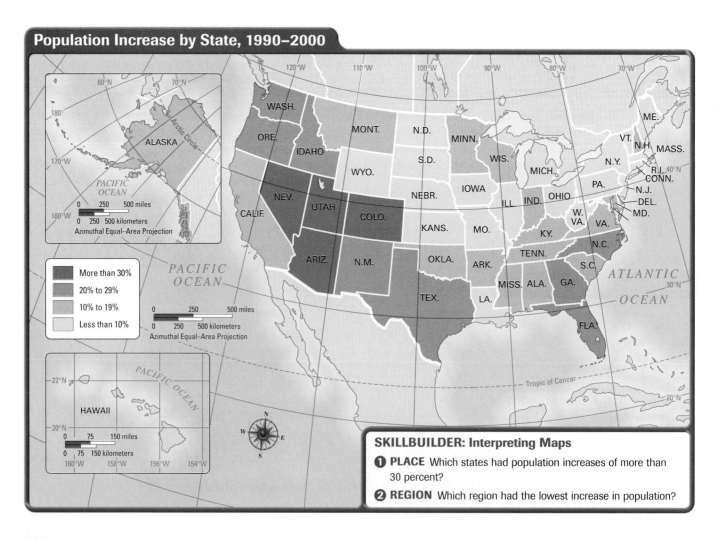

Population Increase by State, 1990–2000

More than 30%
20% to 29%
10% to 19%
Less than 10%

SKILLBUILDER: Interpreting Maps

❶ **PLACE** Which states had population increases of more than 30 percent?

❷ **REGION** Which region had the lowest increase in population?

It stretches through 500 miles of highly urbanized areas from Boston in the north to Washington, D.C., the national capital, in the south. It contains one-sixth of the U.S. population. New York City, the country's cultural and financial center, is located here. Rapid road, rail, and air links have been vital to its economic development and expansion into the South. You will read more about urban growth in Chapter 8.

The Midwest

The subregion that contains the 12 states of the north-central United States is called the **Midwest.** Because of its central location, the Midwest is called the American heartland. It occupies about one-fifth of the nation's land and almost one-fourth of its people live there. Since the Revolutionary War, immigrants from all over the world have made it their destination. Many early settlers came from Britain, Germany, and Scandinavia. Vast, largely flat plains are a distinctive feature of the region. So are numerous waterways, including the Great Lakes and the Mississippi River and its many tributaries.

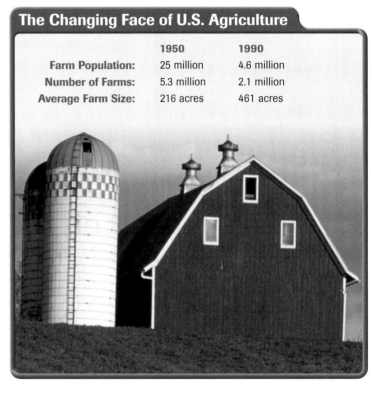

The Changing Face of U.S. Agriculture

	1950	1990
Farm Population:	25 million	4.6 million
Number of Farms:	5.3 million	2.1 million
Average Farm Size:	216 acres	461 acres

AGRICULTURAL AND INDUSTRIAL HEARTLAND The Midwest is the nation's "breadbasket." Fertile soil, adequate rainfall, and a favorable climate enable Midwesterners to produce more food and feed more people than farmers in any comparable area in the world. Among the main products are corn, wheat, soybeans, meat, and dairy goods. Agriculture also is the foundation for many of the region's industries, including meatpacking, food processing, farm equipment, and grain milling. Other traditional industries are steel and automaking.

Its central location and excellent waterways make the Midwest a trade, transportation, and distribution center. Chicago, Illinois, which is located near the southwestern shores of Lake Michigan, is the cultural, financial, and transportation hub of the Midwest. Most of the region's major cities developed near large bodies of water, which were essential for early transportation. Cleveland, Detroit, Chicago, and Milwaukee grew near the Great Lakes, and Cincinnati, St. Louis, Minneapolis, St. Paul, Kansas City, and Omaha developed along rivers. ◄B

CHANGING FACE OF THE MIDWEST Like other regions, the Midwest is changing. The number of farms is declining. More Midwesterners are now employed in providing services than in traditional industries. The region's metropolitan areas are expanding as urban dwellers and businesses leave the central cities for the suburbs. People and industries are also moving to the warmer South and West.

Geographic Thinking

Making Comparisons
B What do the major cities of the Midwest have in common with those of the Northeast?

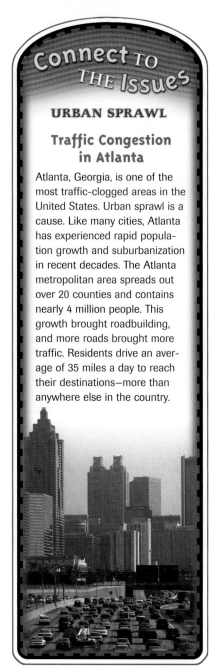
The South

The South is a subregion that covers about one-fourth of the land area of the United States and contains more than one-third of its population. Among its 16 states are 11 that made up the Confederacy during the Civil War. One of these states—Texas—is sometimes included in an area of the West called the Southwest. The South's warm climate, fertile soils, and many natural resources have shaped its development.

THE OLD SOUTH Like the Northeast, the South was also the site of early European settlement. In fact, Virginia was England's first American colony. The South has a mix of cultures that reflects the diversity of its early settlers. In addition to people of British heritage, there are the descendants of Africans brought as slave laborers and Hispanics whose families first migrated from Mexico to Texas. Cajuns of French-Canadian origin and Creoles of French, Spanish, and African descent are found in Louisiana, while Florida is home to many Hispanics who came from Cuba.

Once a rural agricultural area, the South is rapidly changing and its cities growing. Along with the Southwest, it is often referred to as the "sunbelt" because of its climate.

THE NEW SOUTH Agriculture was the South's first economic activity, and cotton, tobacco, fruits, peanuts, and rice are still grown there. Also, livestock production is important in states such as Texas and Arkansas. The South's humid subtropical climate at first hindered industrialization. But the widespread use of air conditioning beginning in the 1950s and the region's vast stores of energy resources—oil, coal, natural gas, and water—gave a boost to industry.

In recent times, the South has attracted many manufacturing and service industries fleeing the harsh weather of the "rust belt." Major industries include petroleum, steel, chemicals, food processing, textiles, and electronics. The South's climate draws millions of tourists and retirees, too. Atlanta, Georgia—a financial, trade, and transportation center—is the hub of the New South. Miami, Tampa-St. Petersburg, New Orleans, Houston, Dallas-Fort Worth, and San Antonio are other rapidly growing **metropolitan areas**—large cities and nearby suburbs and towns.

The West

Look on the map on page 134, and you will see that **the West** is a far-flung subregion consisting of 13 states. It stretches from the Great Plains to the Pacific Ocean and includes Alaska to the north and Hawaii in the Pacific. The West covers about one-half of the land area of the United States but has only about one-fifth of the population. It is a region of dramatic and varied landscapes.

People settle in the West today as they did during its frontier days: wherever landforms and climate are favorable. Some areas, such as its many deserts, are sparsely settled. Nonetheless, California is the

BACKGROUND Washington, Oregon, and Idaho are often called the Northwest. California, Arizona, New Mexico, Nevada, Colorado, Utah, and Texas are called the Southwest.

country's most populous state because of excellent farmland, good harbors, and a mild climate. The West is the most rapidly growing region in the United States. Los Angeles, the country's second largest city, is the West's cultural and commercial center.

BACKGROUND
According to the 2000 census, the population of the West grew by 20 percent from 1990.

DEVELOPING THE WEST The West's growth in the 20th century was helped by air conditioning and by irrigation. The map on this page, for example, shows how water from the Colorado River in Arizona has been diverted to serve many areas. Water supply aided development of inland cities such as Las Vegas, Tucson, and Phoenix.

The economic activities of the West are as varied as its climate and landscape. Among them are farming, ranching, food processing, logging, fishing, mining, oil refining, tourism, filmmaking, and the production of computers. Many cities with good harbors, including Seattle, Los Angeles, and Long Beach, make foreign trade—especially with Asia—important.

You read about the subregions of the United States in this section. In the next chapter, you will learn about the human geography of Canada.

Colorado River Basin

Volume of water
(millions of acre–feet)
IN   OUT
2+   2+
1–2   1–2
0–1   0–1
● City   ▮ Dam
.... Aqueduct

0   100   200 miles
0   100   200 kilometers
Albers Equal-Area Projection

IDAHO
WYOMING
UTAH
COLORADO
NEVADA
Las Vegas
Lake Mead
Lake Powell
Green R.
Colorado R.
Gunnison R.
Dolores R.
San Juan R.
CALIFORNIA
Grand Canyon
Colorado R.
Little Colorado R.
Los Angeles
ARIZONA
NEW MEXICO
San Diego
Phoenix
Tucson
UNITED STATES
MEXICO
PACIFIC OCEAN
Gulf of California

US & CANADA

SKILLBUILDER: Interpreting Maps

❶ **PLACE** What area receives the largest volume of water from the Colorado River?

❷ **MOVEMENT** Which states contribute water to the Colorado River?

SECTION

3 Assessment

❶ Places & Terms

Explain the meaning of each of the following terms.

• New England
• megalopolis
• the Midwest
• the South
• metropolitan area
• the West

❷ Taking Notes

REGION Review the notes you took for this section.

The United States → Subregions

• What are the four subregions of the United States?
• Which subregion is the largest in land area?

❸ Main Ideas

a. Why is the Northeast one of the most heavily industrialized and urbanized areas?

b. How is the economy of the Midwest changing?

c. What helped the economy of the West to grow?

❹ Geographic Thinking

Seeing Patterns How has air conditioning changed the economic activities of the subregions of the United States? **Think about:**

• the South and the West
• the "rust belt" and the "sunbelt"

RESEARCH LINKS
CLASSZONE.COM

GeoActivity

MAKING COMPARISONS Use the Internet to find more information on the economies of the four subregions. Create a **database** comparing the top five industries in each of the four subregions.

Disasters!

The Dust Bowl

Years of unrelenting drought, misuse of the land, and the miles-high dust storms that resulted (shown here) devastated the Great Plains in the 1930s. Rivers dried up, and heat scorched the earth. As livestock died and crops withered, farms were abandoned. Thousands of families—more than two million people—fled to the West, leaving behind their farms and their former lives. Most of these "Okies," as they were called (referring to Oklahoma, the native state of many), made their way over hundreds of miles to California. There they tried to find work as migrant farm laborers and restart their lives. The drought lasted nearly a decade, and it took years for this productive agricultural region to recover.

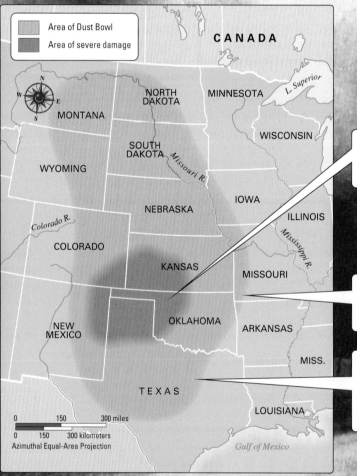

Area of Dust Bowl

Area of severe damage

CANADA

NORTH DAKOTA

MINNESOTA

L. Superior

MONTANA

WISCONSIN

SOUTH DAKOTA

Missouri R.

WYOMING

IOWA

NEBRASKA

ILLINOIS

Colorado R.

Mississippi R.

COLORADO

KANSAS

MISSOURI

NEW MEXICO

OKLAHOMA

ARKANSAS

MISS.

TEXAS

LOUISIANA

0   150   300 miles

0   150   300 kilometers

Azimuthal Equal-Area Projection

Gulf of Mexico

The worst of the devastation was centered in parts of five states—Oklahoma, Kansas, Colorado, New Mexico, and Texas.

Dust from the Great Plains was reported by ships to have blown as far east as 500 miles out into the Atlantic Ocean in 1934.

The most terrible dust storm came on April 14, 1935. A blinding black cloud of swirling dust rolled across the southern plains, blotting out the sun, suffocating animals, and burying machinery.

Thousands of farms like this one in Cimarron County, Oklahoma, were turned into dust-covered wastelands by the drought and dust storms of the 1930s.

Migrants from the Dust Bowl were forced to live any way they could while trying to find jobs picking vegetables or fruit. This mother and her seven children lived in a tent in a California migrant camp, eating vegetables found on the ground and birds they killed.

GeoActivity

REMEMBERING THE VICTIMS
Use the Internet to find personal accounts of Dust Bowl families. Then create a **documentary proposal** about one of them.

- Begin with a brief overview of how the drought affected the family.
- Add a sketch map showing where they lived and copies of any photos available, with captions for each.
- Present your proposal to a panel of student producers.

RESEARCH LINKS
CLASSZONE.COM

GeoData

CAUSES
- Years of poor agricultural practices, such as overplowing and overgrazing, stripped away about 96 million acres of grasslands in the southern plains.
- Seven years of drought, or dry weather, turned the soil to dust.

EFFECTS
- Hundreds of millions of tons of soil were blown away.
- Crops withered and livestock died.
- More than 2 million plains people abandoned their farms.

PREVENTIVE MEASURES
Experts in crop production and soil management proposed the use of scientific farming methods, including

- contour plowing, or plowing across a hill rather than up and down, to stop wind and water erosion
- terracing, or planting crops in stair-stepped rows, to prevent soil erosion
- planting trees to hold the soil in place and to slow the force of the wind

VISUAL SUMMARY
HUMAN GEOGRAPHY OF THE UNITED STATES

History and Government

- The United States was populated by a diverse group of immigrants.
- The United States expanded westward and industrialized.
- The government of the United States is a representative democracy.
- At the start of the 21st century, the United States was the only remaining superpower.

Economy and Culture

- Fertile land, valuable resources, and good location help make the United States an economic leader.
- Much of the U.S. economy is based on service industries.
- Most of the U.S. population lives in urban areas.

Subregions of the United States

- ◎ The Northeast region is heavily populated and industrialized.
- ○ The Midwest produces a variety of agricultural and manufactured goods but is shifting to some service industries.
- ● The South is rapidly becoming more industrialized.
- ◐ The West is a rapidly growing economic region.

Reviewing Places & Terms

A. Briefly explain the importance of each of the following.

1. migration
2. Columbian Exchange
3. suburb
4. representative democracy
5. free enterprise
6. service industry
7. postindustrial economy
8. multinational
9. megalopolis
10. metropolitan area

B. Answer the questions about vocabulary in complete sentences.

11. What role did migration play in populating the United States?
12. What are some examples of items in the Columbian Exchange?
13. Which of the above terms are associated with urban geography?
14. What type of government does the United States have?
15. What is an advantage of free enterprise?
16. How are the service industry and postindustrial economy related?
17. What is an example of a service industry?
18. What makes a business a multinational corporation?
19. In which region is an example of a megalopolis found?
20. How are the terms suburb and metropolitan area related?

Main Ideas

History and Government of the United States (pp. 135-139)

1. Why is the United States called a "nation of immigrants?"
2. How did the Louisiana Purchase change the United States?
3. What factors led the United States to become a superpower?

Economy and Culture of the United States (pp. 140-144)

4. Why is the United States a leader in agricultural production?
5. What are some examples of the cultural diversity of the United States?
6. In what industry do most Americans work?

Subregions of the United States (pp. 145-151)

7. What changes have taken place in the industrial base of the Northeast?
8. What role did water play in the development of the Midwest?
9. What industries are found in the South today?
10. How did California become the nation's most populous state?

Critical Thinking

1. Using Your Notes

Use your completed chart to answer these questions.

a. What resources have been important in the development of the United States?

b. Which subregions make up the "rust belt" and which the "sunbelt"? How are they related?

2. Geographic Themes

a. **REGION** How has the economy of the South changed?

b. **MOVEMENT** How has U.S. population shifted since the country began?

3. Identifying Themes

How did air conditioning and irrigation change the population of the West? Which of the five themes apply to this situation?

4. Determining Cause and Effect

What was the effect of the United States becoming industrialized?

5. Making Generalizations

What has been the result of the United States being populated by many different groups of people?

Additional Test Practice, pp. S1–S37

TEST PRACTICE
CLASSZONE.COM

Geographic Skills: Interpreting Maps

U.S. Population and Geographic Centers

Use the map at right to answer the following questions.

1. **MOVEMENT** In which year did the population center cross the Mississippi River?

2. **MOVEMENT** How would you describe the difference between changes in the geographic center and changes in the population center?

3. **REGION** In which region was the population center from 1790 through 1850?

GeoActivity

Create a series of four maps showing movement of the population center of the United States in 50-year periods. Use the map on this page to help you. Start with the period from 1790 to 1840.

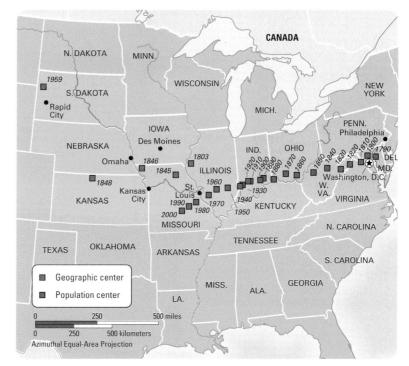

INTERNET ACTIVITY

Use the links at **classzone.com** to do research about the expansion of the United States. Look for the dates when territory was added to the United States.

Writing About Geography Write a report about your findings. Include a map showing the territory acquired to help present the information. List the Web sites that were your sources.

Chapter 7

SECTION 1
History and Government of Canada

SECTION 2
Economy and Culture of Canada

SECTION 3
Subregions of Canada

HUMAN GEOGRAPHY OF CANADA
Developing a Vast Wilderness

Four Subregions of Canada

ARCTIC OCEAN

Beaufort Sea

Baffin Bay

YUKON TERRITORY

NORTHWEST TERRITORIES

NUNAVUT

Great Slave Lake

Davis Strait

Labrador Sea

BRITISH COLUMBIA

Lake Athabasca

Hudson Bay

NEWFOUNDLAND

ALBERTA

MANITOBA

SASKATCHEWAN

Lake Winnipeg

James Bay

QUEBEC

ONTARIO

PRINCE EDWARD ISLAND

Gulf of St. Lawrence

NOVA SCOTIA

NEW BRUNSWICK

L. Superior

L. Michigan

L. Huron

L. Erie

L. Ontario

UNITED STATES

ATLANTIC OCEAN

Gulf of Mexico

Tropic of Cancer

Caribbean Sea

	Atlantic Provinces
	Core Provinces
	Prairie Provinces
	Pacific Province and Territories

0   150   300 miles
0   150   300 kilometers
Azimuthal Equal-Area Projection

GeoFocus

How was such an immense land developed?

Taking Notes Copy the graphic organizer below into your notebook. Use it to record information about the human geography of Canada.

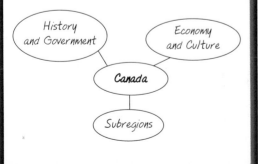

History and Government

Economy and Culture

Canada

Subregions

History and Government of Canada

Main Ideas

- French and British settlement greatly influenced Canada's political development.
- Canada's size and climate affected economic growth and population distribution.

Places & Terms

province

Dominion of Canada

confederation

parliamentary government

parliament

prime minister

CONNECT TO THE ISSUES
DIVERSE SOCIETIES
Conflict between Canadians of French and English ancestry has been a factor throughout much of Canada's history.

US & CANADA

A HUMAN PERSPECTIVE Around A.D. 980, a Viking named Erik the Red sailed to Greenland. Soon after, about 3,000 Vikings colonized the region. About A.D. 1000, Erik's son Leif led an expedition that landed off the Atlantic coast of North America on what is now Newfoundland. Leif called the area Vinland, after the wild grapes that grew there. The Vikings built a settlement but later abandoned it. Five centuries would pass before another European, an Italian navigator named Giovanni Caboto, would come to North America. In 1497, exploring for the English, Caboto (John Cabot in English) landed in Newfoundland and claimed the region for England. European exploration and colonization followed.

The First Settlers and Colonial Rivalry

Canada's vast size and its cold climate significantly affected its development. So did the early migrations of people across its land, the bitter territorial rivalry between the two European nations that colonized it—England and France—and their conflict with the First Nations peoples.

EARLY PEOPLES As you read in Chapter 5, one of the greatest migrations in history took place thousands of years ago, after the last Ice Age. Migrants from Asia began moving into North America across an Arctic land bridge that connected the two continents. Some early peoples remained in what are now the Canadian Arctic and Alaska. These were the ancestors of the Inuit (or Eskimos). Others, the ancestors of the North

LOCATION Quebec City, located on high ground above the St. Lawrence River, was the site of the first permanent French settlement in Canada. **Why was this a desirable location?**

American Indian peoples, gradually moved south, into present-day British Columbia and beyond. When the ice melted, they moved throughout Canada. They settled where they could grow crops.

COLONIZATION BY FRANCE AND BRITAIN During the 16th and 17th centuries, French explorers claimed much of Canada. Their settlements were known as New France. The British, too, were colonizing North America along the Atlantic coast. To both countries, the coastal fisheries and the inland fur trade were important. Soon, the French and British challenged each other's territorial claims. Britain defeated France in the French and Indian War (1754–1763), forcing France to surrender its territory. But French settlers remained.

Steps Toward Unity

By the end of the 18th century, Canada had become a land of two distinct cultures—Roman Catholic French and Protestant English. Conflicts erupted between the two groups, and in 1791, the British government split Canada into two **provinces,** or political units. Upper Canada (later, Ontario), located near the Great Lakes, had an English-speaking majority, while Lower Canada (Quebec), located along the St. Lawrence River, had a French-speaking population. The land to the northwest, called Rupert's Land, was owned by a British fur-trading company.

BACKGROUND
Upper Canada was upriver—on the St. Lawrence—from Lower Canada (Quebec).

ESTABLISHING THE DOMINION OF CANADA Over the next few decades, Quebec City, Montreal, and Toronto developed as major cities in Canada. Population soared as large numbers of immigrants came from Great Britain. Railways and canals were built, and explorers moved across western lands seeking better fur-trading areas.

The conflicts between English-speaking and French-speaking settlers had not ended, however. By the late 1830s, there were serious political and ethnic disputes in both Upper and Lower Canada. The British government decided that major reform was needed. In 1867, it passed the British North America Act creating the **Dominion of Canada.** The Dominion was to be a loose **confederation,** or political union, of Ontario (Upper Canada), Quebec (Lower Canada), and two British colonies on the Atlantic coast—Nova Scotia and New Brunswick. The Dominion

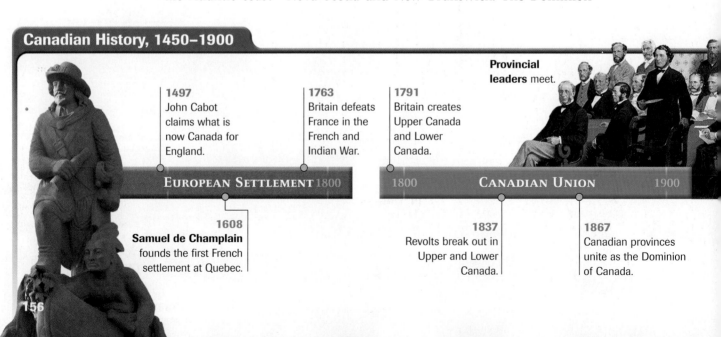

Canadian History, 1450–1900

Provincial leaders meet.

1497
John Cabot claims what is now Canada for England.

1763
Britain defeats France in the French and Indian War.

1791
Britain creates Upper Canada and Lower Canada.

EUROPEAN SETTLEMENT 1800

1800 CANADIAN UNION 1900

1608
Samuel de Champlain founds the first French settlement at Quebec.

1837
Revolts break out in Upper and Lower Canada.

1867
Canadian provinces unite as the Dominion of Canada.

league and its professional ice hockey, baseball, and basketball teams compete in U.S. leagues. The Canadian love of sport goes back to its native peoples, who developed the game we know as lacrosse, and to its early European settlers, who developed ice hockey. Two annual events that are favorites nationwide are the Quebec Winter Carnival, held in Quebec City, and the Calgary Stampede, pictured on page 99.

THE ARTS Not surprisingly, Canada's long history and cultural diversity have given the nation a rich artistic heritage. The earliest Canadian literature was born in the oral traditions of the First Nations peoples. Later, the writings of settlers, missionaries, and explorers lent French and English influences to the literature.

The early visual arts included the realistic carvings of the Inuit and the elaborately decorated totem poles of the First Nations peoples of the West Coast. The artistry of the Inuit carvings has been evident since prehistoric times. Inuit carvers used ivory, whalebone, and soapstone to carve figurines of animals and people in scenes from everyday life. A uniquely Canadian style of painting developed among a group of Toronto-based artists called the Group of Seven early in the 20th century. The performing arts—music, dance, and theater—enjoyed spectacular growth during the last half of the century. The Stratford Festival in Ontario, honoring William Shakespeare, is known worldwide.

In this section, you read about life in Canada today. In the next section, you will learn more about Canada's subregions.

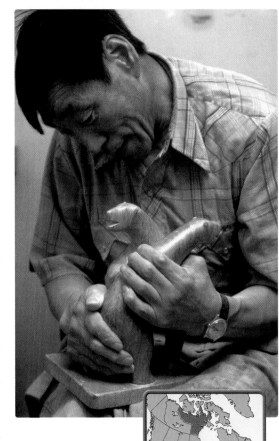

HUMAN–ENVIRONMENT INTERACTION This Inuit artist carves a sculpture of two polar bears from gray soapstone.

Assessment

1 Places & Terms

Identify and explain these terms.

- First Nations
- métis
- reserve

2 Taking Notes

REGION Review the notes you took for this section.

- Which industries drive Canada's economy?
- In which region is the majority of the population located?

3 Main Ideas

a. Why are Canada and the United States close trading partners?

b. How have Canada's urban areas changed?

c. What is Canada's work force like?

4 Geographic Thinking

Drawing Conclusions How have Canada's physical resources contributed to its economic prosperity? **Think about:**

- its location
- its primary industries

S See Skillbuilder Handbook, page R5.

MAKING COMPARISONS Study the information in Chapter 6, Section 2, about the U.S. economy. Create a **Venn diagram** with three circles showing the economic activities Canada and the United States have in common and those that are unique to each.

Comparing Cultures

Transportation

As you have read, one of the five themes of geography is movement—how people move themselves and their goods across the Earth's surface. The earliest humans moved by foot from place to place. Later, they used animals, both to ride and as pack animals. Needing to cross streams, ancient people built primitive boats from available materials, such as wood and reeds. Over the centuries, advances in technology from wheeled vehicles to the steam engine to the construction of lighter-than-air craft has enabled people in different regions to meet the challenges posed by their environments.

Canada

Vietnam

Algeria

Peru

In North African countries like Algeria, camels are often called "ships of the desert" because they can carry freight and people across long distances. The Arabian, or one-humped, camel shown here in the Sahara Desert can cover 40 miles a day for four days carrying 400 pounds.

Flat, smooth roadways crisscrossing Vietnam make it easy for these workers to transport hundreds of fish traps from workshops to customers on the coast by bicycle.

In the northernmost reaches of Canada, roads are scarce. So, vast distances between places are more easily covered by small planes that can touch down on land or water, like this one flying into Cochrane, Ontario.

GeoActivity

RESEARCHING TRANSPORTATION
Working with a partner, use the Internet to research transportation around the world. Then prepare a report that shows the design of a **Web page** highlighting some aspect of world transportation.

- Create text to present the information you have found.
- Select suitable images.
- Locate appropriate links for visitors to your Web site.

RESEARCH LINKS
CLASSZONE.COM

GeoData

LAND TRANSPORTATION
- In the United States, there is one car for every two persons; in Somalia, one for every 500.
- One of the world's longest single rail systems, Russia's Trans-Siberian Railway, covers a distance of 5,867 miles from Moscow to the port of Nakhodka.
- Snowmobiles have replaced dogsleds as transport in remote, cold climates of North America.
- China has more bicycles—about 540,000,000—than any other country.
- Animals, including dogs, horses, donkeys, mules, camels, and elephants, still provide transport for many people around the world.

AIR TRANSPORTATION
- Airliners carried 137 million passengers on more than 1 million flights from the United States to other countries from June 1999 to June 2000.

WATER TRANSPORTATION
- Some modern cruise ships and ocean liners are more than 900 feet long and can carry upwards of 2,000 passengers on a voyage.

This crescent-shaped boat on Lake Titicaca in Peru is made from a reedlike plant. Native peoples of the region have made these boats for centuries.

Subregions of Canada

A HUMAN PERSPECTIVE The Grand Banks, a shallow section of the North Atlantic off the coast of Newfoundland, make up one of the earth's richest fishing grounds. In fact, it was the abundance of fish—including cod, haddock, herring, and mackerel—that first attracted Europeans to the region centuries ago. Today, thousands of hardy Canadians make their living fishing in these coastal waters. One, Alex Saunders of Labrador, remarked that "fishing is a disease. Once you start, you keep at it, do whatever's necessary. I jeopardize my home, all my possessions just to keep this boat going and keep fishing." The Grand Banks are part of the Atlantic Provinces, one of Canada's four subregions.

The Atlantic Provinces

Canada is divided into ten provinces and three territories. Each has a unique population, economy, and resources. Eastern Canada is the location of the four **Atlantic Provinces**—Prince Edward Island, New Brunswick, Nova Scotia, and Newfoundland.

HARSH LANDS AND SMALL POPULATIONS As you can see on the chart below, the Atlantic Provinces are home to just 8 percent of Canada's population. Of these people, most live in coastal cities, such as Halifax, Nova Scotia, and St. John, New Brunswick. The small population is due largely to the provinces' rugged terrain and severe weather.

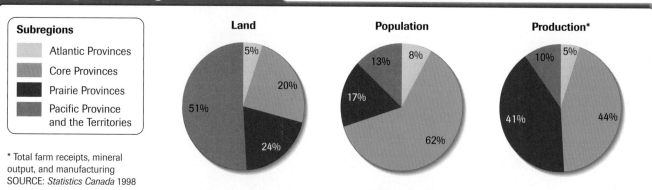

Comparing the Subregions of Canada

Subregions
- Atlantic Provinces
- Core Provinces
- Prairie Provinces
- Pacific Province and the Territories

Land
- 5%
- 20%
- 24%
- 51%

Population
- 8%
- 13%
- 17%
- 62%

Production*
- 5%
- 10%
- 41%
- 44%

* Total farm receipts, mineral output, and manufacturing
SOURCE: *Statistics Canada* 1998

SKILLBUILDER: Interpreting Graphs

❶ **ANALYZING DATA** Which subregion has the highest production?

❷ **MAKING COMPARISONS** How do the Pacific Province and the Territories compare overall to the other three subregions?

The Fight Against Terrorism

How can a country protect itself from terrorism?

Main Ideas

• Terrorism threatens the safety and security of society.

• The United States launched a war against international terrorism after being attacked on September 11, 2001.

Places & Terms

terrorism

global network

coalition

biological weapon

US & CANADA

A HUMAN PERSPECTIVE For Karl Co, a 15-year-old sophomore at Stuyvesant High School in New York City, September 11, 2001, began as "such a normal day." From his classroom, Karl had a clear view of the World Trade Center, just four blocks away. On a normal day, about 50,000 people worked in and 70,000 visited the twin towers. When the north tower burst into flames and smoke, Karl first thought, "It's a bomb. I'm going to die." Then the south tower erupted, and, shortly after, both collapsed. The students soon learned terrorists had crashed airliners into the towers, and the school was evacuated.

The September 11 Attacks

The students at Stuyvesant High had witnessed an act of **terrorism.** Terrorism is the unlawful use of, or threatened use of, force or violence against individuals or property for the purpose of intimidating or causing fear for political or social ends. Like many countries, the United States has been subjected to terrorism, both at home and abroad. But the September 11, 2001, attacks were the most destructive acts of terrorism ever committed on American soil.

On that morning, 19 Arab terrorists hijacked four airliners. They crashed two planes into the World Trade Center towers and one into the Pentagon, the U.S. military headquarters near Washington, D.C. The fourth plane crashed in Pennsylvania without striking its intended target, after some passengers overwhelmed the hijackers.

THE DESTRUCTION The hijacked planes were loaded with fuel. They became destructive missiles as they crashed into their targets. Thousands of workers escaped before the damaged skyscrapers collapsed. Fire and raining debris caused nearby buildings to collapse as well. At the Pentagon, the plane tore a 75-foot hole in the building's west side.

About 3,000 people died in the attacks. The dead included 265 plane passengers and 343 New York City firefighters who had entered the towers to rescue those trapped inside. Nine buildings in the city's financial district were completely destroyed or partly collapsed, and six others suffered major damage. The disaster area covered 16 acres.

THE TERRORISTS Immediately after the attacks, investigators worked to identify both the hijackers and those who directed the attacks. The evidence pointed to a **global network,** or worldwide interconnected group, of extremist Islamic terrorists led by Osama bin Laden, a Saudi Arabian millionaire. The group, known as al-Qaeda, was formed to fight the Soviet invasion of Afghanistan in 1979. Al-Qaeda later began to oppose

BACKGROUND
Osama bin Laden offered to help the Saudi Arabian government when Iraq invaded Kuwait in 1990 and threatened Saudi Arabia. He was angered when the Saudis turned to the United States for military help instead.

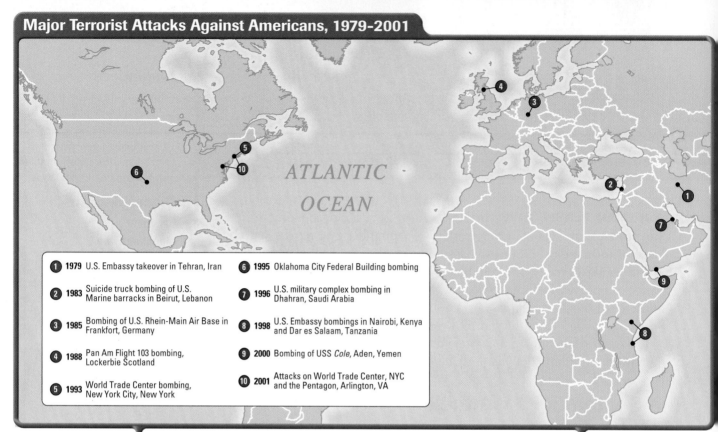

Major Terrorist Attacks Against Americans, 1979–2001

ATLANTIC OCEAN

① 1979 U.S. Embassy takeover in Tehran, Iran

② 1983 Suicide truck bombing of U.S. Marine barracks in Beirut, Lebanon

③ 1985 Bombing of U.S. Rhein-Main Air Base in Frankfort, Germany

④ 1988 Pan Am Flight 103 bombing, Lockerbie Scotland

⑤ 1993 World Trade Center bombing, New York City, New York

⑥ 1995 Oklahoma City Federal Building bombing

⑦ 1996 U.S. military complex bombing in Dhahran, Saudi Arabia

⑧ 1998 U.S. Embassy bombings in Nairobi, Kenya and Dar es Salaam, Tanzania

⑨ 2000 Bombing of USS *Cole*, Aden, Yemen

⑩ 2001 Attacks on World Trade Center, NYC and the Pentagon, Arlington, VA

SKILLBUILDER: Interpreting Maps

LOCATION What region was the site of the most attacks on Americans?

American influence in Muslim lands. It started to target Americans and U.S. allies after the Persian Gulf War in 1991. Since its founding, al-Qaeda has carried out numerous terrorist attacks.

Aftermath of the Attacks

The September 11 attacks shocked and distressed not only Americans but people around the world. President George W. Bush declared war on terrorism and called on other nations to join the United States in fighting global terrorism. He also pushed for new security measures at home and authorized a search for suspected terrorists.

INTERNATIONAL WAR ON TERRORISM The United States organized a **coalition,** or an alliance, of nations to fight the war on terrorism. Canada, China, Great Britain, Pakistan, Russia, and many other nations joined the coalition. They pledged to share information, arrest terrorists in their countries, and seize the financial assets of terrorist groups. The coalition also supported military action in Afghanistan, where al-Qaeda was based. As part of Operation Enduring Freedom, the United States began bombing Afghanistan in October 2001, and later sent in ground forces. By mid-March 2002, Afghanistan's extremist Taliban regime had been removed from power and the al-Qaeda network overthrown.

In March 2003, President Bush expanded the war on terrorism by taking military action against Iraq. The President claimed that Iraqi dictator Saddam Hussein posed a threat to national security. Major combat in Iraq ended in May soon after Hussein's regime had been toppled.

Geographic Thinking

Using the Atlas
Ⓐ Locate Afghanistan on the political map on page A22. What is its location in relation to Saudi Arabia?

HOMELAND SECURITY New airport security measures were enacted after the September 11 attacks. In addition, extra precautions were taken at public places where large numbers of people gather, such as sports stadiums. Other possible targets—nuclear power plants and water supply systems—expanded security. The Department of Homeland Security, led by Secretary Tom Ridge, was established to coordinate antiterrorist efforts.

Facing Terrorist Threats

Terrorism has been a global problem for decades. The prevention of terrorist attacks is one of the most difficult tasks facing the world today.

Geographic Thinking

Making Comparisons
B How does a war on terrorism differ from a conventional war against another country?

TERRORIST OPERATIONS AND WEAPONS Terrorists act in secret and can move from country to country while pursuing their objectives. Some terrorist groups want territory, like Palestinian extremists who use violence trying to gain a homeland in Southwest Asia. Other terrorists, such as the domestic terrorists who bombed the Federal Building in Oklahoma City in 1995, want to attack government policies. ◄**B**

Terrorists can use other weapons besides bombs and fuel-laden planes, including biological, chemical, and nuclear weapons. **Biological weapons** refer to bacteria and viruses that can be used to harm or kill people, animals, or plants. The United States went on an anthrax alert after traces of the anthrax bacteria were found in letters sent to some members of Congress and the news media after the September attacks.

BALANCING SECURITY AND FREEDOM The United States and its allies hope to reduce terrorism by breaking up terrorist groups and by increasing security to make it harder for terrorists to act. But there are many kinds of terrorist threats, and the fight against global terrorism could go on for many years. Democratic countries also have to meet the challenge of providing security for citizens while preserving freedom and individual rights.

SECTION 1 Assessment

① Places & Terms
Identify and explain the following terms.
- terrorism
- global network
- coalition
- biological weapon

② Taking Notes
REGION Review the notes you took for this section.

	Causes	Effects
Issue 1		
Terrorism		

- Why are the United States and its allies so concerned about terrorism?
- How has terrorism affected the policies of the United States and its allies?

③ Main Ideas
a. What happened in the terrorist attacks on the United States on September 11, 2001, and who was believed to be responsible?

b. How did the United States respond to the attacks?

④ Geographic Thinking
Drawing Conclusions What might be some difficulties facing the United States and its allies in fighting terrorism?
Think about:
- terrorists moving from country to country
- the variety of weapons available to terrorists

RESEARCH LINKS
CLASSZONE.COM

GeoActivities

EXPLORING LOCAL GEOGRAPHY Do research to learn how the fight against terrorism is being waged in your state. Write a **press release** describing one of these antiterrorist measures.

Urban Sprawl

How can urban sprawl be controlled?

- Many metropolitan areas in the United States and Canada have sprawled, or spread out, farther and farther.
- Cities are focusing on smart-growth solutions to urban sprawl.

Places & Terms

urban sprawl

infrastructure

smart growth

sustainable community

A HUMAN PERSPECTIVE   Richard Baron is a real estate developer who tried to address the related problems of urban sprawl and inadequate low-income housing. In 1996, he began building Murphy Park, an affordable and attractive housing complex in mid-town St. Louis, Missouri. The development has more than 400 units and contains both apartments and townhouses. It has plenty of green space, art and day-care centers, and an elementary school. More than half of Murphy Park's units are reserved for people with low income. Baron's solution—to bring the attractive features of suburban living to the city—is one of many that are being applied to the problem of urban sprawl.

Growth Without a Plan

Those Americans and Canadians who can afford it often choose to work in a city but live in its suburbs. They are usually attracted by new, upscale housing, better public services, and open space. As suburbs become more numerous, metropolitan areas become larger and more difficult to manage. (See chart to the right.)

URBAN SPRAWL Poorly planned development that spreads a city's population over a wider and wider geographical area is called **urban sprawl.** As outlying areas become more populated, the land between them and the city fills in as well.

In the United States and Canada, urban sprawl is becoming a matter of increasing concern. From 1970 to 2000, people who worked in U.S. cities moved farther and farther from urban centers. The population density of cities in the United States decreased by more than 20 percent as people in cities moved to suburbs and outlying areas. About 30,000 square miles of rural lands were gobbled up by housing developments. For example, the population of the city of Chicago decreased during this period from 3.4 million people to 2.8 million. But the Chicago metropolitan area grew from about 7.0 million persons to 7.3 million.

Canada is less populated than the United States but faces similar problems. In the 1990s, more than 75 percent of all Canadians lived in urban areas.

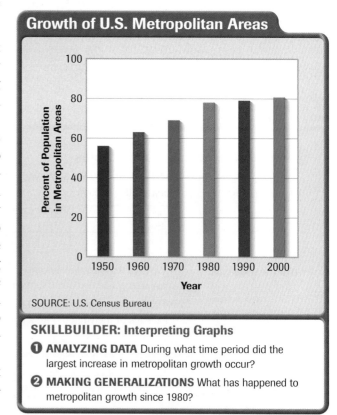

Growth of U.S. Metropolitan Areas

Percent of Population in Metropolitan Areas

(Years: 1950, 1960, 1970, 1980, 1990, 2000)

SOURCE: U.S. Census Bureau

SKILLBUILDER: Interpreting Graphs

❶ **ANALYZING DATA** During what time period did the largest increase in metropolitan growth occur?

❷ **MAKING GENERALIZATIONS** What has happened to metropolitan growth since 1980?

PLACE Las Vegas, Nevada, is a perfect example of urban sprawl. In the 1970s (left), it was a small city. In the 1990s (right), it became the fastest growing city in the country.
What are some of the differences between the photos of Las Vegas above?

CAUSES OF URBAN SPRAWL Sprawl occurs in metropolitan areas that allow unrestricted growth or that have no plans to contain it. Other factors include the widespread use of automobiles and the building of expressways. Autos and relatively cheap gasoline enable Americans to drive many miles to and from their jobs. Despite clogged highways and long commutes, Americans prefer their cars to mass transit. Expressways provide the means for continued reliance on the automobile.

Yet, despite sprawl, there are many reasons why Americans have moved to suburbs. Some people want open spaces or better schools and housing. Still others want to try to recapture the sense of community they experienced while growing up. They want their children to know their neighbors and have a backyard in which to play. Only recently have urban planners started to design big-city neighborhoods to give a sense of community, hoping to slow the flight to the suburbs.

Urban Sprawl's Negative Impact

Urban sprawl has a negative impact on the quality of life in many ways. As suburbs grow, more commuter traffic strains the infrastructure. **Infrastructure** consists of the basic facilities, services, and machinery needed for a community to function. For example, roads and bridges need maintenance. More cars on the road for more time adds to air pollution, too. Also, sources of water, such as rivers or underground aquifers (layers of water-holding rock or soil), become depleted. ◄

Urban sprawl also has other costs. The cost of providing streets, utilities, and other public facilities to suburban communities is often at least 25 per cent higher than for high-density residences in a city. Urban sprawl also separates classes of people. When those in upper-income brackets choose to live in outlying areas, lower-income residents often become isolated in inner-city areas.

Geographic Thinking

Seeing Patterns
A What problems has the automobile caused?

Solutions to Sprawl

More and more cities are developing plans for **smart growth,** which is the efficient use and conservation of land and other resources. Most often this involves encouraging development close to or inside the limits of existing cities. Good public transportation systems help to make smart growth possible by cutting down on auto traffic.

PORTLAND'S GROWTH BOUNDARY In 1979, the city of Portland, Oregon, drew a line around itself to create an urban growth boundary. Building was allowed inside the boundary. The surrounding green space was off limits to developers. This decision caused controversy but has paid off. Portland has contained urban sprawl.

VANCOUVER'S PLAN FOR SUSTAINABLE COMMUNITIES Since 1961, Vancouver, British Columbia, has seen the population of its metropolitan area double. The growth of outlying suburbs often took place at the expense of forests, farms, and flood plains. In 1995, the Greater Vancouver Regional Board adopted a plan to manage growth. It involved turning suburbs into **sustainable communities,** that is, communities where residents could live and work. The same solution was applied to Vancouver's downtown area, where about 40 percent of its residents now walk to work. This has cut down on commuting.

Geographic Thinking

Making Comparisons
B How were the urban growth actions of Portland and Vancouver similar?

GRASSROOTS OPPOSITION In some metropolitan areas, citizens have banded together to offer their own solutions to urban sprawl. For example, citizens in Durham, North Carolina, opposed additional commercial development along a congested area of a nearby interstate highway. They formed CAUSE—Citizens Against Urban Sprawl Everywhere. The organization is working against sprawl through education and political activism.

In this section, you read about the challenge of urban sprawl. In the Case Study that follows, you will learn about challenges increasingly diverse societies bring to the United States and Canada.

SECTION 2 Assessment

❶ Places & Terms

Identify and explain the following places and terms.

- urban sprawl
- infrastructure
- smart growth
- sustainable community

❷ Taking Notes

HUMAN-ENVIRONMENT INTERACTION Review the notes you took for this section.

Issue 2: Urban Sprawl	Causes	Effects

- What are some of the causes of urban sprawl?
- What are some of the effects of urban sprawl?

❸ Main Ideas

a. What happens when metropolitan areas spread farther and farther out?

b. What are some of the ways cities are dealing with urban sprawl?

c. What are some of the ways citizens are dealing with urban sprawl?

❹ Geographic Thinking

Drawing Conclusions
What would happen to the environment if urban sprawl were not controlled? **Think about:**

- the negative effects of urban sprawl
- the quality of life in the United States and Canada

EXPLORING LOCAL GEOGRAPHY Pair with another student and choose a metropolitan area in the United States or Canada to research. Then prepare a **report** on the condition of urban sprawl in that area and present your report to the class. Discuss the effects of urban sprawl and what steps, if any, are being taken to control the sprawl.

Map and Graph Skills

Reading a Bounded-Area Map

Urban growth over time is the theme of this map of the Baltimore, Maryland, and Washington, D.C., areas. Both Baltimore and Washington grew from small cities to important metropolitan areas, spreading outward in all directions. At one time, nearly 30 miles of unsettled area separated them. Today, much of this area has been built up as the Baltimore and Washington metropolitan areas have spread.

THE LANGUAGE OF MAPS **Bounded-area maps** show the distribution of some feature of interest, such as climate, vegetation, precipitation, or, in this case, urban growth in a region. Bounded-area maps use lines, colors, and patterns to communicate information.

Urban Growth in Baltimore and Washington ❶

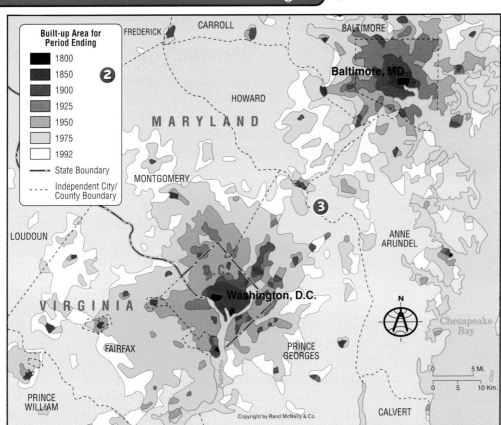

Copyright by Rand McNally & Co.

❶ The title gives you the subject matter of the map.

❷ The key explains the meanings of the colors and symbols.

❸ This map shows the gradual spread of urban areas from two neighboring cities—Baltimore and Washington. The map covers the period of time from 1800 to 1992.

Map and Graph Skills Assessment

1. Seeing Patterns
Like most early settlements, Baltimore and Washington were founded near essential geographic features. What were they?

2. Analyzing Data
During which time period did the greatest expansion take place for the Washington metropolitan area?

3. Drawing Conclusions
At what physical location do the two metropolitan areas seem to have merged?

CaseStudy

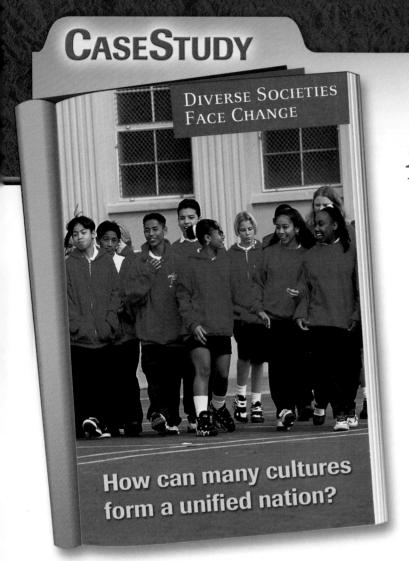

DIVERSE SOCIETIES FACE CHANGE

How can many cultures form a unified nation?

The diverse heritage of the United States is evident in this group of students in California.

As you read earlier in this unit, the first immigrants to North America are believed to have come from Asia. They are thought to have crossed a land bridge that existed in what is now the Bering Strait thousands of years ago. Since that time, millions of people from countries all over the world have immigrated to the United States and Canada. They have come in search of a new life in a new homeland. The challenge for citizens and governments of both the United States and Canada is to make sure that these diverse peoples continue to remain unified.

"Mosaic" or "Melting Pot"

After centuries of immigration, the United States and Canada are culturally diverse. They contain large populations of the world's cultures. Ethnic neighborhoods with populations of Asians, Eastern Europeans, and Latin Americans are found in most large cities of both countries. In New York City alone, immigrant schoolchildren speak more than 100 different languages. The arrival of so many peoples over the years left the United States and Canada with the difficult task of forming a unified society. Each country approached the task of unifying its many cultures differently.

CANADA'S CULTURAL "MOSAIC" Canada's earliest settlers were its native peoples. Its first European settlers came, as you have learned, from two distinct cultural groups—French and English. All of these groups kept their separate identities as the nation developed. Also, Canada encouraged immigration from all over the world. It wanted to fill its vast lands and expand its workforce and its domestic markets. These immigrants also were encouraged to retain their cultural heritage.

As a result, many Canadians have strong ethnic ties. In fact, as you read in Chapter 7, the ethnic identity of French-speaking citizens in Quebec has been so strong that at times they have even considered separating themselves from the Canadian confederation.

The Canadian government has officially recognized the multicultural nature of Canada. In 1988, it enacted the Canadian Multiculturalism Act to protect and promote diversity. Many Canadians believe that this policy ensures equality for people of all origins and enriches their nation. But not all agree. Some Canadians feel that diversity has promoted difference at the expense of "Canadianness."

AMERICA'S "MELTING POT" For many years, people in the United States believed that assimilation was the key. It was thought to be the best way to build one nation from many different peoples. Assimilation occurs when people from a minority culture assume the language, customs, and lifestyles of people from the dominant culture. Native Americans were an example. In the late 19th century, they were encouraged and even forced to learn English, adopt Western dress, and become Christians to assimilate into the dominant white culture.

People expected immigrants to assimilate, too. Those who did not could face prejudice because of their cultural differences. Immigrants soon learned that life would be easier if they adopted the ways of their new country—if they underwent "Americanization." Most of these immigrants had come from Europe. Many wanted to assimilate. They wanted to adopt a common language and culture—to become Americans.

New Immigrants Challenge Old Ways

The immigrants who came to the United States in the late 20th century brought different attitudes. They came mainly from Latin America and Asia. They were culturally or racially unlike earlier immigrant groups, who had come mainly from Europe. These later immigrants were less willing to give up their traditions and beliefs in order to assimilate.

DIVIDED OPINION Some Americans felt that the new immigrants did not understand what made the United States unique. According to this point of view, America's strength has come from blending its diverse cultures to create something new—an American. They also believed that encouraging different languages and customs would promote separation, not unity. In response, they wanted immigration limited and English made the official language.

SEE
PRIMARY SOURCE **C**

Other Americans, including many educators, held different views. They thought that American society would benefit by stressing multiculturalism, as the Canadians do.

As you can see, bringing many cultures together is a continuing challenge both in the United States and in Canada. So, how can cultural diversity be preserved and national unity forged? The Case Study Project and primary sources that follow will help you explore this question.

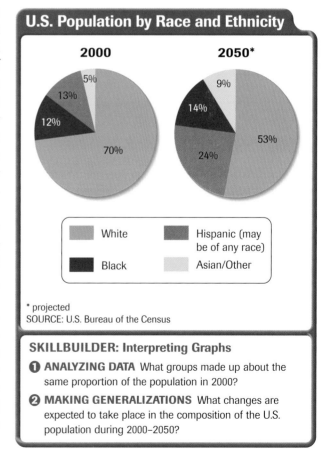

U.S. Population by Race and Ethnicity

2000
5%
13%
12%
70%

2050*
9%
14%
53%
24%

White
Black
Hispanic (may be of any race)
Asian/Other

* projected
SOURCE: U.S. Bureau of the Census

SKILLBUILDER: Interpreting Graphs

❶ **ANALYZING DATA** What groups made up about the same proportion of the population in 2000?

❷ **MAKING GENERALIZATIONS** What changes are expected to take place in the composition of the U.S. population during 2000–2050?

CASE STUDY

PROJECT Talk Show Discussion

Primary sources A, B, C, D, and E offer differing opinions about assimilation and maintaining cultural identity. Use them along with your own research from the library or Internet to prepare for a talk show discussion on the issue of today's cultural diversity.

RESEARCH LINKS
CLASSZONE.COM

Suggested Steps

1. With a group totaling five students, prepare a talk show discussion on the topic, "Can Many Cultures Form a Unified Nation?" One member should act as the discussion leader. Each of the other members should select one of the following positions: for assimilation or against assimilation.

2. Think about the following questions as you prepare for your role. "Must a unified nation have a single culture?" "What are the advantages and disadvantages of assimilation, or the advantages and disadvantages of multiculturalism, in unifying a nation?"

3. Use online and print resources to research your topic.

4. Write an opening statement of your position. Prepare visuals, such as charts or graphs, if you need them to support your position.

5. Present your position as a part of the talk show. Discuss with the leader and other group members the focus question given above.

Materials and Supplies

• posterboard
• colored markers
• reference books, newspapers, and magazines
• Internet access

PRIMARY SOURCE A

Newspaper Article *In 1998, the* Washington Post *published a series of articles titled* The Myth of the Melting Pot. *Staff writer* **William Booth** *offered the following comments about immigration and cultural identity in his piece,* "One Nation, Indivisible: Is It History?"

The immigrants of today come not from Europe but overwhelmingly from the still developing world of Asia and Latin America. They are driving a demographic shift so rapid that within the lifetimes of today's teenagers, no one ethnic group—including whites of European descent—will comprise a majority of the nation's population. . . .

[M]any historians argue that there was a greater consensus in the past on what it meant to be an American, a yearning for a common language and culture, and a desire—encouraged, if not coerced [forced] by members of the dominant white Protestant culture—to assimilate. Today, they say, there is more emphasis on preserving one's ethnic identity, of finding ways to highlight and defend one's cultural roots.

PRIMARY SOURCE B

Social Commentary *Michelle Young is a writer and editor. Much of her work has focused on issues of multiculturalism. In the following excerpt from a 1996 article in the online publication* Career Magazine, *Young contrasts assimilation with multiculturalism.*

The melting pot concept spoke of all Americans being part of the enormous "cultural stew" we call America. . . . Many people . . . saw the United States of America as a place where historical hurts from their homelands could be erased. . . .

But America was not the nation they'd been promised, where the streets were paved with gold. Many newcomers knew that from experience because "they" were doing the paving! As a result, people began to realize that the concept of the melting pot just wasn't realistic. . . .

In contrast to the melting pot, multiculturalism encourages us to take pride in our own roots first, in our ingredients we've added to what has become America's multicultural stew. The nation's promise lies in that multicultural stew, and by appreciating our own cultures, we develop an eagerness to learn about others' origins.

Political Commentary *Patrick Buchanan is a politician who was the presidential candidate of the Reform Party in 2000. Buchanan was a strong supporter of immigration reform and assimilation, as is evident in these words posted on his Web site on August 6, 2000.*

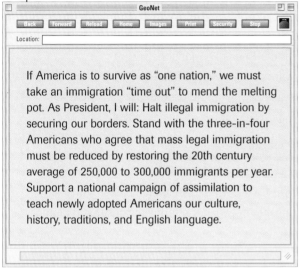

GeoNet

Location:

If America is to survive as "one nation," we must take an immigration "time out" to mend the melting pot. As President, I will: Halt illegal immigration by securing our borders. Stand with the three-in-four Americans who agree that mass legal immigration must be reduced by restoring the 20th century average of 250,000 to 300,000 immigrants per year. Support a national campaign of assimilation to teach newly adopted Americans our culture, history, traditions, and English language.

Government Law *The Canadian Multiculturalism Act was passed by the Canadian parliament in 1988. Its purpose was to make the preservation and enhancement of multiculturalism in Canada the law of the land.*

". . . It is hereby declared to be the policy of the Government of Canada to . . . (b) recognize and promote the understanding that multiculturalism is a fundamental characteristic of the Canadian heritage and identity and that it provides an invaluable resource in the shaping of Canada's future; . . . (c) promote the full and equitable participation of individuals and communities of all origins in the continuing evolution and shaping of all aspects of Canadian society and assist them in the elimination of any barrier to that participation; . . . (f) encourage and assist the social, cultural, economic, and political institutions of Canada to be both respectful and inclusive of Canada's multicultural character; . . . (g) promote the understanding and creativity that arise from the interaction between individuals and communities of different origins."

Government Document *The 2000 census form contained detailed racial and ethnic classifications, showing the diverse peoples that make up the population of the United States.*

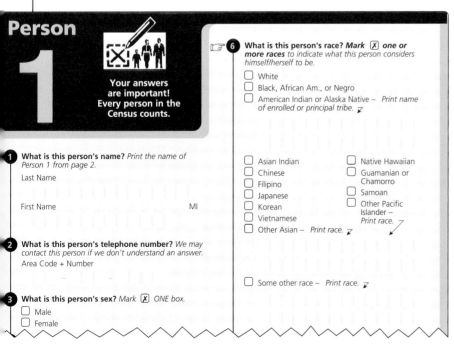

Person

1

Your answers are important! Every person in the Census counts.

1 What is this person's name? *Print the name of Person 1 from page 2.*
Last Name

First Name   MI

2 What is this person's telephone number? *We may contact this person if we don't understand an answer.*
Area Code + Number

3 What is this person's sex? *Mark* ☒ *ONE box.*
☐ Male
☐ Female

☞ **6** What is this person's race? *Mark* ☒ *one or more races to indicate what this person considers himself/herself to be.*
☐ White
☐ Black, African Am., or Negro
☐ American Indian or Alaska Native – *Print name of enrolled or principal tribe.* ↘

☐ Asian Indian   ☐ Native Hawaiian
☐ Chinese   ☐ Guamanian or Chamorro
☐ Filipino
☐ Japanese   ☐ Samoan
☐ Korean   ☐ Other Pacific Islander – *Print race.* ↘
☐ Vietnamese
☐ Other Asian – *Print race.* ↘

☐ Some other race – *Print race.* ↘

PROJECT *CheckList*

Have I . . .

✓ fully researched my topic?

✓ taken into account both sides of an issue?

✓ created informative visuals that make my presentation clear and interesting?

✓ practiced the delivery of my presentation?

VISUAL SUMMARY
TODAY'S ISSUES IN THE UNITED STATES AND CANADA

Conflict

The Fight Against Terrorism
- Terrorists attack the United States on September 11, 2001.
- The United States increases security at home and searches for suspected terrorists within the country.
- A coalition of nations led by the United States launches a war against global terrorism.
- The war begins in Afghanistan, where those responsible for the September attacks—the al-Qaeda terrorists led by Osama bin Laden—are based.

Economics

Urban Sprawl
- Many metropolitan areas in North America have spread out farther and farther.
- This has caused problems such as traffic congestion, air pollution, strains on infrastructure, rising housing costs, and the separation of the well-off from the poor.
- Some governments and citizens are promoting "smart growth" as an answer to urban sprawl.

Government

Case Study: Diverse Societies Face Change
- Centuries of immigration from all parts of the world have given the United States and Canada diverse populations.
- The United States and Canada have approached unifying their many cultures differently.
- Bringing diverse peoples together is a continuing challenge for both countries.

Reviewing Places & Terms

A. Briefly explain the importance of each of the following.

1. terrorism
2. global network
3. coalition
4. biological weapon
5. urban sprawl
6. infrastructure
7. smart growth
8. sustainable community

B. Answer the questions about vocabulary in complete sentences.

9. What is the objective of terrorism?
10. What are the characteristics of a global network?
11. What is the name for an alliance of nations?
12. Which of the terms above might be used to refer to anthrax?
13. How does urban sprawl contribute to air pollution?
14. What are some of the elements that make up infrastructure?
15. Which term involves encouraging development close to or inside city limits?
16. What did Vancouver try to turn into sustainable communities?
17. What is the relationship between the terms terrorism and global network?
18. What is the objective of employing a biological weapon?
19. How does urban sprawl cause housing costs to rise?
20. What system is an important component of smart growth?

Main Ideas

The Fight Against Terrorism (pp. 173-175)

1. What are some of the actions governments can take when faced with terrorism?
2. What are some of the weapons used by terrorists to further their objectives?
3. What might become a problem for democratic governments waging war against terrorism?

Urban Sprawl (pp. 176-179)

4. What are some of the causes of urban sprawl?
5. What are some of the negative effects of urban sprawl?
6. How are governments and concerned citizens trying to find solutions to urban sprawl?

Case Study: Diverse Societies Face Change (pp. 180-183)

7. Why have the United States and Canada become diverse societies?
8. How have Americans reacted to diversity?
9. How have Canadians reacted to diversity?
10. What are some ways suggested for Americans to meet the challenges of the new immigrants?

Critical Thinking

1. Using Your Notes

Use your completed chart to answer these questions.

	Causes	Effects
Issue 1: Terrorism		
Issue 2: Urban Sprawl		

a. How might a negative effect of urban sprawl be halted?

b. What are some of the positive effects of diverse societies?

2. Geographic Themes

a. **MOVEMENT** How have terrorists been able to form global networks?

b. **HUMAN-ENVIRONMENT INTERACTION** How has the spread of urban sprawl affected the environment?

3. Identifying Themes

If you were a government official, how would you promote smart growth? Which of the five themes are reflected in your answer? Explain.

4. Making Decisions

What factors do democratic governments have to consider when waging a war against an enemy such as global terrorism?

5. Making Comparisons

How do the Canadian and American approaches to a diverse society differ?

Additional Test Practice, pp. S1–S37

TEST PRACTICE
CLASSZONE.COM

Geographic Skills: Interpreting Graphs

Region of Last Residence of Legal Immigrants to the United States, 1901–1998

Use the graph to answer the following questions.

1. **ANALYZING DATA** What was the percentage of immigrants from Europe during 1901–1910? during 1991–1998?

2. **MAKING COMPARISONS** Which two regions supplied the largest percentage of immigrants to the United States during the last century?

3. **DRAWING CONCLUSIONS** What significant change took place in the pattern of immigration during the 20th century?

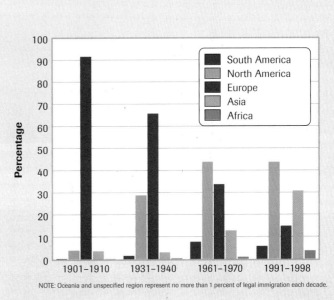

NOTE: Oceania and unspecified region represent no more than 1 percent of legal immigration each decade.

SOURCE: 1998 Statistical Yearbook of the Immigration and Naturalization Service

GeoActivity

Do research to create a chart showing the total number of immigrants from each region during the 20th century. Display the figures for each region on an outline map of the world.

INTERNET ACTIVITY

Use the links at **classzone.com** to research immigration to Canada. Focus on changes in the regions from which immigrants came in the 20th century.

Writing About Geography Write a report on your findings. Combine with a chart listing the regions and the percentages.

Latin America

**Latin America includes parts of
North America, Central America
and the Caribbean, and South
America. The region covers many
latitudes from north to south of
the equator.**

HUMAN–ENVIRONMENT INTERACTION Chacobo
Indians make the dugout canoes they use to explore in
the Amazon River basin in northern Bolivia.

MOVEMENT Villagers from surrounding areas
bring their goods to market in the Aztec city of
Tenochtitlán, depicted in this mural by Diego Rivera.

LOCATION Latin America extends from Mexico southward across the equator to nearly reach Antarctica in the Southern Hemisphere.

REGION It is called "Latin America" because the two main languages spoken there—Spanish and Portuguese—developed from Latin.

REGION This region is bordered by two oceans (Atlantic and Pacific), the Gulf of Mexico, and the Caribbean Sea.

For more information on Latin America . . .

RESEARCH LINKS
CLASSZONE.COM

LATIN AMERICA

PLACE Sugarloaf Mountain is a famous landmark that looks out over Guanabara Bay in Rio de Janeiro, Brazil. The statue of Christ atop the mountain reflects the importance of the Catholic faith to millions of Latin Americans.

187

Today's Issues in Latin America

Three of the most important issues that concern Latin America today are resources, democracy, and the income gap between rich and poor.

As you read Chapters 9 and 10, you will learn helpful background information. You will study the issues themselves in Chapter 11.

In a small group, answer the following questions. Then participate in a class discussion of your ideas.

Exploring the Issues

1. **RESOURCES** What are some resources that are becoming increasingly scarce in the world?

2. **DEMOCRACY** What are some threats to democracy in the world today? What conditions might be necessary for democracy to thrive?

3. **INCOME GAP** Why might an income gap exist in a country? How might a growing gap between rich and poor affect a country?

For more on these issues in Latin America . . .

CURRENT EVENTS
CLASSZONE.COM

RESOURCES

How can we preserve and develop the rain forest?

Agriculture and timber harvesting in Brazil are reducing the size of the rain forests by destroying thriving ecosystems, but are providing food and export products.

DEMOCRACY

How can Latin Americans gain a voice in government?

Demonstrators in Chile rally in support of putting former dictator General Augusto Pinochet on trial. The signs say, "Judgment for Pinochet—truth and justice for Chile."

INCOME GAP

CASE STUDY

How can the economic gulf between rich and poor be bridged?

There is a growing gap between rich and poor in Latin America, with all the problems of slums, homeless children, and street crime. Here, a young girl stands above polluted water in a slum in Belém, Brazil.

189

Unit ATLAS

Patterns of Physical Geography

Use the Unit Atlas to add to your knowledge of Latin America, which stretches from Mexico to the tip of South America. As you look at the maps and graphs, notice geographic patterns and specific details about the region. For example, the graph gives details about two large rivers in the region.

After studying the graphs and physical map on these two pages, jot down answers to the following questions in your notebook.

Making Comparisons

1. Which river systems dominate South America?

2. How are the Andes Mountains of South America similar in location to the Rocky Mountains of the United States?

3. Compare Latin America's landmass and population to those of the United States. Based on that data, how might the overall population densities of the two compare?

For updated statistics on Latin America . . .

DATA UPDATE
CLASSZONE.COM

Comparing Data

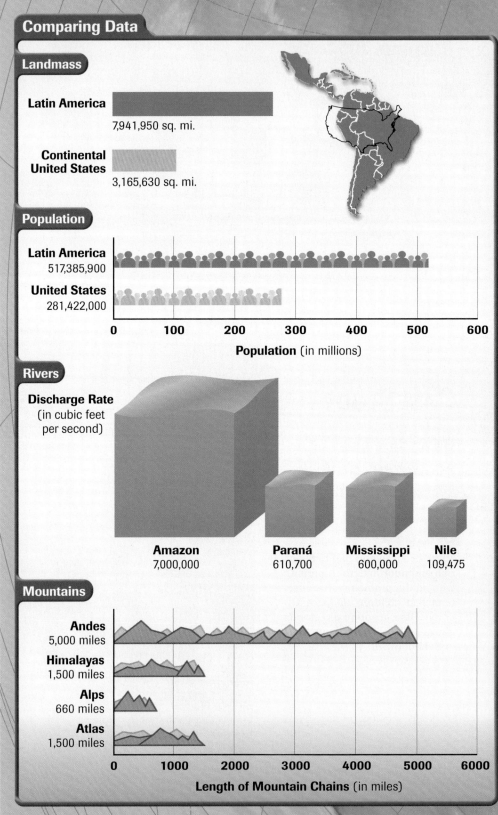

Landmass

Latin America
7,941,950 sq. mi.

Continental United States
3,165,630 sq. mi.

Population

Latin America
517,385,900

United States
281,422,000

Population (in millions)
0   100   200   300   400   500   600

Rivers

Discharge Rate (in cubic feet per second)

Amazon
7,000,000

Paraná
610,700

Mississippi
600,000

Nile
109,475

Mountains

Andes
5,000 miles

Himalayas
1,500 miles

Alps
660 miles

Atlas
1,500 miles

0   1000   2000   3000   4000   5000   6000
Length of Mountain Chains (in miles)

Doctors (per 100,000 pop.) (1992–1997)	GDP[a] (billions $US) (1998–1999)	Import/Export[a] (billions $US) (1998–1999)	Literacy Rate (percentage) (1998)	Televisions (per 1,000 pop.) (1996–1998)	Passenger Cars (per 1,000 pop.) (1991–1998)	Total Area[b] (square miles)
18	1.9	0.620 / 0.574	98	59	34	83,000
8	9.2	0.762 / 0.322	48	5	5	10,714 
83	14.1	2.7 / 1.6	73	90	14	43,433 
140	8.8	2.7 / 1.4	86	323	17	4,471 
186	865.5	142.1 / 136.8	91	261	87	761,600 
86	12.5	1.5 / 0.573	68	190	16	50,193 
167	21.0	6.4 / 4.7	91	187	54	33,659 
110	19.9	3.2 / 3.1	93	101	14	157,047
93	116.0	8.4 / 5.9	89	144	20	496,222 
117	0.2	0.160 / 0.042	90	244	130	139 
47	0.7	0.290 / 0.075	82	211	68	238
88	0.3	0.180 / 0.048	82	162	44	150
25	1.5	0.461 / 0.406	93	217	111	63,251 
79	9.4	3.0 / 2.4	98	331	107	1,980
370	28.0	3.4 / 2.1	97	242	147	68,498 
236	182.8	11.8 / 20.9	92	185	68	352,143 
251	9,255.0	820.8 / 663.0	97	847	489	3,787,319 

Chapter 9

SECTION 1
Landforms and Resources

SECTION 2
Climate and Vegetation

SECTION 3
Human–Environment Interaction

PHYSICAL GEOGRAPHY OF LATIN AMERICA
From the Andes to the Amazon

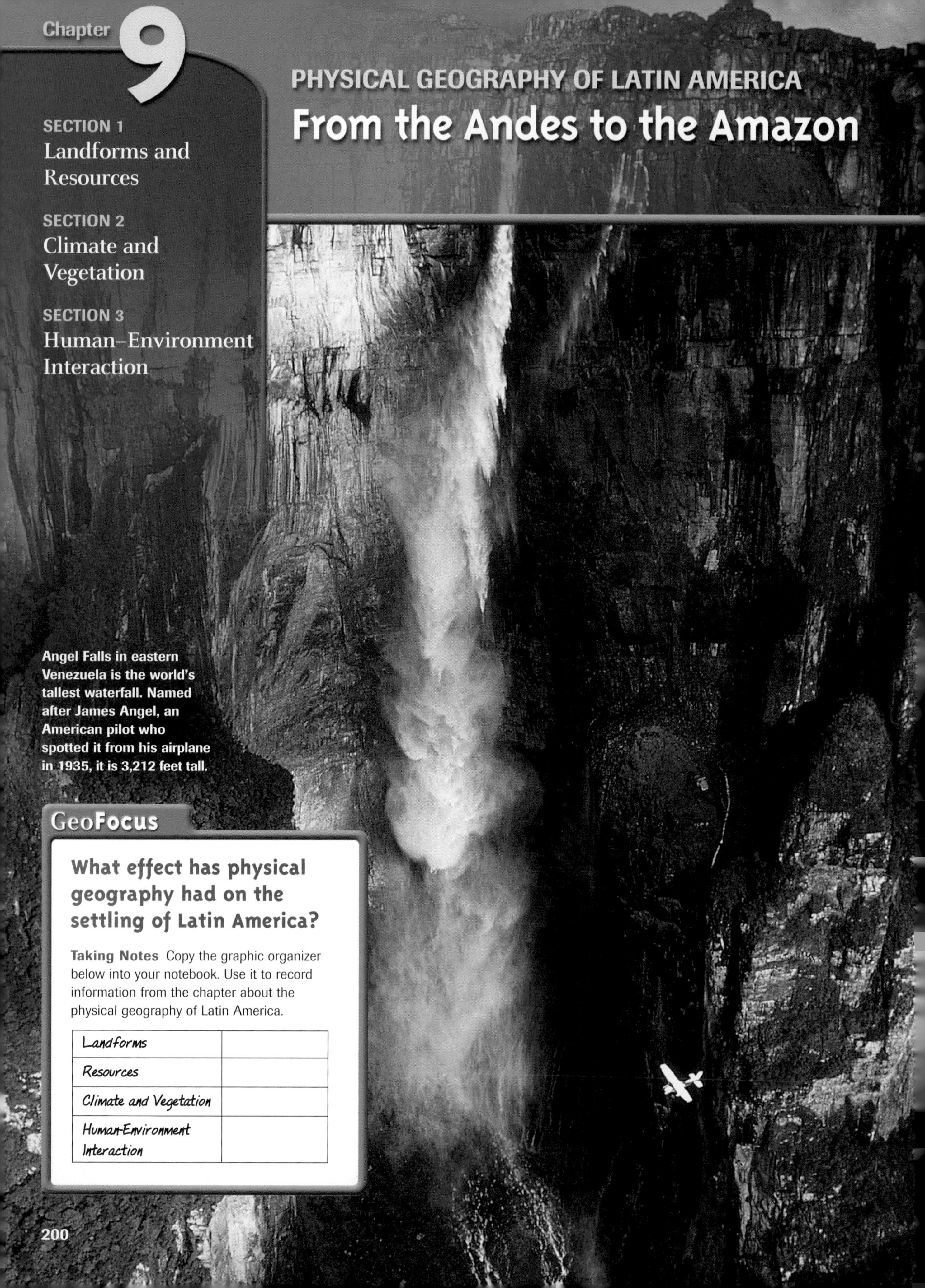

Angel Falls in eastern Venezuela is the world's tallest waterfall. Named after James Angel, an American pilot who spotted it from his airplane in 1935, it is 3,212 feet tall.

GeoFocus

What effect has physical geography had on the settling of Latin America?

Taking Notes Copy the graphic organizer below into your notebook. Use it to record information from the chapter about the physical geography of Latin America.

Landforms	
Resources	
Climate and Vegetation	
Human-Environment Interaction	

Landforms and Resources

Main Ideas

- Latin America's landforms include highlands, lowlands, mountains, and plains.
- The Andes Mountains and the Amazon River are the region's most remarkable physical features.

Places & Terms

Andes Mountains

llanos	Orinoco River
cerrado	Amazon River
pampas	Paraná River

CONNECT TO THE ISSUES

RESOURCES People in Latin America have often struggled over the best way to develop and use natural resources.

A HUMAN PERSPECTIVE Simón Bolívar was a general who led the South American wars of independence against Spain. In August 1819, Bolívar led approximately 2,500 soldiers on a daring march from Venezuela over the mountains into present-day Colombia. Coming from this direction, over the massive barrier of the Andes Mountains, Bolívar and his troops were able to advance unseen. Bolívar's soldiers surprised the Spanish army and won a great victory. Military leaders such as Bolívar were able to use the geography of the region to help the South American republics win their independence from Spain.

Mountains and Highlands

Latin America has an enormous span from north to south, as you can see from the map on page 191. It reaches from the border between the United States and Mexico down to Tierra del Fuego at the southernmost tip of South America, a distance of about 7,000 miles. It covers part of North America, all of Central and South America, and the Caribbean Islands. Its highlands, lowlands, rain forests, and plains are bounded by the Atlantic and Pacific oceans, the Gulf of Mexico, and the Caribbean Sea. The mountains of Latin America form one of the great ranges of the world.

THE ANDES MOUNTAINS The **Andes Mountains** of the South American continent are part of a chain of mountain ranges that run through the western portion of North, Central, and South America. This range is called the Rockies in the United States, the Sierra Madre in Mexico, and the Andes in South America. There are many active volcanoes throughout the region.

All along the west and south coasts of South America, the Andes Mountains are a barrier to movement into the interior. As a result, more settlement in South America has occurred along the eastern and northern coasts.

Even so, the mountain ranges of Latin America were the home of some of the most important civilizations in the hemisphere, including the Inca in Peru.

MOVEMENT Two sure-footed guanacos climb the foothills of the Andes in Patagonia, a region that includes parts of Argentina and Chile.

HIGHLANDS Other ranges in Latin America include the Guiana Highlands in the northeast section of South America. Highlands are made up of the mountainous or hilly sections of a country. The highlands of Latin America include parts of Venezuela, Guyana, Suriname, French Guiana, and Brazil. The Brazilian Highlands (see the map on page 203) are located along the east coast of Brazil.

Plains for Grain and Livestock

South America has wide plains that offer rich soil for growing crops and grasses for grazing livestock.

LLANOS OF COLOMBIA AND VENEZUELA Colombia and Venezuela contain vast plains called **llanos** (LAH·nohs), which are grassy, treeless areas used for livestock grazing and farming. They are similar to the Great Plains in the United States and the pampas of Argentina.

PLAINS OF AMAZON RIVER BASIN Brazil also contains expansive plains in the interior of the country. These are the **cerrado** (seh·RAH·doh), savannas with flat terrain and moderate rainfall that make them suitable for farming. Much of this land is undeveloped. However, the government of Brazil is encouraging settlers to move into the interior and develop the land.

PAMPAS OF ARGENTINA AND URUGUAY In parts of South America, the plains are known as **pampas** (PAHM·puhs), areas of grasslands and rich soil. Pampas are found in northern Argentina and Uruguay. The main products of the pampas are cattle and wheat grain. A culture of the gaucho has grown up in the region, centered on the horsemen of the pampas.

The Amazon and Other Rivers

The countries of Central America and the Caribbean do not have the extensive river systems that are found in South America. In North America, the Rio Grande, which forms part of the border between the United States and Mexico, is longer than any other river in Mexico, Central America, or the Caribbean. However, these areas are all bordered by water. As a result, they are less dependent on river systems for transportation than is South America.

South America has three major river systems. The Orinoco is the northernmost river system, with the Amazon also in the north, and the Paraná in the south of the continent.

ORINOCO RIVER The **Orinoco River** winds through the northern part of the continent, mainly in Venezuela. It flows more than 1,500 miles, partly along the Colombia-Venezuela border, to the Atlantic. The Orinoco River basin drains the interior lands of both Venezuela and Colombia. Some of the areas drained by the Orinoco are home to the few remaining Native American peoples, such as the Yanomamo.

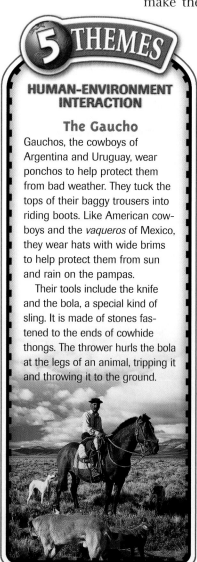

5 THEMES

HUMAN-ENVIRONMENT INTERACTION

The Gaucho

Gauchos, the cowboys of Argentina and Uruguay, wear ponchos to help protect them from bad weather. They tuck the tops of their baggy trousers into riding boots. Like American cowboys and the *vaqueros* of Mexico, they wear hats with wide brims to help protect them from sun and rain on the pampas.

Their tools include the knife and the bola, a special kind of sling. It is made of stones fastened to the ends of cowhide thongs. The thrower hurls the bola at the legs of an animal, tripping it and throwing it to the ground.

Geographic Thinking

Seeing Patterns

A How are the llanos, cerrado, and pampas of South America similar to the Great Plains of the United States?

AMAZON RIVER Farther south, the **Amazon River** flows about 4,000 miles from west to east, emptying into the Atlantic Ocean. Its branches start in the Andes Mountains of South America, close to the Pacific. Yet it flows eastward across the central lowlands toward the Atlantic. The Amazon River is fed by over 1,000 tributaries, some of which are large rivers in themselves. The Amazon carries more water to the ocean than any other river in the world. In fact, it carries more water to the ocean than the next seven largest rivers of the world combined.

PARANÁ RIVER The **Paraná River** has its origins in the highlands of southern Brazil. It travels about 3,000 miles south and west through Paraguay and Argentina, where it is fed by several rivers, and then turns eastward. The last stretch of the river, where it turns into an estuary of the Paraná and Uruguay rivers between Argentina and Uruguay, is called the Río de la Plata. An estuary is the wide lower course of a river where its current is met by the tides.

Landforms and Rivers of Latin America

[Map of Latin America showing landforms and rivers with labels: Sierra Madre Occidental, Sierra Madre Oriental, Gulf of Mexico, Caribbean Sea, ATLANTIC OCEAN, Llanos, Orinoco R., Guiana Highlands, Equator, AMAZON PLAIN, Amazon R., PACIFIC OCEAN, ANDES, Mato Grosso Plateau, BRAZILIAN HIGHLANDS, Atacama Desert, Gran Chaco, Paraná R., Pampas, Patagonia, ATLANTIC OCEAN]

Legend:
- Mountains
- Hills and Plateaus
- Plains

0   500   1,000 miles
0   500   1,000 kilometers
Azimuthal Equal–Area Projection

SKILLBUILDER: Interpreting Maps

❶ **MOVEMENT** Which rivers empty into the Atlantic Ocean?

❷ **REGION** What mountains run along the western edge of South America?

Major Islands of the Caribbean

The Caribbean Islands consist of three major groups: the Bahamas, the Greater Antilles, and the Lesser Antilles. (See the map on page 191.) These islands together are sometimes called the West Indies and were the first land encountered by Christopher Columbus when he sailed to the Western Hemisphere in 1492. They served as a base of operations for the later conquest of the mainland by the Spanish.

The Bahamas are made up of hundreds of islands off the southern tip of Florida and north of Cuba. They extend southeast into the Atlantic Ocean. Nassau is the capital and largest city in the Bahamas.

THE GREATER ANTILLES The Greater Antilles are made up of the larger islands in the Caribbean. These include Cuba, Jamaica, Hispaniola, and Puerto Rico. The island of Hispaniola is divided between the countries of Haiti and the Dominican Republic.

THE LESSER ANTILLES The Lesser Antilles are the smaller islands in the region southeast of Puerto Rico. The Lesser Antilles are divided into the Windward Islands and Leeward Islands. The Windward Islands face winds that blow across them. The Leeward Islands enjoy a more sheltered position from the prevailing northeasterly winds. ◀B

Geographic Thinking

Using the Atlas
B▶ Use the map on page 191. Which of the Antilles are closer to the coast of South America?

Resources of Latin America

Latin America is a treasure house of natural resources. These include mineral resources, such as gold and silver, as well as energy resources, such as oil and natural gas. In addition, the region is rich in agricultural and forest resources, such as timber. These resources have drawn people to the region for centuries.

MINERAL RESOURCES Gold, silver, iron, copper, bauxite (aluminum ore), tin, lead, and nickel—all these minerals are abundant in Latin America. In addition, mines throughout the region produce precious gems, titanium, and tungsten. In fact, South America is among the world's leaders in the mining of raw materials.

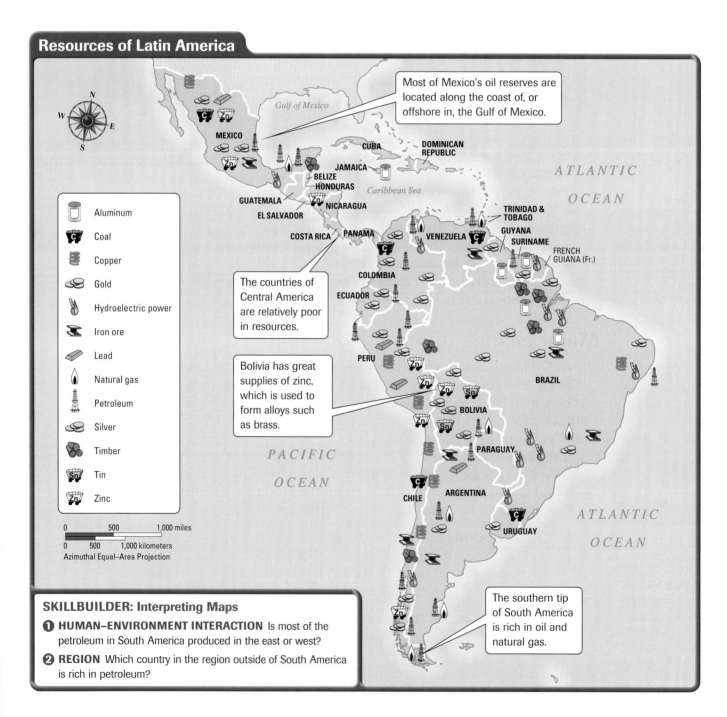

Resources of Latin America

Most of Mexico's oil reserves are located along the coast of, or offshore in, the Gulf of Mexico.

The countries of Central America are relatively poor in resources.

Bolivia has great supplies of zinc, which is used to form alloys such as brass.

The southern tip of South America is rich in oil and natural gas.

Legend:
- Aluminum
- Coal
- Copper
- Gold
- Hydroelectric power
- Iron ore
- Lead
- Natural gas
- Petroleum
- Silver
- Timber
- Tin
- Zinc

0   500   1,000 miles
0   500   1,000 kilometers
Azimuthal Equal–Area Projection

SKILLBUILDER: Interpreting Maps

❶ **HUMAN–ENVIRONMENT INTERACTION** Is most of the petroleum in South America produced in the east or west?

❷ **REGION** Which country in the region outside of South America is rich in petroleum?

Many of these minerals are mined and then exported to other parts of the world, where they are made into valuable goods. For example, Jamaica was originally a plantation economy that depended on the sale of bananas and sugar for its livelihood. Then it turned to the mining and processing of bauxite (aluminum ore) in an attempt to make the country less dependent on agriculture and tourism. Today, this resource is mainly an export that is shipped elsewhere for industrial use.

ENERGY RESOURCES Oil, coal, natural gas, uranium, and hydroelectric power are all plentiful in Latin America. Venezuela and Mexico have major oil reserves. Brazil is rich in hydroelectric power because of its many rivers (including the mighty Amazon) and waterfalls. It is also rich in oil and gas.

Trinidad has discovered vast reserves of natural gas. New factories have turned Trinidad into a major exporter of methanol and ammonia. Natural gas has also attracted developers to the island.

In Mexico and Venezuela, oil has been a very important resource. Venezuela sits on top of major oil deposits. This resource was developed into a significant oil industry. Mexico has huge oilfields centered along the Gulf coast. Because of its reserves, Mexico is able to export oil to other countries. However, changes in the global price of oil have had a great impact on the economies of these countries.

Latin America has great variety in its climate and vegetation. You will read about each in the next section.

CONNECT TO THE ISSUES
RESOURCES
How do the countries of the region make use of their natural resources?

Assessment

1 Places & Terms

Identify and explain where in the region these would be found.

- Andes Mountains
- llanos
- cerrado
- pampas
- Orinoco River
- Amazon River
- Paraná River

2 Taking Notes

PLACE Review the notes you took for this section.

Landforms	
Resources	

- What types of landforms are found in Latin America?
- What is their relative location?

3 Main Ideas

a. How did the Andes Mountains affect settlement along the western coast of South America?

b. How are the landforms of the region both an advantage and disadvantage?

c. What effect did natural resources have on the development of the region?

4 Geographic Thinking

Drawing Conclusions How might the Amazon River have affected movement into the interior of South America?

Think about:

- the network of travel offered by a river system

S **See Skillbuilder Handbook, page R5.**

GeoActivities

SEEING PATTERNS Pair with a partner and draw a **sketch map** of Latin America's rivers and mountains. Use arrows to indicate the directions the rivers flow. Why does the Amazon flow all the way east across the continent even though its headwaters begin in the Andes Mountains along the west coast?

Interpreting a Precipitation Map

This map shows differences in annual precipitation throughout South America. Suppose you have been given a chance to live in either Manaus, Brazil, or Buenos Aires, Argentina, for a year. You don't want to live in a city where it rains a lot. Which city would you choose? To help make your decision, find the two cities on the Unit Atlas map on page 193. Then find their locations on this precipitation map.

THE LANGUAGE OF MAPS A **precipitation map** is a type of thematic map. Many precipitation maps show differences in annual precipitation within a given region.

Precipitation in South America ❶

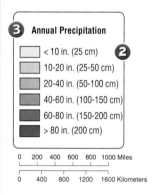

❸ Annual Precipitation

	< 10 in. (25 cm) **❷**
	10-20 in. (25-50 cm)
	20-40 in. (50-100 cm)
	40-60 in. (100-150 cm)
	60-80 in. (150-200 cm)
	> 80 in. (200 cm)

0 200 400 600 800 1000 Miles
0 400 800 1200 1600 Kilometers

Copyright by Rand McNally & Co.

❶ The title gives you the subject matter of the map.

❷ The amount of annual precipitation is shown both in inches and centimeters.

❸ The key shows the colors used on the map and explains their meaning. Each color shows a different range of annual precipitation.

Map and Graph Skills Assessment

1. Making Comparisons
Where are you likely to experience more rain—in Peru or Brazil?

2. Drawing Conclusions
Does Guyana have heavy or light annual precipitation?

3. Making Inferences
Is there heavier annual precipitation in the northern or southern parts of the continent?

Climate and Vegetation

Main Ideas

- Latin America has a variety of climates, from the cold peaks of the Andes to the Amazon rain forest.
- The vegetation of Latin America ranges from grasslands to the largest rain forest in the world.

Places & Terms

rain forest

CONNECT TO THE ISSUES

RESOURCES Latin America's climate and vegetation make up a habitat that is threatened by economic development.

A HUMAN PERSPECTIVE In the 17th century, missionaries and Indians in the area of present-day Paraguay were at times attacked by jaguars, the great cats of Latin America. In 1637, packs of jaguars roamed the countryside, attacking humans. The Indians built barricades for protection from the savage cats. But the jaguars remained a source of fear. The cats were a factor that had to be taken into account in settling and protecting towns and villages. There was no question about it—jaguars and other creatures thrived in the humid climate and thick vegetation of the tropical rain forests.

A Varied Climate and Vegetation

The climate of Latin America ranges from the hot and humid Amazon River basin to the dry and desert-like conditions of northern Mexico and southern Chile. Rain forest, desert, and savanna are all found in the region.

The vegetation varies from rain forests to grasslands and desert scrub. It ranges from the thick trees of the rain forests to mosses of the tundra.

This variety of climate and vegetation is due to several factors. First, Latin America spans a great distance on each side of the equator. Second, there are big changes in elevation because of the massive mountains in the region. Third, the warm currents of the Atlantic Ocean and the cold currents of the Pacific Ocean affect the climate.

Tropical Climate Zones

The tropical climate zones of the region produce both rain forests and the tree-dotted grasslands known as savannas. Rain forests are abundant in Central America, the Caribbean, and South America. Savannas are found in South America.

TROPICAL WET __Rain forests__ are dense forests made up of different species of trees. They form a unique ecosystem—a community of plants and animals living in balance. The climate in these areas is hot and rainy year round. The largest forest is the

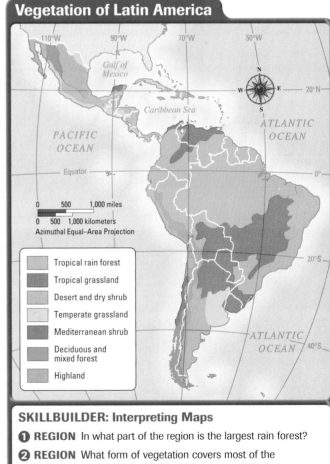

Vegetation of Latin America

- Tropical rain forest
- Tropical grassland
- Desert and dry shrub
- Temperate grassland
- Mediterranean shrub
- Deciduous and mixed forest
- Highland

0 500 1,000 miles
0 500 1,000 kilometers
Azimuthal Equal–Area Projection

SKILLBUILDER: Interpreting Maps

❶ **REGION** In what part of the region is the largest rain forest?

❷ **REGION** What form of vegetation covers most of the southeastern part of South America?

Amazon rain forest, which covers more than two million square miles of South America. Much of this rain forest is located in Brazil.

Rain forests contain many exotic plants and creatures. Scientists have counted more than 2,500 varieties of trees in the Amazon rain forest. These include the Brazil nut tree, which grows 150 feet high. Animals include the anaconda, among the largest snakes in the world, the jaguar, and the piranha, a sharp-toothed, meat-eating fish.

TROPICAL WET AND DRY Wet and dry climates, found primarily in South America, support savannas, which are grasslands dotted with trees common in tropical and subtropical regions. These areas have hot climates with seasonal rain. Savannas are found in Brazil, Colombia, and Argentina.

BACKGROUND
The anaconda lives in and near the rivers of tropical South America. It may grow as long as 25 feet.

Dry Climate Zones

Dry climate zones are found in Mexico on the North American continent and in various countries of South America. Neither Central America nor the Caribbean, though, has dry climate zones.

SEMIARID A semiarid climate is generally dry, with some rain. Vast, semiarid, grass-covered plains are often found in such climates. Desert shrubs also grow in semiarid regions. Such regions are found in Mexico, Brazil, Uruguay, and Argentina.

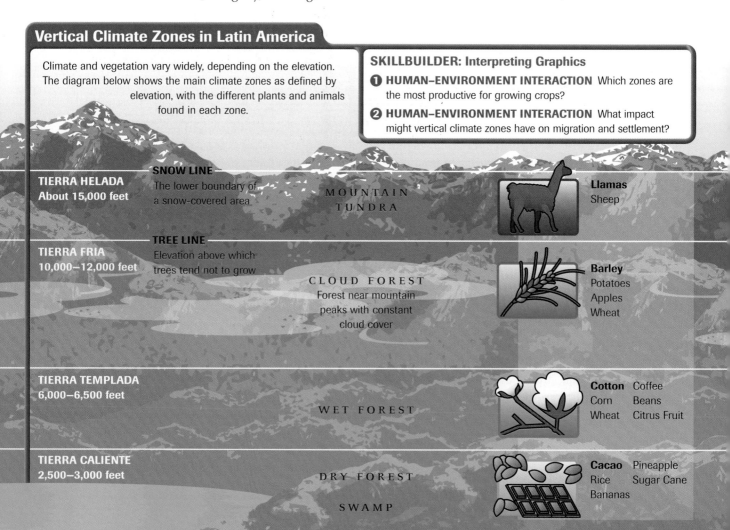

Vertical Climate Zones in Latin America

Climate and vegetation vary widely, depending on the elevation. The diagram below shows the main climate zones as defined by elevation, with the different plants and animals found in each zone.

SKILLBUILDER: Interpreting Graphics

❶ **HUMAN–ENVIRONMENT INTERACTION** Which zones are the most productive for growing crops?

❷ **HUMAN–ENVIRONMENT INTERACTION** What impact might vertical climate zones have on migration and settlement?

TIERRA HELADA
About 15,000 feet

SNOW LINE
The lower boundary of a snow-covered area

MOUNTAIN TUNDRA

Llamas
Sheep

TIERRA FRIA
10,000–12,000 feet

TREE LINE
Elevation above which trees tend not to grow

CLOUD FOREST
Forest near mountain peaks with constant cloud cover

Barley
Potatoes
Apples
Wheat

TIERRA TEMPLADA
6,000–6,500 feet

WET FOREST

Cotton Coffee
Corn Beans
Wheat Citrus Fruit

TIERRA CALIENTE
2,500–3,000 feet

DRY FOREST

SWAMP

Cacao Pineapple
Rice Sugar Cane
Bananas

DESERT Parts of northern Mexico are classified as desert, as is much of the coast of Peru. The Atacama Desert is in northern Chile. Likewise, Argentina's southern zone, Patagonia, contains a desert. The deserts of the region are made up of shrubs growing in gravel or sand.

Mid-Latitude Climate Zones

The mid-latitude, moderate climate zones in the region are located south of the equator, from approximately Rio de Janeiro in Brazil southward.

HUMID SUBTROPICAL Humid subtropical areas have rainy winters and hot, humid summers. Parts of Paraguay, Uruguay, southern Brazil, southern Bolivia, and northern Argentina (including Buenos Aires) are located in humid, subtropical climates. The vegetation is varied.

MEDITERRANEAN Mediterranean climate zones experience hot, dry summers and cool, moist winters. Part of Chile along the west coast is in this zone. You have experienced a similar climate if you have ever been to California. The vegetation in this climate is mainly chaparral.

MARINE WEST COAST Marine west coast climate zones are characterized by cool, rainy winters and mild, rainy summers. One such climate region runs along the coast of southwestern South America. Parts of southern Chile and Argentina have this climate. If you have spent time in Oregon or Washington, you have experienced this type of climate. Forests are the typical vegetation.

Geographic Thinking

Seeing Patterns
A Where are most of the highland climate zones located?

HIGHLANDS Highland climate zones vary from moderate to cold, depending on elevation. Other factors influence highland climates, such as wind, sunlight, and landscape. Highland climates are found in the mountains of Mexico and South America. **A**

In the next section, you will read about how human-environment interaction affects the quality of life in Latin America.

SECTION 2 Assessment

① Places & Terms

Identify and explain where in the region this would be found.

• rain forest

② Taking Notes

PLACE Review the notes you took for this section.

```
Climate and
Vegetation
```

• What vegetation characterizes the Amazon River basin?

• What types of climate zones are found in Latin America?

③ Main Ideas

a. What are two reasons for the variety of climate and vegetation found in Latin America?

b. What effect might elevation have on growing crops and grazing livestock in the region?

c. What are the three main types of moderate climate zones in the region?

④ Geographic Thinking

Making Inferences How might the climate and vegetation of Latin America have affected migration, settlement, and ways of life?
Think about:

• the impact of deserts and rain forests on settlement

RESEARCH LINKS
CLASSZONE.COM

GeoActivity

ASKING GEOGRAPHIC QUESTIONS Research on the Internet the climate and vegetation in your state. Devise three geographic questions, such as "What is the dominant climate zone in my state?" Choose one of your questions and then write a **paragraph** explaining your findings. Be sure to list your sources.

Human–Environment Interaction

3

Main Ideas

- The people of Latin America have altered the land through agriculture and urbanization.
- Tourism is having a growing impact on the environment of Latin America.

Places & Terms

slash-and-burn

terraced farming

push factors

pull factors

infrastructure

CONNECT TO THE ISSUES
INCOME GAP The income gap can be seen in the landless poor, the cities, and the tourist industry.

A HUMAN PERSPECTIVE High in the Andes Mountains, in what is present-day Peru, the ancient Inca needed fields in which to grow crops. By the 1200s, in the highlands around their capital of Cuzco and elsewhere, the Inca carved terraces out of the steep sides of the Andes Mountains. They built irrigation channels to bring water to the terraces. Because of their activity, they were able to grow crops for thousands of people on the slopes of previously barren hillsides. In this way, the Inca altered their environment to meet their needs.

Agriculture Reshapes the Environment

Native peoples were the first in the Western Hemisphere to change their environment to grow food. They burned the forest to clear land for planting and diverted streams to irrigate crops. They built raised fields in swampy areas and carved terraces out of hillsides.

SLASH–AND–BURN To clear fields, native peoples used the **slash-and-burn** technique—they cut trees, brush, and grasses and burned the debris to clear the field. This method was particularly effective in humid and tropical areas.

Today, farmers practice the same method as they move into the Amazon River basin in Brazil and clear land for farming in the rain forest. But the non-landowning poor who are clearing and then settling the

---

Slash-and-Burn Farming

❶ Farmers cut trees, brush, and grasses to clear a field.

❷ They then burn the debris and use the ashes to fertilize the soil.

❸ Farmers plant crops for a year or two, which exhausts the soil.

land sometimes use destructive farming practices. After a few years, they find that the soil is exhausted—all the nutrients have been drained from the land. Then they move on and clear a new patch to farm. This is one of the reasons for the steady shrinking of the rain forests. (For more about the rain forest, see Chapter 11, Section 1, page 245.) ◀A

TERRACED FARMING **Terraced farming** is an ancient technique for growing crops on hillsides or mountain slopes. It is an especially important technique in the mountainous areas of the region. Farmers and workers cut step-like horizontal fields into hillsides and slopes, which allow steep land to be cultivated for crops. The technique reduces soil erosion. As you read earlier, the Inca practiced terraced farming hundreds of years ago in Peru. The Aztecs of Mexico also used terraced farming.

Urbanization: The Move to the Cities

Throughout Latin America, people are moving from rural areas into the cities. They leave farms and villages in search of jobs and a better life. Cities have grown at such a rapid pace in Latin America that today the region is as urban as Europe or North America.

FROM COUNTRY TO CITY Argentina, Chile, and Uruguay are the most highly urbanized countries in South America. In these countries, more than 85 percent of the people live in cities. In Brazil, too, most people live in cities and towns.

People move to the cities in the hope of improving their lives. Many people in rural areas struggle to make a living and feed their families by subsistence farming. With a great deal of effort, they grow barely enough food to keep themselves and their families alive.

Both push and pull factors are at work in moving peasants and farmers off the land and drawing them to the cities. **Push factors** are factors that "push" people to leave rural areas. They include poor medical care, poor education, low-paying jobs, and ownership of the land by a few rich people. **Pull factors** are factors that "pull" people toward cities. They include higher-paying jobs, better schools, and better medical care. ◀B

RAPIDLY GROWING CITIES Six cities in South America rank among the region's largest in population. These include São Paulo and Rio de Janeiro in Brazil, Buenos Aires in Argentina, Lima in Peru, Bogotá in Colombia, and Santiago in Chile. But the most populous city in all of Latin America is Mexico City. Estimates of its population vary from approximately 18 to 20 million people for the city alone to about 30 million for the entire greater metropolitan area.

Similar problems afflict cities throughout the region. Slums spread over larger and larger urban areas. Often unemployment and crime increase. In addition to social problems, there are many environmental problems. These include high levels of air

CONNECT TO THE ISSUES
RESOURCES
Ⓐ What is the impact of slash-and-burn on the rain forest?

CONNECT TO THE ISSUES
INCOME GAP
Ⓑ How might push and pull factors affect the gap between rich and poor?

④ Fields often remain barren or are reclaimed by brush, grass, trees, and scrub.

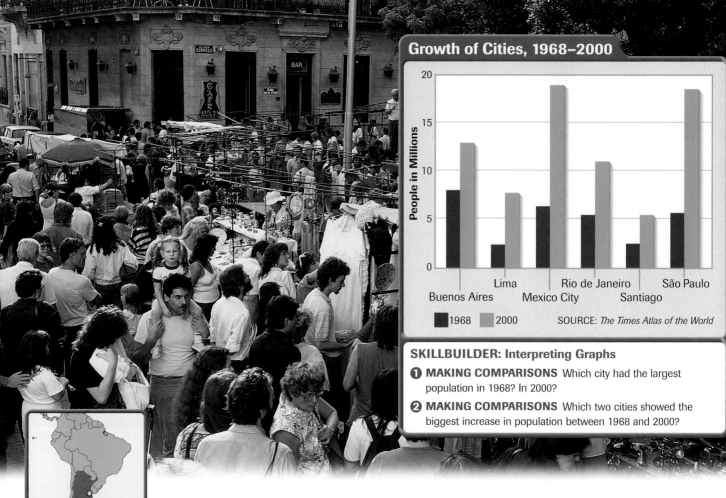

Growth of Cities, 1968–2000

People in Millions

20

15

10

5

0

Buenos Aires   Lima   Mexico City   Rio de Janeiro   Santiago   São Paulo

■ 1968   ■ 2000   SOURCE: *The Times Atlas of the World*

SKILLBUILDER: Interpreting Graphs

❶ **MAKING COMPARISONS** Which city had the largest population in 1968? In 2000?

❷ **MAKING COMPARISONS** Which two cities showed the biggest increase in population between 1968 and 2000?

PLACE Pedestrians crowd a street during a festival in Buenos Aires, Argentina.

pollution from cars and factories. Some cities have shortages of drinkable water as local supplies are used up and underwater supplies are drained.

To make matters worse, local governments cannot afford facilities to handle the population increase. This **infrastructure** includes such things as sewers, transportation, electricity, and housing.

Tourism: Positive and Negative Impacts

Tourism is a growth industry throughout Latin America. It is especially important in Mexico and the Caribbean. But despite the money it brings in to the economies of the region, tourism is a mixed blessing.

ADVANTAGES OF TOURISM Every year millions of tourists visit the resorts of Latin America, spending money and helping to create jobs. New hotels, restaurants, boutiques, and other businesses have sprung up on the islands of the Caribbean and in Mexico to serve the tourist trade. Luxurious cruise ships anchor in the ports of the region. They carry travelers who spend money on souvenirs and trips around the islands. Lavish restaurants serve expensive meals to these tourists. Staffing those ships, hotels, and restaurants are local people who profit from the visitors in their midst.

Resorts offer many activities that provide jobs for local residents. For example, local guides conduct tours of the natural wonders and beautiful scenery. Local companies may offer guided rafting trips down rivers. Sailing and snorkeling expeditions into the waters of the Caribbean and Pacific reveal exotic marine life. All of these activities bring money into the region and employ local people.

In this way, tourism can play a part in reducing the income gap between rich and poor. Jobs in hotels, restaurants, and resorts raise incomes and give the local people a stake in their society.

DISADVANTAGES OF TOURISM

Despite the income and jobs that tourism brings to various places in Latin America, it causes problems as well. As resorts are built in previously unspoiled settings, congestion occurs and pollution increases.

The tourism industry often puts a great strain on the local communities where it builds its resorts. Further, there is an obvious gap between rich tourists and less well-off local residents. This has produced resentment and hostility in places such as Jamaica in the Caribbean and Rio de Janeiro in Brazil.

More important, local governments can run up large public debts by borrowing money to build tourist facilities. Airports and harbors must be constructed. Hotels and resorts must be built. Sewage systems and shopping areas must be expanded.

CONNECT TO THE ISSUES
DEMOCRACY
▶ How might absentee ownership of tourist facilities undermine democracy in a tourist country?

Often the owners of these hotels and airlines do not live in the tourist country. Typically, they send their profits back home. Further, these absentee owners often make decisions that are not in the tourist country's best interest. The owners may be able to influence local elections and business decisions. ◀

In the next chapter, you will read about the human geography of Latin America, including its history, culture, economics, and daily life.

LATIN AMERICA

HUMAN–ENVIRONMENT INTERACTION
A luxury cruise ship is docked in the beautiful harbor of Charlotte Amalie, St. Thomas in the Virgin Islands.
What might be the impact of tourists on the local economy?

SECTION 3 Assessment

1 Places & Terms

Identify and explain the significance of each in the region.

- slash-and-burn
- terraced farming
- push factors
- pull factors
- infrastructure

2 Taking Notes

HUMAN-ENVIRONMENT INTERACTION Review the notes you took for this section.

> Human-Environment Interaction

- What are the steps in slash-and-burn farming?
- What are some of the problems of cities in the region?

3 Main Ideas

a. How have humans changed the environment in Latin America to make it more suitable for agriculture?

b. What factors have drawn people from the countryside into the cities of the region?

c. What are some of the advantages of tourism to the Caribbean?

4 Geographic Thinking

Making Inferences How might the cities of Latin America deal with the increasing demands placed on them by their expanding populations? **Think about:**

- water, sewage, and electricity
- transportation and housing

S See Skillbuilder Handbook, page R4.

GeoActivity

SEEING PATTERNS Pair with a partner and create a **travel poster** about a place in the region that you would like to visit. Show various activities and sports available at the place you choose.

VISUAL SUMMARY
PHYSICAL GEOGRAPHY OF LATIN AMERICA

Landforms

Major Mountain Ranges: Andes, Sierra Madres

Major Rivers: Orinoco, Amazon, Paraná

Major Plains: pampas of Argentina and Uruguay, llanos of Colombia and Venezuela, cerrado of Brazil

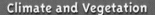

Resources

- Latin America has important mineral and energy resources.
- Venezuela and Mexico have major oil reserves.

Climate and Vegetation

- The variety of climate and vegetation in Latin America is caused by the great distance from north to south, variations in elevation, and ocean currents.
- Latin America has many rain forests.

Human-Environment Interaction

- Two techniques that farmers have used in the region are slash-and-burn and terraced farming.
- Cities in Latin America have grown at a rapid pace, and the region is now highly urbanized.
- Tourism has both advantages and disadvantages for the region.

Reviewing Places & Terms

A. Briefly explain the importance of each of the following.

1. Andes Mountains	**5.** Amazon River	**9.** push factors
2. llanos	**6.** rain forest	**10.** infrastructure
3. pampas	**7.** slash-and-burn	
4. Orinoco River	**8.** terraced farming	

B. Answer the questions about vocabulary in complete sentences.

11. What two countries does the Orinoco River drain?

12. Where are the Andes Mountains located?

13. What agricultural technique involves using ashes to fertilize the soil?

14. What characteristics do the pampas and llanos have in common?

15. What are some of the problems that afflict cities throughout the region?

16. Which river drains the largest rain forest in the region?

17. What are some factors that are pushing farmers off the land and into the cities?

18. Which is the northernmost of the great rivers of South America?

19. What farming technique is especially useful in mountainous regions?

20. What are the main products of the pampas?

Main Ideas

Landforms and Resources (pp. 201-206)

1. How have the Andes Mountains affected settlement in South America?

2. What are the two main purposes for which the plains and grasslands of the region are used?

3. What are the three major island groups of the Caribbean?

4. What Caribbean island is rich in natural gas, and what impact has this had on the economy?

Climate and Vegetation (pp. 207-209)

5. In what part of the region are savannas most common?

6. How do the vertical climate zones of Latin America affect agriculture?

7. What is the dominant vegetation of the Amazon river basin?

Human-Environment Interaction (pp. 210-213)

8. What is the main disadvantage of the slash-and-burn method of growing crops?

9. What factors tend to pull people into the cities from their farms?

10. What are some of the disadvantages of tourism in the region?

Mexican Life Today

The people of Mexico face big challenges in today's world. Jobs, emigration, and education are foremost among their concerns. Many of these issues relate to the income gap between rich and poor.

EMIGRATION Emigration has had an impact on family life in Mexico. Mexico shares a 2,000-mile border with the United States. Many workers leave Mexico and travel to the United States in search of work. This separates families. Nonetheless, most of these workers remain in touch with their families in Mexico. Many send money back to their native villages. Often, after a year or two working in the United States, they return to Mexico with savings to help improve living conditions for their extended families.

WORK AND SCHOOL The rapidly growing population and various government policies have contributed to a shortage of jobs. This has led many Mexicans to migrate to the United States in search of work.

Without education and training, young workers cannot find good jobs. In recent years, attendance of eligible students at school has improved. Today, about 85 percent of school-age children attend school. In the coming years, Mexico will have to invest large sums in education to provide a better life for its young citizens. Education will become even more important as Mexico becomes more industrialized. ◀

In the next section, you will read about Central America and the Caribbean. This subregion links North America and South America.

MOVEMENT
Pedestrians use a walkway in crossing from Nuevo Laredo, Mexico, into Laredo, Texas. Shop signs are in English and Spanish.

CONNECT TO THE ISSUES
INCOME GAP
▷ How might the income gap be narrowed in Mexico?

Assessment

1 Places & Terms

Identify and explain the following terms.

- Spanish conquest
- Tenochtitlán
- Institutional Revolutionary Party (PRI)
- mestizo
- maquiladoras
- NAFTA

2 Taking Notes

REGION Review the notes you took for this section.

- Which two main cultures blended to form modern Mexico?
- Where do most of Mexico's people live today?

3 Main Ideas

a. How might democratic reforms and improved trade agreements contribute to a stronger economy in Mexico?

b. What effect might Mexico's young population have on its development?

c. In what ways have Native American and Spanish influences shaped Mexico?

4 Geographic Thinking

Making Generalizations
How might a shortage of jobs in Mexico affect the movement of its people?
Think about:

- why one might travel to the United States in search of work
- what factors in Mexico might lead people to move

GeoActivity

MAKING COMPARISONS Pair with a partner and make a **chart** of the ten most heavily populated states of Mexico arranged in order from most to least heavily populated. Then compare your chart with a map, and mark those states that are closest to the U.S. border.

Central America and the Caribbean

Main Ideas

- Native peoples, Europeans, and Africans have shaped the culture of this region.
- The economies of the region are based primarily on agriculture and tourism.

Places & Terms

cultural hearth

United Provinces of Central America

Panama Canal

calypso

reggae

informal economy

CONNECT TO THE ISSUES

INCOME GAP The people of Central America and the Caribbean face an uneven distribution of income as one of the effects of colonialism.

A HUMAN PERSPECTIVE Central America forms an isthmus, a land bridge between North and South America. It also divides two oceans. This geographic fact has made the region attractive to the United States and other major world powers and has helped to keep the area fragmented and politically unstable. For example, in the early 20th century, the United States wanted to build a canal across Panama that would connect the Atlantic and Pacific oceans. In 1903, Panama was still a province of Colombia, which did not like the idea. The United States encouraged a revolution in Panama, and when it won its independence, Panama granted the United States a ten-mile-wide zone in which to build a canal. Central America had become a crossroads of world trade.

Native and Colonial Central America

Central America is a cultural hearth as well as a crossroads. A **cultural hearth** is a place from which important ideas spread. Usually, it is the heartland or place of origin of a major culture. The Mayan people built a great civilization in the area that spread throughout the region. The homeland of the Maya stretched from southern Mexico into northern Central America. During the 800s, the Maya began to abandon many of their cities. Why they did so remains a mystery to be solved by archaeologists.

Wooden snake carved by a Taino artist

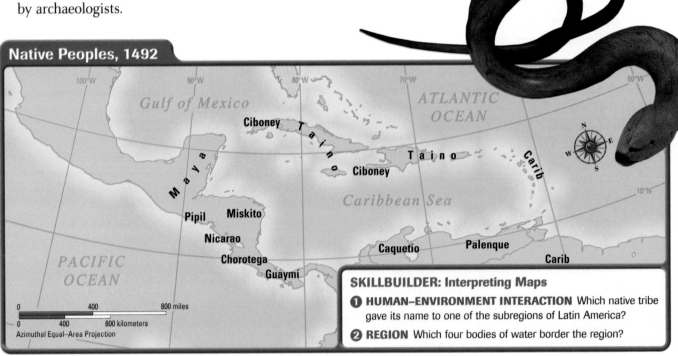

Native Peoples, 1492

SKILLBUILDER: Interpreting Maps

❶ **HUMAN–ENVIRONMENT INTERACTION** Which native tribe gave its name to one of the subregions of Latin America?

❷ **REGION** Which four bodies of water border the region?

MAYAN INFLUENCE The Maya built many cities with temples and palaces in present-day Belize, Guatemala, El Salvador, and Honduras. Each city was an independent state ruled by a god-king and served as a center for religious ceremony and trade. One of their most spectacular cities was Tikal, located in the dense, steamy jungle of northern Guatemala, considered the center of Mayan civilization. The pyramids at Tikal were among the tallest structures in the Americas until the 20th century. The influence of the Maya spread over a region from Mexico to El Salvador. The Mayan culture was carried to other regions through military alliances and trade.

HUMAN–ENVIRONMENT INTERACTION This pyramid at Tikal towers over the great plaza of the ancient city. A temple sits on top of the pyramid.
What might this and similar buildings at Tikal suggest about Mayan civilization?

THE SPANISH IN CENTRAL AMERICA The Spanish conquest of the Aztecs in Mexico opened the door to Spanish control of Central America. Spain ruled Central America until the 19th century. Mexico declared its independence from Spain in 1821. Up to that point, Central America had been governed from Mexico. In 1823, however, the whole region declared its independence from Mexico and took the name of the **United Provinces of Central America.**

By the late 1830s, the United Provinces had split into separate nations. These became El Salvador, Nicaragua, Costa Rica, Guatemala, and Honduras. Later, Panama broke off from Colombia and became an independent country in Central America. Belize, a former British colony, broke away from British Honduras.

Native and Colonial Caribbean

Geographic Thinking

Using the Atlas
A Use the maps on pages 216 and 222. Why might the Caribbean have been settled by more European powers than was Central America?

Although Central America was ruled by Spain, the Caribbean was settled and claimed by many European powers. In addition, Africans who were brought to the Caribbean as slaves played an important role in the settling of the Caribbean.

CARIBBEAN INFLUENCES When Christopher Columbus reached the Caribbean islands in 1492, he thought he had reached the East Indies in Asia. Therefore, he called the natives "Indians." The inhabitants of these islands called themselves the Taino (TY·noh). The Spanish settled some of the islands and established sugar plantations, which were well suited to the climate and soil of the islands. They attempted to use the Taino as forced labor, but many of the natives died from disease and mistreatment.

To replace the Taino, European slave traders brought Africans to the Caribbean by force and put them to work on plantations. As a result, Africans have had a lasting influence on Caribbean life and culture.

Caribbean Colonies

Country	Colony	Major Cultural Influences
Spain	Cuba, Dominican Republic, Puerto Rico	Spanish language Catholic religion
France	Haiti, Guadeloupe, Martinique	French language Catholic religion
Great Britain	Jamaica, Barbados, St. Lucia, St. Vincent, Grenada, Trinidad and Tobago, British Virgin Islands	English language Protestant and Catholic religions
Netherlands	Netherlands Antilles	Dutch language Protestant religion
Denmark	Danish West Indies [1]	Danish language [2] Protestant religion

[1] Became U.S. Virgin Islands in 1917. [2] English is now the official language.

SKILLBUILDER: Interpreting Charts

❶ **REGION** Which European country had the most colonies in the Caribbean?

❷ **PLACE** In the colonies of which European countries was the Catholic religion dominant?

A COLONIAL MOSAIC By the 19th century, the Spanish, French, British, Dutch, and Danish all claimed islands in the Caribbean. Most of the European powers were there to profit from the sugar trade. This trade depended on the forced labor of workers brought in chains from Africa.

CARIBBEAN INDEPENDENCE The first independence movement in Latin America began as a slave revolt in the Caribbean on the island of Haiti. In the 18th century, Haiti was a French colony with an important sugar industry. Africans brought to the island by force worked on the sugar plantations and other plantations. In the 1790s, Toussaint L'Ouverture (too•SAN•loo•vehr•TOOR) led a slave rebellion in Haiti and took over the government of the island. By 1804, Haiti had achieved independence from France. Cuba achieved independence from Spain in 1898 as a result of the Spanish-American War. After an occupation by United States forces, the island became self-governing in 1902. Jamaica and Trinidad and Tobago did not achieve full independence from Great Britain until 1962.

Cultural Blends

Central America and the Caribbean are close to each other geographically, and their cultures show a blending of influences. This mixture affects everything from religion to language.

CULTURE OF CENTRAL AMERICA As you've read, the culture of Central America blends two major elements: Native American influences with those of Spanish settlers. The Spanish were the dominant group of European settlers in Central America—their language remains dominant in the area today. Catholicism is the major religion, although Protestant missionaries are active in the region.

The Spanish took land away from the natives of the region. The conquerors cut down forests, opened up land for grazing livestock, and introduced new crops, such as wheat. They created large farms and ranches, built towns, and moved the native peoples off the land and into the towns. All this altered the way of life in the region.

CULTURE OF THE CARIBBEAN A greater variety of influences was at work in the Caribbean. The Spanish, French, British, Danish, and Dutch existed side by side with the African and Native American. Residents of the islands are of European, African, or mixed ancestry. ▶

African influences were especially important. Most of the people are descendants of the African slaves brought to the islands to work on the

Geographic Thinking

Making Comparisons
Ⓑ How does the culture of the Caribbean differ from the culture of Central America?

This boy is playing baseball, a sport as popular in Cuba as it is in the United States. Baseball traveled from the United States to Cuba in the late 1800s. Baseball is considered the island's national pastime, just as it is in the United States.

Young people in Cuba receive many benefits from the Communist government, including free education and health care. The education system extends from preschool programs through college to graduate programs. However, young people, like all Cubans, live in a police state that limits their economic and political freedoms.

If you lived in Cuba, here are some rights you would enjoy and restrictions you would face:

- You would receive a free education.
- You would receive free medical care.
- You would attend school from age 6 to somewhere between ages 11 and 15.
- You could attend free concerts, ballets, and plays.
- Your freedom of speech and writing would be restricted.
- Your economic opportunities would be very limited.

sugar plantations. They left a lasting mark on all aspects of culture in the islands, including village life, markets, and choice of crops.

The religions of the Caribbean include Catholic and Protestant, as well as Santeria, which combines certain African practices and rituals with Catholic elements. Voodoo is practiced on the island of Haiti. Rastafarianism is a religious and political movement based in Jamaica.

Spanish is spoken on the most populous islands in the Caribbean: Cuba, with a population of about 11 million, and the Dominican Republic, with a population of about 8.5 million. There are also many French speakers (Haiti alone has a population of more than 6 million). English dominates in Jamaica, with a population of almost 3 million. There is a smattering of Dutch and Danish also spoken in the region.

Economics: Jobs and People

In general, most of the people in the countries of the region are poor. This is, in part, a legacy of colonialism. The early success of the sugar crop benefited colonial planters, not the native or African laborers. Also, the region faced competition in the sugar market, and eventually the sugar trade declined. Further, the fact that natural resources were exported and not used locally left the region economically weakened.

FARMING AND TRADE Sugar cane plantations in the Caribbean provide the region's largest export crop. Other important export crops are bananas, citrus fruits, coffee, and spices. All these crops are well adapted to the climate and soil of the region. Many people work on the plantations that grow crops for export. But the pay is poor, and as a result, average per-capita income in the Caribbean is very low.

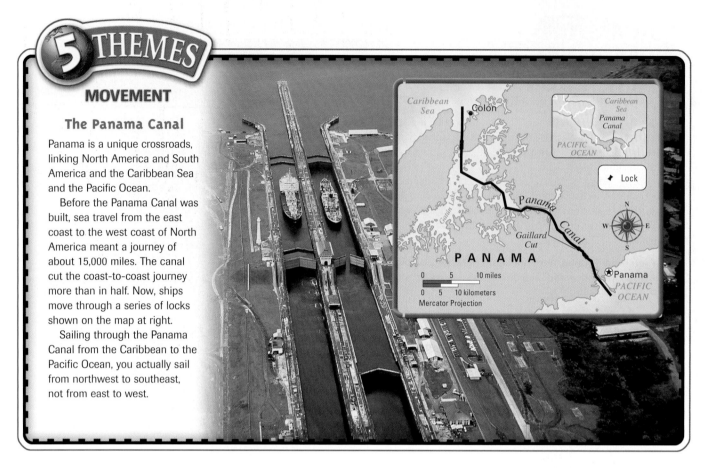

MOVEMENT

The Panama Canal

Panama is a unique crossroads, linking North America and South America and the Caribbean Sea and the Pacific Ocean.

Before the Panama Canal was built, sea travel from the east coast to the west coast of North America meant a journey of about 15,000 miles. The canal cut the coast-to-coast journey more than in half. Now, ships move through a series of locks shown on the map at right.

Sailing through the Panama Canal from the Caribbean to the Pacific Ocean, you actually sail from northwest to southeast, not from east to west.

In Central America, too, the main source of income is the commercial farming practiced on large plantations. These farms produce 10 percent of the world's coffee and 10 percent of the world's bananas. Central America's mines and forests also provide resources for export.

Trade is important because of the **Panama Canal,** which cuts through the land bridge and connects the Atlantic and Pacific oceans. Ships from both hemispheres use the canal, making Panama a crossroads of world trade. The canal made possible the exchange of both goods and ideas.

WHERE PEOPLE LIVE AND WHY Population patterns in Central America and the Caribbean are directly related to their economies. Both Central America and the Caribbean have populations of between 30 and 40 million people. But in Central America, most of the people make their living on farms and, as a result, live in rural areas.

Many of the islands in the Caribbean are densely populated. More than 11 million people live on Cuba, the largest of the islands. Most people live in urban areas, where they hope to find jobs in tourism. The cities attract people who are seeking a better way of life. Unfortunately, many end up living in slums. The region is working to find a way to channel more of the profits from tourism and farming to benefit local communities.

Popular Culture, Tourism, and Jobs

Education and jobs are a major concern to the people of Central America and the Caribbean. Music, heavily influenced and shaped by the African heritage in the region, is an important part of the popular culture of Central America and the Caribbean.

CONNECT TO THE ISSUES
RESOURCES
What resources are exported from Central America and the Caribbean?

MUSIC OF THE CARIBBEAN Both reggae and calypso music started in the Caribbean. **Calypso** music began in Trinidad. Calypso combines musical elements from Africa, Spain, and the Caribbean. Calypso songs are accompanied by steel drums and guitars, and they have improvised lyrics.

Reggae developed in Jamaica in the 1960s. Many reggae songs deal with social problems and religion. African music, Caribbean music, and American music all fed into the roots of reggae. Bob Marley of Jamaica was a pioneer of reggae. The music of the Caribbean is one of the elements that lures tourists to the region, creating jobs for local residents.

BACKGROUND
Bob Marley's son, David "Ziggy" Marley, is carrying on his father's musical legacy.

HUMAN-ENVIRONMENT INTERACTION Many of Bob Marley's songs reflect his faith and political beliefs. **How might popular culture express important ideas and political beliefs?**

LATIN AMERICA

TOURISM AND THE INFORMAL ECONOMY Rapid population growth in the Caribbean is contributing to high unemployment, especially among the young. Many people flee rural areas and move to the cities in search of jobs. Too often, however, they lack job skills. There are schools to help prepare students for jobs in agriculture and tourism.

Tourism is, in fact, an increasingly important industry. Local residents of the islands are able to find jobs working in the hotels, resorts, and restaurants there. In addition, people can make a living working as guides and assistants on fishing excursions, sailing trips, snorkeling adventures, hiking expeditions, and other activities for tourists.

People also find jobs in the **informal economy,** which takes place outside official channels, without benefits or protection for workers. These include jobs such as street vending, shining shoes, and a variety of other activities and services that provide people with a small income.

In Section 3, you will read about Spanish-speaking South America.

SECTION 2 Assessment

① Places & Terms

Identify and explain the following places and terms.

• cultural hearth
• United Provinces of Central America
• Panama Canal
• calypso
• reggae
• informal economy

② Taking Notes

REGION Review the notes you took for this section.

Central America and the Caribbean

Latin America

• What European countries had colonies in the Caribbean?
• Which European country settled most of Central America?

③ Main Ideas

a. What are the major groups that blended to form the culture of this region?

b. What are some major sources of income in the economies of Central America and the Caribbean?

c. What forms of music have evolved in the region?

④ Geographic Thinking

Drawing Conclusions How did the establishment of sugar plantations by Europeans affect the settlement of the Caribbean? **Think about:**

• the people brought in to work on the plantations

RESEARCH LINKS CLASSZONE.COM

GeoActivity

MAKING COMPARISONS Pair with a partner and make a **poster** about the Panama Canal. Do research on the Internet and illustrate your poster with maps and diagrams of the locks in the canal. Provide statistical data about the canal that compares it with other canals, such as the Suez Canal.

Disasters!

Volcano on Montserrat

Montserrat is an island in the Caribbean. One of the outstanding features of the island is its large volcano located in the Soufrière Hills. The volcano had been dormant for approximately five centuries when it began to erupt in 1995. The eruptions continued through 1996 and became particularly severe in 1997. The large map of the island (below) shows the area affected by the eruptions. The dates on the map show the expanding "zones of exclusion." A zone of exclusion is an area too dangerous for people to enter. Two-thirds of the island is now uninhabitable.

ATLANTIC OCEAN

Montserrat

Caribbean Sea

SOUTH AMERICA

Soufrière Hills

October '96

June '97

February '97

Sept '97

July '97

Plymouth

Plymouth, the capital and largest city on the island of Montserrat, lies downwind of Soufrière Hills volcano. As a result, it has been covered with a gray shroud of ash.

The Juggernaut in Puri, India, is a wooden image of the Hindu god Krishna mounted on a cart. The term comes from a Sanskrit word that means "lord of the world." The cart moves on 16 wheels through crowds of Hindu pilgrims on various festival days.

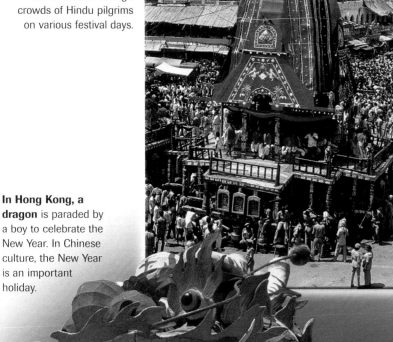

In Hong Kong, a dragon is paraded by a boy to celebrate the New Year. In Chinese culture, the New Year is an important holiday.

GeoActivity

CREATING A POSTER

Working with a partner, use the Internet to research one of the festivals or holidays listed below. Then create a **poster** about the holiday.

* Use visuals and captions to describe the festival or holiday you have chosen.
* Research a different festival and make a second poster to compare festivals from different countries.

RESEARCH LINKS
CLASSZONE.COM

GeoData

FESTIVALS AND HOLIDAYS AROUND THE WORLD

RELIGIOUS

Christianity
Christmas
Easter

Judaism
Rosh Hashanah
Passover

Islam
Feast of Sacrifice
Festival of Breaking Fast
Ashura

Hinduism
Holi
Diwali

OTHER
Independence Day
New Year's Day
Cinco de Mayo
Bastille Day
May Day
Kwanzaa
Thanksgiving

VISUAL SUMMARY
HUMAN GEOGRAPHY OF LATIN AMERICA

Subregions of Latin America

● **Mexico**

- Native peoples and Spanish settlers have shaped the history and culture of Mexico.
- Economic expansion and an increasingly democratic government have developed together.

● **Central America and the Caribbean**

- Native peoples, settlers from many European countries, and Africans have shaped Central America and the Caribbean.
- The economies of the region rely primarily on agriculture and tourism.

○ **Spanish-Speaking South America**

- The countries of South America are developing strategies to improve their economies.
- Among these strategies are wide-ranging trade agreements, including Mercosur.

● **Brazil**

- Brazil is the giant of Latin America.
- Settled originally by the Portuguese, Brazil has welcomed immigrants from all over the world.
- Its economy is among the ten largest in the world.

Reviewing Places & Terms

A. Briefly explain the importance of each of the following.

1. Tenochtitlán
2. Institutional Revolutionary Party (PRI)
3. NAFTA
4. cultural hearth
5. United Provinces of Central America
6. Panama Canal
7. Inca
8. Mercosur
9. Treaty of Tordesillas
10. Carnival

B. Answer the questions about vocabulary in complete sentences.

11. What body of water surrounded Tenochtitlán?
12. Whose election signaled the end of one-party rule in Mexico?
13. Why is the Panama Canal important to world trade?
14. Which two European powers signed the Treaty of Tordesillas?
15. Which countries are associate members of Mercosur?
16. In what city of Brazil is Carnival celebrated in a particularly colorful way?
17. Which countries besides Mexico are members of NAFTA?
18. Why are Central America and the Andes Mountains around Cuzco cultural hearths?
19. Which countries made up the United Provinces of Central America?
20. What language did the Inca speak?

Main Ideas

Mexico (pp. 217–221)

1. What was the Spanish attitude toward Aztec culture?
2. What are the maquiladoras?

Central America and the Caribbean (pp. 222–229)

3. In terms of who settled there, how is the Caribbean different from Mexico and Central America?
4. Which two parts of the economy provide most of the income in Central America and the Caribbean?
5. What are some of the most important export crops in the region?

Spanish-Speaking South America (pp. 230–235)

6. Which countries are full members of Mercosur?
7. Which countries have literacy rates higher than 90 percent?
8. What happened to the Inca language after the Spanish conquest?

Brazil (pp. 236-241)

9. What is the ethnic makeup of Brazil?
10. What are some of the darker aspects of life in Brazil today?

Critical Thinking

1. Using Your Notes

Use your completed chart to answer these questions.

a. Which two European countries colonized the most territory in Latin America?

b. What are some of the ways in which Latin America is developing economically in recent years?

2. Geographic Themes

a. **HUMAN–ENVIRONMENT INTERACTION** How has the Amazon River been used and developed?

b. **MOVEMENT** What has restricted the movement of people from the coast of South America into the interior?

3. Identifying Themes

Interaction between European powers and native peoples occurred throughout the region. What are some of the consequences of this interaction? Which of the five themes are reflected in your answer?

4. Identifying and Solving Problems

What are some of the ways that individual citizens of Latin America are working to improve their economic situation?

5. Making Comparisons

How are Spanish-speaking and Portuguese-speaking South America alike and different?

Additional Test Practice, pp. S1–S37   **TEST PRACTICE** CLASSZONE.COM

Geographic Skills: Interpreting Maps

City of Tenochtitlán

Use the map to answer the questions.

1. **PLACE** This is a Spanish map of the Aztec city of Tenochtitlán. Why did the city require roadway connections to the mainland?

2. **MOVEMENT** Why might this site have been a good location for a city?

3. **HUMAN–ENVIRONMENT INTERACTION** What purpose might the canals within the city have served?

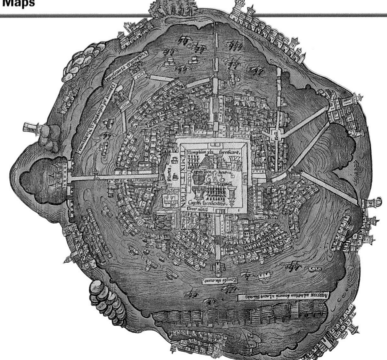

Create a map of a fortress city of your design. Your map should make use of the natural advantages afforded by the site you have chosen.

 INTERNET ACTIVITY

Use the links at **classzone.com** to do research on economic growth in Latin America. Focus on the impact of free-market reforms on the income gap.

Creating Graphs and Charts Present a report of your findings. Include a chart that shows which countries have introduced free-market reforms and what impact these reforms have had on closing the income gap.

Chapter 11

SECTION 1
Rain Forest Resources

SECTION 2
Giving Citizens a Voice

CASE STUDY
THE INCOME GAP

For more on these issues in Latin America . . .

 CURRENT EVENTS
CLASSZONE.COM

TODAY'S ISSUES
Latin America

Timber harvesting (as shown here in Bahia, Brazil) and agriculture have had a devastating effect on the Latin American rain forest.

GeoFocus

How can citizen participation help solve problems?

Taking Notes In your notebook, copy a cause-and-effect chart like the one below. Then take notes on causes and effects of some aspect of each issue.

	Causes	Effects
Issue 1: Resources		
Issue 2: Democracy		
Case Study: Income Gap		

Rain Forest Resources

How can we preserve and develop the rain forest?

Main Ideas

- Special-interest groups make competing demands on the resources of the rain forest.
- As the rain forests are destroyed, the quality of life on Earth is threatened.

Places & Terms

biodiversity

deforestation

global warming

debt–for–nature swap

The Voyageur Experience in World Geography

Costa Rica: Ecotourism and Economic Development

A HUMAN PERSPECTIVE In 1997, biologist Marc van Roosmalen made an incredible discovery. An Amazonian Indian had brought the biologist a tiny monkey huddled inside a tin can. Van Roosmalen realized that the monkey was a kind of pygmy marmoset never before seen by scientists. Over the next three years, Van Roosmalen and his colleagues located the native region of this creature and along the way observed plants and animals unknown to science. These scientists had confirmed the richness of plant and animal life in the Amazon rain forest of Brazil. But for other people, the forest (once cleared) holds the promise of something more—land for farming and timber for sale.

Rain Forest Land Uses

The rain forest is an important global resource. Its vegetation helps to clean the earth's atmosphere, regulate the climate, and shelter several million species of plants, insects, and other wildlife. Scientists have just begun to investigate and understand the rain forest's **biodiversity**—its wide range of plant and animal species. And yet, this variety of life is being destroyed at a rapid rate. At the end of the 20th century, nearly 50 million acres of rain forest worldwide were being destroyed every year.

CLEARING THE RAIN FORESTS The world's demand for timber is great. The Amazon rain forest contains tropical hardwoods, such as mahogany and cedar, that are harvested for export by the timber industry.

Native peoples, living in poverty, travel into the rain forest in search of land on which they can grow crops. They clear the forest,

Major Rain Forests of Latin America

Original rain forest

Rain forest, 2000

SKILLBUILDER: Interpreting Maps

❶ **HUMAN–ENVIRONMENT INTERACTION** Has the rain forest contracted or expanded in recent years? Why might this be so?

❷ **REGION** Around what latitude is most of the rain forest clustered? Why might this be?

not realizing that the soil is not very fertile. Also, cutting down the trees exposes the land to erosion. After a few years, this new farmland becomes less productive, resulting in the need for more timber clearing.

Livestock, too, have been introduced into the rain forest. Ranchers need land on which to graze their cattle, and by clearing the forests for pasture, they can produce a steady supply of beef for the export market.

POPULATION PRESSURES More than half of the Amazon rain forest is located in Brazil. That country's growing population is contributing to the rain forest's decline. The estimated population of Brazil in 2000 was about 173 million people. With an annual growth rate between half a percent and 1 percent, Brazil's population is expected to reach 200 million by 2020. With that many people to shelter, some developers want to build homes on land now covered by the rain forest. ▷

Geographic Thinking
Using the Atlas
◁ Use the map on page 191. In what other countries besides Brazil is the Amazon rain forest located?

The Price of Destruction

There is a cost to pay for **deforestation**—cutting down and clearing away of trees—in the rain forest. The short-term benefits are offset by the high price Latin America and the world are paying in damage to the environment.

ENVIRONMENTAL CONCERNS Rain forests help to regulate the earth's climate. They do this by absorbing carbon dioxide and producing oxygen. As the forests disappear, however, much less carbon dioxide is absorbed. The carbon dioxide that is not absorbed builds up in the atmosphere. This buildup prevents heat from escaping into space. The temperature of the atmosphere begins to rise, and weather patterns start to change. By the beginning of the 21st century, evidence of this **global warming** appeared around the world, causing scientific concern. A common method for clearing the rain forest, known as slash-and-burn (see pages 210–211), produces carbon dioxide and other harmful gases.

PLANTS AND ANIMALS IN DANGER Although the world's rain forests cover about 6 percent of the earth's surface, they are home to an estimated 50 percent of the world's plant and animal species. Medical researchers are developing the processes needed to make use of the many plants that rain-forest dwellers have harvested for thousands of years. The forest dwellers have used these plants to make medicines that heal wounds and cure disease. What is lost as the rain forests disappear is more than biodiversity and a stable environment. The rain forests also hold secrets of nature that might improve and extend the quality of people's lives.

HUMAN–ENVIRONMENT INTERACTION A naturalist and a biologist attach a radio transmitter to a bird to track its movements in the rain forest.
Why might scientists wish to track birds?

BACKGROUND
These gases are referred to as greenhouse gases because they help keep heat in the atmosphere.

Moving Toward Solutions

Saving the rain forests of Latin America is an issue that affects people around the world. Creative solutions will be required to make sure that the forests are not sacrificed to economic development.

A JUGGLING ACT A central problem facing many Latin American countries is how to balance competing interests. Some countries in the region are attempting to restrict economic development until they can find the right balance between economic growth and the preservation of the rain forests.

For example, grassroots organizations are closely observing development projects in the rain forests. Their mission is to educate people about the value of the rain forests and, when necessary, to organize protests against plans that would damage the environment.

Geographic Thinking

Seeing Patterns
B▷ How might the income gap affect the use of the rain forest?

FIGHTING ECONOMICS WITH ECONOMICS Some people think that since economic gain is at the heart of rain forest destruction, the affected governments should be paid to preserve the forests. One such plan is known as a **debt-for-nature swap.** ◀B

Many Latin American nations are burdened by tremendous debt. They've borrowed money to improve living conditions, and now they are struggling to pay it back. In a debt–for–nature swap, an environmental organization agrees to pay off a certain amount of government debt. In return, the government agrees to protect a certain portion of the rain forest. Governments get debt relief; environmentalists get rain forest preservation. This approach was successful in Bolivia. There, an international environmental group paid off some government debt in exchange for the protection of an area of forest and grassland.

The movement to preserve the rain forests has many supporters in the region, as well as around the world. The battle to preserve the rain forests may be one in which everybody wins.

SECTION

Assessment

❶ Places & Terms

Identify and explain the following places and terms.

- biodiversity
- deforestation
- global warming
- debt-for-nature swap

❷ Taking Notes

HUMAN–ENVIRONMENT INTERACTION Review the notes you took for this section.

	Causes	Effects
Issue 1: Resources		

- Why are the rain forests being destroyed?
- What effect might the destruction of the rain forest have on climate?

❸ Main Ideas

a. What are some of the important resources of the rain forest?

b. What are some of the costs of the destruction of the rain forest?

c. What are some factors that might slow destruction of the rain forest?

❹ Geographic Thinking

Making Inferences What might happen to the rain forest in the future? **Think about:**

- economic pressures to destroy the rain forest
- reasons to preserve the rain forest

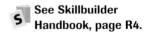
S **See Skillbuilder Handbook, page R4.**

MAKING COMPARISONS Pair with a partner and make a **chart** of the largest rain forests in the world. Then make a copy of a map of the world and color in on the map the rain forests on your chart.

Interpreting Satellite Images

Satellites are orbiting "eyes in the skies." They can give us detailed views of landforms, vegetation, and bodies of water. The satellite image below shows part of the rain forest in the state of Rondônia in Brazil.

THE LANGUAGE OF MAPS A **satellite image** is a visible-light, radar, or infrared picture of land or water taken from space. Depending on the equipment used, satellite images can show land features such as those shown below—ground vegetation and a lake as well as a river that shows some flooding (in the loop below the center of the image). **Landsat satellites** are orbiting satellites that measure reflected light to show features on the earth's surface, including vegetation. A landsat satellite image shows changes in vegetation over time by using a series of images.

A Satellite View of the Rain Forest **❶**

Landsat Image

Sketch Map

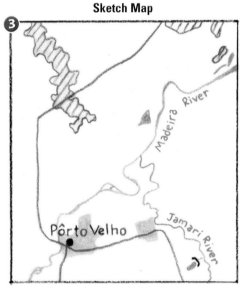

KEY:
- ∿ WATER (RIVER, LAKE, RESERVOIR)
- — ROAD
- ▨ FLOODED AREA, SWAMP
- ⌒ DAM
- ● TOWN

❶ Satellite images are useful in constructing maps, updating maps, and making them more explicit.

❷ In Landsat images, shallow water appears light blue. Thick vegetation appears red. Sparse vegetation appears white.

❸ A researcher made a sketch of the Landsat image, adding names of towns and rivers.

Map and Graph Skills Assessment

1. Making Inferences
The town of Pôrto Velho is near the intersection of what two means of transportation?

2. Making Decisions
In what direction would you travel in going from Pôrto Velho to the Jamari River?

3. Drawing Conclusions
What sort of vegetation predominates in the Landsat image? How can you tell?

Giving Citizens a Voice

How can Latin Americans gain a voice in government?

Main Ideas

• Despite obstacles, democracy is beginning to succeed in Latin America.

• The success of Latin American democracies depends on political, economic, and social reforms.

Places & Terms

oligarchy	caudillo
junta	land reform

A HUMAN PERSPECTIVE From the late 1970s through the early 1980s, the Argentine military waged a campaign of terror against those who supported political reform. As many as 30,000 people mysteriously disappeared. People accused of being terrorists and revolutionaries were kidnapped and questioned. Some were tortured, and then killed or "disappeared"—their bodies were never found. In an effort to learn the truth about their loved ones, a group of women, calling themselves the Mothers of Plaza de Mayo, staged weekly protests in the plaza in Buenos Aires. Their protests were part of the larger attempt by citizens of the region to gain a voice in how their governments were being run.

A Struggle to Be Heard

Latin Americans today seek more democratic governments. Democracy depends on free and fair elections, citizen participation, majority rule with minority rights, and guaranteed freedoms. However, Latin America has shown little support for democratic rule until recently.

THE LEGACY OF COLONIALISM After the Spanish conquest of the region in the 16th century, Native Americans in Central and South America were ruled by governors who took their orders from the king and queen of Spain. Even when Latin American countries won their independence during the 1800s, they continued to be governed mainly by small groups of Spanish colonists.

This government by the few, known as **oligarchy** (AHL·ih·GAHR·kee), was not democratic. The government censored the press, limited free speech, and punished dissent. It also discriminated against all who were not part of the Spanish ruling class. Elections were held, but there was never any doubt who was in charge. If the government was unable to control the people, the military would step in, seize power, and form a new, harsher government known as a **junta** (HOON·tah), which was run by the generals.

THE RULE OF THE *CAUDILLO* Throughout the 20th century, many Latin American countries were ruled by a **caudillo** (kow·DEE·yoh), a military dictator or political boss, such as Juan Perón in Argentina. The caudillo's

"My goodness, if I'd known how badly you wanted democracy I'd have given it to you ages ago."

support came from the military and the wealthy. Surprisingly, the caudillo was sometimes elected directly by the people.

For example, from the 1920s until the end of the 20th century, Mexico was governed by caudillos who were members of the *Partido Revolucionario Institucional* (PRI), or the Institutional Revolutionary Party. For 71 years the PRI dominated Mexican politics.

Opposition parties were legal, but the PRI used fraud and corruption to win elections. Opposition parties made big gains in the 1997 congressional elections. In 2000, Vicente Fox became the first non-PRI president since the adoption of Mexico's constitution in 1917. Finally, it seemed Mexico was ready to fully accept democracy.

BACKGROUND
Caudillo is a Spanish word that means "leader" or "chief."

Establishing Stable Democracies

Creating democracies in Latin America requires political, economic, and land reforms.

THE GOALS OF REFORM One goal of political reform is to establish constitutional government. A freely elected government that respects the law is the basis of democracy. Participation of citizens in political affairs is also critical. This requires that people be well educated and provided with economic security.

Political and economic stability are two sides of the same coin. A lack of prosperity is usually accompanied by social and political unrest. **A**

Argentina in the 1980s was one example of how economic problems damaged a developing democracy. In 1983, Raúl Alfonsín was elected president of Argentina in that nation's first free election in many years. He was faced with a ruined economy after years of military rule.

Argentina suffered from inflation—a rise in the prices of goods and services. To fight inflation, the newly-elected president froze all wages and prices. He issued a new currency to replace the peso. (Later, the peso was brought back.) At first these measures seemed successful, but by 1989, inflation was severe again. In 1989, Argentina elected a new president, Carlos Menem. He introduced a number of capitalist reforms. These included reducing government spending and selling off state-controlled industries and utilities.

Another goal of reform is to recognize and increase the role of women in politics. Throughout the region, women are running for office and taking an active role in government. For example, Marta Suplicy was elected mayor of São Paulo, Brazil, in 2000.

LAND REFORM Latin American countries had been ruled by a wealthy elite. Economic power, as well as land, was in the hands of the few. To spread the wealth more fairly, some governments set up a program of **land reform,** the process of breaking up large landholdings and giving portions of the land to land-poor peasant farmers.

Geographic Thinking

Seeing Patterns
A What effect might the income gap have on political stability in a democracy?

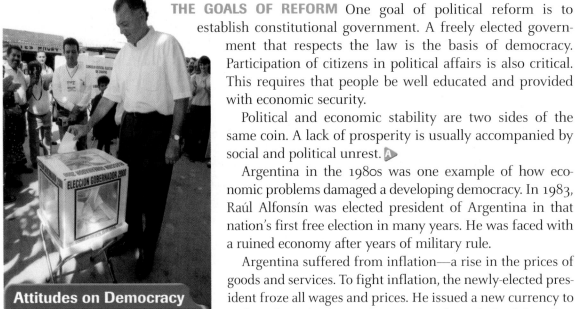

Attitudes on Democracy

Latinobarometro, a Chilean organization, conducts polls asking Latin Americans from a number of countries what they think about different political issues. Recently, the organization asked residents of various countries the following question:

Is democracy the best system of government?

Brazil	50% agree
Central America	49%
Ecuador	41%
Mexico	45%
Paraguay	44%
Peru	21%
Uruguay	86%

SOURCE: *Latinobarometro,* 1998

In Mexico, for example, the process of land reform began with Benito Juarez. He was a Zapotec Indian from a small farm who was elected Mexico's president in 1858. One of his main reform goals was to redistribute the land so that rich landowners could not keep other Mexicans in a cycle of poverty. After the Mexican Revolution in the early part of the 20th century, there was another attempt at land reform. This gave people a better chance at economic equality.

All of these reforms have been aimed at creating stability. With a sound foundation, democracy has a better chance of taking root.

REGION Marta Suplicy holds a press conference after being elected mayor of São Paulo, Brazil, in 2000. **What does her election suggest about the role of women in politics in Brazil?**

Assessment

1 Places & Terms

Identify and explain the following places and terms.

- oligarchy
- junta
- caudillo
- land reform

2 Taking Notes

HUMAN–ENVIRONMENT INTERACTION Review your notes.

	Causes	Effects
Issue 2: Democracy		

- What problems has democracy faced in Latin America?
- What are some of the effects of political reform in the region?

3 Main Ideas

a. How did colonialism affect the development of democracy?

b. What are some of the goals of political reform in the region?

c. Why was land reform necessary, and what was its purpose?

4 Geographic Thinking

Drawing Conclusions
What are the prospects of democracy in the region?
Think about:

- political reforms
- economic reforms

RESEARCH LINKS
CLASSZONE.COM

SEEING PATTERNS Pair with a partner and choose a country in Latin America to research on the Internet. Then prepare a **report** on the condition of democracy in that country and present your report to the class. Discuss what kind of government the country has, the number and names of political parties, and the nature of its legislative and executive functions.

THE INCOME GAP

How can the economic gulf between rich and poor be bridged?

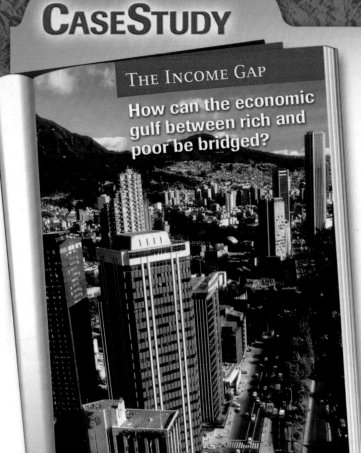

Bogotá, Colombia's glittering financial district

A long the oceanfront in Rio de Janeiro, Brazil, gleaming office buildings and hotels share the boulevards with trendy restaurants and exclusive shops.

Behind all this glitter and glamour, however, is another world, hidden from sight—the *favelas*, or slums, of Rio. Here, the poor live among swamps and garbage dumps, and on barren hillsides.

These contrasting conditions are evidence of what economists call an income gap. This is the difference between the quality of life enjoyed by the rich and the poor. In many Latin American countries, the gap is widening. Some solutions have been proposed for this problem.

The Nature of the Problem

As you've learned in this unit, the income gap in Latin America has many causes, some of which reflect the impact of colonialism in the region. There are three angles to exploring the income gap: it is a moral issue, an economic dilemma, and a political problem.

A MORAL ISSUE Some people argue that Latin America's income gap raises important ethical questions. How can any caring society, they ask, justify vast wealth in the hands of a few while most people live in poverty from which they will likely never escape? The Catholic Church and other religious faiths in Latin America have argued that narrowing the gap between rich and poor is more than just an economic necessity; it is a matter of social justice.

AN ECONOMIC DILEMMA Most Latin American countries now have free-market economies with a minimum of government rules. A free-market economy offers many people the freedom and rewards they need to create wealth. However, in Latin America the poor often lack the basic skills that would make taking part in the economy possible.

Often, the poor have little education. Many cannot read. Most cannot find jobs. Those who find work may end up sweeping streets or shining shoes. Conditions in the slums breed disease and encourage crime. In fact, the life spans of slum dwellers are shorter than those of the middle and upper classes. To the poor of Latin America, the doors to economic equality appear shut.

A POLITICAL PROBLEM Poverty can make people desperate. Those who think they have nothing to lose are sometimes willing to take great risks.

SEE
PRIMARY SOURCE C

Throughout history, battles have been waged and governments have been overthrown by citizens protesting what they regard as an unjust society in which a few have too much while the many have too little.

Argentina, Bolivia, Brazil, Colombia, El Salvador, and Guatemala have all seen bloody rebellions put down by harsh military measures. In the process, human rights and human dignity have been violated. The story is usually the same. The rebels seek economic justice, and the military protects the wealthy. Clearly, attitudes will have to change before the poor in Latin America will be able to participate fully in their nations' economies. Some attitudes are already changing as, for example, more money is going to education.

Possible Solutions

The income gap in Latin America varies from one country to another. For example, according to a recent report issued by the United Nations' Development Program, nearly 45 percent of all Brazilians live in poverty, existing on less than two dollars a day. In Ecuador, Paraguay, and Uruguay, on the other hand, the income gap is much narrower than it is in Brazil.

SEE
PRIMARY SOURCE A

EDUCATION, POLITICS, AND ECONOMICS Many of the countries of the region have put in place free-market economies that they hope will eventually help to narrow the gap by providing economic opportunity and stability for all citizens.

Along with market economies, democracy is now seen by many countries as an essential part of the equation needed to achieve widespread prosperity. Democracy provides an outlet for protest and opposition so that policies can be adjusted to reflect the will of the majority of the people.

Finally, education is an important part of the mix. A literate, well-educated population will be needed to fill the jobs that will become available in an increasingly complex economy. A case study project on the income gap follows on the next two pages.

REGION A girl plays amid garbage and polluted water in Belém, Brazil.
What do the photographs on these pages suggest about the distribution of money in the region?

CASESTUDY

PROJECT

Multimedia Report

Primary sources A, B, C, and D offer information about the income gap in Latin America. Use these resources along with your own research to prepare a multimedia report. The report should define the income gap, personalize it with accounts from the very poor, and identifiy possible solutions.

RESEARCH LINKS
CLASSZONE.COM

Suggested Steps

1. Research possible solutions or initiatives to deal with the income gap in Latin America.
2. Use video, audio, online, and print resources to research your topic.
3. Think about the following questions during your research:
   - What are the roots of the income gap?
   - How does the income gap hinder the participation of the poor in national economies?
   - What are some possible solutions to the problem?

4. Create charts and graphs and use videotapes, audio CDs, and other electronic media to make your report clear and convincing.
5. Prepare a brief talk to introduce and explain your topic.

Materials and Supplies

- Reference books, newspapers, and magazines
- Computer with Internet access
- Printer
- VCR and television
- CD player

PRIMARY SOURCE A

Graph This pie graph shows income distribution in Latin America. The gap was wider at the end of the 1990s than at the end of the 1970s.

Income Distribution in Latin America

☐ **Poorest 40%**
Poorest 40 percent of the population has only 8 percent of the income.

■ **Wealthiest 20%**
The wealthiest 20 percent controls 62 percent of the income.

■ **Middle 40%**
The middle 40 percent of the population has 30 percent of the income.

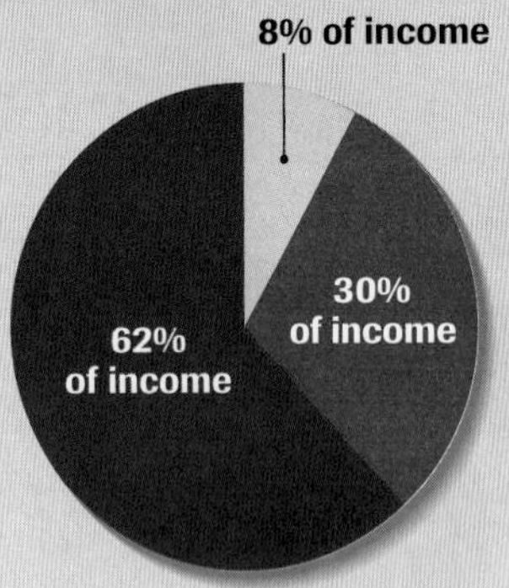

8% of income

62% of income

30% of income

SOURCE: UNICEF, *State of the World's Children, 1999*

PRIMARY SOURCE B

Cable News Story For the homeless children of Rio de Janeiro, the income gap is more than just an economic hardship. It is a matter of life and death, as detailed in this report filed by CNN correspondent Marina Marabella.

April 29, 1996—Four men, including three police officers, went on trial in Rio Monday for the 1993 slaying of eight street children. The murder, the worst massacre of children on record in Brazil, took place outside Candelaria Cathedral in the city center. . . .

Of all the dangers faced by Rio's homeless children, the one they fear the most is being murdered by death squads while they sleep. "When we can, we sleep during the day," said Ricardo, 13. "It's too risky at night". . .

Yvonne Bezerra de Mello has spent years helping Brazil's estimated 2,000 to 3,000 street children. "Until now, no policemen were ever convicted for killing street kids. This is a very good step for Brazilian justice," she said.

She and other human rights activists say the death squads that murder Brazil's homeless children are hired by shopkeepers and others to get rid of those suspected of stealing. . . .

[O]fficial police estimates say about 500 of Rio's homeless children are murdered each year.

Newspaper Report *On September 5, 2000, Steven Gutkin filed this story from Caracas, Venezuela, to* The Times of India Online. *It shows clearly that the consequences of the income gap can be found throughout Latin America.*

Caracas—The Sambil shopping mall in eastern Caracas is Latin America's largest. It boasts 450 stores, two movie theatres, an amusement park, a 30,000-gallon aquarium—and a McDonald's where Big Macs cost a half day's pay for the average Venezuelan worker.

A slum just a few miles to the west has open sewers running alongside tin shacks perched on unstable hillsides, flies buzzing in uncollected garbage and idle young men nursing bullet wounds. Blanca Vera, 65, lifts her baby granddaughter's blouse to reveal blotches on her tiny stomach. "This is from the pollution," she says.

[I]nequality of wealth and opportunity is a huge obstacle to development in Latin America. The existence of so many have-nots threatens to undermine the success of the region's two great experiments of recent years: democracy and free markets.

In Chile, the highest-paid 6 percent of workers get 30 percent of salaries, while 75 percent of workers get just 4 percent, according to the United Nations' Economic Commission for Latin America and the Caribbean.

Some blame the growing inequality on globalization. . . . Yet most economists say the real culprit is not globalization but misguided state policies that deprive the poor of a decent education, fail to collect taxes, and encourage corruption.

There's another factor that's harder to define but likely is just as real: a culture of elitism that regards poor people as unworthy. "You can't operate in a globalized economy with a narrow, tiny elite sector that has absolutely no connection or appreciation of the vast majority of people in society," says Michael Shifter, a Latin America specialist at the Washington-based Inter-American Dialogue.

Magazine Article *There are some initiatives to deal with the consequences of poverty. A reporter for the British magazine,* The Economist, *wrote about a program in Pôrto Alegre, Brazil, to help street children.*

"Is it true that in your country parents can be jailed for beating their children?" 16-year-old Jose asks your correspondent. Clearly there is no need to ask what made him run away from home, to become, briefly, one of Brazil's "street children." Luckily for him, the city on whose streets he ended up sleeping is Pôrto Alegre. Its municipal council this year, for the second year running, won an award given by the Abrinq Foundation, a Brazilian children's rights charity, to the local authority with the best social services for children. After only a short while on the streets, Jose now sleeps in a council-run dormitory and spends most of his days in the city's "Open School," which allows current and former street children to come and go as they please, aiming gradually to draw them back to something like a normal life and perhaps to an education. . . .

Pôrto Alegre is one of a handful of cities . . . that are trying. The services they offer are modest: a shelter where the children can sleep, eat, and wash; a day center staffed with a few teachers, drug counsellors, and so on; and some staff to patrol the streets at night looking for children in need.

PROJECT *CheckList*

Have I . . .

✔ fully researched my topic?

✔ searched for a mix of media sources from which to build my report?

✔ created informative visuals that make my report clear and convincing?

✔ practiced the delivery of my presentation?

✔ made sure that I am familiar with the video and audio equipment I plan to use?

VISUAL SUMMARY
TODAY'S ISSUES IN LATIN AMERICA

Environment

Rain Forest Resources
- There are a number of competing demands on the resources of the rain forest.
- Farmers, ranchers, environmentalists, the timber industry, and pharmaceutical companies all have their own interests in the rain forests of the region.
- Intelligent management and development of the rain forests depend on careful balancing of these competing interests.

Government

Giving Citizens a Voice
- After a long struggle to overcome the legacy of colonialism, most countries in Latin America are struggling toward more democratic forms of government.
- Political stability and economic progress often go hand in hand.

Economics

Case Study: The Income Gap
- The gap between rich and poor in Latin America presents a challenging problem.
- It is likely that a widening income gap will undermine political stability in the region.
- For this reason, government, businesses, and education must all work together to try to narrow the gap.

Reviewing Places & Terms

A. Briefly explain the importance of each of the following.

1. biodiversity
2. deforestation
3. global warming
4. debt-for-nature swap
5. oligarchy
6. junta
7. caudillo
8. land reform

B. Answer the questions about vocabulary in complete sentences.

9. Why is the biodiversity of the rain forest important?
10. What are some examples of the kinds of trees being harvested in the rain forest?
11. What is one byproduct of slash-and-burn clearing of the rain forest that is harming the atmosphere?
12. Why is it in the interest of governments to participate in debt-for-nature swaps?
13. Why is an oligarchy undemocratic?
14. Why is a junta undemocratic?
15. From where does the caudillo gain support?
16. Who benefits from land reform?
17. Is the biodiversity of the region increasing or decreasing?
18. In a debt-for-nature swap, what does the government agree to do?
19. Who loses in a program of land reform?
20. Which of the eight terms listed above represent the negative impact of colonialism on the politics of the region?

Main Ideas

Rain Forest Resources (pp. 245–248)

1. Why is the rain forest an important global resource?
2. What are some of the reasons the rain forest is being cleared?
3. What is one mission of the grassroots organizations in the rain forest?

Giving Citizens a Voice (pp. 249–251)

4. Who are some democratically elected leaders in the region?
5. What are some of the elements upon which democracy depends?
6. What sorts of reforms are essential to stable democracy in the region?

Case Study: The Income Gap (pp. 252–255)

7. Which groups have argued that the income gap presents a moral issue?
8. What is the basic economic dilemma confronted by poor people in Latin America?
9. Why is the income gap a political issue?
10. Do all countries of Latin America have a similar income gap? Explain.

Critical Thinking

1. Using Your Notes

Use your completed chart to answer these questions.

	Causes	Effects
Issue 1: Resources		
Issue 2: Democracy		

a. How might the income gap undermine democracy?

b. What effect might the exploitation of rain forest resources have upon the income gap in the region?

2. Geographic Themes

a. **MOVEMENT** What effect has the movement of people had on the rain forest?

b. **REGION** What are some of the major historical facts that have hindered the development of democracy in Latin America?

3. Identifying Themes

How might the use and development of the region's resources be connected to the gap between rich and poor? Which of the five themes apply to this situation?

4. Making Decisions

If you were a government official in the region, how might you try to balance competing demands on rain forest resources?

5. Drawing Conclusions

How might democratic government in the region promote economic prosperity?

Additional Test Practice, pp. S1–S37

TEST PRACTICE
CLASSZONE.COM

Geographic Skills: Interpreting Graphs

Poverty in Latin America

Use the graph to answer the following questions.

1. **REGION** In which three countries of Latin America is the percentage of people living in poverty the lowest?

2. **REGION** In which three countries is the poverty rate highest?

3. **PLACE** Brazil is the largest country in the region, in terms of both area and population. What is its poverty rate?

GeoActivity

Create a poster showing the effects of poverty and the income gap in one or more countries in the region. Include a map, as well as photographs and diagrams.

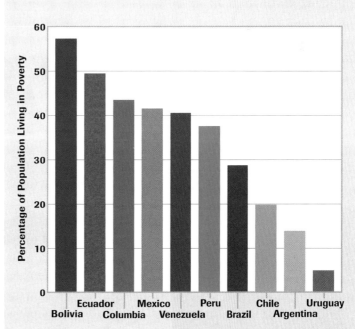

SOURCE: *Social Panorama of Latin America*, 1998

 INTERNET ACTIVITY

Use the links at **classzone.com** to do research about the Amazon rain forest. Focus on solutions and strategies to slow the dwindling of the rain forest.

Creating Multimedia Presentations Combine charts, maps, or other visual images in an electronic presentation showing strategies for preserving the rain forest.

Unit 4

Europe

Europe is the world's second smallest continent. Located in the Northern Hemisphere, Europe has great diversity of landforms and cultures.

LOCATION The dazzling White Cliffs of Dover in England face the English Channel. The cliffs are made of soft chalk and are slowly eroding.

GeoData

REGION Many people view the Ural Mountains as the eastern border of Europe, but for historic and cultural reasons, Russia and other former republics of the Soviet Union are in Unit 5.

PLACE Europe's coastline is longer than that of Africa, the world's second largest continent.

HUMAN-ENVIRONMENT INTERACTION Historically, Europeans used the oceans and seas to make voyages for exploration and trade. Their culture spread around the world.

For more information on Europe . . .

RESEARCH LINKS
CLASSZONE.COM

PLACE The Eiffel Tower stands 984 feet above the Paris skyline. It was completed in 1889 for an International Exposition celebrating the French Revolution.

MOVEMENT Europeans are wildly enthusiastic about soccer. Teams and fans travel to matches held all over the world. In this game, held during the 1998 World Cup matches, Germany played Italy.

259

Unit PREVIEW
4

Today's Issues in Europe

Today, Europe faces the issues previewed here. As you read Chapters 12 and 13, you will learn helpful background information. You will study the issues themselves in Chapter 14.

In a small group, answer the questions below. Then participate in a class discussion of your answers.

Exploring the Issues

1. **CONFLICT** Search a print or online newspaper for articles about ethnic or religious conflicts in Europe today. What do these conflicts have in common? How are they different?

2. **POLLUTION** Make a list of possible pollution problems faced by Europe and those faced by the United States. How are these problems similar? Different?

3. **UNIFICATION** To help you understand the issues involved in unifying Europe, compare Europe to the United States. Imagine what might occur if each U.S. state were its own country. List five problems that might result.

For more on these issues in Europe . . .

CURRENT EVENTS
CLASSZONE.COM

CONFLICT

How can people resolve their differences?

In central Bosnia, a child stands near the ruins of a Muslim mosque. Bosnian Croats destroyed the mosque during an "ethnic cleansing" campaign to drive out Muslims during the 1992–1995 Bosnian war.

off the Balkan Peninsula from the rest of Europe. Historically, they also have isolated the peninsula's various ethnic groups from each other.

UPLANDS Mountains and uplands differ from each other in their elevation. **Uplands** are hills or very low mountains that may also contain mesas and high plateaus. Some uplands of Europe are eroded remains of ancient mountain ranges. Examples of uplands include the Kjølen (CHUR·luhn) Mountains of Scandinavia, the Scottish highlands, the low mountain areas of Brittany in France, and the central plateau of Spain called the **_Meseta_** (meh·SEH·tah). Other uplands border mountainous areas, such as the Central Uplands of Germany, which are at the base of the Alps. About one-sixth of French lands are located in the uplands called the **_Massif Central_** (ma·SEEF sahn·TRAHL).

BACKGROUND
Brittany is a region located on a peninsula in northwest France.

Rivers: Europe's Links

Traversing Europe is a network of rivers that bring people and goods together. These rivers are used to transport goods between coastal harbors and the inland region, aiding economic growth. Historically, the rivers also have aided the movement of ideas.

Two major castle-lined rivers—the Danube and the Rhine—have served as watery highways for centuries. The Rhine flows 820 miles from the interior of Europe north to the North Sea. The Danube cuts through the heart of Europe from west to east. Touching 9 countries over its 1,771-mile length, the Danube River links Europeans to the Black Sea.

Many other European rivers flow from the interior to the sea and are large enough for ships to traverse. Through history, these rivers helped connect Europeans to the rest of the world, encouraging both trade and travel. Europeans have explored and migrated to many other world regions.

Rivers of Europe

SKILLBUILDER: Interpreting Maps

❶ MOVEMENT Which rivers empty into the North Sea? Into the Mediterranean Sea?

❷ PLACE What port is at the mouth of the Rhine?

Geographic Thinking

Seeing Patterns
How does the direction in which European rivers flow aid in linking Europeans to the world?

Fertile Plains: Europe's Bounty

One of the most fertile agricultural regions of the world is the Northern European Plain (see the map on page 263), stretching in a huge curve across parts of France, Belgium, the Netherlands, Denmark, Germany, and Poland. Relatively flat, this plain is very desirable agricultural land that has produced vast quantities of food over the centuries. However, the plain's flatness has also allowed armies and groups of invaders to use it as an open route into Europe. Smaller fertile plains used for farming also exist in Sweden, Hungary, and Lombardy in northern Italy.

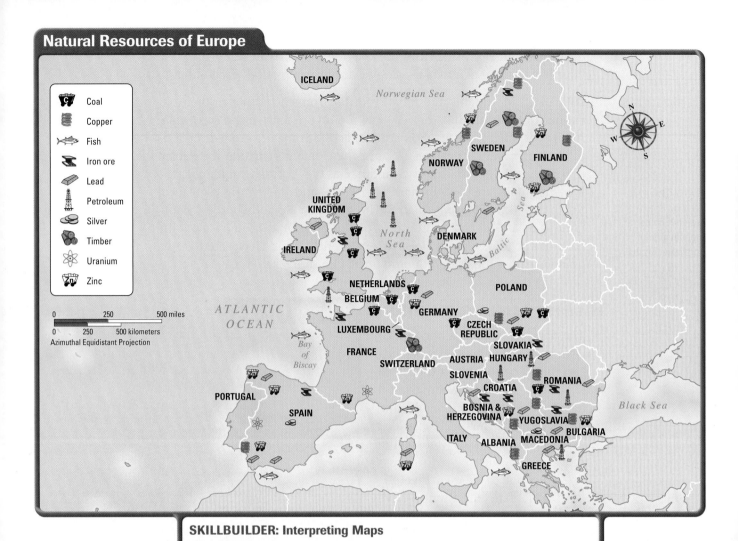

SKILLBUILDER: Interpreting Maps

❶ **LOCATION** Where are major petroleum deposits found in Europe?

❷ **REGION** Which countries in Europe have relatively few natural resources?

Resources Shape Europe's Economy

Europe has abundant supplies of two natural resources—coal and iron ore—needed for an industrialized economy. The map above shows a band of coal deposits stretching from the United Kingdom across to Belgium and the Netherlands and from there to France, Germany, and Poland. Near many of these coal deposits are iron ore deposits. Having both of these resources makes it possible to produce steel. The Ruhr (roor) Valley in Germany, the Alsace-Lorraine region of France, and parts of the United Kingdom are heavily industrialized because these minerals are found there and good transportation exists. But as a result, these regions have suffered from industrial pollution. (See Chapter 14 for more on pollution.) ▷

ENERGY Oil and natural gas were found beneath the North Sea floor in 1959. Energy companies began to tap gas fields between the United Kingdom and the Netherlands. In 1971, new technologies made it possible to construct offshore oil rigs in the North Sea despite its deep, stormy waters. Norway, the Netherlands, the United Kingdom, and Denmark now pump oil from rigs as far as 400 miles out in the ocean. The North Sea oil fields are major sources of petroleum for the world.

CONNECT TO THE ISSUES
POLLUTION
◁ What types of pollution might industry create?

AGRICULTURAL LAND About 33 percent of Europe's land is suitable for agriculture. The world average is 11 percent, so Europe is especially well off. The land produces a variety of crops: grains, grapes, olives, and even cork. Timber is cut from vast forests on the Scandinavian Peninsula and in the Alps.

BACKGROUND
Cork is the outer bark of the cork oak tree.

Resources Shape Life

As is true of every region, the resources available in Europe help shape the lives of its people. Resources directly affect the foods people eat, the jobs they hold, the houses in which they live, and even their culture. For example, traditional European folk tales often take place in deep, dark forests that were a major part of the European landscape centuries ago.

The distribution of resources also creates regional differences within Europe. For instance, because Ireland lacks energy sources, the Irish cut peat from large beds and burn it as fuel. **Peat** is partially decayed plant matter found in bogs. In contrast, coal is plentiful in other parts of Europe and has been mined for centuries. For example, generations of Polish miners have worked the mines that modern-day Poles work.

Just as landforms and resources influence the lives of people, so does climate. In Section 2, you will learn that the climates of Europe are mild near the Atlantic Ocean and grow harsher inland. You will also learn about the climates of the Mediterranean and the Arctic regions.

PLACE Harvesting peat is common in Ireland because other fuel sources are scarce. **Why is it cut in blocks?**

Assessment

① Places & Terms

Identify and explain where in the region these would be found.

- fjord
- uplands
- *Meseta*
- *Massif Central*
- peat

② Taking Notes

PLACE Review the notes you took for this section.

Landforms	
Resources	

- What types of landforms are found in Europe?
- What resources help with farming?

③ Main Ideas

a. Why is Europe called a "peninsula of peninsulas"?

b. How are the landforms of Europe both an advantage and a disadvantage to life in Europe?

c. How did natural resources help Europe to become industrialized?

④ Geographic Thinking

Drawing Conclusions
What role did the waterways of Europe play in the development of its economy?
Think about:

- the nearness to seas and oceans
- the network of rivers

RESEARCH LINKS
CLASSZONE.COM

EXPLORING LOCAL GEOGRAPHY Do research to learn the top three natural resources in your state. Then study the map on page 276 to determine which European country has the most resources in common with your state. Create a **Venn Diagram** showing the resources your state has in common with that country and the resources that are different.

Climate and Vegetation

Main Ideas

• Much of Europe has a relatively mild climate because of ocean currents and warm winds.

• Eastern Europe has a harsher climate because it is farther from the Atlantic Ocean.

Places & Terms

North Atlantic Drift

sirocco

mistral

CONNECT TO THE ISSUES
POLLUTION Industrial air pollution leads to acid rain, which kills trees and other vegetation.

A HUMAN PERSPECTIVE Because of Greece's mild climate, the ancient Greeks spent much time outdoors. Greek men liked to talk with their friends in the marketplace. They also enjoyed sports. Large crowds gathered for athletic contests that were held during religious festivals. The most important of these was a footrace held every four years in the town of Olympia, a contest called the Olympic Games. In time, these games came to include other sports such as wrestling. In this form, they were the model for our modern Olympics. If ancient Greece had had a cold climate, we might not have Olympic Games today.

Westerly Winds Warm Europe

A marine west coast climate exists in much of Europe—from northern Spain across most of France and Germany to western Poland. It also exists in the British Isles and some coastal areas of Scandinavia. With warm summers and cool winters, the region enjoys a milder climate than do most regions at such a northern latitude.

The nearby ocean and the dominant winds create this mild climate. The **North Atlantic Drift,** a current of warm water from the tropics, flows near Europe's west coast. The prevailing westerlies, which blow west to east, pick up warmth from this current and carry it over Europe. No large mountain ranges block the winds, so they are felt far inland. They also carry moisture, giving the region adequate rainfall.

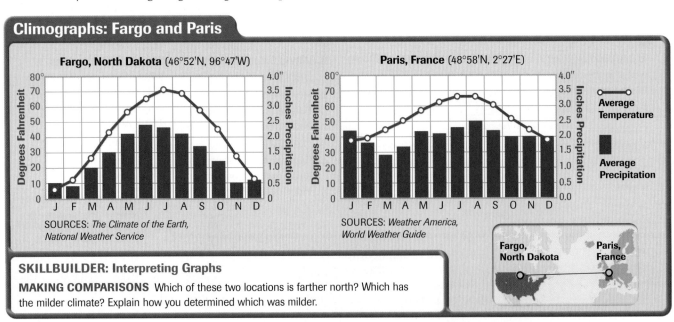

Climographs: Fargo and Paris

Fargo, North Dakota (46°52'N, 96°47'W)

SOURCES: *The Climate of the Earth, National Weather Service*

Paris, France (48°58'N, 2°27'E)

SOURCES: *Weather America, World Weather Guide*

Average Temperature

Average Precipitation

Fargo, North Dakota   Paris, France

SKILLBUILDER: Interpreting Graphs

MAKING COMPARISONS Which of these two locations is farther north? Which has the milder climate? Explain how you determined which was milder.

The Alps create a band of harsher conditions next to this climate zone. Because of their high elevation, the Alps have a much colder climate. Above 5,000 feet, snow can reach a depth of 33 feet in winter. 🔺

FORESTS TO FARMS Originally, mixed forests covered much of the marine west coast climate region. Over the centuries, people cleared away most of the forest so they could settle and farm the land. Today, farmers in the region grow grains, sugar beets, livestock feed, and root crops such as potatoes.

Harsher Conditions Inland

People who live far from the Atlantic Ocean do not benefit from the moderating influence of the westerlies. As a result, much of Sweden and Finland and the eastern parts of Poland, Slovakia, and Hungary have a humid continental climate, as does all of Romania. These places have cold, snowy winters and either warm or hot summers (depending upon their latitude). In general, the region receives adequate rainfall, which helps agriculture.

Like most of Europe, the region has suffered much deforestation, but the forests that do survive tend to be coniferous. The region also has broad fertile plains that were originally covered with grasses. Today, farmers grow grains such as wheat, rye, and barley on these plains. Other major crops include potatoes and sugar beets.

The Sunny Mediterranean

A mild climate lures people to live and vacation in the region bordering the Mediterranean Sea. This Mediterranean climate extends from southern Spain and France through Italy to Greece and other parts of the Balkan Peninsula. Summers are hot and dry with clear, sunny skies, while winters are moderate and wet. One reason for the climate is that mountain ranges block cold north winds from reaching the Iberian, Italian, and Balkan peninsulas.

SPECIAL WINDS An exception to this pattern is the Mediterranean coast of France, which is not protected by high mountains. In winter, this coast receives the **mistral** (MIHS•truhl), a cold, dry wind from the north. 🅱

Most Mediterranean countries experience a wind called the sirocco. The **sirocco** (suh•RAHK•oh) is a hot, steady south wind that blows from North Africa across the Mediterranean Sea into southern Europe. Some siroccos pick up moisture from the sea and produce rain; others carry dust from the desert.

EUROPE

REGION In some Mediterranean fields, such as this one in southern France, olive trees and grape vines are grown side by side. **Why might farmers choose to plant a field with two crops instead of one?**

PLACE In the village of Jukkasjärvi, Sweden, the Ice Hotel is built every winter out of 10,000 tons of ice and 30,000 tons of snow. **How does climate make this possible?**

THE CLIMATE ATTRACTS TOURISTS

The Mediterranean region has primarily evergreen shrubs and short trees that grow in climates with hot, dry summers. The region's major crops are citrus fruits, olives, grapes, and wheat. The sunny Mediterranean beaches also attract thousands of people, making tourism a major industry in the region.

Land of the Midnight Sun

In far northern Scandinavia, along the Arctic Circle, lies a band of tundra climate. As explained in Chapter 3, the land in such a climate is often in a state of permafrost, in which the subsoil remains frozen year-round. No trees grow there—only mosses and lichens. To the south of this lies the subarctic climate, which is cool most of the time with very cold, harsh winters. Little grows there but stunted trees. Because of the climate, agriculture is limited to southern Scandinavia.

This far northern region witnesses sharp variations in the amount of sunlight received throughout the year. Winter nights are extremely long, as are summer days. North of the Arctic Circle, there are winter days when the sun never rises and summer days when the sun never sets. The region is often called the Land of the Midnight Sun.

In the next section, you will read about ways in which Europeans have altered their environment—both positively and negatively.

BACKGROUND
A lichen is an organism made of a fungus and an alga growing together.

SECTION 2

Assessment

1 Places & Terms

Identify these terms and explain how they affect climate.

- North Atlantic Drift
- mistral
- sirocco

2 Taking Notes

REGION Review the notes you took for this section.

> Climate and Vegetation

- Which regions of Europe have the harshest, coldest climates?
- Which climate zones produce the richest variety of vegetation?

3 Main Ideas

a. How do the North Atlantic Drift and the prevailing westerlies affect Europe's climate?

b. How are a mistral and a sirocco different?

c. Why is northern Scandinavia sometimes called the Land of the Midnight Sun?

4 Geographic Thinking

Making Decisions If you wanted to attract tourists to far northern Scandinavia, how would you advertise the region? **Think about:**

- recreational activities suitable for such a climate

RESEARCH LINKS
CLASSZONE.COM

GeoActivity

MAKING COMPARISONS Choose a place in Europe, and then find a place in North America at about the same latitude. Do Internet research to learn about the climate and vegetation of the two places. Create a **chart** comparing the two.

Interpreting a Bar Graph

How much rain and snow does your area receive in a year? Average yearly precipitation varies widely throughout the United States, with extremes ranging from a low of less than 2 inches a year in Death Valley, California, to as much as 151.25 inches a year in Yakutat, Alaska. The figures for average yearly precipitation don't reveal how much rain or snow falls in a given month, but they can provide a general indication of a place's suitability for agriculture or other activities.

THE LANGUAGE OF GRAPHS A **bar graph** is a visual way of showing quantities. On a bar graph, it is easy to see how different examples in a category compare; the longer the bar, the greater the quantity. Depending on the subject, the quantities are expressed using measurements such as inches, dollars, or tons. The categories vary from graph to graph. Time periods and places are common categories. Below, a bar graph shows annual precipitation for several European cities.

Average Annual Precipitation in Europe

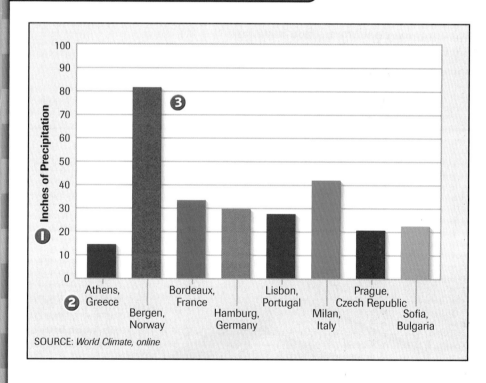

SOURCE: *World Climate, online*

❶ The vertical axis tells you that on this graph, the precipitation is expressed in inches.

❷ The horizontal axis tells you that the category is selected cities of Europe.

❸ A quick glance at this bar graph tells you which cities have high and low amounts of precipitation. By examining the bars more carefully and measuring their heights against the horizontal lines, you can estimate actual amounts of precipitation.

Map and Graph Skills Assessment

1. Analyzing Data
Which cities on this graph have the lowest and highest amounts of annual precipitation?

2. Drawing Conclusions
To which city would you move if your doctor advised you to live in a dry climate?

3. Analyzing Data
What is the average annual precipitation for these eight cities?

Human–Environment Interaction

A HUMAN PERSPECTIVE "1800 DIE IN WIND-WHIPPED FLOOD WATERS!" February 1, 1953, witnessed a disaster in the Netherlands. Winds estimated at 110 to 115 miles per hour piled up gigantic waves that ripped through **dikes**—earthen banks—holding back the North Sea. When the storm was over, 4.5 percent of the Netherlands was flooded, and thousands of buildings were destroyed. The Netherlands is prone to floods because much of its land is below sea level.

Polders: Land from the Sea

An old saying declares, "God created the world, but the Dutch created Holland." (Holland is another name for the Netherlands.) Because the Dutch needed more land for their growing population, they reclaimed land from the sea. At least 40 percent of the Netherlands was once under the sea. Land that is reclaimed by diking and draining is called a **polder.**

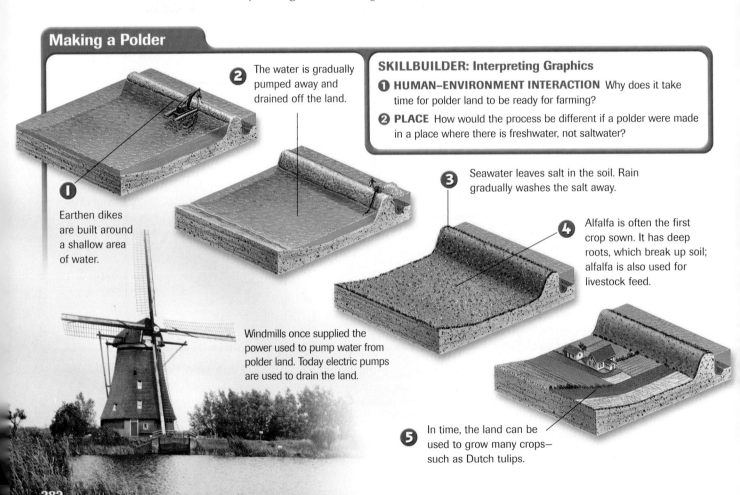

Making a Polder

2 The water is gradually pumped away and drained off the land.

SKILLBUILDER: Interpreting Graphics

1 HUMAN–ENVIRONMENT INTERACTION Why does it take time for polder land to be ready for farming?

2 PLACE How would the process be different if a polder were made in a place where there is freshwater, not saltwater?

3 Seawater leaves salt in the soil. Rain gradually washes the salt away.

1 Earthen dikes are built around a shallow area of water.

4 Alfalfa is often the first crop sown. It has deep roots, which break up soil; alfalfa is also used for livestock feed.

Windmills once supplied the power used to pump water from polder land. Today electric pumps are used to drain the land.

5 In time, the land can be used to grow many crops— such as Dutch tulips.

SEAWORKS The Dutch erected **seaworks,** structures that are used to control the sea's destructive impact on human life. Those seaworks include dikes and high earthen platforms called **terpen.** The dikes hold back the sea, while the terpen provide places to go for safety during floods and high tides.

Over the centuries, the Dutch found ways to reinforce the dikes and to control water in the low-lying areas the dikes protected. In the 1400s, the Dutch began using their windmills to power pumps that drained the land. When the French conqueror Napoleon viewed a site with 860 windmills pumping an area dry, he reportedly said, "Without equal." Today the pumps use electric motors instead of windmills. ◁

Geographic Thinking

Making Comparisons
▷ What are possible disadvantages of windmills and of electric pumps?

TRANSFORMING THE SEA Another remarkable Dutch alteration of their environment was the transformation of the **Zuider Zee** (ZEYE•duhr ZAY). It was an arm of the North Sea and is now a freshwater lake. The idea was originally proposed in 1667. But it was not until the late 1800s and early 1900s that the Dutch perfected a plan to build dikes all the way across the entrance to the Zuider Zee. Since no saltwater flowed into that body of water, it eventually became a freshwater lake. It is now called **Ijsselmeer** (EYE•suhl•MAIR). The land around the lake was drained, creating several polders that added hundreds of square miles of land to the Netherlands.

Waterways for Commerce: Venice's Canals

BACKGROUND
A land link to Venice was built in 1846. A railway bridge connected Venice to the mainland.

Like the Netherlands, Venice, Italy, is a place where humans created a unique environment. About 120 islands and part of the mainland make up the city of Venice. Two of the largest islands are San Marco and Rialto. A broad waterway called the Grand Canal flows between them.

Moving people or goods in Venice depends upon using the more than 150 canals that snake around and through the islands. Consequently, to get from one place to another in Venice, you generally have two choices: take a boat or walk. Almost anything that is moved on wheels elsewhere is moved by water in Venice.

AN ISLAND CITY GROWS Venice began when people escaping invaders took shelter on inhospitable islands in a lagoon. They remained there and established a settlement that eventually became Venice. The city is located at the north end of the Adriatic Sea, a good site for a port. As a result, trade helped Venice grow.

BUILDING ON THE ISLANDS Building Venice required construction techniques that took into account the swampy land on the islands. Builders sunk wooden pilings into the ground to help support the structures above. So many pilings were required that oak forests in the northern Italian countryside and in Slovenia were leveled to supply the wood. The weight of the buildings is so great that it has compressed the underlying ground. This is one of the reasons that Venice is gradually sinking. Other reasons include rising sea levels and the removal of too much groundwater by pumping.

PROBLEMS TODAY Severe water pollution threatens historic Venice. Industrial waste, sewage, and saltwater are combining to eat away the

Canals of Venice

Venice

Lagoon of Venice

San Michele

Grand Canal

St. Mark's Canal

Sacca Fisola

Giudecca Canal

La Giudecca

San Giorgio Maggiore

La Grazia

0 .4 .8 miles
0 .4 .8 kilometers
Transverse Mercator Projection

Street

SKILLBUILDER: Interpreting Maps

❶ **MOVEMENT** What is the advantage of having both streets and canals in Venice?

❷ **HUMAN-ENVIRONMENT INTERACTION** How have humans altered the environment of the islands of Venice?

MOVEMENT Waterbuses, motorboat taxis, small river boats, and gondolas move people and goods on the canals of Venice. Gondolas generally are too expensive for local people to hire. Instead, tourists use them for sightseeing.

foundations of buildings and damage the buildings themselves. Erosion has allowed increased amounts of seawater into the lagoon. Because of this, floods also endanger the city. In November 1966, six feet of flood-water engulfed the city and ruined many of its buildings and the art-work that they housed. Agricultural runoff flowing into Venice's harbor creates conditions that promote algae growth, sometimes called "killer algae." These algae grow rapidly and, after they die, decay. The decaying process uses up oxygen in the water, so that fish also die. Dead fish attract insects and create a stench, especially in warm weather.

A Centuries-Old Problem: Deforestation

Throughout history, humans have damaged and destroyed Europe's forests. The term deforestation means the clearing of forests from an area. Often when we think of deforestation, we think of losing the great rain forests of the world, such as those in South America, which you learned about in Unit 3. But people have also been clearing the forests of Europe since ancient times. Forests provided wood to burn for fuel and to use as building material for ships and houses. When Europeans began to develop industry in the 1700s and 1800s, they needed even

HUMAN–
ENVIRONMENT
INTERACTION A
forest in Bohemia in
the Czech Republic is
dying from the effects
of acid rain.
**Why would restoring
the forest be a slow
process?**

more wood to make charcoal for blast furnaces. Eventually, they used coal as a fuel in place of wood, but not before huge areas of Europe had lost their native forests.

ACID RAIN STRIPS FORESTS In the 1960s, people noticed that many trees of the Black Forest in Germany were discolored, losing needles and leaves, and dying. In time, scientists identified one cause of the tree deaths as acid rain. Europe's factories produce high amounts of sulfur dioxide and nitrogen oxide emissions. These combine with water vapor and oxygen to form acid rain or snow. Winds carry the emissions to other parts of Europe, affecting an estimated one-fourth of all European forests. This problem has hit Scandinavia particularly hard, since the prevailing winds blow in that direction. As mentioned earlier, the Black Forest in Germany also has suffered extreme damage. To save the remaining forests, nations must work together to reduce air pollution. You can read more about this in Chapter 14. **B**

As you will read in Chapter 13, the ways people live upon the land and interact with each other make up the human geography of Europe.

**CONNECT TO
THE ISSUES
UNIFICATION
B** How might a
union of nations
affect the clean-up
effort?

EUROPE

SECTION 3 Assessment

① Places & Terms

Identify and explain where in the region these would be found.

- dike
- polder
- seaworks
- terpen
- Zuider Zee
- Ijsselmeer

② Taking Notes

**HUMAN-ENVIRONMENT
INTERACTION** Review your notes for this section.

Human-Environment
Interaction

- What are examples of human adaptation to the environment?
- What are examples of an environment changed by humans?

③ Main Ideas

a. How have the people of the Netherlands been able to create more land for their country?

b. How has pollution affected the city of Venice?

c. How has industrialization hurt the forests of Europe?

④ Geographic Thinking

Making Comparisons
What is similar about the ways that the people of the Netherlands and the people of Venice interact with their environments? **Think about:**

- seaworks in the Netherlands
- canals in Venice

 **See Skillbuilder
Handbook, page R3.**

SEEING PATTERNS Pollution has affected both Venice and the forests of Scandinavia. Create two **cause-and-effect charts** outlining the causes and effects of pollution in each place. Then write a sentence or two summarizing the similarities.

Chapter 12 Assessment

VISUAL SUMMARY
PHYSICAL GEOGRAPHY OF EUROPE

Landforms

Major Peninsulas:
Scandinavian, Jutland, Iberian, Italian, Balkan

Major Mountain Ranges:
Alps, Pyrenees, Carpathians, Apennines, Balkans

Major Rivers: Danube, Rhine, Seine, Loire, Elbe, Oder

Resources

• Oil from North Sea oil rigs is an important energy source for Europe.

• Coal and iron ore are found in abundance, making heavy industry possible.

Climate and Vegetation

• The North Atlantic Drift and the prevailing westerlies moderate much of Europe's climate.

• Lands bordering the Mediterranean Sea have a climate that encourages large-scale commercial agriculture.

Human-Environment Interaction

• Polders are an example of how Europeans have altered their environment.

• The canals of Venice demonstrate how Europeans have adapted to their environment.

• Deforestation of the land is a long-standing environmental problem in Europe.

Reviewing Places & Terms

A. Briefly explain the importance of each of the following.

1. fjord
2. uplands
3. *Meseta*
4. *Massif Central*
5. peat
6. mistral
7. polder
8. seaworks
9. terpen
10. Zuider Zee

B. Answer the questions about vocabulary in complete sentences.

11. What are fjords and where are they found?
12. Which of the above terms are examples of uplands?
13. What is France's highland area called?
14. How does the North Atlantic Drift influence climate?
15. In what part of Europe would you find the mistral?
16. How is peat used?
17. Which of the above terms is a type of seaworks?
18. How did the Zuider Zee become Ijsselmeer?
19. What are polders and where are they found?
20. Which of the above terms are associated with human-environment interaction?

Main Ideas

Landforms and Resources (pp. 273–277)

1. How do the mountain ranges of Europe impact the lives of the people who live near them?
2. Why are the rivers of Europe an important aspect of its geography?
3. Where are the most important oil fields of Europe located, and which countries pump oil from them?

Climate and Vegetation (pp. 278–281)

4. How do the prevailing westerlies affect the climate of Europe? Explain which part of Europe is most affected.
5. In which climate area of Europe would you find citrus fruits growing? Explain why.
6. What types of vegetation are found on the Scandinavian Peninsula?

Human-Environment Interaction (pp. 282–285)

7. Why did the Dutch build seaworks?
8. In what ways have the people of the Netherlands changed the physical geography of their land?
9. What kinds of pollutants are found in the Venice canals?
10. Why were forests chopped down in Europe?

Critical Thinking

1. Using Your Notes

Use your completed chart to answer these questions.

Landforms	
Resources	

a. Which of the human-environment interactions try to make the best use of landforms?

b. Which interactions focus on problems with resources?

2. Geographic Themes

a. **PLACE** In what ways has the physical geography of the Balkan Peninsula affected the people who live there?

b. **LOCATION** How would you describe Europe's location relative to bodies of water and to other regions?

3. Identifying Themes

Considering the climate and landforms, evaluate which areas of Europe would be the most agriculturally productive. Which of the five themes apply to this situation?

4. Identifying and Solving Problems

What factors must the people of Venice consider when dealing with the water pollution in their city?

5. Making Comparisons

How are the Scandinavian Peninsula and the Italian Peninsula alike and how are they different? Discuss landforms, resources, and climates.

Additional Test Practice, pp. S1–S37

TEST PRACTICE
CLASSZONE.COM

Geographic Skills: Interpreting Maps

Mountain Ranges of Europe

Use the map to answer the following questions.

1. **MOVEMENT** Which mountains hinder travel between Spain and France?

2. **REGION** Which mountain ranges are in Eastern Europe?

3. **LOCATION** What is the relative location of the Alps?

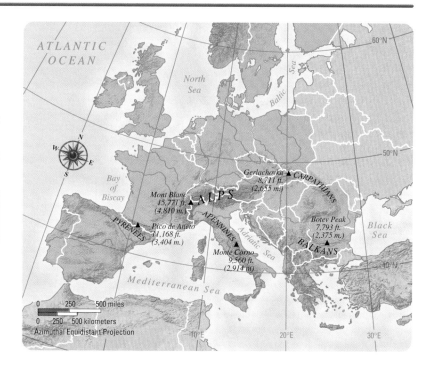

GeoActivity

Create your own sketch map of the physical geography of Europe. Combine the information from this map with the information from the rivers map on page 275 and the peninsulas map on page 273.

INTERNET ACTIVITY

Use the links at **classzone.com** to do research about acid rain in Europe. Focus on one aspect of acid rain, such as how the European Union is fighting acid rain or how European students learn about acid rain.

Writing About Geography Write a report of your findings. Include a map or a chart that visually presents information on acid rain. List the Web sites that you used in preparing your report.

Chapter

13

HUMAN GEOGRAPHY OF EUROPE
Diversity, Conflict, Union

Four Subregions of Europe

Legend:
- Mediterranean
- Western
- Northern
- Eastern

GeoFocus

How have cultural differences in Europe caused conflict?

Taking Notes In your notebook, copy a cluster diagram like the one shown below. As you read, take notes about the history, economics, culture, and modern life of each subregion of Europe.

Mediterranean Europe

Main Ideas

- The ancient Greek and Roman civilizations and the Renaissance all began in Mediterranean Europe.
- In the 20th century, the region has seen economic growth and political turmoil.

Places & Terms

city-state	**Renaissance**
republic	**aqueduct**
Crusades	

CONNECT TO THE ISSUES

UNIFICATION Membership in the European Union has helped the economies of the Mediterranean nations.

The Voyageur Experience in World Geography

Italy: Natural Hazards and Disasters

A HUMAN PERSPECTIVE Have you ever heard the saying, "All roads lead to Rome"? The Mediterranean region was home to the two great civilizations of ancient Europe—ancient Greece and ancient Rome. The city of Rome was founded in about 753 B.C., and Rome conquered a huge empire by about A.D. 100. To aid communication and make it possible for the army to march quickly to distant locations, Rome built a large network of well-paved roads. In ancient Europe, most roads did indeed lead to Rome, enabling that city to control a vast region.

A History of Ancient Glory

Two geographic advantages helped the Mediterranean to become the region where European civilization was born. First, the mild climate made survival there easier than in other areas. So societies had time to develop complex institutions such as government. Second, the nearby Mediterranean Sea encouraged overseas trade. When different societies trade with each other, they also exchange ideas. The spread of ideas often leads to advances in knowledge.

GREECE: BIRTHPLACE OF DEMOCRACY Beginning about 2000 B.C., people from the north moved onto the Balkan Peninsula. They built villages there. The region is mountainous, so those villages were isolated from each other and developed into separate city-states. A **city-state** is a political unit made up of a city and its surrounding lands.

Ancient Greece left a lasting legacy to modern civilization. The city-state of Athens developed the first democracy, a government in which the people rule. In Athens, all free adult males were citizens who had the right to serve in the law-making assembly. Athenian democracy helped inspire the U.S. system of government. And Greek science, philosophy, drama, and art helped shape modern culture.

In the 400s B.C., conflict weakened Greece. Several city-states fought a costly series of wars with Persia, an empire in southwest Asia. Then Athens fought a ruinous

PLACE In Athens, ancient ruins such as the Parthenon, shown here, stand near modern buildings.

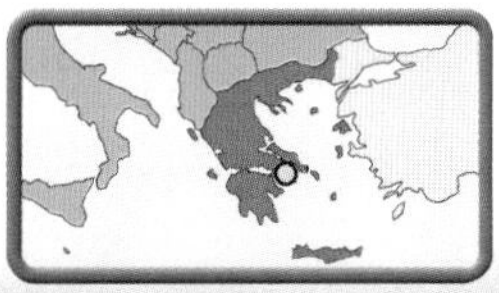

Cultural Legacy of the Roman Empire

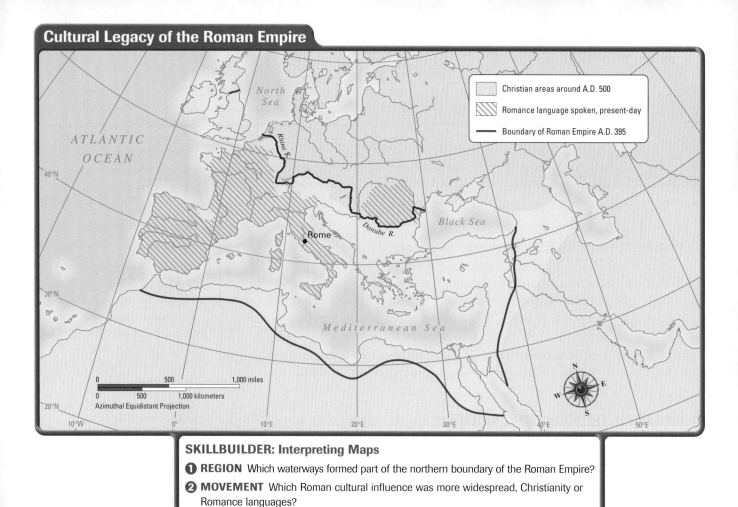

Christian areas around A.D. 500

Romance language spoken, present-day

Boundary of Roman Empire A.D. 395

SKILLBUILDER: Interpreting Maps

❶ **REGION** Which waterways formed part of the northern boundary of the Roman Empire?

❷ **MOVEMENT** Which Roman cultural influence was more widespread, Christianity or Romance languages?

war with Sparta, a rival Greek city-state. Finally, in 338 B.C., Macedonia (a kingdom to the north) conquered Greece. Beginning in 336 B.C., the Macedonian general Alexander the Great conquered Persia and part of India. His empire spread Greek culture but broke apart after his death.

THE ROMAN EMPIRE As Greece lost power, a state to the west was rising. That state, Rome, ruled most of the Italian Peninsula by 275 B.C. At the time, Rome was a **republic,** a government in which citizens elect representatives to rule in their name.

The Roman Empire grew by conquering territory overseas, including the Iberian and Balkan peninsulas. At home in Italy, unrest over inequalities led to decades of turmoil that caused Romans to seek strong leaders. Rome began to be ruled by an emperor, ending the republic.

One of Rome's overseas territories was Palestine, the place where Jesus was born. Christianity spread from there across the empire, and by the late 300s, Christianity was Rome's official religion.

By A.D. 395, the empire was too big for a single government, so it split into a western and an eastern half. The Western Roman Empire grew weak, in part because of German invaders from the north, and fell in A.D. 476. The Eastern Roman Empire lasted nearly 1,000 years longer.

BACKGROUND
The Roman republic was a model for modern governments such as those of France and the United States.

Moving Toward Modern Times

After 476, the three Mediterranean peninsulas had very different histories. The Balkan Peninsula stayed part of the Eastern Roman Empire (also called the Byzantine Empire) for nearly 1,000 years. Beginning in the 1300s, Italy saw the birth of the Renaissance, and in the 1400s, Portugal and Spain launched the Age of Exploration.

ITALIAN CITY-STATES The invaders who overran the Italian Peninsula had no tradition of strong central government. Italy eventually became divided into many small states and remained so for centuries.

In 1096, European Christians launched the **Crusades,** a series of wars to take Palestine from the Muslims. Italians earned large profits by supplying the ships that carried Crusaders to the Middle East. Italian cities such as Florence and Venice became rich from banking and foreign trade. This wealth helped them grow into powerful city-states.

The **Renaissance,** which began in the Italian city-states, was a time of renewed interest in learning and the arts that lasted from the 14th through 16th centuries. It was inspired by classical art and writings. Renaissance ideas spread north to the rest of Europe.

But the wealth of Italy did not protect it from disease. In 1347, the bubonic plague reached Italy from Asia and in time killed millions of Europeans. (See pages 294–295.)

SPAIN'S EMPIRE In the 700s, Muslims from North Africa conquered the Iberian Peninsula. Muslims controlled parts of the Iberian Peninsula for more than 700 years. Spain's Catholic rulers, Ferdinand and Isabella, retook Spain from the Muslims in 1492.

Also in 1492, Queen Isabella paid for Christopher Columbus's first voyage. Portugal had already sent out many voyages of exploration. Both Spain and Portugal established colonies in the Americas and elsewhere. Their empires spread Catholicism and the Spanish and Portuguese languages throughout the world.

A Rich Cultural Legacy

Mediterranean Europe's history shaped its culture by determining where languages are spoken and where religions are practiced today. And the people of the region take pride in the artistic legacy of the past.

ROME'S CULTURAL LEGACY Unlike many areas of Europe that Rome conquered, Greece retained its own language. Greek was, in fact, the official language of the Byzantine Empire. In contrast, Portuguese, Spanish, and Italian are Romance languages that evolved from Latin, the language of Rome.

The two halves of the Roman Empire also developed different forms of Christianity. The majority religion in Greece today is Eastern Orthodox Christianity. Roman Catholicism is strong in Italy, Spain, and Portugal. ◀

CENTURIES OF ART This region shows many signs of its past civilizations. Greece and Italy have ancient ruins, such as the Parthenon, that reveal what classical

REGION Italian Renaissance paint–ings often show the Virgin Mary and baby Jesus. Muslim art, like the Spanish wall design below (*bottom*), often uses calligraphy to praise God.

The Virgin and Child Surrounded by Five Angels, Sandro Botticelli

Alhambra Palace, Granada, Spain

Geographic Thinking

Making Comparisons
A▷ What is similar about the cultural legacies left by the Roman and Spanish empires?

architecture was like. Spain has Roman **aqueducts,** structures that carried water for long distances, and Muslim mosques, places of worship.

The region also has a long artistic legacy, which includes classical statues, Renaissance painting and sculpture, and modern art produced by such artists as Pablo Picasso of Spain. The pictures on page 291 contrast Renaissance Italian art with Muslim Spanish art.

Economic Change

Because of the Mediterranean region's sunny climate and historic sites, tourism has long been a large part of its economy. In other ways, the economy has been changing rapidly since World War II.

AGRICULTURE TO INDUSTRY In general, the Mediterranean nations are less industrial than those of Northern and Western Europe. For centuries, the region's economy was based on fishing and agriculture. Fishing remains important, and olives, grapes, citrus, and wheat are still major agricultural crops.

But in the late 20th century, the region's economy grew and changed. Today, manufacturing is increasing. The making of textiles is Portugal's biggest industry. Spain is a leading maker of automobiles, and Italy is a major producer of clothing and shoes. Service industries, such as banking, also make up a much larger part of the economy than before.

In the 1980s, Greece, Portugal, and Spain joined the European Union (EU). This aided growth by promoting trade with other EU nations and by making financial aid from the EU available.

BACKGROUND
The EU is an economic and political alliance of 15 nations. Italy was one of the founding members.

ECONOMIC PROBLEMS The region still faces economic challenges. For example, Italy's northern region is much more developed than its southern half. The reasons for this include the following:

- The north is closer to other industrial countries of Europe, such as Germany and France.
- The south has poorer transportation systems.
- The government tried to promote growth in the south but made bad choices. It started industries that did not benefit the local people.

Another problem is that the entire Mediterranean region is poor in energy resources and relies heavily on imported petroleum. This makes the region vulnerable because trade problems or wars could halt oil supplies and prevent industries from functioning.

Modern Mediterranean Life

Mediterranean Europe saw political turmoil in the 20th century. Two dictators, Benito Mussolini in Italy

Economic Activity*

Greece

1952: 31%, 35%, 10%, 24%
1995: 14%, 21%, 14%, 51%

Italy

1952: 24%, 25%, 12%, 39%
1995: 3%, 26%, 19%, 52%

Portugal

1952: 27%, 29%, 8%, 36%
1995: 4%, 33%, 16%, 47%

■ Trade   ■ Agriculture
■ Industry   ■ Other

SOURCE: *United Nations Statistical Yearbooks,* 1955, 1997

NOTES: *Industry* includes mining and construction; *other* includes government, utilities, transportation, and service industries.
* as percentage of GDP

SKILLBUILDER: Interpreting Graphs

❶ **SEEING PATTERNS** From 1952 to 1995, which economic activities increased and which decreased?

❷ **MAKING INFERENCES** Why do you think the category "other" changed so significantly? Give possible reasons.

and Francisco Franco in Spain, ruled for long periods. After Franco died in 1975, Spain set up a constitutional government. After World War II, Italy became a republic but has had dozens of governments since then. Greece has also experienced political instability.

THE BASQUES Spain has had an ongoing conflict with a minority group. The Basque people live in the western foothills of the Pyrenees. Their language is the only pre-Roman language still spoken in southwestern Europe. In the late 1970s, Spain granted the Basque region self-rule. But some Basques want complete independence and have used violence to fight for it. The conflict remains unresolved.

CITY GROWTH The transition from agriculture to manufacturing and service industries has encouraged people to move from the country to the city. Urban growth has created housing shortages, pollution, and traffic jams. The people of Mediterranean Europe want to preserve their historic cities, so they are trying to solve these problems. For example, Athens is expanding its subway system to reduce traffic and pollution.

Despite their problems, Mediterranean cities give intriguing insight into the past. In Rome and Athens, classical ruins stand near modern buildings. Florence has glorious works of Renaissance art. Granada, Spain, has Catholic cathedrals and a Muslim palace. In Section 2, you will read about Western Europe, a region that also has a rich history.

Geographic Thinking

Using the Atlas
B Locate the Basque language on the map on page 267. What other country besides Spain has Basque speakers?

PLACE Pamplona, Spain, holds a festival in which young men run through the streets before a herd of stampeding bulls. **What might this activity show about Spanish culture?**

EUROPE

Assessment

① Places & Terms

Identify these terms and explain their importance in the region's history or culture.

- city-state
- republic
- Crusades
- Renaissance
- aqueduct

② Taking Notes

REGION Review the notes you took for this section.

- What are the two ancient civilizations of this region?
- What type of movement is the result of recent economic change?

③ Main Ideas

a. How was the Renaissance an example of the movement of ideas?

b. What is Rome's cultural legacy in Mediterranean Europe today?

c. How has Mediterranean Europe's economy changed since World War II?

④ Geographic Thinking

Identifying and Solving Problems What might help preserve the historic cities of Mediterranean Europe?
Think about:

- how to provide housing and reduce both pollution and traffic

RESEARCH LINKS
CLASSZONE.COM

GeoActivity

ASKING GEOGRAPHIC QUESTIONS Review the paragraph about the Crusades on page 291. Write three to five geographic questions about the Crusades, such as "Why did many Crusaders purchase supplies for their ships in Italy?" Do research to answer as many of your questions as possible. Then create a set of **quiz show questions and answers.**

Disasters!

Bubonic Plague

By the 1300s, Italian merchants were growing rich from the trade in luxury goods from Asia. Then in October 1347, trading ships sailed into the port of Messina, Sicily, carrying a terrifying cargo—the disease we now call bubonic plague. Over the next four years, the plague spread along trade routes throughout Europe. An estimated 25 million Europeans died, about one-fourth to one-third of the population. In terms of its death toll, the plague (also called the Black Death) was the worst disaster Europe ever suffered.

Spread of the Bubonic Plague

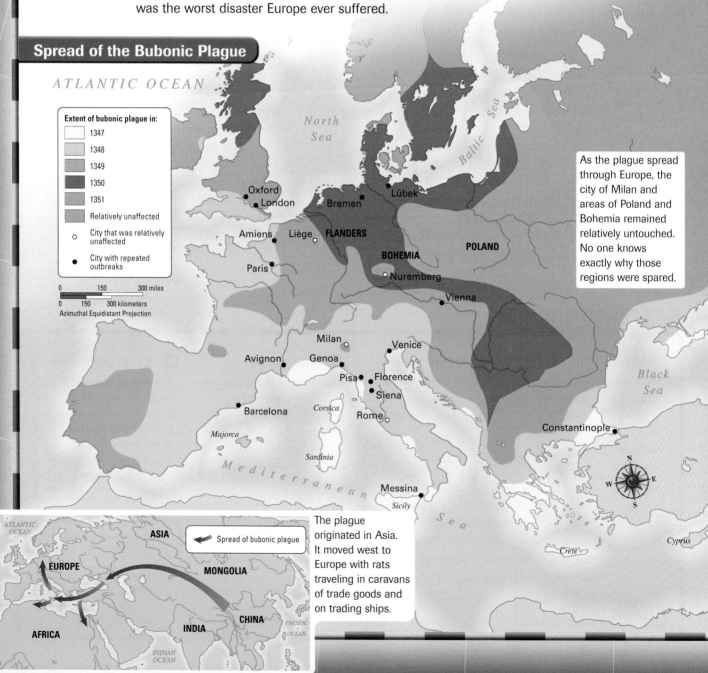

ATLANTIC OCEAN

Extent of bubonic plague in:
- 1347
- 1348
- 1349
- 1350
- 1351
- Relatively unaffected
- ○ City that was relatively unaffected
- ● City with repeated outbreaks

0   150   300 miles
0   150   300 kilometers
Azimuthal Equidistant Projection

North Sea

Baltic Sea

Oxford
London
Bremen
Lübek
Amiens   Liège   FLANDERS
Paris
BOHEMIA
POLAND
Nuremberg
Vienna

Milan
Venice
Avignon   Genoa
Pisa   Florence
Siena
Barcelona   *Corsica*
Rome
Majorca
Sardinia

Mediterranean

Messina
Sicily

Sea

Black Sea

Constantinople

As the plague spread through Europe, the city of Milan and areas of Poland and Bohemia remained relatively untouched. No one knows exactly why those regions were spared.

Crete

Cyprus

ATLANTIC OCEAN

ASIA   → Spread of bubonic plague

EUROPE   MONGOLIA

AFRICA   INDIA   CHINA   PACIFIC OCEAN

INDIAN OCEAN

The plague originated in Asia. It moved west to Europe with rats traveling in caravans of trade goods and on trading ships.

Transmission of the Plague

① The bacterium that causes bubonic plague, *Yersinia pestis,* lives in the guts of fleas. The fleas bite rats and feed on their blood, infecting them with the disease.

② Sometimes, an infected rat comes into contact with humans. Because the rat is dying, the fleas jump onto the humans to feed off them.

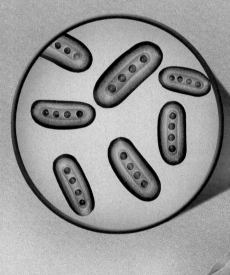

③ People catch bubonic plague from flea bites. In some, the plague enters their lungs, becoming pneumonic plague. These victims cough, sneeze, and spit up infected blood and saliva—spreading the disease more quickly.

GeoActivity

UNDERSTANDING EPIDEMICS

Working with a partner, use the Internet to research an epidemic on the time line below and create a **presentation** about it.

* Create a diagram showing the symptoms of the disease and the methods of treating it.
* Add a map of the region affected by this epidemic.
* Last, write a report explaining how the epidemic affected society.

RESEARCH LINKS
CLASSZONE.COM

GeoData

PREVENTIVE MEASURES

In the 1300s, most doctors recommended these methods of purifying the air to prevent plague:

* Burn richly scented incense.
* Fill the house with flowers.
* Sprinkle the floors with vinegar.
* Have doctors wear a bird mask with perfume in the beak.

OTHER DISASTROUS EPIDEMICS

1507–1518
Smallpox killed one-third to one-half of the people of Cuba, Haiti, and Puerto Rico.

1918–1919
About 30 million people died from an influenza outbreak that spread around the world.

2000
A UN report said that AIDS had killed 19 million people worldwide. Seven African countries had 20 percent of their population infected.

Western Europe

Main Ideas
- France and the Germanic countries developed very different cultures.
- These cultural differences led to conflicts that shaped the history of Western Europe.

Places & Terms

Benelux	**nationalism**
Reformation	**Holocaust**
feudalism	**Berlin Wall**

CONNECT TO THE ISSUES
UNIFICATION France and Germany have resolved their past conflicts and now cooperate in the European Union.

A HUMAN PERSPECTIVE Today, the French call Émile Durkheim the father of French sociology (the study of society). But he wasn't always honored. During World War I, some French patriots considered him a disloyal foreigner. Why? Perhaps it was because he had a German last name and came from Lorraine, a region that had switched between French and German rule many times. France and Germany have long had a deep rivalry, based in part on cultural differences.

A History of Cultural Divisions

France and Germany are the dominant countries in Western Europe. They are the two largest countries, and their access to resources, ports, and trade routes helped them to build productive economies.

French culture is strong in France and Monaco; German culture is strong in Germany, Austria, and Liechtenstein. Switzerland and the **Benelux** countries of Belgium, the Netherlands, and Luxembourg have their own cultures—but also have been influenced by Germany and France. Western Europe's cultural divisions have historic roots.

ROME TO CHARLEMAGNE One cultural division, language, dates from ancient times. By 50 B.C., the Roman Empire had conquered the Celtic tribes in what is now France. French is one of the Romance languages that evolved from Latin (Rome's language). But Rome never fully conquered the Germanic tribes that migrated into the lands east of France. Germanic languages are still spoken there. (See the chart on page 297.)

Western European History, 800–2000

800
Charlemagne unites much of Europe.

1347
Bubonic plague starts to sweep through Europe.

MIDDLE AGES

1099
Crusaders from Europe capture Jerusalem from Muslims.

1455
First printing of Gutenberg Bible

1516
Leonardo da Vinci moves to France, bringing Renaissance ideas.

RENAISSANCE AND REFORMATION

1517
Martin Luther criticizes the Catholic Church. The Reformation begins.

of the Nordic countries, which provide many welfare services for their citizens. For example, Finland, Norway, and Sweden give families a yearly allowance to help raise their children. The Nordic governments help fund national health insurance programs. Britain also has a national health insurance program. To pay for the programs, the people in those countries have very high taxes.

DISTINCTIVE CUSTOMS Some social customs of Northern Europe have gained worldwide fame. For example, the British are known for afternoon tea, a small meal of sandwiches, breads, cakes, and tea. Swedes developed the smorgasbord. It is a large assortment of hot and cold dishes served buffet style. Finns are famous for their sauna, in which people sit in a hot room to work up a sweat that cleans the skin's pores. Afterward, they plunge into a cold bath or icy lake.

LEISURE Even though the Nordic countries have some of the coldest climates in Europe, outdoor sports remain popular there. Some of the sports in the winter Olympics developed in Norway and the other Nordic countries. They include cross-country skiing and ski jumping.

Many British enjoy horseback riding, horse jumping, and fox hunting. These traditionally were pastimes for the wealthy upper classes on their large country estates. In addition, the British developed two sports that are unique. Rugby is a form of football, and cricket is played with a ball, a bat, and wickets. Spread by British colonialism, cricket is played around the world.

In Section 4, you will read about Eastern Europe, a region that continues to be torn apart by ethnic conflicts.

BACKGROUND
Because of Sweden's closeness to the sea, smorgasbords feature a variety of seafood such as salmon and herring.

5 THEMES

MOVEMENT

Tea Time

Nothing seems more English than tea, but it is really an import from Asia. Dutch traders introduced tea to Europe, and it was sold for the first time in England in 1657. Tea soon became Britain's national beverage.

Perhaps one reason for its popularity is that clean water was scarce; boiling water for tea purified it. Tea also had caffeine, giving tea drinkers energy during the long stretch between the midday meal and supper. The custom of taking food with afternoon tea began in the 1800s.

SECTION 3 Assessment

❶ Places & Terms

Identify these terms and explain their importance in the region's history, culture, or economy.

- Nordic countries
- parliament
- Silicon Glen
- euro

❷ Taking Notes

REGION Review the notes you took for this section.

- Where did the Industrial Revolution begin and to where did it spread?
- What are some characteristics of governments in Northern Europe?

❸ Main Ideas

a. How did conquest influence the languages spoken in Northern Europe?

b. How did the Industrial Revolution spur the growth of Britain's empire?

c. How did the Reformation affect Northern Europe?

❹ Geographic Thinking

Determining Cause and Effect Why is there conflict in Northern Ireland? **Think about:**

- the history of Britain's relationship with Ireland
- religious differences
- arguments for and against a union of the Republic of Ireland and Northern Ireland

---

SEEING PATTERNS Compare the map on page 304 with the world map on pages A4 and A5 to learn the present names of former British colonies. Then do research to learn which former colonies still use English as an official language. Present this information on a **chart**.

Comparing Cultures

Geographic Sports Challenges

Over time, humans have found ways to enjoy even the most forbidding climates and terrains. Some popular sports evolved from activities that people used to overcome geographic challenges, such as mountains or snowy climates. Other sports were created to take advantage of special geographic features, such as recurring winds or waves. On these two pages, you will learn about geographically inspired sports from around the world.

Norway

United States

Mexico

Australia

Surfing, shown here off the coast of Australia, dates back to prehistoric times. It may have originated when Polynesian sailors of the Pacific Islands needed to reach land from large canoes floating offshore.

Skiing originated as a means of travel in northern Europe, and ski jumping probably evolved in hilly Norway. In 1924, ski jumping became an Olympic sport. Competitors are judged not only on how far they jump but also on the technique they use.

Acapulco, Mexico, is famous for its cliff diving. This dangerous sport often involves diving from heights nearly three times higher than those used in Olympic platform diving. Cliff divers have been killed by hitting their heads on rocks.

The Iditarod Sled Dog Race is held in Alaska. Susan Butcher, shown here, was the first person to have won it three years in a row. The Inuit people first used sled dogs to travel across snow-covered terrain; racing evolved later.

GeoActivity

EXPLORING MOUNTAIN CLIMBING

Working with a small group, use the Internet to research mountain climbing, another geographic sports challenge. Then create a **presentation** about the sport.

• Draw a world map, label popular mountains to climb, and give their altitudes.

• Make a chart listing the dangers of mountain climbing.

RESEARCH LINKS
CLASSZONE.COM

GeoData

Skiing
• Skis that are more than 4,000 years old have been found in Scandinavian bogs.
• Skiing was once a military skill. Norwegian troops skied in the Battle of Oslo in 1200.

Surfing
• The explorer James Cook first reported seeing surfing in 1778.
• European missionaries banned surfing in 1821. It was revived in 1920 by a Hawaiian, Duke Kahanamoku.

Cliff Diving
• Women did not compete at Acapulco until 1996.
• Divers enter the water at speeds of up to 65 mph.

Sled Dog Racing
• The Iditarod honors a 1925 emergency mission to deliver medicine to Nome, Alaska.
• During the 1985 race, a moose charged across Susan Butcher's path. The collision that resulted killed 2 dogs and wounded 13 other dogs.

Eastern Europe

Main Ideas
- Eastern Europe has great cultural diversity because many ethnic groups have settled there.
- Many empires have controlled parts of the region, leaving it with little experience of self-rule.

Places & Terms

cultural crossroads

balkanization

satellite nation

market economy

folk art

anti-Semitism

CONNECT TO THE ISSUES
CONFLICT Nationalism and ethnic differences have fueled conflicts that have torn apart the Balkans in recent times.

A HUMAN PERSPECTIVE Eastern Europe has many plains that allow invaders to move from east to west and vice versa. In World War II, Germany invaded the Communist Soviet Union, killing millions. After the war, the Soviet Union decided to protect itself from invasion by setting up a political barrier. So it established Communist governments in the nations of Eastern Europe, which lay between the Soviet Union and its enemies to the west. Soviet dictator Joseph Stalin wanted Eastern Europe to "have governments whose relations to the Soviet Union are loyal." For decades, the Soviet Union crushed political reform and free trade in Eastern Europe. The region is still recovering.

History of a Cultural Crossroads

Eastern Europe's location between Asia and the rest of Europe shaped its history. Many groups migrated into the region, creating great diversity. Strong empires ruled parts of Eastern Europe, delaying the rise of independent nation-states there. Today the region includes Albania, Bosnia and Herzegovina, Bulgaria, Croatia, the Czech Republic, Hungary, Macedonia, Poland, Romania, Slovakia, Slovenia, and Yugoslavia.

CULTURES MEET Eastern Europe is a **cultural crossroads**, or a place where various cultures cross paths. Since ancient times, people moving between Europe and Asia—traders, nomads, migrants, and armies—have passed through this region. Because the region is an important crossroads, many world powers have tried to control it.

Eastern European History, 1389–2000

1389	1566	1686	1867
The Ottoman Empire defeats the Serbs at the Battle of Kosovo.	**Suleiman I**, the Ottoman ruler, dies during a siege in Hungary.	The Austrians drive the Ottomans out of Hungary.	Hungary demands equal status with Austria. The empire becomes Austria-Hungary.

CONFLICT AMONG EMPIRES

1618	1795
Bohemia (now the Czech Republic) revolts against its Austrian ruler, starting the Thirty Years' War.	The Russian ruler **Catherine the Great** divides Poland among Russia, Prussia, and Austria.

BACKGROUND
Romania means "land of the Romans."

EMPIRES AND KINGDOMS By about A.D. 100, ancient Rome held the Balkan Peninsula, Bulgaria, Romania, and parts of Hungary. After the Roman Empire was split, the Byzantine Empire held onto those lands for centuries. In the 1300s and 1400s, the Ottoman Empire of Turkey (see Unit 7) gradually took over the southern part of Eastern Europe.

Various Slavic groups moved into Eastern Europe from the 400s through the 600s. Several kingdoms, such as Poland in the north and Serbia on the Balkan Peninsula, formed. In the late 800s, a non-Slavic group called the Magyars swept into what is now Hungary and in time established a kingdom. The Ottomans later conquered it.

Beginning in the 1400s, the nation of Austria became a great power. Austria drove the Ottomans out of Hungary and took control of that state. In the late 1700s, Austria, Prussia (a German state), and Russia divided up Poland among themselves. Poland ceased to exist.

Turmoil in the 20th Century

CONNECT TO THE ISSUES
CONFLICT
A What might happen if two different ethnic groups wanted to establish a nation on the same land?

Responding to centuries of foreign rule, most ethnic groups in Eastern Europe fiercely guarded their identities. Many wanted their own nation-states, even though few had a history of self-rule. These characteristics sparked many conflicts in Eastern Europe during the 20th century. **A**

WAR AFTER WAR By 1908, the Balkan nations of Bulgaria, Greece, Montenegro, Romania, and Serbia had broken free from the Ottoman Empire. In 1912, Greece, Bulgaria, and Serbia went to war against the Ottomans, who lost most of their remaining European territory. In 1913, the Balkan countries fought over who should own that territory. Their actions led to a new word, **balkanization.** The term refers to the process of a region breaking up into small, mutually hostile units.

The Slavic nation of Serbia also wanted to free the Slavs in Austria-Hungary. In 1914, a Serb assassinated an Austrian noble, sparking World War I. Austria-Hungary and Serbia each pulled their allies into the conflict until most of Europe was involved. After the war, Austria and Hungary split apart. Albania, Bulgaria, Czechoslovakia, Poland, and Yugoslavia gained independence. The Ottoman Empire ended and was replaced by the nation of Turkey.

EUROPE

1914
A Serb kills Austrian **Archduke Francis Ferdinand.** World War I erupts.

1939
Germany overruns Poland. World War II starts.

TWO WORLD WARS

1918
The Kingdom of Serbs, Croats, and Slovenes (now Yugoslavia) is created.

1946–1948
Communist governments are set up in Eastern Europe.

1989
Czechoslovakia, Hungary, Romania, and Poland end Communist rule.

COMMUNISM AND DEMOCRACY

1945
Josip Broz Tito becomes dictator of Yugoslavia.

2000
Yugoslavia elects a reform leader, Vojislav Kostunica, as president.

311

In 1939, Germany seized Poland, starting World War II. Near the end of that war, the Soviet Union advanced through Eastern Europe as part of an Allied strategy to crush Germany from two sides. The Soviet Union later refused to withdraw from Eastern Europe until it had set up Communist governments there. Eastern Europe became a region of **satellite nations**—nations dominated by another country.

RECENT CHANGES The Soviet Union controlled Eastern Europe for four decades. But by the late 1980s, the Soviet Union had severe economic problems, and a new leader, Mikhail Gorbachev, was making reforms. As one reform, he gave Eastern Europe more freedom.

The impact was dramatic. Eastern Europeans demanded political and economic reforms. In 1989, Czechoslovakia, Hungary, Poland, and Romania ended Communist control of their governments and held free elections. In 1990, Bulgaria and Yugoslavia followed suit.

Instability followed. The old governments had taught people to be loyal only to the Communist Party. After those governments fell, people

MOVEMENT In 1989, the desire for democracy swept Eastern Europe. Country after country saw demonstrations like this one in Budapest, Hungary.

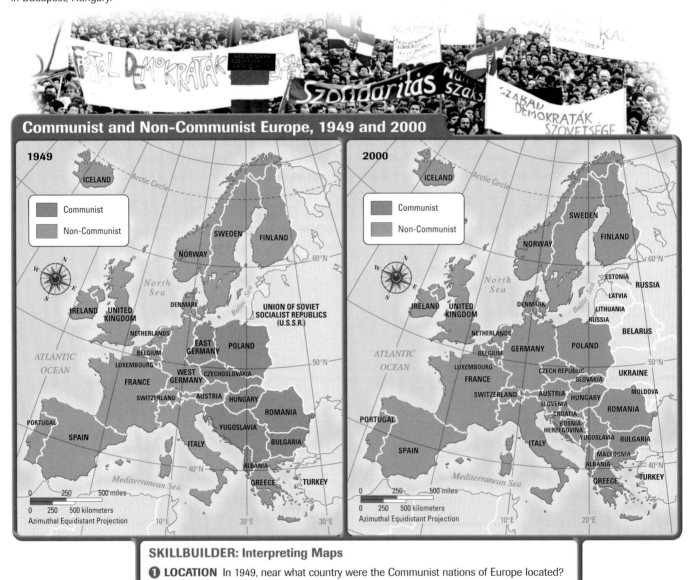

Communist and Non-Communist Europe, 1949 and 2000

SKILLBUILDER: Interpreting Maps

❶ **LOCATION** In 1949, near what country were the Communist nations of Europe located?

❷ **REGION** How would you describe the political change that happened in the region?

returned to ethnic loyalties. That was especially true in Yugoslavia, a nation consisting of six republics. In the early 1990s, four of the six Yugoslav republics voted to become separate states. Serbia objected, leading to civil war. (See Chapter 14 for details.) In contrast, Czechoslovakia peacefully split into the Czech Republic and Slovakia.

Developing the Economy

Because of its fertile plains, Eastern Europe has traditionally been a farming region. After 1948, the Soviet Union promoted industry there.

INDUSTRY Under communism, the government owned all factories and told them what to produce. This system was inefficient because industries had little motive to please customers or to cut costs. Often, there were shortages of goods. Eastern European nations traded with the Soviet Union and each other, so they didn't keep up with the technology of other nations. As a result, they had difficulty selling goods to nations outside Eastern Europe. And their outdated factories created heavy pollution.

After 1989, most of Eastern Europe began to move toward a **market economy**, in which industries make the goods consumers want to buy. Many factories in Eastern Europe became privately owned instead of state owned. The changes caused problems, such as inflation, the closing of factories, and unemployment. Since then, however, many factories have cut their costs and improved production. As a result, the Czech Republic, Hungary, and Poland have all grown economically. ◀B

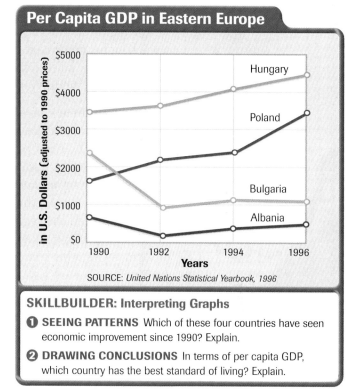

Per Capita GDP in Eastern Europe

SOURCE: *United Nations Statistical Yearbook, 1996*

SKILLBUILDER: Interpreting Graphs

❶ **SEEING PATTERNS** Which of these four countries have seen economic improvement since 1990? Explain.

❷ **DRAWING CONCLUSIONS** In terms of per capita GDP, which country has the best standard of living? Explain.

CONNECT TO
THE ISSUES
UNIFICATION
B Do you think the nations of Eastern Europe will want to join the European Union? Why or why not?

LINGERING PROBLEMS Some Eastern European nations have had trouble making economic progress—for many different reasons.

- Albania's economic growth is slowed by old equipment, a lack of raw materials, and a shortage of educated workers.
- Few of Romania's citizens have money to invest in business. In addition, the Romanian government still owns some industries. Foreigners don't want to invest their money in those industries.
- The civil wars of the 1990s damaged Yugoslavia and its former republics of Bosnia and Herzegovina and Croatia. Equipment and buildings were destroyed; workers were killed or left the country.

In general, it will take years for Eastern Europe to overcome the damage caused, in part, by decades of Communist control.

A Patchwork Culture

Because Eastern Europe contains a variety of ethnic groups, the region as a whole is a patchwork of different languages and religions.

CULTURAL DIVERSITY The map on page 267 shows the languages of Eastern Europe. The number of languages makes it difficult to unify the region. In some places, the national language is most closely related to a language spoken in a different region. For example, Hungarian is related to Finnish, and Romanian is related to Italian, French, and Spanish. Neither are related to the Slavic languages of the countries around them. This pattern was created by long-ago migrations.

Similarly, many different religions can be found in Eastern Europe. The Roman Empire introduced Catholicism, and after Rome fell, the Byzantine Empire spread Eastern Orthodox Christianity. Some countries also have a Protestant minority. And under the Ottoman Empire, some Eastern Europeans converted to Islam.

The region also has a small Jewish minority. Jews once made up a much higher percentage of Eastern Europeans, but in the Holocaust, Nazi Germany killed 6 million Jews. About half of them were from Poland. After World War II, many surviving Jews migrated to Israel.

FOLK ART Religious belief, rural customs, and Byzantine art have all influenced Eastern European folk art. In general, **folk art** is produced by rural people with traditional lifestyles instead of by professional artists. Eastern European folk artists create items such as pottery, woodcarving, and embroidered traditional costumes.

Many Eastern European ethnic groups also have their own folk music. This music influenced the region's classical musicians. Frédéric Chopin based some of his piano music on Polish dances. Anton Dvořák wove Czech folk music into his compositions.

Geographic Thinking

Seeing Patterns
Why do you think folk art has remained important in Eastern Europe?

Moving Toward Modern Life

Since their Communist governments fell, many Eastern Europeans have expressed a longing for more economic growth and political freedom. These goals provide the region with some major challenges.

LESS URBAN DEVELOPMENT Eastern Europe has several large cities, such as Prague in the Czech Republic. More than 1,000 years old, Prague is one of Europe's most interesting cities, with quaint buildings, a rich history of music and culture, and thriving industries.

In general, though, Eastern Europe is much less urban than the rest of Europe. For example, the percentage of city dwellers is only 40 percent in Bosnia and Herzegovina and only 37 percent in Albania.

As Eastern Europe develops more industry, its cities will grow. That will have both positive and negative effects. Cities are often places of culture, learning, and modern technology. But urban growth creates problems such as pollution, traffic jams, and housing shortages.

CONFLICT As you read earlier, many Eastern Europeans have fierce loyalties to their own ethnic groups. One result of that has been conflict. For example, many Serbs hate Croats (KROH•ATS) because they believe the Croats betrayed them in World War II by working with the Nazis.

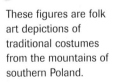

These figures are folk art depictions of traditional costumes from the mountains of southern Poland.

PLACE Crossing the Vltava River in Prague is the famous 650-year-old Charles Bridge. The bridge is now reserved for pedestrians. **Why do you think cars are banned from this bridge?**

Eastern European minority groups have often faced discrimination. Throughout history, Jews have suffered from **anti-Semitism**, which is discrimination against Jewish people. Another minority that experiences prejudice is the Romany, or Gypsy, people who are scattered across Eastern Europe. Traditionally, the Romany have moved from place to place. Because of this, other groups often look down on them.

DEMOCRACY To obtain true democracy, Eastern Europeans need to overcome old hatreds and work together. They also need to accept democratic ideals such as the rule of law—which means that government officials must obey the law. The dictators that ruled Eastern Europe in the past did not do so. But in recent years, Eastern Europeans have often held their leaders accountable. For example, in 2000, the Yugoslav people forced a dictator to accept election results that turned him out of office. You will read about this event in Chapter 14, along with other major issues of European life today.

 Assessment

1 **Places & Terms**

Identify these terms and explain their importance in the region.

- cultural crossroads
- balkanization
- satellite nation
- market economy
- folk art
- anti-Semitism

2 **Taking Notes**

REGION Review the notes you took for this section.

- What country dominated Eastern Europe after World War II?
- What problems did the move toward a market economy cause?

3 **Main Ideas**

a. Why is Eastern Europe considered a cultural crossroads?

b. What role did the Soviet Union play in the rise and fall of communism in Eastern Europe?

c. What are some important ways that Eastern Europe is different from Western Europe?

4 **Geographic Thinking**

Making Inferences The Balkan region has been called the "powder keg of Europe." Why do you think it earned that name? **Think about:**

- the wars in 1912 and 1913
- World War I

S **See Skillbuilder Handbook, page R4.**

EXPLORING LOCAL GEOGRAPHY Like Eastern Europe, most places in the United States have been controlled by various cultural groups or nations over time. Research the history of your area and create a **time line,** like the one on pages 310-311, listing changes in control.

VISUAL SUMMARY
HUMAN GEOGRAPHY OF EUROPE

Subregions of Europe

○ **Mediterranean Europe**

- The influence of ancient Greece, ancient Rome, and Renaissance Italy on art, philosophy, religion, and language shaped modern life.
- In the late 1900s, Mediterranean Europe began to have more manufacturing and service industries.

○ **Western Europe**

- Germany and France developed very different cultures, and throughout history, conflicts between them involved much of Europe.
- Western Europe has a highly developed economy. It is a leader in the economic and political alliance known as the European Union.

● **Northern Europe**

- This region was a leader in the Industrial Revolution and the rise of representative government.
- The region has a history of seafaring conquerors. Great Britain established an empire that spread British culture and the English language worldwide.

● **Eastern Europe**

- Because it is a cultural crossroads, Eastern Europe has a diverse culture with many ethnic groups.
- Domination by outside powers, most recently the Soviet Union, has characterized the region's history.

Reviewing Places & Terms

A. Briefly explain the importance of each of the following.

1. city-state
2. republic
3. Benelux
4. nationalism
5. Berlin Wall
6. Nordic countries
7. euro
8. cultural crossroads
9. balkanization
10. satellite nation

B. Answer the questions about vocabulary in complete sentences.

11. Which of the terms above are the names of regions?
12. Would a supporter of nationalism want to adopt the euro? Explain.
13. Which of the terms above have to do with conflict?
14. In which part of Europe did the countries become satellite nations of the Soviet Union?
15. How does the geographic theme of movement relate to a cultural crossroads?
16. Which ancient civilization was organized into city-states and which was a republic?
17. In what part of Europe is Benelux found?
18. What is the origin of the term *balkanization*?
19. Which of the terms above can also be applied to the United States? Explain.
20. Which two major peninsulas are found in the Nordic countries?

Main Ideas

Mediterranean Europe (pp. 289–295)

1. What legacy did ancient Athens leave for modern governments?
2. What effect did the empires of Spain and Portugal have on the rest of the world?
3. Why does Spain have a conflict with the Basque people?

Western Europe (pp. 296–301)

4. How did the Reformation create new cultural divisions?
5. How did nationalism lead to conflicts?
6. For what artistic legacy are Germany and Austria famous?

Northern Europe (pp. 302–309)

7. Who were the Vikings, and what did they do?
8. What geographic advantages helped Great Britain build its empire?

Eastern Europe (pp. 310–315)

9. Why did independent nation-states develop later in Eastern Europe than in Western Europe?
10. What problems existed in the Eastern European economy under Communist rule?

Critical Thinking

1. Using Your Notes

Use your completed chart to answer these questions.

a. What similarities exist between the ways the Roman Empire and the British Empire influenced other regions of the world?

b. In what ways are Eastern Europe and Northern Europe different?

2. Geographic Themes

a. **LOCATION** Do you think the location of France and Germany relative to the rest of Europe is a geographic advantage or disadvantage? Explain.

b. **MOVEMENT** What geographic reason might account for the fact that Spain and Great Britain colonized much of the Americas?

3. Identifying Themes

Explain which countries were the first to develop industry and which developed industry later. If you identify those countries on a map, what spatial patterns do you see? Which geographic themes relate to your answer?

4. Seeing Patterns

How did ancient migrations affect the pattern of where certain languages are spoken in Europe today? Give examples.

5. Making Inferences

Millions of Europeans have migrated to other parts of the world. What are some geographic factors that you think might have encouraged this?

Additional Test Practice, pp. S1–S37

TEST PRACTICE CLASSZONE.COM

Geographic Skills: Interpreting Maps

A Divided Germany

Use the map to answer the following questions.

1. **PLACE** How did the size of West Germany compare with that of East Germany?

2. **LOCATION** In which of the two countries was the city of Berlin located?

3. **LOCATION** Which of the two Germanys was closer to the Soviet Union?

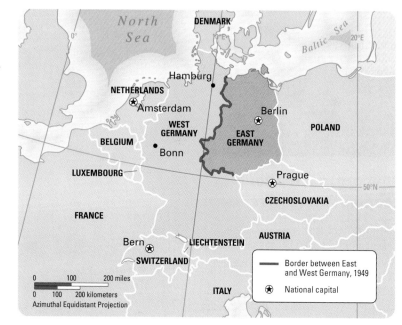

GeoActivity

West Germany was divided into several zones after World War II. Use a history book or historical atlas to learn which three countries controlled those zones. Create a historical map showing the zones.

INTERNET ACTIVITY

Use the links at **classzone.com** to do research about the population of a single society in Europe. Look for such information as age distribution, religions, ethnic or minority groups, and literacy rates.

Constructing a Population Pyramid Use the information you have gathered to construct a population pyramid describing the population characteristics of the European society you have chosen.

TODAY'S ISSUES
Europe

SECTION 1
Turmoil in the Balkans

SECTION 2
Cleaning Up Europe

CASE STUDY
THE EUROPEAN UNION

For more on these issues in Europe . . .

 CURRENT EVENTS
CLASSZONE.COM

Throughout the 1990s, ethnic conflict tore apart the Balkan region.

GeoFocus

How can international cooperation resolve issues?

Taking Notes In your notebook, copy a cause-and-effect chart like the one shown below. Then take notes on the causes and effects of the issues.

	Causes	Effects
Issue 1: Conflict		
Issue 2: Pollution		
Case Study: Unification		

Cleaning Up Europe

How can Europeans clean up their environment?

Main Ideas

• Pollution has many complex causes and results. It often spreads across borders, contaminating a region.

• The nations of Europe are cooperating to try to clean up their environment.

Places & Terms

cyanide

European Environmental Agency

particulates

smog

ozone

The Voyageur Experience in World Geography
Greece: Urbanization and the Environment

EUROPE

A HUMAN PERSPECTIVE In January 2000, a gold mine in Romania released cyanide into local streams. The **cyanide,** a deadly poison, flowed into the Tisza River in Hungary. Before the accident, the river held some of Europe's rarest fish. The poison killed an estimated 80 percent of the fish in the Tisza. Balazs Meszaros, whose family has commercially fished the Tisza for generations, said, "Now I don't know how I am going to live." Even worse than the loss of jobs was the threat to health. Experts feared that the poison would seep into wells and contaminate crops and livestock. The damage will take years to undo.

Pollution is a complex example of human-environment interaction. People damage the environment, which in turn affects human lives. For instance, pollution is thought to cause 1 out of every 17 deaths in Hungary. Because cleaning up pollution is time-consuming, difficult, and costly, it remains a serious issue in Europe—and around the world.

Saving Europe's Water

As the story of the Tisza demonstrates, pollution rarely remains at its point of origin but often spreads to neighboring regions. As a result, water pollution is a problem that concerns almost all of Europe.

CAUSES OF WATER POLLUTION Mines and factories create much of Europe's water pollution. Industries often discharge chemicals into streams and rivers. Factories sometimes bury solid waste. Poisons from this waste seep into ground water and contaminate wells and rivers. And, as you read in Chapter 12, the burning of coal and other fuels causes acid rain. Acid rain changes the chemistry of lakes and rivers, often killing fish.

The link between industry and pollution creates a dilemma. Most countries want to develop industry, and some accept environmental damage as the price they must pay for progress. Other nations force industry to use pollution controls, but these are usually expensive.

HUMAN-ENVIRONMENT INTERACTION A cyanide spill poisoned Eastern Europe's streams and rivers. These dead fish are from the Tisza River in Hungary.

323

Industry is not the only source of water pollution. Other sources include the following:

- **Sewage** Ideally, cities should have treatment plants that remove harmful substances from sewage before it is released into bodies of water. But in Poland, for example, from 1988 to 1990, 44 percent of the cities had no sewage treatment plants. The water in most of Poland's rivers is unsafe to drink. It has also contaminated the soil so that some crops are toxic.

- **Chemical fertilizers** Rain washes fertilizers from fields into bodies of water, where they cause algae and plants to grow faster than fish can eat them. The plants and algae die and decay, a process that uses up oxygen. The lack of oxygen kills fish—which then decay, using more oxygen. In time, these bodies of water can no longer support life.

- **Oil spills** For example, in December 1999, a tanker sank off the west coast of France and spilled 10,000 tons of oil that spread along 250 miles of coastline. The oil killed tens of thousands of shorebirds.

CLEANING UP THE WATER Because water pollution spreads so easily, nations must cooperate to solve the problem. For example, pollution levels in the Rhine River rose sharply in the mid-1900s. To correct this, representatives from France, Germany, Luxembourg, the Netherlands, and Switzerland formed the International Commission for the Protection of the Rhine. Since it began meeting in 1950, the commission has recommended programs such as the treatment of sewage before it enters the Rhine. As a result, pollution of the Rhine has decreased.

In addition, the European Union has passed environmental laws that its member nations must obey. The EU also set up the **European Environmental Agency,** which provides the EU with reliable information about the environment.

HUMAN–ENVIRONMENT INTERACTION This mill in Nowa Huta, Poland, is making coke—a byproduct of coal. The smokestacks cause heavy air pollution.
What else besides the smokestacks might be causing air pollution?

Improving Europe's Air Quality

Although they are often considered separately, the different types of pollution are connected. For example, water pollution can be caused by air pollution—because rain washes chemicals out of dirty air and into bodies of water.

CAUSES OF AIR POLLUTION Air pollution is made up of harmful gases and **particulates,** very small particles of liquid or solid matter. Many human activities create air pollution by expelling these gases and particulates into the atmosphere.

- **Using fossil fuels** The burning of petroleum, gas, and coal causes much air pollution. It contributes

324

BACKGROUND
The word *smog* was formed by combining the words *smoke* and *fog*.

to the formation of **smog**—a brown haze that occurs when the gases released by burning fossil fuels react with sunlight to create hundreds of harmful chemicals. One such chemical is **ozone,** a form of oxygen that causes health problems.

- **Fires** Forest fires caused by careless human behavior and the burning of garbage release smoke and particulates into the atmosphere.

- **Chemical use** Dry cleaning, refrigeration, air conditioning, and the spraying of pesticides are among the human activities that release harmful chemicals into the air.

- **Industry** Factories discharge chemicals such as sulfur into the air. The factories of former Communist countries have been especially heavy polluters. Because of this, air pollution levels are much higher in the former East Germany than in the United States.

Geographic Thinking

Seeing Patterns You learned in Chapter 13 that Eastern Europe used old technology. How might this relate to pollution?

RESULTING PROBLEMS Breathing polluted air can contribute to respiratory diseases such as asthma, bronchitis, and emphysema. Air pollution is also suspected to be one of the causes of lung cancer. In addition, air pollution harms livestock and stunts plant growth. It also causes acid rain, which kills forests and damages buildings, such as the famous Parthenon in Athens, Greece.

CLEANING UP THE AIR Individual European countries are passing laws to make their air safer to breathe. France, for example, now requires improved thermal insulation of new buildings. This reduces the need to burn fossil fuels for heat. Other European governments are also passing laws to protect the air.

Nations are also cooperating to clean the air. For example, in 1998, the members of the European Union agreed that, starting in 2000, they would require reduced emissions from cars and vans. As that example indicates, a leader in the effort to restore Europe's environment will be the European Union—which is discussed in the following Case Study.

EUROPE

SECTION 2 Assessment

1 Places & Terms

Identify these terms and explain their relationship to the issue.

- cyanide
- European Environmental Agency
- particulates
- smog
- ozone

2 Taking Notes

HUMAN-ENVIRONMENT INTERACTION Review the notes you took for this section.

	Causes	Effects
Issue 2: Pollution		

- What river has an international group been trying to save?
- What diseases are linked to air pollution?

3 Main Ideas

a. What dilemma is faced by countries that are developing industry?

b. What is a harmful result of burning fossil fuels?

c. Why is the European Union a leader in the fight against pollution?

4 Geographic Thinking

Seeing Patterns How are the different types of pollution interrelated?

Think about:

- how air pollution, water pollution, and buried waste cause other types of pollution

 See Skillbuilder Handbook, page R8.

GeoActivity

EXPLORING LOCAL GEOGRAPHY Find out how your community deals with pollution. Learn about laws passed by your local government, environmental safeguards used by industry, or water treatment facilities. Then write a **news article** on the subject.

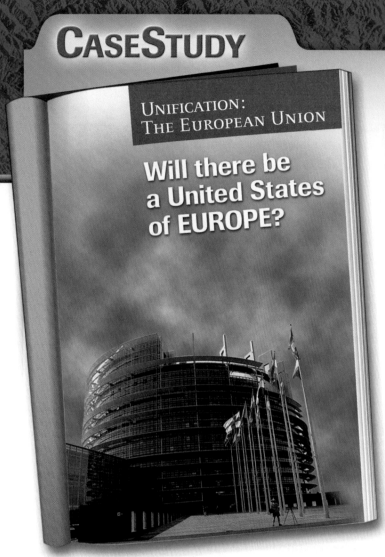

UNIFICATION: THE EUROPEAN UNION

Will there be a United States of EUROPE?

EU headquarters in Brussels, Belgium

Europe's long history of conflict reached a crisis in World War II (1939-1945). In the wake of that destructive war, two goals emerged: to rebuild the nations' shattered economies and to prevent new conflict. Some people believed the best way to achieve both goals was to unify Europe. As you read the Case Study, consider the pros and cons of that idea.

Steps Toward Unity

In 1951, France and West Germany began the process of unification by signing a treaty that gave control of their coal and steel resources to a multinational group, the European Coal and Steel Community (ECSC). Italy and the Benelux countries also joined the ECSC. The six countries' leaders thought this alliance would have many positive results. Because the nations would depend on each other for industrial resources, their economies would suffer if they fought again. No country could prepare for war secretly because each knew what the others were manufacturing. In addition, the ECSC would set a tone of cooperation that would help Europe rebuild its economy.

The next step toward unity came in 1957 with the formation of the European Economic Community (EEC), also called the Common Market. This alliance removed trade barriers, set common economic goals, and allowed people to live and work in any member country. Between 1958 and 1968, trade among the EEC nations quadrupled.

In 1967, the EEC merged with the ECSC and another European alliance to become the European Community (EC). In 1973, the EC began to admit other European nations. (See the map on page 327.)

The Road to European Unity

1957
European Economic Community (EEC) forms.

1999
Euro introduced in banking and finance.

1950   1975   2000

1951
European Coal and Steel Community (ECSC) forms.

1967
EEC, ECSC, and European Atomic Energy Community (Euratom) become the European Community (EC).

1993
Maastricht Treaty establishing European Union (EU) takes effect.

2002
Euro adopted as standard currency.

The European Union Today

In 1993, the Maastricht Treaty took effect, and the European Union (EU) replaced the EC. With 15 member nations, the EU faces many issues. They include settling political and economic differences, replacing national currencies with the euro, and expanding EU membership.

ECONOMICS AND POLITICS EU members wonder how the union will affect their national economies. For example, workers may move to areas with higher wages, creating shifts in national populations.

SEE
PRIMARY SOURCE Ⓐ

Further, some countries believe that switching to the euro will mean losing control of economic factors such as interest rates. Others don't want to give up the national identities associated with having their own currencies. But many people believe the euro has benefits. These include greater business efficiency and increased international trade. In 1999, financial institutions began to calculate transactions in euros. Euros will begin to be used in everyday life in 2002.

SEE
PRIMARY SOURCE Ⓑ

The EU also affects politics. For example, on February 4, 2000, Joerg Haider and his Freedom Party became part of a coalition government in Austria. (In a coalition government, several parties share power.) In the past, Haider had made statements that were sympathetic to Nazi Germany, so EU nations criticized Austria for its support of Haider. Haider did step down, but some observers fear that a controversial leader like Haider could some day tear the EU apart.

EXPANDING THE EU One of the complex issues facing the EU is growth. In time, it might expand to 28 countries that presently have about 475 million people. Running such a huge alliance could be difficult. Many of the proposed members were once Communist nations. In general, they are less prosperous and have little experience of democracy. Such differences may create tensions that the EU will have to resolve.

SEE
PRIMARY SOURCE Ⓒ

On the following pages, you will find primary sources about the EU. Use them to form your own opinion.

EUROPE

The European Union, 2000

Legend:
- Original European alliance
- Later European Union members
- European Union candidates
- Non-member countries
- 1973 Year of joining European alliance

ICELAND

SWEDEN 1995
FINLAND 1995
NORWAY

ATLANTIC OCEAN

IRELAND 1973
UNITED KINGDOM 1973
DENMARK 1973
ESTONIA
LATVIA
LITHUANIA
NETHERLANDS 1951
BELGIUM 1951
(1990) (WEST) GERMANY 1951
POLAND
CZECH REPUBLIC
LUXEMBOURG 1951
SLOVAKIA
SWITZERLAND
AUSTRIA 1995
HUNGARY
FRANCE 1951
SLOVENIA
CROATIA
ROMANIA
YUGOSLAVIA
Black Sea
PORTUGAL 1986
SPAIN 1986
BOSNIA-HERZEGOVINA
BULGARIA
ALBANIA
MACEDONIA
ITALY 1951
GREECE 1981
TURKEY
Mediterranean Sea
MALTA
CYPRUS

Bay of Biscay
North Sea
Baltic Sea

0   250   500 miles
0   250   500 kilometers
Azimuthal Equidistant Projection

SKILLBUILDER: Interpreting Maps

❶ **MOVEMENT** In what years did the largest expansion of the European alliance occur? What countries joined?

❷ **REGION** Which region of Europe is not part of the EU? Why?

CASESTUDY

PROJECT *Panel Discussion*

Primary sources A to E on these two pages present differing opinions on expansion of the EU. Use these sources and your own research to prepare for a panel discussion on EU expansion. You might use the Internet and the library for research.

RESEARCH LINKS
CLASSZONE.COM

Suggested Steps

1. Choose a European (EU or non-EU) country to represent.

2. Research your country's position on EU expansion. Use encyclopedias, books, or the Internet to help you find the right information.

3. Consider the following questions during your research.
   - Why do certain countries want to join the EU?
   - What do current EU members have to gain and lose in expansion?
   - Why do certain countries want to remain independent?

4. Create a visual to be shown during the panel discussion.

5. Give a 2–3 minute speech that introduces your country's position.

Materials and Supplies

- Writing paper
- Posterboard
- Felt-tip markers
- Encyclopedias and reference books
- Computer
- Internet access

PRIMARY SOURCE A

Political Commentary *Global Britain, a conservative group in the United Kingdom, gave this view of the euro on January 25, 1999. Although the United Kingdom belongs to the EU, it is reluctant to adopt the euro, a central issue for EU expansion.*

The Single Currency is a political project designed to hasten the creation of a Single European State in which nation-states like Britain would be provinces. . . . In joining the Single Currency, a nation hands over control of its interest rate, exchange rate and gold and currency reserves, as well as control over tax and spending, to [the EU]. All of this is set out in the Maastricht Treaty which Britain signed in 1992. . . . There are 43 nation-states in Europe, of which only 11 have joined the "single" European currency. Those 11 countries, unlike Britain, are in varying degrees economic satellites of Germany and France. . . . A single currency eliminates the interest rate and exchange rate safety valves, which allow changing national economies to adjust to each other. . . .
Preparations for the "single" currency have already helped to cause mass unemployment in Germany, France, and Italy, where real jobless rates are at least three times as high as in Britain.

PRIMARY SOURCE B

Speech *Günter Verheugen of Germany, the European Commissioner for Enlargement, expressed his views on EU expansion in speeches in the United States on April 4–6, 2000. Germany is an original member of the EU and its predecessors.*

Enlargement is the biggest challenge the Union is facing at the dawn of the new millennium. . . . We are committed to this historical mission: to integrate the Central and East European countries which can and want to participate in our common achievements. . . . Our objective is to promote political and economic stability—and make this process irreversible. . . . What are the political benefits? First and foremost, the enlargement process is vital for securing political stability, democracy and respect of human rights on the European continent as a whole. . . . Political stability and freedom will be increased throughout Europe. Against the background of many years of crisis . . . the only way to achieve lasting stability in Europe is further integration.

Data *Eurobarometer is a company that surveys public opinion. It asked people living in the 15 EU countries how they felt about various countries joining the EU. This chart lists the various countries and the support for them.*

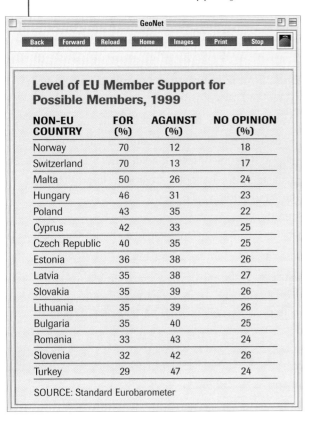

GeoNet

Back | Forward | Reload | Home | Images | Print | Stop

Level of EU Member Support for Possible Members, 1999

NON-EU COUNTRY	FOR (%)	AGAINST (%)	NO OPINION (%)
Norway	70	12	18
Switzerland	70	13	17
Malta	50	26	24
Hungary	46	31	23
Poland	43	35	22
Cyprus	42	33	25
Czech Republic	40	35	25
Estonia	36	38	26
Latvia	35	38	27
Slovakia	35	39	26
Lithuania	35	39	26
Bulgaria	35	40	25
Romania	33	43	24
Slovenia	32	42	26
Turkey	29	47	24

SOURCE: Standard Eurobarometer

Political Analysis *Edmund L. Andrews published this article in the* New York Times *on June 21, 1999. He examined some of the problems and issues of EU expansion into Central Europe.*

By becoming [EU] members, the Central European nations would eventually gain full access to European markets. Their citizens would be free to live and work throughout Western Europe. . . .

As more detailed negotiations loom . . . between the European Union and the Poles, Czechs, and Hungarians . . . both sides face the need for painful change. . . .

Central Europeans have the added burden of history. Many of them remain suspicious of Germany, the European Union's largest power and Central Europe's neighbor. And most adults, reared under Communism, are still adjusting to . . . the Western way of doing things.

As for the European Union, the prospect of a flood of labor from the East raises irrational fears among Westerners already grumbling about too many immigrants. . . . Another big fear [in Central Europe] is that foreigners—by which most people mean Germans—will buy up their land, which is another basic right accorded to anybody living within the European Union.

Political Cartoon *Pat Oliphant, a political cartoonist, shows the leaders of the EU trying to navigate stormy seas on a euro.*

PROJECT CheckList

Have I . . .

✓ researched my country's perspective?

✓ answered all relevant questions?

✓ created an interesting, colorful visual for the discussion?

✓ practiced my speech?

VISUAL SUMMARY
TODAY'S ISSUES IN EUROPE

Conflict

Turmoil in the Balkans
- Yugoslavia was a nation of many ethnic groups distributed among six republics.
- Serbia tried to dominate Yugoslavia, causing several republics to declare independence. Brutal wars followed. The UN and the United States negotiated peace.
- Seeking to re-establish control over Kosovo, Serbia tried to drive Albanians from Kosovo. NATO intervened to stop the violence.

Environment

Cleaning Up Europe
- Industry, sewage, agriculture, and other activities have caused water and air pollution in Europe.
- Pollution has caused disease, damaged buildings, and harmed livestock.
- Both national and international efforts are being made to clean up Europe.

Economics

The European Union
- After the destruction of World War II, France, Germany, Italy, and the Benelux countries joined in an economic alliance to foster cooperation.
- In time, this alliance began to admit other nations and to pursue more general goals.
- The alliance became the European Union (EU) in 1993. The EU faced the issues of adopting a common currency, settling political and economic differences, and expanding EU membership.

Reviewing Places & Terms

A. Briefly explain the importance of each of the following.

1. Slobodan Milošević
2. South Slavs
3. ethnic cleansing
4. KLA
5. Vojislav Kostunica
6. cyanide
7. European Environmental Agency
8. particulates
9. smog
10. ozone

B. Answer the questions about vocabulary in complete sentences.

11. What is the relationship between ozone and smog?
12. What effect did cyanide have on the rivers of Europe?
13. How are Slobodan Milošević and Vojislav Kostunica different?
14. What do Milošević and Kostunica have in common?
15. Which of the terms listed above might appear in a report by the European Environmental Agency?
16. Who were the South Slavs?
17. Who was the leader associated with the policy of ethnic cleansing?
18. Which groups were targets of ethnic cleansing?
19. Can Slobodan Milošević and the KLA best be described as allies or enemies? Explain.
20. Which type of pollution is associated with particulates? Explain.

Main Ideas

Turmoil in the Balkans (pp. 319–322)

1. How did historic events contribute to the conflict over Kosovo?
2. How did the diversity of Bosnia and Herzegovina's population contribute to the conflict there?
3. What did international officials discover after Serbian forces withdrew from Kosovo?
4. What are possible sources of future conflict in the Balkans?

Cleaning Up Europe (pp. 323–325)

5. What are the effects of acid rain?
6. Which region became heavily polluted under Communist rule?
7. Why is pollution such a difficult issue to resolve?

The European Union (pp. 326–329)

8. What organizations were forerunners of the European Union?
9. Why did European leaders believe that an economic alliance would help prevent war?
10. What are some possible problems associated with admitting formerly Communist countries to the EU?

Critical Thinking

1. Using Your Notes

Use your completed chart to answer these questions.

	Causes	Effects
Issue 1: Conflict		
Issue 2: Pollution		

a. Which of these issues has caused physical damage to Europe? Explain.

b. Do you think the issues are linked? Explain.

2. Geographic Themes

a. **REGION** In what way is the European Union creating a new region?

b. **MOVEMENT** What natural processes spread pollution from its point of origin?

3. Identifying Themes

Reread the story about the Tisza River on page 323. How do the five themes of geography relate to that story?

4. Making Inferences

What factors do you think led the Yugoslav people to vote Slobodan Milošević out of office?

5. Drawing Conclusions

How important is international cooperation in solving Europe's problems? Explain using specific examples.

Additional Test Practice, pp. S1–S37

TEST PRACTICE
CLASSZONE.COM

Geographic Skills: Interpreting Graphs

Trade with EU, 1995
(as percentage of total trade)

Use the graph to answer the following questions.

1. **PLACE** Which country does the highest percentage of its trade within the EU?

2. **PLACE** Which two countries do the lowest percentage of trade with the EU?

3. **MOVEMENT** Judging by the countries shown here, is there more trade within the EU or between the EU and non-member countries? Explain.

GeoActivity

Research trade statistics for the eight EU countries not shown here. Create an expanded graph showing data for all 15 EU members.

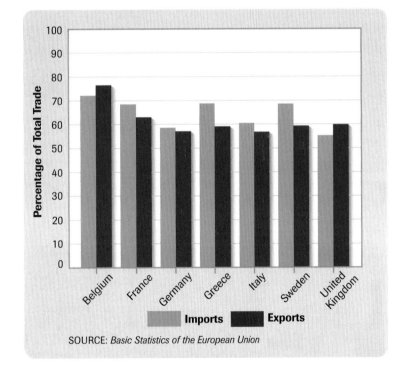

SOURCE: *Basic Statistics of the European Union*

INTERNET ACTIVITY

Use the links at **classzone.com** to do research about pollution in Europe. Learn about the "Green" political parties and their views on what should be done.

Writing About Geography Write a summary of your findings. Include a chart listing the programs proposed by the "Green" political parties. List the Web sites that were your sources.

Unit
5

Russia and the Republics

Between 1922 and 1991, Russia and
most of the Republics were part
of the Union of Soviet Socialist
Republics (USSR), also known
as the Soviet Union.

LOCATION Shoppers stroll around Russia's famous
State Department Store. The mall, which opened in 1893,
is located in Moscow, the capital of Russia.

REGION Russia and the Republics cross over 11 time zones and cover nearly one-sixth of the earth's land surface.

LOCATION Most of the region is hundreds of miles from the open sea.

HUMAN-ENVIRONMENT INTERACTION Freezing temperatures can continue so long that people use frozen rivers as roadways.

For more information on Russia and the Republics . . .

RESEARCH LINKS
CLASSZONE.COM

RUSSIA & REP.

PLACE The Caucasus Mountains stretch between the Black and Caspian seas. A great variety of peoples have settled in the region surrounding the mountains.

MOVEMENT Invaders from Arabia brought Islam to the southern areas of the region by the 8th century. Beautiful mosques adorn many of the region's cities.

Today's Issues in Russia and the Republics

Today, Russia and the Republics face the issues previewed here. As you read Chapters 15 and 16, you will learn helpful background information. You will study the issues themselves in Chapter 17.

In a small group, answer the questions below. Then participate in a class discussion of your answers.

Exploring the Issues

1. **CONFLICT** Search a newspaper for articles about conflicts in Russia and the Republics today. What do these conflicts have in common? How are they different?

2. **ECONOMIC CHANGE** Think about the different economic systems you learned about in Chapter 4. How might changing from a command economy to a market economy be difficult?

3. **NUCLEAR LEGACY** What impact could Soviet nuclear programs have on the region's economy?

For more on these issues in Russia and the Republics . . .

CURRENT EVENTS
CLASSZONE.COM

CONFLICT

How do new nations establish law and order?

After the collapse of the Soviet Union in 1991, groups in different parts of the region took up arms to fight for independence. This photo shows a woman and child from a region of Russia called Chechnya. Russia invaded Chechnya twice in the 1990s to end an independence movement in the region.

ECONOMIC CHANGE

How does a nation change its economic system?

For more than 70 years, the Soviet government made all the important economic decisions in the region. This cartoon illustrates a major economic challenge faced by the region's new leaders, as demonstrated here by former Russian president Boris Yeltsin. The leaders are trying to move their nations from a command economy to a market economy.

CASESTUDY

How have Soviet decisions affected new leaders?

In 1965, Soviet officials exploded a nuclear bomb to create this lake in Kazakhstan. The blast exposed nearby residents to harmful radiation. The region's new leaders inherited many problems caused by Soviet nuclear programs.

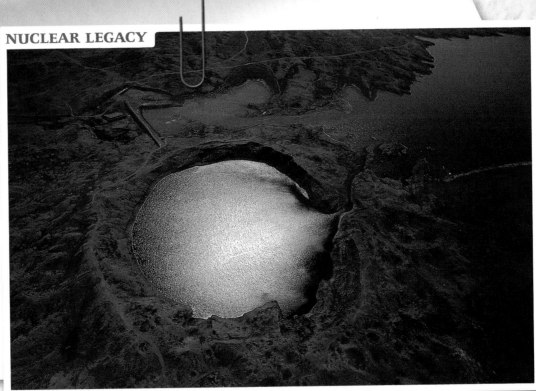

NUCLEAR LEGACY

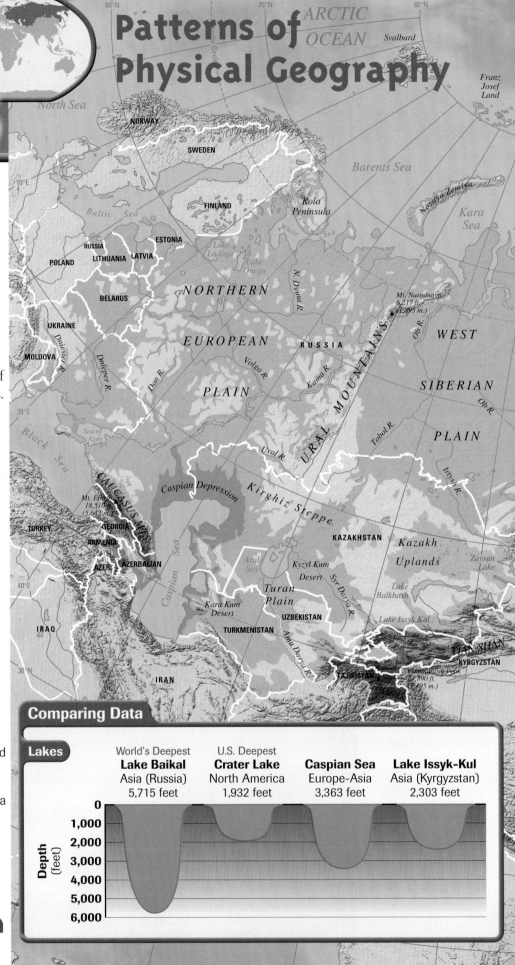
Patterns of Physical Geography

Russia and the Republics span two continents. The part of the region that lies to the west of the Ural Mountains is part of Europe. The part of the region that lies to the east of the Urals is part of Asia.

Use the Unit Atlas to add to your knowledge of Russia and the Republics. As you study the maps and charts, notice geographic patterns and specific details about the region.

Jot down answers to the following questions in your notebook.

Making Comparisons

1. What ocean lies to the north of Russia and the Republics? How might this ocean affect the region's climate?

2. How much deeper is Lake Baikal than the deepest lake in the United States?

3. Based on these maps and charts, which region do you think has the higher population density: Russia and the Republics or the United States? Why?

For updated statistics on Russia and the Republics . . .

DATA UPDATE
CLASSZONE.COM

Comparing Data

Lakes

	World's Deepest **Lake Baikal** Asia (Russia) 5,715 feet	U.S. Deepest **Crater Lake** North America 1,932 feet	**Caspian Sea** Europe-Asia 3,363 feet	**Lake Issyk-Kul** Asia (Kyrgyzstan) 2,303 feet

Depth (feet)

0
1,000
2,000
3,000
4,000
5,000
6,000

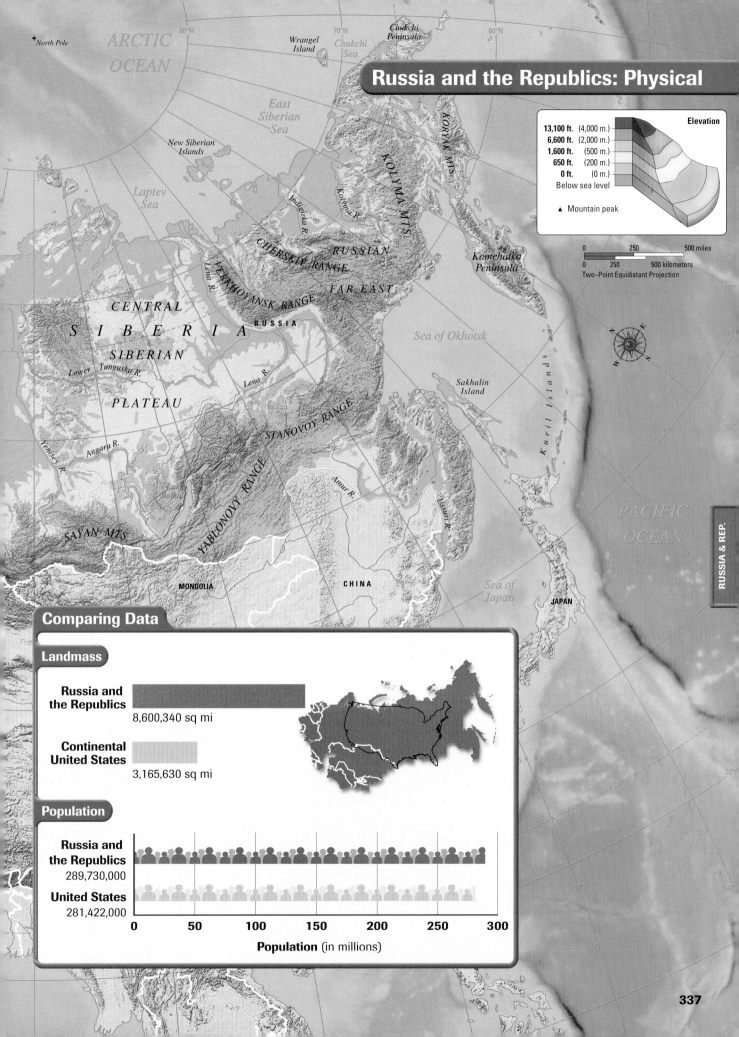

ARCTIC OCEAN

+ North Pole

Wrangel Island

Chukchi Peninsula

Chukchi Sea

East Siberian Sea

New Siberian Islands

Laptev Sea

KORYAK MTS.

Indigirka R.

Kolyma R.

KOLYMA MTS.

CHERSKIY RANGE

RUSSIAN

VERKHOYANSK RANGE

FAR EAST

Kamchatka Peninsula

CENTRAL SIBERIA

SIBERIAN PLATEAU

Lower Tunguska R.

RUSSIA

Lena R.

Lena R.

Sea of Okhotsk

Sakhalin Island

Yenisey R.

Angara R.

STANOVOY RANGE

YABLONOVY RANGE

Kuril Islands

PACIFIC OCEAN

SAYAN MTS.

Amur R.

Ussuri R.

MONGOLIA

CHINA

Sea of Japan

JAPAN

Elevation

13,100 ft.	(4,000 m.)
6,600 ft.	(2,000 m.)
1,600 ft.	(500 m.)
650 ft.	(200 m.)
0 ft.	(0 m.)

Below sea level

▲ Mountain peak

0 250 500 miles
0 250 500 kilometers
Two-Point Equidistant Projection

RUSSIA & REP.

Comparing Data

Landmass

Russia and the Republics
8,600,340 sq mi

Continental United States
3,165,630 sq mi

Population

Russia and the Republics
289,730,000

United States
281,422,000

| 0 | 50 | 100 | 150 | 200 | 250 | 300 |

Population (in millions)

Patterns of Human Geography

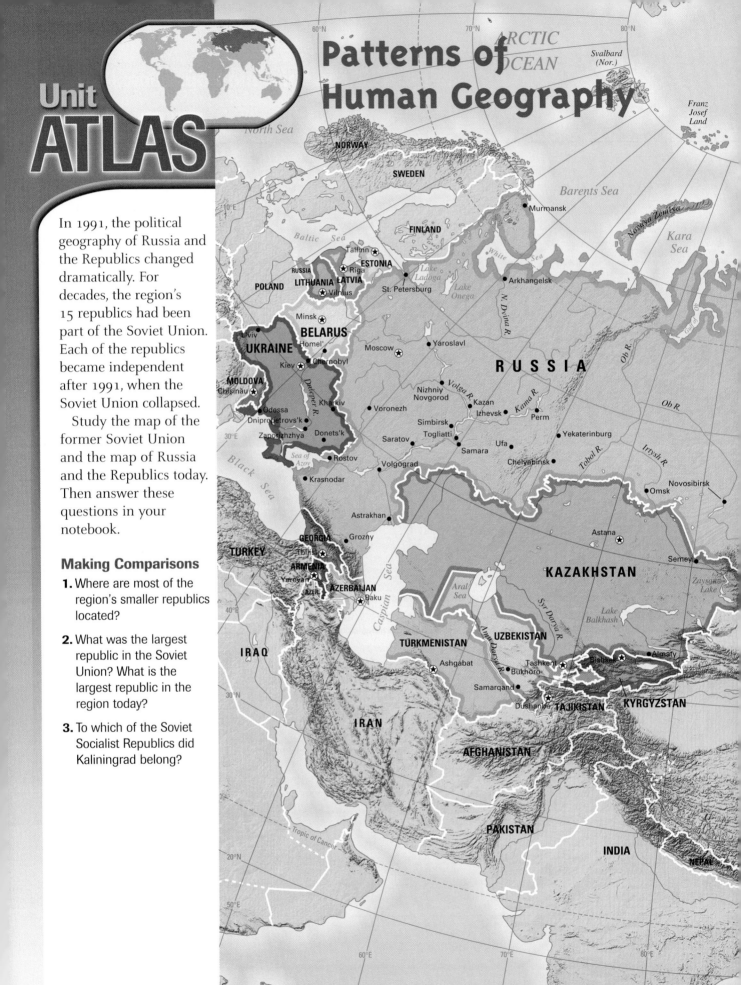

In 1991, the political geography of Russia and the Republics changed dramatically. For decades, the region's 15 republics had been part of the Soviet Union. Each of the republics became independent after 1991, when the Soviet Union collapsed.

Study the map of the former Soviet Union and the map of Russia and the Republics today. Then answer these questions in your notebook.

Making Comparisons

1. Where are most of the region's smaller republics located?

2. What was the largest republic in the Soviet Union? What is the largest republic in the region today?

3. To which of the Soviet Socialist Republics did Kaliningrad belong?

ARCTIC OCEAN

+ North Pole

80°N

Wrangel Island

Chukchi Sea

70°N

60°N

Bering Sea

Severnaya Zemlya

New Siberian Islands

East Siberian Sea

Laptev Sea

Indigirka R.

Kolyma R.

Lena R.

Oymyakon

RUSSIA

Petropavlovsk-Kamchatski

★ National capital
● Other city

0   250   500 miles
0   250   500 kilometers
Two–Point Equidistant Projection

Norilsk

Yakutsk

Sea of Okhotsk

Lower   Tunguska R.

Lena R.

Sakhalin Island

Kuril Islands

Yenisey R.

Angara R.

Krasnoyarsk

Irkutsk

Amur R.

Khabarovsk

PACIFIC OCEAN

RUSSIA & REP.

CHINA

Vladivostok

MONGOLIA

Sea of Japan

JAPAN

NORTH KOREA

Former Soviet Union, 1989

0   500   1,000 miles
0   500   1,000 kilometers
Two-Point Equidistant Projection

60°N

ARCTIC OCEAN

80°N

80°N

60°N

40°N

Bering Sea

Arctic Circle

KALININGRAD (R.S.F.S.R.)

FINLAND

LITHUANIAN S.S.R.

Baltic Sea

ESTONIAN S.S.R.

SLOV.

POLAND

LATVIAN S.S.R.

BELORUSSIAN S.S.R.

ROMANIA

UKRAINIAN S.S.R.

MOLDAVIAN S.S.R.

RUSSIAN SOVIET FEDERATED SOCIALIST REPUBLIC (R.S.F.S.R.)

Sea of Okhotsk

TURKEY

Black Sea

GEORGIAN S.S.R.

Caspian Sea

Aral Sea

KAZAKH S.S.R.

Lake Baikal

PACIFIC OCEAN

ARMENIAN S.S.R.

Lake Balkhash

Sea of Japan

AZERBAIJAN S.S.R.

TURKMEN S.S.R.

UZBEK S.S.R.

MONGOLIA

N. KOREA

JAPAN

IRAN

KIRGHIZ S.S.R.

S. KOREA

20°E

40°E

60°E

TADZHIK S.S.R.

CHINA

100°E

120°E

NOTE: S.S.R. is the abbreviation for Soviet Socialist Republic.

Unit
ATLAS

Regional Patterns

These two pages contain a pie graph and three thematic maps. The pie graph shows the religions of Russia and the Republics. The maps show other important features of the region: its different climates, numerous ethnic groups, and population density. After studying these two pages, answer the questions below in your notebook.

Making Comparisons

1. Where is the population of Russia and the Republics most dense? Which climate do those areas have? How might climate affect population density?

2. How would you describe the ethnic and religious populations of Russia and the Republics? Which is the most wide-spread ethnic group in the region?

Religions of Russia and the Republics

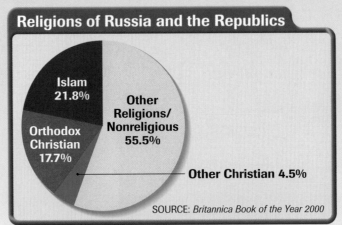

Islam
21.8%

Other Religions/ Nonreligious
55.5%

Orthodox Christian
17.7%

Other Christian 4.5%

SOURCE: *Britannica Book of the Year 2000*

Climates of Russia and the Republics

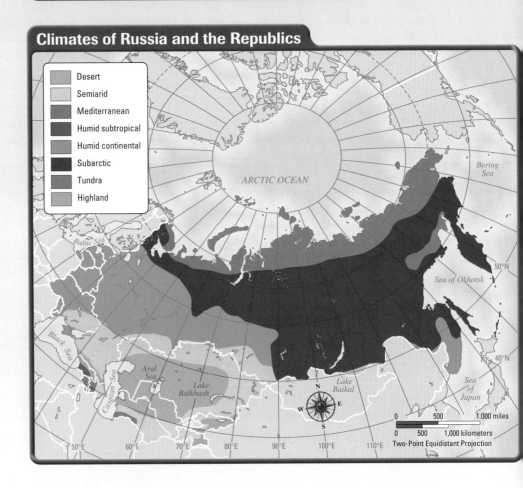

Desert
Semiarid
Mediterranean
Humid subtropical
Humid continental
Subarctic
Tundra
Highland

ARCTIC OCEAN

Baltic Sea

Black Sea

Aral Sea

Caspian Sea

Lake Balkhash

Lake Baikal

Bering Sea

Sea of Okhotsk

Sea of Japan

50°N

40°N

50°E 60°E 70°E 80°E 90°E 100°E 110°E

0 500 1,000 miles
0 500 1,000 kilometers
Two-Point Equidistant Projection

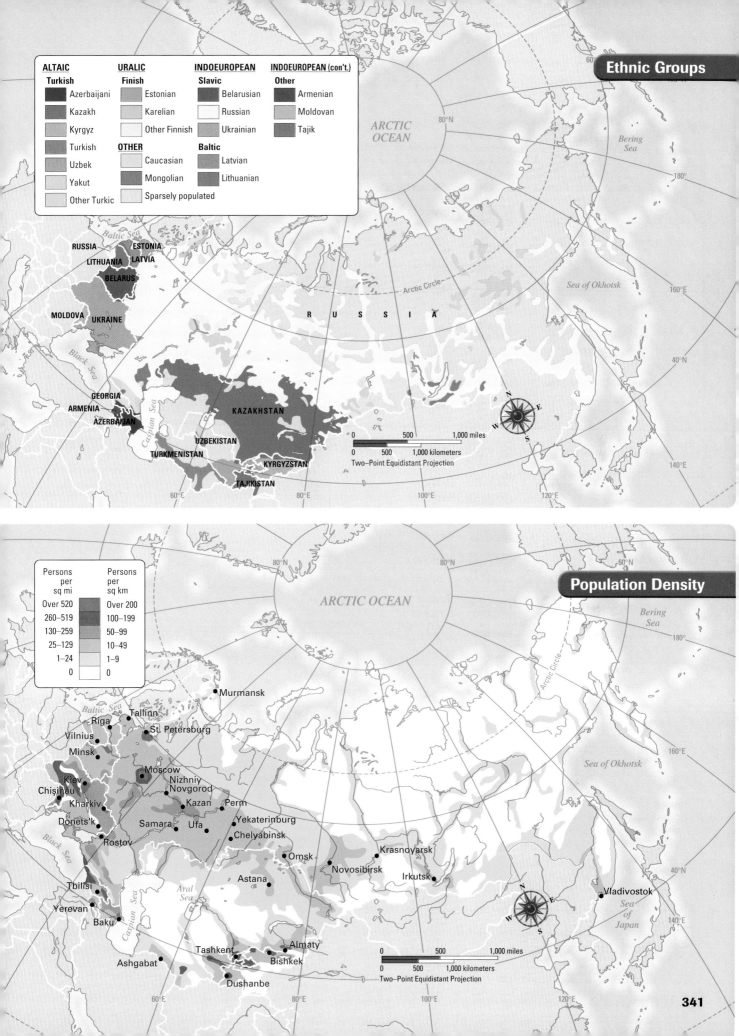

Ethnic Groups

ALTAIC
Turkish
- Azerbaijani
- Kazakh
- Kyrgyz
- Turkish
- Uzbek
- Yakut
- Other Turkic

URALIC
Finish
- Estonian
- Karelian
- Other Finnish

OTHER
- Caucasian
- Mongolian
- Sparsely populated

INDOEUROPEAN
Slavic
- Belarusian
- Russian
- Ukrainian

Baltic
- Latvian
- Lithuanian

INDOEUROPEAN (con't.)
Other
- Armenian
- Moldovan
- Tajik

ARCTIC OCEAN
Bering Sea
Baltic Sea
RUSSIA
ESTONIA
LITHUANIA
LATVIA
BELARUS
MOLDOVA
UKRAINE
Arctic Circle
R U S S I A
Sea of Okhotsk
Black Sea
GEORGIA
ARMENIA
AZERBAIJAN
Caspian Sea
KAZAKHSTAN
UZBEKISTAN
TURKMENISTAN
KYRGYZSTAN
TAJIKISTAN

0   500   1,000 miles
0   500   1,000 kilometers
Two–Point Equidistant Projection

Population Density

Persons per sq mi	Persons per sq km
Over 520	Over 200
260–519	100–199
130–259	50–99
25–129	10–49
1–24	1–9
0	0

ARCTIC OCEAN
Bering Sea
Baltic Sea
Arctic Circle
Sea of Okhotsk
Murmansk
Tallinn
Riga
St. Petersburg
Vilnius
Minsk
Moscow
Nizhniy Novgorod
Kiev
Chişinău
Kharkiv
Kazan
Perm
Donets'k
Samara
Ufa
Yekaterinburg
Rostov
Chelyabinsk
Omsk
Krasnoyarsk
Astana
Novosibirsk
Irkutsk
Vladivostok
Tbilisi
Yerevan
Baku
Aral Sea
Caspian Sea
Black Sea
Sea of Japan
Tashkent
Almaty
Ashgabat
Bishkek
Dushanbe

0   500   1,000 miles
0   500   1,000 kilometers
Two–Point Equidistant Projection

341

Regional Data File

Study the charts on the countries of Russia and the Republics. In your notebook, answer these questions.

Making Comparisons

1. Which five republics have the highest infant mortality rates? Do you notice any pattern?

2. Examine the literacy rates for the region. What do the figures tell you about the value placed on education in the region?

Sources:
CIA World Factbook 2000 online
Europa World Year Book 2000
Human Development Report 2000,
  United Nations
International Data Base (IDB), 2000
  updates, U.S. Census Bureau online
Merriam-Webster's Geographical
  Dictionary, 3d ed., 1998
Statesman's Yearbook 2001
WHO Estimates of Health Personnel,
  1998, World Health Organization
  online
World Almanac and Book of Facts 2001
World Education Report 2000,
  UNESCO online
2000 World Population Data Sheet,
  Population Reference Bureau online

Notes:
[a] Life expectancy figures for Russia
  and several other republics in the
  former USSR declined significantly in
  the 1990s.
[b] A comparison of the prices of the
  same items in different countries
  is used to figure these data.
[c] Includes land and water, when
  figures are available.

For updated statistics on Russia and the Republics . . .

DATA UPDATE
CLASSZONE.COM

Country Flag	Country/Capital	Population (2000 estimate)	Life Expectancy[a] (years) (2000)	Birthrate (per 1,000 pop.) (2000)	Infant Mortality (per 1,000 live births) (2000)
	Armenia Yerevan	3,809,000	75	11	41
	Azerbaijan Baku	7,734,000	72	18	83
	Belarus Minsk	10,004,000	68	9	15
	Estonia Tallinn	1,433,000	70	8	13
	Georgia Tbilisi	5,454,000	73	11	53
	Kazakhstan Astana	14,865,000	65	17	59
	Kyrgyzstan Bishkek	4,929,000	67	26	77
	Latvia Riga	2,416,000	70	8	16
	Lithuania Vilnius	3,697,000	72	10	15
	Moldova Chişinău	4,276,000	67	13	43
	Russia Moscow	145,231,000	67	9	20
	Tajikistan Dushanbe	6,374,000	68	34	117
	Turkmenistan Ashgabat	5,239,000	66	29	73
	Ukraine Kiev	49,509,000	68	9	22
	Uzbekistan Tashkent	24,760,000	69	26	72
	United States Washington, D.C.	281,422,000	77	15	7

Doctors (per 100,000 pop.) (1998)	GDP[b] (billions $US) (1999 est.)	Import/Export[b] (billions $US) (1999)	Literacy Rate (percentage) (1998)	Televisions (per 1,000 pop.) (1996–1998)	Passenger Cars (per 1,000 pop.) (1996–1997)	Total Area[c] (square miles)	
316	9.9	0.782 / 0.24	98	217	2	11,506	
360	14.0	1.46 / .885	99	254	36	33,436	
443	55.2	5.76 / 6.0	99	314	111	80,154	
297	7.9	3.4 / 2.5	99	480	294	17,413	
436	11.7	0.84 / 0.33	99	472	80	26,911	
353	54.5	4.8 / 5.2	99	234	61	1,048,300	
301	10.3	0.59 / 0.515	97	44	32	76,641	
282	9.8	2.8 / 1.9	99	593	174	24,595	
395	17.3	4.5 / 3.3	99	376	242	25,174	
400 (1995)	9.7	0.56 / 0.47	98	297	46	13,012	
421	620.3	48.2 / 75.4	99	420	120	6,592,812	
201	6.2	0.77 / 0.634	99	285	31	55,251	
300 (1997)	7.7	1.25 / 1.1	98	201	N/A	188,455	
299	109.5	11.8 / 11.6	99	490	97	233,089	
309	59.3	3.1 / 2.9	88	273	37	173,591	
251	9,255.0	820.8 / 663.0	97	847	489	3,787,319	

Chapter 15

SECTION 1
Landforms and Resources

SECTION 2
Climate and Vegetation

SECTION 3
Human–Environment Interaction

PHYSICAL GEOGRAPHY OF
RUSSIA AND THE REPUBLICS
A Land of Extremes

Russia's Lake Baikal is the world's deepest lake and holds over 20 percent of the earth's fresh water. Russians treasure Lake Baikal as much as Americans treasure the Grand Canyon.

GeoFocus

How do the extremes of physical geography in Russia and the Republics affect the lives of the region's people?

Taking Notes Copy the graphic organizer below into your notebook. Use it to record information from the chapter about the physical geography of Russia and the Republics.

Landforms	
Resources	
Climate and Vegetation	
Human-Environment Interaction	

1

Landforms and Resources

Main Ideas

- Flat plains stretch across the western and central areas of the region. In the south and east, the terrain is more mountainous.

- Many resources in Russia and the Republics are in hard-to-reach regions with brutal climates.

Places & Terms

chernozem   Transcaucasia

Ural Mountains   Central Asia

   Siberia

Eurasia

CONNECT TO THE ISSUES
ECONOMIC CHANGE
Leaders must strike a balance between environmental protection and economic growth.

A HUMAN PERSPECTIVE Russia and the Republics occupy a tremendous expanse of territory—approximately three times the land area of the United States. The region sprawls across the continents of both Europe and Asia and crosses 11 time zones. When laborers in the western city of Kaliningrad are leaving their jobs after a day's work, herders on the region's Pacific coast are just beginning to awaken their animals for the next day's grazing.

Northern Landforms

The geography of Russia and the Republics is the geography of nearly one-sixth of the earth's land surface—over eight and a half million square miles. In spite of this huge size, the region's landforms follow a simple overall pattern. You can divide the northern two-thirds of the region into four different areas. Moving from west to east, they are the Northern European Plain, the West Siberian Plain, the Central Siberian Plateau, and the Russian Far East. (See the physical map on pages 336–337 of the Unit Atlas.)

THE NORTHERN EUROPEAN PLAIN The Northern European Plain is an extensive lowland area. It stretches for over 1,000 miles from the western border of Russia and the Republics to the Ural Mountains.

One of the world's most fertile soils—**chernozem,** or black earth—is abundant on this plain. It sometimes occurs in layers three feet deep or more. Because of the high quality of its soil, many of the region's agricultural areas are located on this plain.

Nearly 75 percent of the region's 290 million people live on this plain. Three of the region's largest cities are located there: Moscow, Russia's capital; St. Petersburg; and Kiev, the capital of Ukraine.

PLACE Ukraine, which lies on the Northern European Plain, has been called the region's breadbasket because of the enormous grain crops produced on its farms.

WEST SIBERIAN PLAIN The **Ural Mountains** separate the Northern European and West Siberian plains. Some geographers recognize the Urals as a dividing line between Europe and Asia. Others consider Europe and Asia to be a single continent, which they call **Eurasia.**

The West Siberian Plain lies between the Urals and the Yenisey River and between the shores of the Arctic Ocean and the foothills of the Altay Mountains. Because the plain tilts northward, its rivers flow toward the Arctic Ocean.

CENTRAL SIBERIAN PLATEAU AND RUSSIAN FAR EAST Although extensive plains lie east of the Yenisey River, uplands and mountains are the dominant landforms. High plateaus—with average heights of 1,000 to 2,000 feet—make up the Central Siberian Plateau, which lies between the Yenisey and Lena rivers.

East of the Lena River lies the Russian Far East and its complex system of volcanic ranges. The Kamchatka Peninsula alone contains 120 volcanoes, 20 of which are still active. The Sakhalin and Kuril islands lie south of the peninsula. Russia seized the islands from Japan after World War II. Japan still claims ownership of the Kuril Islands.

BACKGROUND
Russia and Japan never signed a formal peace treaty after World War II ended in 1945. Technically, they are still at war.

Southern Landforms

LOCATION The Tian Shan, which is Chinese for "Heavenly Mountains," stretch for nearly 1,500 miles, mainly between China and Kyrgyzstan.
Why might a river be flowing at the base of these mountains?

The southern areas of Russia and the Republics feature towering mountains, barren uplands, and semiarid grasslands.

THE CAUCASUS AND OTHER MOUNTAINS The Caucasus Mountains stretch across the land that separates the Black and Caspian seas. The mountains form the border between Russia and **Transcaucasia**—a region that consists of the republics of Armenia, Azerbaijan, and Georgia. Farther east, along the southern border of Russia and the Republics, rises a colossal wall of mountains, including the Tian Shan, shown below.

Some of these mountains are located along the southeastern border of **Central Asia**—a region that includes the republics of Kazakhstan,

346

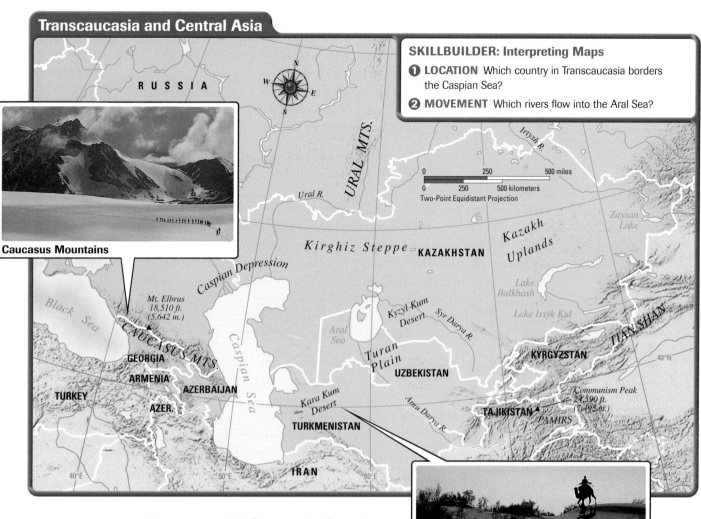

RUSSIA

Caucasus Mountains

URAL MTS.

Ural R.

Caspian Depression

Kirghiz Steppe **KAZAKHSTAN**

Kazakh Uplands

Irtysh R.

Zaysan Lake

Lake Balkhash

Lake Issyk Kul

Mt. Elbrus 18,510 ft. (5,642 m.)

Black Sea

Aral Sea

Kyzyl Kum Desert

Syr Darya R.

Turan Plain

CAUCASUS MTS.

Caspian Sea

GEORGIA

ARMENIA

AZERBAIJAN

AZER.

TURKEY

Kara Kum Desert

Amu Darya R.

UZBEKISTAN

KYRGYZSTAN

TIAN SHAN

Communism Peak 24,590 ft. (7,495 m.)

TAJIKISTAN

PAMIRS

40°N

TURKMENISTAN

IRAN

40°E   50°E   60°E

N

W   E

S

0   250   500 miles
0   250   500 kilometers
Two-Point Equidistant Projection

SKILLBUILDER: Interpreting Maps

❶ **LOCATION** Which country in Transcaucasia borders the Caspian Sea?

❷ **MOVEMENT** Which rivers flow into the Aral Sea?

Kara Kum Desert

RUSSIA & REP.

⊕ **Geographic Thinking**

Using the Atlas

Ⓐ Examine the climate map on page 340. What is the relationship between landforms and climate zones in Central Asia?

Kyrgyzstan, Tajikistan, Turkmenistan, and Uzbekistan. These ranges are so high that they prevent moist air from entering the region from the south, contributing to the arid climate of Central Asia.

THE TURAN PLAIN An extensive lowland called the Turan Plain lies between the Caspian Sea and the mountains and uplands of Central Asia. Although two major rivers, the Syr Darya and Amu Darya, cross the plain, much of the lowland is very dry. Two large deserts stretch across the plain—the Kara Kum and the Kyzyl Kum.

Rivers and Lakes

Some of the world's longest rivers flow through the vast plains of Russia and the Republics. The region also boasts some of the largest and deepest lakes in the world.

DRAINAGE BASINS AND RIVERS The region's rivers flow through a number of large drainage basins. You may recall from Chapter 2 that a drainage basin is an area drained by a major river and its tributaries. The main drainage basins in Russia and the Republics are the Arctic Ocean, Caspian Sea, Pacific Ocean, Baltic Sea, Black Sea, and Aral Sea basins.

The Arctic basin is the region's largest. The basin's three powerful rivers—the Ob, the Yenisey, and the Lena—drain an area of more than three million square miles. These rivers deliver water to the Arctic Ocean at a combined rate of nearly 1,750,000 cubic feet per second. **B**

The Volga River, the longest river on the European continent, drains the Caspian Sea basin. The Volga begins near Moscow and flows southward for about 2,300 miles until it arrives at the Caspian. This important waterway carries about 60 percent of Russia's river traffic.

LAKES In addition to some of the world's longest rivers, Russia and the Republics also boast some of the largest lakes on our planet. Two of them, the Caspian and Aral seas, are located in Central Asia.

The Caspian Sea, which is actually a saltwater lake, stretches for nearly 750 miles from north to south, making it the largest inland sea in the world. The Aral Sea, which lies east of the Caspian, is also a saltwater lake. Since the 1960s, the Aral has lost about 80 percent of its water volume. This enormous loss is the result of extensive irrigation projects that have diverted water away from the rivers that feed the lake. Unless drastic action is taken, the Aral Sea could vanish within 20 to 30 years.

LAKE BAIKAL The crown jewel among the region's lakes is Lake Baikal—the deepest lake in the world. At its deepest point, Baikal is more than a mile from the surface to the bottom. From north to south, the lake stretches for nearly 400 miles. It holds 20 percent of the world's fresh water.

Though it has some pollution, most of Lake Baikal is remarkably clean. Thousands of species of plants and animals live in the lake. Twelve hundred species, including the world's only freshwater seal, are unique to Lake Baikal.

Regional Resources

Russia and the Republics have a great wealth of natural resources. Regional leaders have found it difficult to properly manage these resources. One challenge has been how to transport resources from harsh and distant regions. Another has been how to use the resources without damaging the environment in the process.

Geographic Thinking

Seeing Patterns
B Examine the map on pages 336–337. Why might many of the region's rivers flow toward the north?

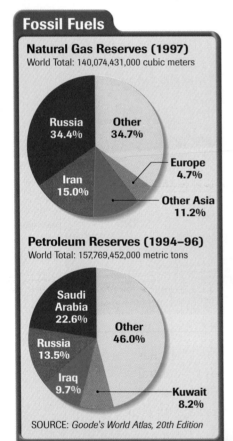

Fossil Fuels

Natural Gas Reserves (1997)
World Total: 140,074,431,000 cubic meters

- Russia 34.4%
- Other 34.7%
- Europe 4.7%
- Other Asia 11.2%
- Iran 15.0%

Petroleum Reserves (1994–96)
World Total: 157,769,452,000 metric tons

- Saudi Arabia 22.6%
- Other 46.0%
- Russia 13.5%
- Iraq 9.7%
- Kuwait 8.2%

SOURCE: *Goode's World Atlas, 20th Edition*

SKILLBUILDER:
Interpreting Graphs

❶ **ANALYZING DATA** What country had the largest reserves of natural gas in 1997?

❷ **ANALYZING DATA** About how many cubic meters of natural gas did Russia have in 1997?

LOCATION Workers adjust machinery at the Samotlor oil field in Russia.

Swamps form when northward flowing rivers, swollen by spring rains, run into still-frozen water further north. Soon, enormous black clouds of insects are attacking Siberia's residents.

The climate also affects construction in Siberia. Permafrost makes the ground in Siberia iron-hard. However, a heated building will thaw the permafrost. As the ground thaws, buildings sink, tilt, and eventually topple over. To prevent such problems, builders raise their structures a few feet off the ground on concrete pillars.

WAR AND "GENERAL WINTER" Russia's harsh climate has caused difficulties for its inhabitants, but it has also, at times, come to their aid.

In the early 1800s, the armies of the French leader Napoleon Bonaparte were taking control of Europe. In the spring of 1812, Napoleon decided to extend his control over Russia. He gathered his army together in Poland and from there began the march on Moscow.

But as his troops advanced, so did the seasons. When Napoleon arrived in Moscow in September, the Russian winter was not far behind. Moreover, the citizens of Moscow had set fire to their city before fleeing, so there was no shelter for Napoleon's troops.

Napoleon had no choice but to retreat during the bitter Russian winter. He left Moscow with 100,000 troops. But by the time his army arrived back in Poland, the cruel Russian winter had helped to kill more than 90,000 of his soldiers. Some historians believe that Russia's "General Winter" succeeded in defeating Napoleon where the armies of Europe had failed. **B**

Geographic Thinking

Seeing Patterns
B Besides climate, what other geographic factors might an army invading Russia have to consider?

Crossing the "Wild East"

At the end of the 19th century, Siberia was similar to the "Wild West" of the United States. Travel through the region was dangerous and slow. For these reasons, Russia's emperor ordered work to start on a **Trans-Siberian Railroad** that would eventually link Moscow to the Pacific port of Vladivostok.

Rail Routes Across Russia

Rail Routes
— Trans-Siberian Railroad
— Trans-Manchurian Railroad
— Baikal-Amur Mainline
+1 hr. Time difference from Moscow
200 Miles between cities
⭐ Capital   ● City

SKILLBUILDER: Interpreting Maps

❶ LOCATION As a train moves eastward after passing over the Ural Mountains, what is the first major stop?

❷ MOVEMENT What railroad route would you take if you wanted to pass north of Lake Baikal?

355

MOVEMENT A train from Ukraine travels on Trans-Siberian tracks on its journey toward Vladivostok. **What impact do railroads have on commerce?**

AN ENORMOUS PROJECT The project was a massive undertaking. The distance to be covered was more than 5,700 miles, and the tracks had to cross seven time zones. Between 1891 and 1903, approximately 70,000 workers moved 77 million cubic feet of earth, cleared more than 100,000 acres of forest, and built bridges over several major rivers.

RESOURCE WEALTH IN SIBERIA Russian officials did not undertake this massive project simply to speed up travel. They also wanted to populate Siberia in order to profit from its many resources.

Ten years after the completion of the line in 1904, nearly five million settlers, mainly peasant farmers, had taken the railway from European Russia to settle in Siberia.

As migrants streamed into Siberia, resources, such as coal and iron ore, poured out. Siberia, one author wrote, began to yield riches that "she has under guard of eternal snow and ice, so long held in trust for future centuries." In the years that followed, the railroad would aid the political and economic development of Russia and the Republics, which you will read about in the next chapter.

SECTION 3 Assessment

❶ Places & Terms

Explain the importance of the following terms.

• runoff

• Trans-Siberian Railroad

❷ Taking Notes

REGION Review the notes you took for this section.

Human-Environment Interaction

• What precautions must builders take in Siberia? Why?

• How did the construction of the Trans-Siberian Railroad affect the region's landscape?

❸ Main Ideas

a. Why is the Aral Sea shrinking?

b. How has the region's harsh climate helped its inhabitants?

c. What were the main reasons for the construction of the Trans-Siberian Railroad?

❹ Geographic Thinking

Making Decisions If you were a regional leader, what steps would you take to end the Aral Sea disaster? **Think about:**

• how your solutions will affect people in the region

RESEARCH LINKS CLASSZONE.COM

MAKING COMPARISONS Do more research on the Trans-Siberian Railroad. Then do research on the construction of the transcontinental railroad in the United States. Use a **Venn diagram** to compare and contrast the two projects.

Understanding Time Zones

In 1884, international officials agreed to divide the map of the earth's surface into 24 time zones, one for each hour of the day. Because the earth rotates 360° each day, each zone was to represent 15° longitude (360° ÷ 24 hours = 15°). Officials used the prime meridian (0°) as the starting point for the time zones. They named this base time Greenwich Mean Time (GMT). The International Date Line was set at 180° longitude. To the east of this line, the calendar date is one day earlier than to the west.

THE LANGUAGE OF MAPS A **time zone map** shows the time zones that are in use around the world today. Officials have adjusted the boundaries of many time zones to keep political units, such as countries, within a single time zone.

> Non-standard time zones
> Time varies from the standard
> time zone by less than an hour.

Time Zones of the World

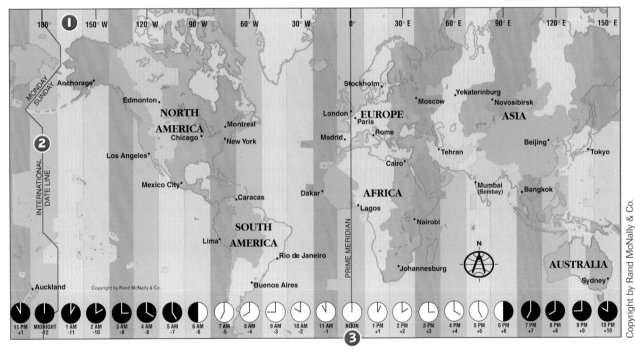

① Each band of color represents one time zone.

② Officials set the International Date Line at 180°, but the line moves east or west of it in places to avoid dividing countries.

③ Positive and negative numbers show the difference between local time and Greenwich Mean Time.

Copyright by Rand McNally & Co.

Map and Graph Skills Assessment

1. Drawing Conclusions
How many time zones are there in the continental United States?

2. Making Comparisons
What is the current time in the time zone in which you live? What is the current time in Greenwich, England?

3. Drawing Conclusions
If it is 6:00 Sunday morning in New York, what are the day and time in Auckland, New Zealand?

VISUAL SUMMARY
PHYSICAL GEOGRAPHY OF RUSSIA AND THE REPUBLICS

Landforms

Major Geographical Areas: Northern European Plain, West Siberian Plain, Central Siberian Plateau, Russian Far East, Turan Plain

Important Mountain Ranges: Urals, Caucasus, Tian Shan

Important Rivers and Lakes: Caspian and Aral seas, Lake Baikal, Volga, Ob, Yenisey, and Lena rivers

Resources

• Regional resources include huge coal reserves and deposits of iron ore and other metals.

• The Caspian Sea region has enormous reserves of oil and gas.

Climate and Vegetation

• Continentality and a wall of high southeastern mountains have a major impact on the climate of Russia and the Republics.

Human-Environment Interaction

• The shrinking of the Aral Sea is an example of the dramatic impact that agricultural policies can have on the environment.

• The hardships faced by Napoleon's army show how Russia's environment influenced human events.

• The construction of the Trans-Siberian Railway changed the population distribution and economic geography of the region.

Reviewing Places & Terms

A. Briefly explain the importance of each of the following.

1. chernozem
2. Ural Mountains
3. Eurasia
4. Transcaucasia
5. Central Asia
6. Siberia
7. continentality
8. taiga
9. runoff
10. Trans-Siberian Railroad

B. Answer the questions about vocabulary in complete sentences.

11. What is the name of the region crossed by the Trans-Siberian Railroad?
12. Which region is located south of the Caucasus Mountains?
13. What is chernozem and where is it found?
14. How can runoff affect the environment?
15. What are the five republics located in Central Asia?
16. Why are the Ural Mountains important for geographers?
17. What is the name of the non-European part of Russia?
18. Which vegetation region allows Russia to boast one-fifth of the world's timber resources?
19. Why do Russia and the Republics receive limited precipitation?
20. Which landmass is named after the continents of Asia and Europe?

Main Ideas

Landforms and Resources (pp. 345–349)

1. What facts could you provide to give an idea of the enormous size of Russia and the Republics?
2. How does the tilt of the West Siberian Plain affect the region's physical geography?
3. How is the region's use of its resources affected by climate?

Climate and Vegetation (pp. 350–352)

4. What are major influences on the region's climate?
5. How does latitude affect the type of vegetation found in Russia's forests?
6. Where is the steppe located in Russia and the Republics?

Human-Environment Interaction (pp. 353–357)

7. What effect have irrigation projects had on the Aral Sea?
8. How has the shrinking of the Aral Sea affected public health in the surrounding region?
9. What factors contribute to the formation of swamps in Siberia, and how do the swamps affect people living in the region?
10. How long did it take to complete the main line of the Trans-Siberian Railway?

Critical Thinking

1. Using Your Notes

Use your completed chart to answer these questions.

Landforms	
Resources	

   a. Which region contains a large number of volcanoes?

   b. Who was "General Winter"?

2. Geographic Themes

   a. **HUMAN-ENVIRONMENT INTERACTION** How has Siberia's climate affected transportation in the region?

   b. **MOVEMENT** What impact did the Trans-Siberian Railway have on Russia's population?

3. Identifying Themes

What factor might explain why Russia and the Republics receive relatively little precipitation and frequently experience extreme temperatures? Which of the five themes applies to this situation?

4. Determining Cause and Effect

What is a major factor contributing to the large subtropical climate zone in Transcaucasia?

5. Drawing Conclusions

Given what you have read about the dependency of Central Asian farmers on the water from the Amu Darya and Syr Darya rivers, how likely do you think it is that the Aral Sea will eventually recover?

Additional Test Practice, pp. S1–S37

TEST PRACTICE
CLASSZONE.COM

Geographic Skills: Interpreting Maps

Mineral Resources and Pollution

Use the map to answer the following questions.

1. **REGION** This map shows how close mining sites are to polluted areas. Why might the two be related?

2. **MOVEMENT** How might locating a mining site near a river affect the spread of pollution?

3. **PLACE** Why might the areas around Moscow and St. Petersburg be polluted even though there seem to be few mining sites nearby?

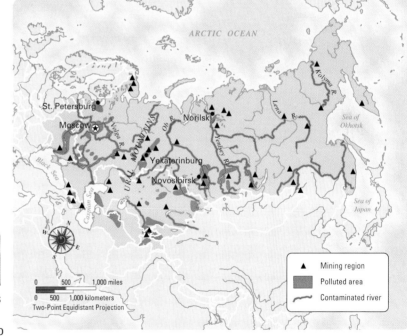

GeoActivity

Do more research on mining pollution's impact on public health in one area of the region. Use presentation software to share your results.

INTERNET ACTIVITY

Use the links at **classzone.com** to do research on Siberia. Focus on how people cope with the region's low temperatures. For example, investigate the kinds of clothing people wear or how they move about in the winter.

Writing About Geography Write a report of your findings. Include photos or illustrations that visually present information about life in the region. List the Web sites that you used in preparing your report.

Chapter

16

SECTION 1
Russia and the
Western Republics

SECTION 2
Transcaucasia

SECTION 3
Central Asia

HUMAN GEOGRAPHY OF RUSSIA AND THE REPUBLICS

A Diverse Heritage

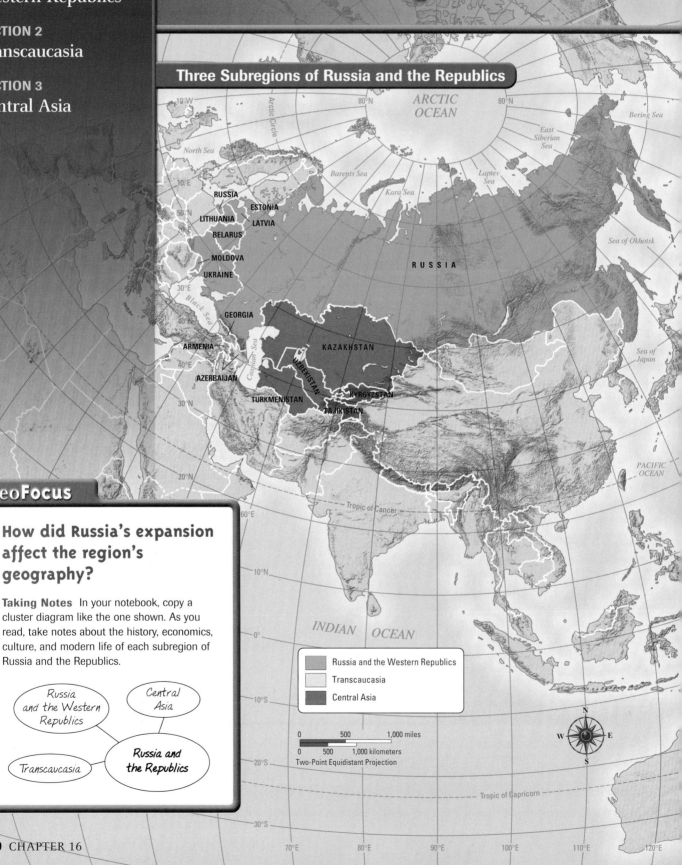

Three Subregions of Russia and the Republics

Russia and the Western Republics

Transcaucasia

Central Asia

0   500   1,000 miles
0   500   1,000 kilometers
Two-Point Equidistant Projection

GeoFocus

How did Russia's expansion affect the region's geography?

Taking Notes In your notebook, copy a cluster diagram like the one shown. As you read, take notes about the history, economics, culture, and modern life of each subregion of Russia and the Republics.

Russia and the Western Republics

Central Asia

Transcaucasia

Russia and the Republics

Russia and the Western Republics

Main Ideas
- From modest beginnings, Russia expanded to become the largest country in the world.
- The rise and fall of the Soviet Union affected the world's political geography.

Places & Terms
Baltic Republics
czar
Russian Revolution
USSR
Cold War
command economy
collective farm

CONNECT TO THE ISSUES
ECONOMIC CHANGE The region is struggling to move from a command economy to a market economy.

A HUMAN PERSPECTIVE Early in the 1500s, the Russian leader Ivan the Great put an end to two centuries of foreign rule in his homeland. Russia then entered a period of explosive growth. From its center in Moscow, Russia expanded at a rate of about 55 square miles a day for the next four centuries. During the expansion, Russians made so much progress toward the east that they swallowed up a future U.S. state, Alaska. Russia had taken control of the territory by the late 18th century but did not sell it to the United States until 1867.

A History of Expansion

Russia's growth had lasting effects on nearby lands and peoples. You can see these effects even today in the republics to its west: Belarus, Moldova, Ukraine and the **Baltic Republics** of Estonia, Latvia and Lithuania. But Russian expansion not only affected its neighbors. It also had an impact on the entire world's political geography.

BIRTH OF AN EMPIRE The Russian state began in the region between the Baltic and Black seas. In the ninth century, Vikings from Scandinavia came to the region to take advantage of the river trade between the two seas. They established a settlement near what is now Kiev, a city near the Dnieper River. In time, the Vikings adopted the customs of the local Slavic population. Soon the settlement began to expand.

Expansion was halted in the 13th century with the arrival of invaders from Mongolia, called Tatars. The ferocity of those Mongol warriors is legendary. It is said that "like molten lava, they destroyed everything in their path." The Tatars sacked Kiev between 1237 and 1240.

The Mongols controlled the region until the 1500s, when Ivan the Great, the powerful prince of Moscow, put an end to their rule. Russia continued once again to expand to the east. By the end of the 17th century, it had built an empire that extended to the Pacific Ocean. As the leaders of Russia added more territory to their empire, they also added more people. Many of these people belonged to different ethnic groups, spoke different languages, and practiced different religions.

MOVEMENT This Mongol armor from the 14th or 15th century includes a case for bow and arrows. Mongol warriors were skilled archers, even on horseback.

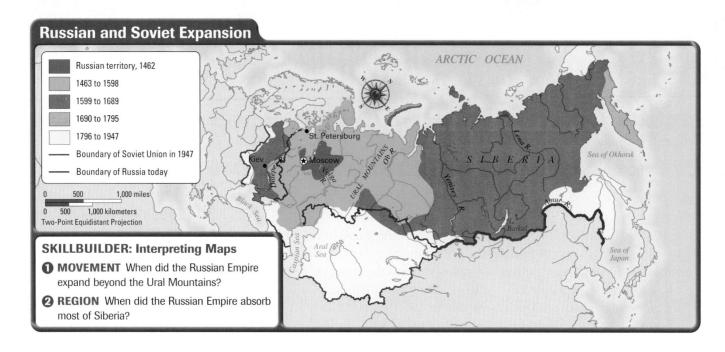

Russian and Soviet Expansion

Russian territory, 1462
1463 to 1598
1599 to 1689
1690 to 1795
1796 to 1947
—— Boundary of Soviet Union in 1947
—— Boundary of Russia today

0   500   1,000 miles
0   500   1,000 kilometers
Two-Point Equidistant Projection

ARCTIC OCEAN

St. Petersburg
Kiev
Moscow
SIBERIA
URAL MOUNTAINS
Ob R.
Lena R.
Yenisey R.
Volga R.
Dnieper R.
Amur R.
L. Baikal
Sea of Okhotsk
Sea of Japan
Black Sea
Caspian Sea
Aral Sea

SKILLBUILDER: Interpreting Maps

❶ MOVEMENT When did the Russian Empire expand beyond the Ural Mountains?

❷ REGION When did the Russian Empire absorb most of Siberia?

RUSSIA LAGS BEHIND WESTERN EUROPE Russia's territorial growth was rapid, but its progress in other ways was less impressive. Russian science and technology lagged behind that of its European rivals. Peter the Great, who was **czar**—or emperor—of Russia from 1682 to 1725, tried to change this. For example, he moved Russia's capital from Moscow to a city on the Baltic Sea. The new capital, named St. Petersburg, provided direct access by sea to Western Europe. Russians called St. Petersburg their "window to the West."

Peter the Great made impressive strides toward modernizing Russia, but the empire continued to trail behind the West. While the Industrial Revolution swept over many Western European countries in the first half of the 1800s, Russia did not even begin to industrialize until the end of the century. When industry did come to Russia, it resulted in harsh working conditions, low wages, and other hardships. These problems contributed to the people's anger at the czars who ruled Russia.

BACKGROUND
The word *czar* comes from the Latin for *Caesar*, the title of address for Roman emperors.

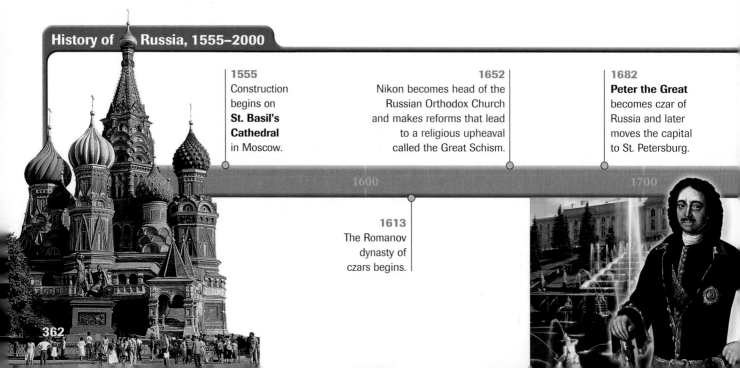

History of Russia, 1555–2000

1555
Construction begins on **St. Basil's Cathedral** in Moscow.

1652
Nikon becomes head of the Russian Orthodox Church and makes reforms that lead to a religious upheaval called the Great Schism.

1682
Peter the Great becomes czar of Russia and later moves the capital to St. Petersburg.

1600

1700

1613
The Romanov dynasty of czars begins.

THE RISE AND FALL OF THE SOVIET UNION During World War I (1914–1918), the Russian people's anger exploded into revolt. In 1917, the **Russian Revolution** occurred, ending the rule of the czars. The Russian Communist Party, led by V. I. Lenin, took control of the government. The Party also took charge of the region's economy and gave Communist leaders control over all important economic decisions.

By 1922, the Communist Party had organized the different peoples absorbed during the centuries of Russia's imperial expansion. This new nation was called the Union of Soviet Socialist Republics (**USSR**), or the Soviet Union for short. The leaders of the Soviet Union ruled the nation from its new capital in Moscow.

By the time World War II broke out in 1939, Joseph Stalin had taken over the leadership of the USSR. In 1941, he led the Soviet Union in the fight against Nazi Germany. However, as the war dragged on, relations between the Soviet Union and its allies—including the United States—began to worsen.

After the war, Stalin installed pro-Soviet governments in the Eastern European countries that his armies had liberated from Germany. U.S. leaders feared that a new stage of Russian expansion was beginning and that Stalin would spread communism all over the globe. By the late 1940s, tensions between the United States and the Soviet Union led to conflict. Diplomats called this conflict the **Cold War** because it never grew into open warfare between the two nations. ◀Ⓐ

The rivalry between the two superpowers continued into the mid-1980s. At that time, Soviet leader Mikhail Gorbachev started to give more economic and political freedom to the Soviet people. This began a process that led to the collapse of the Communist government and the Soviet Union in 1991—and the end of the Cold War.

After the fall of the Soviet Union, the region was divided into 15 independent republics. Of these, Russia, formally known as the Russian Federation, is the largest and most powerful. Today, Russia has a popularly elected president. Its legislature, the Federal Assembly, is divided into two chambers—the Federation Council and the State Duma.

BACKGROUND
Soviet comes from the Russian word for a governmental council, or assembly.

🌐 **Geographic Thinking**◀

Using the Atlas
Ⓐ Examine the map on page 339. Why do you think the Soviet Union had so much power over Eastern Europe?

1812
The French general, **Napoleon** (below), retreats during winter after a failed attack on Russia.

1861
Czar Alexander II issues an order freeing all Russian serfs, that is, peasants held in servitude.

1991
The Soviet Union falls, and Russia becomes independent.

2000
The Russians elect **Vladimir Putin** president.

1800

1900

1917
The Russian Revolution occurs. **Lenin** (above left), then **Stalin** (above right), take power.

1922
Russia becomes part of the Soviet Union.

REGION The Soviet government's control over the economy was often inefficient. Citizens had to wait in line for hours, even to buy basic consumer goods, such as a handbag.

Building a Command Economy

The Communists who overthrew czarist Russia in 1917 had strong ideas about the future. When they put their ideas into practice, they drastically transformed the economic geography of the region.

AN ECONOMIC DREAM The Communists had been inspired by the work of Karl Marx, a German philosopher who had examined the history of economic systems. Marx believed that the capitalist system was doomed because it concentrated wealth in the hands of a few and left everyone else in poverty. He predicted that a Communist system would replace capitalism. In a Communist society, he argued, citizens would own property together, and everyone would share the wealth.

A HARSH REALITY To move their society toward communism, Soviet leaders adopted a **command economy**—one in which the central government makes all important economic decisions. The government took control of the major sources of the state's wealth, including land, mines, factories, banks, and transportation systems. Government planners decided what products factories would manufacture, what crops farms would grow, and even what prices merchants would charge for their goods.

Rapid industrialization became a major goal of Soviet planning. Even farming became an industry under Stalin. The Soviet government created enormous **collective farms** on which large teams of laborers were gathered to work together. People were moved to the farms by the thousands. By 1939, nearly nine out of ten farms were collectives. The Soviets had firmly established their power over the countryside.

Although industrial and agricultural production increased, the region's people had to make great sacrifices for this rapid transformation. Millions of citizens starved to death in famines caused, in part, by the creation of collective farms. Those who survived soon realized that only a small number of individuals had benefited from the economic changes.

Many people tried to do something about this betrayal, but at great risk. Under Stalin's rule, the police swiftly punished any form of protest. Some historians estimate that Stalin was responsible for the deaths of more than 14 million people.

Since the fall of the Soviet Union in 1991, leaders in Russia and the Republics have tried to reduce the state's monopoly on economic power and return some control to private individuals and businesses. You will learn more about these changes in Chapter 17. B▷

CONNECT TO THE ISSUES
ECONOMIC CHANGE
◀B Considering how long the Soviet command economy lasted, why might the change to a market economy be hard for the region's citizens?

A Rich Culture

Russia and the Western Republics faced hard times under the czars and the communists. But these leaders could not destroy the cultural and spiritual traditions of the region's people.

ETHNICITY AND RELIGION The region has a rich variety of ethnic groups because of the many peoples absorbed during the centuries of Russian expansion. Russia has the greatest ethnic diversity of the region's republics. Russians make up the largest ethnic group there, with about 80 percent of the total. But nearly 70 other peoples live in Russia, including Finnish, Turkic, and Mongolian peoples. (See the map on page 341 of the Unit Atlas.)

Russia and the Western Republics are home to a great number of religions. Most Russians follow Orthodox Christianity, a religion Russia adopted in the 10th century. But the region is home to many other religions, including Buddhism and Islam. Judaism is also an important religion in the region. However, persecution has led large numbers of Jews to emigrate, especially to Israel and the United States.

ARTISTIC GENIUS Religion and art are closely related in Russia and the Western Republics. The art and architecture of Orthodox Christian churches, for example, are among the region's earliest artistic achievements. Even today, citizens adore the beautiful onion-shaped domes and the icons—images of sacred Christians—that ornament the churches.

Regional culture went through great change after Peter the Great began to promote communication with Western Europe. As Russian artists combined artistic ideas from the West with their own experiences, a truly golden age of culture began.

Geographic Thinking

Seeing Patterns
How did the expansion of the Russian Empire affect the ethnic and religious makeup of the region?

HUMAN-ENVIRONMENT INTERACTION The two churches in this photo are on Kizhi Island, in Karelia, Russia. The churches' onion domes help to prevent the accumulation of snow during the winter.
How can architecture reflect the influence of geography?

REGION The Communist Party recruited artists to help promote Soviet industry. The poster above, from 1947, and the one below, from 1931, promise punishment for laziness and rewards for hard work.

In the 18th and 19th centuries, audiences around the world marveled at the work of writers such as Aleksandr Pushkin and Feodor Dostoyevsky. Their dramatic scenes and colorful psychological studies give an important portrait of Czarist Russia.

Great composers such as Peter Tchaikovsky and Igor Stravinsky also earned worldwide attention, as did the Russian ballet. Russian ballet companies, such as the Kirov and Bolshoi, are famous for producing magnificent dancers and creative choreographers, such as Mikhail Baryshnikov.

Art underwent another major change after the Communist Party began to outlaw artists who did not work in the official style. This style, called socialist realism, promoted Soviet ideals by optimistically showing citizens working to create a socialist society. In spite of the censorship, many artists took great risks to continue producing original work. Since the collapse of the Soviet Union, artistic expression has begun to gain strength.

Tradition and Change in Russian Life

Since the collapse of the Soviet Union, the region is more open to the influence of other countries—especially those in the West. At the same time, the region's people continue to honor their traditions and work hard to preserve them.

A MORE OPEN SOCIETY The region's people—especially in larger cities—have begun to enjoy more social and cultural opportunities. Large cities, such as Moscow and St. Petersburg, now resemble major cities in the West. City dwellers can read books, magazines, and newspapers from all over the world. They are able to keep up with new movies, music, and clothing trends. They can also experience a wide variety of foods and cuisines.

Although the variety of social and cultural opportunities has increased, native traditions have survived. For example, in spite of the many cuisines now available in Russian cities, many Russians still favor their traditional foods. Many of the foods, such as rye bread,

Geographic Thinking

Seeing Patterns

D How do Russian foods reflect regional geography?

reflect the large crops of grain produced on the region's steppes. Kasha is another popular food made from grain. It is cooked and eaten with butter. Even Russia's national drink, vodka, is made from rye or wheat grains. ◁)

DACHAS AND BANYAS Only a quarter of Russia's population lives in rural areas. Even so, many Russians cherish the nation's countryside. Nearly 30 percent of the population own homes in the country, where they spend weekends and vacations. These homes, called *dachas,* are usually small, plain houses and often have gardens in which to grow vegetables.

One of the customs that Russians enjoy both in the countryside and the cities is visiting a *banya.* A *banya* is a bathhouse in which Russians perform a cleaning ritual that combines a dry sauna, steam bath, and often a plunge into ice-cold water.

Russians begin the ritual by warming up in a sauna heated to around 200°F. They then move into a steam room, where they use birch twigs to ease the muscles and perfume the body. After spending time in the steam room, many bathers plunge into an icy-cold pool—which might be a hole cut in the ice of a river or a lake. The ice bath is followed by hot tea, and the process is repeated. A visit to the *banya* can sometimes last for two to three hours.

The preservation of such customs and traditions by the Russian people has played an important role since the fall of the Soviet Union. It has helped to make the change from the isolated Soviet past to the more open society of the present less difficult.

PLACE A man enjoys an ice-cold dip during his trip to the *banya.*

RUSSIA & REP.

SECTION 1 Assessment

1 Places & Terms

Explain the importance of each of the following terms and places.

- Baltic Republics
- czar
- Russian Revolution
- USSR
- Cold War
- command economy
- collective farm

2 Taking Notes

REGION Review the notes you took for this section.

- How did the Russian Empire lag behind its European rivals?
- How did the Communist Party control artistic expression?

3 Main Ideas

a. What was the extent of the Russian Empire's expansion between the 9th century and the end of the 17th century?

b. What were the origins of the Soviet Union?

c. How did the Soviet Union come to an end?

4 Geographic Thinking

Making Inferences How did the economic policies of the Soviet Union affect its human geography? **Think about:**

- industrialization
- collective farms

S **See Skillbuilder Handbook, page R4.**

EXPLORING LOCAL GEOGRAPHY You read in this section how Russia's traditional food is related to its geography. Do research on the traditional foods where you live, and try to determine how they might be related to the physical or human geography of your region. Explain the connections that you find in an **oral report.**

Disasters!

Nuclear Explosion at Chernobyl

On April 28, 1986, engineer Cliff Robinson arrived at Sweden's Forsmark nuclear power plant. He was startled when a radiation detector went off as he entered his office. When he checked the radiation levels on his clothing, he could not believe his eyes. "My first thought," said Robinson, "was that a war had broken out and that somebody had blown up a nuclear bomb." What Forsmark had detected was a radioactive cloud from the city of Chernobyl—site of a Soviet nuclear power plant nearly 800 miles away.

One of Chernobyl's nuclear reactors had exploded, spewing radioactive dust across the region. It took two days for Soviet officials to admit that the explosion had occurred. The blast killed 31 people. No one is certain what toll accident-related diseases will take on the region's population in the future.

The Spread of Radiation from Chernobyl

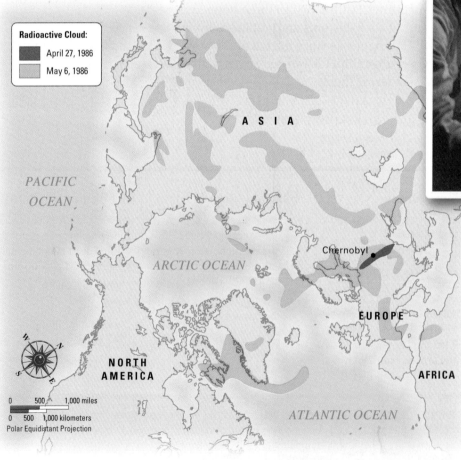

Radioactive Cloud:
- April 27, 1986
- May 6, 1986

ASIA

PACIFIC OCEAN

ARCTIC OCEAN

Chernobyl

EUROPE

NORTH AMERICA

AFRICA

0   500   1,000 miles
0   500   1,000 kilometers
Polar Equidistant Projection

ATLANTIC OCEAN

Workers test radiation levels from a helicopter. After the explosion, hundreds of thousands of workers helped in cleanup operations. Many were exposed to radiation and required emergency medical treatment.

The radioactive cloud from Chernobyl eventually spread over the entire Northern Hemisphere.

Serious health problems, such as thyroid cancer, have increased dramatically among children since the accident at Chernobyl.

A close-up of the damage at Chernobyl's Unit 4 reactor (left). The color image below shows the concrete and steel "sarcophagus," or enclosure, later built around the contaminated reactor.

GeoActivity

PLANNING A PRESENTATION

With a partner, use the Internet to research Chernobyl today. Plan a **multimedia presentation** about the disaster's legacy.

- Design charts, graphs, and maps that show the disaster's impact on public health and the environment.
- Include personal stories from individuals whose lives have been affected by the explosion.

RESEARCH LINKS
CLASSZONE.COM

GeoData

DAMAGE REPORT

- The Chernobyl plant is located about 80 miles north of Kiev, Ukraine's capital.
- The plant once employed nearly 9,200 people.
- On April 26, 1986, a poorly planned safety experiment led to the explosion at Chernobyl, which was made worse by a faulty reactor design.
- The reactor explosion was the world's worst civilian nuclear accident.
- The explosion contaminated around 100,000 square miles of land in Ukraine, Russia, and Belarus.
- Officials evacuated and resettled approximately 250,000 people from different towns around Chernobyl.
- Chernobyl continued to produce electricity until December 15, 2000, when officials finally shut down its last operating reactor.
- Costs related to the disaster have been estimated at over $300 billion.

Transcaucasia

Main Ides

Main Ideas

- Transcaucasia has been a gateway between Europe and Asia.
- The Caspian Sea's oil and gas reserves have given the region great economic potential.

Places & Terms

Red Army

supra

CONNECT TO THE ISSUES
CONFLICT Ethnic tensions in Transcaucasia erupted in conflict after the fall of the Soviet Union.

A HUMAN PERSPECTIVE Throughout history, human beings have migrated through Transcaucasia, which today consists of the republics of Armenia, Azerbaijan, and Georgia. Recent discoveries have shown just how early such migrations began. In the summer of 1999, a team of scientists discovered two 1.7-million-year-old human skulls in the Transcaucasian republic of Georgia. They were the oldest human fossils found outside Africa. Reports suggest that the skulls could belong to the first people to have migrated from Africa.

A Gateway of Migration

People have long used Transcaucasia as a migration route, especially as a gateway between Europe and Asia. Trade routes near the Black Sea led to the thriving commercial regions of Mediterranean Europe. And trade routes leading to the Far East began on the shores of the Caspian Sea.

A VARIETY OF CULTURES Because of the presence of so many trade routes, Transcaucasia has been affected by many different peoples and cultures. Today, more than 50 different peoples live in the region.

Migrants brought a great variety of languages to the region. Arab geographers called the region *Jabal Al-Alsun,* or the "Mountain of Language." The Indo-European, Caucasian, and Altaic language families are the region's most common.

MIGRATION BRINGS RELIGIONS The people of Transcaucasia follow a number of different religions. However, most of the region's people belong to either the Christian or the Islamic faith.

Languages Around the Caucasus

INDO-EUROPEAN
- **Armenian** Armenian
- **Iranian** Kurdish, Ossetic
- **Slavic** Russian

CAUCASIAN
- **Abkhazo-Adyghian** Abkhaz, Adyghian
- **Kartvelian** Georgian, Mingrelian/Laz
- **Nakho-Dagestanian** Avar, Chechen, Dargin, Lak, Lezgi, Tabasaran

ALTAIC
- **Turkic** Azerbaijani, Karachay, Kumyk

OTHER LANGUAGES

- - - Transcaucasia

RUSSIA

Black Sea

GEORGIA
Tbilisi

ARMENIA
Yerevan

TURKEY

AZERBAIJAN
NAGORNO-KARABAKH   Baku

AZER.

Caspian Sea

IRAN

0   100   200 miles
0   100   200 kilometers
Lambert Azimuthal Equal-Area Projection

SKILLBUILDER: Interpreting Maps

❶ **REGION** Which is the most common language group in Azerbaijan?

❷ **PLACE** Which language is spoken in the Nagorno-Karabakh region of Azerbaijan?

These faiths arrived in the region at an early date, because Transcaucasia is close to the areas in Southwest Asia where the two religions began. Armenia and Georgia, for example, are among the oldest Christian states in the world. Armenia's King Tiridates III converted to Christianity in A.D. 300. A year later, he made his state the first in the world to adopt Christianity.

Not long after the 7th-century beginnings of Islam in Southwest Asia, Muslim invaders stormed into the southern Caucasus and converted many Transcaucasians to Islam. Today, the great majority of Azerbaijan's people are Muslim.

CONFLICT The region's diverse population has not always lived together in harmony. Tensions seldom erupted into open hostility under the rigid rule of the Soviets. However, after the collapse of the USSR in 1991, tensions among different groups have resulted in violence. Civil war broke out in Georgia, and Armenia fought a bitter war with Azerbaijan over a disputed territory called Nagorno-Karabakh.

The story of conflict is not new to Transcaucasia. Its history of conflict, as you will read below, can be explained, in part, by its location.

CONNECT TO THE ISSUES
CONFLICT
Why did ethnic tensions seldom erupt into violence during the Soviet era?

PLACE The beautiful Karmravor Church is located in the Armenian village of Ashtarak. It dates from the 7th century. **How long after Armenia adopted Christianity was the church built?**

A History of Outside Control

Over the centuries, Transcaucasia has been a place where the borders of rival empires have come together. Imperial armies have repeatedly invaded the region to protect and extend those borders.

CZARIST AND SOVIET RULE In the 18th century, the troops of the Russian Empire joined the list of invaders. Russia's southward expansion had begun as early as the 1500s, but it was only in the 1700s that the czar's army began making progress south of the Caucasus Mountains.

The inhabitants of the region resisted the Russians, but the czar's troops prevailed. By 1723, Peter the Great's generals had taken control of Baku, the capital of Azerbaijan. In 1801, Russia annexed Georgia. In 1828, Russian armies took control of a large stretch of Armenian territory, including the plain of Yerevan. By the late 1870s, the czar's troops had added Transcaucasia to the Russian Empire.

After the Russian Revolution in 1917, the Transcaucasian republics enjoyed a brief period of independence. By the early 1920s, however, the **Red Army**—the name of the Soviet military—had taken control of the region.

In the decades following the Soviet takeover, the people of Transcaucasia experienced the same painful economic and political changes as the rest of the Soviet Union. Many people lost their lives in famines triggered by the shift to collective farming or were killed because of their political beliefs. The republics of Transcaucasia regained their political independence in 1991 after the fall of the Soviet Union. Since then, the region's leaders have struggled to rebuild their nations' economies.

BACKGROUND
Stalin was especially harsh on Transcaucasia, even though he was from the Georgian town of Gori.

Economic Potential

Today, economic activity in the Transcaucasian republics ranges from the tourism and wine industries of subtropical Georgia to large-scale oil production in Azerbaijan.

AGRICULTURE AND INDUSTRY Although much of Transcaucasia's terrain is mountainous, each of the republics has a significant agricultural output. Transcaucasians have taken advantage of the region's climate and the potential of the limited amount of land fit for farming.

The humid subtropical lowlands and foothills of the region are ideal for valuable crops such as tea and fruits. Grapes are one of the most important fruit crops. Georgians use the grapes cultivated along their Black Sea coast to produce their famous wines. Georgia's mild climate also once fueled a profitable tourist industry.

There was little industry in Transcaucasia before the Soviet Union took control of the region. Soviet planners transformed Transcaucasia from a largely agricultural area into an industrial and urban region.

PLACE

Trouble in Georgia

In the late 1980s, more than 3.6 million tourists visited Georgia each year. But tourism slowed to a trickle after ethnic conflict broke out in the region in the early 1990s.

One conflict took place in Abkhazia—a resort area that stretches for more than 100 miles along Georgia's Black Sea coast. Ethnic Abkhazians sought independence and rebelled against Georgia, which sent troops to prevent the uprising. The conflict remained unresolved at the beginning of 2001.

In this photo, from 1993, soldiers help a boy flee from street fighting in Sokhumi, the capital of Abkhazia.

A number of industrial centers built by the Soviets continue to produce iron, steel, chemicals, and consumer goods for the region's economy. But today, the oil industry is most important. The oil industry has an impact not only on oil-rich republics, such as Azerbaijan. It also affects Armenia and Georgia because oil producers want to build pipelines across their territory to bring the oil to market.

LAND OF FLAMES The significance of oil in the region has a long history. In fact, the name Azerbaijan means "land of flames." The republic's founders chose the name because of the fires that erupted seemingly by magic from both the rocks and the waters of the Caspian Sea. The fires were the result of underground oil and gas deposits.

DIVIDING THE CASPIAN SEA Since the breakup of the Soviet Union, Azerbaijan and the other four countries bordering the Caspian Sea have argued about whether the Caspian is an inland sea or a lake. The resolution of this argument will decide how resources are divided among the five countries.

If the Caspian is a sea, then each country has legal rights to the resources on its own part of the sea bed. If it is a lake, the law says that most of the resource wealth must be shared equally among each of the countries. Azerbaijan, with large reserves off its coast, says the Caspian is an inland sea. Russia, with few offshore reserves, insists that the Caspian is a lake. **◀B**

The oil industry has given the region's people hope for a better life. But oil revenue has benefited few Transcaucasians. Many continue to live in poverty.

Dividing the Caspian

As Lake

KAZAKHSTAN
RUSSIA
Caspian Sea
Joint area
UZBEKISTAN
GEORGIA
AZERBAIJAN
Baku
TURKMENISTAN
AZER.
ARMENIA
IRAN

As Inland Sea

KAZAKHSTAN
RUSSIA
Caspian Sea
UZBEKISTAN
GEORGIA
AZERBAIJAN
Baku
TURKMENISTAN
AZER.
ARMENIA
IRAN

SKILLBUILDER: Interpreting Maps

REGION Which are the five countries that border the Caspian Sea?

⊕ **Geographic Thinking◀**

Seeing Patterns
B How can the geographic definition of a body of water affect economic relationships?

Modern Life in Transcaucasia

Although times are tough for many, the region has much to offer, including a well-educated population and a reputation for hospitality.

AN EDUCATED PEOPLE The educational programs of the Soviet Union had a largely positive impact on its people. At the time of the Russian Revolution, only a small percentage of Transcaucasia's population was literate. Communist leaders decided to train a new generation of skilled workers who would be prepared to undertake the tasks of industrial development and modernization. They succeeded, as literacy rates in Transcaucasia rose to nearly 99 percent, among the highest in the world. Today, high quality educational systems remain a priority for Transcaucasians.

HOSPITALITY In their quest for a modern system of education, Transcaucasians have not forgotten the value of their traditions. Among the most important are the region's mealtime celebrations.

RUSSIA & REP.

PLACE At a dinner party held in the Georgian town of Kutol, a woman raises her glass to deliver a toast. **How do the foods you see in the image reflect what you have read about Georgia's climate?**

The Georgian ***supra,*** or dinner party, is one of the best examples of such gatherings. The word *supra* means tablecloth but also refers to any occasion at which people gather to eat and drink.

A *supra* involves breathtaking quantities of food and drink. Meals begin at a table spread with a great number of cold dishes. Two or three hot courses and fruit and desserts follow those. Georgians add locally grown foods, such as grated walnuts, garlic, and an array of herbs and spices to their recipes. And they are able to serve meals with remarkable freshness, thanks to the region's mild climate.

In addition to food and drink, a *supra* is accompanied by a great number of toasts, short speeches given before taking a drink. Georgians take the toasts very seriously because they show a respect for tradition, eloquence, and the value of bringing people together—a goal of great importance for the future of the region.

SECTION 2 Assessment

1 Places & Terms

Explain the importance of each of the following terms.

• Red Army

• *supra*

2 Taking Notes

REGION Review the notes you took for this section.

Russia and the Republics

Transcaucasia

• How do Transcaucasia's republics differ in terms of religion?

• What sorts of activities take place during a Georgian *supra*?

3 Main Ideas

a. How would you describe the ethnic and linguistic makeup of Transcaucasia?

b. What roles did Russia and the Soviet Union play in Transcaucasia?

c. How has the oil industry affected the people of Transcaucasia?

4 Geographic Thinking

Determining Cause and Effect How did the economic goals of the Soviet Union affect educational values in Transcaucasia? **Think about:**

• Transcaucasia's economy before the 1920s

• the impact of economic changes on the region's workers

MAKING COMPARISONS Carry out more research on the religions of Transcaucasia. Then write a **script** for a five minute documentary that compares the architectural styles used in two different houses of worship.

Because they are always on the move and must carry what they own, nomads have few possessions. They usually carry what is most useful. Even so, many of the possessions of Central Asia's nomads are both useful and beautiful.

YURTS Among the most valuable of the nomads' possessions are their tents—called **yurts.** Yurts are light and portable. They usually consist of several layers of felt stretched around a wooden frame, often made of willow. The outermost layer of felt is coated with the waterproof fat of sheep.

As the photo on page 378 shows, the inside of a yurt can be stunningly beautiful. To block the wind, nomads hang reed mats, intricately woven with the grasses of the steppe. For storage, they suspend woven bags on their tent walls. The inlaid wooden saddles of their horses and their carved daggers also ornament the yurt.

Perhaps the most beautiful and useful of all the yurt's furnishings are the handwoven carpets. Their elaborate designs, colored with natural plant and beetle dyes, have made the carpets famous. Nomads use them for sleeping, or as floor coverings, wall linings, and insulation.

PRESERVING TRADITIONS The nomadic lifestyle of the peoples of Central Asia is not nearly as widespread as it once was. But many people are working hard to preserve the tradition. One group has organized a network of shepherds' families in Kyrgyzstan who are willing to take in guests. In this way, tourists can experience the daily life of the shepherds, who, in turn, receive a source of income for their families. Central Asians will benefit greatly from such imaginative and productive uses of their traditions.

HUMAN-ENVIRONMENT INTERACTION Two Kyrgyz men wrestle on horseback. This sport, in which contestants try to unseat their opponents, requires strength, skill, and good horsemanship.

Assessment

1 Places & Terms

Explain the importance of each of the following terms.

- Silk Road
- Great Game
- nomad
- yurt

2 Taking Notes

REGION Review the notes you took for this section.

Russia and the Republics — Central Asia

- What were some of the objects traded or transported over the Silk Road?
- Why have some people suggested that a new Great Game is beginning in Central Asia?

3 Main Ideas

a. What was the cause of the Great Game?

b. What impact has Soviet nuclear testing had in Central Asia?

c. What are two important unifying forces in Central Asia?

4 Geographic Thinking

Drawing Conclusions How did the Soviet Union use the human geography of Central Asia to establish control of the region? **Think about:**

- ethnic groups in the region
- how Soviet planners drew borders

RESEARCH LINKS
CLASSZONE.COM

SEEING PATTERNS Carry out more research on the lives of nomads in Central Asia. Focus on the period before the Soviet Union took control of the region. Then make up a **diary entry** that describes the daily activity of a typical nomadic family.

Comparing Cultures

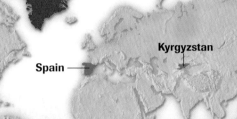

Greenland

Kyrgyzstan

Spain

Indonesia

Homes and Shelters

The geography of the region in which people live influences the nature of their homes and shelters. People who live in forested areas, for example, might build log cabins. People living in grasslands, on the other hand, may use thatch—plant stalks and leaves—to build their homes. On these two pages, you will learn how homes in different parts of the world reflect local geographic possibilities and limitations.

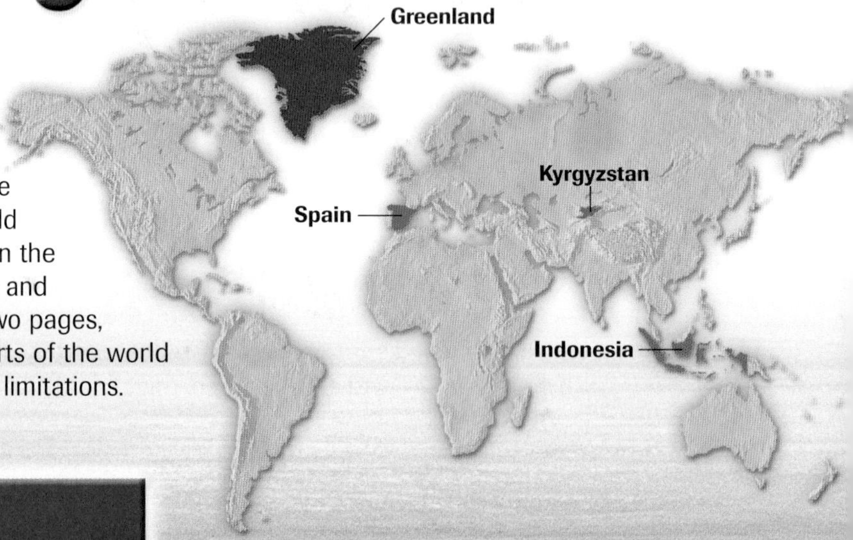

Arctic peoples in Canada and Greenland take advantage of their environment by using blocks of snow to build dome-shaped winter shelters called igloos. They sometimes add windows made with sheets of ice or seal intestines.

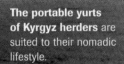

The portable yurts of Kyrgyz herders are suited to their nomadic lifestyle.

The Korowai of Irian Jaya, Indonesia, live in tree houses that protect them from rival tribes, as well as the insects, scorpions, and snakes of the rain forest.

People in the Spanish town of Guadix have turned underground caves into homes to protect against the region's extreme temperatures.

GeoActivity

CREATING AN EXHIBIT
Working with a partner, use the Internet to do research on homes in a region other than those shown on these two pages. Create an **exhibit** that shows the relationship between the region and its homes.

- Construct a model of the homes you are researching.
- Add a world map to the exhibit that shows where the homes are located.

RESEARCH LINKS
CLASSZONE.COM

GeoData

Igloos
- The blocks of snow in an igloo are about 2 feet high, 4 feet long, and 8 inches thick.
- An experienced builder can finish an igloo in one to two hours.

Caves
- About 50 percent of Guadix's inhabitants live underground.
- Some of Guadix's caves are quite luxurious, with marble floors, modern kitchens, fax machines, and Internet connections.

Tree Houses
- The Korowai people build tree houses as high as 150 feet above ground.
- Korowai tree houses have separate areas for men and women, each with its own entrance.

Yurts
- A nomadic family can set up their yurt in approximately a half-hour.
- Felt—the material used to cover yurts—is a fabric of compressed animal fibers, such as wool or fur.

VISUAL SUMMARY
HUMAN GEOGRAPHY OF RUSSIA AND THE REPUBLICS

Subregions of Russia and the Republics

○ **Russia and the Western Republics**

- The explosive growth of the Russian Empire and the following decades of Soviet rule have had a lasting impact on both the physical and human geography of the region.
- The dramatic economic changes that accompanied the rise and fall of the Soviet Union affected both Russia and the Republics and the world.

○ **Transcaucasia**

- Migrating peoples have created a mosaic of languages and ethnicities in Transcaucasia.
- Today, leaders in Transcaucasia are struggling to maintain harmony among the region's different cultural groups and bring stability to the region's three newly-independent republics.

● **Central Asia**

- Central Asia's fractured cultural geography still reflects the political goals of the old Soviet government.
- Powerful unifying forces, such as Islam, may help the region's new republics as they continue to rebuild their social and economic systems.

Reviewing Places & Terms

A. Briefly explain the importance of each of the following.

1. Baltic Republics
2. czar
3. Russian Revolution
4. USSR
5. Cold War
6. command economy
7. collective farm
8. Red Army
9. Silk Road
10. yurt

B. Answer the questions about vocabulary in complete sentences.

11. What were the emperors of the Russian Empire called?
12. What are the names of the three former Soviet republics located on the Baltic Sea?
13. What event ended the Russian Empire and the rule of the czars?
14. What is another name for the Soviet Union?
15. What was the name of the 20th-century conflict between the United States and the Soviet Union?
16. In what type of system are all major economic decisions made by the central government?
17. How did the Soviet Union turn agriculture into an industry?
18. What was the name of the Soviet military?
19. What caravan route contributed to the growth of magnificent trading cities such as Samarqand?
20. What is the name for the felt-covered dwellings of Central Asia's nomads?

Main Ideas

Russia and the Western Republics (pp. 361–369)

1. What former Soviet republics are located west of Russia?
2. What event delayed the growth of Russia before the 16th century?
3. What were the origins of the Cold War?
4. What is the largest religious group in Russia and the Western Republics?

Transcaucasia (pp. 370–374)

5. Of what republics does Transcaucasia consist?
6. Transcaucasia's location between which two seas made it an ideal migration route?
7. What factors may have contributed to instability in Transcaucasia?

Central Asia (pp. 375–381)

8. Of what republics does Central Asia consist?
9. Why did the Silk Road cross over Central Asia?
10. How did Islam become a major religion in Central Asia?

Critical Thinking

1. Using Your Notes

Use your completed chart to answer these questions.

a. What percentage of the Russian population lives in rural areas?

b. What are the main religions in Transcaucasia?

2. Geographic Themes

a. **PLACE** Why did Azerbaijan's founders call it the "land of flames"?

b. **LOCATION** Where was the Soviet nuclear test site called "the Polygon" located?

3. Identifying Themes

Which country in Russia and the Western Republics has the greatest ethnic diversity, and what is its largest ethnic group? Which of the five themes applies to this situation?

4. Making Comparisons

How did the rise of the Soviet Union affect Transcaucasia and Central Asia?

5. Making Generalizations

How can the type of government that a country has affect the kind of work the country's artists create?

Additional Test Practice, pp. S1–S37

TEST PRACTICE
CLASSZONE.COM

Geographic Skills: Interpreting Maps

Central Moscow

Use the map to answer the following questions.

1. **LOCATION** On what river does Russia's capital lie?

2. **MOVEMENT** Which of the ring roads would you take to visit Gorky Park?

3. **MOVEMENT** In which direction would you walk to get from Lenin's tomb to the State Department Store?

GeoActivity

Choose one of the buildings shown on this map, and carry out further research on that building. Create a poster that includes a sketch of the site's floor plan and a paragraph about the history of the building.

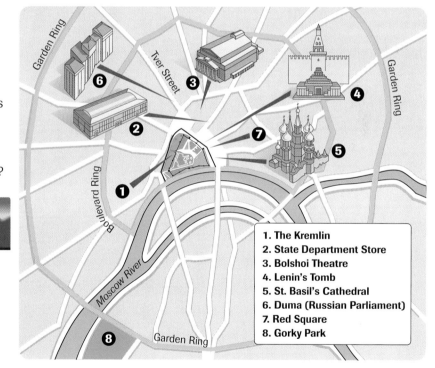

1. The Kremlin
2. State Department Store
3. Bolshoi Theatre
4. Lenin's Tomb
5. St. Basil's Cathedral
6. Duma (Russian Parliament)
7. Red Square
8. Gorky Park

INTERNET ACTIVITY

Use the links at **classzone.com** to do research on two of the former Soviet republics to the west of Russia. Focus on the characteristics of the republics' geography and people.

Creating Charts and Graphs Use your research to create charts and graphs that compare the two republics that you have chosen. List the Web sites that you used in preparing your report.

RUSSIA & REP.

SECTION 1
Regional Conflict

SECTION 2
The Struggle for Economic Reform

CASE STUDY
THE SOVIET UNION'S NUCLEAR LEGACY

For more on these issues in
Russia and the Republics . . .

CURRENT EVENTS
CLASSZONE.COM

This woman and child are
from the Russian Republic
of Chechnya. Russia
invaded Chechnya twice
in the 1990s to prevent
the republic from
becoming independent.

TODAY'S ISSUES
Russia and the Republics

GeoFocus

What impact has the fall of the Soviet Union had on the region?

Taking Notes In your notebook, copy a
cause-and-effect chart like the one below.
Then take notes on causes and effects of
some aspect of each issue.

	Causes	Effects
Issue 1: Conflict		
Issue 2: Economy		
Case Study: Nuclear Legacy		

far below the poverty line. Some people even wondered whether things had been better under the Soviet Union.

Obstacles to Economic Reform

Russians have made slow, if painful, strides toward capitalism. Even so, many obstacles remain. Russia's enormous size and the rise of organized crime are among the most important.

DISTANCE DECAY A major obstacle facing economic reformers is **distance decay.** This means that long distances between places make communication and transportation difficult. Russia is an enormous nation, stretching across 11 time zones. Spread over this vast area are 89 different regional governments. The interaction and cooperation of these regional leaders with Moscow is crucial if the government's economic reforms are to be successful. But because the central government in Moscow has been weak, officials far from the capital sometimes refuse to carry out the government's reform programs.

In the spring of 2000, Russian President Vladimir Putin created seven large federal districts to gain more control over regional leaders. Each has its own governor-general. Putin hopes that the heads of the new federal districts will force regional officials to carry out the economic reforms that Moscow wants. ◢

Geographic Thinking

Seeing Patterns
Ⓐ Some Russians have objected to the creation of new federal districts. Why might there be disagreement over the districts?

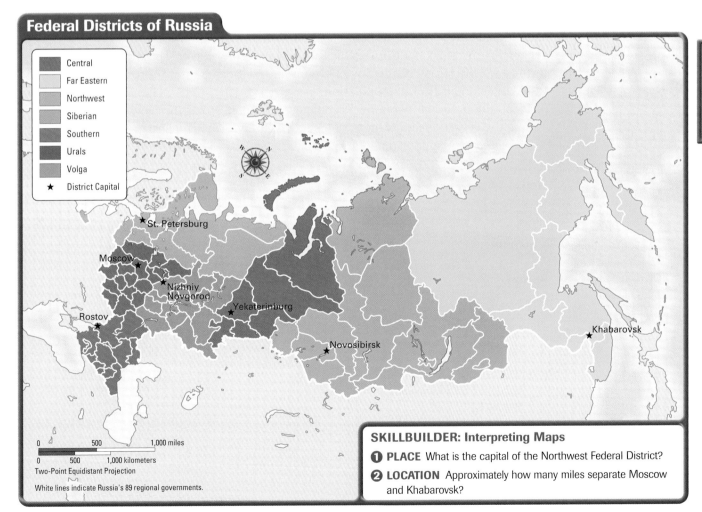

Federal Districts of Russia

Legend:
- Central
- Far Eastern
- Northwest
- Siberian
- Southern
- Urals
- Volga
- ★ District Capital

St. Petersburg
Moscow
Nizhniy Novgorod
Rostov
Yekaterinburg
Novosibirsk
Khabarovsk

0   500   1,000 miles
0   500   1,000 kilometers
Two-Point Equidistant Projection

White lines indicate Russia's 89 regional governments.

SKILLBUILDER: Interpreting Maps
❶ **PLACE** What is the capital of the Northwest Federal District?
❷ **LOCATION** Approximately how many miles separate Moscow and Khabarovsk?

ORGANIZED CRIME As the government tries to improve the economy, it must also face a powerful enemy—organized crime. The "Russian mafia," as criminal organizations in the republic are sometimes labeled, grew rapidly during the 1990s.

By the end of the decade, the mafia had created its own economy. In 1998, the government estimated that organized criminals controlled 40 percent of private companies and 60 percent of state-owned enterprises. Russian criminal activity also expanded outside of Russia. The mafia even tried to sell a Russian submarine to drug barons in Colombia.

PLACE Officers from a special police force in Moscow arrest a suspected mafia car thief in August 1997.
Why might organized crime present a special problem for the Russian government?

The growth of organized crime has slowed economic reform by rewarding illegal activity over honest business. And because illegal activities often go undetected, the government cannot collect taxes on them. Russian officials have taken initiatives to combat organized crime, including the addition of more officers to a special tax police.

FUTURE PROSPECTS In February 2001, Russia's prime minister reported increases in tax and customs revenues. Government officials said the increases are a sign that the Russian economy is on track. If the growth in revenues continues, Russia will be better able to come to terms with the legacy of the Soviet Union and will be able to improve the living standards of its population.

In addition to the economic problems inherited from the Soviet Union, this legacy includes the problems created by Soviet nuclear programs, which you will read about in the next section.

SECTION 2 Assessment

❶ Places & Terms
Explain the importance of each of the following terms.
- privatization
- distance decay

❷ Taking Notes
REGION Review the notes you took for this section.

	Causes	Effects
Issue 2: Economy		

- Why did the Russian government issue vouchers in 1992?
- What impact might organized crime have on government revenue?

❸ Main Ideas
a. What is one of the toughest issues facing Russia's economic reformers?

b. How has Russia moved toward a capitalist system?

c. What are some of the obstacles to economic reform?

❹ Geographic Thinking
Drawing Conclusions Why did President Putin establish seven new federal districts in Russia? **Think about:**
- the number of its regional governments
- Russia's size

S **See Skillbuilder Handbook, page R5.**

SEEING PATTERNS Do research on a U.S. company doing business in Russia. Create a set of **guidelines** that the company might follow in conducting business in Russia.

Reading Line and Pie Graphs

Russia's economy has changed dramatically since the fall of the Soviet Union. To keep track of these changes and plan for the future, economists gather statistics. Presenting statistical data visually in graph form makes the data easier to read.

THE LANGUAGE OF GRAPHS **Line graphs** show the relation between two variables. The line graph below shows changes in Russia's unemployment rate. The vertical axis lists rates of unemployment. The horizontal axis shows the passage of time.

   Pie graphs use percentages to show the relationship of parts to a whole. The pie represents the whole, and each slice of the pie represents a part. The pie graph below shows the distribution of income in Russia.

Economic Conditions in Russia

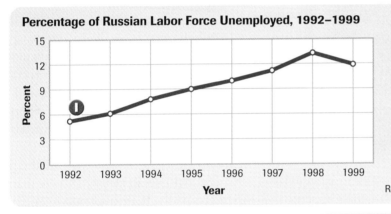

Percentage of Russian Labor Force Unemployed, 1992–1999

SOURCE: IMF Staff Country Report No. 00/150, 2000

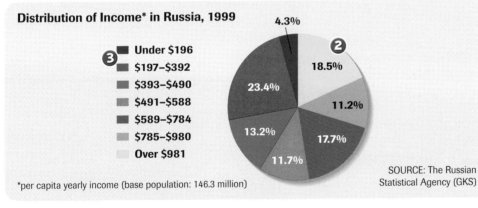

Distribution of Income* in Russia, 1999

■	Under $196
■	$197–$392
■	$393–$490
■	$491–$588
■	$589–$784
■	$785–$980
■	Over $981

4.3% 18.5% 11.2% 23.4% 13.2% 17.7% 11.7%

*per capita yearly income (base population: 146.3 million)

SOURCE: The Russian Statistical Agency (GKS)

1 After the Soviet Union and its command economy collapsed in 1991, Russia became a separate republic. As it struggled to reform its economy, the rate of unemployment began to rise.

2 In 1999, only 18.5 percent of Russia's population earned more than $981.

3 The color key shows the income ranges that correspond to the different slices of the pie graph.

Map and Graph Skills Assessment

1. Seeing Patterns
What was the trend in Russia's unemployment rate after 1992? When did it begin to change?

2. Analyzing Data
What was Russia's unemployment rate in 1998?

3. Analyzing Data
What percentage of Russians earned less than $491 in 1999?

THE SOVIET UNION'S NUCLEAR LEGACY

How have Soviet decisions affected new leaders?

In 1965, Soviet officials used a nuclear bomb to create this reservoir in Semey, Kazakhstan.

A s you have read, the breakup of the Soviet Union sparked regional conflicts and economic hardship. Equally serious were the problems caused by the Soviet Union's nuclear programs. These included nuclear warheads atop ballistic missiles, poorly constructed and maintained nuclear power stations, and decaying nuclear waste dumps. All threatened the region's people and environment.

An Unwelcome Legacy

When the USSR fell apart in the early 1990s, leaders around the world had serious concerns about the fate of the region's nuclear weapons. The Soviet Union, which had once controlled those weapons, was now separated into 15 independent republics. World leaders wanted to know who was in control of the weapons, where they were located, and how well they were protected. They also wondered what would become of the nuclear scientists who had worked on the weapons systems.

The weapons industry was just part of the problem. As the 1986 disaster at Chernobyl had so clearly shown, many of the region's nuclear reactors were badly built and poorly managed. Many reactors of the same design as the one that exploded at Chernobyl still exist. Observers fear another disaster may occur in the region.

Selected Nuclear Reactors in Former Soviet Republics

Chernobyl-type nuclear reactor

2 Number of reactors

* Last reactor in operation was closed Dec. 2000

SWEDEN

Baltic Sea

FINLAND

ESTONIA
St. Petersburg
RUSSIA LATVIA 4
POLAND LITHUANIA
SLOVAKIA Ignalina
2
HUNGARY Smolensk
BELARUS 1
Chernobyl*
4
ROMANIA UKRAINE
Kursk
MOLDOVA 4 RUSSIA

BULGARIA

Black Sea

0   250   500 miles
0   250   500 kilometers
Two-Point Equidistant Projection

SKILLBUILDER: Interpreting Maps

❶ **PLACE** How many Chernobyl-type nuclear reactors are operating in Lithuania?

❷ **REGION** Why might Finland worry about nuclear power in Russia?

The Consequences of Collapse

The nuclear legacy of the USSR has had serious political, economic, and environmental consequences.

POLITICAL TENSIONS When the communist government could no longer keep the USSR together, the security of the region's nuclear materials became uncertain. This has caused political tension between the region's leaders and other nations, especially the United States.

SEE
PRIMARY SOURCE A

In January 2000, a task force of former U.S. officials issued a report that suggested just how important the issue is. The report said that the possibility of Russian nuclear materials being stolen or misused is "the most urgent unmet national security threat" facing the United States. The task force recommended a $30 billion program to help ensure the safety of Russia's nuclear weapons.

ECONOMIC HEALTH The Soviet Union's nuclear legacy also affects the economic health of Russia and the former Soviet Republics. For example, many regional leaders have been reluctant to shut down aging Soviet reactors because of the expense of building new plants that run on other kinds of fuel, such as natural gas.

Some republics have taken questionable steps to revive their economies. For instance, Russian lawmakers recently approved plans to make their country the world's nuclear dump. In January 2001, the Duma, or legislature, gave preliminary approval to a plan to import, store, and treat nuclear waste from other countries. Officials hope the project will earn Russia as much as $21 billion over the next ten years.

ENVIRONMENTAL PROSPECTS Plans for the disposal of other nations' nuclear waste angered Russian environmentalists. But other developments have given some hope that the region's environmental prospects might improve. In December 2000, the government of Ukraine finally shut down the last active reactor at Chernobyl. Officials there pledged to spend millions of dollars on a new protective dome for the site.

PLACE A Ukrainian official examines a nuclear missile just before it is to be dismantled as part of a U.S.-sponsored program.
Why would the United States sponsor this program in Ukraine?

Help has also come from overseas. In October 2000, a U.S.-funded treatment plant opened near the White Sea. The 17-million-dollar facility will treat radioactive waste from Russia's fleet of nuclear submarines—waste that used to be dumped in the sea.

SEE
PRIMARY SOURCE D

You will learn more about these developments as you examine the primary sources and complete the Case Study Project on the following pages.

CASE STUDY

Damage Assessment Report

Primary sources A to D on these two pages offer different views of the Soviet Union's nuclear legacy. Use these resources and your own research to prepare a damage assessment report of the region's nuclear situation today.

RESEARCH LINKS
CLASSZONE.COM

Suggested Steps

1. Choose a nuclear threat to investigate and examine its political, economic, and environmental consequences.

2. Use online and print resources to research your topic.

3. Be sure your damage assessment includes both causes and effects. Also, explain the steps being taken by regional officials to address the problems.

4. Search for interesting statistics, compelling stories, and first-person accounts to enliven your assessment.

5. Provide maps, charts, graphs, and photos to add visual interest to the assessment.

6. Prepare a brief oral introduction that introduces and explains your topic.

Materials and Supplies

- posterboard
- colored markers
- computer with Internet access
- reference books, newspapers, and magazines
- printer

PRIMARY SOURCE A

Political Cartoon *This cartoon by* **Nick Anderson** *illustrates the frightening prospect of a collapsing nuclear superpower.*

Copyright © 2000 Nick Anderson, The Washington Post Writers Group. Reprinted with permission.

Editorial Commentary On January 21, 1999, the New York Times *offered its comments on Russia's nuclear legacy.*

There is no longer any threat of Russia's deliberately attacking the United States. But Moscow's still-formidable stocks of nuclear bombs, nuclear ingredients, and biological and chemical warfare agents pose a different kind of danger. Much of this material is inadequately secured, and the workers guarding it are paid poorly or not at all. That creates an unacceptably high risk that some material could be sold to potential aggressors like Iraq, Libya, North Korea, or Serbia. Many Russian weapons scientists are also unemployed or unpaid and vulnerable to foreign recruitment.

During the Cold War, the United States spent trillions of dollars to deter Russia from using its nuclear, biological, and chemical weapons. It would not take much more than $10 billion to eliminate most of the risks from those weapons today.

News Report In his dispatch of September 30, 1997, London Daily Telegraph *reporter Christopher Lockwood relates yet another terrible tale from Russia's nuclear legacy of an environmental disaster waiting to happen.*

Nothing on the outside indicates what lies within the retired Russian supply ship *Lepse* except the presence of a Kalashnikov-armed guard and the fact that the vessel is moored at the farthest possible point of the Atomplot shipyard in Murmansk.

In fact, *Lepse* may well be the most terrifying vessel on Earth, loaded with a deadly cargo of warped nuclear-reactor parts and spent fuel rods that would be sufficient to poison the world's population. . . .

For the past six years, Norway and Finland have been negotiating with Russia in an attempt to clear up the mess left by Russia's Northern Fleet, which had its headquarters in Severomorsk, near Murmansk. About 200 disused nuclear reactors and tens of thousands of fuel rods are haphazardly stored at its bases around Murmansk, in the Kola Peninsula.

"If there is a catastrophe in the Kola Peninsula, it can affect the whole of Europe's climate, perhaps for hundreds of years," said Norwegian Defense Minister Joergen Kosmo.

News Report On December 15, 2000, 14 years and 7 months after the reactor explosion at Chernobyl, Ukraine's president ordered the plant closed. The excerpt below, written by New York Times *reporter Michael Wines, outlines the economic impact that the shutdown will have.*

The closing of Unit 3 [the plant's last working reactor] will cut off 5 percent of the electricity supply in a nation already deeply in [debt] to Russia for natural gas and dogged by shortages in its shoddily run power grid.

The closing will also gradually eliminate jobs of thousands of Ukrainians whose work depends, directly or indirectly, on Chernobyl's continued operation as a power plant. Beyond the layoffs at the plant itself, thousands of Ukrainians provide goods or services to Chernobyl workers.

Ukraine also faces immense costs in the future—$750 million to cover the disaster site with a new [protective dome], hundreds of additional millions of dollars to remove 180 tons of lethal melted fuel and steel from the damaged core of Unit 4 and to store it safely, millions to build a new heating system and other necessities for the crews that will permanently care for the idle reactor site and millions for solid and liquid waste-processing plants to handle the fuel from the closing of Unit 3.

PROJECT CheckList

Have I . . .

✓ fully researched my topic?

✓ balanced my report by discussing both sides of the issue?

✓ created informative visuals that make my report clear and interesting?

✓ practiced explaining my report?

✓ anticipated questions others might ask and prepared answers?

VISUAL SUMMARY
TODAY'S ISSUES IN RUSSIA AND THE REPUBLICS

Conflict

Regional Conflict

• Since the fall of the Soviet Union in 1991, a number of ethnic and religious groups have sought more control over their own affairs. Their demands have frequently led to conflict.

• Regional leaders who are trying to end these conflicts face a dilemma. How can they maintain order without resorting to the undemocratic rule of the past?

Economics

The Struggle for Economic Reform

• Another dilemma facing leaders in Russia and the former Soviet republics is how to move away from the old Soviet command economy toward a market economy.

• Leaders are struggling to make reforms without causing too much turmoil for citizens.

Government

The Soviet Union's Nuclear Legacy

• The impact of Soviet nuclear programs did not end with the fall of the Soviet Union in 1991. Russia and the Republics inherited the former state's nuclear weapons, power plants, and waste.

• This legacy has had serious political, economic, and environmental consequences.

Reviewing Places & Terms

A. Briefly explain the importance of each of the following.

1. Caucasus
2. Chechnya
3. Nagorno-Karabakh
4. privatization
5. distance decay

B. Answer the questions about vocabulary in complete sentences.

6. In which nation is Chechnya located?
7. Which region is the subject of a dispute between Armenia and Azerbaijan?
8. How might a nation move from a command economy to a market economy?
9. What is another name for Caucasia?
10. What is the name for the decreasing interaction between places as the distance between them increases?

Main Ideas

Regional Conflict (pp. 385–387)

1. What is the connection between the fall of the Soviet Union and the outbreak of ethnic conflicts in Russia and the Republics?
2. Why might ethnic tensions in the Caucasus be stronger than in other regions?
3. In the Russian part of Caucasia, where has the most serious conflict taken place?

The Struggle for Economic Reform (pp. 388–391)

4. Over the past decade, what has been one of the major goals of Russian economic reformers?
5. How have reformers moved Russia toward a market economy?
6. What are some of the problems faced by economic reformers?

Case Study: The Soviet Union's Nuclear Legacy (pp. 392–395)

7. Why were world leaders concerned about the security of nuclear weapons in Russia and the Republics after 1991?
8. What other aspect of the Soviet nuclear legacy concerned observers?
9. How has the United States assisted Russia in dealing with the nuclear legacy of the Soviet Union?
10. How are the nuclear policies of Russia related to its economic problems?

Critical Thinking

1. Using Your Notes

Use your completed chart to answer these questions.

	Causes	Effects
Issue 1: Conflict		
Issue 2: Economy		

a. What caused several ethnic groups in the Caucasus to believe they might successfully demand independence in the 1990s?

b. What is the intended effect of Russia's new federal districts?

2. Geographic Themes

REGION Why did the division of the USSR into 15 independent republics concern observers of the region's nuclear programs?

3. Identifying Themes

Why did the United States fund a nuclear waste treatment plant near the White Sea? Which of the five themes applies to this situation?

4. Making Inferences

Why might Russian economic reformers worry about causing too much hardship for citizens?

5. Drawing Conclusions

Why do you think Russian legislators want to import, store, and treat nuclear waste from other countries in spite of the environmental risks involved?

Additional Test Practice, pp. S1–S37

TEST PRACTICE
CLASSZONE.COM

Geographic Skills: Interpreting Graphs

Global Male Life Expectancy

Use the graph to answer the following questions.

1. **PLACE** How does male life expectancy in Russia differ from world trends?

2. **PLACE** What was the life expectancy of Russian men in 1990? In 2000?

3. **PLACE** What might account for the dip in life expectancy for Russian men?

Create another line graph that shows how the population of Russia changed during the same period of time.

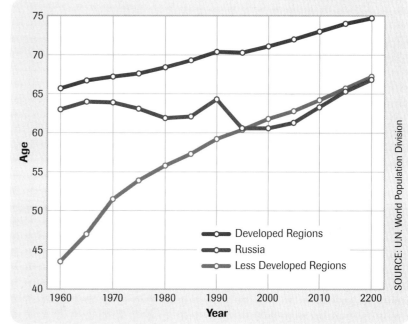

SOURCE: U.N. World Population Division

INTERNET ACTIVITY

Use the links at **classzone.com** to do research on current economic conditions in Russia. Compare the statistics you find on the Russian economy, such as inflation and poverty rates, with statistics on the U.S. economy.

Creating a Multimedia Presentation Create a multimedia presentation of your findings. Include maps and graphs that visually present the information that you discovered.

Africa is the world's second largest continent. Its unique location—almost centered over the equator—affects its vegetation, climate, and population patterns.

LOCATION
A man prays in front of the pyramids at Giza in Egypt.

MOVEMENT
People travel to a market outside of Mali's Great Mosque in Djenné. The mosque is one of the world's largest mud-brick buildings.

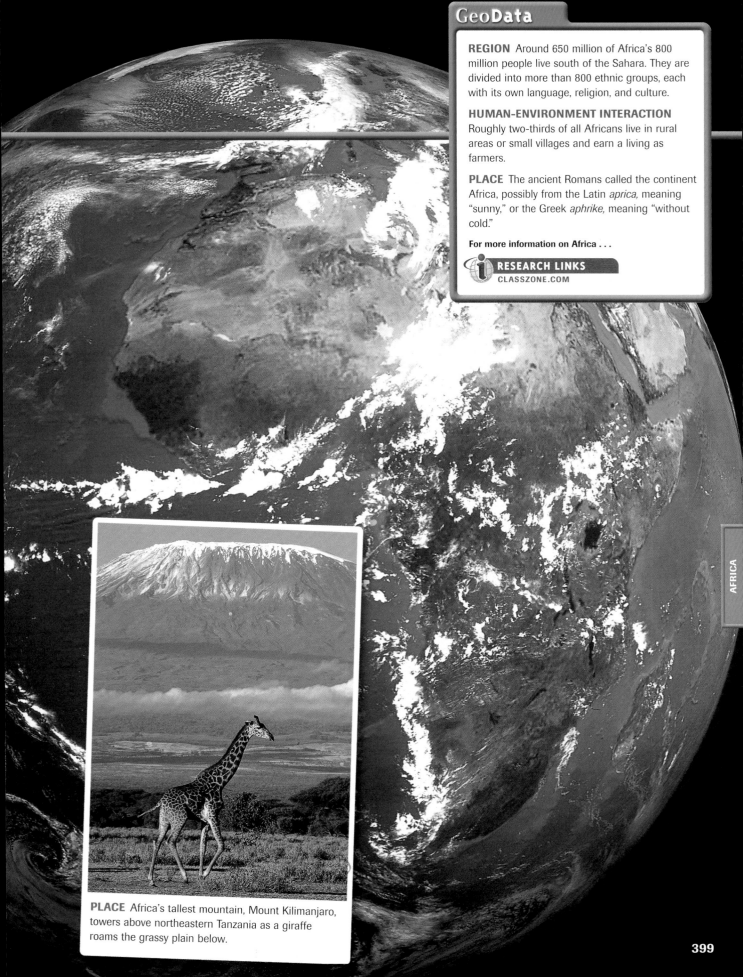

GeoData

REGION Around 650 million of Africa's 800 million people live south of the Sahara. They are divided into more than 800 ethnic groups, each with its own language, religion, and culture.

HUMAN-ENVIRONMENT INTERACTION Roughly two-thirds of all Africans live in rural areas or small villages and earn a living as farmers.

PLACE The ancient Romans called the continent Africa, possibly from the Latin *aprica,* meaning "sunny," or the Greek *aphrike,* meaning "without cold."

For more information on Africa . . .

RESEARCH LINKS
CLASSZONE.COM

AFRICA

PLACE Africa's tallest mountain, Mount Kilimanjaro, towers above northeastern Tanzania as a giraffe roams the grassy plain below.

399

Unit PREVIEW

6

Today's Issues in Africa

Africa faces the issues previewed here. As you read Chapters 18 and 19, you will learn background information. You will study the issues themselves in Chapter 20. In a small group, answer the questions below. Then have a class discussion of your answers.

Exploring the Issues

1. **ECONOMIC DEVELOPMENT** Make a list of some of the pros and cons of economic development. How would economic development benefit people living in Africa?

2. **HEALTH CARE** Search the Internet for information about how African nations are trying to slow the spread of various diseases. What strategies and actions are being employed by these countries?

3. **EFFECTS OF COLONIALISM** Find one news story about political or ethnic violence. How might colonialism be a cause or have contributed to the problem?

For more on these issues in Africa . . .

CURRENT EVENTS
CLASSZONE.COM

ECONOMIC DEVELOPMENT

How can African nations develop their economies?

African nations rely too much on the exportation of natural resources. These miners in Johannesburg, South Africa, mine gold, one of the country's main exports.

HEALTH CARE

How can African countries eliminate the diseases that threaten their people and cultures?

A health clinic in Nairobi, Kenya, attempts to slow the spread of AIDS through various education programs.

← CLINIC

CASESTUDY

How can African nations bring peace and stability to their people?

Many African countries are still suffering from the effects of colonialism. Africa's problems after colonialism are shown in this cartoon about the Democratic Republic of the Congo (formerly known as Zaire). This cartoon shows that there are no easy solutions.

EFFECTS OF COLONIALISM

APATHY

NEGLECT

ZAIRE

AMBIVALENCE

INTERNATIONAL COMMUNITY

INTERVENTION

DESTINATION UNKNOWN...

KING
30.10.96
THE CITIZEN

Unit ATLAS

Patterns of Physical Geography

Use the Unit Atlas to add to your knowledge of Africa. As you look at the maps and charts, notice geographic patterns and specific details about the region. After studying the graphs and physical map on these two pages, jot down in your notebook the answers to the questions below.

Making Comparisons

1. Compare Africa's size and population to that of the United States. How much larger in terms of population and size is Africa compared to the United States?

2. Compare Africa's longest river, the Nile, to the Mississippi. How much difference is there in the lengths?

3. How much bigger is the Sahara than the largest desert in the United States? What is the difference in size between the Sahara and the continental United States?

For updated statistics on Africa . . .

DATA UPDATE
CLASSZONE.COM

Comparing Data

Landmass

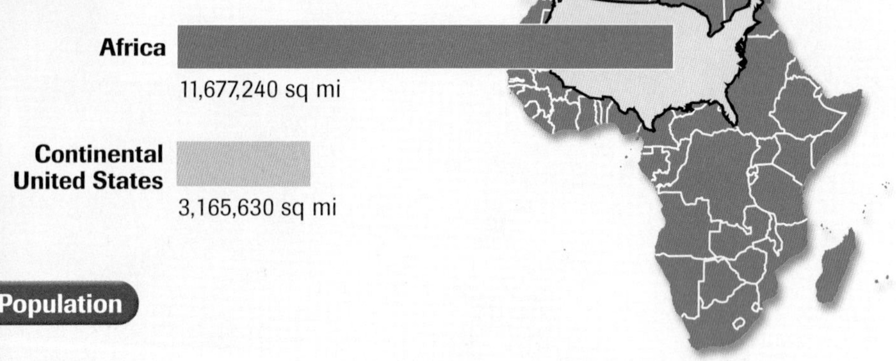

Africa
11,677,240 sq mi

Continental United States
3,165,630 sq mi

Population

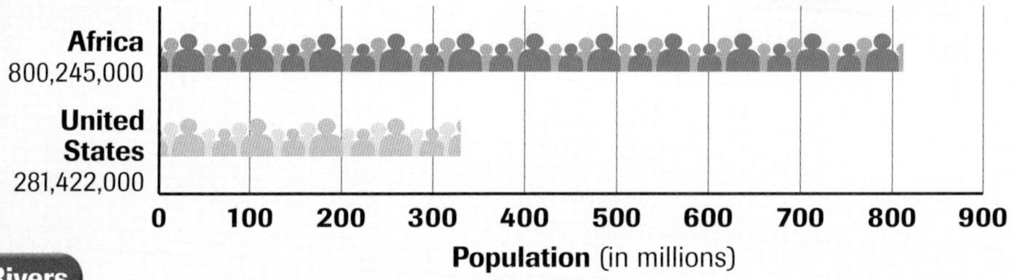

Africa 800,245,000

United States 281,422,000

| 0 | 100 | 200 | 300 | 400 | 500 | 600 | 700 | 800 | 900 |

Population (in millions)

Rivers

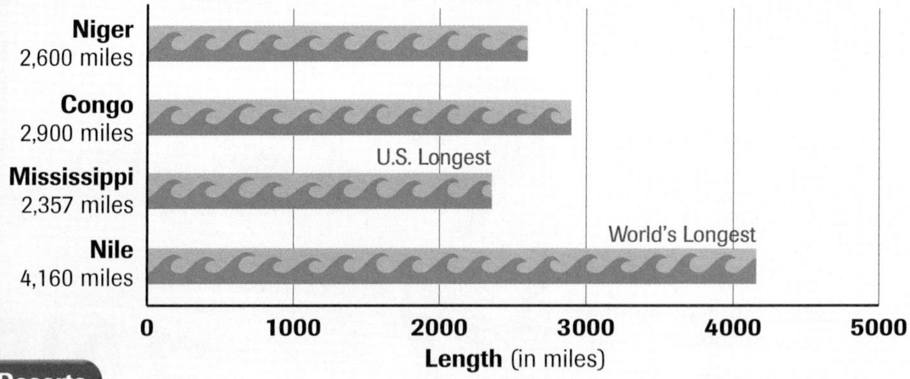

Niger 2,600 miles

Congo 2,900 miles

Mississippi 2,357 miles — U.S. Longest

Nile 4,160 miles — World's Longest

| 0 | 1000 | 2000 | 3000 | 4000 | 5000 |

Length (in miles)

Deserts

World's Largest
Sahara
Africa
3,500,000 sq. miles

U.S. Largest
Mojave
United States
25,000 square miles

Namib
Africa
102,248 square miles

Kalahari
Africa
about 100,000 square miles

UNITED KINGDOM

GERMANY

FRANCE
SWITZ.
HUNGARY
CROATIA
ROMANIA
YUGO.
ITALY
BULGARIA
ALB.
GREECE
TURKEY

SPAIN

PORTUGAL

CYPRUS
LEBANON
SYRIA
ISRAEL
IRAQ
IRAN

Strait of Gibraltar

*Madeira Is.
(Port.)*

ATLAS MOUNTAINS

TUNISIA

Mediterranean Sea

Gulf of Sidra

Suez Canal

JORDAN
KUWAIT

MOROCCO

Canary Islands

WESTERN SAHARA
(MOROCCO)

ALGERIA

LIBYA

EGYPT

SAUDI ARABIA

QATAR

U.A.E.

S   A   H   A   R   A

AHAGGAR MOUNTAINS

Tropic of Cancer

LIBYAN DESERT

Nile R.

L. Nasser

Red Sea

Cape Verde

MAURITANIA

MALI

NIGER

Tibesti Mountains

Nubian Desert

ERITREA

YEMEN

Gulf of Aden

Senegal R.

CAPE VERDE

SENEGAL

S   A   H   E   L

Niger R.

CHAD

L. Chad

SUDAN

White Nile
Blue Nile

DJIBOUTI

GAMBIA

GUINEA-BISSAU

GUINEA

BURKINA FASO

NIGERIA

Benue R.

ETHIOPIAN PLATEAU

ETHIOPIA

HORN OF AFRICA

SIERRA LEONE

CÔTE D'IVOIRE

GHANA

TOGO

BENIN

Mt. Cameroon
13,451 ft.
(4,100 m.) ▲

CENTRAL AFRICAN REPUBLIC

SOMALIA

LIBERIA

CAMEROON

Gulf of Guinea

EQUATORIAL GUINEA

SÃO TOMÉ AND PRÍNCIPE

Ubangi R.

Congo R.

C O N G O

UGANDA

KENYA

Mt. Kenya
17,058 ft.
(5,199 m.) ▲

L. Turkana

Equator

GABON

REP. OF CONGO

B   A   S   I   N

DEMOCRATIC REPUBLIC OF CONGO

RWANDA

BURUNDI

Lake Victoria

Mt. Kilimanjaro
19,341 ft.
(5,895 m.) ▲

SEYCHELLES

ATLANTIC OCEAN

ANGOLA

TANZANIA

L. Tanganyika

Pemba Is.
Zanzibar Is.

St. Helena

Katanga Plateau

L. Nyasa

COMOROS

ANGOLA

ZAMBIA

MALAWI

MAURITIUS

Victoria Falls

Zambezi R.

MOZAMBIQUE

MADAGASCAR

Réunion

ZIMBABWE

NAMIB DESERT

NAMIBIA

BOTSWANA

Limpopo R.

Mozambique Channel

Tropic of Capricorn

KALAHARI DESERT

Orange R.

LESOTHO

SWAZILAND

SOUTH AFRICA

Drakensberg

Karroo

INDIAN OCEAN

Cape of Good Hope

N
W   E
S

Elevation

13,100 ft.	(4,000 m.)
6,600 ft.	(2,000 m.)
1,600 ft.	(500 m.)
650 ft.	(200 m.)
0 ft.	(0 m.)
Below sea level	

▲ Mountain peak

0   400   800 miles
0   400   800 kilometers
Lambert Azimuthal Equal-Area Projection

AFRICA

Unit ATLAS

Patterns of Human Geography

In the years preceding World War I (1914–1918), the political map of Africa changed dramatically. European colonial powers had replaced traditional African states and empires. Study the political maps of Africa in 1913 and Africa today to see how the continent changed by the end of the 20th century. Then answer these questions in your notebook.

Making Comparisons

1. What independent nations appear on the map of Africa in 1913 and also appear on the map of Africa today?

2. Which two European powers controlled the most land in Africa in 1913? Which country controlled the least amount?

3. Which countries in Africa today formed French West Africa in 1913?

4. Which three African countries emerged from colonialism with the most territory?

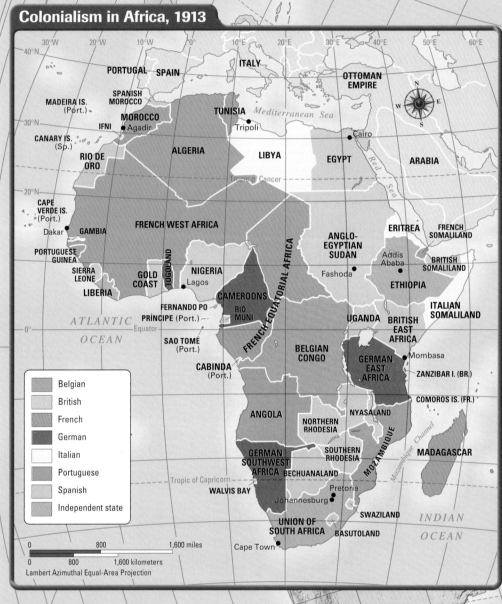

Colonialism in Africa, 1913

Legend:
- Belgian
- British
- French
- German
- Italian
- Portuguese
- Spanish
- Independent state

0 — 800 — 1,600 miles
0 — 800 — 1,600 kilometers
Lambert Azimuthal Equal-Area Projection

Map labels: PORTUGAL, SPAIN, ITALY, OTTOMAN EMPIRE, MADEIRA IS. (Port.), SPANISH MOROCCO, TUNISIA, Mediterranean Sea, MOROCCO, IFNI, Agadir, Tripoli, Cairo, CANARY IS. (Sp.), ALGERIA, LIBYA, EGYPT, ARABIA, RIO DE ORO, Tropic of Cancer, Red Sea, CAPE VERDE IS. (Port.), FRENCH WEST AFRICA, ERITREA, FRENCH SOMALILAND, Dakar, GAMBIA, ANGLO-EGYPTIAN SUDAN, Addis Ababa, BRITISH SOMALILAND, PORTUGUESE GUINEA, SIERRA LEONE, GOLD COAST, TOGOLAND, NIGERIA, Lagos, Fashoda, ETHIOPIA, LIBERIA, CAMEROONS, FRENCH EQUATORIAL AFRICA, ITALIAN SOMALILAND, FERNANDO PO, PRÍNCIPE (Port.), RIO MUNI, UGANDA, BRITISH EAST AFRICA, ATLANTIC OCEAN, Equator, SAO TOMÉ (Port.), BELGIAN CONGO, GERMAN EAST AFRICA, Mombasa, CABINDA (Port.), ZANZIBAR I. (BR.), COMOROS IS. (FR.), ANGOLA, NYASALAND, NORTHERN RHODESIA, MOZAMBIQUE, Mozambique Channel, GERMAN SOUTHWEST AFRICA, SOUTHERN RHODESIA, MADAGASCAR, BECHUANALAND, Tropic of Capricorn, WALVIS BAY, Pretoria, Johannesburg, SWAZILAND, INDIAN OCEAN, UNION OF SOUTH AFRICA, BASUTOLAND, Cape Town

404 UNIT 6

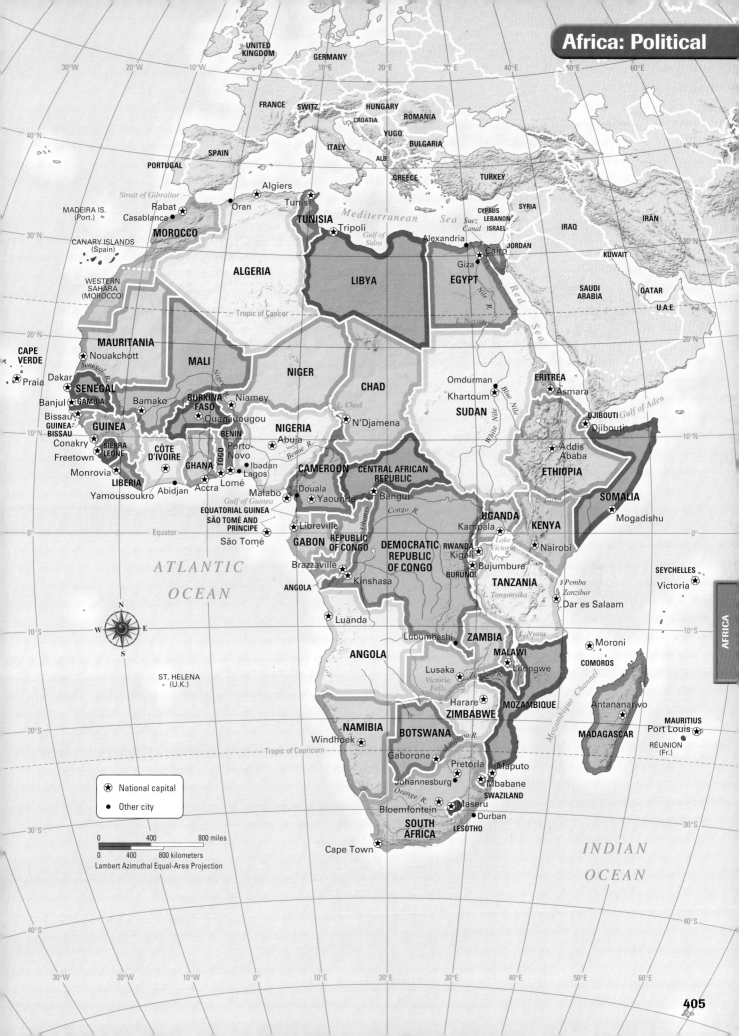

UNITED KINGDOM
GERMANY

FRANCE
SWITZ.
HUNGARY
CROATIA
ROMANIA
YUGO.
BULGARIA

PORTUGAL
SPAIN
ITALY
ALB.
GREECE
TURKEY

MADEIRA IS. (Port.)
Rabat
Casablanca
Algiers
Oran
Tunis
TUNISIA
Tripoli
Mediterranean Sea
Gulf of Sidra
Suez Canal
Alexandria
Cairo
Giza
CYPRUS
LEBANON
ISRAEL
SYRIA
IRAQ
JORDAN
IRAN
KUWAIT

Strait of Gibraltar

CANARY ISLANDS (Spain)
MOROCCO

WESTERN SAHARA (MOROCCO)

ALGERIA
LIBYA
EGYPT
Nile R.
L. Nasser
Red Sea
SAUDI ARABIA
QATAR
U.A.E.

Tropic of Cancer

CAPE VERDE
Praia

MAURITANIA
Nouakchott
Senegal R.
Dakar
SENEGAL
Banjul
GAMBIA
Bissau
GUINEA-BISSAU
Conakry
SIERRA LEONE
Freetown
Monrovia
LIBERIA
Yamoussoukro
Abidjan

MALI
Bamako
Niger R.

BURKINA FASO
Niamey
Ouagadougou

NIGER

CHAD

CÔTE D'IVOIRE
GHANA
BENIN
TOGO
Accra
Lomé
Porto-Novo
NIGERIA
Abuja
Ibadan
Lagos
Benue R.

N'Djamena

Omdurman
Khartoum
SUDAN
Blue Nile
White Nile
L. Tana

ERITREA
Asmara
DJIBOUTI
Djibouti
Gulf of Aden

Addis Ababa
ETHIOPIA

SOMALIA
Mogadishu

CAMEROON
Malabo
Douala
Yaoundé
EQUATORIAL GUINEA
SÃO TOMÉ AND PRÍNCIPE
São Tomé
Gulf of Guinea

CENTRAL AFRICAN REPUBLIC
Bangui

Congo R.

UGANDA
Kampala
KENYA
Nairobi

Libreville
GABON
REPUBLIC OF CONGO
Brazzaville
Kinshasa
DEMOCRATIC REPUBLIC OF CONGO
RWANDA
Kigali
BURUNDI
Bujumbura
Lake Victoria

Equator

ATLANTIC OCEAN

Luanda
ANGOLA

TANZANIA
Dar es Salaam
L. Tanganyika
Pemba
Zanzibar

SEYCHELLES
Victoria

Lubumbashi
ZAMBIA
Lusaka
Victoria Falls
Zambezi R.
MALAWI
Lilongwe
L. Nyasa
COMOROS
Moroni

Harare
ZIMBABWE
MOZAMBIQUE

Mozambique Channel

MADAGASCAR
Antananarivo
MAURITIUS
Port Louis
RÉUNION (Fr.)

ST. HELENA (U.K.)

ANGOLA

NAMIBIA
Windhoek
BOTSWANA
Gaborone

Limpopo R.

Tropic of Capricorn

Pretoria
Maputo
Johannesburg
Mbabane
SWAZILAND
Orange R.
Bloemfontein
Maseru
Durban
SOUTH AFRICA
LESOTHO

Cape Town

INDIAN OCEAN

★ National capital
• Other city

0   400   800 miles
0   400   800 kilometers
Lambert Azimuthal Equal-Area Projection

AFRICA

Unit ATLAS

Regional Patterns

These two pages contain a graph and two thematic maps. The graph shows the religions of Africa. The maps show other important features of Africa: its diversity of languages and its population distribution. After studying these two pages, jot down in your notebook the answers to the questions below.

Making Comparisons

1. Where are most of the people in Africa living? In what areas of Africa are the fewest people living?

2. What geographic factors may account for these population patterns?

3. What do you notice about the number of languages in Africa? Do they belong to one language group or several?

Religions of Africa

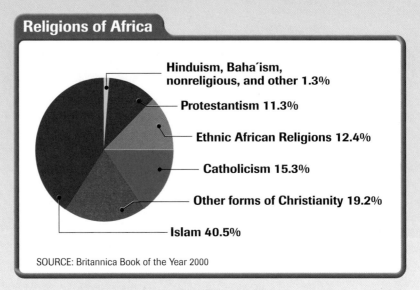

Hinduism, Baha´ism, nonreligious, and other 1.3%

Protestantism 11.3%

Ethnic African Religions 12.4%

Catholicism 15.3%

Other forms of Christianity 19.2%

Islam 40.5%

SOURCE: Britannica Book of the Year 2000

Population Distribution of Africa

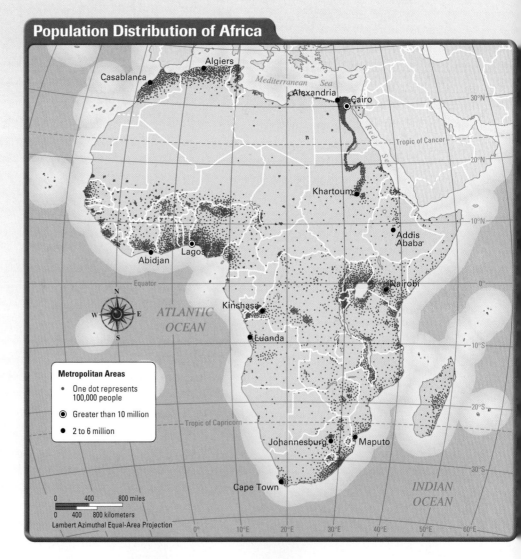

Algiers
Casablanca
Mediterranean Sea
Alexandria
Cairo — 30°N
Red Sea
Tropic of Cancer
20°N
Khartoum
10°N
Addis Ababa
Lagos
Abidjan
Equator
Nairobi 0°
ATLANTIC OCEAN
Kinshasa
Luanda
10°S

Metropolitan Areas
• One dot represents 100,000 people
◉ Greater than 10 million
● 2 to 6 million

Tropic of Capricorn — 20°S
Johannesburg   Maputo

30°S
Cape Town
INDIAN OCEAN

0   400   800 miles
0   400   800 kilometers
Lambert Azimuthal Equal-Area Projection
0°   10°E   20°E   30°E   40°E   50°E   60°E

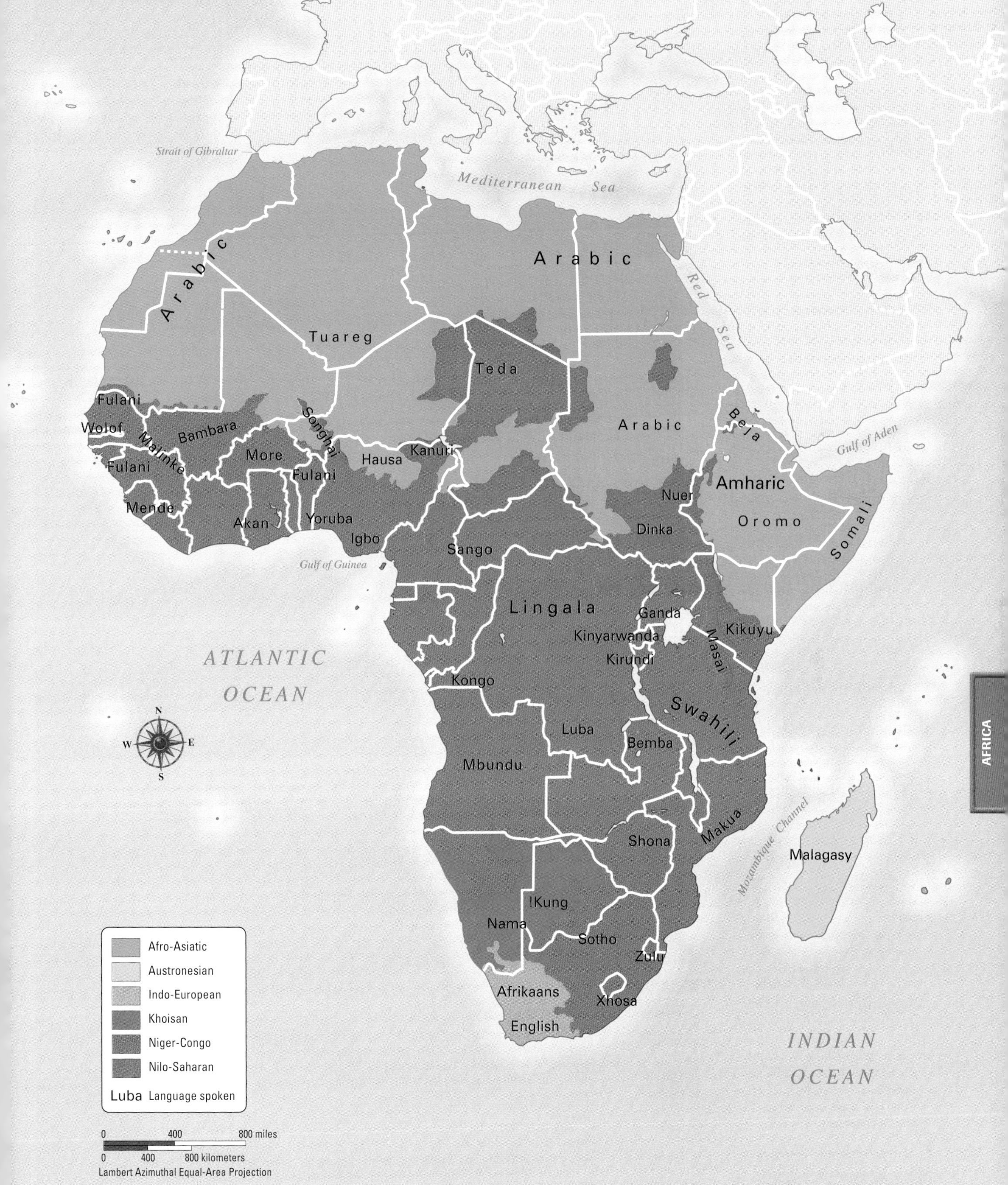

Strait of Gibraltar

Mediterranean Sea

Arabic

Arabic

Red Sea

Tuareg

Teda

Arabic

Beja

Gulf of Aden

Fulani

Wolof

Bambara

Songhai

Kanuri

More

Hausa

Malinke

Fulani

Nuer

Amharic

Fulani

Mende

Akan

Yoruba

Dinka

Oromo

Igbo

Somali

Sango

Gulf of Guinea

Lingala

Ganda

Kikuyu

Kinyarwanda

Kirundi

Masai

ATLANTIC

OCEAN

Kongo

Swahili

Luba

Mbundu

Bemba

Makua

N
W E
S

Shona

Mozambique Channel

Malagasy

AFRICA

!Kung

Nama

Sotho

Zulu

Afrikaans

Xhosa

English

INDIAN

OCEAN

	Afro-Asiatic
	Austronesian
	Indo-European
	Khoisan
	Niger-Congo
	Nilo-Saharan
Luba	Language spoken

0   400   800 miles
0   400   800 kilometers
Lambert Azimuthal Equal-Area Projection

Unit ATLAS

Regional Data File

Country Flag	Country/ Capital	Population (2000)	Life Expectancy[a] (years) (2000)	Birthrate (per 1,000 pop.) (2000)	Infant Mortality (per 1,000 live births) (2000)
	Algeria Algiers	31,471,000	69	29	44.0
	Angola Luanda	12,878,000	47	48	125.0
	Benin Porto-Novo	6,396,000	50	45	93.9
	Botswana Gaborone	1,576,000	44	32	57.2
	Burkina Faso Ouagadougou	11,946,000	47	47	105.3
	Burundi Bujumbura	6,054,000	47	42	74.8
	Cameroon Yaoundé	15,422,000	55	37	77.0
	Cape Verde Praia	401,000	68	37	76.9
	Central African Republic, Bangui	3,513,000	45	38	96.7
	Chad N'Djamena	7,977,000	48	50	109.8
	Comoros Moroni	578,000	59	38	77.3
	Congo, Democratic Republic of, Kinshasa	51,965,000	49	48	108.6
	Congo, Republic of, Brazzaville	2,831,000	48	40	108.6
	Côte d'Ivoire Yamoussoukro	15,980,000	47	38	112.2
	Djibouti Djibouti	638,000	48	39	115.0
	Egypt Cairo	68,344,000	65	26	52.3
	Equatorial Guinea Malabo	453,000	50	41	108.0
	Eritrea Asmara	4,142,000	55	43	81.8

Study the charts on the countries of Africa. In your notebook, answer these questions.

Making Comparisons

1. Which three African countries have the most people? Locate them on the map. Are they also the largest countries in terms of total area?

2. Which three African countries have the fewest people? Locate them on the map. Are they the smallest countries in terms of total area?

3. Look at Angola's life expectancy, infant mortality, and number of doctors. Judging from these statistics, does Angola have good health care?

(continued on page 410)

Notes:
[a] Life expectancy figures for many African countries are declining significantly, mainly due to poverty, politics, and the spread of AIDS.
[b] Doctors are defined as graduates of a school of medicine in any medical field.
[c] A comparison of the prices of the same items in different countries is used to figure these data.
[d] Includes land and water, when figures are available.

For updated statistics on Africa . . .

DATA UPDATE
CLASSZONE.COM

Doctors[b] (per 100,000 pop.) (1992–1998)	GDP[c] (billions $US) (1999)	Import/Export[c] (billions $US) (1997–1999)	Literacy Rate (percentage) (1998–1999)	Televisions (per 1,000 pop.) (1996–1998)	Passenger Cars (per 1,000 pop.) (1996–1997)	Total Area[d] (square miles)	
85	147.6	9.3 / 13.7	66	68	17	919,590	
8	11.6	3.0 / 5.0	42	124	21	481,351	
6	8.1	0.566 / 0.396	38	91	6	43,483	
24	5.7	2.05 / 2.36	76	27	53	231,804	
3	12.4	0.572 / 0.311	22	6	3	105,869	
6	4.2	0.108 / 0.056	46	10	2	10,759	
7	31.5	1.5 / 2.0	74	81	7	183,591	
17	0.618	0.225 / 0.038	73	45	29	1,557	
4	5.8	0.17 / 0.195	44	5	3	240,534	
3	7.6	0.359 / 0.288	39	2	1	495,752	
7	0.41 (1998)	0.05 / 0.009	59	4	18	719	
7	35.7	0.46 / 0.53	59	43	7	905,365	
25	4.15	0.77 / 1.7	78	8	10	132, 047	
9	25.7	2.6 / 3.9	45	70	11	124,503	
14	0.55	0.44 / 0.26	62	73	31	8,958	
202	200.0	15.8 / 4.6	54	127	20	386,900	
25	0.96	0.3 / 0.555	81	162	9	10,830	
3	2.9	0.44 / 0.053	52	14	2	47,320	

Making Comparisons

(continued)

4. Use the map on page 405 to choose a country in East Africa. How many televisions and cars does it have per 1,000 people? How does that compare to the United States?

5. Make a list of the top three African countries in GDP. Where are these countries located? Do you notice any pattern?

6. Use the map on page 405 to identify two countries in Southern Africa. For each of those countries, calculate per capita GDP by dividing total GDP by population. Which country has the higher per capita GDP?

(continued on page 412)

Country Flag	Country/ Capital	Population (2000)	Life Expectancy[a] (years) (2000)	Birthrate (per 1,000 pop.) (2000)	Infant Mortality (per 1,000 live births) (2000)
	Ethiopia Addis Ababa	64,117,000	46	45	116.0
	Gabon Libreville	1,226,000	52	38	87.0
	Gambia Banjul	1,305,000	45	43	130.0
	Ghana Accra	19,534,000	58	34	56.2
	Guinea Conakry	7,466,000	45	42	98.0
	Guinea-Bissau Bissau	1,213,000	45	42	130.0
	Kenya Nairobi	30,340,000	49	35	73.7
	Lesotho Maseru	2,143,000	53	33	84.5
	Liberia Monrovia	3,164,000	50	50	139.1
	Libya Tripoli	5,114,000	75	28	33.3
	Madagascar Antananarivo	14,858,000	52	44	96.3
	Malawi Lilongwe	10,385,000	39	41	126.8
	Mali Bamako	11,234,000	53	47	122.5
	Mauritania Nouakchott	2,670,000	54	41	92.0
	Mauritius Port Louis	1,189,000	70	17	19.4
	Morocco Rabat	28,778,000	69	23	37.0
	Mozambique Maputo	19,105,000	40	41	133.9
	Namibia Windhoek	1,771,000	46	36	68.3

Notes:

[a] Life expectancy figures for many African countries are declining significantly, mainly due to poverty, politics, and the spread of AIDS.

[b] Doctors are defined as graduates of a school of medicine in any medical field.

[c] A comparison of the prices of the same items in different countries is used to figure these data.

[d] Includes land and water, when figures are available.

Doctors[b] (per 100,000 pop.) (1992–1998)	GDP[c] (billions $US) (1999)	Import/Export[c] (billions $US) (1997–1999)	Literacy Rate (percentage) (1998–1999)	Televisions (per 1,000 pop.) (1996–1998)	Passenger Cars (per 1,000 pop.) (1996–1997)	Total Area[d] (square miles)	
4	33.3	1.25 / 0.42	36	5	0.8	471,776	
19	7.9	1.2 / 2.4	63	136	21	103,346	
4	1.4	0.201 / 0.132	35	4	7	4,127	
6	35.5	2.5 / 1.7	69	115	5	92,100	
13	9.2	0.56 / 0.695	36	41	2	94,925	
17	1.1	0.023 / 0.027	37	N/A	3	13,948	
13	45.1	3.3 / 2.2	81	21	10	224,960	
5	4.7 (1998 est.)	0.7 / 0.235	82	24	3	11,720	
2	2.85	0.142 / 0.039	38	27	9	43,000	
128	39.3	7.0 / 6.6	78	143	126	679,358	
11	11.5	0.793 / 0.6	65	46	4	226,658	
2	9.4	0.512 / 0.51	58	2	3	47,747	
5	8.5	0.65 / 0.64	38	11	3	478,764	
14	4.9	0.444 / 0.425	41	91	7	397,955	
85	12.3	2.1 / 1.7	84	228	61	790	
46	108.0	9.5 / 7.1	47	160	39	172,413	
4	18.7	1.44 / 0.3	42	4	4	302,328	
30	7.1	1.5 / 1.4	81	32	38	318,000	

Regional Data File

Country Flag	Country/ Capital	Population (2000)	Life Expectancy[a] (years) (2000)	Birthrate (per 1,000 pop.) (2000)	Infant Mortality (per 1,000 live births) (2000)
	Niger Niamey	10,076,000	41	54	123.1
	Nigeria Abuja	123,338,000	52	42	77.2
	Rwanda Kigali	7,229,000	39	43	120.9
	São Tomé and Príncipe São Tomé	160,000	64	43	50.8
	Senegal Dakar	9,481,000	52	41	67.7
	Seychelles Victoria	82,000	71	18	8.5
	Sierra Leone Freetown	5,233,000	45	47	157.1
	Somalia Mogadishu	7,253,000	46	47	125.8
	South Africa, Pretoria/ Cape Town/Bloemfontein	43,421,000	55	25	45.4
	Sudan Khartoum	29,490,000	51	33	69.5
	Swaziland Mbabane	1,004,000	38	41	107.7
	Tanzania Dodoma	35,306,000	53	42	98.8
	Togo Lomé	5,019,000	49	42	79.7
	Tunisia Tunis	9,619,000	69	22	35.0
	Uganda Kampala	23,318,000	42	48	81.3
	Zambia Lusaka	9,582,000	37	42	109.0
	Zimbabwe Harare	11,343,000	40	30	80.0
	United States Washington, D.C.	281,422,000	77	15	7.0

Making Comparisons
(continued)

7. Calculate the GDP per capita for Sierra Leone, Zambia, and Eritrea by dividing GDP by population. Where do those countries rank in life expectancy? What might be the relationship between a country's GDP and its life expectancy?

Sources:
ABC-CLIO
CIA World Factbook 2000 online
Columbia Gazetteer
Population Reference Bureau 2000 online
Statesman's Yearbook 2001
UN Human Development Report 2000
U.S. Census Bureau online
World Almanac 2000
World Health Organization online
N/A = not available

Notes:
[a] Life expectancy figures for many African countries are declining significantly, mainly due to poverty, politics, and the spread of AIDS.
[b] Doctors are defined as graduates of a school of medicine in any medical field.
[c] A comparison of the prices of the same items in different countries is used to figure these data.
[d] Includes land and water, when figures are available.

Doctors[b] (per 100,000 pop.) (1992–1998)	GDP[c] (billions $US) (1999)	Import/Export[c] (billions $US) (1997–1999)	Literacy Rate (percentage) (1998–1999)	Televisions (per 1,000 pop.) (1996–1998)	Passenger Cars (per 1,000 pop.) (1996–1997)	Total Area[d] (square miles)	
4	9.6	0.266 / 0.269	15	26	4	489,189	
19	110.5	10.0 / 13.1	61	67	5	356,669	
4	5.9	0.242 / 0.071	64	N/A	2	10,169	
47	.169	0.02 / 0.005	73	227	30	372	
8	16.6	1.2 / 0.925	36	41	12	76,124	
132	.59	0.363 / 0.091	84	190	85	178	
7	2.5	0.166 / 0.041	31	26	4	27,699	
4	4.3	0.327 / 0.187	24	13	2	246,200	
56	296.1	26.0 / 28.0	85	125	102	471,445	
9	32.6	1.26 / 0.58	56	141	1	967,494	
15	4.2	1.05 / 0.825	78	107	29	6,705	
5	23.3	1.44 / 0.828	74	21	2	364,898	
8	8.6	0.45 / 0.4	55	20	17	21,853	
70	52.6	7.47 / 5.8	69	198	28	63,378	
4	24.2	1.1 / 0.471	65	26	1	91,134	
7	8.5	1.15 / 0.9	76	137	16	290,585	
14	26.5	2.0 / 2.0	87	29	3	150,820	
251	9,255.0	820.8 / 663.0	97	847	489	3,787,319	

Chapter 18

PHYSICAL GEOGRAPHY OF AFRICA

The Plateau Continent

SECTION 1
Landforms and Resources

SECTION 2
Climate and Vegetation

SECTION 3
Human–Environment Interaction

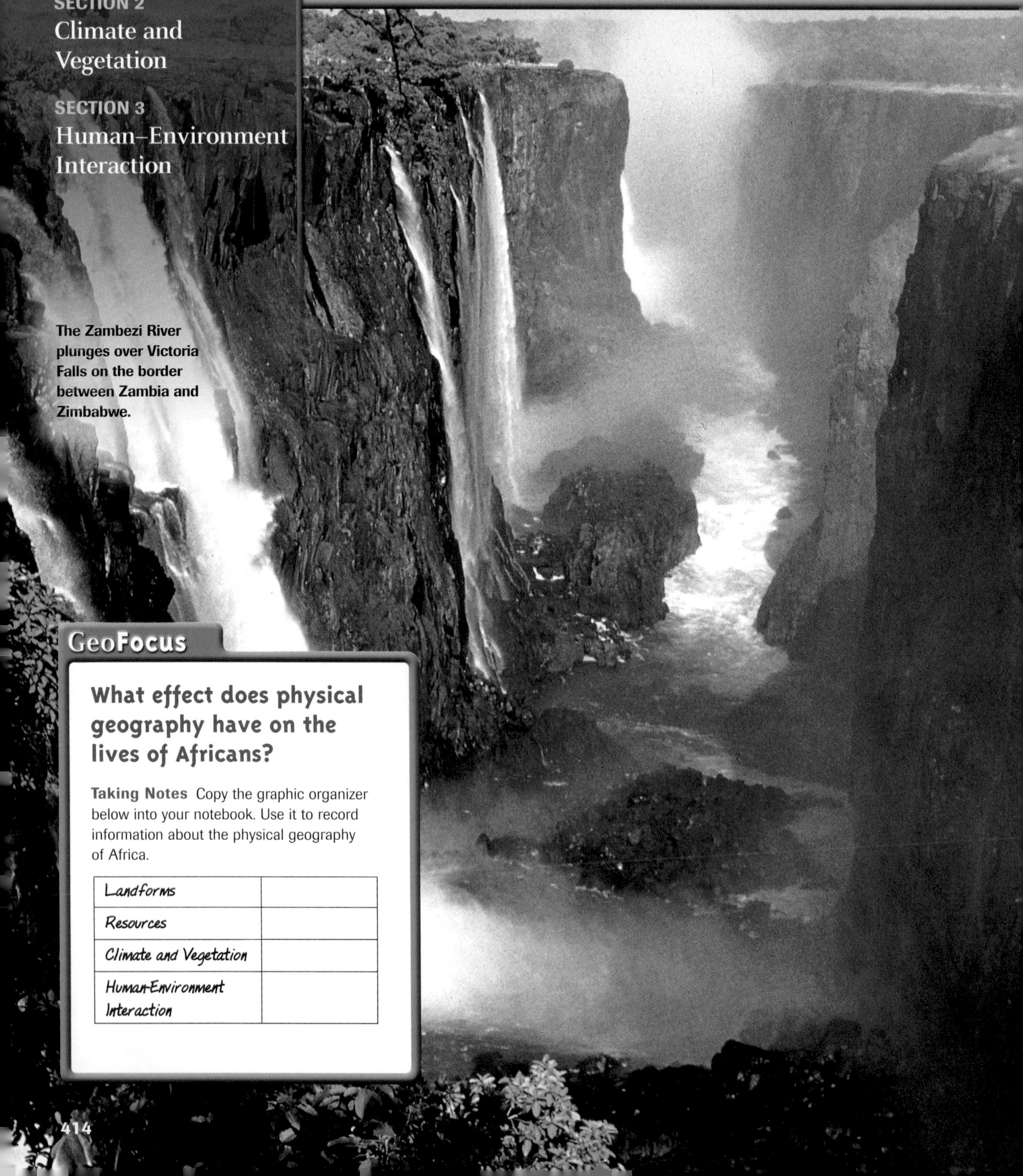

The Zambezi River plunges over Victoria Falls on the border between Zambia and Zimbabwe.

GeoFocus

What effect does physical geography have on the lives of Africans?

Taking Notes Copy the graphic organizer below into your notebook. Use it to record information about the physical geography of Africa.

Landforms	
Resources	
Climate and Vegetation	
Human-Environment Interaction	

Reading an Economic Activity Map

Subsistence farming and nomadic herding are the primary economic activities in large sections of Africa. Even though African nations have a wealth of natural resources, mining and drilling for these resources are not evenly distributed throughout the continent. The thematic map below shows a wide variety of economic activities in Africa.

THE LANGUAGE OF MAPS An **economic activity map** is a thematic map that shows the location of economic activities over a large area such as a continent.

Economic Activities in Africa

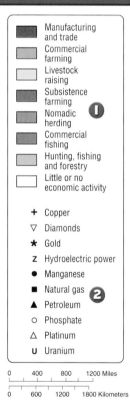

Manufacturing and trade
Commercial farming
Livestock raising
Subsistence farming
Nomadic herding **❶**
Commercial fishing
Hunting, fishing and forestry
Little or no economic activity

+ Copper
▽ Diamonds
★ Gold
Z Hydroelectric power
● Manganese
■ Natural gas **❷**
▲ Petroleum
○ Phosphate
△ Platinum
U Uranium

0   400   800   1200 Miles
0   600   1200   1800 Kilometers

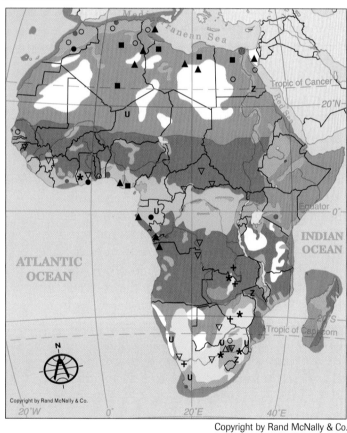

Copyright by Rand McNally & Co.

❶ Each color represents the economic activity in which the majority of people are engaged.

❷ The black symbols represent major drilling and mining for natural resources.

❸ The symbols and colors show the combination of economic activities and natural resources in a particular location. For example, this map shows that Southern Africa contains livestock raising, mining, commercial fishing, and commercial farming.

Map and Graph Skills Assessment

1. Making Generalizations
In what region of Africa does most of the livestock raising take place?

2. Making Inferences
Why do you think so many manufacturing and trade centers are located near rivers or on the coasts?

3. Drawing Conclusions
What is the most common type of farming done in Africa?

Climate and Vegetation

A HUMAN PERSPECTIVE In 1352, 48-year-old Ibn Battuta, a great traveler from Morocco, set out for the empire of Mali in West Africa. His most challenging obstacle was the Sahara, a desert nearly the same size as the continental United States. Battuta and his caravan set out in February. They traveled only in early morning and early evening to avoid the midday heat. Even so, they still battled temperatures of near-ly 100 degrees during the day and freezing temperatures at night. Reaching Mali around April, Batutta covered more than 1,000 miles, all on foot. The Sahara today remains just as hazardous—fewer than 2 million of Africa's approximately 800 million people live in it.

A Warm Continent

You can see from the map on page 421 that Africa lies almost entirely between the tropic of Cancer and the tropic of Capricorn. This location gives most of Africa warm, tropical temperatures.

THE DESERTS The **Sahara** is the largest desert in the world. Sahara actually means "desert" in Arabic. It stretches about 3,000 miles across the continent, from the Atlantic Ocean to the Red Sea, and also runs 1,200 miles from north to south. Temperatures can rise as high as 136.4°F in the summer, hot enough to fry an egg on the sand. But tem-peratures can also fall below freezing at night in winter.

Only about 20 percent of the Sahara consists of sand. Towering mountains, rock formations, and gravelly plains make up the rest. For instance, the Tibesti Mountains, located mostly in northwestern Chad, rise to heights of more than 11,000 feet. Other African deserts include the Kalahari and the Namib.

Travel in the Sahara is risky because of the extreme conditions. Many travelers rely on the camel as desert transportation. A camel can go for

PLACE Rolling sand dunes are only a small part of the Sahara's varied landscape.
How might an expanding desert affect the lives of the people living near it?

up to 17 days without water. In addition, wind-blown sand has little effect on a camel. It closes its nostrils and just keeps walking.

Ironically, as much as 6,000 feet under this hottest and driest of places lie huge stores of underground water called **aquifers.** In some places, this water has come to the surface. Such a place is called an **oasis.** It supports vegetation and wildlife and is a critical resource for people living in the desert.

THE TROPICS Africa has a large tropical area—the largest of any continent. In fact, nearly 90 percent of the continent lies within the tropics of Cancer and Capricorn, as you can see on the map to the right. Temperatures run high most of the year. The hottest places are in the parts of the Sahara that lie in the nation of Somalia. July temperatures average between 110°F and 115°F almost every day. Differences in temperature between winter and summer in the Tropics are barely noticeable. Differences in temperature between night and day actually tend to be greater than any difference between seasons. A saying in Africa says that nighttime is the "winter" of the tropics.

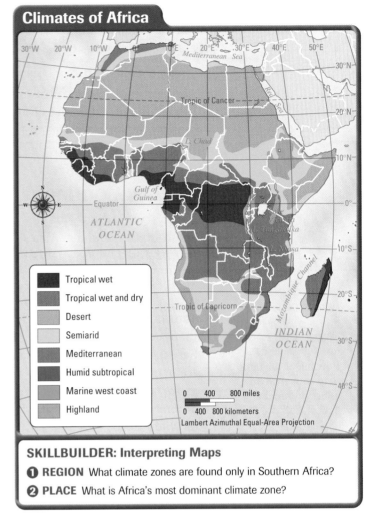

Climates of Africa

- Tropical wet
- Tropical wet and dry
- Desert
- Semiarid
- Mediterranean
- Humid subtropical
- Marine west coast
- Highland

0   400   800 miles
0   400   800 kilometers
Lambert Azimuthal Equal-Area Projection

SKILLBUILDER: Interpreting Maps

❶ **REGION** What climate zones are found only in Southern Africa?

❷ **PLACE** What is Africa's most dominant climate zone?

Geographic Thinking

Using the Atlas
Ⓐ Which climate regions of Africa do the tropics of Cancer and Capricorn pass through?

Sunshine and Rainfall

Rainfall in Africa is often a matter of extremes. Some parts get too much rain, while other parts receive too little. The amount of rainfall can also vary greatly from year to year as well as season to season. These variations have had a tremendous impact on East Africa, which endured several droughts in the 1980s and 1990s.

RAINFALL PATTERNS The rain forest in Central Africa receives the most precipitation, as rain falls throughout the year. Most of the rest of Africa, however, has one or two rainy seasons. Africa's tropical savanna stretches through the middle of the continent. It covers nearly half the total surface area of Africa. Rainy seasons in this area can last up to six months. The closer an area is to the equator, the longer the rainy season. The closer an area is to the desert, the longer the dry season.

Africa's west coast also receives a great deal of rain. The region around Monrovia, Liberia, experiences an average annual rainfall of more than 120 inches. In contrast, many parts of Africa barely get 20 inches of rain over the course of a year. In the Sahara and other deserts, rain may not fall for years. Children living in those areas may not see rain until they are five or six years old!

AFRICA'S MODERATE AREAS A Mediterranean climate exists on the northern and southern tips of the continent. Clear, blue skies in these places are normal. Rain falls usually only in the winter—December and January in North Africa and June and July in Southern Africa. Summer temperatures in Johannesburg, South Africa, average around 68°F.

A Grassy Continent

Africa's vegetation—like its climate—is almost mirrored north and south of the equator. Africa's vegetation consists of grasslands, rain forests, and a wide variety of other plant life.

BACKGROUND
Serengeti means "endless plains" in the Masai language.

TROPICAL GRASSLAND Tropical grassland covers most of the continent. One example of this grassland is the **Serengeti Plain** in northern Tanzania. Its dry climate and hard soil prevent the growth of trees and many crops, but these conditions are perfect for growing grass. Serengeti National Park, located within the Serengeti Plain, contains some of the best grasslands in the world. Some of these grasses can grow taller than the average person. The abundance of grass makes Serengeti National Park an ideal place for grazing animals. Huge herds of wildebeests, gazelles, and zebras roam there. It is the place where the largest numbers of land mammals still make annual migrations.

PLACE These wildebeests live in Serengeti National Park, which was founded in 1951. **How might the park help conservation efforts in Africa?**

Africa's Extremes

An enormous tropical rain forest stretches across Central Africa.

RAIN FOREST The major rain forests of Africa sit on the equator in the area of the Congo Basin. One square acre of rain forest can contain almost 100 different kinds of trees. It may also be home to hundreds of species of birds. The massive number of plants, leaves, and trees block out much of the sunlight that would otherwise hit the floor of the rain forest. Beneath this umbrella of vegetation, the air is hot and filled with moisture. As a result, plants and other vegetation quickly decompose, or decay. For example, a fallen leaf in Europe decomposes in about a year. A leaf on the jungle floor in Africa decomposes in about six weeks. **B**

Most animals in a rain forest live in the canopy. The **canopy** refers to the uppermost layer of branches, about

Geographic Thinking

Seeing Patterns
B How do rain forests maintain such a high level of moisture?

150 feet above the ground. Birds, monkeys, and flying foxes move from tree to tree and enjoy the bounty of the rain forest. A large number of snakes live in these rain forests, too. The Gaboon viper, the largest African viper, can weigh as much as 18 pounds and have fangs more than two inches long. Another snake, the black-necked cobra, can shoot its venom more than eight feet through the air.

BACKGROUND
The National Cancer Institute estimates that 70 percent of the plants found useful for cancer treatment are found only in rain forests.

However, farmers using slash-and-burn agricultural methods are endangering the existence of the rain forest. As you read in Chapter 9, slash-and-burn farming is a method in which people clear fields by cutting and burning trees and other vegetation, the ashes of which fertilize the soil. After farmers have exhausted the soil, they burn another patch of forest. Slash-and-burn farming is responsible for the nearly complete destruction of Madagascar's rain forest. Experts estimate that over half of Africa's original rain forest has been destroyed.

MOVEMENT
Animals, such as this chimpanzee, move about the Ituri rain forest in the Democratic Republic of the Congo.

VARIETIES OF PLANTLIFE All of Africa's regions contain a variety of vegetation. North Africa contains sizable oak and pine forests in the upper reaches of the Atlas Mountains. The mangrove tree of West Africa sprouts up along river banks in swamps and river deltas. Mangrove tree roots are breeding grounds for fish. They also help to build up dry land by holding silt. In the next section, you will read about different ways that people in Africa have interacted with their environment.

AFRICA

SECTION 2 Assessment

1 Places & Terms

Explain the meaning of each of the following terms.

- Sahara
- aquifer
- oasis
- Serengeti Plain
- canopy

2 Taking Notes

PLACE Review the notes you took for this section.

Climate and Vegetation	

- What are the different climates found in Africa?
- How does climate affect the vegetation of Africa?

3 Main Ideas

a. What is the largest climatic feature in Africa?

b. Why does most of Africa have high temperatures?

c. What are the different kinds of vegetation growing in Africa?

4 Geographic Thinking

Making Comparisons What are the similarities between the climates of Africa north and south of the equator?
Think about:

- the Sahara
- where the equator cuts across Africa

RESEARCH LINKS
CLASSZONE.COM

MAKING COMPARISONS Choose a place in Africa and a place in the United States at about the same latitude. Use encyclopedias or the Internet to compare the climate and vegetation of the two places. Create a **chart** comparing the two locations.

Human–Environment Interaction

Main Ideas
- The Sahara's expansion is causing problems for Africa's farmers.
- The Nigerian oil industry has caused serious environmental damage in the Niger delta.

Places & Terms
Niger delta

Sahel

desertification

Aswan High Dam

silt

CONNECT TO THE ISSUES
COLONIALISM European colonialism has caused political, economic, and environmental problems in Africa today.

A HUMAN PERSPECTIVE Akierou Awe lives in a mud-brick house in Nigeria's **Niger delta,** a region that contains most of Nigeria's oil. On the morning of July 10, 2000, Awe's four sons had been collecting fuel from a leaking pipeline to help scrape out a living in this poverty-stricken region. They hoped to resell the fuel for more than the going rate of 21 cents a quart. Suddenly, an explosion shook the area, and a fire spread along a mile-long stretch of the pipeline. The blast killed more than 300 people, including three of Awe's sons. This accident is one of many in the recent past that have claimed the lives of hundreds of Nigerians. Nigeria has become one of the top oil producers in the world, but at the cost of thousands of lives and major environmental ruin in the region.

Desertification of the Sahel

Sahel means "shore of the desert" in Arabic. You can see from the physical map on page 403 that the Sahel is a narrow band of dry grassland that runs east to west along the southern edge of the Sahara. People use the Sahel for farming and herding. Since the 1960s, the desert has spread into the Sahel. This shift of the desert is called desertification. **Desertification** is an expansion of dry conditions into moist areas that are next to deserts. Normally, it results from nature's long-term cycle, but as you can see in the illustration below, human activity is speeding up the process.

HUMAN CAUSES OF DESERTIFICATION Geographers and other scientists have identified several human activities that increase the pace of desertification. For example, allowing overgrazing of vegetation by

The Process of Desertification

1 The Sahel receives little rainfall. The vegetation lives in a fragile state, having barely enough water and food to survive.

2 Farming, overgrazing by livestock, and burning wood for fuel all contribute to desertification.

livestock exposes the soil. Animals also trample the soil, making it more vulnerable to erosion.

Farming also increases the pace of desertification. When farmers clear the land to plant crops, they expose the soil to wind, which can cause erosion. In addition, when farmers drill for water to irrigate crops, they put further stress on the Sahel. Widespread drilling and more irrigation increase salt levels in the soil, which prevent the growth of vegetation.

Increasing population levels are an indirect cause of desertification. More people require more food. As a result, farmers continue to clear more land for crops, burn more wood for fuel, and overfarm the land they already have.

RESULTS OF DESERTIFICATION Desertification has affected many parts of Africa. For example, large forests once existed around Khartoum, Sudan. In addition, desertification is slowly destroying a tropical rain forest around Lake Chad in the southern edge of the Sahel. Slowing desertification is difficult. Some African countries have increased tree planting and promoted more efficient use of forests and farmland in hopes of slowing the process.

Using the Atlas
Refer to the physical map on page 403. What countries are probably most affected by desertification of the Sahel?

Harming the Environment in Nigeria

Another environmental issue concerns the discovery of oil in Nigeria in 1956. Rich oil deposits in the Niger delta made Nigeria one of Africa's wealthiest countries. However, in drilling for oil, the Nigerian government and foreign oil companies have often damaged the land and harmed the people living in the Niger delta.

A MAJOR OIL PRODUCER Nigeria is the sixth leading oil exporter in the world. Two million barrels are extracted each day, much of it shipped to the United States. Oil accounts for 80 to 90 percent of Nigeria's income.

During the 1970s, high oil prices made Nigeria one of the wealthiest nations in Africa. As a result, the government borrowed heavily against the future sale of its oil. However, oil prices eventually fell, and the Nigerian government owed millions of dollars to other nations, including the United States. Mismanagement, poor planning, corruption, and a decline in oil prices left Nigeria poorer than before the oil boom.

AFRICA

3 During desertification, dry grasses die and are replaced by tougher plants like shrubs. These plants do not cover the soil as well as grass.

4 With less vegetation covering the soil, any rain that falls evaporates quickly. Over the years, the wind then blows the dry soil into a desert-like state.

425

DESTROYING THE LAND AND PEOPLE The damage caused by oil companies and the Nigerian government has been severe. More than 4,000 oil spills have occurred in the Niger delta over the past four decades. Cleanup operations have been slow and sometimes non-existent. Fires often resulted, causing acid rain and massive deposits of soot, and people in the region contracted respiratory diseases. In addition, between 1998 and 2000, oil pipeline explosions killed more than 2,000 people—including three of Akierou Awe's sons.

Many of these explosions were not accidents but were caused intentionally. Bandits, in cooperation with corrupt government officials and the military, drain fuel from the pipelines and then resell it. In 1999, these bandits damaged about 500 pipelines. Once the bandits finish draining oil, local villagers arrive. They use small cans to collect any spilled oil and then sell it. **B**

A NEW START In May 1999, Olusegun Obasanjo became Nigeria's new president. Although a former Nigerian military leader himself, he has distanced himself from the armed forces. He has started many economic reforms and fired corrupt government officials. Now he faces the task of finding ways for Nigeria to benefit from oil.

Geographic Thinking

Seeing Patterns
B Why did bandits and corrupt government officials drain fuel from the pipelines?

Controlling the Nile

Egypt faces environmental challenges caused by another resource—water. Throughout history, the Egyptians have tried to control the floodwaters of the Nile River. Ancient Egyptians built canals and small dams. In spite of these efforts, though, the people still experienced cycles of floods and droughts. To solve these problems, Egyptians completed the first Aswan Dam on the Nile in 1902, which quickly became outmoded.

THE ASWAN HIGH DAM Four miles upriver from the first Aswan Dam, the Egyptians cut a huge channel through the land beside the Nile River. The builders used the rocks from the channel as a base for their new creation—the **Aswan High Dam**—which was completed in 1970. Lake Nasser, which Egypt shares with Sudan, is the artificial lake created behind the dam. It stretches for nearly 300 miles.

Aswan High Dam

Mediterranean Sea

Flood plain

Alexandria
Cairo
El Minya
SINAI PEN.
LIBYA
EGYPT
Asyut

0   150   300 miles
0   150   300 kilometers
Lambert Azimuthal
Equal-Area Projection

Aswan High Dam
Aswan
First Cataract
Lake Nasser
Tropic of Cancer

SUDAN
NUBIAN DESERT

Red Sea

30°N
25°N
25°E
35°E
40°E

HUMAN–ENVIRONMENT INTERACTION The Aswan High Dam has helped Egypt control the flooding of the Nile River.
What are some of the benefits of the Aswan High Dam?

The dam gives farmers a regular supply of water. It holds the Nile's floodwaters, releasing them as needed so that farmers can use the water effectively for irrigation. As a result of the dam, farmers can now have two or three harvests per year rather than one. Irrigation canals even keep some fields in continuous production through the use of artificial fertilizers. The dam has increased Egypt's farmable land by 50 percent. The dam has also helped Egypt avoid droughts and floods.

PROBLEMS WITH THE DAM Though the dam has provided Egypt with many benefits, it has also created some problems. During the dam's construction, many people had to be relocated, including thousands of Nubians, whose way of life was permanently changed. In addition, one of ancient Egypt's treasures, the temples at Abu Simbel, had to be moved. Other smaller ancient treasures could not be saved and now lie at the bottom of Lake Nasser.

The dam also decreased the fertility of the soil around the Nile. First, the river no longer deposits its rich **silt**, or sediment, on the farmland. Farmers must now rely on expensive artificial fertilizers to enrich the soil. Second, this year-round irrigation has resulted in a rising water table in Egypt. As a result, salts from deep in the earth have decreased the fertility of the soil. Before the dam was built, floodwaters flushed out the salt. Now expensive field drains have to be installed.

Rates of malaria and other diseases have increased due to greater numbers of mosquitos, which thrive in the still waters of Lake Nasser and the irrigation canals. Furthermore, because Lake Nasser holds the floodwaters, Egyptians lose millions of gallons of fresh water every year to evaporation. Measuring the success of the Aswan High Dam is difficult. For all the ways it has helped Egyptians, it has also created new problems.

Geographic Thinking

Seeing Patterns
▷ How do farmers fertilize their land now that the dam traps all the silt?

Assessment
SECTION 3

① Places & Terms

Explain the meaning or identify the location of each of the following terms.
- Niger delta
- Sahel
- desertification
- Aswan High Dam
- silt

② Taking Notes

HUMAN–ENVIRONMENT INTERACTION Review the notes you took for this section.

Human-Environment Interaction	

- Which activities illustrate human control of the environment?
- Which examples illustrate an environment changed by humans?

③ Main Ideas

a. What are some of the causes of desertification?

b. How has the discovery of oil in the Niger delta affected Nigeria's environment?

c. What were some of the reasons that the Egyptian government built the Aswan High Dam?

④ Geographic Thinking

Drawing Conclusions Do you think that the benefits of the Aswan High Dam have outweighed its problems?
Think about:
- the dam's effect on Egypt's food supply and farmers

S See Skillbuilder Handbook, page R5.

GeoActivity

ASKING GEOGRAPHIC QUESTIONS Study the map of the Aswan High Dam on page 426. Write three geographic questions about the map, such as one concerning the location of the dam. Write a **report** answering one of your three questions. Then present your findings to the class.

AFRICA

VISUAL SUMMARY
PHYSICAL GEOGRAPHY OF AFRICA

Landforms and Resources

- A large plateau covers most of Africa.
- Long, thin valleys, called rift valleys, stretch along East Africa.
- Africa contains many valuable resources including oil, diamonds, and gold.

Climate and Vegetation

- The Sahara, the largest desert in the world, stretches across northern Africa.
- Nearly 90 percent of Africa lies within the Tropics.
- A large, grassy area called the Serengeti Plain provides an ideal natural habitat for Africa's wild animals.

Human-Environment Interaction

- Desertification results from nature's cycle, farming, overgrazing, and clearing too much land for crops.
- People and the environment in Nigeria have suffered as a result of the country's poor management and corruption of the oil industry.
- The Aswan High Dam helped to increase Egypt's food supply but has also caused environmental problems.

Reviewing Places & Terms

A. Briefly explain the importance of each of the following.

1. Nile River
2. rift valley
3. escarpment
4. Sahara
5. oasis
6. Serengeti Plain
7. Sahel
8. desertification
9. Aswan High Dam
10. silt

B. Answer the questions about vocabulary in complete sentences.

11. What is sediment that is deposited on farmland by rivers and also acts as a fertilizer?
12. What is the longest river in the world?
13. What does the pulling apart of continental plates create?
14. What is the largest desert in the world?
15. What is the process in which dry conditions spread into areas that are moist?
16. Where do Africa's large mammal migrations take place?
17. What supports vegetation and is a critical resource for people living in the desert?
18. What is a narrow region of grassland on the southern edge of the Sahara?
19. What marks the edge of Africa's plateau in Southern Africa?
20. Which of the terms above is an example of how humans have adapted to the environment?

Main Ideas

Landforms and Resources (pp. 415–419)

1. In what ways does the Nile River support life?
2. What are some of the abundant resources in Africa?
3. Why does oil in Angola not always benefit Angolans?

Climate and Vegetation (pp. 420–423)

4. What is the physical geography of the Sahara?
5. What is the general pattern of rainfall in Africa?
6. How does the Serengeti Plain help support much of Africa's wildlife?
7. What are some of the benefits of rain forests?

Human-Environment Interaction (pp. 424–427)

8. How might desertification affect people's lives in the Sahel?
9. What are some problems created by the Nigerian oil industry?
10. What are some of the problems created by the Aswan High Dam?

Critical Thinking

1. Using Your Notes

Use your completed chart to answer these questions.

Landforms	
Resources	

a. Why has Africa not been able to take advantage of its abundant resources?

b. What are some of the problems facing Africa's rain forests?

2. Geographic Themes

a. **REGION** What are some of the aspects of Africa's physical geography that make interior transportation difficult?

b. **LOCATION** In what way does Africa's location impact its climate?

3. Identifying Themes

How does desertification alter Africa's surrounding environment? Which of the five themes of geography apply to this situation?

4. Making Generalizations

How has the Aswan High Dam affected the lives of Egyptians?

5. Seeing Patterns

Has the Nigerian oil industry and the Aswan High Dam had positive or negative effects on the surrounding environment? Explain.

Additional Test Practice, pp. S1–S37

TEST PRACTICE
CLASSZONE.COM

Geographic Skills: Interpreting Maps

Profile of Africa

Use the profile to answer the following questions.

1. **PLACE** What is the tallest landform on the map?

2. **MOVEMENT** How many feet would a person have to climb to reach the peak of the Ahaggar Mountains from the lowest point at this latitude?

3. **REGION** How does this profile illustrate Africa's nickname as the plateau continent?

GeoActivity

Sketch your own profile of Africa. Use this profile as a model for your own map. Examine the physical map on page 403 and choose a latitude from which to draw your profile.

AFRICA

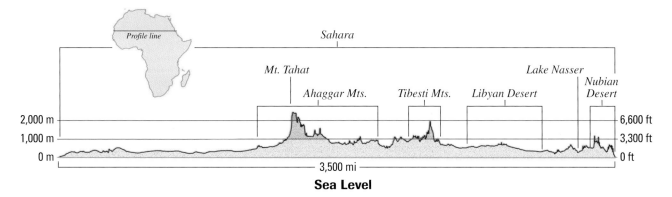

Profile line

Sahara

Mt. Tahat

Ahaggar Mts.

Tibesti Mts.

Libyan Desert

Lake Nasser

Nubian Desert

2,000 m — 6,600 ft

1,000 m — 3,300 ft

0 m — 0 ft

3,500 mi

Sea Level

INTERNET ACTIVITY

Use the links at **classzone.com** to do research about desertification. Focus on determining the long-term effects of desertification in the Sahel.

Creating a Multimedia Presentation Combine charts, maps, or other visual images in an electronic presentation showing how the Sahel will be affected by desertification.

Chapter

19

SECTION 1
East Africa

SECTION 2
North Africa

SECTION 3
West Africa

SECTION 4
Central Africa

SECTION 5
Southern Africa

HUMAN GEOGRAPHY OF AFRICA
From Human Beginnings to New Nations

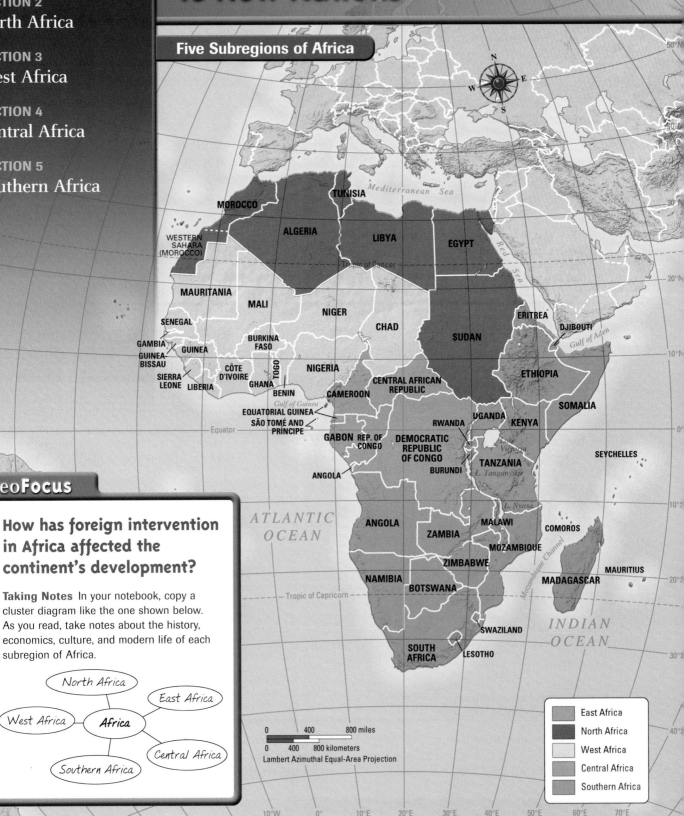

Five Subregions of Africa

0   400   800 miles
0   400   800 kilometers
Lambert Azimuthal Equal-Area Projection

East Africa
North Africa
West Africa
Central Africa
Southern Africa

GeoFocus

How has foreign intervention in Africa affected the continent's development?

Taking Notes In your notebook, copy a cluster diagram like the one shown below. As you read, take notes about the history, economics, culture, and modern life of each subregion of Africa.

North Africa

West Africa — Africa — East Africa

Southern Africa — Central Africa

East Africa

Main Ideas
- East Africa is known as the "cradle of humanity."
- East Africa's location has made it a trading center.

Places & Terms

Olduvai Gorge

Aksum

Berlin Conference

cash crop

Masai

pandemic

CONNECT TO THE ISSUES
ECONOMIC DEVELOPMENT
East Africa's political conflicts have limited its economic development.

A HUMAN PERSPECTIVE East Africa is called the "cradle of humanity" because of the large number of prehistoric human remains found in the region. In 1931, Louis Leakey, an English archaeologist, began doing research in **Olduvai Gorge,** located in northern Tanzania. Olduvai Gorge has contained the most continuous known record of humanity. The gorge has yielded fossils from 65 individual hominids, or humans that walk upright. In 1959, Leakey and his wife, Mary Leakey, discovered a fossil there of a species called *Homo habilis,* the first human creatures to make stone tools. They lived about two million years ago. Throughout history, East Africa has been a crossroads of humanity because of its geographic position near seas and oceans.

Continental Crossroads

Bounded on the east by the Red Sea and Indian Ocean, East Africa includes Burundi, Djibouti, Eritrea, Ethiopia, Kenya, Rwanda, Seychelles, Somalia, Tanzania, and Uganda. Scientists believe that the world's first humans lived there.

A TRADING COAST East Africa was also a place where early civilizations developed. An important civilization was **Aksum,** which emerged in present-day Ethiopia in the A.D. 100s. Its location on the Red Sea and the Indian Ocean made it an important trading center and contributed to its expansion and power. People from Aksum regularly traded with the people of Egypt and the eastern Roman Empire.

During the sixth century, however, Aksum lost many trading partners, and several geographic factors weakened the empire. Traders on routes between the eastern Mediterranean region and Asia began passing through the Persian Gulf rather than

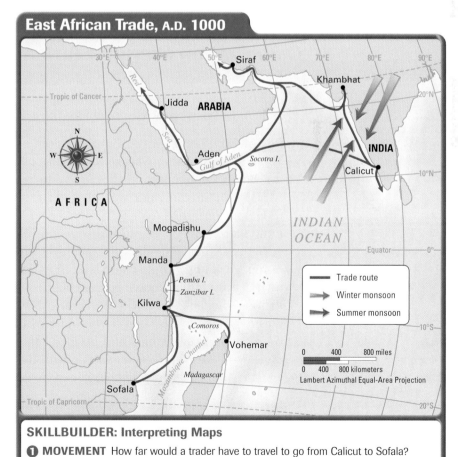

East African Trade, A.D. 1000

Legend:
- Trade route
- Winter monsoon
- Summer monsoon

0   400   800 miles
0   400   800 kilometers
Lambert Azimuthal Equal-Area Projection

SKILLBUILDER: Interpreting Maps

❶ **MOVEMENT** How far would a trader have to travel to go from Calicut to Sofala?

❷ **HUMAN–ENVIRONMENT INTERACTION** Which monsoon would a trader rely on to sail from Africa to India?

East Africa **431**

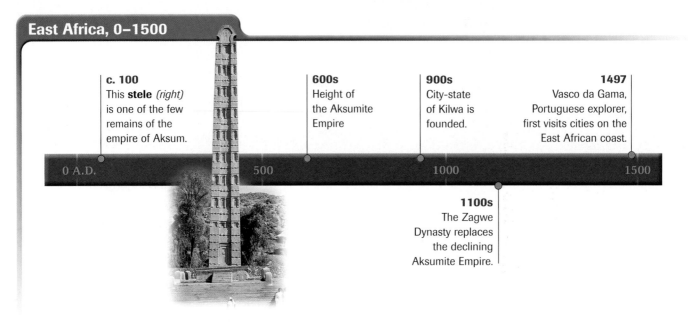

c. 100
This **stele** *(right)* is one of the few remains of the empire of Aksum.

600s
Height of the Aksumite Empire

900s
City-state of Kilwa is founded.

1497
Vasco da Gama, Portuguese explorer, first visits cities on the East African coast.

0 A.D.　　　500　　　1000　　　1500

1100s
The Zagwe Dynasty replaces the declining Aksumite Empire.

the Red Sea. In addition, the cutting down of forests and overuse of the soil led to a population decline, which reduced the empire's power.

Around the seventh century, Arab, Persian, and Indian traders once again made East Africa an international trading center. By 1300, many trading cities dotted the eastern coast of Africa. The trading city of Kilwa emerged as one of the most important cities of the time. Kilwa flourished on the southern coast of what is now Tanzania. All this movement of goods, ideas, and people made East Africa a cultural crossroads.

Colonization Disrupts Africa

In the 19th century, Europe's industrialized nations became interested in Africa's raw materials. Those European nations wanted to colonize and control parts of Africa to obtain those resources.

SCRAMBLE FOR AFRICA Europeans did not want to fight over Africa. To prevent European wars over Africa, 14 European nations convened the **Berlin Conference** in 1884–1885 to lay down rules for dividing Africa. No African ruler was invited to attend this conference, even though it concerned Africa's land and people. By 1914, only Liberia and Ethiopia remained free of European control.

Nations that attended the Berlin Conference decided that any European country could claim land in Africa by telling other nations of their claims and by showing they could control the area. The European nations divided Africa without regard to where African ethnic or linguistic groups lived. They set boundaries that combined peoples who were traditional enemies and divided others who were not. Europe's division of Africa is often cited as one of the root causes of the political violence and ethnic conflicts in Africa in the 20th century.

ETHIOPIA AVOIDS COLONIZATION Ethiopia is one country that escaped European colonization. Ethiopia's emperor, Menelik II, skillfully protected his country from the Italian invasion with weapons from France and Russia. In addition, the Ethiopian army had a greater knowledge of the area's geography than did the Italians. As a result, Ethiopia defeated Italy in 1896.

Geographic Thinking

Seeing Patterns
Ⓐ Which group of nations participated in the Berlin Conference? Which group did not?

BACKGROUND
The Ethiopian victory was the first time native Africans successfully defended themselves against a colonial power.

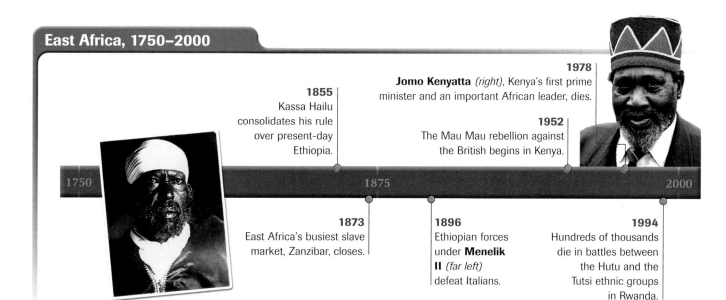

1855
Kassa Hailu consolidates his rule over present-day Ethiopia.

1978
Jomo Kenyatta *(right)*, Kenya's first prime minister and an important African leader, dies.

1952
The Mau Mau rebellion against the British begins in Kenya.

1750   1875   2000

1873
East Africa's busiest slave market, Zanzibar, closes.

1896
Ethiopian forces under **Menelik II** *(far left)* defeat Italians.

1994
Hundreds of thousands die in battles between the Hutu and the Tutsi ethnic groups in Rwanda.

CONFLICT IN EAST AFRICA By the 1970s, most of East Africa had regained its independence from Europe. However, internal disputes and civil wars became a serious problem. For example, colonialism inflamed the peoples of Rwanda and helped to cause a bloody conflict in the 1990s.

One cause of those problems was that European colonial powers had not prepared East African nations for independence. Furthermore, the ethnic boundaries created by the Europeans forced cultural divisions that had not existed before colonialism. Those cultural divisions often caused internal conflicts among native groups. Colonialism also greatly affected the economy of East Africa, which today centers around tourism and farming.

Farming and Tourism Economies

Agriculture forms the economic foundation of East Africa. In addition, East Africa's world-famous wildlife parks generate millions of dollars of revenue.

FARMING IN EAST AFRICA East Africa is more than 70 percent rural. Since European colonization in the 19th century, countries have relied more on **cash crops** such as coffee, tea, and sugar, which are grown for direct sale. They bring in much-needed revenue but reduce the amount of farmland that otherwise could be devoted to growing food for use in the region. Relying on cash crops for revenue can be risky because the price of crops varies according to the world market. ◀B

East Africa's agricultural balance is changing, however, because people are leaving farms for greater economic opportunities in cities. For example, Addis Ababa, the capital of Ethiopia, has grown by more

Geographic Thinking

Seeing Patterns
B How does growing cash crops both help a country's economy and hurt the people living in the country?

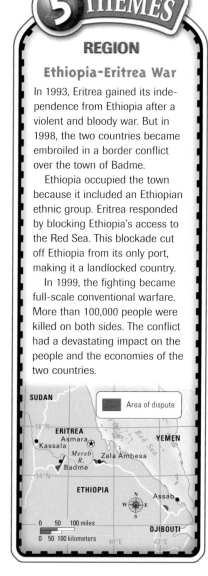

5 THEMES

REGION

Ethiopia-Eritrea War

In 1993, Eritrea gained its independence from Ethiopia after a violent and bloody war. But in 1998, the two countries became embroiled in a border conflict over the town of Badme.

Ethiopia occupied the town because it included an Ethiopian ethnic group. Eritrea responded by blocking Ethiopia's access to the Red Sea. This blockade cut off Ethiopia from its only port, making it a landlocked country.

In 1999, the fighting became full-scale conventional warfare. More than 100,000 people were killed on both sides. The conflict had a devastating impact on the people and the economies of the two countries.

AFRICA

The Masai are members of an ethnic group that live in Kenya. All Masai children address adults as either "mother" or "father."

A typical Masai girl *(pictured at the right)* takes on responsibilities that include:

- household chores
- child care
- the processing and distribution of milk

Each boy is assigned to a group called an age-set. Boys at the bottom of the age-set do the following:

- herd young animals
- learn to protect the herd from predatory animals

Between the ages of 14 and 18, boys receive a new name in a ceremony marking their transition from youth to manhood.

Around eight years of age, boys and girls have the upper part of their ears pierced. Two years later, the lower lobes are pierced. Wooden plugs are inserted into the holes to increase their size. Masai consider large ear lobes to be beautiful.

than one million people since 1991. However, such rapid population growth can put a strain on a city's resources and a country's agricultural production.

TOURISM CREATES WEALTH AND PROBLEMS One of the main economic activities in East Africa is tourism. The region's vast wildlife parks in Kenya, Uganda, and Tanzania are world famous. In 1938, Europeans created the game reserves because they were killing animals for sport at a high rate. Most African peoples did not need the parks because they hunted only for survival. However, the wild animal parks, which are no longer used for hunting, have now become important sources of income for Africans, generating millions of dollars each year from tourists.

Competing demands for the parkland exist, though. For example, Kenya's increasing population requires more food. As a result, some groups want to eliminate or reduce the size of the wildlife reserves to create more farmland. Some desperate farmers have even begun to plow the land around the parks.

BACKGROUND
Serengeti National Park in Tanzania covers nearly 6,000 square miles and contains 35 species of plains animals and 350 species of birds.

Maintaining Traditional Cultures

East Africa's position as a major trading region has given it a diverse culture. It is a melting pot of more than 160 different ethnic groups.

CULTURES OF EAST AFRICA Two major ethnic groups in East Africa are the **Masai** and the Kikuyu. The Masai, whom you read about above, are an East African ethnic group that lives on the grasslands of the rift valleys in Kenya and Tanzania. Most of the Masai herd livestock and farm the land.

Typical Masai dress includes clothes made from calfskin or buffalo hide. Women wear long skirt-like robes, while men wear a shorter version of the robe. They often grease their clothes with cow fat to protect themselves from the sun and rain. The Masai are also known for making intricate beadwork and jewelry.

The Kikuyu are the largest ethnic group in Kenya, numbering around 6.6 million. Their homeland is centered around Mount Kenya. Like the Masai, the Kikuyu traditionally were herders. However, today the Kikuyu live throughout the country and work in a variety of jobs. During British colonial rule, the Kikuyu organized a society called the Mau Mau, which fought against the British. The British killed around 11,000 Africans—mostly Kikuyu—during the Mau Mau rebellion between 1952 and 1960. ◀️

Seeing Patterns
▷ How have the lives of the Kikuyu changed during the last century?

Health Care in Modern Africa

The people of East Africa face many health care problems. The most critical is acquired immune deficiency syndrome (AIDS), which spread throughout Africa in the 1980s and 1990s.

HEALTH CARE IN AFRICA AIDS has become a pandemic and is having a devastating effect on the continent. A **pandemic** is an uncontrollable outbreak of a disease affecting a large population over a wide geographic area.

AIDS is caused by the human immunodeficiency virus (HIV). People infected with HIV do not necessarily have AIDS and can carry HIV for years without knowing it. As a result, AIDS statistics can be misleading. The number of people who have AIDS lags behind the number of those infected with HIV. Though AIDS education is increasing, some governments hide the scope of the disease. Many doctors in Africa say that more AIDS cases exist than are reported.

Some medical geographers predict that the populations of Africa's worst affected countries could decline by 10 to 20 percent.

You will read more about AIDS and other major health issues in Chapter 20. In the next section , you will learn about North Africa.

Assessment

① Places & Terms

Identify these terms and explain their importance in the region's history or culture.

- Olduvai Gorge
- Aksum
- Berlin Conference
- cash crop
- Masai
- pandemic

② Taking Notes

PLACE Review the notes you took for this section.

- How did Aksum's location help the empire grow?
- What are some of the problems created by tourism?

③ Main Ideas

a. Why did East Africa become an international trading center early in its history?

b. How did the Berlin Conference change Africa?

c. How is AIDS affecting the population of Africa?

④ Geographic Thinking

Making Generalizations
In what way has colonialism affected East Africa? **Think about:**

- the Berlin Conference
- problems in the 20th century

S See Skillbuilder Handbook, page R6.

SEEING PATTERNS Do research to learn about two ethnic groups other than the Masai and the Kikuyu in East Africa. Create a **time line** tracing the origins of those ethnic groups to the present day. Examine the groups' history, movement patterns, and evolution of their lifestyles.

Disasters!

Famine in Somalia

Famine—an extreme and long-term shortage of food—causes widespread hunger and sometimes death to millions of people. Natural causes, such as weather, plant diseases, and massive insect infestations, can cause famine. Drought is the most common natural cause. In addition, human beings can cause famine. Wars and political violence often destroy crops and prevent the adequate distribution of food. The worst famines usually involve a combination of both human and natural causes. The Horn of Africa, which includes Ethiopia and Somalia, has been the site of recent famines in the 1980s and 1990s.

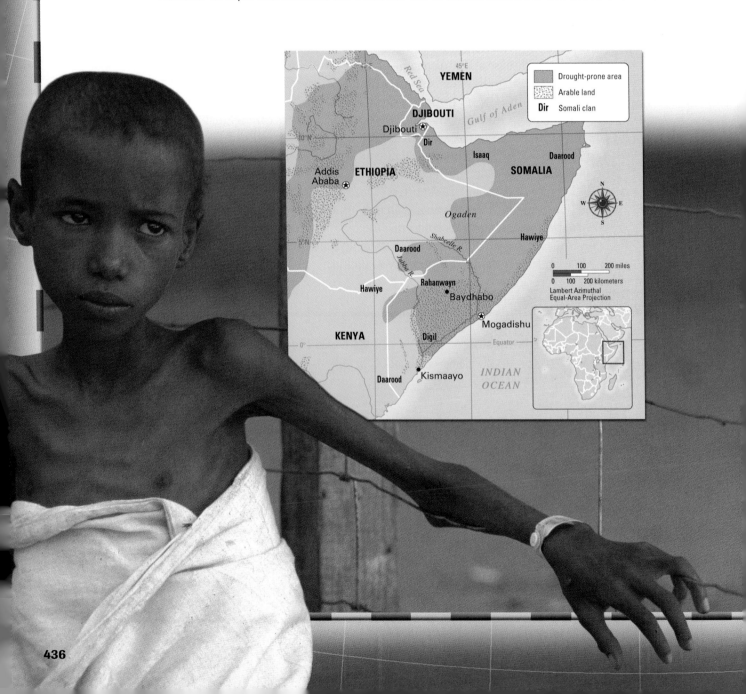

Natural Causes

A lack of rain in successive seasons resulted in drought. Drought prevented the growth of enough food to feed the country's population.

Human Causes

Somali gunmen often looted relief shipments and then extracted payment for protecting relief workers. Other political causes, such as disagreements between warring factions, also prevented the delivery of food supplies.

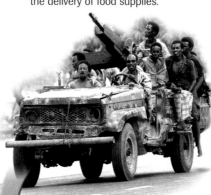

Results

Thirsty Somalis plead for water delivered by the International Red Cross in Baidoa, Somalia, in 1992. Aid agencies estimated that famine killed between 25 and 50 people a day in this town in 1992.

GeoActivity

UNDERSTANDING FAMINE

Working with a partner, use the Internet to research different international aid organizations. Then write a **news report** about those organizations.

- Create a visual aid comparing the various groups.
- Include information about how the groups are funded.

RESEARCH LINKS
CLASSZONE.COM

GeoData

FAMINE IN SOMALIA

In the early 1990s, more than 300,000 Somalis died of famine, and another 30,000 died in a related civil war.

- Principal causes included drought, desertification, and civil war.
- Underlying causes, such as increased growth of cash crops and reliance on livestock, stemmed from a history of foreign intervention dating back to Italian and British colonization in the 19th century.

OTHER FAMINES

1876–1878 India
Drought caused famine that killed about five million people.

1932–1934 Soviet Union
Between six and eight million peasants died because of actions by the government.

1958–1960 China
Around 20 million people died during government reforms.

North Africa

Main Ideas
• The Nile River valley and ancient Egypt, one of the world's great civilizations, formed a cultural hearth.
• North Africa shares the Arabic language and the Islamic religion and culture with Southwest Asia.

Places & Terms
Carthage

Islam

rai

CONNECT TO THE ISSUES
ECONOMIC DEVELOPMENT
The discovery of oil in North Africa has helped the region's economy to grow.

A HUMAN PERSPECTIVE According to legend, around 814 B.C. a Phoenician queen founded **Carthage,** one of the great cities of ancient Africa. She located it on a peninsula on the Gulf of Tunis. The location was ideal. The Lake of Tunis protected the rear of the peninsula from invasion. In addition, because Carthage was on the coast of the Mediterranean Sea, it had access to trading routes. Consequently, it became a trading and commercial force in the ancient world for hundreds of years. Carthage's history shows that a city's or a civilization's geographic position always plays an important part in its ability to thrive and grow.

Roots of Civilization in North Africa

North Africa includes Algeria, Egypt, Libya, Morocco, Sudan, and Tunisia. Egypt and the Nile River valley formed a cultural hearth, a place where ideas and innovations come together to change a region. Those ideas and innovations reached other regions through cultural diffusion.

EGYPT BLOSSOMS ALONG THE NILE The Nile River made possible the existence of the great civilization of ancient Egypt. The river flooded at roughly the same time every year, providing the people with water and rich soil for their crops. The ancient Greek historian Herodotus remarked in the fifth century B.C. that Egypt was the "gift of the Nile."

Egyptians had been living in farming villages around the Nile River since 3300 B.C. Each village followed its own customs and rituals. Around 3100 B.C., a strong king united all of Egypt and established the first Egyptian dynasty. The history of ancient Egypt would span 2,600 years and around 30 dynasties. During the Middle Kingdom, Egyptian god-kings, called Pharaohs, ruled Egypt. Egyptians believed that those kings ruled even after death, and they built pyramids to house the Pharaohs' remains.

Movement influenced ancient Egypt and the Nile valley. Egyptian ideas about farming, the building of their cities, and their system of

HUMAN-ENVIRONMENT INTERACTION An irrigation ditch from the Nile River nourishes the fields outside Al Fayyam, Egypt. **Why has Egypt been called the "gift of the Nile"?**

writing may have come from the Mesopotamians, who lived in what is now Southwest Asia. Egyptians pioneered the use of geometry in farming to set boundaries after the Nile's annual flood. Furthermore, Egyptian medicine was famous throughout the ancient world. Egyptians could make splints for broken bones and effectively treat wounds and fevers. Trade and travel on the Nile River, the Mediterranean and Red seas, and overland trade routes helped spread those practices.

ISLAM IN NORTH AFRICA North Africa lies close to Southwest Asia and across the Mediterranean Sea from Europe. As a result, it has been invaded and occupied by many people and empires from outside Africa. Greeks and Romans from Europe and Phoenicians and Ottoman Turks from Southwest Asia all invaded North Africa.

Islam, however, remains the major cultural and religious influence in North Africa. Islam, a monotheistic religion, is based on the teachings of the prophet Muhammad, whom you will read about in Chapter 22. Muslim invaders from Southwest Asia brought their language, culture, and religion to North Africa. Beginning in A.D. 632, the successors of Muhammad began to spread Islam through conquest and through trade. Around 634, Muslim armies swept into lower Egypt, which was then part of the Byzantine Empire. By 750, Muslims controlled most of North Africa. Muslims bound their territory together with a network of sea-linked trading zones. They used the Mediterranean Sea and the Indian Ocean to connect North Africa and Europe with Southwest Asia. ◁**A**

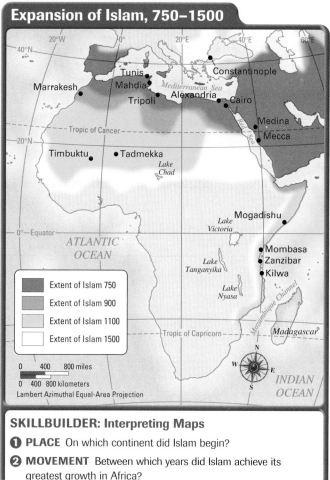

Expansion of Islam, 750–1500

Extent of Islam 750
Extent of Islam 900
Extent of Islam 1100
Extent of Islam 1500

0   400   800 miles
0   400   800 kilometers
Lambert Azimuthal Equal-Area Projection

SKILLBUILDER: Interpreting Maps

❶ PLACE On which continent did Islam begin?

❷ MOVEMENT Between which years did Islam achieve its greatest growth in Africa?

🌐 **Geographic Thinking** ◁

Using the Atlas
▷ Using the map on this page and the unit atlas on page 405, identify the first Islamic countries in Africa.

Economics of Oil

North Africa began with an economy based on agriculture. Over the course of its history, it evolved into an economy based on the growth of cash crops and mining. Today, the economy revolves around the discovery of oil in the region.

BLACK GOLD Oil has transformed the economies of some North African countries, including Algeria, Libya, and Tunisia. In Algeria, oil has surpassed farm products as the major export and source of revenue. Furthermore, oil makes up about 99 percent of Libya's exports. Libya and Algeria supply the European Union with much of its oil and gas.

Although oil has helped the economies of those countries, it has also caused some problems. For example, Libya's labor force cannot meet the demands of the oil industry because of a lack of training and education. Oil companies therefore are forced to give many high-paying jobs to foreign workers. Despite the oil industry, overall unemployment is still a problem. As a result, large numbers of North Africans have migrated to Europe in search of jobs.

A Culture of Markets and Music

North African culture is a combination of Arabic influences and traditional African ethnic groups.

NORTH AFRICAN *SOUKS* *Souks,* or marketplaces, are common features of life in North Africa. A country *souk* opens early in the morning. Tents are erected, and storytellers, musicians, and fortunetellers entertain the crowds. A typical city *souk* is located in the *medina,* or old section, of a North African town or city. A *medina* has narrow, winding streets. Some of the best *souks* in North Africa can be found in Marrakesh, Morocco. The markets are known for high-pressure sales, and shoppers must be prepared to bargain fiercely for the lowest price. ▶

In both the city and the country, people fill the *souks* throughout the day. All kinds of bartering and haggling take place for a range of products, including brightly colored clothes, spices, and a variety of foods. The aroma of lamb, spices, and animals fills the air. It is also a place where one can eat traditional foods such as couscous, a kind of steamed grain.

PROTEST MUSIC Algeria is home to **_rai,_** a kind of music developed in the 1920s by poor urban children. *Rai* was at first carefree and centered around topics for youths. The music is fast paced and contains elements of popular Western music.

Before Algerian independence in 1962, however, performers began using *rai* to communicate Algerian resentment toward their French

Geographic Thinking ◀

Making Comparisons
◀ⓑ How are country and city *souks* alike? Different?

MOVEMENT
Moroccans flood this typical market in Marrakesh.
What role do markets play in the movements of goods and people?

440

colonizers. After independence, the Algerian government tried to ban *rai*. In the 1990s, Islamic fundamentalists have criticized *rai* for its Western-style qualities. *Rai* is now used as a form of rebellion against Islamic fundamentalists, especially by women.

MOVEMENT Two women in Western-style clothing pause outside a popular marketplace in Marrakesh, Morocco.

Changing Roles of Women

Modern life in North Africa is in a constant state of change. The role of women, especially, has shifted during the past several years.

WOMEN AND THE FAMILY North African households tend to be centered around males. Men go out to work in offices or on farms. Few women hold jobs after they marry. Men and women also generally eat and pray separately.

Women's roles, however, are changing, especially in Tunisia, where having more than one wife at a time has been abolished. It has also increased the penalty for spousal abuse. Moreover, either spouse can now seek a divorce. In addition, Tunisia no longer permits preteen girls in arranged marriages and requires equal pay for equal jobs.

Women in North Africa have also made gains outside the home, particularly in cities. Growing numbers of them, for instance, have professional jobs. Women hold seven percent of Tunisia's parliamentary seats and manage nearly nine percent of the businesses in Tunis, the capital of Tunisia.

In the next section, you will read about how trade formed the foundation of ancient civilizations in West Africa.

Assessment

❶ Places & Terms

Identify these terms and explain their importance in the region's history or culture.

- Carthage
- Islam
- *rai*

❷ Taking Notes

PLACE Review the notes you took for this section.

- What is the single biggest cultural influence in North Africa?
- Which commodity supports some of North Africa's economies?

❸ Main Ideas

a. How did the Nile help support the growth of ancient Egypt?

b. Where did Islam spread after its beginnings in Southwest Asia?

c. In which ways have women's roles changed in North Africa?

❹ Geographic Thinking

Drawing Conclusions
How has Islam influenced life in North Africa? **Think about:**

- its impact on women
- the religion that people practice

RESEARCH LINKS
CLASSZONE.COM

GeoActivity

SEEING PATTERNS Use the Internet or encyclopedias to learn about all the economic and recreational activities supported by the Nile River. Then create an **illustration** of the Nile River with those activities taking place.

AFRICA

West Africa

Main Ideas

- Wealth from the gold and salt trades supported a series of West African empires.
- West Africa has a rich cultural tradition that has influenced many parts of the world.

Places & Terms

Gorée Island

stateless society

Ashanti

CONNECT TO THE ISSUES

COLONIALISM European nations took raw materials from West Africa. Today many West African countries rely on exports to support their economies.

A HUMAN PERSPECTIVE A visit to **Gorée Island,** off the coast of Senegal, can be a moving experience. This island served as one of the busiest points for exporting slaves during the slave trade. From the mid-1500s to the mid-1800s, Europeans transported about 20 million Africans through Gorée Island. The island has a slave house, a dark, damp building that housed captive Africans. Europeans packed these captives onto slave ships bound for plantations in the Americas. Approximately 20 percent of all Africans died on the transatlantic voyage—and the rest never saw their West African homes or families again. Slavery had a profound effect on West Africa that is still being felt there today.

A History of Rich Trading Empires

West Africa includes Benin, Burkina Faso, Cape Verde, Chad, Côte d'Ivoire, Gambia, Ghana, Guinea, Guinea-Bissau, Liberia, Mali, Mauritania, Niger, Nigeria, Senegal, Sierra Leone, and Togo. West Africa is a cultural hearth, and its ideas and practices spread to North America and Europe.

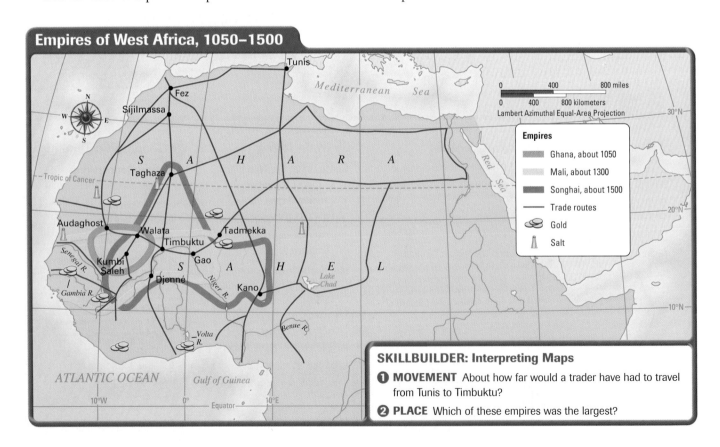

Empires of West Africa, 1050–1500

Empires

- Ghana, about 1050
- Mali, about 1300
- Songhai, about 1500
- Trade routes
- Gold
- Salt

SKILLBUILDER: Interpreting Maps

❶ **MOVEMENT** About how far would a trader have had to travel from Tunis to Timbuktu?

❷ **PLACE** Which of these empires was the largest?

THREE TRADING EMPIRES The empires of Ghana, Mali, and Songhai thrived in West Africa because of their location on trade routes across the Sahara. Gold and salt were the main products traded. By A.D. 200, trade across the Sahara had existed for many years.

Many of the trade routes crossed an area farmed by the Sonike people. They called their leader *ghana,* or war chief. Traders began to refer to this area as Ghana, which grew rich from taxing the traders who passed through its territory. Traders exchanged mostly gold and salt. Ghana became an empire around A.D. 800 but began to decline in power by the end of the 11th century. ◀A

By 1235, the kingdom of Mali emerged. Mali's first great leader, Sundiata, conquered Ghana. He promoted agriculture and reestablished the gold and salt trade. Some experts estimate that until 1350, about two-thirds of the world's gold came from West Africa. Around 1400, Mali declined because of a lack of leadership and the discovery of new gold fields farther east.

Around 1400, the empire of Songhai replaced Mali. Sunni Ali ruled for 28 years, beginning in 1464. In 1591, a Moroccan army invaded Songhai and defeated it, destroying the empire.

STATELESS SOCIETIES West Africa is filled with many different cultures and peoples. Before colonialism, some of these people lived in what are called stateless societies.

A **stateless society** is one in which people rely on family lineages to govern themselves, rather than an elected government or a monarch. A lineage is a family or group that has descended from a common ancestor. Members of a stateless society work through their differences to cooperate and share power.

One example of a stateless society is the Igbo of southeast Nigeria. Relying on family lineages worked well for the Igbo and other African societies. However, many stateless societies faced challenges from 18th- and 19th-century European colonizers, who expected one ruler to govern the society.

West Africa Struggles Economically

Trade is as important to West Africa today as it was in the past. The economic well-being of West Africa is based on the sale of its products to industrialized countries in Europe, North America, and Asia. The economies of West Africa range in strength from the relatively solid economy of Ghana to the weak economy of Sierra Leone.

Geographic Thinking

Seeing Patterns
A▷ Why did three empires prosper and grow in this area of West Africa?

BACKGROUND
One stateless society, the Nuer of southern Sudan, organized thousands of people without an official ruler.

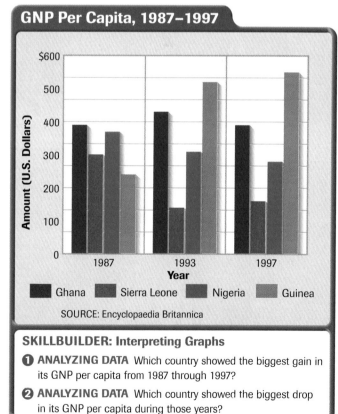

GNP Per Capita, 1987–1997

Amount (U.S. Dollars) vs. Year (1987, 1993, 1997)

Legend: Ghana, Sierra Leone, Nigeria, Guinea

SOURCE: Encyclopaedia Britannica

SKILLBUILDER: Interpreting Graphs

❶ **ANALYZING DATA** Which country showed the biggest gain in its GNP per capita from 1987 through 1997?

❷ **ANALYZING DATA** Which country showed the biggest drop in its GNP per capita during those years?

AFRICA

GHANA'S STABILITY Ghana's economy relies primarily on the export of gold, diamonds, magnesium, and bauxite to the industrialized world. Ghana has a per capita income of $1,900, the second highest in West Africa. However, its per capita income is low compared to the industrialized world.

Ghana's transition from colonialism to democracy has had setbacks, including military rule and civil war. However, in 1992, 1996, and 2000, Ghana held free and fair elections. As a result of this new political stability, the economy is growing at a healthy rate. But other West African countries have not been as fortunate. **B**

PROBLEMS IN SIERRA LEONE The worst economic conditions in West Africa exist in Sierra Leone, which once produced some of the world's highest-quality diamonds. However, years of political instability and civil wars have left the economy in shambles. In addition, a relatively uneducated population—with a 31 percent literacy rate—leaves a shortage of skilled workers. Finally, the road and transportation system contains few highways and only 800 miles of roads. In contrast, Benin, another West African country close to Sierra Leone in size, has about 5,000 miles of roads.

Geographic Thinking

Seeing Patterns
B How is the economy of Ghana similar to those of the ancient West African kingdoms?

HUMAN-ENVIRONMENT INTERACTION A West African weaver makes *kente* cloth. **What are some skills that a weaver might need?**

Cultural Symbols of West Africa

West African cultures, such as the Ashanti and Benin, have produced elaborate craftwork and colorful textiles.

ASHANTI CRAFTS The <u>**Ashanti,**</u> who live in what is now Ghana, are known for their work in weaving colorful *asasia*—what Westerners usually call *kente* cloth. The designs of *kente* cloth contain colorful woven geometric figures with specific meanings. Only royalty were allowed to wear *kente* cloth.

Other crafts include making masks and carving wooden stools. An Ashanti stool symbolizes the unity between ancestral spirits and the living members of a family. Fathers often give their sons a stool as their first gift. In the case of kings, the stool represents the unity of the state with its people.

BENIN ART The kingdom of Benin, which has no direct connection with the current

444

country of Benin, arose in what is now Nigeria in the 1200s. Benin artists made beautiful objects of metal and terra cotta. However, their most important works were fashioned from brass and are called Benin "bronzes." They include statues, masks, and jewelry. A common subject of Benin "bronzes" was that of the queen mother.

Music in Daily Life

Music is a large part of life in West Africa. West African music has become an important influence on world music.

WEST AFRICAN MUSIC West African popular music involves a blend of traditional African music with American forms of jazz, blues, and reggae— which also had their origins in West Africa because of the slave trade and the contact between the two regions. Over the years, West African musicians used French and English lyrics to attract an international audience. West African music is played on a wide variety of drums and other instruments such as the kora, a cross between a harp and a lute. The kora originated in what is now Guinea-Bissau.

MOVEMENT King Sunny Adé's music blends sounds from North America and West Africa. **How do you think music moved from West Africa to North America?**

King Sunny Adé, also known as the "minister of enjoyment," is a popular musician from Nigeria. King Sunny and his band, the African Beats, play an informal type of music characterized by tight vocals, complex guitar work, traditional talking drums, percussion instruments, and the pedal steel guitar and accordion.

In Section 4, you can read more about culture and life in Central Africa.

SECTION 3 Assessment

① Places & Terms

Identify these terms and explain their importance in the region's history or culture.

- Gorée Island
- stateless society
- Ashanti

② Taking Notes

HUMAN-ENVIRONMENT INTERACTION Review the notes you took for this section.

West Africa ⎯ Africa

- How did natural resources affect the ancient empires in West Africa?
- How do stateless societies differ from those with a centralized government?

③ Main Ideas

a. What three empires flourished because of trade in West Africa?

b. What are some of the roadblocks to economic development in West Africa?

c. What is the significance of the stool in Ashanti society?

④ Geographic Thinking

Making Comparisons How do the economics of Sierra Leone and Ghana differ?
Think about:
- Ghana's political stability
- the state of infrastructure in Sierra Leone

RESEARCH LINKS
CLASSZONE.COM

GeoActivity

MAKING COMPARISONS Review the information about the West African economies on pages 443–444. Using the Internet or encyclopedias, find the per capita income of four other West African countries during the last ten years. Then create a **chart** comparing their growth or decline during that time.

Comparing Cultures

Feasts

All over the world, people celebrate certain events by holding a ritual feast. The autumn harvest, when the season's crops are gathered, is an important time in most cultures. As a result, many people have a special meal to celebrate the earth's bounty. Most harvest feasts are accompanied by a legend or story that tells of the feast's origins. For example, the American harvest feast began in 1621 when Pilgrims invited Native Americans to join them in a three-day celebration marking the harvest.

People in India celebrate *Sankranti* by eating on traditional banana leaf plates. *Sankranti* celebrates the end of the year's harvest. Rice is a staple of this meal.

Chinese celebrate the moon festival in Hong Kong. Throughout history, the Chinese have planted and harvested according to the moon. The Chinese eat moon-shaped pastries filled with red bean and lotus seed paste.

Critical Thinking

1. Using Your Notes

Use your completed chart to answer these questions.

a. How were the precolonial kingdoms of West Africa similar to or different from the precolonial kingdoms of Southern Africa?

b. How did colonialism change Africa and its people?

2. Geographic Themes

a. **MOVEMENT** How did the movement of Islam from Southwest Asia to North Africa affect the African continent?

b. **HUMAN-ENVIRONMENT INTERACTION** What role did natural resources play in the colonization of Africa?

3. Identifying Themes

How did natural resources affect the formation of ancient African kingdoms and empires? Which of the five themes apply to this situation?

4. Making Comparisons

How did stateless societies in Africa differ from centralized governments?

5. Determining Cause and Effect

What prompted the Berlin Conference, and what effects did it have on Africa's culture and economy?

Additional Test Practice, pp. S1–S37

TEST PRACTICE
CLASSZONE.COM

Geographic Skills: Interpreting Graphs

Languages of Nigeria

Use the graph below to answer the following questions.

1. **ANALYZING DATA** What percentage of Nigerians speak English?

2. **MAKING GENERALIZATIONS** Which language group is the most commonly spoken?

3. **MAKING INFERENCES** How might the number of languages in Nigeria affect a newly formed democratic government?

GeoActivity

Choose another country in Africa. Then using the library, encyclopedias, or other reference books, create your own language pie chart.

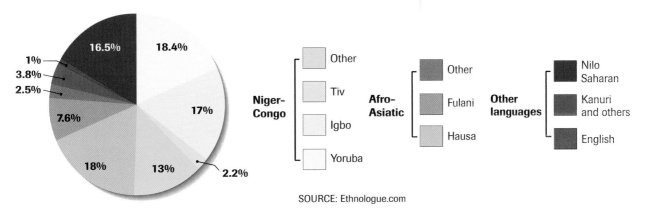

SOURCE: Ethnologue.com

INTERNET ACTIVITY

Use the links at **classzone.com** to do research on the people of one African country. Look for such information as age range, religions, ethnic groups, literacy rates, and per capita income.

Constructing a Population Pyramid Using the information you have gathered, construct a population pyramid describing the population characteristics of the society you have chosen.

Chapter 20

TODAY'S ISSUES
Africa

SECTION 1
Economic Development

SECTION 2
Health Care

CASESTUDY
EFFECTS OF COLONIALISM

For more on these issues in Africa . . .

CURRENT EVENTS
CLASSZONE.COM

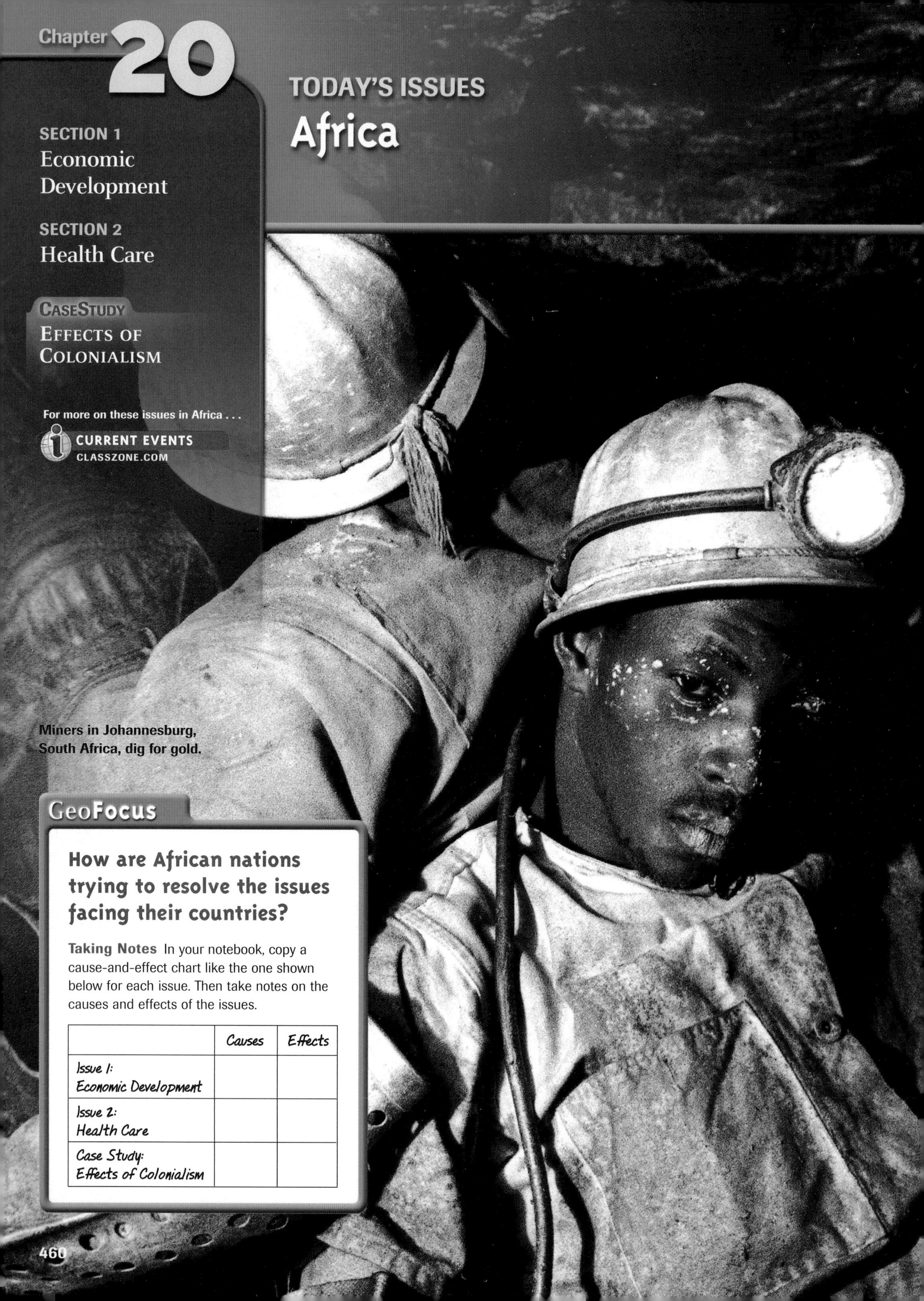

Miners in Johannesburg, South Africa, dig for gold.

GeoFocus

How are African nations trying to resolve the issues facing their countries?

Taking Notes In your notebook, copy a cause-and-effect chart like the one shown below for each issue. Then take notes on the causes and effects of the issues.

	Causes	Effects
Issue 1: Economic Development		
Issue 2: Health Care		
Case Study: Effects of Colonialism		

Economic Development

How can African nations develop their economies?

Main Ideas
- Africa's history of colonization has had long-term effects on its economy.
- Barriers to African economic development include illiteracy, foreign debt, and a lack of manufacturing industries.

Places & Terms

"one-commodity" country

commodity

diversify

A HUMAN PERSPECTIVE Mauwa Funidi wonders about the future of her country, the Democratic Republic of the Congo, as she looks around the rundown university library where she works. She has not been paid her salary of 12 dollars per month in many months. Classes at the university have been suspended because of a lack of funds. Funidi survives only by selling little bags of charcoal on the streets of Kisangani. Funidi, like many other Africans, is trying to scrape out a living on a continent where people's standard of living has gotten worse over the last 30 years. Nevertheless, many African countries have vowed to change their fortunes with better government, better relations with neighbors, more investment in education, and a diverse economy.

Africa's Economy Today

Most African nations have little manufacturing of their own. Their economies are based on providing raw materials—oil, minerals, or agricultural products—to the world's industrialized countries.

A HISTORY OF PROBLEMS As you learned in the previous chapter, European colonizers exploited Africa's resources and people during the past few centuries. Millions of Africans were sold into slavery, and countless others have died in Africa from harsh working conditions while obtaining raw materials for foreign interests. In addition, the land has been mined and drilled with little regard for the environment. This history of exploitation has limited Africa's economic growth and fostered political instability. Without political stability, consistent economic growth is difficult.

AFRICA'S ECONOMIC STATUS Today, most African countries are worse off economically than they were in the 1960s, just after many of them gained independence from European colonizers. In the last 30 years, average incomes in Africa have decreased, while they have increased in most of the rest of the world. Africa accounts for only 1 percent of total world GNP and 1.5 percent of total dollar value of world exports—both

HUMAN-ENVIRONMENT INTERACTION Much of Africa's infrastructure is undeveloped. With few paved roads, trucks get bogged down on muddy roads, such as this one in the Democratic Republic of the Congo.

AFRICA

small numbers compared to Africa's population and natural resources. The whole of Africa's economy is about as large as that of Argentina's.

Furthermore, the economic infrastructure needed for substantial growth is not in place. Roads, airports, railroads, and ports are not adequate to help African nations further their economic growth.

In addition, most Africans don't have access to computers or other aspects of high technology. High technology has fueled economic growth in other parts of the world such as North America, Europe, and Asia.

Geographic Thinking

Seeing Patterns
A Why do you think good roads are important to the functioning of an economy?

On the Road to Development

Despite this legacy of exploitation, African nations are struggling to build economies based on the careful use of natural and human resources.

REDUCING DEBT AND INCREASING COOPERATION When the colonial nations pulled out of Africa, they often left the newly independent nations without money for transportation, education, and businesses. To build their economies, African countries borrowed heavily. By 1997, total public debt of sub-Saharan African governments—about 227 billion dollars—was strangling them. As a result, many Western leaders have urged their countries to forgive Africa's debts so that it has more money to build its economies.

Another way that Africa seeks to improve its economy is through regional cooperation. The Economic Community of West African States (ECOWAS) and the Southern African Development Community (SADC) are both striving to promote trade. For example, ECOWAS is working toward removing duties and creating a common currency. Efforts of SADC include working to improve the transportation and communication infrastructures.

BACKGROUND
In 1998, over 57 percent of Africa's total GNP went to repaying its debts.

BUILDING INDUSTRIES The economy of many African nations is based on the export of raw materials. Furthermore, several of Africa's countries rely on just one or two principal commodities for much of their earnings. These are called **"one-commodity" countries.** A **commodity** is an agricultural or mining product that can be sold. The value of a commodity varies from day to day based on worldwide supply and demand. That makes the economies of the producing nations—especially "one-commodity" countries—unstable. Economists believe African nations must **diversify,** or create variety in, their economies and promote manufacturing to achieve economic growth and stability.

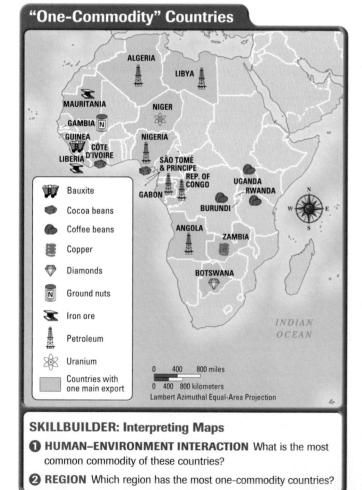

"One-Commodity" Countries

ALGERIA
LIBYA
MAURITANIA
GAMBIA
GUINEA
LIBERIA
CÔTE D'IVOIRE
NIGER
NIGERIA
SÃO TOMÉ & PRINCIPE
REP. OF CONGO
GABON
ANGOLA
UGANDA
RWANDA
BURUNDI
ZAMBIA
BOTSWANA

INDIAN OCEAN

Bauxite
Cocoa beans
Coffee beans
Copper
Diamonds
Ground nuts
Iron ore
Petroleum
Uranium
Countries with one main export

0   400   800 miles
0   400   800 kilometers
Lambert Azimuthal Equal-Area Projection

SKILLBUILDER: Interpreting Maps
❶ **HUMAN–ENVIRONMENT INTERACTION** What is the most common commodity of these countries?
❷ **REGION** Which region has the most one-commodity countries?

BACKGROUND
The Highland's Water Project of Lesotho will eventually allow the country to generate its own electricity, instead of having to buy it from South Africa.

Some African nations are making strides toward that goal. In East Africa, Djibouti is using its location on the Gulf of Aden to establish a major international shipping center.

Educating Workers

A key to developing Africa's economies is improving its education system to provide people with a high level of skills. African nations must also find ways to prevent their educated citizens from leaving the continent.

Geographic Thinking

Seeing Patterns
Why do you think education is important to Africa's economy?

IMPROVING EDUCATION A large barrier to economic development in Africa is an unschooled population. For example, the average length of school attendance for African women has increased only by 1.2 years in the last 40 years. In some countries, such as Angola and Somalia, civil wars have all but destroyed the school systems.

Some African countries, however, are making progress. For example, in Algeria, 94 percent of the country's school-age population receive a formal education. Mauritius has also made huge gains. Currently, 83 percent of Mauritians over the age of 15 are literate.

REVERSING THE BRAIN DRAIN Another priority is slowing the departure of African professionals to Western countries. In 1983, the International Organization for Migration began a campaign to encourage these professionals to return home.

As Africa moves into the 21st century, efforts to improve education, invest in industry, and create stable governments provide hope for the future.

Geography TODAY

Hi-Tech Tracking

As the scorching sun of the Kalahari Desert beats down on Karel Kleinman, he puts information into a palm-sized computer. Kleinman tracks animals in the same desert as his Koi grandfather did. But now Kleinman uses high technology to follow the animals.

Louis Liebenberg of South Africa developed this application to collect information about animals more efficiently. For example, some of the data will help protect certain species from drought. Liebenberg's idea both protects Africa's resources and shows how well-educated people can solve problems.

AFRICA

SECTION 1 Assessment

1 Places & Terms

Explain the meaning of each of the following terms.

- "one-commodity" country
- commodity
- diversify

2 Taking Notes

PLACE Review the notes you took for this section.

	Causes	Effects
Issue 1		

- What are some of the causes of economic problems of African countries?
- What impact is Africa's debt having on its ability to build its economy?

3 Main Ideas

a. What has happened to people's incomes over the last half-century?

b. What is one problem for "one-commodity" countries?

c. Why is improving education important to Africa's economy?

4 Geographic Thinking

Making Generalizations
What actions should African nations take to form a solid economic foundation? **Think about:**

- economic cooperation
- education

S **See Skillbuilder Handbook, page R6.**

GeoActivity

EXPLORING LOCAL GEOGRAPHY Find out how your city or state promotes economic development. Learn about laws passed to promote growth or tax breaks given to certain industries. Then write a **news article** on the topic.

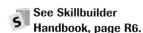

Reading a City Map

Johannesburg, South Africa, is one of the youngest major cities in the world. It grew rapidly following the discovery of gold in 1886. Today, it is South Africa's largest city and the country's financial and industrial center. Looking at the city map below, you can see that the streets of the city center are laid out in a grid. A grid is something resembling a framework of crisscrossing parallel bars.

THE LANGUAGE OF MAPS A **city map** is essentially another kind of road map. However, it is usually set at a larger scale than a state road map in order to show greater detail to guide both visitors and residents. Many city maps show the names of streets, major tourist attractions, bus and train stations, and other useful buildings.

Johannesburg•

Johannesburg, South Africa

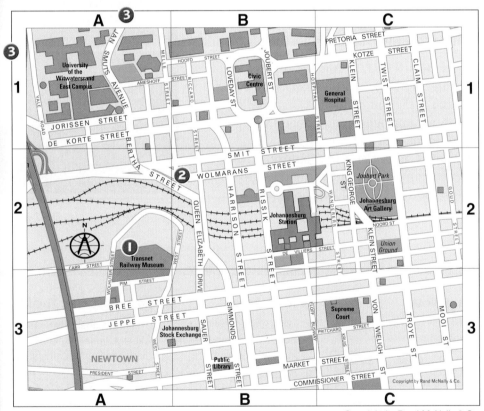

Copyright by Rand McNally & Co.

Legend:
- Parks
- Principal buildings
- Rail/Bus stations
- Major highways
- Streets and roads
- Railways

0   0.1   0.2 Mi.
0   0.1   0.2   0.3 Km.

❶ Points of interest are shown to help tourists plan their visit.

❷ Labeling major streets is necessary to guide people around the city.

❸ Letters at the top and bottom and numbers on the sides identify the grid sections created by the black lines. The grid sections help readers find places on the map. Most city maps have an index listing places on the map and the grid sections where they appear.

Map and Graph Skills Assessment

1. Making Generalizations
If you took the train into Johannesburg, how would you get to the railway museum by foot and about how long would it take?

2. Drawing Conclusions
Where is the most likely place for a picnic in this part of the city? What are the place's index coordinates?

3. Making Inferences
If you spent a weekend in Johannesburg, what are at least five activities available to you in the city?

Health Care

How can African countries eliminate the diseases that threaten their people and cultures?

Main Ideas
• Epidemic diseases are killing Africa's people in huge numbers.
• African nations and countries around the world are using a variety of methods, including education, to eradicate disease.

Places & Terms
AIDS
cholera
malaria
tuberculosis
UNAIDS

A HUMAN PERSPECTIVE On June 1, 2001, Nkosi Johnson died from the human immunodeficiency virus (HIV)—the virus that causes **acquired immune deficiency syndrome (AIDS).** He was the longest living South African child born with HIV. In February, he celebrated his 12th birthday—but weighed just 27 pounds. Living with a foster mother, the child had become a symbol of hope in a nation suffering from AIDS. He frankly discussed the problems of the disease and received cheers at the world's largest AIDS conference in Durban, South Africa, in July 2000. His plight was typical of many on the continent, as African nations struggle to deal with this and other diseases.

Disease and Despair

Controlling AIDS and other diseases is essential if Africans are to improve their quality of life and live a normal lifespan.

SERIOUS DISEASES African nations are threatened by a variety of diseases. Inadequate sanitation and lack of a clean water supply can lead to **cholera,** an infection that is often fatal if not treated. In 2000–2001, widespread flooding caused some cases of cholera in Mozambique, but international relief efforts prevented a widespread outbreak.

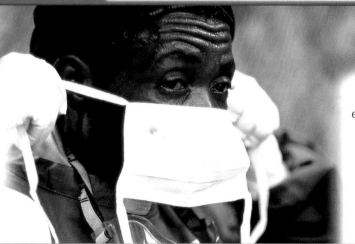

Diseases in Africa, 1900 and 2000

Leading Diseases 1900

Malaria
First reference in Greece around 400 B.C.

Sleeping Sickness
First described around A.D. 1300 in present-day Mali.

Smallpox
First evidence c. 1156 B.C. in Egypt. Eradicated A.D. 1977.

Leading Diseases 2000

Malaria
Ninety percent of world's estimated 250 million malaria cases occur in Africa.

Sleeping Sickness
Affects 60 million people annually in Africa.

AIDS
Origins of HIV traced to Central Africa in 1959.

SKILLBUILDER: Interpreting Charts
❶ **MAKING COMPARISONS** What was a leading disease in Africa in 1900 but not in 2000?
❷ **DRAWING CONCLUSIONS** What disease occurs in Africa in 2000 but not in 1900?

Mosquitos carrying **malaria**—an infectious disease marked by chills and fevers that is often fatal—are common in African countries. The disease has become resistant to standard drugs because of overuse of those drugs in treating the disease during the past several decades. AIDS and HIV, however, create the most severe problems. Seventy percent of the world's adult AIDS cases and 80 percent of the world's children with AIDS are in African nations. AIDS is often accompanied by **tuberculosis,** a respiratory infection spread between humans.

AIDS Stalks the Continent

In 2000, AIDS took the lives of three million people worldwide. Of these, 2.4 million lived in sub-Saharan Africa. In Swaziland, three of every four deaths were attributed to AIDS. The AIDS epidemic in Swaziland has caused life expectancy there to drop from 58 years to 39 years. In 2000, nearly 26 million people in Africa were living with either HIV or AIDS.

A HIGH PRICE TO PAY Widespread disease has economic consequences. People who are sick work less or not at all, earn less, and thus are pushed further into poverty. Economists project that by 2010, the GDP of South Africa will be 17 percent lower than it would have been if not for AIDS. Furthermore, AIDS patients' medical care is also expensive. **UNAIDS,** the UN program that studies the world's AIDS epidemic, estimates that $4.63 billion will be needed to fight AIDS in Africa.

BACKGROUND
According to the U.S. Agency for International Development, by 2010, nearly 30 million children will have lost at least one parent to AIDS.

Nations Respond

Response to these epidemics comes both from African nations and from countries around the world.

A VARIETY OF ANSWERS To fight malaria and other insect-borne diseases, African nations have used spraying programs since the 1930s to reduce the number of insects. In 2000, the Global Fund for Children's Vaccines pledged more than $250 million for use over the next five years for immunization programs in Africa, Asia, Latin America, and Europe.

Some African countries are fighting disease by improving their health care systems. Gabon, for example, has used oil revenues to improve its health care system substantially. In addition, the African Development Fund approved a loan of nearly 12.3 million dollars to enable Mozambique to upgrade its public health facilities. Ⓐ▸

Geographic Thinking◂

Using the Atlas
Ⓐ Using the atlas and the map on the left, which other countries could rely on revenues from oil to improve their health care systems?

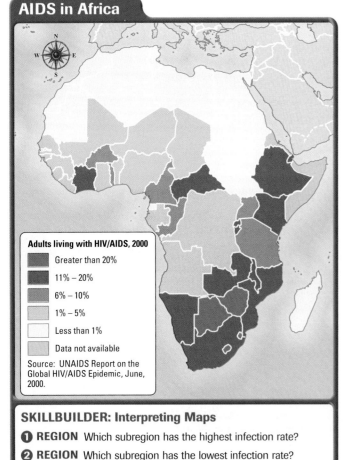

AIDS in Africa

Adults living with HIV/AIDS, 2000
- Greater than 20%
- 11% – 20%
- 6% – 10%
- 1% – 5%
- Less than 1%
- Data not available

Source: UNAIDS Report on the Global HIV/AIDS Epidemic, June, 2000.

SKILLBUILDER: Interpreting Maps
❶ **REGION** Which subregion has the highest infection rate?
❷ **REGION** Which subregion has the lowest infection rate?

STRATEGIES AGAINST AIDS
Fighting and preventing AIDS is being done on many levels. In December 2000, South Africa and Brazil reached an agreement to work together on AIDS prevention and care. Brazil's public health policies to combat AIDS and other diseases are considered a model for developing countries.

SUCCESS STORIES Two countries, Uganda and Senegal, have had success in reducing the spread of HIV. Uganda's government has spearheaded efforts to combat AIDS. For example, in 1997, Uganda began to offer same-day HIV tests and education programs. Infection rates among 15 to 24 year olds have dropped by 50 percent. On the other hand, Senegal has controlled the spread of the disease from the outset through an intensive education program. Infection rates have remained below two percent since the mid-1980s.

UNAIDS says that HIV infection rates in 2000 in sub-Saharan Africa dropped by 200,000 cases from 1999. However, UNAIDS cautions that the drop in HIV infection rates could mean that almost as many people are dying of AIDS as are being infected with HIV. Nevertheless, many African nations are taking action. With these efforts, African countries can build an effective health care system and make progress against the epidemics that threaten its peoples and cultures.

Geographic Thinking

Making Comparisons
B▷ What are the differences in the ways that Uganda and Senegal have tried to slow the spread of AIDS?

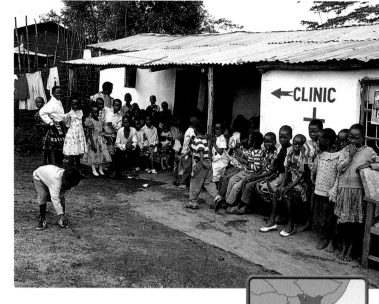
PLACE Kenyans gather outside a health clinic near Nairobi to learn about the dangers of AIDS.

AFRICA

SECTION 2 Assessment

1 Places & Terms

Explain the meaning of each of the following terms.
- AIDS
- cholera
- malaria
- tuberculosis
- UNAIDS

2 Taking Notes

PLACE Review the notes you took for this section.

	Causes	Effects
Issue 2		

- What are some of the serious diseases affecting African countries?
- How has AIDS affected Swaziland?

3 Main Ideas

a. What are some of the causes of cholera and malaria?

b. How is AIDS affecting the population of Africa's countries?

c. Why might the drop in HIV infection rates not indicate progress in slowing AIDS?

4 Geographic Thinking

Identifying and Solving Problems How have African nations slowed the spread of the continent's diseases?
Think about:
- the different programs
- international cooperation

RESEARCH LINKS
CLASSZONE.COM

GeoActivity

MAKING COMPARISONS Using encyclopedias or the Internet, find out what the leading diseases are in the United States and identify their primary causes. Compare your findings to the leading diseases in Africa in a **chart** on the topic.

EFFECTS OF COLONIALISM

How can African nations bring peace and stability to their people?

OUR FREE NATION - NAMIBIA

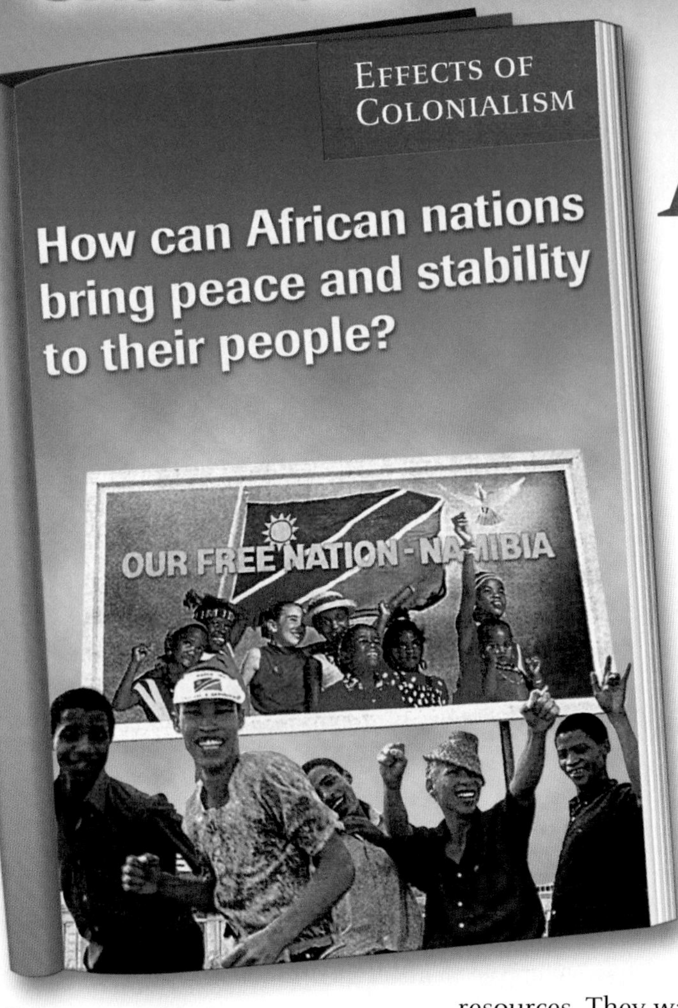

Young people celebrate Namibia's independence from South Africa in 1990.

The Voyageur Experience in World Geography
Kenya: National Identity and Unity

Africa, at the beginning of the 19th century, was home to great empires and rich cultures such as the Zulu, the Ashanti, and the Hausa. At the end of the 19th century, Africa was a place of European colonial power and oppression. European governments and financial agents based in such places as French West Africa, Belgian Congo, and British East Africa controlled much of the continent. Africa has not been the same since. Much of the poverty and violence of the 20th century is the direct result of colonialism. As you read the Case Study, consider how Africa might overcome the legacy of European colonialism.

Colonizing Africa

During the 15th century, Portuguese ships, looking for trade routes to Asia, landed in Africa. Soon other European countries established coastal trading stations there.

EUROPEANS IN AFRICA By the mid-1800s, Europeans knew of Africa's rich natural resources. They wanted these raw materials to fuel their own industrial economies and to establish markets to sell and trade their goods. In 1884–1885, the Berlin Conference, which you read about in Chapter 19, set down rules for dividing up Africa. European colonial control of Africa began to end in the early 20th century, but most African countries gained their independence in the 1960s. The Europeans did long-term damage to Africa, affecting its cultural and ethnic boundaries, and ruining its economy.

Challenges of Independence

When the European colonial powers were forced to leave Africa, the newly independent African countries did not have stable governments in place. For the next 40 years, many of the newly established African nations and their peoples suffered through dictatorships and civil wars. Many of these conflicts had lasting consequences for the continent's economy and the people's well-being.

COLONIAL TRANSITION European governments did not understand the incredible ethnic diversity in Africa. Certain African ethnic groups are living together today only because European colonizers established national borders that grouped them together. Examine the map on page 469 and you will see the ethnic and cultural complexity in

SEE
PRIMARY SOURCE B

Africa. Each area marked by a red line is an ethnic group. Many of these groups now reside together in the present-day countries created by Europeans. Many groups living in the same country are historical enemies. For example, German and Belgian colonial governments aggravated historically tense relations between the Hutu and Tutsi ethnic groups in present-day Rwanda and Burundi. In the early 1990s, the ethnic violence between these two groups resulted in a war that led to the deaths of hundreds of thousands of people.

Because of the way these colonial borders were drawn, many African governments had difficulty getting different ethnic groups to cooperate in building stable democracies. Dictators, such as Mobutu Sese Seko of what is now the Democratic Republic of the Congo, became common. In addition, many Africans had no experience living in democratic governments.

Traditional Ethnic Boundaries of Africa

— Ethnic group

0   400   800 miles
0 400 800 kilometers
Lambert Azimuthal Equal-Area Projection

SKILLBUILDER: Interpreting Maps

1 PLACE Which modern-day country probably contains the most traditional ethnic groups?

2 REGION Which subregion of Africa was probably the least affected by the postcolonial reconstruction of Africa's boundaries?

CAUSE FOR HOPE Establishing a democratic tradition is a primary goal for many African nations. Only through political stability can a nation bring peace and prosperity to its people. In the past decade, some African nations have been making progress. In 1994, the white minority government in South Africa finally yielded power to the black majority, ending decades of government-sanctioned racial discrimination and social injustice.

Furthermore, in 2001, Ghana swore in a new president in a peaceful transfer of power, unlike the coups and assassinations that had occurred during previous changes of government. These events are promising in a continent that is hoping for radical progress in the 21st century. Complete the Case Study Project on the following two pages to learn more about how Africa is dealing with the effects of colonialism.

SEE
PRIMARY SOURCE D

MOVEMENT Voters line up during elections in South Africa.
How will elections help improve the lives of people in Africa?

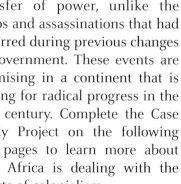

469

CaseStudy

PROJECT *News Report*

Primary sources A, B, C, D, and E on these two pages are about colonial and postcolonial Africa. Use these resources along with your own research to prepare a news report on postcolonial Africa.

RESEARCH LINKS
CLASSZONE.COM

Suggested Steps

1. Select one African country to study.
2. Use online and print resources to research your country's precolonial, colonial, and postcolonial history.
3. Highlight its people, resources, colonizers, and postcolonial activity.
4. Prepare a news report on the current status of your country, covering such topics as conflicts, the health and welfare of its people, the economy, and prospects for the future.
5. Practice your news report in front of a small audience. Ask them for ways to improve it.
6. Use a tape recorder or video recorder to tape your broadcast.

Materials and Supplies

- computer with Internet access
- reference books, newspapers, magazines, and encyclopedias
- tape recorder or video recorder

PRIMARY SOURCE A

Eyewitness Account *In his desire for more and more rubber from the Congo, Belgian King Leopold II adopted terrorism as his preferred method of persuasion. In 1899, the British vice consul offered this eyewitness account.*

An example of what is done was told me up the Ubangi [River]. This officer's method was to arrive in canoes at a village, the inhabitants of which invariably bolted on their arrival; the soldiers were then landed, and commenced looting, taking all the chickens, grain, etc. out of the houses; after this, they attacked the natives until able to seize their women; these women were kept as hostages until the chief of the district brought in the required number of kilograms of rubber. The rubber having been brought, the women were sold back to their owners for a couple of goats apiece, and so he continued from village to village until the requisite amount of rubber had been collected.

PRIMARY SOURCE B

Statement of Principle *Kwame Nkrumah was the leader of post-colonial Ghana until he was overthrown in 1966. In his book,* I Speak of Freedom, *published in 1961, he wrote about his hopes for postcolonial Africa.*

For centuries, Europeans dominated the African continent. The white man [claimed] the right to rule and to be obeyed by the non-white; his mission, he claimed, was to "civilize" Africa. Under this cloak, the Europeans robbed the continent of vast riches and inflicted unimaginable suffering on the African people.

All this makes a sad story, but now we must be prepared to bury the past with its unpleasant memories and look to the future. All we ask of the former colonial powers is their goodwill and cooperation to remedy past mistakes and injustices and to grant independence to the colonies in Africa.

It is clear that we must find an African solution to our problems, and that this can only be found in African unity. Divided we are weak; united, Africa could become one of the greatest forces for good in the world.

News Analysis *Ron Daniels, writing in the magazine* Black World Today, *offered this analysis of the Trade and Development Act of 2000. This law recognizes the need to promote economic growth and reduce poverty in Africa, but the law only helps a small number of countries.*

How ironic, tragic even, that as we prepare to enter a new century and millennium, Africa, the motherland, is so afflicted by poverty, underdevelopment, hunger, disease, corruption, and debt that African leaders, out of desperation . . . , are in effect begging to be recolonized. How ironic that the continent whose historical underdevelopment under slavery and colonialism, whose vast human and material resources contributed mightily to the enrichment and development of Europe and America must now turn to the former slave-masters and colonizers for a "bail-out."

Political Cartoon *Cartoonist Alan King drew this cartoon in 1996. The cartoon appeared in the* Ottawa Citizen *in Ottawa, Canada. King shows the unending cycle of indecisive attitudes on the part of the international community. The Democratic Republic of the Congo, which was formerly known as Zaire, suffers from these indecisive attitudes.*

Editorial Commentary *On January 8, 2001, the* New York Times *editorial page included this essay on the changes that have taken place in Ghana. The editorial was titled "An African Success Story."*

In its first two decades of independence, the West African nation of Ghana was an archetypal political disaster, brought low by successive coups and dictatorships, corruption and near total economic collapse. Today, Ghana is a welcome African example of legitimate democracy and successful economic reform. In an unusually peaceful transfer of power, a civilian government that grew out of a military regime has accepted an election defeat and surrendered power to the opposition.

John Kufuor, an Oxford-trained lawyer and businessman, and the leader of Ghana's opposition New Patriotic Party, was sworn in as president yesterday. He defeated John Atta Mills, the incumbent vice president, in an election widely viewed as free and fair. President Jerry Rawlings, the charismatic former flight lieutenant who has dominated Ghana for nearly 20 years, stepped down after reaching a constitutional two-term limit as elected president.

DESTINATION UNKNOWN...

PROJECT CheckList

Have I . . .

✓ fully researched the country I chose to investigate?

✓ included information about the current status of the country?

✓ taped my broadcast in the form of an actual news report?

VISUAL SUMMARY
TODAY'S ISSUES IN AFRICA

Economics

Economic Development
- Africa's economy suffered because many European nations exploited Africa for its resources.
- African nations are concentrating on economic cooperation and economic diversification to build their economies.
- Many African nations are improving their educational systems to produce skilled workers.

Environment

Health Care
- Diseases are killing millions in Africa. They include cholera, malaria, tuberculosis, and AIDS.
- AIDS is threatening the continent's population and reducing life expectancies in many countries.
- Many African nations are improving efforts to educate citizens about AIDS.

Government

Effects of Colonialism
- European nations began colonizing Africa once exploration revealed a vast amount of valuable natural resources.
- Colonialism caused much political and ethnic violence because it disrupted many long-standing political and ethnic boundaries.

Reviewing Places & Terms

A. Briefly explain the importance of each of the following.

1. "one-commodity" country
2. commodity
3. diversify
4. AIDS
5. cholera
6. malaria
7. tuberculosis
8. UNAIDS

B. Answer the questions about vocabulary in complete sentences.

9. What is a nation called when it relies on one product for its economic well-being?
10. What is the name of the disease that is carried by a mosquito and was also a leading disease in both 1900 and 2000?
11. What is the name of the respiratory disease that often accompanies AIDS?
12. What is the process whereby countries employ many different ways to help their economies grow?
13. What disease is spread by poor sanitation and a polluted water supply?
14. What is a product called that is bought and sold and has value in a worldwide market?
15. Which organization tracks the world's AIDS problem?

Main Ideas

Economic Development (pp. 461–464)

1. How has Africa's economic status changed during the past 40 years?
2. What is one of the main problems preventing Africa from spending money on economic development?
3. What is a danger with a country's having only one valuable product that it relies on for its economic well-being?

Health Care (pp. 465–467)

4. How are African nations fighting some of the diseases afflicting their continent?
5. What are some of the economic implications of disease in Africa?
6. What do Uganda's and Senegal's AIDS programs have in common?
7. Why might the drop in HIV infection rates be misleading?

Effects of Colonialism (pp. 468–471)

8. What are some of the empires and peoples that controlled areas of Africa at the beginning of the 19th century?
9. What was one of the main reasons that European countries wanted to control Africa?
10. Why did colonization cause so much political and ethnic violence in the 20th century?

Critical Thinking

1. Using Your Notes

Use your completed chart to answer these questions.

	Causes	Effects
Issue 1: Economic Development		
Issue 2: Health Care		

a. What is the primary foundation for most African nations' economies?

b. How might disease and economic development be related?

2. Geographic Themes

a. **MOVEMENT** How are diseases such as malaria and cholera spread?

b. **REGION** In what way is the modern map of Africa not a true reflection of the continent's people?

3. Identifying Themes

How would you relate one of the five themes of geography to the primary way in which African nations support their economies?

4. Making Inferences

How do you think Africa's economic health affects the spread of diseases such as cholera and AIDS?

5. Drawing Conclusions

How important do you think regional cooperation is in building Africa's economy? Why?

Additional Test Practice, pp. S1–S37

TEST PRACTICE
CLASSZONE.COM

Geographic Skills: Interpreting Maps

Dates of African Independence

Use the map at right to answer the following questions.

1. **PLACE** Which two countries remained free of European control?

2. **PLACE** Which country most recently gained its independence?

3. **REGION** Which decade saw the most countries gain independence?

GeoActivity

Choose one country in West Africa that was once controlled by France. Then using the library, encyclopedias, or other reference books, research how France's influence is still felt today in that country's economy, government, schools, and language.

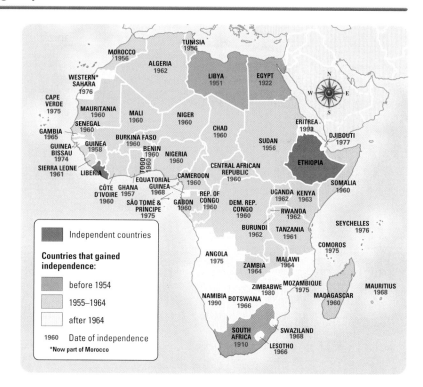

TUNISIA 1956
MOROCCO 1956
ALGERIA 1962
LIBYA 1951
EGYPT 1922
WESTERN* SAHARA 1976
CAPE VERDE 1975
MAURITANIA 1960
MALI 1960
NIGER 1960
CHAD 1960
ERITREA 1993
DJIBOUTI 1977
SENEGAL 1960
GAMBIA 1965
GUINEA-BISSAU 1974
GUINEA 1958
BURKINA FASO 1960
BENIN 1960
NIGERIA 1960
SUDAN 1956
SIERRA LEONE 1961
LIBERIA
TOGO 1960
CENTRAL AFRICAN REPUBLIC 1960
ETHIOPIA
SOMALIA 1960
CÔTE D'IVOIRE 1960
GHANA 1957
GUINEA 1968
EQUATORIAL
CAMEROON 1960
SÃO TOMÉ & PRINCIPE 1975
GABON 1960
REP. OF CONGO 1960
DEM. REP. CONGO 1960
UGANDA 1962
KENYA 1963
RWANDA 1962
BURUNDI 1962
TANZANIA 1961
SEYCHELLES 1976
COMOROS 1975
ANGOLA 1975
ZAMBIA 1964
MALAWI 1964
ZIMBABWE 1980
MOZAMBIQUE 1975
MAURITIUS 1968
NAMIBIA 1990
BOTSWANA 1966
MADAGASCAR 1960
SOUTH AFRICA 1910
SWAZILAND 1968
LESOTHO 1966

Independent countries

Countries that gained independence:

before 1954

1955–1964

after 1964

1960 Date of independence

*Now part of Morocco

INTERNET ACTIVITY

Use the links at **classzone.com** to do research on the postcolonial economy of one African country. Look for attempts to diversify the economy, education programs, growth in per capita income, and amount of manufacturing.

Writing About Geography Write a report of your findings. Include charts, pie graphs, and other visuals to help present the information. List the Web sites that you used as sources.

Unit 7

Southwest Asia

Southwest Asia, sometimes called a cradle of civilization, is the home of oil rich lands, vast deserts, and difficult political problems.

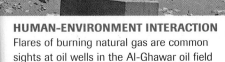

HUMAN-ENVIRONMENT INTERACTION
Flares of burning natural gas are common sights at oil wells in the Al-Ghawar oil field in Saudi Arabia.

PLACE Two holy places in Jerusalem, Israel, can be seen in this photograph: a shrine known as the Dome of the Rock, which is sacred to Muslims, and the Western Wall, a spot sacred to Jews.

GeoData

REGION More than half of the world's oil reserves are found in this region.

HUMAN-ENVIRONMENT INTERACTION Some experts believe that the freshwater supplies of the Arabian peninsula will be exhausted in the next 25 to 30 years.

LOCATION Southwest Asia connects three continents: Europe, Asia, and Africa.

For more information on Southwest Asia . . .

RESEARCH LINKS
CLASSZONE.COM

MOVEMENT Crossing the desert areas of Southwest Asia by land would be almost impossible without oases to provide water and a resting place. This oasis was on the caravan route from Yemen to Palestine.

SW ASIA

Unit PREVIEW 7

Today's Issues in Southwest Asia

Today, Southwest Asia faces the issues previewed here. As you read Chapters 21 and 22, you will learn helpful background information. You will study the issues themselves in Chapter 23.

In a small group, answer the questions below. Then participate in a class discussion of your answers.

Exploring the Issues

1. **POPULATION RELOCATION** Think about why a group of people may leave a place they call home. What problems might relocation cause for the group? Then make a list of the reasons people relocate and the problems that are caused by moving.

2. **ECONOMIC DEVELOPMENT** Make a list of major rivers found in the region and a list of major rivers found in the United States. How do the lists compare? What does this suggest about scarce resources in the region?

3. **RELIGIOUS CONFLICT** Study the cartoon on page 477. Who are the figures in the cartoon?

For more on these issues in Southwest Asia . . .

CURRENT EVENTS
CLASSZONE.COM

POPULATION RELOCATION

What kind of population movement is taking place in Southwest Asia?

This nomadic Kurdish family rests in the hills of eastern Turkey. The Kurds claim a homeland that crosses the boundaries of five countries: Turkey, Iraq, Iran, Syria, and Armenia.

ECONOMIC DEVELOPMENT

How can oil wealth help develop the region's economies?

Wealth from oil wells, like this one located at Al Ghawar in Saudi Arabia, may be used to develop economic activities that do not depend on oil.

CaseStudy

Who should control Jerusalem?

In this cartoon, the dove symbolizes peace between Arabs and Israelis in Southwest Asia. Jerusalem plays a vital role in the peace process.

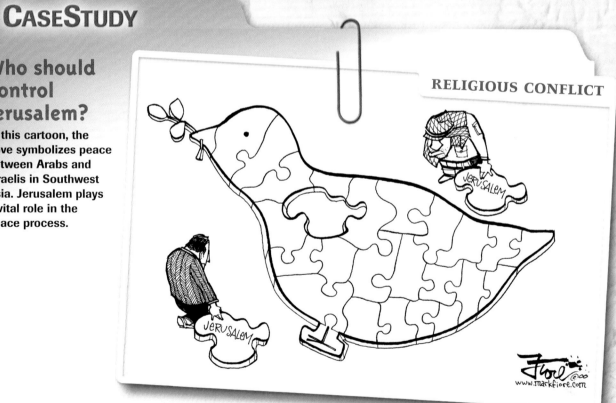

RELIGIOUS CONFLICT

Unit ATLAS

Use the Unit Atlas to add to your knowledge of Southwest Asia. As you look at the maps and charts, notice geographic patterns and specific details about the region. For example, the chart gives details about the mountains and deserts of Southwest Asia.

After studying the graphics and physical map on these two pages, jot down answers to the following questions in your notebook.

Making Comparisons

1. Which of Southwest Asia's deserts is about the same size as the Mojave Desert of the United States?

2. How do the tallest mountains of Southwest Asia compare to the tallest U.S. mountain?

3. Which mountain chains cut off Turkey and Iran from the rest of the region? How might isolation affect the way a country develops economically?

For updated statistics on Southwest Asia . . .

DATA UPDATE
CLASSZONE.COM

Comparing Data

Landmass

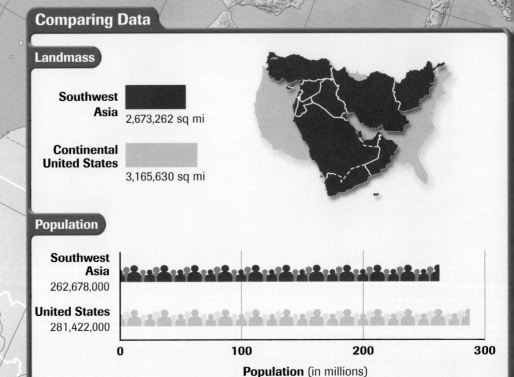

Southwest Asia
2,673,262 sq mi

Continental United States
3,165,630 sq mi

Population

Southwest Asia
262,678,000

United States
281,422,000

| 0 | 100 | 200 | 300 |

Population (in millions)

Deserts

World's Largest
Sahara
Africa
3,500,000
square miles

U.S. Largest
Mojave
United States
25,000
square miles

Rub al-Khali
Arabian
Peninsula
250,000
square miles

An-Nafud
Arabian
Peninsula
25,000
square miles

Negev
Israel
4,700
square miles

Mountains

World's Tallest
Mt. Everest
Nepal-Tibet
29,035 feet

U.S. Tallest
Mt. McKinley
United States
20,320 feet

Damavand
Iran
18,606 feet

Mt. Ararat
Turkey
16,945 feet

Mt. Hermon
Lebanon-Syria
9,232 feet

MOLDOVA
UKRAINE
ROMANIA
RUSSIA
BULGARIA
KAZAKHSTAN

30°E 40°E 50°E 60°E 70°E

Black Sea
Sea of Marmara
Bosporus
KYRGYZSTAN

GEORGIA
Caspian Sea
UZBEKISTAN
Anatolia
Pontic Mountains
ARMENIA AZERBAIJAN
TAJIKISTAN
Amu Darya
TURKEY
Mt. Ararat 16,945 ft. (5,165 m.) AZER.
Hindu Kush
TURKMENISTAN
Lake Van
Taurus Mountains
Lake Urmia
Khyber Pass 3,518 ft. (1,072 m.)

Crete (Gr.)
N. CYPRUS
CYPRUS
Euphrates River
Tigris River
Diyala R.
Elburz Mountains
Mt. Damavand 18,934 ft. (5,771 m.)
AFGHANISTAN

Mediterranean Sea
LEBANON
SYRIA
Mt. Hermon 9,232 ft. (2,814 m.)
Dasht-e Kavir
Plateau Of Iran

ISRAEL
Dead Sea -1,312 ft. (-400 m.)
Syrian Desert
IRAQ
ZAGROS MOUNTAINS
IRAN
Dasht-e Lut

30°N

JORDAN
PAKISTAN
EGYPT
An-Nafud
KUWAIT
Persian Gulf

Red Sea
Hejaz
N a j d
BAHRAIN
Strait of Hormuz
OMAN
Gulf Of Oman
QATAR

SAUDI ARABIA
UNITED ARAB EMIRATES
Tropic of Cancer
Arabian Sea

70°E
20°N

OMAN
ARABIAN PENINSULA
Rub Al-Khali

INDIAN OCEAN

YEMEN
Socotra (Yemen)

Inset map:
0 25 50 miles
0 25 50 kilometers
LEBANON
Golan Heights
Mediterranean Sea
ISRAEL
Jordan R.
West Bank
Gaza Strip
Negev
Suez Canal
Sinai
EGYPT
Peninsula
JORDAN
SAUDI ARABIA

SUDAN

Gulf of Suez
DJIBOUTI
Gulf Of Aden
ETHIOPIA

SOMALIA

Equator
0°
10°N

Elevation
13,100 ft. (4,000 m.)
6,600 ft. (2,000 m.)
1,600 ft. (500 m.)
650 ft. (200 m.)
0 ft. (0 m.)
Below sea level

▲ Mountain peak

0 250 500 miles
0 250 500 kilometers
Lambert Conformal Conic Projection

SW ASIA

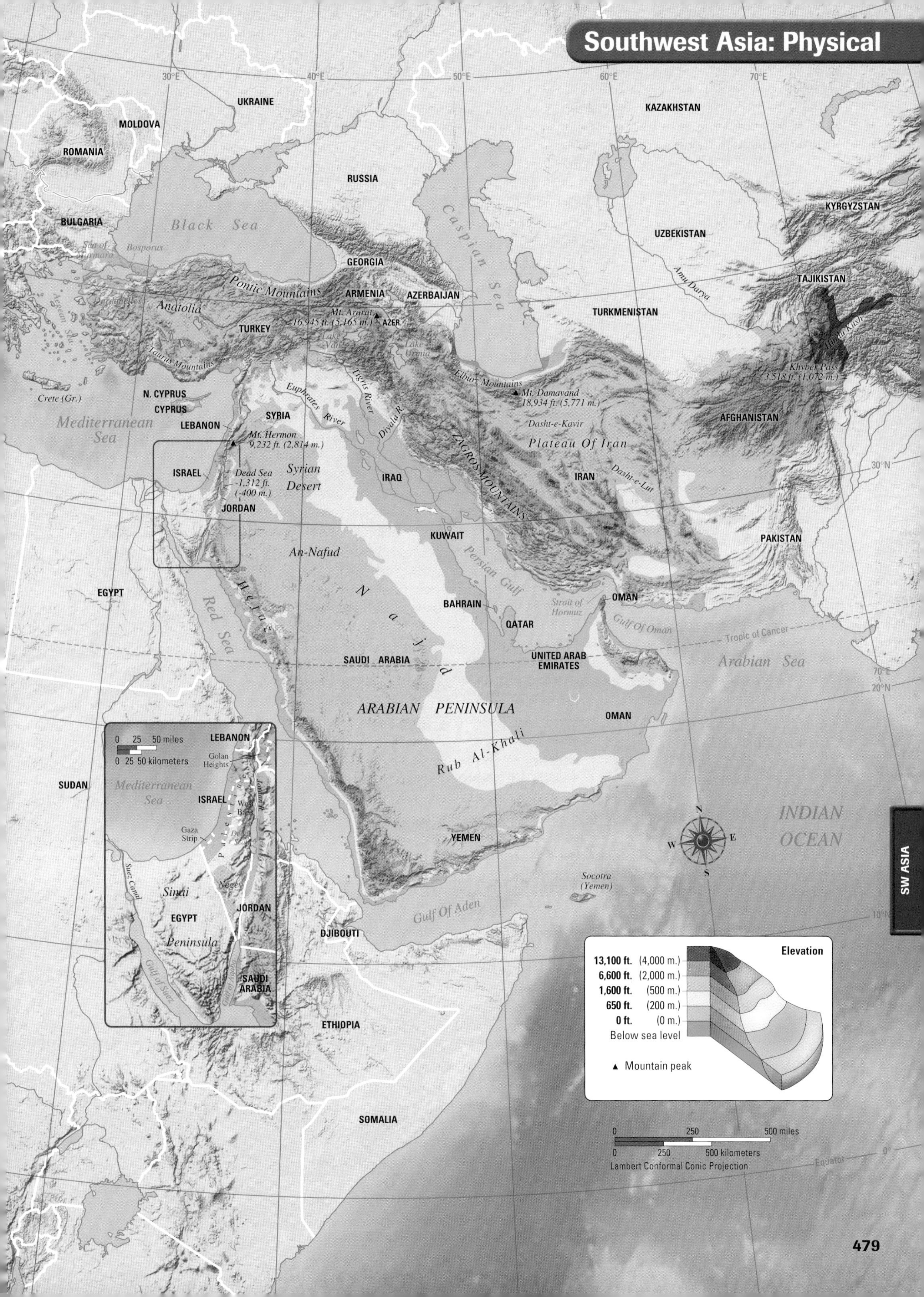

Patterns of Human Geography

After World War II (1939-1945), the nation of Israel was created in 1948. Since that time, the peoples and nations of the region have been in conflict with one another.

Study the political map of Southwest Asia and the Israel maps at the right to see how possession of the lands changed. Then write the answers to these questions in your notebook.

Making Comparisons

1. Which areas did Israel occupy in 1967?

2. Study both maps of Israel and the political map and write a sentence describing the changes in land possession from 1948 to the present.

3. What nation is in possession of the Sinai Peninsula today?

4. Which four nations surround the Golan Heights? Who controls the area?

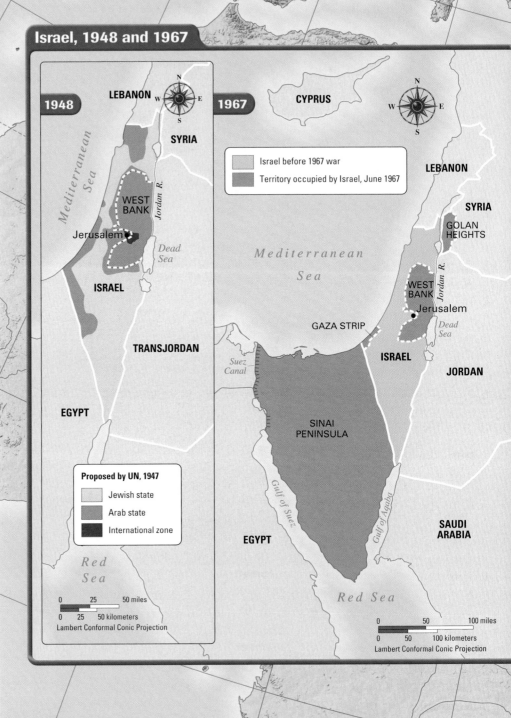

Israel, 1948 and 1967

1948

LEBANON
SYRIA
Mediterranean Sea
WEST BANK
Jordan R.
Jerusalem
Dead Sea
ISRAEL
TRANSJORDAN
EGYPT
Red Sea

Proposed by UN, 1947
- Jewish state
- Arab state
- International zone

0   25   50 miles
0   25   50 kilometers
Lambert Conformal Conic Projection

1967

CYPRUS

Israel before 1967 war
Territory occupied by Israel, June 1967

LEBANON
SYRIA
GOLAN HEIGHTS
Mediterranean Sea
WEST BANK
Jordan R.
Jerusalem
Dead Sea
GAZA STRIP
ISRAEL
JORDAN
Suez Canal
SINAI PENINSULA
Gulf of Suez
Gulf of Aqaba
EGYPT
SAUDI ARABIA
Red Sea

0   50   100 miles
0   50   100 kilometers
Lambert Conformal Conic Projection

MOLDOVA

ROMANIA

BULGARIA

UKRAINE

Black Sea

Sea of Marmara

Istanbul

Ankara

Bosporus

Dardanelles

Samsun

RUSSIA

GEORGIA

ARMENIA

AZERBAIJAN

Caspian Sea

KAZAKHSTAN

UZBEKISTAN

KYRGYZSTAN

TURKMENISTAN

TAJIKISTAN

Izmir

Aegean Sea

T U R K E Y

Erzurum

AZER.

Tabriz

Adana

Aleppo

N. CYPRUS

Nicosia

CYPRUS

LEBANON

Beirut

Euphrates River

Tigris River

Mosul

Arbil

SYRIA

Damascus

Mashhad

Tehran

Qom

I R A N

Kabul

AFGHANISTAN

Crete (Gr.)

Mediterranean Sea

ISRAEL

Jerusalem

Amman

JORDAN

Baghdad

I R A Q

Esfahan

Basra

KUWAIT

Kuwait

Shiraz

PAKISTAN

EGYPT

Red Sea

Manama

BAHRAIN

Persian Gulf

OMAN

Doha

QATAR

Abu Dhabi

UNITED ARAB
EMIRATES

Muscat

Strait of Hormuz

Gulf Of Oman

Tropic of Cancer

Arabian Sea

Medina

Riyadh

SAUDI ARABIA

Jiddah

Mecca

OMAN

Gulf of Aden

YEMEN

Sanaa

Aden

Socotra
(Yemen)

INDIAN OCEAN

N
W   E
S

SW ASIA

0   25   50 miles

0   25 50 kilometers

LEBANON

Golan Heights

Mediterranean Sea

Haifa

ISRAEL

Tel Aviv-Yafo

Jerusalem

*Gaza Strip

West Bank

Jordan R.

Dead Sea

Amman

SUDAN

Port Said

Suez

Suez Canal

EGYPT

JORDAN

Gulf of Suez

Gulf of Aqaba

SAUDI
ARABIA

ETHIOPIA

DJIBOUTI

SOMALIA

* The status of the West Bank and Gaza
Strip is under negotiation

⊛ National capital

● Other city

0   250   500 miles

0   250   500 kilometers

Lambert Conformal Conic Projection

Dashed border indicates disputed boundary

481

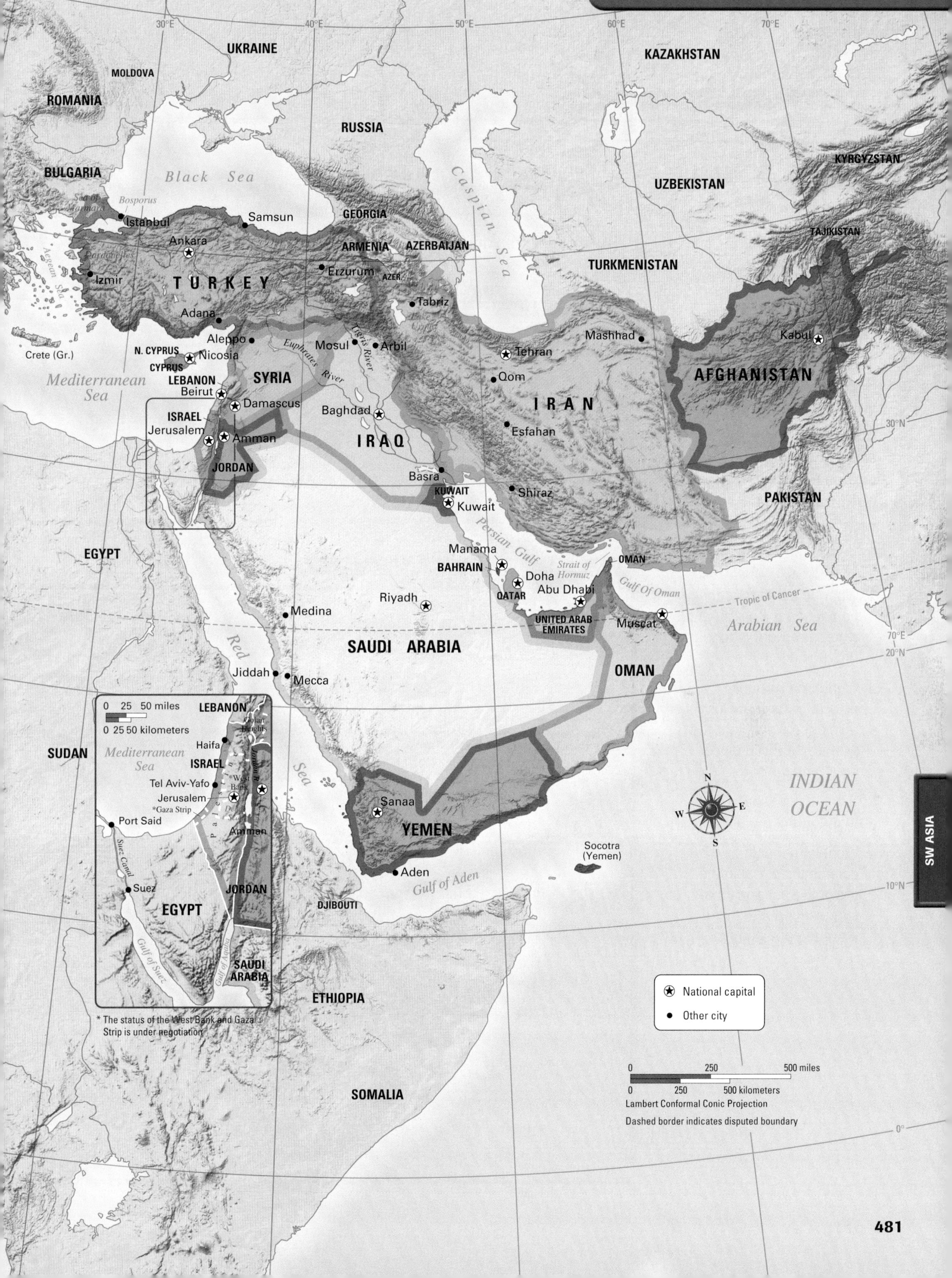

Regional Patterns

These two pages contain a graph and three thematic maps. The graph and two of the maps show the ethnic and religious diversity of Southwest Asia. The third map shows you how people in the region earn a living. After studying these two pages, answer the questions below in your notebook.

Making Comparisons

1. What percentage of the population is Kurdish and where are Kurds found in the region?

2. What area has holy places for three major religions? Why might the location of these places be a problem?

3. What energy sources are found in the region?

4. What is the main economic activity in the region? What does that suggest about the land and the population on it?

Ethnic Groups of Southwest Asia*

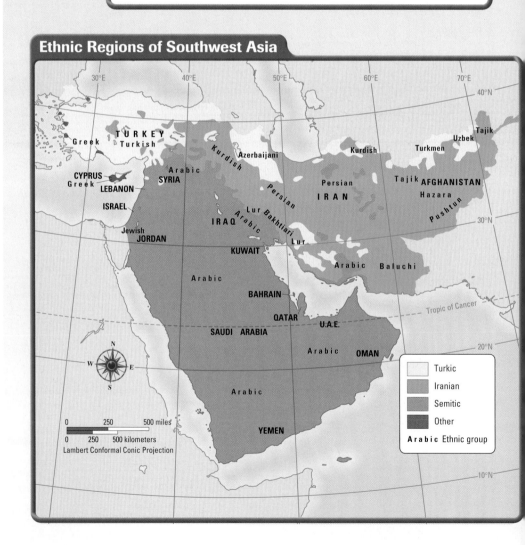

Azerbaijani 7%
Kurd 9%
Arab 31%
Persian 14%
Turk 23%
Other 16%*

* Includes Jews, who are of different ethnic groups.
SOURCE: *Britannica Book of the Year 2000;* U.S. Census Bureau, International Data Base; CIA *World Factbook 2000*

Ethnic Regions of Southwest Asia

Greek
TURKEY
Turkish
Kurdish
Azerbaijani
Kurdish
Tajik
Uzbek
Turkmen
CYPRUS
Greek
Arabic
SYRIA
LEBANON
ISRAEL
Persian
Persian
IRAN
Tajik AFGHANISTAN
Hazara
Pushtun
Jewish
JORDAN
Lur Bakhtiari
Arabic
IRAQ
Lur
KUWAIT
Arabic
Baluchi
Arabic
BAHRAIN
QATAR
U.A.E.
SAUDI ARABIA
Arabic
OMAN
Tropic of Cancer
Arabic
Arabic
YEMEN

0   250   500 miles
0   250   500 kilometers
Lambert Conformal Conic Projection

Turkic
Iranian
Semitic
Other
Arabic Ethnic group

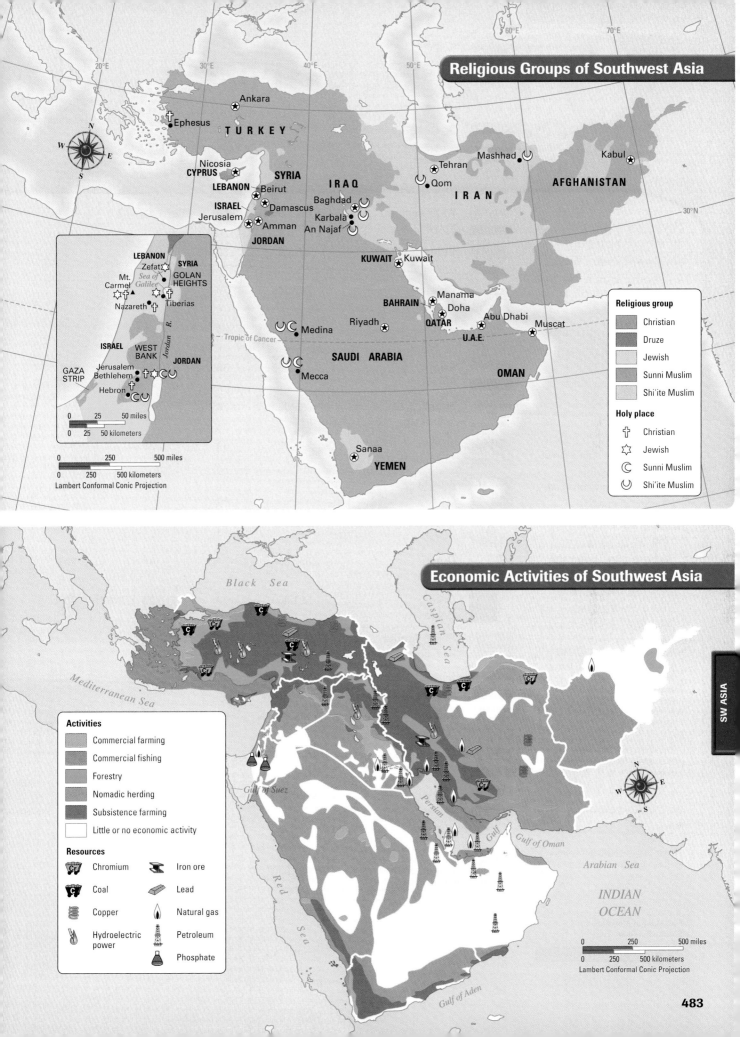

Religious Groups of Southwest Asia

Ankara
Ephesus
TURKEY
Nicosia
CYPRUS
SYRIA
LEBANON
Beirut
Damascus
ISRAEL
Jerusalem
Amman
JORDAN
IRAQ
Baghdad
Karbala
An Najaf
Mashhad
Tehran
Qom
IRAN
Kabul
AFGHANISTAN
30°N
KUWAIT Kuwait
BAHRAIN
Manama
Doha
QATAR
Abu Dhabi
Muscat
U.A.E.
Medina
Riyadh
SAUDI ARABIA
OMAN
Mecca
Tropic of Cancer
Sanaa
YEMEN

20°E 30°E 40°E 50°E 60°E 70°E

N W E S

Inset map:
LEBANON
Zefatt
SYRIA
GOLAN HEIGHTS
Mt. Carmel
Sea of Galilee
Nazareth
Tiberias
Jordan R.
ISRAEL
WEST BANK
JORDAN
GAZA STRIP
Jerusalem
Bethlehem
Hebron

0 25 50 miles
0 25 50 kilometers

0 250 500 miles
0 250 500 kilometers
Lambert Conformal Conic Projection

Religious group
- Christian
- Druze
- Jewish
- Sunni Muslim
- Shi'ite Muslim

Holy place
- ✝ Christian
- ✡ Jewish
- ☪ Sunni Muslim
- ☽ Shi'ite Muslim

Economic Activities of Southwest Asia

Black Sea
Caspian Sea
Mediterranean Sea
Gulf of Suez
Persian Gulf
Gulf of Oman
Red Sea
Gulf of Aden
Arabian Sea
INDIAN OCEAN

N E W S

SW ASIA

Activities
- Commercial farming
- Commercial fishing
- Forestry
- Nomadic herding
- Subsistence farming
- Little or no economic activity

Resources
- Chromium
- Coal
- Copper
- Hydroelectric power
- Iron ore
- Lead
- Natural gas
- Petroleum
- Phosphate

0 250 500 miles
0 250 500 kilometers
Lambert Conformal Conic Projection

483

Regional Data File

Country Flag	Country/ Capital	Population (2000 estimate)	Life Expectancy (years) (2000)	Birthrate (per 1,000 pop.) (2000)	Infant Mortality (per 1,000 live births) (2000)
	Afghanistan Kabul	26,668,000	46	43	149.8
	Bahrain Manama	691,000	69	22	8.1
	Cyprus Nicosia	882,000	77	14	7.8
	Iran Tehran	67,411,000	69	21	30.8
	Iraq Baghdad	23,115,000	59	38	127.0
	Israel Jerusalem	6,227,000	78	22	6.0
	Jordan Amman	5,083,000	69	33	34.0
	Kuwait Kuwait	2,190,000	72	24	12.5
	Lebanon Beirut	4,202,000	70	23	34.5
	Oman Muscat	2,353,000	71	44	25.0
	Qatar Doha	591,000	72	20	20.0
	Saudi Arabia Riyadh	21,607,000	70	35	46.4
	Syria Damascus	16,482,000	67	33	34.6
	Turkey Ankara	65,311,000	69	22	37.9
	United Arab Emirates Abu Dhabi	2,835,000	74	24	16.0
	Yemen Sanaa	17,030,000	59	39	75.3
	United States Washington, D.C.	281,422,000	77	15	7.0

Study the information on the countries of Southwest Asia. In your notebook, answer these questions.

Making Comparisons

1. Which nations have more doctors per 100,000 population than the United States?

2. Study the information to determine which nation after Afghanistan is the poorest. On which categories did you base your judgment?

3. Using the map on page 479, make a list of the nations that border the Persian Gulf. How many of those nations have more exports than imports?

Sources:
Human Development Report 2000, UN
International Data Base, U.S. Census Bureau online
Merriam-Webster's Geographical Dictionary, 3d ed., 1997
World Education Report 2000, UNESCO online
World Population Data Sheet 2000, Population Reference Bureau online
WHO Estimates of Health Personnel, online
World Almanac and Book of Facts 2001
World Factbook 2000, CIA online

Notes:
[a] A comparison of the prices of the same items in different countries is used to figure these data.
[b] Includes land and water, when figures are available.

For updated statistics on Southwest Asia . . .

DATA UPDATE
CLASSZONE.COM

Doctors (per 100,000 pop.) (1996–1998)	GDP[a] (billions $US) (1998–1999)	Import/Export[a] (billions $US) (1998–1999)	Literacy Rate (percentage) (1998–1999)	Televisions (per 1,000 pop.) (1998)	Passenger Cars (per 1,000 pop.) (1996–1997)	Total Area[b] (square miles)	
11	21.0	0.2 / 0.1 1996	32	10	2	250,775	
100	8.6	3.5 / 3.3	87	467	242	255	
255	Greek Cypriot 9.0 Turkish Cypriot 0.8	GrkCyp 3.5 / 1.1 TrkCyp 0.4 / 0.1	97	322	316	3,572	
85	347.6	13.8 / 12.2	75	63	26	635,932	
55	59.9	8.9 / 12.7	54	80	32	168,927	
385	105.4	30.6 / 23.5	96	290	224	7,992	
166	16.0	3.0 / 1.8	89	80	40	34,575	
189	44.8	8.1 / 13.5	81	370	318	6,880	
210	16.2	5.7 / 0.9	85	366	325	3,949	
133	19.6	5.4 / 7.2	69	657	108	82,000	
126	12.3	4.2 / 6.7	80	401	151	4,400	
166	191.0	28.0 / 48.0	75	257	89	865,000	
144	42.2	3.2 / 3.3	73	67	9	71,498	
121	409.4	36.0 / 26.0	84	189	53	301,380	
181	41.5	27.5 / 34.0	75	104	144	32,278	
23	12.7	2.3 / 2.0	44	28	15	203,849	
251	9,255.0	820.8 / 663.0	97	847	489	3,787,319	

Chapter 21

SECTION 1
Landforms and Resources

SECTION 2
Climate and Vegetation

SECTION 3
Human–Environment Interaction

PHYSICAL GEOGRAPHY OF SOUTHWEST ASIA

Harsh and Arid Lands

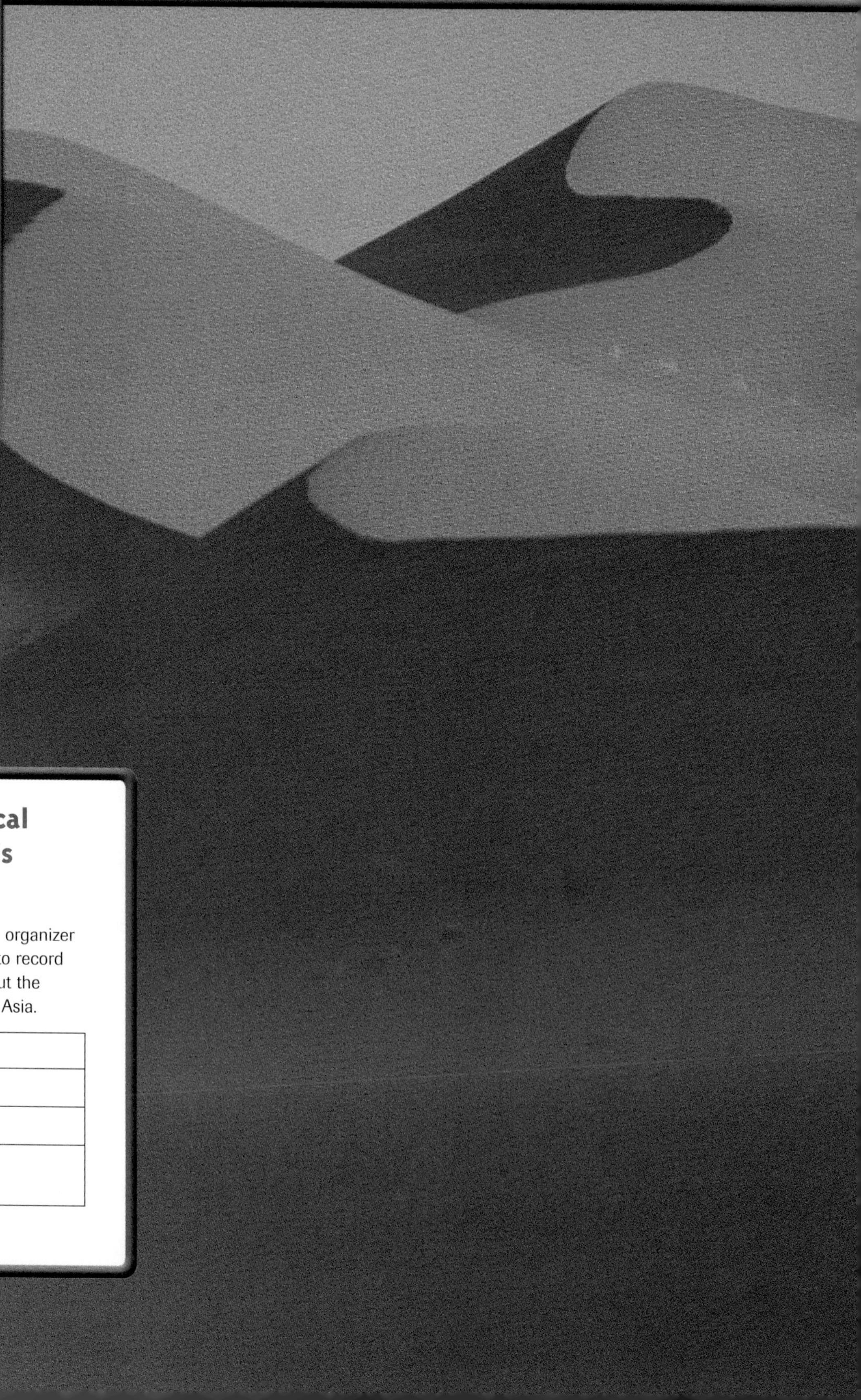

Wind-shaped sand dunes in Arabia's An-Nafud Desert sometimes reach a height of 600 feet.

GeoFocus

Why does the physical geography make this a vital region?

Taking Notes Copy the graphic organizer below into your notebook. Use it to record information from the chapter about the physical geography of Southwest Asia.

Landforms	
Resources	
Climate and Vegetation	
Human-Environment Interaction	

Landforms and Resources

A HUMAN PERSPECTIVE Artillery shells and sniper fire rained down on the lands below a small plateau in southwestern Syria. Airplanes bombed the military positions on the plateau itself. Families in nearby villages huddled in their homes, hoping for the shelling to stop. Israeli Army engineers struggled to build a road to enable tanks to reach the top. Thousands died in the 1967 war when Syria and Israel fought for control of the **Golan Heights,** also called Al Jawlan, a hilly plateau overlooking the Jordan River and the Sea of Galilee. This landform's strategic location has made it the site of conflict in Southwest Asia for decades. It is one of many landforms that divide the region.

Landforms Divide the Region

People sometimes picture Southwest Asia as a region of rippling sand dunes and parched land occasionally interrupted with an oasis. But the lands of Southwest Asia actually range from green coastal plains to snow-peaked mountains. Southwest Asia forms a land bridge connecting Asia, Africa, and Europe. As you can see on the map on page 37, the region is situated at the edge of a huge tectonic plate. Parts of the Arabian Peninsula are pulling away from Africa, and parts of the Anatolian Peninsula are sliding past parts of Asia. Still other plates are pushing up mountains in other areas of the Asian continent.

PENINSULAS AND WATERWAYS The most distinctive landform in Southwest Asia is the Arabian Peninsula, which is separated from the continent of Africa by the Red Sea on the southwest and from the rest of Asia by the Persian Gulf on the east. The Red Sea covers a rift valley created by the movement of the Arabian plate. The Zagros, Elburz, and Taurus mountains at the north side of the plate cut off part of the region from the south. Another important landform in the region is the Anatolian Peninsula, which is occupied by the country of Turkey. It marks the beginning of the Asian continent. (See the map on page 479.)

Both peninsulas border on strategic waterways. On the southwest side of the Arabian Peninsula are the Red Sea and a strategic opening to the Mediterranean Sea—the Suez Canal. Goods from Asia flow through this canal to ports in Europe and North Africa.

Main Ideas
- The Southwest Asian landforms have had a major impact on movement in the region.
- The most valuable resources in Southwest Asia are oil and water.

Places & Terms
Golan Heights

wadi

Tigris River

Euphrates River

Jordan River

Dead Sea

CONNECT TO THE ISSUES
RESOURCES Enormous oil reserves have brought changes to the economic and political standing of this region.

PLACE The Golan Heights are a strategic location near the source of water in the region. **How will control of this area affect those who live on lands below the top of the plateau?**

SW ASIA

487

The Anatolian Peninsula is located between the Black Sea and the Mediterranean Sea. Two narrow waterways, the Bosporus Strait and the Dardenelles Strait, are situated at the west end of the peninsula. Both straits have always been highly desirable locations for controlling trade and transportation to Russia and the interior of Asia.

Farther south is a narrow passageway leading from the Arabian Sea to the Persian Gulf called the Straits of Hormuz. These straits are the only waterway to the huge oilfields of Kuwait, Saudi Arabia, and Iraq. Because access to oil is essential to the world-wide economy, this waterway is very important.

BACKGROUND
The Persian Gulf is also called the Arabian Gulf.

PLAINS AND HIGHLANDS Much of the Arabian Peninsula is covered by plains. Because of the dry, sandy, and windy conditions, few activities using the land take place here. Most of the land is barren with some low hills, ridges, and **wadis,** which are riverbeds that remain dry except during the rainy seasons. On the southwestern corner of the peninsula, a range of mountains—the Hejaz Mountains—pokes out of the land. People living on the Arabian Peninsula have adapted to the harsh conditions by living nomadic lives in search of water.

The heart of Iran is a plateau surrounded by mountains. Isolated and very high, the land is a stony, salty, and sandy desert. The foothills surrounding the plateau are able to produce some crops. Much of the Anatolian Peninsula is also a plateau. Some areas are productive for agriculture, while other areas support flocks of grazing animals such as sheep and goats. The Northern Plain of Afghanistan, a well-watered agricultural area, is surrounded by high mountains that isolate it from other parts of the region. Ⓐ

MOUNTAINS Rugged mountains divide the land and countries. As you study the map on page 479, you will see that the Hindu Kush Mountains of Afghanistan are linked with other ranges of mountains that frame southern Asia. Afghanistan is landlocked and mountainous, so contact with the outside world is difficult.

The Zagros Mountains on the western side of Iran help isolate that country from the rest of Southwest Asia. The Elburz Mountains south of the Caspian Sea cut off easy access to that body of water by Iran. Finally, the Taurus Mountains separate Turkey from the rest of Southwest Asia. In spite of these physical barriers, people, goods, and ideas move through the entire region. One of the ways they move is by water.

Geographic Thinking

Making Comparisons
Ⓐ How are the plateaus of Iran and Anatolia different?

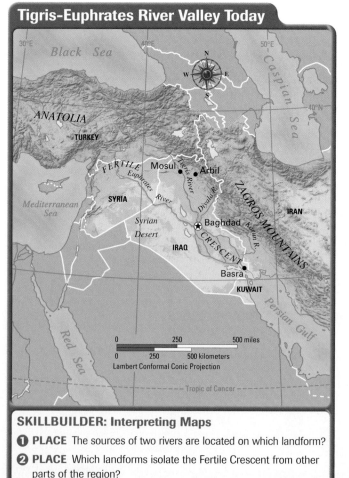

Tigris-Euphrates River Valley Today

SKILLBUILDER: Interpreting Maps

❶ PLACE The sources of two rivers are located on which landform?

❷ PLACE Which landforms isolate the Fertile Crescent from other parts of the region?

Human–Environment Interaction

> **Main Ideas**
> • Water is critical to regional physical survival and economic development.
> • Discovery of oil increased the global economic importance of Southwest Asia.

Places & Terms

drip irrigation   crude oil

desalinization   refinery

fossil water

CONNECT TO THE ISSUES

RESOURCES Southwest Asian nations face the challenge of how to use the income from oil resources to develop their economies.

A HUMAN PERSPECTIVE Icebergs for fresh water? As you have seen, fresh water is in short supply in Southwest Asia. In 1977, a Saudi prince, Muhammad ibn Faisal, formed a company to investigate the possibility of towing icebergs from Antarctica to the port of Jidda on the Red Sea. The icebergs would then be melted to release huge quantities of fresh water. It cost one million dollars to find out that no ship was powerful enough to tow an enormous iceberg, and there was no way to keep the iceberg from breaking up on the way. In 1981, the iceberg project was suspended. This story illustrates just how precious fresh water is in Southwest Asia. For centuries, people living in the region have struggled to find fresh water for themselves and for crops.

Providing Precious Water

Water has been a valuable resource since life began on earth. Even though oil brings a great deal of money into Southwest Asia, the most critical resource in this dry region is water. Fresh water supplies are available only in small amounts and not consistently. Ancient civilizations constantly faced the problem of finding and storing water in order to survive and prosper. Today, the same challenge exists for modern nations. To find reliable water supplies, nations today use both ancient and modern practices. The pictures on page 496 include examples of both ancient and modern techniques for providing water.

DAMS AND IRRIGATION SYSTEMS Ancient practices for providing water work well for small fields but are not efficient for large-scale farming. To meet the needs of large farms and for growing populations, countries must construct dams and irrigation systems. Turkey is building a series of dams and a man-made lake on the upper Euphrates River. The dams and lake will provide water and hydroelectricity for parts of the country. But the project is controversial—countries downstream from the dam will lose the use of the water for irrigation or hydroelectricity.

The National Water Carrier project in Israel carries water from the northern part of the country to sites in the nation's center and south. The water comes from mountain areas, including the Golan Heights, the Jordan River, and Lake Kinneret (Sea of Galilee). Some of the water is used in agricultural projects in the Negev Desert, and some for drinking

PLACE Date palms thrive in this oasis in the arid country of Oman.
Where does the water for an oasis come from?

SW ASIA

Water Systems

① **Drip irrigation** places water just at the root zone, reducing evaporation of precious water. This system is located in the Negev Desert in Israel.

② A bag of water is collected by using this pump. It is a part of a **qanat**—a system of underground brick-lined tunnels and wells that collect runoff water from the mountains.

③ This **irrigation canal** in Oman has delivered water for over a thousand years. The canals are carefully maintained to provide water for agriculture.

④ A **noria**—or waterwheel run by the flow of water or by animal power—is used to lift water from the river to the fields. These two are located in Syria on the Orontes River.

water. Because the water sources flow through several countries and access to the water is restricted, the National Water Carrier Project is a source of international conflict.

MODERN WATER TECHNOLOGY Several countries in the region use **drip irrigation.** This is the practice of using small pipes that slowly drip water just above ground to conserve water used for crops. Other nations are developing ways to use ocean water. **Desalinization,** the removal of salt from ocean water, is done at technically sophisticated water treatment plants. However, the desalinated water may be too salty to use for irrigation so it is used in sewage systems. Desalinization plants are very expensive and cannot provide adequate quantities of water to meet all the needs of people in Southwest Asia. Another alternative source of water, especially for agriculture, is the treatment of wastewater. Wastewater treatment plants constructed in the region fail to generate enough water to meet all the needs.

Water pumped from underground aquifers is called **fossil water,** because it has been in the aquifer for very long periods of time. Fossil water has very little chance of being replaced because this region has too little rainfall to recharge the aquifers. It is estimated that at the current rate water is being pumped, only about 25 to 30 years of water usage remain. Finding ways to conserve or even reuse water must be a top priority for the nations of this region.

Geographic Thinking

Making Comparisons
A How are the water projects of Turkey and Israel different?

Oil From the Sand

The oil fields discovered in the sands of Southwest Asia have been a bonanza for the region. These fields contain about one-half of all of the petroleum reserves in the world. Petroleum is the source of gasoline for automobiles, heating oil, and the basis of many chemicals used to make everything from fertilizers to plastics. Thus, petroleum products are an important part of the world economy. Having huge oil resources makes Southwest Asia a very important region economically.

FORMING PETROLEUM Oil and natural gas deposits were formed millions of years ago when an ancient sea covered the area of Southwest Asia. Microscopic plants and animals lived and died in the waters. Their remains sank and became mingled with the sand and mud on the bottom of the sea. Over time, pressure and heat transformed the material into hydrocarbons, which form the chemical basis of oil and natural gas.

Oil and natural gas do not exist in large pools beneath the ground, but are trapped inside rocks. You could hold a rock containing oil in your hand and not be able to see the oil because it is trapped in the microscopic pores of the rock. The more porous the rock, the more oil can be stored. A barrier of nonporous rock above the petroleum deposit prevents the gas or oil from moving out of the rock and to the surface.

Engineers use sophisticated equipment to extract, or remove, the oil. It also takes technical skill and special equipment to find deposits of oil. For this reason, oil was not discovered in some parts of the region until the 1920s and 1930s.

EARLY EXPLORATION Industrialization and the increasing popularity of automobiles made petroleum a highly desired resource. Beginning in the late 1800s, oil companies searched all over the world for oil resources. The first Southwest Asia oil discovery was in 1908 in Persia, now known as Iran. In 1938, oil companies found more oil fields in the Arabian Peninsula and Persian Gulf. Then, World War II interrupted further exploring. In 1948, oil companies discovered portions of what would become one of the world's largest oil fields at *al-Ghawar,* just on the eastern edge of the Rub al-Khali. This field contains more than one-quarter of all Saudi Arabia's reserves of oil. ◄B

TRANSPORTING OIL Petroleum that has not been processed is called **crude oil.** Crude oil pumped from the ground must be moved to a **refinery.** The job of a refinery is to convert the crude oil into useful products. Pipelines transport the crude oil either to refineries or to ports where the oil is picked up by tankers and moved to other places for processing. Study the diagram on page 498 to learn how oil is processed and moved.

Geographic Thinking ◄

Seeing Patterns
B▷ Why did oil companies continue to search for oil deposits in Southwest Asia?

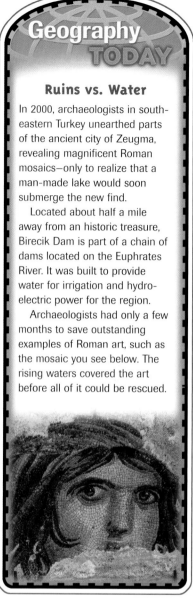

Geography TODAY

Ruins vs. Water

In 2000, archaeologists in southeastern Turkey unearthed parts of the ancient city of Zeugma, revealing magnificent Roman mosaics—only to realize that a man-made lake would soon submerge the new find.

Located about half a mile away from an historic treasure, Birecik Dam is part of a chain of dams located on the Euphrates River. It was built to provide water for irrigation and hydro-electric power for the region.

Archaeologists had only a few months to save outstanding examples of Roman art, such as the mosaic you see below. The rising waters covered the art before all of it could be rescued.

Processing Petroleum

Oil Field

Storage

Oil Refinery

Pumping Station

1. Drilling rigs cut through nonporous rock above the trapped crude oil and natural gas and pump it to storage tanks.

2. Natural gas, water, and sediments are removed from the crude oil. Oil is sent to the pumping station.

3. The crude oil is pumped to tankers or sent to a refinery to be processed. Some pipelines carry as much as a million barrels a day.

4. At the oil refinery, the crude oil is converted into useful products like gasoline.

5. The products are transported to markets all over the world. Ocean-going tankers carry more than a million barrels; railroad tank cars carry about 1,500 barrels; tank trucks hold about 300 barrels.

Oil Pipelines in Southwest Asia

RUSSIA

Black Sea

GEORGIA

ARMENIA AZERBAIJAN
Baku

TURKEY

Dörtyol

CYPRUS Tartus
LEBANON SYRIA
Mediterranean Sidon
Sea Haifa
Suez ISRAEL
Canal

JORDAN

EGYPT

Strait of Tiran

IRAQ

Basra
Kuwait
KUWAIT
Al Ahmadi
Ras Tanura
Manama
Doha
BAHRAIN
Umm Said
QATAR U.A.E.
SAUDI ARABIA

Abadan

Khark Is.

Lavan Is.

Persian Gulf

Gulf of Oman

Bandar-e 'Abbas
Strait of Hormuz

Matrah Tropic of Cancer

OMAN

IRAN

AFGHANISTAN

Caspian Sea

Arabian Sea

Yanbu al Bahr

Red Sea

Salif YEMEN Bab el Mandeb Gulf of Aden

0   250   500 miles
0   250   500 kilometers
Lambert Conformal Conic Projection

Legend:
- Oil field
- Gas field
- ▲ Major oil port
- Pipeline

Oil Production

Millions of barrels per day

■ Saudi Arabia   ■ United States

SOURCE: U.S. Department of Energy

SKILLBUILDER: Interpreting Maps

1. **REGION** Where are the largest number of oil and gas fields located?

2. **MOVEMENT** In what direction is much of the oil moved?

Using the Atlas

Use the maps on pages 481 and 498. What countries would likely receive oil shipments from ports located on the Mediterranean Sea?

Placement of pipelines depends on the location of existing ports or access to worldwide markets. Study the map on page 498. Notice that in this region, the pipelines move the crude oil to ports on the Persian Gulf, the Red Sea, and the Mediterranean Sea. From these locations, oil tankers carry the petroleum to markets in the rest of the world.

In some places, refineries process the crude oil near ports. Tanks to hold the oil products are located at port facilities. Many Southwest Asian nations have updated and outfitted their ports to service the very large ocean-going tankers.

RISKS OF TRANSPORTING OIL Moving oil from one location to another always involves the risk of oil spills. The largest oil spill ever recorded occurred in January 1991, during the Persian Gulf War. A series of tankers and oil storage terminals in Kuwait and on islands off its coast were blown up. More than 240 million gallons of crude oil were spilled into the water and on land.

Buried pipelines in Southwest Asia help reduce the danger of above-ground accidents. However, oil spills on land do happen. Because oil is such a valuable commodity, the pipelines are carefully monitored for any drop in pressure that might signal a leak in the line. Any leaks are quickly repaired.

On the other hand, ocean-going tankers transporting oil are at a much higher risk for causing pollution. Many tankers operate in shallow and narrow waterways such as the Red Sea, the Suez Canal, the Persian Gulf, and the Straits of Hormuz. Here, there is danger of oil spills due to collisions or running aground. Most modern tankers have double hulls so that minor accidents will not result in oil spills. In addition, oil-producing nations in Southwest Asia have taken legal steps to protect their environments.

In the next chapter, you will learn more about the people and cultures of the subregions of Southwest Asia.

SECTION 3 > Assessment

1 Places & Terms

Identify and explain where in the region these would be found.

- drip irrigation
- desalinization
- fossil water
- crude oil
- refinery

2 Taking Notes

HUMAN–ENVIRONMENT INTERACTION Review the notes you took for this section.

Human–Environment Interaction	

- What are some ways water is supplied in this region?
- In what ways is oil moved from the source to the market place?

3 Main Ideas

a. Why must both ancient and modern water supply methods be used in the region?

b. Why might water projects in Southwest Asia cause controversy?

c. What are some of the risks in transporting oil?

4 Geographic Thinking

Making Inferences What impact has technology had on the supply of oil and water in the region? **Think about:**

- finding large reserves of oil or water
- environmental hazards

RESEARCH LINKS
CLASSZONE.COM

ASKING GEOGRAPHIC QUESTIONS Study the map of oil pipelines on page 498. Devise three geographic questions about the map, such as "What problems might there be in choosing locations for these pipelines?" Choose one of your questions and write several **paragraphs** answering the question. Present your findings to the class. Be sure to identify your data sources.

SW ASIA

VISUAL SUMMARY
PHYSICAL GEOGRAPHY OF SOUTHWEST ASIA

Landforms

Peninsulas: Anatolian, Arabian

Mountain Ranges: Hindu Kush, Elburz, Zagros, Taurus

Major Waterways: Tigris, Euphrates, Jordan, Red Sea-Suez Canal, Bosporus Strait, Straits of Hormuz

Resources

• Water is a scarce resource.

• Oil is an abundant resource that shapes the region's economy.

Climate and Vegetation

Deserts:

• Rub al-Khali, An-Nafud, Syrian, and Negev are mostly sandy.

• Dasht-e Kavir and Dasht-e Lut are salt flat deserts.

Human-Environment Interaction

• Water is provided through both old and new technologies.

• Oil is pumped from the ground, processed, and transported out of Southwest Asia.

Reviewing Places & Terms

A. Briefly explain the importance of each of the following.

1. Golan Heights
2. wadi
3. Tigris River
4. Euphrates River
5. oasis
6. salt flat
7. drip irrigation
8. desalinization
9. crude oil
10. refinery

B. Answer the questions about vocabulary in complete sentences.

11. Where would you most likely find a wadi?
12. The Golan Heights are an example of which type of landform?
13. Where were several ancient river valley civilizations located?
14. Which terms above deal with water usage?
15. Why are refineries needed?
16. Where might you find a refinery?
17. Why is drip irrigation used?
18. Where would you find a salt flat desert in Southwest Asia?
19. What is the source of water for an oasis?
20. What are drawbacks to using water from a desalinization plant?

Main Ideas

Landforms and Resources (pp. 487–490)

1. How do the landforms of the region restrict movement?
2. What are the most valuable resources in the region and why are they valuable?
3. How large are the oil reserves in the region?

Climate and Vegetation (pp. 491–494)

4. What types of deserts are found in the region?
5. Why is extensive irrigation needed in the region?
6. Where in the region are well-watered lands found?

Human-Environment Interaction (pp. 495–499)

7. What are some examples of the ways in which water is provided in the region?
8. In what ways do major water projects cause political problems?
9. Where are the major oil fields in the region located?
10. What are some dangers in transporting oil?

Critical Thinking

1. Using Your Notes

Use your completed chart to answer these questions.

Landforms	
Resources	

a. How are landforms and desert climate connected?

b. How is oil production related to the economy of the region?

2. Geographic Themes

a. **LOCATION** Why is the relative location of Southwest Asia important to world trade of oil?

b. **PLACE** Why is the Persian Gulf considered a strategic location?

3. Identifying Themes

Why are the Tigris and Euphrates rivers so important to Southwest Asia? Which of the five themes applies to this situation?

4. Making Generalizations

In what ways do oil and water shape the lives of the people of Southwest Asia?

5. Making Inferences

How does climate affect the distribution of population in the region?

Additional Test Practice, pp. S1–S37

TEST PRACTICE
CLASSZONE.COM

Geographic Skills: Interpreting a Cartogram

Estimated Worldwide Oil Reserves

Use the cartogram at the right to answer the following questions. (See page 22 or page 733 for more on cartograms.)

1. **PLACE** Which nations have the greatest oil reserves?

2. **PLACE** What is the approximate amount of reserves for the United States?

3. **REGION** How does this cartogram help to explain the importance of the region?

Create a three-dimensional model to show the information on the cartogram. Be sure to label each of the countries and give an approximate total amount of oil reserves.

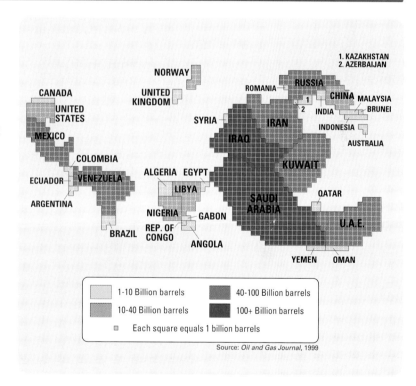

1-10 Billion barrels	40-100 Billion barrels
10-40 Billion barrels	100+ Billion barrels

□ Each square equals 1 billion barrels

Source: *Oil and Gas Journal,* 1999

INTERNET ACTIVITY

Use the links at **classzone.com** to do research about oil production. Find out what products are made from crude oil.

Creating Graphs and Charts Create an illustrated chart showing the types of products that are produced from petroleum. List the Web sites that you used in preparing your report.

HUMAN GEOGRAPHY OF SOUTHWEST ASIA
Religion, Politics, and Oil

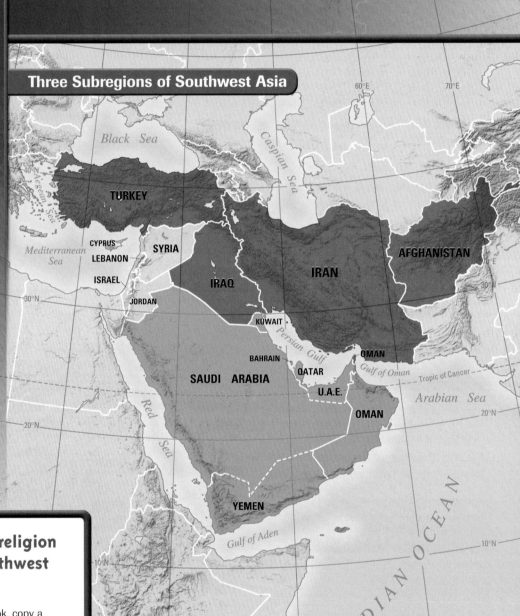

Three Subregions of Southwest Asia

Black Sea

Caspian Sea

TURKEY

Aegean Sea

Mediterranean Sea

CYPRUS
LEBANON
SYRIA

ISRAEL

JORDAN

IRAQ

IRAN

AFGHANISTAN

KUWAIT

Persian Gulf

BAHRAIN
QATAR
U.A.E.

OMAN
Gulf of Oman

SAUDI ARABIA

Tropic of Cancer

Arabian Sea

OMAN

Red Sea

YEMEN

Gulf of Aden

INDIAN OCEAN

Equator

	Arabian Peninsula
	Eastern Mediterranean
	Northeast

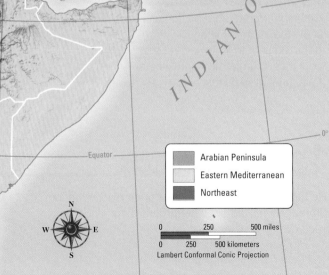

0   250   500 miles
0   250   500 kilometers
Lambert Conformal Conic Projection

GeoFocus

What impact have religion and oil had on Southwest Asia?

Taking Notes In your notebook, copy a cluster diagram like the one below. As you read, take notes about the history, culture, and modern life of each subregion of Southwest Asia.

Northeast

Southwest Asia

Arabian Peninsula

Eastern Mediterranean

The Arabian Peninsula

Main Ideas
- The Arabian Peninsula is heavily influenced by the religious principles of Islam.
- Oil production dominates the economy of the region.

Places & Terms

Mecca	mosque
Islam	theocratic
Muhammad	OPEC

CONNECT TO THE ISSUES
RELIGIOUS CONFLICT
Muslim claims to land in the region laid the foundation for future conflict.

A HUMAN PERSPECTIVE Two million people pour into the Saudi Arabian city of Mecca for a few weeks each year. They come from all over the world. In the past, the trip to Mecca involved a difficult journey across oceans and over miles of desert. Today, pilgrims arrive on airplanes. These people are fulfilling the Islamic religious duty of hajj, which is a pilgrimage to the holiest city of Islam—**Mecca.** For five or more days, all are dressed in simple white garments and all perform special activities, rituals, and ceremonies. It is a powerful example of spiritual devotion by the followers of one of the three major religions that claim a home in Southwest Asia.

Islam Changes Desert Culture

The modern nations in this subregion are Bahrain, Kuwait, Oman, Saudi Arabia, Qatar, United Arab Emirates, and Yemen. They are located at the intersection of three continents: Africa, Asia, and Europe. Because of this location, there were many opportunities for trade, and exchange of culture and religion.

TOWN AND DESERT In the past, some towns in the subregion served as trade centers for caravans moving across the deserts. Other cities were ports where goods were exchanged from the Silk Roads in East Asia, Indian Ocean trade from South Asia, and Mediterranean Sea trade from Europe. Still other towns were near oases and fertile lands along major rivers.

Nomadic desert dwellers called Bedouins moved across the peninsula from oasis to oasis. They adapted to the harsh conditions of the desert and built a culture based on strong family ties. They often fought against other families and clans for pasturelands for their livestock. Their fighting skills would eventually help to spread a new religion that developed in the region—Islam.

Islam is a monotheistic religion based on the teachings of its founder, the Prophet **Muhammad.** Muhammad lived part of his life in the city of Mecca.

PLACE Thousands of Muslim pilgrims gather at the holy site of the Ka'aba in Mecca. The Ka'aba is the black box at the right in the picture.

ISLAM BRINGS A NEW CULTURE The new religion united the people of the Arabian Peninsula in a way that had not been done previously. Islam requires certain religious duties of all who follow its teachings. The basic duties are called the Five Pillars. By performing these religious duties, all converts to Islam, called Muslims, practiced a similar culture. The Five Pillars are:

- **Faith** All believers must testify to the following statement of faith: "There is no God but Allah, and Muhammad is the Messenger of Allah."

- **Prayer** Five times a day, Muslims face toward the holy city of Mecca to pray. They may do this at a place of worship called a **mosque** or wherever they find themselves at the prayer times.

- **Charity** Muslims believe they have a responsibility to support the less fortunate by giving money for that purpose.

- **Fasting** During the Islamic holy month of Ramadan, Muslims do not eat or drink anything between sunrise and sunset. This action reminds Muslims that there are things in life more important than eating. It is also a sign of self-control and humility.

- **Pilgrimage** All able Muslims are expected to make a pilgrimage (hajj) to Mecca at least once during their lifetime.

THE SPREAD OF ISLAM As more and more people on the Arabian Peninsula began to convert to Islam, they spread its teachings. Armies of Bedouin fighters moved across the desert, conquered lands, and put Muslim leaders in control. Arabic language and Islamic teachings and culture spread across Southwest Asia. Muslim armies spread across three continents—Asia, Africa, and Europe. By the Middle Ages, a large area of the world was controlled by Muslim empires.

BACKGROUND Ramadan is the ninth month of the 12-month lunar year calendar used by Muslims. It does not match the calendar used by most Americans.

Governments Change Hands

The governments of lands controlled by Muslims were **theocratic.** This means religious leaders control the government. Rulers relied on religious law and consulted with religious scholars on running the country.

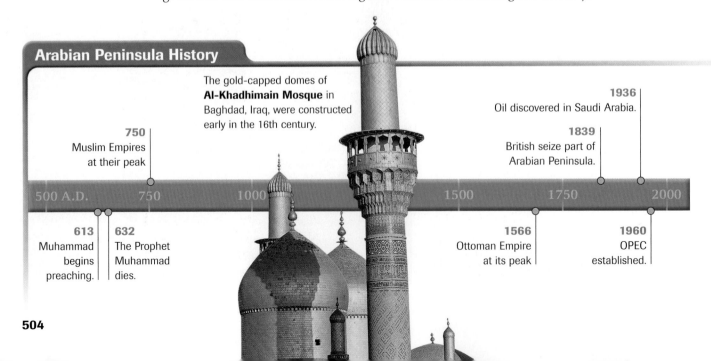

Arabian Peninsula History

The gold-capped domes of **Al-Khadhimain Mosque** in Baghdad, Iraq, were constructed early in the 16th century.

1936
Oil discovered in Saudi Arabia.

750
Muslim Empires at their peak

1839
British seize part of Arabian Peninsula.

500 A.D.   750   1000   1500   1750   2000

613
Muhammad begins preaching.

632
The Prophet Muhammad dies.

1566
Ottoman Empire at its peak

1960
OPEC established.

504

In some of the modern nations of this region—Iran, for example—religious leaders are in control of the government.

COLONIAL POWERS TAKE CONTROL Toward the end of the 1600s, the leaders of Muslim nations were weak. At the same time, countries like Britain and France were growing in power and establishing empires throughout the world. Much of Southwest Asia fell under the control of those two nations, especially after World War I and the breakup of the Muslim-held Ottoman Empire. The region was valuable to colonial powers for two reasons: because of the Suez Canal, a vital link between colonial holdings in the rest of Asia and European ports, and because oil was discovered there after 1932.

However, only a part of the region was colonized. On the Arabian Peninsula, a new power was rising. It was Abdul al-Aziz Ibn Saud. A daring leader, Abdul al-Aziz consolidated power over large areas of the Arabian Peninsula in the name of the Saud family. By the end of the 1920s, only small countries on the Arabian Gulf and parts of Yemen remained free of his control. The whole area became known as Saudi Arabia in 1932. Descendants of Abdul al-Aziz still rule Saudi Arabia today. Ⓐ

Geographic Thinking
Using the Atlas
Ⓐ Using the map on page 479, make a list of the countries that were not under the control of Abdul al-Aziz.

Oil Dominates the Economy

The principal resource in the economy of the Arabian Peninsula is oil. The region grew in global importance as oil became more important to the economies of all nations. Arabian Peninsula nations make almost all of their export money and a large share of GDP from oil, so oil prices are very important to them. Large increases in oil prices allow the oil-producing nations to funnel money into development of other parts of their economies, especially water development projects.

In 1960, a group of oil-producing nations, including Saudi Arabia and Kuwait, established an organization to coordinate policies on selling petroleum products. The group is the Organization of Petroleum Exporting Countries, also known as **OPEC.** The purpose of OPEC is to help members control worldwide oil prices by adjusting oil prices and production quotas. OPEC is a powerful force in international trade. Other Southwest Asian members include Qatar, the United Arab Emirates, Iran, and Iraq.

BACKGROUND
Other members of OPEC include Algeria, Gabon, Indonesia, Libya, Nigeria, and Venezuela.

Modern Arabic Life

Changes in the nations of the Arabian Peninsula during the 20th century were dramatic. The region is developing quickly with an emphasis on modernizing. Use of Western technology and machines undermined traditional ways of life. Camels, which used to be the mainstay of life in

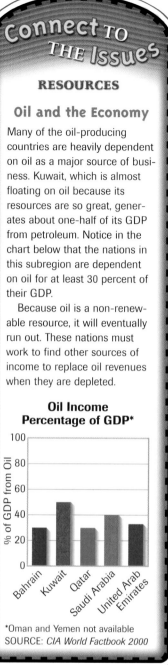
Connect TO THE Issues

RESOURCES

Oil and the Economy

Many of the oil-producing countries are heavily dependent on oil as a major source of business. Kuwait, which is almost floating on oil because its resources are so great, generates about one-half of its GDP from petroleum. Notice in the chart below that the nations in this subregion are dependent on oil for at least 30 percent of their GDP.

Because oil is a non-renewable resource, it will eventually run out. These nations must work to find other sources of income to replace oil revenues when they are depleted.

Oil Income Percentage of GDP*

*Oman and Yemen not available
SOURCE: *CIA World Factbook 2000*

SW ASIA

the Arabian Peninsula, are no longer used as extensively as they once were. Pick-up trucks, automobiles, and motorcycles have replaced them.

Gone, too, are some of the traditional marketplaces called bazaars or souks (sooks). These open-air markets brought together buyers and sellers with a great variety of merchandise, food, and entertainment. The market was a place to meet neighbors or friends, or to conduct business. Today, Western-style supermarkets or malls may be the shopping location of choice instead of the traditional bazaar.

THE CHANGE TO URBAN LIFE Cities were always a part of life in Southwest Asia. However, because of changes in the economy, the entire area is much more urbanized. Millions of people abandoned their lives as villagers, farmers, and nomads and moved into cities. In 1960, the region was about 25 percent urbanized. By the 1990s, this number had risen to about 58 percent. According to estimates, 70 percent of the population will live in cities by 2015. Saudi Arabia has an urban population of 83 percent. About 4 million people jam the capital, Riyadh. B▷

As the economy switched to providing petroleum and petroleum products, the types of jobs available in cities changed as well. Workers who could read and write and had technical skills were in great demand. Arabic nations on the peninsula scrambled to upgrade educational systems to meet the needs of the technological age. When those needs could not be fully met, foreign workers were brought in to work at jobs the native population could not fill. As a result, a large number of foreign workers now live in peninsula countries. In some cases, such as Qatar, only one in five workers is a native of the land.

RELIGIOUS DUTIES SHAPE LIVES Despite its rapid modernization, some aspects of Muslim culture have remained the same for centuries. If you traveled to Southwest Asia, one of the first things you would likely notice is that women cover their heads, hair, and sometimes faces with a scarf or veil. This is in keeping with the belief that covering those parts of the body is pleasing to God. Women's roles have gradually expanded during the 20th century. More Arabic women are becoming educated and are able to pursue careers in other nations. Because

Geographic Thinking◀

Making Comparisons
◀B How does the percentage of people living in cities of the Arabian Peninsula compare to that of the United States?

PLACE Camels are transported to pasture land by truck.
How does this photograph illustrate the change oil production has made in the region?

CONNECT TO THE ISSUES

▶ **RESOURCES**

Why might it be important for women to become more educated?

family is viewed as very important, many women stay at home to manage household affairs. **C**

As you read earlier in this section, all Muslims are expected to perform certain activities. One of the duties, prayer, is performed at prescribed times—dawn, noon, mid-afternoon, sunset, and before bed. Faithful Muslims stop the activities they are engaged in to carry out this responsibility. In some countries, traffic stops during prayer time. If a person is not near a place of worship, he or she may unroll a small prayer rug on which to kneel to pray. On Fridays, the day for congregational prayer, Muslims assemble for prayers at a mosque.

PLACE The female doctor above shows a blend of traditional and modern lifestyles. **How does this photograph illustrate changes in the roles of women in the region?**

Fasting in the month of Ramadan is another duty that shapes the lives of Muslims. During this month, adult Muslims do not eat or drink from before dawn until sunset. Fasting is a way of reminding Muslims of the spiritual part of their lives. After sunset, Muslims may eat a light meal of lentil or bean soup, a few dates, yogurt, and milky tea. A festival, 'Id al-Fitr, marks the end of Ramadan. New clothes, gifts, and elaborate dinners, along with acts of charity, are part of the celebration.

Since the Muslim culture is found throughout Southwest Asia, many of the same activities of modern life on the Arabian peninsula take place in other areas of Southwest Asia as well. However, as you will learn in the next section, other groups with different religions and lifestyles also live in the region.

SECTION Assessment

❶ Places & Terms

Explain the meaning of each of the following terms.

- Mecca
- Islam
- Muhammad
- mosque
- theocratic
- OPEC

❷ Taking Notes

REGION Review the notes you took for this section.

- How have Islamic beliefs affected this region?
- Why did this region grow in economic importance?

❸ Main Ideas

a. What are the Five Pillars of Islam?

b. Why was the region of Southwest Asia important to colonial powers?

c. What is the purpose of OPEC?

❹ Geographic Thinking

Drawing Conclusions How has the presence of large deposits of oil changed the lives of the people of the Arabian peninsula? **Think about:**

- where people live
- the types of jobs available

▶ **SW ASIA**

RESEARCH LINKS
CLASSZONE.COM

MAKING COMPARISONS Use the Internet to find more information on the increase in oil production over the last 25 years for the nations shown in the graph on page 505. Create a **line graph** showing the increases in oil production for the five nations.

Comparing Cultures

Religious Architecture

Throughout the world and across time, people have created spaces in their communities for the worship of God. Sometimes the space is reserved for only a special few, such as priests. Other times the space is designed to bring many worshippers together to create a sense of community. Religious requirements, available building materials, and artistic expression come together in the "houses of god."

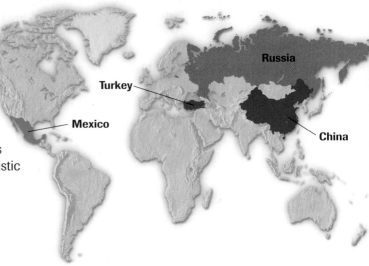

A Buddhist temple, such as this one located in Chufu, China, is sometimes called a pagoda. The temple itself usually is a wooden hall with several tiled roofs that curve up on the edges.

The Sultan Ahmed Cami Mosque in Turkey is considered one of the finest examples of Muslim religious architecture. Most mosques feature a minaret, a slender tower from which believers are called to prayer. This mosque is unusual because it has six minarets.

St. Basil's Cathedral in Moscow, Russia, is really eight smaller churches around a main one. The basic plan of the church forms a cross. The exterior was originally white. The colorful domes are covered with roof tiles that were added in the 17th century.

The Pyramid of the Sun in Mexico is the largest Meso-American religious structure. Its size was designed to inspire awe in the worshipper. A small temple on the top was usually visited only by priests.

GeoActivity

CREATING A MODEL

Choose one of the major religions of the world. With a small group, use the Internet to research more about the religious architecture of that religion.

• Create a model of a worship space showing the unique aspects of that religion's architecture.

• Create a brochure explaining your model.

RESEARCH LINKS
CLASSZONE.COM

GeoData

PLACES OF WORSHIP

THE MOSQUE

• Muslims are instructed to face toward Mecca when they pray. Inside the mosque, a special recess in the wall—*mihrab*—marks the direction of Mecca.

• The Sultan Ahmed Cami Mosque is also called the Blue Mosque because of the bluish haze given to the interior by 21,043 blue-glazed tiles on the walls.

THE PYRAMID

• Standing 216 feet high and 720 by 760 feet at the base, the Pyramid of the Sun is one of the largest structures of its type in the Western Hemisphere.

ST. BASIL'S CATHEDRAL

• St. Basil's was built by Ivan IV, also called Ivan the Terrible, as an offering to God for military victories over Tatar armies.

• Legend has it that the architect of St. Basil's was blinded so that he could never create anything similar to St. Basil's.

The Eastern Mediterranean

Main Ideas

- The holy places of three religions are found in this subregion.
- There is a great deal of political tension among nations in this subregion.

Places & Terms

Western Wall

Dome of the Rock

Zionism

Palestine Liberation Organization (PLO)

CONNECT TO THE ISSUES
RELIGIOUS CONFLICT
Creation of the nation of Israel led to conflict in the region.

A HUMAN PERSPECTIVE On September 28, 2000, riots broke out in the city of Jerusalem. The cause was a visit by an Israeli political leader to a Jewish holy place at a location on the Temple Mount. Muslims also have a holy place on the Temple Mount. They viewed the visit by the Israeli leader as disrespectful to Muslims. Hundreds of people died in the civil unrest that followed.

To understand why a simple visit to a holy place would cause such problems, it is necessary to understand the deep-seated hostility Arabs and Jews feel for each other. They have an enormous disagreement over the control of the city of Jerusalem and of the land called the Occupied Territories. (See the map on page 480.) In fact, the relations between Arabs and Jews affect the entire region of the Eastern Mediterranean.

Religious Holy Places

Three major monotheistic religions—Judaism, Christianity, and Islam—were founded in Southwest Asia. All three claim Jerusalem as a holy city. The City of Jerusalem, which covers 42 square miles, has Jewish, Christian, Armenian Christian, and Muslim sections. Followers of all three religions come to the Old City to visit locations with strong spiritual meaning.

JEWISH PRESENCE For Jews, Jerusalem, the capital of Israel, is the center of their modern and ancient homeland. Located in the old part of the city, the Temple Mount once housed the religion's earliest temples. There, King Solomon built the First Temple. The Second Temple was constructed after the Jews returned to their homeland in 538 B.C. Modern Jews come to pray at the holiest site in Jerusalem, a portion of the Second Temple known as the **Western Wall**—also called the Wailing Wall. It is the only remaining piece of the Second Temple, which was destroyed in A.D. 70 by the Romans.

CHRISTIAN HERITAGE For Christians, Jerusalem is the sacred location of the final suffering and crucifixion of Jesus. Towns and villages important in the life of Jesus are found near Jerusalem. Every year, Christians visit places like the Mount of Olives and the Church of the Holy Sepulchre by the thousands. When Jerusalem was under Muslim control, Christians launched the Crusades to regain the lands and place them under the

PLACE Christian pilgrims walk on the road to the Mount of Olives on a holy day—Palm Sunday.

control of Christians. Eventually, the lands returned to the control of Muslims and remained that way until the nation of Israel was established in May of 1948.

ISLAMIC SACRED SITES After Mecca and Medina, Jerusalem is considered the third most holy city to Muslims. A shrine there, called **Dome of the Rock,** houses the spot where Muslims believe the Prophet Muhammad rose into heaven. Jews believe it is the site where Abraham, a Jewish forefather, prepared to sacrifice his son Isaac to God. The Dome of the Rock and a nearby mosque, Al-Aqsa, are located on the Temple Mount next to the Western Wall. Because these most holy sites are so close together, they have been the site of clashes between Jews and Muslims. ◀

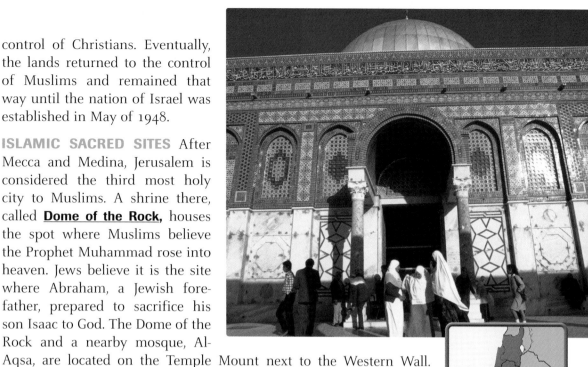

PLACE Muslim visitors gather at the Dome of the Rock, a holy site in the city of Jerusalem. **How did control of Jerusalem change over many centuries?**

CONNECT TO THE ISSUES
RELIGIOUS CONFLICT
Ⓐ What problems might emerge when three different religious groups claim the same area as a holy place?

A History of Unrest

The nations of the Eastern Mediterranean have been plagued with a history of political tension and unrest. The Ottoman Empire, a Muslim government based in Turkey, ruled the Eastern Mediterranean lands from 1520 to 1922. But the Ottoman Empire grew weaker and less able to solve problems with groups seeking independence. By the beginning of the 20th century, its collapse was not far away. The Ottoman Empire sided with Germany during World War I. At the end of the war, the Ottoman Empire fell apart. Britain and France received the lands in the Eastern Mediterranean as part of the war settlement.

BACKGROUND
The League of Nations gave the Ottoman lands to France and Britain.

THE LEGACY OF COLONIALISM After World War I, Britain and France divided the Ottoman lands in the Eastern Mediterranean region. France took the northern portion, including the present-day countries of Lebanon and Syria. Britain controlled the southern section, which included the present-day nations of Jordan and Israel. Britain and France were supposed to rule these lands until they were ready for independence. During the time of their control, the French frequently played different religious groups against each other. Those tensions remain in the region today. The Syrians hated the French and in the 1920s and 1930s rebelled against them. Lebanon became independent in 1943, and Syria gained independence in 1946.

BRITISH CONTROL PALESTINE The land controlled by Britain was known as Palestine. In the 19th century, a movement called **Zionism** began. Its goal was to create and support a Jewish homeland in Palestine. Jewish settlers started buying land and settling there. By 1914, just before World War I, about 12 percent of the population in Palestine was Jewish. After the war, the British took command of the region and continued to allow Jewish immigration to Palestine. Early

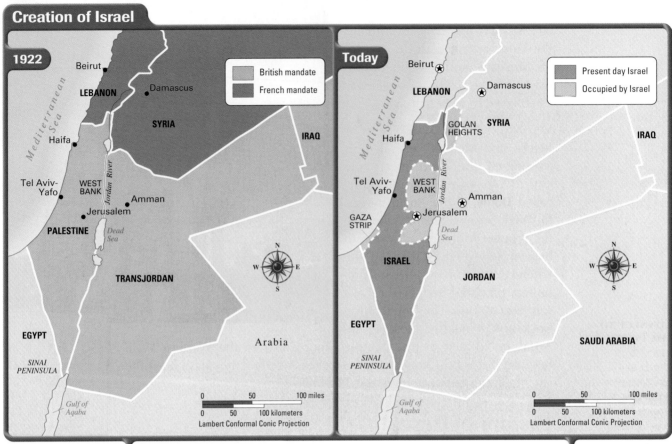

Creation of Israel

1922

British mandate
French mandate

LEBANON
Beirut
Damascus
SYRIA
IRAQ
Mediterranean Sea
Haifa
Jordan River
Tel Aviv-Yafo
WEST BANK
Amman
Jerusalem
PALESTINE
Dead Sea
TRANSJORDAN
EGYPT
Arabia
SINAI PENINSULA
Gulf of Aqaba

0 50 100 miles
0 50 100 kilometers
Lambert Conformal Conic Projection

Today

Present day Israel
Occupied by Israel

Beirut
Damascus
LEBANON
SYRIA
GOLAN HEIGHTS
Haifa
IRAQ
Mediterranean Sea
Jordan River
Tel Aviv-Yafo
WEST BANK
Amman
Jerusalem
GAZA STRIP
Dead Sea
ISRAEL
JORDAN
EGYPT
SAUDI ARABIA
SINAI PENINSULA
Gulf of Aqaba

0 50 100 miles
0 50 100 kilometers
Lambert Conformal Conic Projection

SKILLBUILDER: Interpreting Maps

❶ **PLACE** Which bodies of water form a natural boundary between Jordan and Israel?

❷ **PLACE** Which three areas are occupied by Israel?

on, Arabs and Jews in the region cooperated. But as more and more Jews poured into Palestine to escape persecution in Germany, the Arabs resisted the establishment of a Jewish state. In 1939, to reduce tensions the British halted Jewish immigration to Palestine.

As you study the map on this page, you will see that the area controlled by the British was divided into two sections—Transjordan and Palestine. The land was divided to relieve tensions between Arabs and Jews. An Arab government jointly ruled Transjordan with the British. Britain controlled Palestine, along with local governments that included both Jews and Arabs.

CREATING THE STATE OF ISRAEL At the end of World War II, thousands of Jewish survivors of the Holocaust wanted to settle in Palestine. Palestine was considered the Jewish homeland. World opinion supported the establishment of a Jewish nation-state. Britain eventually referred the question of a Jewish homeland to the United Nations. In 1947, the United Nations developed a plan to divide Palestine into two states—one for Arabs and one for Jews.

Arabs in the region did not agree with the division. However, the nation of Israel was established on May 14, 1948. Immediately, the surrounding Arab nations of Egypt, Syria, Lebanon, Jordan, Iraq, Saudi Arabia, and Yemen invaded Israel to prevent the establishment of the state. Jewish troops fought back. By the 1950s, Israel was a firmly established nation. The 1948 war was the beginning of hostilities that continue to this day.

Geographic Thinking

Using the Atlas

Ⓑ Use the Atlas on page 480. How was the land Israel occupied in 1967 different from the land it held in 1948?

BACKGROUND
A refugee is a person who leaves home or country to find safety in another location.

Caught in the middle of this turmoil were Palestinian Arabs and Christians. Many of these people had roots in Palestine that went back for centuries. They either fled their homes or were forced into UN-sponsored refugee camps just outside Israel's borders. The land designated for the Palestinians on the West Bank and Gaza Strip is under Israeli control. In the 1960s, the **Palestine Liberation Organization (PLO)** was formed to regain the land for Palestinian Arabs. Over the years, the PLO has pursued political and military means to take possession of Arab land in Israel and allow refugees to return to their homes.

In 2003, the PLO named Mahmoud Abbas (also known as Abu Mazen) as the first Palestinian prime minister. Abu Mazen agreed to support a peace plan for the Middle East called the "road map." The goals of this step-by-step plan included establishing a Palestinian state and resolving the rights of the Palestinian refugees.

Modernizing Economies

The nations in the Eastern Mediterranean subregion are young. Most became independent shortly after World War II. Cyprus received its independence from Britain in 1960. Political divisions, refugees, and a weak infrastructure make it difficult to develop healthy economies.

REFUGEES AND CIVIL WARS The creation of Israel produced a large number of Palestinian refugees. Today, those refugees and their descendants total almost 3.6 million people. They are scattered across many of the countries in the region. Some still live in UN-sponsored camps.

Many of the refugees have struggled to find food, shelter, and employment. Providing education and other services for them is difficult for nations such as Jordan, one of the poorest in the region—and the one with the largest Palestinian refugee population.

Civil wars in Lebanon and Cyprus have also caused huge economic problems. Lebanon, a more developed nation, was hard hit by a civil war that lasted from 1975 to 1976. The conflict widened to include other nations, and in 1982 Israel invaded Lebanon. Some Israeli troops remained in Lebanon until 2000.

MODERN INFRASTRUCTURE All of the nations of the Eastern Mediterranean subregion have great potential for developing agriculture, tourism, and trade. What many of them lack, however, is an infrastructure that would support a growing economy. Roads in war-torn areas, for example, must be rebuilt. Especially needed are irrigation systems to make the area bloom. Better communication systems and power sources are needed for developing high-tech industries in the region. Israel has been able to build sophisticated industries such as computer software development.

BACKGROUND
The island of Cyprus has two countries. One is controlled by Greek Cypriots and one by Turkish Cypriots. Only Turkey officially recognizes the Turkish republic.

Connect TO THE Issues

POPULATION

Palestinian Refugee Camps

In 1949, the UN authorized the creation of 53 Palestinian refugee camps. The camps were supposed to be used only for a short time until the Palestinians were resettled. That was over 50 years ago. Today, most of the Palestinians living in the camps were actually born there and have never been to the lands designated for the Palestinian state.

The camps house upwards of 35,000 people and some as many as 50,000 people. The UN and other nations provide money for education and health care needs. Since the Israeli government restricts all travel for work, economic opportunities are very limited.

SW ASIA

growing up in...Israel

This young woman is a member of the Israel Defense Forces.
Unmarried Jewish young women are required to serve for two years.
They serve in various parts of the armed forces, in jobs such as tank
instructors, helicopter pilots, military police, rescue workers, and
office workers. They are not permitted to serve in active combat units.
Service in the armed forces helps build unity and identity for Israelis.

If you lived in Israel, you would pass these milestones:

- You would go to school from age 5 to age 15.
- At age 14, you would choose between going to a technical school or a more academic school.
- You could begin working at age 15.

- You could drive at age 17.
- You could get married at age 17.
- You would enter the armed forces at age 18: men for 3 years, women for 2 years.

Modern Life

Modern life in the Eastern Mediterranean is a curious blend of old and new. Strong cultural traditions exist but they are combined with changes that were brought about by modern innovations. Cell phones, computers, and Internet access are increasingly common. One aspect of life here that remains quite traditional, however, is the dining experience.

EATING OUT, EATING IN Eating in restaurants in Eastern Mediterranean countries is not as common as in the United States. Some restaurants have separate sections for men and women. Cafes serving coffee and tea are generally for men only. Most meals are eaten in the home. Families and sometimes friends gather to have meals. The last meal of the day is usually served between 8 and 11 P.M.

Typically, a meal begins with small portions of hummus, ground chickpeas mixed with lemon juice and parsley, and baba ganouzh, an eggplant dip served with pita, a flat bread with a pocket. A salad called tabbouleh, made of bulgur (cracked wheat), parsley, onions, mint, tomatoes, and lemon juice, is common. Chicken or lamb is more likely to be served as a main course than beef. Many meals are finished with fresh fruit or sweets such as kolaicha, a sweet cake made of barley flour, sugar, oil, and cardamom seed. Thick coffee or tea is also served. The host of a dinner may not eat with the guests so that he can attend to all their needs during the meal. ⓒ

A VARIETY OF CULTURES Muslim Arabs make up the majority of people who live in the countries of the Eastern Mediterranean. However, in several nations, especially Lebanon and Israel, there is a variety of cultures.

Geographic Thinking

Making Comparisons
ⓒ In what ways is the dining experience in this region different from that of the United States?

The quake destroyed 85,000 buildings. Many of the buildings were poorly constructed with inferior building materials. Floors of buildings "pancaked" and crushed the residents.

About 40,000 families were made homeless by the quake. Survivors were housed in 168 tent cities. Unfortunately, few were winterized, and thousands of people shivered through Turkey's winter.

GeoActivity

MAKING A DEMONSTRATION

Working with a small group, use the Internet to research the causes and effects of earthquakes. Then create a **demonstration** about earthquakes.

- Build a model or create a diagram showing how an earthquake occurs.
- Create a chart showing the type of damage caused by earthquakes.
- Add a world map showing the major fault lines.

RESEARCH LINKS
CLASSZONE.COM

GeoData

THE MERCALLI INTENSITY SCALE

- The Mercalli Intensity Scale measures an earthquake's effect on people and buildings.
- Mercalli ranges from I to XII. Here are some examples.

  I. No damage

  VI. Pictures fall off the wall

  VII. Slight damage to structures

  X. Most masonry structures destroyed; landslides; ground cracked

  XII. Total damage

RICHTER SCALE

- The Richter Scale measures the magnitude of energy released during an earthquake.
- Here are some examples of Richter Scale measurements:

  2 Just felt

  4.5 Damage newsworthy

  7 A major quake

  8 Great damage

  8.9 Largest quake ever recorded

VISUAL SUMMARY
HUMAN GEOGRAPHY OF SOUTHWEST ASIA

Subregions of Southwest Asia

○ **The Arabian Peninsula**

- The teachings of Islam shape the lives of the people of the region.
- Oil forms the basis of the economy of the region.
- The subregion has experienced rapid modernization.

○ **The Eastern Mediterranean**

- The region has holy places of three religions: Judaism, Christianity, and Islam.
- The Jewish nation-state of Israel was created in 1948.
- Political unrest in the region has disrupted life and created problems with refugees and the economy.

● **The Northeast**

- The region has a variety of ethnic groups, most of whom practice Islam.
- The region has economies that range from developed to one of the poorest nations in the world—Afghanistan.
- There are divisions among the people of this region over modern and traditional lifestyles.

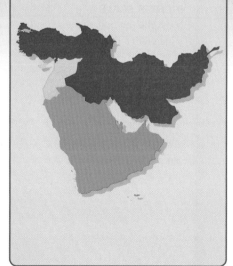

Reviewing Places & Terms

A. Briefly explain the importance of each of the following.

1. Mecca
2. Islam
3. OPEC
4. Western Wall
5. Dome of the Rock
6. Zionism
7. Palestine Liberation Organization
8. Sunni
9. Shi'ite
10. Taliban

B. Answer the questions about vocabulary in complete sentences.

11. Why is Mecca an important site to Muslims?
12. How are Islam, Sunni, and Shi'ite related to each other?
13. Which branch of Islam has the largest number of followers?
14. Where are the Western Wall and the Dome of the Rock located?
15. With which religion is the Dome of the Rock associated?
16. Why is the Western Wall important to Jews?
17. Which of the terms above is associated with international oil trade?
18. What is the goal of the Palestine Liberation Organization?
19. How is Zionism connected to the formation of the state of Israel?
20. In which country were members of the Taliban harboring terrorists?

Main Ideas

The Arabian Peninsula (pp. 503–509)

1. How did the teachings of Islam unite the people of the Arabian Peninsula?
2. Why is oil so important to the economies of the Arabian Peninsula?
3. How has modern Arabic life changed in the past 50 years?

The Eastern Mediterranean (pp. 510–515)

4. For which religions is Jerusalem a holy city?
5. Why was the state of Israel created?
6. What factors have made it difficult to build healthy economies in the Eastern Mediterranean countries?
7. How are populations of Lebanon and Israel different from other countries in the region?

The Northeast (pp. 516-521)

8. How are language, ethnic groups, and religion in the Northeast region different from other parts of Southwest Asia?
9. What steps need to be taken to improve the economies of the Northeast region?
10. Why are there internal struggles in some of the nations of the Northeast region?

Critical Thinking

1. Using Your Notes

Use your completed chart to answer these questions.

- a. How is Israel different from the other nations in the region?
- b. How must infrastructure be changed in the region?

2. Geographic Themes

- a. **HUMAN-ENVIRONMENT INTERACTION** What impact does the presence of oil in the region have on the economies of the countries in Southwest Asia?
- b. **LOCATION** How would Israel's relative location be described?

3. Identifying Themes

Which nations are dealing with large numbers of refugees or immigrants? Which of the five themes applies to this situation?

4. Making Inferences

How has the presence of many different ethnic groups in this region caused political unrest?

5. Making Generalizations

In what ways has oil production changed life in Southwest Asia?

Additional Test Practice, pp. S1–S37

TEST PRACTICE
CLASSZONE.COM

Geographic Skills: Interpreting Maps

Ottoman Empire, 1683

Use the map at the right to answer the following questions.

1. **LOCATION** What is the relative location of the Ottoman Empire?
2. **PLACE** On which continents was the Ottoman Empire located?
3. **PLACE** Which large bodies of water are within the Ottoman Empire?

GeoActivity

On a current map showing the same area as in the map at the right, outline the Ottoman Empire. Make a list of the modern countries that were once a part of the Ottoman Empire.

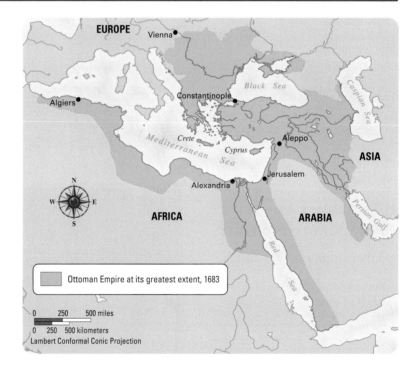

Ottoman Empire at its greatest extent, 1683

0   250   500 miles
0   250   500 kilometers
Lambert Conformal Conic Projection

INTERNET ACTIVITY

Use the links at **classzone.com** to do research about OPEC. Make a list of the current members of the organization. Focus on the impact on the price of oil as a result of actions taken by the group.

Analyzing Data Study the data you collected on oil prices and the actions of OPEC. Create charts or graphs to illustrate the information. Then write a generalization about the information you found.

CASESTUDY

**RELIGIOUS CONFLICT
OVER LAND**

For more on these issues in
Southwest Asia . . .

 CURRENT EVENTS
CLASSZONE.COM

TODAY'S ISSUES

Southwest Asia

A Kurdish family rests
at its camp in eastern
Turkey. Many Kurds
are nomadic and move
across lands in several
countries.

GeoFocus

Can Southwest Asia solve long-standing problems?

Taking Notes Copy the cause-and-effect
chart below into your notebook. Use it to
record information about solving economic
and political problems in Southwest Asia.

	Causes	Effects
Issue 1: Population Relocation		
Issue 2: Economic Development		
Case Study: Religious Conflict		

Population Relocation

What kind of population movement is taking place in Southwest Asia?

Main Ideas
- Economic growth brings foreign workers to the region.
- Political factors have shifted the region's population.

Places & Terms
guest workers

stateless nation

Palestinians

West Bank

Gaza Strip

A HUMAN PERSPECTIVE In the 1980s, Kurds living in Turkey were attacked by the Turkish military. The parents of 10-year-old Garbi Yildirim feared for their son's safety. Reluctantly they sent him from Turkey to live with relatives in Germany. When Garbi reached his 18th birthday, he was notified by the German government that he would have to return to Turkey. Upon his return, he knew that he would have to serve in the Turkish military. This meant he would have to use weapons against his own people—the Kurds. He refused to return to Turkey and was placed in a deportation prison to await the recommendation of a German court on the case. Garbi's case is an example of the problems some ethnic groups face in Southwest Asia.

New Industry Requires More Workers

Life in Southwest Asia in 1900 seemed only slightly different from life there in 1100. Some people lived in villages or cities while others moved livestock from one source of water to another.

Then, in the early years of the 20th century, everything changed. Geologists discovered huge deposits of petroleum and natural gas under the sands and seas of Southwest Asia. Western oil companies quickly leased land in the region and supplied the technology and the workers to pump the fuel from the ground.

Many countries in Southwest Asia grew enormously wealthy from oil profits. The oil boom set off decades of rapid urbanization. Extensive road construction made cities and towns more accessible. Many thousands of people migrated to the cities in search of jobs and a chance to share in the region's newfound riches. So many jobs were available that some were left unfilled.

FOREIGN WORKERS To fill the job openings, companies recruited people, mostly from South and East Asia. These **"guest workers"** are largely unskilled laborers. They fill jobs that the region's native peoples find culturally or economically unacceptable. In parts of the Arabian Peninsula, the immigrant workers actually outnumber the native workers. For example, in 1999, nearly 90 percent of the United Arab Emirates (UAE) work force was made up of immigrants.

PLACE Great wealth makes this United Arab Emirates golf club possible. In the middle of the desert, it features green fairways, a pool, and a freshwater lake. Guest workers fill jobs at sites like this.

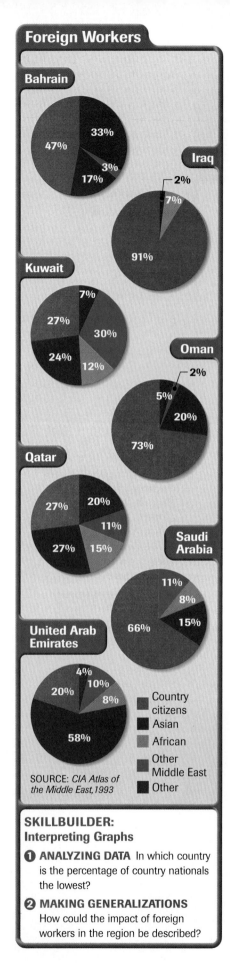

Bahrain

33%
47%
3%
17%

Iraq

2%
7%
91%

Kuwait

7%
27%
30%
24%
12%

Oman

2%
5%
20%
73%

Qatar

20%
27%
11%
27%
15%

Saudi Arabia

11%
8%
66%
15%

United Arab Emirates

4%
10%
20%
8%
58%

- Country citizens
- Asian
- African
- Other Middle East
- Other

SOURCE: *CIA Atlas of the Middle East,* 1993

SKILLBUILDER:
Interpreting Graphs

❶ **ANALYZING DATA** In which country is the percentage of country nationals the lowest?

❷ **MAKING GENERALIZATIONS** How could the impact of foreign workers in the region be described?

PROBLEMS OF GUEST WORKERS The presence of so many guest workers has led to problems. Cultural differences often exist between the guest workers and their employers. Misunderstandings over certain customs can result in severe penalties. For example, a Filipino man was given six months in jail and expelled from the UAE for brushing past a woman on a bus. Arabs viewed his behavior as insulting to the woman.

Sometimes the workers must live in special districts apart from the Arab population. Some workers have been abandoned. Others receive no wages for months at a time. Many immigrants find themselves unemployed and without money to get back home.

The large number of guest workers is a concern to the governments of Southwest Asia. Some government officials worry that depending on these workers will prevent their nation's own workers from developing their skills. Others worry about the intolerance and even violence that these workers face. And, finally, some fear the immigrants could weaken their country's sense of national identity. Solving the cultural and economic issues over guest workers will be a challenge to the governments of the region. Ⓐ

Political Refugees Face Challenges

Rapidly changing economic conditions have caused population shifts in Southwest Asia. Political conflict in the region has also caused relocation.

STATELESS NATION One of the longest conflicts has been over the ethnic group known as the Kurds. After World War I, the Allies recommended creating a national state for the group. Instead, the land intended for the Kurds became part of Turkey, Iraq, and Syria. The Kurds became a **stateless nation**—a nation of people without a land to legally occupy. Turkey, Iraq, Iran, and Syria tried to absorb the Kurds into their populations but were not successful. The Kurds resisted control in each of the countries. Governments forcibly moved thousands of Kurds in an attempt to control them.

In Iraq, this forced migration ruined Kurdish homes, settlements, and farms. As you read in Chapter 22, the Iraqi government used deadly chemical weapons on settlements of Kurds to kill them or force them to leave the area. In the year 2000, as many as 70,000 Kurds had been displaced from areas they called home. Many of the Kurds have been forced to live in crowded relocation camps.

Geographic Thinking

Seeing Patterns
Ⓐ How did changes in the economy of the region change the make-up of the population?

Geographic Thinking

Making Comparisons

B▶ How is the diversification of Oman's economy different from that of Saudi Arabia?

Other nations are making efforts to develop other mineral resources. Oman revived its copper industry and chromium mines. Chromium is used in steel production for jet aircraft. Expanding these industries allowed the Omani economy to reduce its dependence on oil profits. ◀B

HUMAN RESOURCES People are a valuable resource in any nation. Southwest Asian nations are developing their **human resources**—the skills and talents of their people. Many of those nations also realize that they must invest in all their people, including women. Providing education and technology training is critical. Nations are expanding the opportunities for their citizens to gain an education. For example, Kuwait has established free education for all children through the university level. For students who wish to study outside the country, the government pays the fees and provides money to cover living expenses.

Many societies in Southwest Asia have strict rules concerning women's roles in society. Often it is difficult for women to get an education and find employment. However, the shortage of workers in the region has opened economic opportunities for women. Important economic and political changes are taking place in Southwest Asia. As the nations work to develop their physical and human resources, opportunities for all who live there will expand. A successful economy is built on the efforts of all its people working together toward the goal of diversification.

PLACE Muslim girls in Tehran, Iran, discuss a lesson.
Why is it important for all citizens to be educated?

Assessment

① Places & Terms

Identify and explain the meaning of these terms.

• strategic commodity

• human resources

② Taking Notes

REGION Review the notes you took for this section.

	Causes	Effects
Issue 2: Economic Development		

• Why are this region's resources so valuable?

• What changes need to be made to the region's infrastructure?

③ Main Ideas

a. What effect have unpredictable oil prices had on the economy of the region?

b. What steps have nations in the region taken to diversify their economic base?

c. Why must the human resources of the region be developed?

④ Geographic Thinking

Determining Cause and Effect How has oil wealth changed the economy of the region? **Think about:**

• the cost of modernizing the infrastructure

• the need for a diverse economy

S See Skillbuilder Handbook, page R9.

SW ASIA

MAKING COMPARISONS Do some research to find information about the projected freshwater supplies for the nations in the region. Create a set of symbols to represent the projected water supply figures. Then draw a **map** of the region, and place the appropriate water supply symbol on each nation.

CaseStudy

Religious Conflict Over Land

Who should control Jerusalem?

Jerusalem checkpoints deepen Palestinian resentment.

C onflict between Jews and Arabs over land and statehood in Southwest Asia disrupts life in the region. One aspect of this conflict centers around Jerusalem. The city is sacred to Jews, Christians, and Muslims. Control of Jerusalem is a deeply emotional issue that affects the region's politics and population.

Control of Jerusalem

After World War II, the UN recommended that the city of Jerusalem become an international city. It would be under the control of an international body rather than an Arab or a Jewish government. But by the end of the Arab-Israeli war in 1948, Jerusalem was divided between Arabs and Israelis. Arabs took the Old City and East Jerusalem located in the West Bank sector. The Israelis took control of West Jerusalem. During the Six-Day War of 1967, the Israelis captured the rest of Jerusalem.

Control of the holy sites within the Old City also became an issue. Although the Israelis captured the city, the Muslims retained control of their holy site, *Haram ash-Sharif,* called the Temple Mount by the Jews.

As the Israelis gained control of the entire city of Jerusalem, they began adding Arab lands to the city. They placed Jewish settlements on those lands. Palestinian Arabs fled or were forced to leave the settlement lands. The Palestinians in Jerusalem and elsewhere have maintained they should have the "right of return" to the lands in Israel. Their claims are supported by United Nations Resolution 194, which states that Palestinians have the "right of return" to former homelands.

1978
Camp David Accords set up Palestinian self-rule in West Bank.

| 1940 | ARAB-ISRAELI CONFLICT | 1960 | | 1970 | | 1980 | | 1990 | | 2000 | | 2010 |

1948
The **State of Israel** is created; war with Arabs follows immediately.

1967
Israel takes control of Jerusalem, West Bank, and Gaza Strip at the end of the Six-Day War.

1993
Oslo Accords allow Palestinians to establish self-rule in West Bank and Gaza Strip.

2003
A "road map" for peace is launched, calling for the establishment of a Palestinian state and security guarantees for Israel.

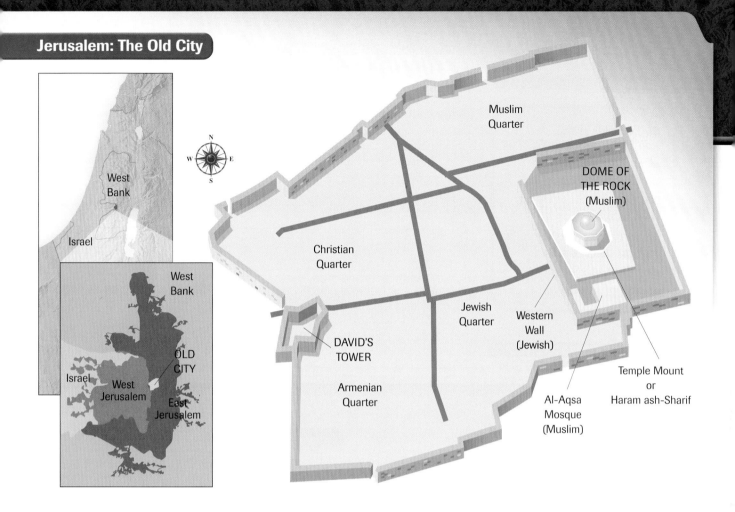

West Bank

Israel

West Bank

OLD CITY

Israel

West Jerusalem

East Jerusalem

Muslim Quarter

DOME OF THE ROCK (Muslim)

Christian Quarter

Jewish Quarter

Western Wall (Jewish)

DAVID'S TOWER

Armenian Quarter

Al-Aqsa Mosque (Muslim)

Temple Mount or Haram ash-Sharif

Proposed Solutions to the Conflict

SEE
PRIMARY SOURCE D

The emotional and political issue of who should control Jerusalem makes it a very difficult diplomatic problem to solve. Because both the Israelis and the Palestinians claim Jerusalem as the capital of their nation, neither is willing to give it up to the other group. The following solutions have been proposed for control of Jerusalem:

- Palestinians retain control of certain parts of East Jerusalem while Israel annexes several Jewish settlements near Jerusalem. This would enlarge Israeli territory in the area.

- Israel retains control of West Jerusalem and the Jewish Quarter of the Old City, but the Palestinians control the Old City and East Jerusalem. This is basically how the city is controlled today.

SEE
PRIMARY SOURCE C

- Palestinians control the Temple Mount but give up the right of return to Israel. The Israeli government fears that the sheer numbers of returning Palestinians would overwhelm Israel.

- An international agency has control of all holy sites.

On the following pages, you will find primary sources that present different views on the control of the city of Jerusalem. Use them to help you form an opinion about the best way to solve the problem.

SW ASIA

CASESTUDY

PROJECT

A Peace Conference

Primary sources A, B, C, D, and E on these two pages offer differing views about control of Jerusalem. Use these resources along with your own research to prepare a peace conference that presents both Israeli and Arab solutions for control of Jerusalem.

RESEARCH LINKS
CLASSZONE.COM

Suggested Steps

1. Choose one of the proposed solutions to the control of Jerusalem to investigate.

2. Use online and print resources to research the positions of Israelis, Palestinians, and Americans.

3. Create visuals—maps, charts, graphs—to make the conference discussion clearer.

4. Select two or three representatives from each group to take part in the conference. The rest of the class should act as journalists, take notes on the presentation, and be prepared to ask questions of the representatives.

Materials and Supplies

- Posterboard
- Markers
- Reference books, newspapers, and magazines
- Video monitor with VCR or DVD capability
- Computer with Internet access/printer

PRIMARY SOURCE A

United Nations Resolution *UN Resolution 181, adopted on November 29, 1947, declared that Jerusalem would become an international city with both Jewish and Muslim inhabitants.*

Part III City of Jerusalem

A. The City of Jerusalem shall be established as a *corpus separatum* [separate body] under a special international regime and shall be administered by the United Nations. The Trusteeship Council shall be designated to discharge the responsibilities of the Administering Authority on behalf of the United Nations.

*   *   *

C. 1(a) To protect and to preserve the unique spiritual and religious interests located in the city of the three great monotheistic faiths throughout the world, Christian, Jewish, and Moslem; to this end to ensure that order and peace, and especially religious peace, reign in Jerusalem.

(b) To foster co-operation among all the inhabitants of the city in their own interests as well as in order to encourage and support the peaceful development of the mutual relations between the two Palestinian peoples throughout the Holy Land.

PRIMARY SOURCE B

Official Statement *This statement was made December 31, 2000, by the Palestinian cabinet, which opposed President Clinton's plan for resolving the issue of "right of return" and control of the holy sites in Jerusalem.*

The Palestinian leadership confirms its commitment to the full right of refugees to return to their lands and homes in accordance with Resolution 194, the cabinet said, referring to the United Nations resolution adopted in December 1948.

Our people will never, under any circumstances, concede one inch from our Jerusalem and our Islamic and Christian holy sites.

PRIMARY SOURCE C

Personal Observation *Yossi Sarid, head of the Meretz party in Israel, is a leading advocate of peace in the region. On December 31, 2000, he expressed his opinion on the central issue of the Palestinian right of return.*

There is only one issue that could, God forbid, make this [Clinton peace proposal] fail, and that is the right of return. It is important for the Palestinians to understand and internalize this. Realization of the right of return means—how should I put it?—the suicide of Israel.

If we open the gates to hundreds of thousands of refugees, that means the state of Israel as created by the Zionist dream will be bankrupt.

PRIMARY SOURCE D

Editorial Commentary *Kenneth L. Woodward, religion editor for* Newsweek *magazine, expresses an opinion about why any solution for the Jerusalem question is one that is important not just to Jews and Arabs but to millions of others.*

Thus, for billions of believers who may never see it, Jerusalem remains a city central to their sacred geography. This is why the future of the city is not just another Middle Eastern conflict between Arabs and Jews. . . . Both Israel and the Palestinians have real roots in the Holy Land, and both want to claim Jerusalem as their capital. The United Nations, supported by the Vatican, would have the city internationalized under its jurisdiction. The issue, however, is not merely one of geopolitics. There will be no enduring solution to the question of Jerusalem that does not respect the attachments to the city formed by each faith. Whoever controls Jerusalem will always be constrained by the meaning the city has acquired over three millenniums of wars, conquest and prophetic utterance.

PRIMARY SOURCE E

Political Cartoon *Mark Fiore drew this cartoon about the situation in Jerusalem. What message is the cartoonist sending about prospects for peace between Israelis and Palestinians?*

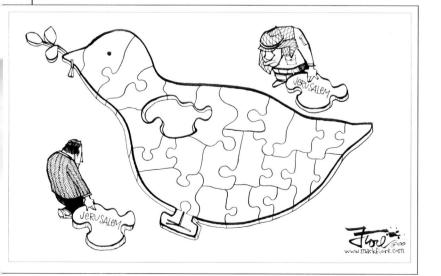

PROJECT CheckList

Have I . . .

✓ looked at all sides of the issue?

✓ identified the key players and their points of view?

✓ created informative visuals that make my presentation clear and interesting?

✓ practiced the delivery of my presentation?

VISUAL SUMMARY
TODAY'S ISSUES IN SOUTHWEST ASIA

Population

Population Relocation
- Urban areas in the region have grown significantly since the 1960s.
- Thousands of foreign workers fill jobs in the region.
- Kurds claim homelands in four countries: Turkey, Iraq, Iran, and Syria.
- Arab Palestinians claim lands in Israel.

Economics

Oil Wealth Fuels Change
- Huge oil resources shape the region's economy.
- The infrastructure needs to be updated.
- The region's economy must be diversified.
- Human resources need to be developed.

Conflict

Religious Conflict Over Land
- Jerusalem is holy to three major religions: Judaism, Christianity, and Islam.
- Israel controls Jerusalem.
- Control of holy sites in Jerusalem is one aspect of the conflict.
- Arab Palestinians claim the "right of return" to Jerusalem, after leaving it as the result of wars.

Reviewing Places & Terms

A. Briefly explain the importance of each of the following.

1. guest workers
2. stateless nation
3. Palestinians
4. West Bank
5. Gaza Strip
6. strategic commodity
7. human resources

B. Answer the questions about vocabulary in complete sentences.

8. Why is it necessary to have guest workers in Southwest Asia?
9. Which terms above refer to land areas in Israel?
10. Which of the above terms includes the location of Jerusalem?
11. Why might Palestinians and Kurds be considered stateless nations?
12. Which group claims the right of return to the Gaza Strip?
13. Why is oil considered a strategic commodity?
14. In what way could water be considered a strategic commodity?
15. What groups make up human resources?

Main Ideas

Population Relocation (pp. 525–528)

1. What concerns have been raised about foreign workers in the region?
2. Why don't the Kurds have a homeland?
3. Which lands are claimed by Arab Palestinians?
4. Where do a large majority of Palestinians live?

Oil Wealth Fuels Change (pp. 529–531)

5. Why must nations stop depending solely on oil wealth?
6. Which areas of the region's economy need to be developed and diversified?
7. Why is providing education and technology training an important aspect of developing human resources?

Religious Conflict Over Land (pp. 532–535)

8. How did the Israelis gain control of Jerusalem?
9. What is the "right of return"?
10. What are some proposed solutions to the issue of control of Jerusalem?

Critical Thinking

1. Using Your Notes

Use your completed chart to answer these questions.

	Causes	Effects
Issue 1: Population Relocation		
Issue 2: Economic Development		

a. How did the Kurds become a stateless nation?

b. What effect has an expanding economy had on population relocation in the region?

2. Geographic Themes

a. **MOVEMENT** How are Palestinian refugee camps and the "right to return" related?

b. **REGION** Why is this region considered to be a strategic location?

3. Identifying Themes

Why is the control of Jerusalem such a difficult issue to resolve? Which of the five themes applies to this situation?

4. Making Inferences

Why must some oil wealth be used to develop water resources in the region?

5. Making Decisions

Which of the proposed solutions for the control of Jerusalem do you favor and why?

Additional Test Practice, pp. S1–S37

TEST PRACTICE
CLASSZONE.COM

Geographic Skills: Interpreting Graphs

Availability of Water Resources*

Use the graph at the right to answer the following questions.

1. **MAKING COMPARISONS** Which country is projected to have the greatest available water supplies by 2050? Which country will have the least?

2. **MAKING INFERENCES** What are some reasons why the availability of water resources will decrease?

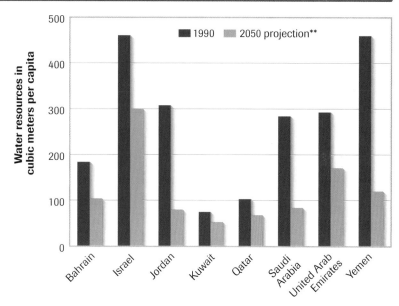

* Freshwater resources of below 1,000 cubic meters per year per capita will likely cause chronic water shortages.

**Estimates are based on a high projection of population increase.

SOURCE: Adapted from Robert Engelman and Pamela LeRoy, *Sustaining Water: An Update,* Population Action International, Washington D.C. 1995

GeoActivity

Use the Regional Data File to create a chart showing the population of the nations listed in the Water Stress Index. Create a second bar graph showing water stress by placing the countries in order by population.

INTERNET ACTIVITY

Use the links at **classzone.com** to do research on water scarcity in the region and proposed solutions to the problem. Focus on finding a solution that would be environmentally friendly.

Identifying and Solving Problems Using the information you gathered, propose a solution to the need for fresh water in Southwest Asia. Support your proposal with charts or graphs illustrating both the need for water and the sources of fresh water.

Unit 8

South Asia

South Asia includes the Indian subcontinent and its nearby islands. It is a region of ancient cultures, spectacular landforms, and rapidly growing populations.

PLACE The Taj Mahal, at Agra, India, is said to be one of the world's most beautiful buildings. Constructed of marble, it was built in the 17th century by Emperor Shah Jahan as a tomb for his wife.

GeoData

LOCATION South Asia is mainly a triangular peninsula that juts out from the Asian mainland into the Indian Ocean.

REGION The seven countries of South Asia have great cultural and religious diversity.

HUMAN-ENVIRONMENT INTERACTION Life in South Asia is greatly influenced by its varied landforms and its extreme weather, especially the seasonal monsoons.

For more information on South Asia . . .

RESEARCH LINKS CLASSZONE.COM

LOCATION Elephants wearing richly decorated cloth coverings are central figures in the 14-night Esala Perahera festival in Kandy, Sri Lanka. It is one of many religious festivals held in South Asia.

REGION The world's highest mountains, the majestic snow-capped Himalayas, form the northern border of the Indian subcontinent. Mt. Everest, to the left, is the world's tallest peak at 29,035 feet.

SOUTH ASIA

Today's Issues in South Asia

Today, South Asia faces the issues previewed here. As you read Chapters 24 and 25, you will learn helpful background information. You will study the issues themselves in Chapter 26.

In a small group, answer the questions below. Then have a class discussion of your answers.

Exploring the Issues

1. **POPULATION** What might be some of the effects of rapid population growth on both humans and the environment?

2. **EXTREME WEATHER** Consider news stories that you have heard or read about that refer to extreme weather in various parts of South Asia. Make a list of the types of extreme weather that affect South Asians.

3. **TERRITORIAL DISPUTE** Search the Internet for the latest information about the dispute over Kashmir. What position does each side hold?

For more on these issues in South Asia . . .

CURRENT EVENTS
CLASSZONE.COM

POPULATION EXPLOSION

How can South Asia's population growth be managed?

Many problems come with rapid population growth, including crowded cities. Kolkata, pictured here, had a population of more than 4 million in the 1990s, and a population density of more than 61,900 persons per square mile.

EXTREME WEATHER

How do people cope with extreme weather?

People find a way to continue with their lives despite the severe flooding that plagues South Asia during the summer monsoons. Residents of Dhaka, Bangladesh, shown here, navigate flooded streets as best they can.

CASESTUDY

How can India and Pakistan resolve their dispute over Kashmir?

India and Pakistan have spent millions of dollars to develop nuclear weapons in their continuing dispute over Kashmir. This has left less money to spend on improving the lives of their citizens.

TERRITORIAL DISPUTE

INDIA'S POOR

PAKISTAN'S POOR

Patterns of Physical Geography

Use the Unit Atlas to add to your knowledge of South Asia. As you look at the maps and charts, notice geographic patterns and specific details about the region. For example, the chart to the right gives details about the rivers and mountains of South Asia.

After studying the illustrations, graphs, and physical map on these two pages, jot down in your notebook the answers to the following questions.

Making Comparisons

1. How much longer is the Nile than each of the three major rivers of South Asia?

2. Compare the size and population of South Asia to that of the United States. Which is larger in terms of size? Which is larger in terms of population?

3. How do the tallest mountains of South Asia compare to the tallest U.S. mountain?

For updated statistics on South Asia . . .

DATA UPDATE
CLASSZONE.COM

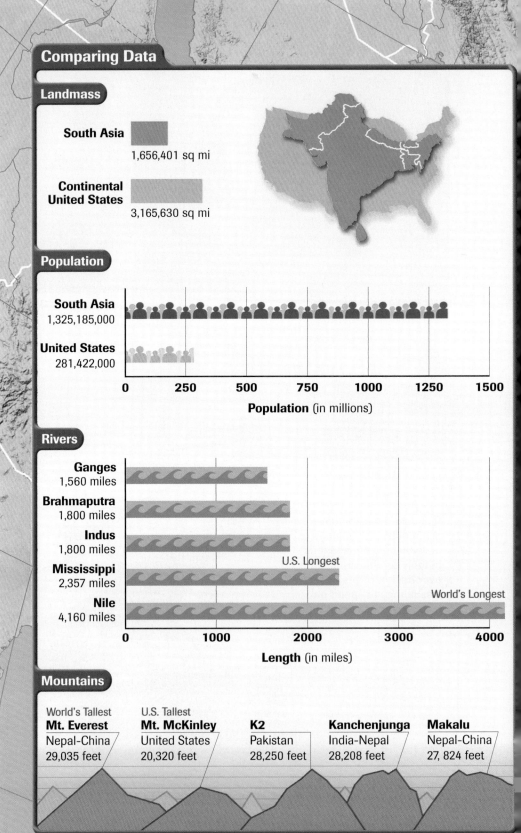

Comparing Data

Landmass

South Asia
1,656,401 sq mi

Continental United States
3,165,630 sq mi

Population

South Asia
1,325,185,000

United States
281,422,000

Population (in millions)
0   250   500   750   1000   1250   1500

Rivers

Ganges
1,560 miles

Brahmaputra
1,800 miles

Indus
1,800 miles

Mississippi
2,357 miles   U.S. Longest

Nile
4,160 miles   World's Longest

Length (in miles)
0   1000   2000   3000   4000

Mountains

World's Tallest Mt. Everest	U.S. Tallest Mt. McKinley	K2	Kanchenjunga	Makalu
Nepal-China	United States	Pakistan	India-Nepal	Nepal-China
29,035 feet	20,320 feet	28,250 feet	28,208 feet	27, 824 feet

AFGHANISTAN

CHINA

HINDU KUSH

Khyber Pass

Karakoram Range

K2
28,250 ft.
(8,611 m.)

Sulaiman Range

Indus R.

HIMALAYA MTS.

PAKISTAN

Indus R.

Thar Desert

INDO-GANGETIC PLAIN

Mt. Everest
29,035 ft.
(8,850 m.)

NEPAL

BHUTAN

Brahmaputra R.

Ganges R.

Rann of
Kutch

INDIA

BANGLADESH

Tropic of Cancer

Arabian
Sea

Gulf of
Khambhat

Vindhya Range

Narmada R.

Chota Nagpur
Plateau

Ganges Delta

MYANMAR

Godavari R.

WESTERN GHATS

Krishna R.

DECCAN

PLATEAU

EASTERN GHATS

Bay of
Bengal

Andaman
Is.

Andaman Sea

Laccadive Is.

Laccadive Sea

Pak Str.

Gulf of
Mannar

SRI LANKA

Nicobar
Is.

MALDIVES

SOUTH ASIA

Equator

INDIAN OCEAN

N
W   E
S

0   250   500 miles
0   250   500 kilometers
Two-Point Equidistant Projection

Elevation

13,100 ft.	(4,000 m.)
6,600 ft.	(2,000 m.)
1,600 ft.	(500 m.)
650 ft.	(200 m.)
0 ft.	(0 m.)
Below sea level	

▲ Mountain peak

Patterns of Human Geography

The first great civilization of South Asia developed along the banks of the Indus River more than 4,000 years ago. Study the historical map of the Indus Valley civilization and the political map of South Asia on these two pages. In your notebook, jot down the answers to these questions.

Making Comparisons

1. In which countries of modern South Asia was the Indus Valley civilization located? Which of these countries is the larger country?

2. What might have been some of the reasons for a civilization developing at that location?

3. What modern city or cities are closest to the locations of ancient Mohenjo-Daro, Harappa, Kalibangan, and Lothal? (In some cases, more than one city will be an acceptable answer.)

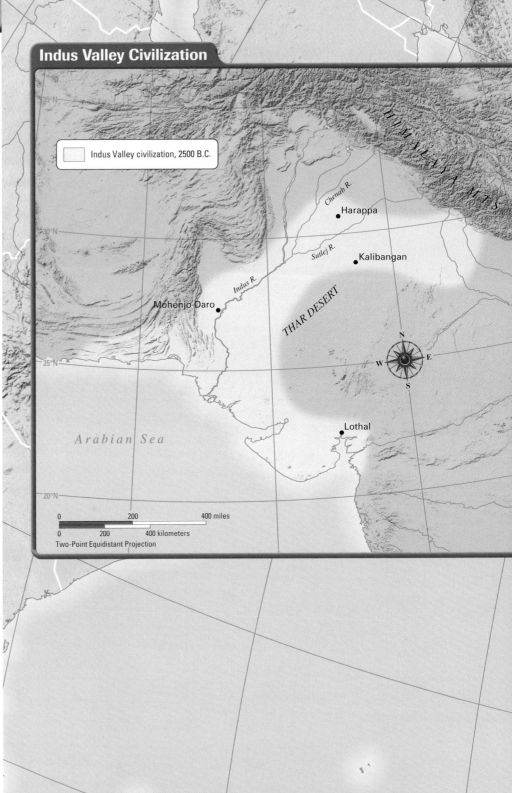

Indus Valley Civilization

Indus Valley civilization, 2500 B.C.

HIMALAYA MTS.

Chenab R.
Harappa

Sutlej R.
Kalibangan

Indus R.

THAR DESERT

Mohenjo-Daro

Lothal

Arabian Sea

0   200   400 miles
0   200   400 kilometers
Two-Point Equidistant Projection

Doctors (per 100,000 pop.) (1992–1999)	GDP[a] (billions $US) (1999 estimate)	Import/Export[a] (billions $US) (1998–1999)	Literacy Rate (percentage) (1998)	Televisions (per 1,000 pop.) (1996–1998)	Passenger Cars (per 1,000 pop.) (1996–1999)	Total Area[b] (square miles)	
20	187	8.01 / 5.1	40	7	1	55,126	
16	2.1	0.122 / 0.111	42 (1995)	19	1	16,000	
48	1,805	50.2 / 36.3	56	69	4	1,195,063	
40	0.54	0.312 / 0.098	96	39	3	115	
4	27.4	1.2 / 0.485	39	4	N/A	54,362	
57	282	9.8 / 8.4	44	88	8	310,403	
37	50.5	5.3 / 4.7	91	92	12	25,332	
251	9,255	820.8 / 663.0	97	847	489	3,787,319	

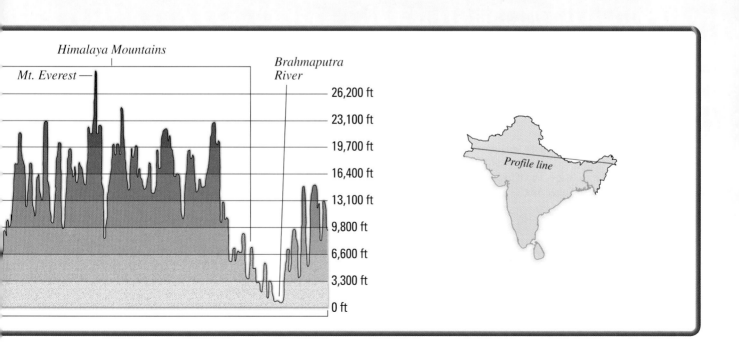

Himalaya Mountains

Mt. Everest

Brahmaputra River

26,200 ft
23,100 ft
19,700 ft
16,400 ft
13,100 ft
9,800 ft
6,600 ft
3,300 ft
0 ft

Profile line

PHYSICAL GEOGRAPHY OF SOUTH ASIA
The Land Where Continents Collided

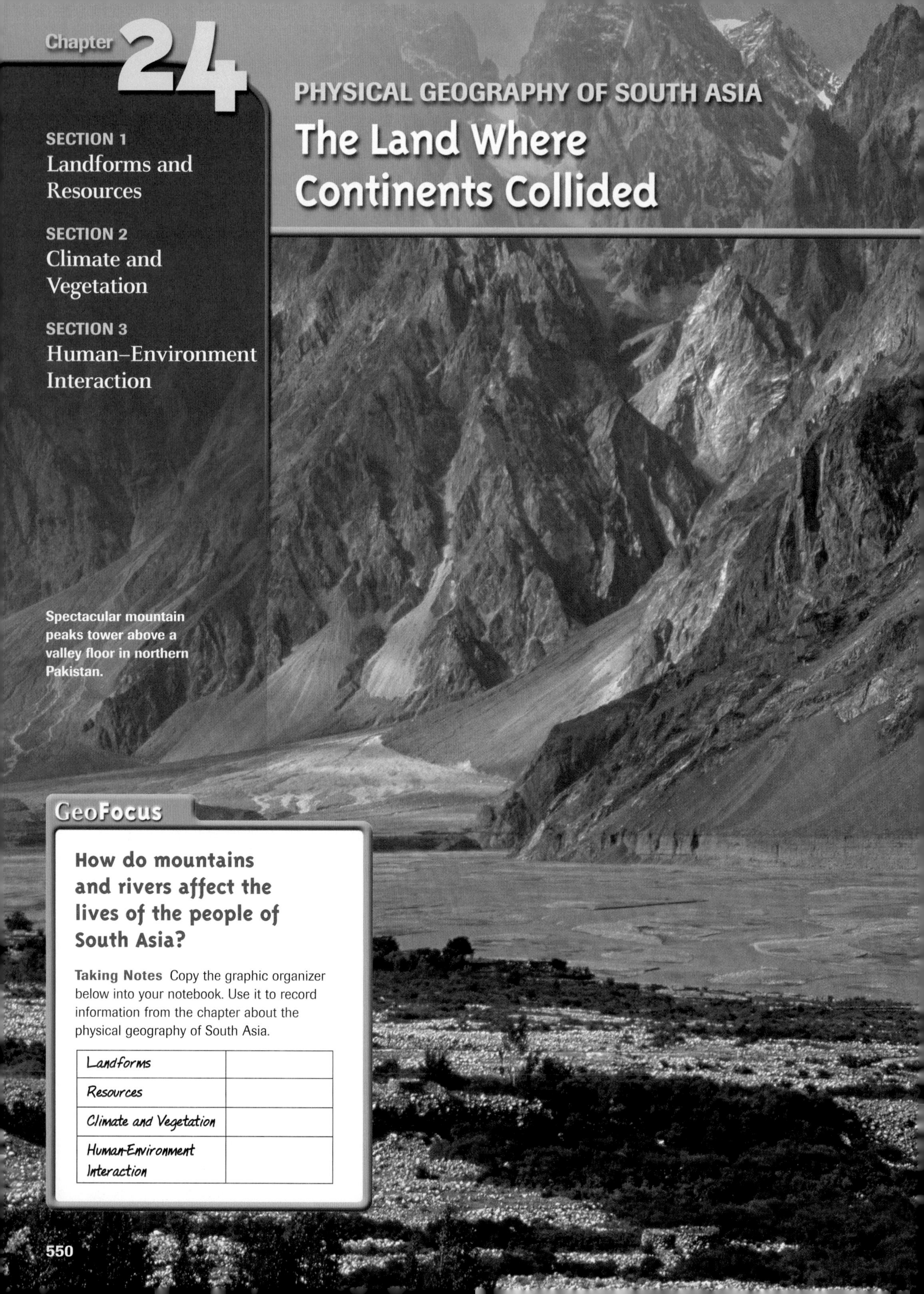

SECTION 1
Landforms and Resources

SECTION 2
Climate and Vegetation

SECTION 3
Human–Environment Interaction

Spectacular mountain peaks tower above a valley floor in northern Pakistan.

GeoFocus

How do mountains and rivers affect the lives of the people of South Asia?

Taking Notes Copy the graphic organizer below into your notebook. Use it to record information from the chapter about the physical geography of South Asia.

Landforms	
Resources	
Climate and Vegetation	
Human-Environment Interaction	

Landforms and Resources

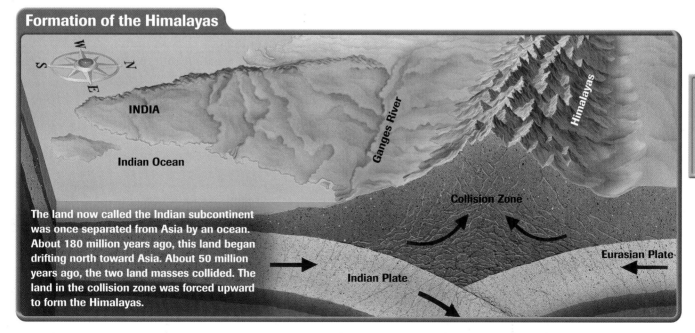

> **Main Ideas**
> • South Asia is a subcontinent of peninsulas bordered by mountains and oceans.
> • A wide variety of natural resources helps sustain life in the region.
>
> **Places & Terms**
> **Himalaya Mountains**
> **subcontinent**
> **alluvial plain**
> **archipelago**
> **atoll**
>
> ---
>
> **CONNECT TO THE ISSUES**
> **TERRITORIAL DISPUTE**
> Kashmir is an area in the western Himalayas on the border of India and Pakistan. It has been a source of dispute between the two countries.

A HUMAN PERSPECTIVE Thousands of years ago, the Hindus of what is now north India imagined a gigantic mountain reaching more than 80,000 miles into the sky. They believed that this enormous peak, called Mt. Meru, was the center of the physical and spiritual world. In their writings, they described "rivers of sweet water" flowing down the sides of the mountain. While Mt. Meru exists only in myth, it did have a real-life inspiration—Mt. Everest, the world's tallest mountain peak at 29,035 feet above sea level. Mt. Everest and the other towering peaks of the **Himalaya Mountains** have been a lure to mountain climbers around the world. Many climbers had died on Everest's icy slopes before Sir Edmund Hillary and Tenzing Norgay, his Sherpa guide, became the first people to reach its summit in 1953.

Mountains and Plateaus

The Himalayas are part of South Asia, a region that includes seven countries—India, Pakistan, Bangladesh, Bhutan, Nepal, Sri Lanka, and the Maldives. South Asia is sometimes called a **subcontinent,** a large landmass that is smaller than a continent. In fact, it is often referred to as the Indian subcontinent because India dominates the region. Although South Asia is about half the size of the continental United States, it has more than one billion inhabitants—one-fifth of the world's population.

Formation of the Himalayas

INDIA

Indian Ocean

Ganges River

Himalayas

Collision Zone

Eurasian Plate

Indian Plate

The land now called the Indian subcontinent was once separated from Asia by an ocean. About 180 million years ago, this land began drifting north toward Asia. About 50 million years ago, the two land masses collided. The land in the collision zone was forced upward to form the Himalayas.

SOUTH ASIA

As you saw on the map on page 543, natural barriers help to separate the South Asian subcontinent from the rest of Asia. The Himalayas and other mountain ranges form the northern border, while water surrounds the rest of the region. The South Asian peninsula, which extends south into the Indian Ocean, is bordered by the Arabian Sea to the west and the Bay of Bengal to the east.

NORTHERN MOUNTAINS Millions of years ago, the land that is now South Asia was actually part of East Africa. About 50 million years ago, it split off and drifted northward. As the illustration on page 551 shows, it collided with Central Asia. The gradual collision of these two large tectonic plates forced the land upward into enormous mountain ranges. These mountains, which are still rising, now form the northern edge of the South Asian subcontinent.

The magnificent Himalayas are a system of parallel mountain ranges. They contain the world's highest mountains, with nearly two dozen peaks rising to 24,000 feet or above. The Himalayas stretch for 1,500 miles and form a giant barrier between the Indian subcontinent and China. Mt. Everest, the world's tallest peak, sits at the heart of the Himalayas. Nestled high up within these mountains are the remote, landlocked kingdoms of Nepal and Bhutan.

The Hindu Kush are mountains that lie at the west end of the Himalayas. They form a rugged barrier separating Pakistan from Afghanistan to the north. For centuries, the Hindu Kush stood in the way of Central Asian tribes trying to invade India. Bloody battles have been fought over control of major land routes through these mountains, including the Khyber Pass. The mighty Karakoram Mountains rise in the northeastern portion of the chain. They are the home of the world's second highest peak, K2.

SOUTHERN PLATEAUS The collision of tectonic plates that pushed up the Himalayas also created several smaller mountain ranges in central India, including the Vindhya (VIHN•dyuh) Range. To the south lies the Deccan Plateau. This large tableland tilts east, toward the Bay of Bengal, and covers much of southern India. Two mountain ranges, the Western Ghats and the Eastern Ghats, flank the plateau, separating it from the coast. These mountains also block most moist winds and keep rain from reaching the interior. As a result, the Deccan is a largely arid region.

Rivers, Deltas, and Plains

The Northern Indian Plain, or Indo-Gangetic Plain, lies between the Deccan Plateau and the northern mountain ranges. This large lowland region stretches across northern India and into Bangladesh. It is formed by three great river systems: the Indus, the Ganges, and the Brahmaputra.

GREAT RIVERS The three great rivers of South Asia have their origins among the snowcapped peaks of the high

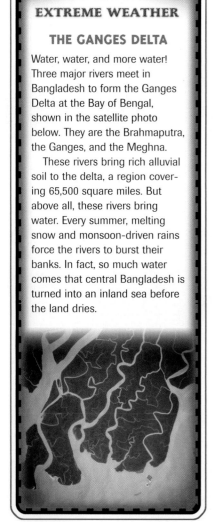

Connect TO THE Issues

EXTREME WEATHER

THE GANGES DELTA

Water, water, and more water! Three major rivers meet in Bangladesh to form the Ganges Delta at the Bay of Bengal, shown in the satellite photo below. They are the Brahmaputra, the Ganges, and the Meghna.

These rivers bring rich alluvial soil to the delta, a region covering 65,500 square miles. But above all, these rivers bring water. Every summer, melting snow and monsoon-driven rains force the rivers to burst their banks. In fact, so much water comes that central Bangladesh is turned into an inland sea before the land dries.

Geographic Thinking

Seeing Patterns
🔲 What role have the Himalayas played in the development of Nepal and Bhutan?

BACKGROUND
The name *Himalayas* is Sanskrit for "abode of snow."

Himalayas. The Indus flows west and then south through Pakistan to the Arabian Sea. The Ganges drops down from the central Himalayas and flows eastward across northern India. The Brahmaputra winds its way east, then west and south through Bangladesh. The Ganges and Brahmaputra eventually meet to form one huge river delta before entering the Bay of Bengal.

FERTILE PLAINS These rivers play a key role in supporting life in South Asia. Their waters provide crucial irrigation for agricultural lands. They also carry rich soil, called alluvial soil, on their journey down from the mountains. When the rivers overflow their banks, they deposit this soil on **alluvial plains,** lands that are rich farmlands. As a result, the Indo-Gangetic Plain is one of the most fertile farming regions in the world.

The Indo-Gangetic Plain is also the most heavily populated part of South Asia. In fact, the area contains about three-fifths of India's population. Many of the subcontinent's largest cities, including New Delhi and Kolkata in India, and Dakha in Bangladesh, are located there. Population densities at the eastern end of the plain, particularly in the Ganges-Brahmaputra delta, are especially high, as you can see on the map on page 547. To the west, in the area between the Indus and Ganges rivers, the plain becomes drier and requires more irrigation. To the south lies one of the world's most arid regions—the Thar, or Great Indian Desert. ◀B

Geographic Thinking ◀

Using the Atlas
B▷ Use the map on page 543. Locate the Thar Desert. What two countries share its land?

Offshore Islands

Two island groups are also countries of South Asia—Sri Lanka and the Maldives. Sri Lanka is located in the Indian Ocean just off India's southeastern tip. The Maldives island group is situated farther off the Indian coast to the southwest.

SRI LANKA: THE SUBCONTINENT'S "TEAR DROP"
Sri Lanka (sree LAHNG·kuh) is a large, tear-shaped island country. It is a lush tropical land of great natural beauty. Dominating the center of the island is a range of high, rugged mountains that reach more than 8,000 feet in elevation. Many small rivers cascade from these mountains to the lowlands below. The northern side of the island consists of low hills and gently rolling farmland. Circling the island is a coastal plain that includes long, palm-fringed beaches.

THE MALDIVES ARCHIPELAGO The Maldives comprise an **archipelago,** or island group, of more than 1,200 small islands. These islands stretch north to south for almost 500 miles off the Indian coast near the equator. The islands (shown at right) are the low-lying tops of submerged volcanoes, surrounded by coral reefs and shallow lagoons. This type of island is called an **atoll.** The total land area of the Maldives is 115 square miles (roughly twice the size of Washington, D.C.). Only about 200 of the islands are inhabited.

PLACE Not one of the more than 1,200 small coral islands that make up the Maldives rises more than six feet above the Indian Ocean.
How might global warming affect these islands?

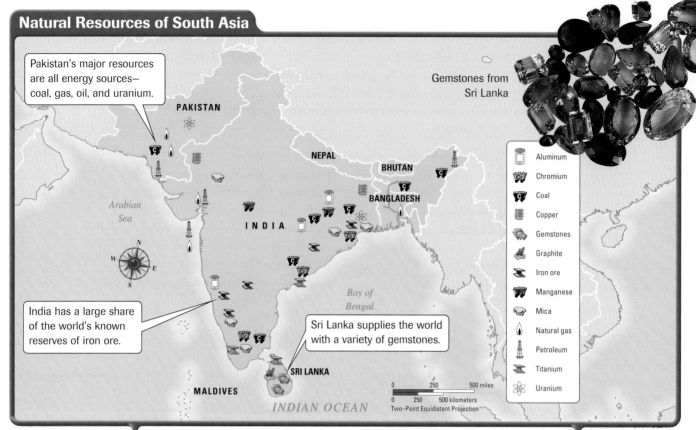

Natural Resources of South Asia

Pakistan's major resources are all energy sources—coal, gas, oil, and uranium.

Gemstones from Sri Lanka

India has a large share of the world's known reserves of iron ore.

Sri Lanka supplies the world with a variety of gemstones.

	Aluminum
	Chromium
	Coal
	Copper
	Gemstones
	Graphite
	Iron ore
	Manganese
	Mica
	Natural gas
	Petroleum
	Titanium
	Uranium

PAKISTAN
NEPAL
BHUTAN
BANGLADESH
Arabian Sea
INDIA
Bay of Bengal
SRI LANKA
MALDIVES
INDIAN OCEAN

0   250   500 miles
0   250   500 kilometers
Two–Point Equidistant Projection

SKILLBUILDER: Interpreting Maps

❶ **LOCATION** How would you describe the distribution of petroleum resources in South Asia?

❷ **REGION** Why might terrain be a reason no major mineral resources are shown in Nepal?

Natural Resources

The natural beauty of the southern islands is just one of the many physical assets of South Asia. In fact, the subcontinent boasts a wide variety of natural resources that support human life. At the same time, South Asia's rapidly growing population puts great pressure on its land and resources.

WATER AND SOIL South Asia relies heavily on its soil and water resources to provide food through farming and fishing. The great river systems that bring alluvial soil down from the mountains help enrich the land. They also bring the water necessary for crops to grow. Both small- and large-scale irrigation projects divert the water to the farmlands that need it. Many types of fish are also found in South Asian rivers and coastal waters, including mackerel, sardines, carp, and catfish.

South Asian waters also provide a means of transportation and power. Boats travel the rivers and coastlines, carrying goods and people from town to town. Governments also are working to harness hydroelectric energy from the waters. For example, India and Pakistan have a number of hydroelectric and irrigation projects underway.

FORESTS Timber and other forest products are another important resource in South Asia. Rain forests in India produce hardwoods like sal and teak, along with bamboo and the fragrant sandalwood. Highland forests in Bhutan and Nepal have thick stands of pine, fir, and other softwood trees. Deforestation is a severe problem, however. It causes

BACKGROUND
Only one-tenth of India's original forest cover remains uncut.

soil erosion, flooding, landslides, and loss of wildlife habitats. Over-cutting has devastated formerly dense forests in India, Bangladesh, and Sri Lanka.

MINERALS Much of South Asia's energy is still generated from mineral resources. For example, India ranks fourth in the world in coal production and has enough petroleum to supply about half its oil needs. India, Pakistan, and Bangladesh also have important natural gas resources. Uranium deposits in India provide fuel for nuclear energy.

South Asia also has large iron-ore deposits, particularly in India's Deccan Plateau. India is one of the world's leading exporters of iron ore, which is also used in that country's large steel industry. Other South Asian minerals include manganese, gypsum, chromium, bauxite, and copper.

India supplies most of the world's mica, a key component in electrical equipment. This is one of the reasons that India has a growing computer industry. Mica is also found in Nepal. India and Sri Lanka both have substantial gemstone deposits. India is traditionally known for its diamonds, while Sri Lanka produces dozens of types of precious and semi-precious stones. The island is most famous for its beautiful sapphires and rubies.

In this section, you read about the landforms and resources of South Asia. In the next section, you will learn about climate and vegetation.

HUMAN-ENVIRONMENT INTERACTION
These Nepalese are harvesting timber from depleted forests in southern Nepal.
What are some ways deforestation might affect the lives of South Asians?

SECTION

Assessment

① Places & Terms

Identify and explain where in the region these would be found.

- Himalaya Mountains
- subcontinent
- alluvial plain
- archipelago
- atoll

② Taking Notes

PLACE Review the notes you took for this section.

Landforms	
Resources	

- What mountain ranges separate the subcontinent from the rest of Asia?
- Why might South Asia have a large steel industry?

③ Main Ideas

a. When and how was South Asia formed?

b. What are South Asia's three largest rivers, and what is their source?

c. How do the island countries that lie off the subcontinent's coast differ from one another?

④ Geographic Thinking

Seeing Patterns How do the Himalayas contribute to South Asia's resource wealth? **Think about:**

- river systems
- agriculture

S **See Skillbuilder Handbook, page R8.**

GeoActivity

MAKING COMPARISONS Do research on one of the mountain climbing expeditions to the peak of Mt. Everest. Write a **news article** about the expedition and present it to the class. Use standard sentence structure, spelling, grammar, and punctuation.

SOUTH ASIA

Climate and Vegetation

2

Main Ideas

- Climate conditions in South Asia range from frigid cold in the high mountains to intense heat in the deserts.
- Seasonal winds affect both the climate and vegetation of South Asia.

Places & Terms

monsoon

cyclone

CONNECT TO THE ISSUES
EXTREME WEATHER
Seasonal droughts and flooding take a heavy toll in lives and property in South Asia each year.

A HUMAN PERSPECTIVE Every April and May, much of South Asia bakes in the heat. People endure temperatures that regularly top 100°F. Dust fills the air, and streams dry up. People walk for miles looking for water. Then—when it seems that no one can survive another day—the clouds roll in. The skies open up, and the rains come. People celebrate when the land turns green.

But their celebration is short-lived, as the downpour continues. Soon, the ground can hold no more water. Rivers overflow their banks. Families are forced from their homes as towns and cities are flooded. Thousands may die before the waters eventually recede, and the land dries out. South Asians see this cycle repeat itself each year.

Climate—Wet and Dry, Hot and Cold

Half of the climate zones that exist on Earth can be found in South Asia. This means that South Asians must adapt to widely varying conditions.

CLIMATE ZONES South Asia has six main climate zones, as you can see on the map on page 557. The highland zone has the coldest climate. This is the area of the Himalayas and other northern mountains, where snow exists year-round. The lower elevations, which include the lush foothills and valleys of Nepal, Bhutan, and northern India, are much warmer. They are in the humid subtropical zone that stretches across South Asia. The Indo-Gangetic Plain also occupies much of this region.

The semiarid zone—a region of high temperatures and light rainfall—is found at the western end of the Plain and in parts of the Deccan Plateau. The desert zone covers much of the lower Indus Valley, in the borderlands of western India and southern Pakistan. The driest part of

MOVEMENT Camels, who can go days without water, are used to move goods and people across the sands of the Thar Desert, which straddles northwest India and southeast Pakistan.
What does this photo show about the climate and vegetation of the Thar Desert?

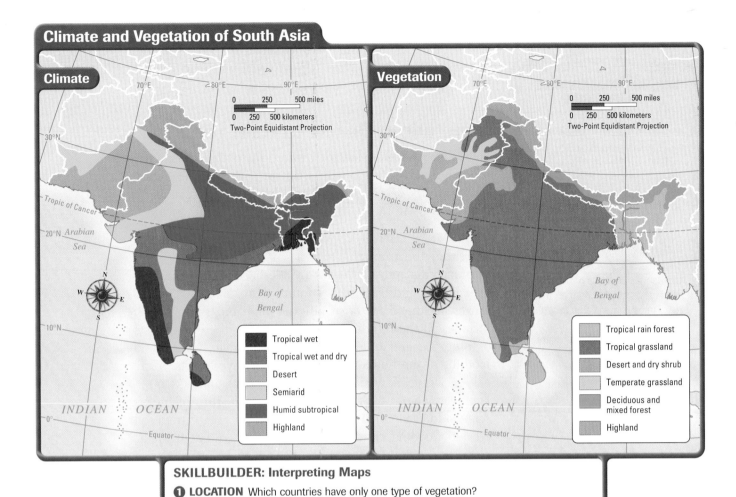

Climate and Vegetation of South Asia

Climate

0 250 500 miles
0 250 500 kilometers
Two-Point Equidistant Projection

- Tropical wet
- Tropical wet and dry
- Desert
- Semiarid
- Humid subtropical
- Highland

Vegetation

0 250 500 miles
0 250 500 kilometers
Two-Point Equidistant Projection

- Tropical rain forest
- Tropical grassland
- Desert and dry shrub
- Temperate grassland
- Deciduous and mixed forest
- Highland

SKILLBUILDER: Interpreting Maps

❶ **LOCATION** Which countries have only one type of vegetation?

❷ **REGION** Which areas of South Asia receive the most rainfall?

this area, the Thar Desert, gets very little rain—averaging 10 inches a year. The tropical wet zone is found along the western and eastern coasts of India and in Bangladesh. Temperatures are high, and rainfall is heavy. In fact, Cherrapunji in northeastern India holds the world's record for rainfall in a month—366 inches. Southern Sri Lanka also has a tropical wet climate, while the north is tropical wet and dry.

MONSOONS AND CYCLONES Although climate varies in South Asia, the region as a whole is greatly affected by **monsoons,** or seasonal winds. Each year, from October through February, dry winds blow across South Asia from the northeast. From June through September, the winds blow in from the southwest, bringing moist ocean air. Heavy rains fall, especially in the southwestern and Ganges Delta portions of South Asia. The illustration on page 598 shows how the monsoons blow across the region.

This rainfall is crucial to life on the subcontinent. Yet, the monsoons can cause severe hardship for millions, especially those living in the lowlands of India and Bangladesh. The monsoons also are highly unpredictable. Some areas may get too little rain, while others get too much. The monsoons are a sometimes beneficial, sometimes difficult feature of life in South Asia. Ⓐ

The most extreme weather pattern of South Asia is the **cyclone,** a violent storm with fierce winds and heavy rain. Cyclones are most destructive in Bangladesh, a low-lying coastal region where high waves can swamp large parts of the country. A severe cyclone can cause

Geographic Thinking

Seeing Patterns
Ⓐ How are the monsoons both beneficial and destructive to South Asia?

SOUTH ASIA

Climate and Vegetation **557**

widespread damage and kill thousands of people. In the Disasters! feature on pages 578–579, you will read about a cyclone that killed more than 300,000 in 1970.

Vegetation: Desert to Rain Forest

Plant life in South Asia varies according to climate and altitude. As you can see on the map on page 557, vegetation ranges from desert shrub and temperate grasslands to dense forests in the wettest areas.

VEGETATION ZONES The most forested parts of South Asia lie within the tropical wet zone, particularly the western coast of India and southern Bangladesh. Lush rain forests of teak, ebony, and bamboo are found there, along with mangroves in the delta areas. In the highland zone, which includes northern India, Nepal, and Bhutan, there are forests of pine, fir, and other evergreens. The river valleys and foothills of the humid subtropical zone have forests of sal, oak, chestnut, and various palms. But deforestation is a problem everywhere. For example, less than one-fifth of India's original forests remain. Cutting down forests has caused soil erosion, flooding, climate changes, and lost wildlife habitats.

In the semiarid areas of South Asia, such as the Deccan Plateau and the Pakistan-India border, there is less vegetation. The main plant life is desert shrubs and grasses. The driest areas, like the Thar Desert, have little plant life, and as a result, few people live there. The tropical wet and dry climate of northern Sri Lanka produces both grasses and trees. How South Asians interact with their environment will be discussed in the next section.

Geographic Thinking

Using the Atlas
B Use the atlas on pages 543 and 547. What is the average population density in the Thar Desert?

SECTION 2 Assessment

1 Places & Terms

Explain the importance of each of the following places and terms.

• monsoon

• cyclone

2 Taking Notes

PLACE Review the notes you took for this section.

Climate and Vegetation	

• How many different climate zones does South Asia have?

• What percentage of India's original forest remains today?

3 Main Ideas

a. In what part of South Asia is there a desert climate?

b. What are monsoons, and when do they affect South Asia?

c. Where are South Asia's tropical rain forests located?

4 Geographic Thinking

Making Inferences What might be some of the long-term effects of deforestation on life in South Asia? **Think about:**

• soil erosion and flooding

• climate changes

• lost wildlife habitats

RESEARCH LINKS CLASSZONE.COM

GeoActivity

SEEING PATTERNS Do more research on the different trees that grow in South Asia, such as teak, ebony, and bamboo. Create a **sketch map** of the region that shows where these various trees grow.

Reading a Weather Map

Suppose you have decided to take a trip to South Asia and want to know what the weather in the area you are going to visit will be like. To see what the weather is predicted to be for the next several days, you would look at a weather map. Most daily newspapers and news broadcasts show weather maps for a region or a country every day.

THE LANGUAGE OF MAPS A **weather map** shows weather conditions and patterns for a specific area at a point in time. Weather maps show temperatures, precipitation, weather fronts (rapid changes in weather), and air pressure. The weather map below shows weather conditions in South Asia on a typical day during the winter monsoon season—February 21, 2001.

Weather Map of South Asia

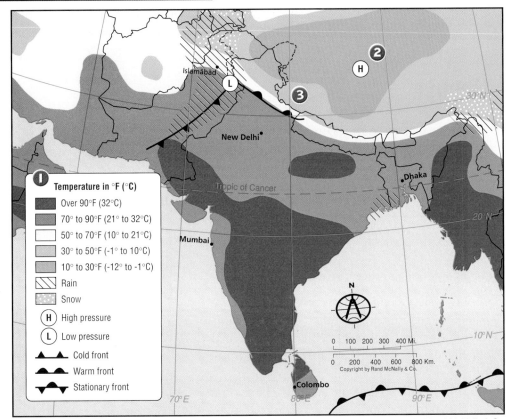

Temperature in °F (°C)

- Over 90°F (32°C)
- 70° to 90°F (21° to 32°C)
- 50° to 70°F (10° to 21°C)
- 30° to 50°F (-1° to 10°C)
- 10° to 30°F (-12° to -1°C)
- Rain
- Snow
- (H) High pressure
- (L) Low pressure
- Cold front
- Warm front
- Stationary front

0 100 200 300 400 Mi.
0 200 400 600 800 Km.
Copyright by Rand McNally & Co.

Copyright by Rand McNally & Co.

❶ The key shows colors and patterns that are used to indicate temperatures and precipitation. The temperatures are shown in both Fahrenheit and Celsius.

❷ Letter symbols on the map represent high and low pressure systems. Air pressure is the force of the air pressing down on the earth's surface.

❸ These symbols show weather fronts. In addition to showing whether a warm front or a cold front is approaching an area, the symbols show in which direction the front is moving.

Map and Graph Skills Assessment

1. Drawing Conclusions
Which South Asian cities are having temperatures over 70°F?

2. Making Comparisons
Which area of South Asia would have the most pleasant weather conditions for a visitor at the time?

3. Making Inferences
Judging from the map, will the weather in northwestern India stay the same or change?

Human–Environment Interaction

Main Ideas
- Rivers play a central role in the lives of South Asians.
- Water pollution and flooding pose great challenges to South Asian countries.

Places & Terms

Hinduism	storm surge
Ganges River	estuary

CONNECT TO THE ISSUES
POPULATION The large population of South Asia is in danger of using up the region's water resources.

A HUMAN PERSPECTIVE <u>Hinduism</u> is the religion of most Indians. During one Hindu religious festival, millions of Indians gather near the city of Allahabad, where the Ganges and Yamuna rivers meet. A temporary tent city goes up, complete with markets, temples, and teahouses. People visit the market stalls and pray at the temples. They also watch plays based on Hindu myths and legends.

Mainly, though, the Hindus wait for the appointed moment when they will wade into the Ganges and wash their sins away in its holy waters. To Hindus, the **Ganges River** is not only an important water resource, but it is also a sacred river. It is the earthly home of the Hindu goddess Ganga.

Living Along the Ganges

The Ganges is the most well-known of all the South Asian rivers. It flows more than 1,500 miles from its source in a Himalayan glacier to the Bay of Bengal. Along the way, it drains a huge area nearly three times the size of France. This area is home to about 350 million people. Although it is shorter than both the Indus and Brahmaputra rivers, the impact of the Ganges on human life in the region is enormous.

A SACRED RIVER The Ganges is extremely important for the livelihood of Indians. It provides water for drinking, farming, and transportation. Just as important, though, is the spiritual significance of the river. The Ganges is known in India as *Gangamai,* which means "Mother Ganges." In Bangladesh, where the Ganges joins the Brahmaputra, the river is called the Padma. According to Hindu beliefs, the Ganges is a sacred river that brings life to its people. As you read above, the Hindus worship the river as a goddess, and they believe its waters have healing powers.

Many temples and sacred sites line the banks of the Ganges. In some places, wide stone steps lead down to the water. Pilgrims come from all parts of the world to drink and bathe in its waters. They also come to scatter the ashes of deceased family members on the river.

At Varanasi (shown at right), one of the most sacred sites on the Ganges, thousands of people gather every day. As the sun rises, Hindu pilgrims enter the water for purification and prayer. They float baskets of flowers and burning candles on the water, as bells ring and trumpeters blow on conch shells. It is a daily celebration of their faith in the Ganges and its sacred waters.

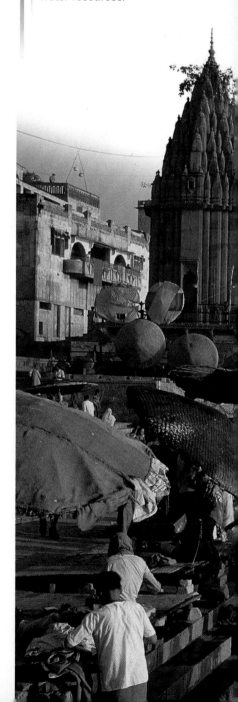

Pakistan and Bangladesh

Main Ideas

• Pakistan and Bangladesh are Muslim countries formed as a result of the partition of British India.

• Both Pakistan and Bangladesh have large populations and face great economic challenges.

Places & Terms

Indus Valley civilization

partition

Kashmir

microcredit

entrepreneur

Ramadan

CONNECT TO THE ISSUES

EXTREME WEATHER Bangladesh is severely affected by seasonal monsoons and cyclones.

A HUMAN PERSPECTIVE Some workers in the port of Chittagong, Bangladesh, have an unusual job. They are ship breakers. When ocean-going ships reach the end of their useful life, they take their last voyage to Chittagong. There, ship breakers wait on the beach with sledgehammers, crowbars, torches, and wrenches. They attack each ship, tearing it apart piece by piece. Within weeks, they can dismantle a ship. Then, they sell its scrap metal for recycling purposes. The job doesn't pay very well, but it is necessary work for the shipping industry, the workers, and the Bangladeshi economy.

New Countries, Ancient Lands

Like India, Pakistan and Bangladesh are young countries with an ancient history and with rapidly growing populations. They, too, are striving to make their way in the modern world.

EARLY HISTORY The largest of the world's first civilizations arose in what is now Pakistan. The **Indus Valley civilization** began around 2500 B.C. It featured well-planned cities like Harappa and Mohenjo-Daro, which had brick buildings (shown below) and sophisticated sanitation systems. The map on page 544 depicts the extent of the civilization at the height of its power. It fell around 1500 B.C., and the Aryans invaded soon after. Later on, the Mauryan, Gupta, and Mughal empires ruled the territory that included modern Pakistan and Bangladesh. The British were the next to take control of the region.

PLACE The ruins of Mohenjo-Daro, one of the great cities of the ancient Indus Valley civilization, lie on the Indus River in south-central Pakistan.

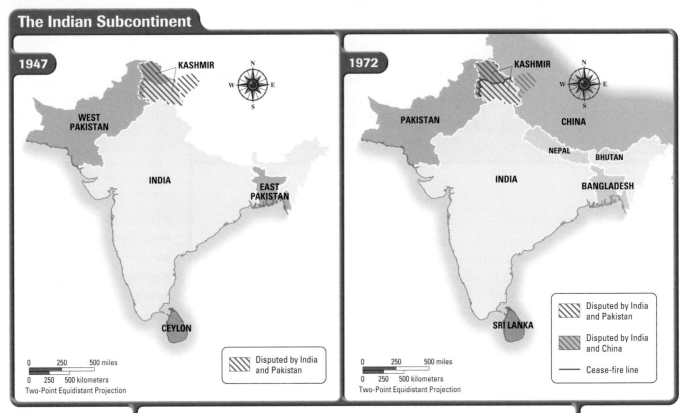

The Indian Subcontinent

1947

KASHMIR

WEST PAKISTAN

INDIA

EAST PAKISTAN

CEYLON

| 0 | 250 | 500 miles |
| 0 | 250 | 500 kilometers |

Two-Point Equidistant Projection

Disputed by India and Pakistan

1972

KASHMIR

PAKISTAN

CHINA

NEPAL

BHUTAN

INDIA

BANGLADESH

SRI LANKA

| 0 | 250 | 500 miles |
| 0 | 250 | 500 kilometers |

Two-Point Equidistant Projection

Disputed by India and Pakistan

Disputed by India and China

Cease-fire line

SKILLBUILDER: Interpreting Maps

❶ **PLACE** What had happened to the territory of Kashmir by 1972?

❷ **REGION** What other changes had taken place in South Asia from 1947 to 1972?

PARTITION AND WAR The end of British rule in 1947 brought the **partition,** or division, of British India. Two new countries were created— India (predominantly Hindu) and mainly Muslim Pakistan (separated into West Pakistan and East Pakistan). Partition led to much violence between Muslims and Hindus. About one million people died in the conflict. Another 10 million fled across national borders. Muslims in India moved to Pakistan, while Hindus in Pakistan crossed into India.

West Pakistan and East Pakistan shared a religious bond, but ethnic differences and their 1,100-mile separation eventually drove them apart. The people of East Pakistan began to call for their own state. But the government in West Pakistan opposed such a move. Civil war broke out in 1971. That year, with help from India, East Pakistan won its independence as Bangladesh.

MILITARY RULE Both Pakistan and Bangladesh have had political struggles since independence. Short periods of elected government have alternated with long periods of military rule. Political corruption has plagued both countries. Pakistan also has fought several destructive wars with India over the territory of **Kashmir.** These wars are discussed in the Case Study in Chapter 26. In the 1990s, both Bangladesh and Pakistan had women prime ministers, a rarity in the Muslim world.

BACKGROUND
Bangladesh means "land of the Bangla (or Bengal)-speaking people."

Struggling Economies

Pakistan and Bangladesh have large, rapidly growing populations. In fact, Bangladesh is the eighth most populous country in the world. Both have economies that depend primarily on agriculture. As in India, per capita incomes are low, and much of the population lives in poverty. Bangladesh is one of the poorest countries in the world.

SUBSISTENCE FARMING Most farmers in Pakistan and Bangladesh work small plots of land and struggle to grow enough crops to feed their families. The government has tried to help modernize farming methods, but many farmers continue to follow less productive traditional ways. Climate also hinders crop yields. Large areas of Pakistan are arid, while Bangladesh is severely affected by seasonal monsoons and cyclones.

The most productive farming areas of Pakistan are the irrigated portions of the Indus Valley. Here, farmers grow enough cotton and rice to allow for export. The farmers also produce substantial amounts of wheat for domestic consumption. The moist delta lands of Bangladesh are ideal for the cultivation of rice, the country's principal food crop. The main export crop is jute (a plant used in the production of rope, carpets, and industrial-quality sacks). Fishing, mainly for freshwater fish, is also vital to the economy of Bangladesh.

SMALL INDUSTRY Neither Pakistan nor Bangladesh is highly industrialized. Most factories are relatively small and lack the capital, resources, and markets required for expansion. Even so, both countries are trying to increase their industrial base. They have growing textile industries that provide an important source of revenue and employment. Both countries export cotton garments, and Pakistan also exports wool carpets and leather goods. ◀🅐

Geographic Thinking

Making Comparisons
🅐 How do the economies of Pakistan and Bangladesh compare with each other?

An important economic development has been the introduction of **microcredit.** This policy makes small loans available to poor **entrepreneurs,** people who start and build a business. Businesses that are too small to get loans from banks can often join forces to apply for these microloans. They then accept joint responsibility for repaying the loan. This program, begun in Bangladesh, has helped small businesses grow in South Asia and has raised living standards for many producers, especially women.

Economic Activity in Pakistan and Bangladesh

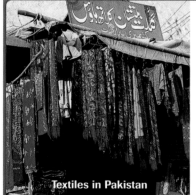

Textiles in Pakistan

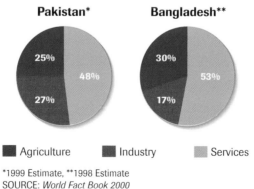

Pakistan*

25%
48%
27%

Bangladesh**

30%
53%
17%

■ Agriculture   ■ Industry   ■ Services

*1999 Estimate, **1998 Estimate
SOURCE: *World Fact Book 2000*

Fishing in Bangladesh

SKILLBUILDER: Interpreting Graphs

❶ **MAKING COMPARISONS** Which of the two countries is more industrialized?

❷ **ANALYZING DATA** In both Pakistan and Bangladesh, which economic sector employs the most people?

One Religion, Many Peoples

Most of the people of Pakistan and Bangladesh are Muslims. In both countries, Islam is an important unifying force. At the same time, ethnic differences promote cultural diversity, particularly in Pakistan.

ISLAMIC CULTURE Islam has long played an important role in Pakistan and Bangladesh. Both lands were key parts of the Muslim Mughal Empire that ruled the Indian subcontinent for centuries, and their cultures bear the stamp of Islam. The faithful observe Islamic customs. These include daily prayer and participation in **Ramadan,** a month-long period of fasting from sunrise to sunset. Mosques in both countries are often large and impressive structures.

The two countries differ somewhat in their Islamic practices, however. In general, Pakistan is stricter in imposing Islamic law on its citizens. For example, many Pakistanis follow the custom of *purdah,* the seclusion of women. This custom prevents women from having contact with men who are not relatives. When women appear in public, they must wear veils. In Bangladesh, purdah is much less common and religious practices are less strict.

REGION Most Pakistanis are Sunni Muslims. Here, men attend a Muslim prayer service in a mosque in Karachi.

ETHNIC DIVERSITY Pakistan is also more ethnically diverse than Bangladesh. Pakistan has five main ethnic groups—Punjabis, Sindhis, Pathans, Muhajirs, and Balochs. Each group has its own language. The Punjabis make up more than half of the population. Each group has its own regional origins within the country except for the Muhajirs, who migrated from India as a result of the partition in 1947. To avoid favoring one region or group over another, the government chose Urdu—the language of the Muhajirs—as the national language. Today, most Pakistanis understand Urdu, even though they may use another language as their primary language.

In contrast, the people of Bangladesh are mainly Bengalis. Bengal is the historic region that includes Bangladesh (once known as East Bengal) and the Indian state of West Bengal. Bengalis speak a language based on Sanskrit, the ancient Indo-Aryan language. Bangladesh also has a small population of Urdu-speaking Muslims and various non-Muslim tribal groups. About 10 percent of the population are Hindus.

BACKGROUND Punjabi is the principal spoken language of Pakistan because the majority of Pakistanis are Punjabis. Arabic is a secondary language for Muslim Pakistanis.

Modern Life and Culture

As in India, life in Pakistan and Bangladesh revolves around the family. Arranged marriages are common, and families tend to be large. Most people live in small villages, in simple homes made of such materials as sun-baked mud, bamboo, or wood. The large cities are busy places, crowded with traffic and pedestrians. People in both countries enjoy sports such as soccer and cricket, and also enjoy going to see movies.

A LOVE OF POETRY Poetry is a special interest in both Pakistan and Bangladesh, where the tradition of oral literature is strong. Many Pakistanis memorize long poems and can recite them by heart. Poets are popular figures, and poetry readings—called *mushairas*—can draw thousands of people, much like a rock concert does in some countries.

The greatest literary figure in Bangladesh is the poet Rabindranath Tagore, who won the Nobel Prize for Literature in 1913. Although Tagore was born in Calcutta (now Kolkata), India, he wrote about the Ganges and his Bengal homeland. Bangladesh adopted his song, "My Golden Bengal," as its national anthem.

MUSIC AND DANCE Music and dance are also important forms of expression in Bangladesh and Pakistan. Both countries share music traditions similar to those of India. Folk music of various types is popular in cities and in rural areas. *Qawwali*—a form of devotional singing performed by Muslims known as Sufis—is famous not only in South Asia but also in parts of Europe and the United States. Bangladesh also has a long tradition of folk dances, in which elaborately costumed dancers act out Bengali myths, legends, and stories. ◀**B**

You have been reading about Pakistan and Bangladesh, India's western and eastern neighbors. Next, you will learn about India's northern neighbors, Nepal and Bhutan.

Geographic Thinking ◀

Seeing Patterns
B What roles do music and dance play in the lives of the people of Pakistan and Bangladesh?

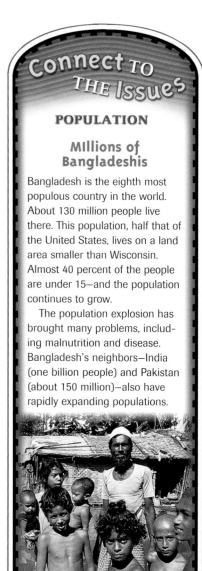

Connect TO THE Issues

POPULATION

Millions of Bangladeshis

Bangladesh is the eighth most populous country in the world. About 130 million people live there. This population, half that of the United States, lives on a land area smaller than Wisconsin. Almost 40 percent of the people are under 15—and the population continues to grow.

The population explosion has brought many problems, including malnutrition and disease. Bangladesh's neighbors—India (one billion people) and Pakistan (about 150 million)—also have rapidly expanding populations.

Assessment

1 Places & Terms

Identify each of the following places and terms.

- Indus Valley civilization
- partition
- Kashmir
- microcredit
- entrepreneur
- Ramadan

2 Taking Notes

PLACE Review the notes you took for this section.

Pakistan and Bangladesh

South Asia

- How were the countries of Pakistan and Bangladesh formed?
- What role does farming play in the economies of Pakistan and Bangladesh?

3 Main Ideas

a. What have been some of the problems for Pakistan and Bangladesh since they were formed?

b. What role does Islam play in Pakistan and Bangladesh?

c. How would you describe Pakistan's ethnic makeup?

4 Geographic Thinking

Making Comparisons How do Pakistan and Bangladesh differ in their Islamic practices? **Think about:**

- the treatment of women
- how much of Pakistan follows strict Islamic law

RESEARCH LINKS
CLASSZONE.COM

SOUTH ASIA

GeoActivities

MAKING COMPARISONS Review the information about Islam on page 576. Then use the Internet or an encyclopedia to compare Islam in Pakistan or Bangladesh with a Muslim country in either Africa or Southwest Asia. Create a **chart** comparing the two countries using such topics as treatment of women, eating practices, and how strictly a country enforces Islamic law.

Disasters!

The Cyclone of 1970

On November 13, 1970, a violent tropical storm struck Bangladesh, bringing death and destruction in its wake. Hundreds of thousands of people and their homes, crops, and animals were swept away in the fury of the 20th century's worst tropical storm. The cyclone's winds, rains, and floods claimed an estimated 300,000 to 500,000 lives. Also, approximately one million were left homeless, roughly 80 percent of the rice crop was lost, and about 70 percent of the country's fishing boats were wrecked. More than any other South Asian country, Bangladesh—with its low-lying coastal plain—suffers from these frequently occurring storms.

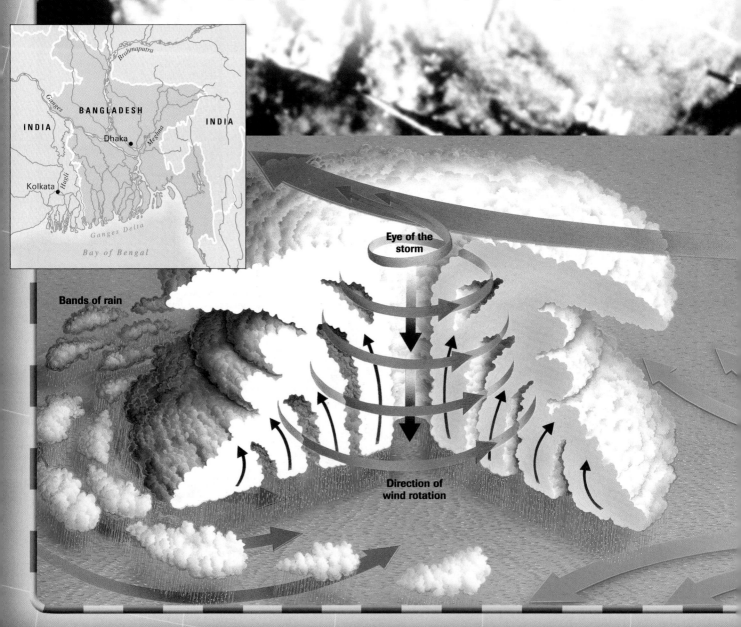

Eye of the storm

Bands of rain

Direction of wind rotation

The damage inflicted on this village in Bangladesh in 1991 is typical of the destructive force of a cyclone's winds and the torrential rains and floods that are a part of this weather system.

Concrete shelters constructed on stilts, as shown here, and reinforced school buildings are refuges from high floodwaters and winds that can knock down all but the strongest buildings.

GeoData

TROPICAL STORMS
Violent tropical storms are called cyclones in the Indian Ocean, typhoons in the northwestern Pacific Ocean, and hurricanes in the Atlantic Ocean. These storms:

* develop over tropical waters in the late summer and fall when ocean temperatures are warmest
* usually begin as a cluster of thunderstorms that start to spiral and then form a single violent storm
* may be as wide as 675 miles
* have winds that range from 75 to 150 miles per hour
* generally last a week but some may take two or three weeks to die out
* produce heavy flooding that is the cause of most of the destruction and deaths
* inflict most of their damage along coastlines

OTHER BANGLADESHI STORMS
* May 28–29, 1963—22,000 deaths
* May 11–12, 1965—17,000 deaths
* June 1–12, 1965—30,000 deaths
* April 30, 1991—139,000 deaths

Nepal and Bhutan

A HUMAN PERSPECTIVE In the novel *Lost Horizon,* James Hilton described an imaginary mountain valley called Shangri-La, hidden high in the Himalayas. He wrote, "The floor of the valley, hazily distant, welcomed the eye with greenness; sheltered from winds . . . completely isolated by the lofty and sheerly unscalable ranges on the further side." Shangri-La was an earthly paradise: a land of peace, harmony, and beauty, where hunger, disease, and war did not exist. Hilton located this mythical land somewhere in Tibet, but it could just as easily have been in Nepal or Bhutan. Although neither of these countries is a paradise, both are remote lands of great beauty and peace.

Mountain Kingdoms

Nepal and Bhutan share a number of important characteristics. Both are located in the Himalayas, a factor that has had a great impact on their history and economic development. Both also are kingdoms with strong religious traditions.

GEOGRAPHIC ISOLATION The main geographic feature of Nepal and Bhutan is their mountainous landscape. Each country consists of a central upland of ridges and valleys leading up to the high mountains, with a small lowland area along the Indian border. The towering, snow-capped Himalayas run along the northern border with China. They are craggy and forbidding and have steep mountain passes and year-round ice fields. The world's tallest mountain peak, Mt. Everest, is located there.

The rugged landscape of Nepal and Bhutan has isolated the two countries throughout their histories. Their mountainous terrain and landlocked location—neither country has access to the sea—made them hard to reach and difficult to conquer and settle. China controlled Bhutan briefly in the 18th century. In the 19th century, Great Britain had influence over both countries because of its control of neighboring India. But Nepal and Bhutan generally remained independent and isolated. In fact, until the past few decades, foreigners rarely entered either country.

EVOLVING MONARCHIES For much of their history, Nepal and Bhutan were split into small religious kingdoms or ruling states. Hindu kings ruled in Nepal, while Buddhist priests controlled Bhutan. In time, unified kingdoms emerged in both countries, led by hereditary monarchs who passed the throne on to their heirs.

Today, the governments of both Nepal and Bhutan are **constitutional monarchies**—kingdoms in which the ruler's powers are limited by a

REGION Richly decorated cloths that display Buddhist religious symbols, such as the cloth shown below, have covered the thrones of Bhutanese rulers. **Why might there be religious symbols on a throne cloth used by secular rulers?**

PLACE A blend of
the old and the new is
evident in the
architecture of this
square in Kathmandu,
Nepal's capital city.
**Why might this rich
cultural tradition
make Kathmandu
attractive to tourists?**

constitution. In Bhutan, the king is still the supreme ruler, while in Nepal the king shares power with an elected parliament. Both governments face difficult political challenges, including the need to balance the interests of their two powerful neighbors, China and India. Both countries also face difficult economic challenges.

Developing Economies

Decades of isolation and difficult topography have limited economic development in Nepal and Bhutan. Now each country is trying to find effective ways to promote economic growth.

LIMITED RESOURCES Nepal and Bhutan are poor countries with economies based mainly on agriculture. Because of the mountainous terrain, neither country has much land suitable for cultivation. Most farm plots are small, soils are poor, and erosion is a problem. Farmers create terraces on the mountainsides to increase the amount of farmland and limit soil loss, a process you read about in Chapter 9. Common farm products include rice, corn, potatoes, and wheat. Common livestock are cattle, sheep, and yaks—longhaired animals related to the ox. In Bhutan, the government has promoted the growing of fruit for export and has tried to improve farming practices.

The timber industry is very important to both countries, although deforestation is a problem. The forests of Nepal are being cut down at a rate of about 1 percent a year. But some valuable timberlands remain. Around 70 percent of Bhutan is still forested. A growing manufacturing sector of the economy includes wood products, food processing, and cement production. Most trade for both countries is with India. ◀A

INCREASING TOURISM One of the fastest growing industries in Nepal is tourism. Tourists come from around the world to visit the valley of Kathmandu, the capital, and to climb the Himalayas. Hotels and restaurants, transportation, and other services have grown to meet the needs of the tourist industry. But tourism is a mixed blessing. It has

**Geographic
Thinking**◀

**Making
Comparisons**
▶A What activities
are important to
the economies
of Nepal and
Bhutan?

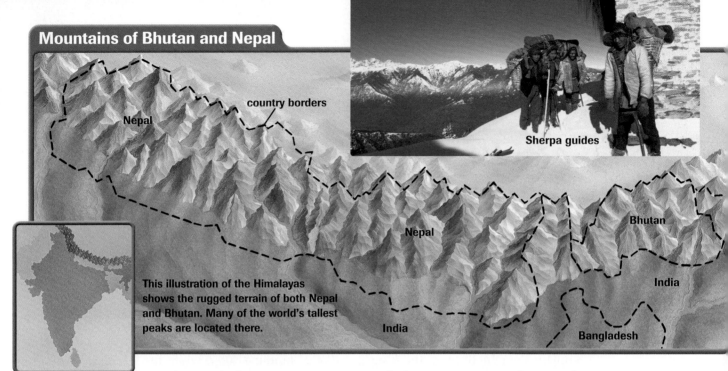

Mountains of Bhutan and Nepal

Nepal

country borders

Sherpa guides

Nepal

Bhutan

India

This illustration of the Himalayas shows the rugged terrain of both Nepal and Bhutan. Many of the world's tallest peaks are located there.

India

Bangladesh

HUMAN–ENVIRONMENT INTERACTION The Sherpa are known for their mountaineering skills and their ability to carry heavy loads at high altitudes.
Why might mountain climbers seek out the Sherpas as guides and porters?

damaged the environment, particularly on mountain slopes, where increased trash and pollution have been most noticeable.

Bhutan, which offers many of the same natural attractions as Nepal, has taken a different approach to tourism. Concerned about the impact of tourists on national life, Bhutan regulates the tourist industry. It allows only limited numbers of visitors and keeps some areas of the country off-limits. Even so, tourism is providing increasing revenues to Bhutan and offers significant economic potential for the future.

Rich Cultural Traditions

Visitors to Nepal and Bhutan come not only for the spectacular mountain scenery but also for a glimpse of the rich cultural traditions of the Himalayan people.

A MIX OF PEOPLES Various ethnic groups inhabit the Himalayan region. In Nepal, the majority of the people are Indo-Nepalese Hindus whose ancestors came from India many centuries ago. These groups speak Nepali, a variation of Sanskrit, an ancient Indo-Aryan language. Nepal also has a number of groups of Tibetan ancestry. Among them are the **Sherpas.** These people from the high Himalayas are the traditional mountain guides of the Everest region.

The main ethnic group in Bhutan is the Bhote, who also trace their origins to Tibet. Most Bhotes live in two-story houses made of wood and stone. The families live on the second floor, while the first floor is reserved for livestock. Bhutan also has a sizable Nepalese minority in the southern lowlands. The Nepalese have preserved their language and customs, even though the government of Bhutan has tried to assimilate them into national life.

RELIGIOUS CUSTOMS Religion is a powerful force in both Nepal and Bhutan. Although the great majority of Nepalese are Hindus, Buddhism also has deep roots in Nepal. The founder of Buddhism, **Siddhartha Gautama,** known as the Buddha, was born on the borders of present-day

BACKGROUND Another Nepalese people, the Gurkhas from the valleys west of Kathmandu, are known as fierce fighters. They have been recruited since the mid-19th century to serve in the British and Indian armies.

declining, Sri Lanka is still one of the world's leading tea-producing countries. Although manufacturing is increasing, other sectors of the Sri Lankan economy are less important. Overcutting has damaged the timber industry, and the fishing and mining industries are relatively small. One exception is gem mining. Sri Lanka is famous for its gemstones—including sapphires, rubies, and topaz.

The economy of the Maldives is different from the economies of the rest of South Asia. Farming is limited by a lack of land, and most food has to be imported. Fishing—for tuna, marlin, and sharks—was long the main economic activity. It still provides one-fourth of the jobs and a large share of the country's export earnings. But it has been replaced in importance by tourism. The islands' beautiful beaches, coral reefs, and impressive marine life draw visitors from around the world.

TOUGH CHALLENGES Until the 1980s, tourism was also growing in Sri Lanka. Then civil war began, and the tourist industry collapsed. Warfare has also disrupted other economic activities and damaged the country's infrastructure—its roads, bridges, power systems, and other services. Until peace returns to Sri Lanka, the economy is likely to struggle. While the Maldives is at peace, it faces a challenge of a different kind: global warming. The islands lie very low in the water, and any rise in sea level—caused by melting of the polar icecaps—could flood them completely. Scientists say this could happen by the end of the 21st century.

In this chapter, you read about modern life in South Asia. In the next chapter, you will read about issues facing South Asians.

LOCATION Tourist resorts in the Maldives are built only on previously uninhabited islands.
What might be a reason for locating the resorts on such islands?

Assessment

1 Places & Terms

Identify these terms and explain their importance in the region.

- Sinhalese
- Tamils
- sultan

2 Taking Notes

MOVEMENT Review the notes you took for this section.

- How were the islands of the Maldives settled?
- Where do the different ethnic groups in Sri Lanka live?

3 Main Ideas

a. What happened between the Sinhalese and the Tamils after Sri Lanka gained independence?

b. What are some of the aspects of cultural life in the Maldives?

c. What are some of the economic strengths of Sri Lanka and the Maldives?

4 Geographic Thinking

Seeing Patterns How do the Maldives's 1,200 islands affect its economy? **Think about:**

- fishing for food or sport
- the number of beaches

S See Skillbuilder Handbook, page R8.

SEEING PATTERNS Review the information about tourism in the Maldives on this page. Do research on different activities for tourists and different places to visit in the country. Then create a **travel poster** advertising the Maldives as an ideal tourist destination.

SOUTH ASIA

Comparing Cultures

Musical Instruments

No one is exactly certain when or where music began to be made or what the first musical instrument was. But scholars believe that music has been part of all cultures, possibly even from earliest times. The first musical instrument may have been the human voice, which may have been used to mimic the sound of birds. Next, the human body was used to make rhythms, by clapping hands or stomping feet. When instruments began to be made, early musicians adapted available materials, such as wood and animal skins. Eventually, four basic types of instruments were developed: percussion, wind, string, and keyboard. Today, thousands of different instruments are played worldwide.

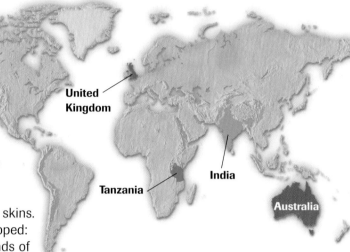

United Kingdom

India

Tanzania

Australia

The bagpipe is a wind instrument that is associated with Scotland, although it is played in other countries. It consists of an animal skin or rubberized cloth bag fitted with one or more pipes that produce a continuous flow of sound when blown.

The drum is a percussion instrument from Africa that probably was made first from wood or stone. It is played by striking with hands or other objects. These drums are made of skins stretched over a frame. Also shown is another percussion instrument—the xylophone.

The didgeridoo is a wind instrument played by aboriginal people in Australia. Made of bamboo or a hollow sapling, it can be as long as five feet. It is generally painted and used in ritual ceremonies.

The sitar is a stringed instrument from India. It has a wooden body and is used mainly to play classical music. Anoushka Shankar, shown here with her sitar, is the daughter of famed sitarist Ravi Shankar, who brought the instrument to the world's attention with his playing in the 1960s.

GeoActivity

FORMING A BAND
With a small group, research other musical instruments. Plan a band that includes at least one of each of the four types of instruments. Then create **a multimedia presentation.**

• Provide visuals of each instrument.

• Write a description of each instrument's sound.

• Make an audiotape that has the sound of each instrument and play it in class.

 **RESEARCH LINKS**
CLASSZONE.COM

GeoData

OTHER INSTRUMENTS

ASIA
• Empty conch shells with broken tips give off a loud sound when blown and have been used in ceremonies for centuries in many regions, including the islands of Polynesia.

EUROPE
• The organ is the oldest keyboard instrument and was found in ancient Greece more than 2,000 years ago. It gave birth to other keyboard instruments such as the harpsichord, clavichord, and piano.

THE AMERICAS
• Native American cultures have strongly emphasized the voice in making music.

AFRICA
• Wall paintings in 4,000-year-old tombs in Egypt show musicians playing lutes.

• Some African cultures still use a stone gong—a hanging stone that gives off a sound when struck.

VISUAL SUMMARY
HUMAN GEOGRAPHY OF SOUTH ASIA

Subregions of South Asia

○ India
- India is the largest country in South Asia and dominates the region.
- India is the world's largest democracy; Hinduism is its principal religion.

● Pakistan and Bangladesh
- Pakistan and Bangladesh were both eventually formed after the partition of India.
- Farming is the main source of people's livelihoods.
- Islam is the primary cultural force in those countries.

○ Nepal and Bhutan
- Nepal and Bhutan developed in relative isolation because of the Himalaya Mountains.
- Nepal has a religious mix of both Hindus and Buddhists, while Bhutan is a predominantly Buddhist country.

● Sri Lanka and the Maldives
- Sri Lanka contains a variety of ethnic and religious groups, including Sinhalese Buddhists, Tamil Hindus, and Muslims.
- Sri Lanka's economy is based on farming and gem mining, while the Maldives relies on fishing and tourism.

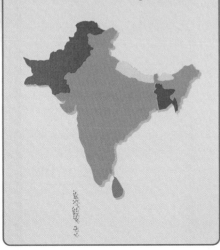

Reviewing Places & Terms

A. Briefly explain the importance of each of the following.

1. Mughal Empire
2. nonviolent resistance
3. caste system
4. partition
5. Kashmir
6. microcredit
7. Sherpa
8. mandala
9. Sinhalese
10. Tamils

B. Answer the questions about vocabulary in complete sentences.

11. How did the great Indian leader Mohandas Gandhi protest British control of India?
12. Over which territory have India and Pakistan fought several wars?
13. What financial aid do poor South Asian entrepreneurs seek?
14. How was the country of West and East Pakistan formed after Indian independence?
15. What did the Muslims establish in India during the 16th century?
16. What people arrived in Sri Lanka from southern India and occupied the northern portion of the island?
17. Who created an advanced civilization in Sri Lanka and built sophisticated irrigation systems?
18. Who guides mountain climbers in the Everest region?
19. What are geometric designs that are symbols of the universe and aid in meditation?
20. What is a Hindu system of social classes?

Main Ideas

India (pp. 567–572)

1. How did Britain gain control of India?
2. What are the major economic activities in India?
3. What are the major languages of India?

Pakistan and Bangladesh (pp. 573–579)

4. What are some of the characteristics of the Indus Valley civilization?
5. What manufactured products are produced in Pakistan and Bangladesh?
6. What type of literature is important in Pakistan and Bangladesh?

Nepal and Bhutan (pp. 580–583)

7. What are some of the groups of people that live in Nepal?
8. What are some important religious customs in Bhutan?

Sri Lanka and the Maldives (pp. 584–589)

9. What are the two major ethnic groups in Sri Lanka and where did they come from?
10. What are some of the challenges facing the economies of Sri Lanka and the Maldives?

Critical Thinking

1. Using Your Notes

Use your completed chart to answer these questions.

a. What role does agriculture play in the economies of the South Asian countries?

b. What are the major religions practiced in the region?

2. Geographic Themes

a. **HUMAN-ENVIRONMENT INTERACTION** How did the mountainous terrain and the landlocked location of Nepal and Bhutan affect their development?

b. **LOCATION** How do the landforms and location of the Maldives ensure that its economy is different from other South Asian countries?

3. Identifying Themes

What groups of people first populated the Indian subcontinent and eventually helped to populate all of South Asia? Which of the five themes apply to this situation?

4. Making Comparisons

How do Pakistan and Bangladesh differ in their practice of Islam?

5. Determining Cause and Effect

What are some of the reasons for the ongoing violence between the Tamils and the Sinhalese in Sri Lanka?

Additional Test Practice, pp. S1–S37

TEST PRACTICE
CLASSZONE.COM

Geographic Skills: Interpreting Maps

Languages of South Asia

Use the map at right to answer the following questions.

1. **LOCATION** How many major languages are spoken in South Asia?

2. **REGION** Which language group is the most commonly spoken?

3. **MOVEMENT** How might the number of languages in South Asia affect its developing economies?

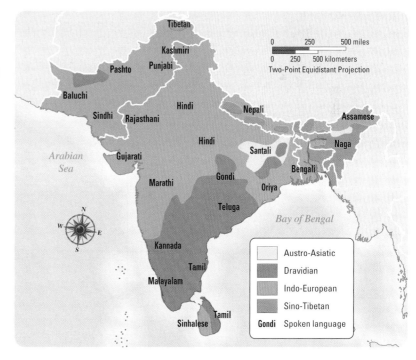

Choose a country in South Asia in which more than one language is spoken, and prepare a chart showing the number of people speaking each language. Use library references or the Internet for your research.

INTERNET ACTIVITY

Use the links at **classzone.com** to do research on the people of one South Asian country. Look for such information as life expectancy, religions, ethnic groups, literacy rates, and per capita income.

Writing About Geography Write a report about your findings. Use standard grammar, spelling, sentence structure, and punctuation in your report. List the Web sites that you used as sources.

SOUTH ASIA

SECTION 1
Population Explosion

SECTION 2
Living with Extreme Weather

CASESTUDY

TERRITORIAL DISPUTE

For more on these issues in
South Asia . . .

CURRENT EVENTS
CLASSZONE.COM

**Kolkata is one of
India's most densely
populated cities.**

GeoFocus

How can people and governments work together to solve problems?

Taking Notes In your notebook, copy a
cause-effect chart like the one shown below
for each issue. Then take notes on the causes
and effects of some aspect of each issue.

	Causes	Effects
Issue 1: Population		
Issue 2: Extreme Weather		
Case Study: Territorial Dispute		

Population Explosion

How can South Asia's population growth be managed?

Main Ideas

• Explosive population growth in South Asia has contributed to social and economic ills in the region.

• Education is key to controlling population growth and improving the quality of life in South Asia.

Places & Terms

basic necessities

illiteracy

The Voyageur Experience in World Geography

India: Population and Resources

A HUMAN PERSPECTIVE On May 11, 2000, at 5:05 A.M., a baby girl was born in a New Delhi hospital. Her parents named her Astha, which means "faith" in the Hindi language. Ordinarily, Astha's birth would not have made news. After all, an estimated 42,000 babies are born in India every day—15,330,000 each year. Astha, however, was special. With this child's birth, the population of India officially hit 1 billion. It was the second country to reach a billion in population; China was the first.

Growing Pains

India's milestone was a mixed blessing. Its population at the beginning of the 21st century is growing so quickly that many of its citizens lack life's **basic necessities**—food, clothing, and shelter. The question for India, and for South Asia as a whole, is how to manage population growth so that economic development can continue.

POPULATION GROWS When India gained its independence from Britain in 1947, the population stood at 300 million. By 2000, the population had more than tripled. India's population is so large that even an annual growth rate of less than 2 percent is producing a population explosion. Unless that growth slows down, in 2045, India will be home to more than 1.5 billion people—all living in a land about one-third the size of the United States. India will be the most populous country in the world, surpassing China.

India is not alone in its skyrocketing population. In fact, of the 10 most populous countries in the world in 1998, three were located in South Asia: India, Pakistan, and Bangladesh. South Asia is home to 22 percent of the world's population. But these people live on less than 3 percent of the world's land area.

INADEQUATE RESOURCES As South Asia's population has increased, regional governments have found it more and more difficult to meet the needs of their people. Widespread poverty and **illiteracy,** the inability to read or

REGION The homeless poor are a common sight in many of India's large cities, such as Mumbai, pictured below. **What might be some ways in which the homeless can be helped?**

SOUTH ASIA

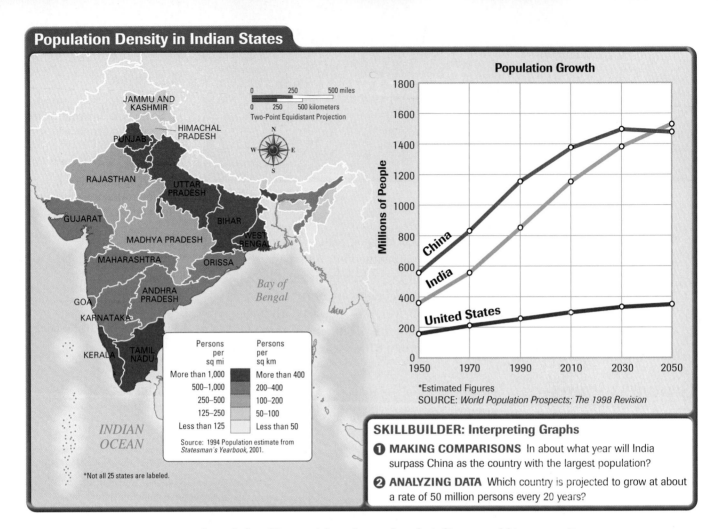

Population Density in Indian States

Population Growth

*Estimated Figures
SOURCE: *World Population Prospects; The 1998 Revision*

Persons per sq mi | Persons per sq km
More than 1,000 | More than 400
500–1,000 | 200–400
250–500 | 100–200
125–250 | 50–100
Less than 125 | Less than 50

Source: 1994 Population estimate from *Statesman's Yearbook*, 2001.

*Not all 25 states are labeled.

SKILLBUILDER: Interpreting Graphs

❶ **MAKING COMPARISONS** In about what year will India surpass China as the country with the largest population?

❷ **ANALYZING DATA** Which country is projected to grow at about a rate of 50 million persons every 20 years?

write, have left millions without hope that their lives would improve. Poor sanitation and the lack of health education have led to outbreaks of disease, which have overwhelmed the region's limited health care systems.

Officials estimate that in order to keep pace with population growth, India will have to do the following *every year:* build 127,000 new village schools, hire nearly 400,000 new teachers, construct 2.5 million new homes, create 4 million new jobs, and produce an additional 6 million tons of food.

Managing Population Growth

South Asia has struggled for decades to find solutions to its population explosion. But efforts have met with only limited success.

SMALLER FAMILIES Today, India spends much of its nearly $1 billion annual health-care budget encouraging Indians to have smaller families. "Let's have small families for a stronger India" is one of the slogans of the campaign. For many reasons, however, these programs have had only limited success. Indian women usually marry before age 18 and start having babies early. Also, for the very poor, children are a source of income. They can beg for money in the streets as early as their third birthday and can work the fields not too many years later. Ⓐ

For many Indians, children represent security in old age. The more children a family has, the more likely someone will be around to take care of the parents when they are elderly. Also, the infant mortality rate

Geographic Thinking

Seeing Patterns
Ⓐ How might smaller families affect India's economic development?

is very high in South Asia—around 75 per 1,000 live births compared to 7 per 1,000 in the United States. As a result, parents try to have many children to ensure that at least some will reach adulthood.

EDUCATION IS A KEY Many factors that affect population growth can be changed through education. However, South Asia's governments have a difficult task ahead of them because education funds are limited. For example, India spends less than $6 per pupil annually on primary and secondary education. (Only a small fraction of this sum is spent on girls.) By contrast, annual per pupil spending on education in the United States is $6,320—more than 1,000 times as much.

BACKGROUND
Statistics for 1997-1998 showed that about 85 percent of Indian boys aged 6 to 12 are in school, compared to about 70 percent of girls.

Education is essential to break the cycle of poverty and provide South Asians with the means to raise their standard of living. It also helps to improve the status of females by giving them job opportunities outside the home. Better health education also can reduce the need for large families by ensuring that more babies reach adulthood. The future development of South Asia depends on the success of such efforts to control population growth.

In the next section, you will learn how the people of South Asia are coping with another problem—the region's extreme weather.

HUMAN-ENVIRONMENT INTERACTION The rural poor build settlements on unused land in many cities, such as these in Ahmadabad, India.
Why might the rural poor be attracted to urban areas?

 Assessment

① Places & Terms

Explain the importance of each of the following terms and places.

• basic necessities

• illiteracy

② Taking Notes

PLACE Review the notes you took for this section.

	Causes	Effects
Issue 1: Population		

• How much did India's population grow in the second half of the 20th century?

• If this growth rate continues, what will India's population be in 2045?

③ Main Ideas

a. Why is the size of India's population a problem?

b. How has the government of India addressed population issues?

c. Why have government programs had mixed success?

④ Geographic Thinking

Making Inferences How does the population density in India compare to that in the United States? **Think about:**

• population size

• territorial size

SOUTH ASIA

RESEARCH LINKS
CLASSZONE.COM

MAKING COMPARISONS Carry out further research focused on comparing 20th-century population growth in a city in India and one in the United States. Use the data that you gather to create a **line graph** that compares population growth in these two cities.

Reading a Population Pyramid

Every nation has a certain distribution of population by age group. India, for instance, has a young population; the majority of people are under the age of 30. To show how the population of a country is distributed by age, a population pyramid is a very useful tool.

THE LANGUAGE OF GRAPHS A **population pyramid** is a type of bar graph. It shows the number or percentage of people that fall into specific age groups. It may also compare the distribution of age groups by sex, ethnic group, or some other category. The population pyramid below shows the distribution of age groups by sex in India.

Population of India, 2000

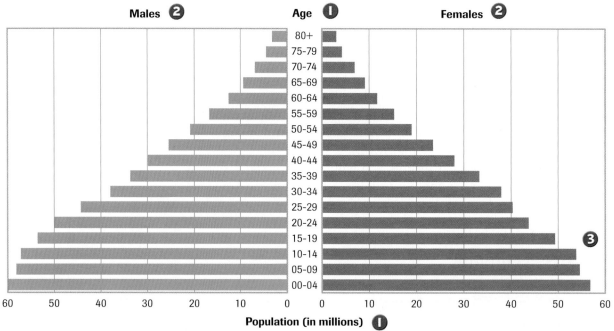

SOURCE: U.S. Census Bureau, International Data Base

❶ The horizontal axis shows population in millions. The vertical axis lists age groups.

❷ The left side of the pyramid shows the population distribution of males in India. The right side shows females.

❸ Notice that there is a steady drop in population as Indians reach their late teens. This indicates that the life expectancy of Indians is relatively short.

Map and Graph Skills Assessment

1. Analyzing Data
Find the bar on the pyramid that would be your age and sex. How many millions of persons fall into that group in India?

2. Making Comparisons
What age group is the largest? What is the largest age group by sex?

3. Making Inferences
What might be said about the male/female composition of the population of India?

2 Living with Extreme Weather

How do people cope with extreme weather?

Main Ideas
- South Asia experiences a yearly cycle of floods, often followed by drought.
- The extreme weather in South Asia leads to serious physical, economic, and political consequences.

Places & Terms

summer monsoon

winter monsoon

A HUMAN PERSPECTIVE In May 1996, a fierce tornado tore through northern Bangladesh, leaving more than 700 people dead and 30,000 injured. Winds reached speeds of 125 mph. Within 30 minutes, nearly 80 villages had been destroyed. In the town of Rampur, Reazuddin Ahmed and his family sought shelter behind a concrete wall. All the while, houses were tossed into the air around them. Babul Ahmed, Reazuddin's 10-year-old son, described his family's terror: "It was dust and wind everywhere. We prayed to God: 'Save us.'" The tornado that terrorized the family was not unusual. It was just one of many types of extreme weather that plague South Asia and make life both difficult and dangerous.

The Monsoon Seasons

South Asia is home to an annual cycle of powerful, destructive weather, including the monsoon. The monsoon is a wind system, not a rainstorm. There are two monsoon seasons—the moist summer monsoon and the dry, cool winter monsoon. (The illustrations on the next page show the monsoon pattern in winter and summer.)

The **summer monsoon** is a wind system that blows from the southwest across the Indian Ocean toward South Asia from June through September. These winds stir up powerful storms that release vast amounts of rain and cause severe flooding.

The **winter monsoon** is a wind system that blows from the northeast across the Himalayas toward the sea from October through February. Unlike the summer monsoon, the winter winds carry little moisture. A drought can result if the summer monsoon has failed to bring normal levels of moisture. From March through May, there are no strong prevailing wind patterns.

Impact of the Monsoons

The monsoon winds shape the rhythms of life for South Asia's people and also affect relations between its countries.

PHYSICAL IMPACT The rains that accompany the summer monsoons are critical to the agriculture of

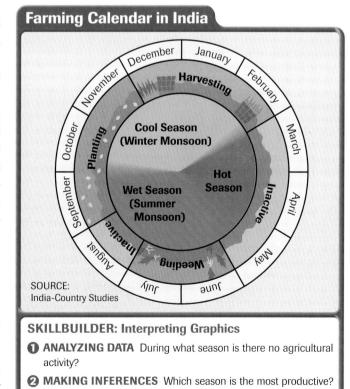

Farming Calendar in India

December · January · February · March · April · May · June · July · August · September · October · November

Harvesting

Cool Season (Winter Monsoon)

Hot Season

Wet Season (Summer Monsoon)

Planting · Inactive · Weeding · Inactive

SOURCE: India-Country Studies

SKILLBUILDER: Interpreting Graphics

❶ **ANALYZING DATA** During what season is there no agricultural activity?

❷ **MAKING INFERENCES** Which season is the most productive?

SOUTH ASIA

Summary and Winter Monsoons

Summer

Himalayas

heavy rain

wet winds from
the southwest

Indian
Ocean

Winter

Himalayas

little moisture

dry winds from
the northeast

Indian
Ocean

SKILLBUILDER: Interpreting Maps

① **MOVEMENT** What direction do the summer monsoon winds follow?

② **MOVEMENT** What direction do the winter monsoon winds follow?

South Asia, as the farming calendar on page 597 shows. They help nourish the rain forests, irrigate crops, and produce the floodwaters that deposit layers of rich sediment to replenish the soil. However, heavy flooding can also damage crops.

At the same time, the summer monsoon can cause tremendous devastation. Cyclones are common and deadly companions to the summer monsoon. (These storms are called hurricanes in North America.) Cyclones destroy farmland, wipe out villages, and cause massive flooding. Their fury is legendary. As you read in the Disasters! feature on pages 578–579, the 1970 cyclone that struck the southern coast of Bangladesh killed more than 300,000 people. It left hundreds of thousands homeless and destitute. In fact, because of the monsoons, Bangladesh was the site of some of the worst natural disasters of the 20th century.

The droughts that come with the dry winter monsoon bring their own problems. Lush landscapes can become arid wastelands almost overnight. These droughts—along with storms and floods—cause havoc for the people and economies of South Asia.

ECONOMIC IMPACT The climate of South Asia makes agriculture difficult. Crops often disappear under summer floodwaters or wither in drought-parched soil. With so many mouths to feed, the countries of South Asia must buy what they cannot grow, and the threat of famine is ever present. But the people suffer from more than just crop failures. They may also lose their homes and families to weather-related catastrophes. Most people are too poor to rebuild their homes and lives, and

Geographic Thinking

Using the Atlas
Ⓐ Use the maps on page 545 and this page. What country of South Asia seems least affected by the summer and winter monsoons?

governments often lack the necessary resources to provide significant help. However, the people of South Asia have taken some steps to prevent or lessen damage. These include building houses on stilts, erecting concrete cyclone shelters, and building dams to control floodwaters.

The region also receives international aid. Other governments and international agencies have lent billions of dollars to South Asian nations. But often this aid does not go far because of the frequency of disasters. Also, the aid burdens these countries with heavy debts. ◀B

POLITICAL TENSIONS Conditions caused by the weather patterns in South Asia have also caused political disputes. For instance, to bring water to the city of Kolkata, India constructed the Farakka dam across the Ganges at a point just before it enters Bangladesh. (See map on page 545.) Because India and Bangladesh share the Ganges, the dam left little water for drinking and irrigation in southern Bangladesh. Many Bangladeshi farmers lost farmland, and some illegally fled to India.

The two countries finally settled the dispute in 1997, when they signed a treaty giving each country specific water rights to the Ganges. Still, the dispute provided a graphic example of the role weather plays in both the politics and economics of South Asia. In the Case Study that follows, you will read about another political conflict—a territorial dispute between India and Pakistan.

Geographic Thinking◀
Seeing Patterns
B▶ How might the governments of South Asia use foreign aid?

REGION Dams on the Ganges divert water to irrigate Indian farms. But the dams decrease water downstream in Bangladesh.
Why might such a result cause conflict between India and Bangladesh?

Section 2 Assessment

1 Places & Terms

Explain the importance of each of the following terms and places.

• summer monsoon

• winter monsoon

2 Taking Notes

PLACE Review the notes you took for this section.

	Causes	Effects
Issue 2: Extreme Weather		

• What are cyclones called in the United States?

• What kind of devastation can cyclones cause?

3 Main Ideas

a. Why do some people mistake monsoons, which are actually wind systems, for rainstorms?

b. What problems are associated with the winter monsoon?

c. What are some of the economic effects of monsoons?

4 Geographic Thinking

Identifying and Solving Problems How have attempts to address the challenges of South Asian weather patterns sometimes resulted in political disputes? How might disputes be avoided in the future?

Think about:

• the importance of water to the region

• who owns rivers

SOUTH ASIA

ASKING GEOGRAPHIC QUESTIONS Do research on the issue of water distribution in one South Asian country. Then, come up with a geographic question about the issue, perhaps one considering how geography can be used to improve the situation. Answer the question and write a **newspaper article** about the issue.

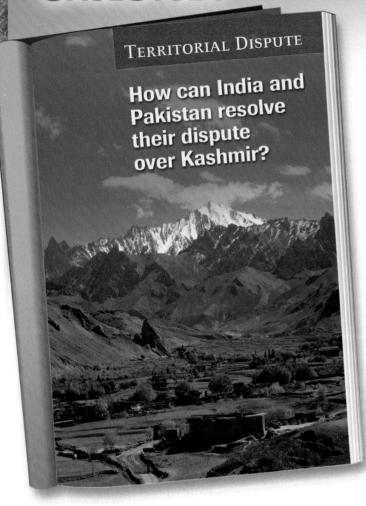

TERRITORIAL DISPUTE

How can India and Pakistan resolve their dispute over Kashmir?

Snowcapped mountains tower over a village in the valley of the Suru River in the disputed territory of Kashmir.

Kashmir is a territory of towering mountains, dense forests, and fertile river valleys. It is strategically located at the foot of the Himalayas and is surrounded by India, Pakistan, and China. Since 1947, India and Pakistan have fought to control this territory of 12 million people. The territorial dispute has caused three Indo-Pakistani wars and, in just the last decade alone, cost up to 75,000 lives. It poses a threat to the political stability of South Asia and the economic well-being of the countries involved. And, because both India and Pakistan have nuclear weapons, the Kashmir conflict has the potential to lead to nuclear war.

A Controversy Over Territory

In 1947, the British government formally ended its colonial rule over the Indian subcontinent after 90 years. It partitioned, or divided, the subcontinent into two independent countries. India had a predominantly Hindu population. Pakistan was mostly Muslim. Britain gave each Indian state the choice of joining either country or remaining independent. Muslim states joined with Pakistan, while Hindu states remained part of India. Kashmir, however, had a unique problem.

POLITICS AND RELIGION Kashmir was mainly Muslim, but its leader, the Maharajah of Kashmir, was a Hindu. Faced with a difficult decision, the maharajah tried to keep Kashmir independent. But the plan failed. The maharajah then ceded Kashmir to India in 1947, but Pakistani soldiers invaded Kashmir. After a year's fighting, India still controlled much of the territory. Since then, India and Pakistan have fought two

KASHMIR CONFLICT

1940   1960   1970   1980   1990   2000

1947
India and Pakistan gain independence.

1948
Year-long **Indo-Pakistani war** over Kashmir ends.

1965
Second Indo-Pakistani war is fought.

1971
India and Pakistan fight again over Kashmir.

1972
Cease-fire agreement signed by India and Pakistan.

1974
India explodes its first nuclear device.

1998
Pakistan begins nuclear testing.

1999
Indian and Pakistani military clash at cease-fire line.

more wars, in 1965 and in 1971. Although a cease-fire was signed in 1972, the situation remains unresolved. As you can see on the map below, India and Pakistan each control part of the disputed territory. Even China controls a portion, having seized a remote northern mountain area in 1962.

A QUESTION OF ECONOMICS There's more to this conflict than just politics and religion. The Indus River flows through Kashmir, and many of its tributaries originate in the territory. The Indus is a critical source of drinking and irrigation water for all of Pakistan. As a result, the Pakistanis are unwilling to let India control such a vital resource. Kashmir has become a strategic prize that neither country is willing to give up.

A Nuclear Nightmare

SEE PRIMARY SOURCE B

In 1998, India and Pakistan each tested nuclear weapons. The rest of the world was horrified by the thought that the 50-year-old dispute over Kashmir might finally end with vast areas of South Asia destroyed by nuclear bombs. After the tests, both nations vowed to seek a political solution to the conflict. But the possibility of a nuclear war has made the dispute even more dangerous. Despite frequent cease-fires, the border clashes have continued. Also, Pakistan is supporting Muslims in Kashmir who have been fighting Indian rule since the late 1980s.

SEE PRIMARY SOURCE A

A QUESTION OF PRIORITIES Both India and Pakistan have large populations and widespread poverty. The money that they have spent on troops, arms, and nuclear programs might have been used to educate millions of children and to address many social problems.

SEE PRIMARY SOURCE E

Resolving the status of Kashmir would offer the people of India, Pakistan, and Kashmir the peace they need to begin improving the quality of their lives. It would also reduce political tensions in the region. The Case Study Project and primary sources that follow will help you to explore the Kashmir question.

Kashmir

AFGHANISTAN
Kashmir
CHINA
PAKISTAN
Arabian Sea

0   100   200 miles
0   100   200 kilometers
Two-Point Equidistant Projection

TAJIKISTAN
AFGHANISTAN
CHINA
KASHMIR
PAKISTAN
INDIA

Controlled by China, also claimed by India	—	Cease-fire line
Controlled by Pakistan, also claimed by India	—	International border
Controlled by India, also claimed by Pakistan	•••••	Disputed border
	-----	State boundary

SKILLBUILDER: Interpreting Maps

❶ **REGION** Which countries does Kashmir border?

❷ **LOCATION** Where was the cease-fire line drawn?

SOUTH ASIA

CASESTUDY

PROJECT

A Newspaper Feature

Primary sources A, B, C, D, and E offer different views of the dispute over Kashmir. Use these resources along with your own research to write a newspaper feature on how the people of Kashmir, India, and Pakistan have suffered in this conflict. Include their own words.

RESEARCH LINKS
CLASSZONE.COM

Suggested Steps

1. Divide into small groups representing ordinary Kashmiris (such as women, farmers, and rebel soldiers), as well as Indian and Pakistani officials or soldiers. Then begin gathering personal accounts about the conflict from newspapers, magazines, and Internet sites.

2. Search for visuals—illustrations, maps, photographs, political cartoons, charts, and graphs—that help illustrate the points you are making.

3. When everyone in the class has collected enough material, work together to plan the feature story.

4. When you have finished planning, prepare the feature.

5. Share your project with other groups at your school or in your community.

Materials and Supplies

- Reference books, newspapers, and magazines
- Computer with Internet access and printer

PRIMARY SOURCE Ⓐ

Government Document *The Ministry of Foreign Affairs of Pakistan published this policy statement on Kashmir in 1999, after a visit to the United States by the **Pakistani prime minister, Nawaz Sharif.***

In order to find an early and just solution to the 50-year old . . . Kashmir dispute, Pakistan has welcomed offers of good offices and third-party mediation. It has encouraged the international community to play an active role and facilitate the peaceful settlement of disputes between Pakistan and India.

While Pakistan is committed to a peaceful settlement of the Jammu and Kashmir dispute, adequate measures have been taken to safeguard the country's territorial integrity and national sovereignty. Pakistan will continue to extend full political, diplomatic and moral support to the legitimate Kashmiri struggle for their right to self-determination as enshrined in the relevant United Nations resolutions. In the context of the bilateral dialogue, it calls on India to translate its commitments into reality.

PRIMARY SOURCE Ⓑ

Government Policy Declaration *At a state dinner in India for President Bill Clinton in March 2000, **Indian President Kocheril Raman Narayanan** warned that India would fight to protect its interests in Kashmir.*

It has been suggested that the Indian subcontinent is the most dangerous place in the world today, and Kashmir is a nuclear flashpoint. These alarmist descriptions will only encourage those who want to break the peace and indulge in terrorism and violence. The danger is not from us who have declared solemnly that we will not be the first to use nuclear weapons, but rather it is from those who refuse to make any such commitment.

We are publicly committed to the abolition of nuclear weapons together with other nuclear powers who possess them in awesome stockpiles capable of destroying the world many times over. India does not threaten any other country and will not engage in an arms race, but India will maintain a minimum credible nuclear deterrent—no more, no less—for her own security.

PRIMARY SOURCE C

Political Speech *Mehbooba Mufti is a leader of the Jammu and Kashmir People's Democratic Party, a political party in Kashmir. In 1999, she spoke about the conflict and her hope that the dispute will be peacefully resolved.*

Everything has changed, mostly for the worse. Take just the physical destruction of whatever we had, the schools, the colleges, the roads, the bridges, the buildings, everything we had for the last 50 years, that has been more or less destroyed. We used to have a very good education system, with very good teachers, but now that has gone. . . .

I think Kashmir finally has to become a bridge between India and Pakistan. Finally. Maybe not today, maybe not tomorrow, but after some years, it is finally going to become a bridge. Have an open relationship. It's a dream!

PRIMARY SOURCE D

Personal Story *Kashmiri native Mohammed Aziz lives in Kargil, a city on the border between the Pakistani- and Indian-controlled regions of Kashmir. In 1999, he described how the conflict had affected his hometown.*

We never know when the shell will come. . . . For the last three years, no one sleeps well there. Whoever flees leaves everything there. He takes nothing with him. The cattle are left on their own. Nobody cares for them, so we don't know what happens to them. . . .

Before, tourism was OK. Before the shelling there used to be 25 hotels, but now I don't think any hotel is open. We can't calculate the damage. . . .

The children's education is stopped, and whoever is ill dies because there is no medication nor anyone to care for them. Whoever resides in Kargil, does so at his own risk.

PRIMARY SOURCE E

Political Cartoon *This 1998 political cartoon shows how the development of nuclear weapons by India and Pakistan has caused economic suffering among the people of both countries.*

PROJECT *CheckList*

Have I . . .

✓ fully researched my topic?

✓ located primary source quotations to tell my story?

✓ taken into account both sides of the issue?

✓ arranged the quotations so that they tell a coherent, interesting story?

✓ created informative visuals that make my story clear and interesting?

VISUAL SUMMARY
TODAY'S ISSUES IN SOUTH ASIA

Economics

Population Explosion
- Though only about one-third the size of the United States, India has over three times as many people.
- India's government has taken steps to control population growth but has had only mixed success.
- Many parents continue to have large numbers of children because of India's high infant mortality rates, the extra income brought in by children, and the need for caregivers as parents age.

Environment

Living with Extreme Weather
- The physical damage caused by extreme weather patterns in South Asia, such as cyclones, can be devastating to the region's people.
- The impact of extreme weather is not limited to physical damage. These forces can also disrupt the economy and cause serious political tensions.

Government

Case Study: Territorial Dispute
- Kashmir is a strategically located territory, surrounded by Pakistan, India, and China.
- India and Pakistan have fought three wars over this territory since 1947.
- Money spent by India and Pakistan for armaments, including nuclear weapons, has not been available to help improve the lives of the people of these countries.

Reviewing Places & Terms

A. Briefly explain the importance of each of the following.

1. basic necessities
2. illiteracy
3. summer monsoon
4. winter monsoon

B. Answer the questions about vocabulary in complete sentences.

5. Which winds stir up powerful storms in South Asia that release vast amounts of rain and cause severe flooding?
6. Which winds blow from the southwest across the Indian Ocean toward South Asia from June through September?
7. Food, shelter, and clothing are all examples of what?
8. Which winds blow from the northeast across the Himalayas from October through February?
9. What is the term for the inability to read or write?
10. What was the Indian government finding difficult to provide for its people?

Main Ideas

Population Growth (pp. 593–596)

1. Currently, about how many babies are born in India every day? Annually?
2. Why might the lack of basic necessities in a region concern demographers—people who study population?
3. Why might a high rate of infant mortality affect the size of families?
4. What percentage of the world's population is found in South Asia?
5. How would education play an important role in slowing population growth?

Living with Extreme Weather (pp. 597–599)

6. What are South Asia's two monsoon seasons? How do they differ?
7. When do these wind seasons occur?
8. What are some of the precautions that people in South Asia have taken to lessen the damage caused by cyclones?
9. What type of international aid have the countries of South Asia received?
10. What political tensions have resulted from the effects of extreme weather?

Case Study: Territorial Dispute over Kashmir (pp. 600–603)

11. Where is Kashmir located?
12. What countries have fought three wars over control of Kashmir?
13. When and why did the dispute over Kashmir begin?
14. Why are world leaders particularly concerned about the dispute?
15. What might happen if the dispute were resolved?

Critical Thinking

1. Using Your Notes

Use your completed chart to answer these questions.

	Causes	Effects
Issue 1: Population		
Issue 2: Extreme Weather		

a. Why might parents in India want a large family?

b. Why is Kashmir economically important to Pakistan?

2. Geographic Thinking

a. **REGION** How is the religious make-up of Kashmir related to conflict over the territory?

b. **MOVEMENT** Why might people in India and the other heavily populated countries in South Asia move to other parts of the world?

3. Identifying Themes

Why is Bangladesh especially vulnerable to the cyclones that occasionally devastate the region? Which of the five themes applies to this situation?

4. Making Comparisons

Why might India and Bangladesh fear the weather that can arrive during the summer?

5. Determining Cause and Effect

How might the dispute over Kashmir affect the social and educational programs in the region?

Additional Test Practice, pp. S1–S37

TEST PRACTICE
CLASSZONE.COM

Geographic Skills: Interpreting Graphs

Ethnic Indian Population Outside of India

Use the graph at right to answer the following questions.

1. **PLACE** On what continent outside of South Asia do most Indians live?

2. **PLACE** About how many Indians live in South America?

3. **LOCATION** Why do you think most ethnic Indians living outside of India live in South Asian countries?

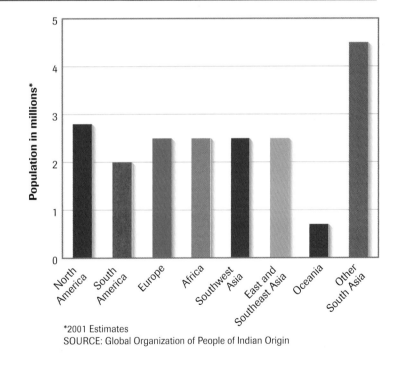

*2001 Estimates

SOURCE: Global Organization of People of Indian Origin

GeoActivity

Carry out research on people from India who live in the United States. Create a table of the five cities with the largest populations of people from India.

 INTERNET ACTIVITY

Use the links at **classzone.com** to continue research on population growth in India. Focus on how the limited availability of basic necessities has affected the daily life of the country's people.

Creating a Multimedia Presentation Use your research to create an electronic presentation. Combine charts, maps, images, objects, and written accounts to provide your audience with a picture of daily life in India.

Unit 9

East Asia

East Asia is made up of a vast mainland area and a number of important islands off the eastern coast.

PLACE The Forbidden City is a walled enclosure in Beijing, China. Inside is a complex of palaces where 24 emperors ruled. Once closed to the public, it is now a museum and tourist attraction.

PLACE East Asia includes huge mountains and large deserts.

LOCATION The region is called "East Asia" because it is on the eastern edge of the Asian continent, bordered by the Pacific Ocean to the east, Russia to the north, and the countries of south and southeast Asia to the south.

REGION This area is bordered by a number of bodies of water, including the Pacific Ocean, the Sea of Japan, the East China Sea, and the South China Sea.

For more information on East Asia . . .

 RESEARCH LINKS
CLASSZONE.COM

HUMAN–ENVIRONMENT INTERACTION
The traditional pagoda, or temple, sits amidst the bustle of economic activity in the city center of Seoul, South Korea.

LOCATION Mount Fuji, the highest peak in Japan at 12,388 feet, is a volcano that last erupted in 1707. It is considered a sacred mountain.

EAST ASIA

Today's Issues in East Asia

Today, East Asia faces the issues previewed here. As you read Chapters 27 and 28, you will learn helpful background information. You will study the issues themselves in Chapter 29.

In a small group, answer the questions below. Then participate in a class discussion of your answers.

Exploring the Issues

1. PHYSICAL FORCES What might be some of the effects of earthquakes and volcanoes on daily life in the region? How might the effects be similar or different in an urban and a rural area?

2. TRADE What are some items you or your family have bought that were made in East Asia?

3. POPULATION Parts of East Asia are very crowded. What might be some of the advantages and challenges of living around so many people?

For more on these issues in East Asia . . .

CURRENT EVENTS
CLASSZONE.COM

PHYSICAL FORCES

How might people in East Asia prepare for earthquakes and volcanoes?

A bus teeters on the edge of a highway torn apart by an earthquake in Kobe, Japan, in 1995.

TRADE

What are some benefits of global trade?

Hong Kong is a thriving center of trade and economic activity. Once a colony of Britain, it is now a part of China. Its wealth and trading expertise are helping China compete with leading industrial nations.

POPULATION

CaseStudy

What pressures does population put on the environment?

Subway attendants in Tokyo push people into crowded subway trains. Japan has a large number of people living on a small amount of land.

Patterns of Physical Geography

Use the Unit Atlas to add to your knowledge of East Asia. As you look at the maps and charts, notice geographic patterns and specific details about the region. For example, the charts on pages 610–611 give details about the rivers and mountains of East Asia.

After studying the pictures, graphs, and physical map on these two pages, jot down in your notebook the answers to the following questions.

Making Comparisons

1. What three main river systems run from west to east in China?

2. Which of the bodies of water surrounding Japan is the largest?

3. Compare East Asia's size and population to those of the United States. Based on that data, how might the population densities of the two compare?

For updated statistics on East Asia . . .

DATA UPDATE
CLASSZONE.COM

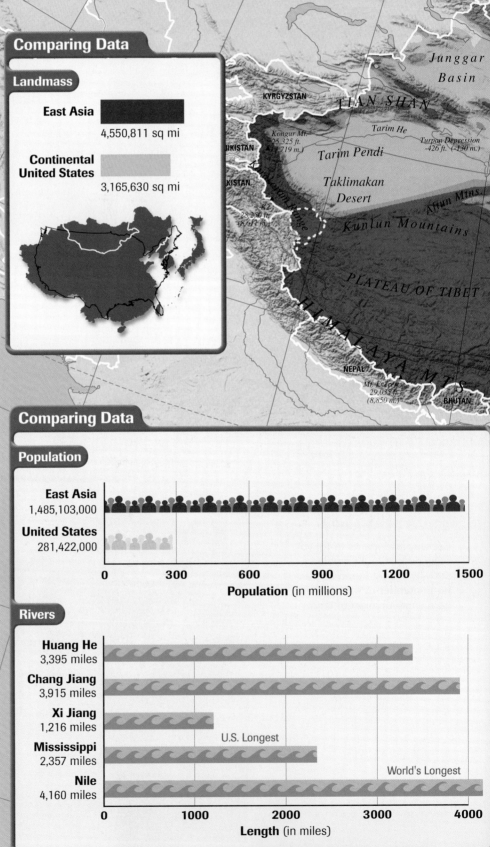

KAZAKHSTAN

KYRGYZSTAN

Junggar Basin

TIAN SHAN

Kongur Mt.
25,325 ft.
(7,719 m.)

Tarim He

Turpan Depression
-426 ft. (-130 m.)

Tarim Pendi

Taklimakan Desert

Altun Mtns.

K2
28,250 ft.
(8,611 m.)

Kunlun Mountains

Karakoram Range

PLATEAU OF TIBET

HIMALAYA MTS

NEPAL

Mt. Everest
29,035 ft.
(8,850 m.)

BHUTAN

Comparing Data

Landmass

East Asia
4,550,811 sq mi

Continental United States
3,165,630 sq mi

Comparing Data

Population

East Asia
1,485,103,000

United States
281,422,000

| 0 | 300 | 600 | 900 | 1200 | 1500 |

Population (in millions)

Rivers

Huang He
3,395 miles

Chang Jiang
3,915 miles

Xi Jiang
1,216 miles

U.S. Longest

Mississippi
2,357 miles

World's Longest

Nile
4,160 miles

| 0 | 1000 | 2000 | 3000 | 4000 |

Length (in miles)

Sakhalin I.

RUSSIA

Hokkaido

MONGOLIA

Great Khingan Mountains

Amur R.

Songhua R.

ALTAI MTS.

MONGOLIAN PLATEAU

Honshu

G O B I

Liao He

Manchurian Plain

Sea of Japan

Mt. Fuji 12,388 ft. (3,776 m.)

Yalu Jiang

NORTH KOREA

JAPAN

Huang He (Yellow R.)

Mu Us Desert

Korea Bay

Bo Hai

SOUTH KOREA

Shikoku

Korea Strait

Huang He (Yellow R.)

Yellow Sea

Cheju I.

Kyushu

Wei He

Grand Canal

Qinling Shandi

North China Plain

PACIFIC OCEAN

CHINA

Chang Jiang (Yangtze R.)

East China Sea

Gongga Shan 24,790 ft. (7,556 m.)

Dongting Hu

Poyang Hu

Ryukyu Islands

Mekong R.

Xi Jiang (West R.)

Salween R.

Taiwan Strait

TAIWAN

Elevation

13,100 ft. (4,000 m.)
6,600 ft. (2,000 m.)
1,600 ft. (500 m.)
650 ft. (200 m.)
0 ft. (0 m.)
Below sea level

▲ Mountain peak

MYANMAR

VIETNAM

Luzon Strait

Gulf of Tonkin

LAOS

South China Sea

Hainan

PHILIPPINES

THAILAND

0   250   500 miles
0   250   500 kilometers
Two-Point Equidistant Projection

EAST ASIA

Comparing Data

Mountains

World's Tallest	U.S. Tallest			
Mount Everest	**Mount McKinley**	**Mount Kongur**	**Mount Paektu**	**Mount Fuji**
Nepal-Tibet	United States	China	Korea	Japan
29,035 feet	20,320 feet	25,325 feet	9,022 feet	12,388 feet

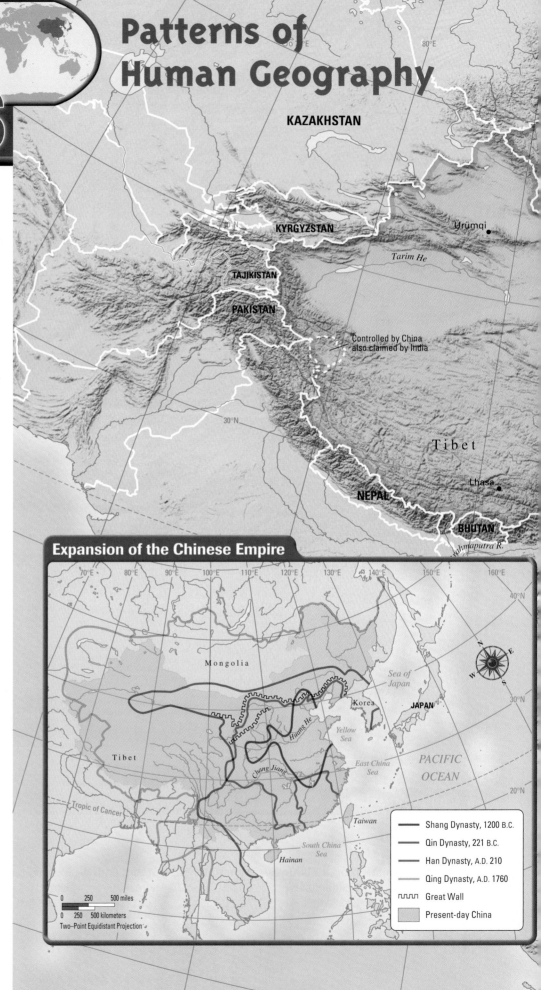

Unit ATLAS

Patterns of Human Geography

Over the course of centuries, the political map of East Asia has changed. The Chinese empire expanded over thousands of years, absorbing much of the region. Study the historical and political maps of East Asia on these two pages. In your notebook, answer these questions.

Making Comparisons

1. What differences do you notice when you compare the historical map of the Chinese empire to the map of East Asia today?

2. What are some of the similarities between the historical map and the contemporary map of East Asia?

3. What countries in the region used to be a part of the Chinese empire but are now independent? Which country in the region was never a part of the empire?

KAZAKHSTAN

KYRGYZSTAN

TAJIKISTAN

PAKISTAN

Tarim He

Ürümqi

Controlled by China also claimed by India

30°N

Tibet

Lhasa

NEPAL

BHUTAN

Brahmaputra R.

Expansion of the Chinese Empire

70°E 80°E 90°E 100°E 110°E 120°E 130°E 140°E 150°E 160°E

40°N

Mongolia

Sea of Japan

Korea

JAPAN

30°N

Huang He

Tibet

Yellow Sea

Chang Jiang

East China Sea

PACIFIC OCEAN

20°N

Tropic of Cancer

Taiwan

South China Sea

Hainan

0 250 500 miles
0 250 500 kilometers
Two–Point Equidistant Projection

— Shang Dynasty, 1200 B.C.
— Qin Dynasty, 221 B.C.
— Han Dynasty, A.D. 210
— Qing Dynasty, A.D. 1760
ᴨᴨᴨ Great Wall
▢ Present-day China

RUSSIA

MONGOLIA

Ulaanbaatar

Manchuria

Amur R.

Songhua R.

Harbin

Changchun

Liao He

Shenyang

Yalu Jiang

NORTH KOREA

Pyongyang

Korea Bay

Sea of Japan

Sakhalin I.

Hokkaïdo

Sapporo

Honshu

JAPAN

Tokyo

Nagoya

Yokoha

Kyoto

Osaka

Huang He (Yellow R.)   *Great Wall*

Beijing

Tianjin

Bo Hai

Seoul

SOUTH KOREA

Pusan

Yellow Sea

Cheju I.

Hiroshima

Fukuoka

Shikoku

Kyushu

PACIFIC OCEAN

Great Wall

Lanzhou

Wei He   Xi'an

Huang He (Yellow R.)   *Grand Canal*

CHINA

Chengdu

Chang Jiang (Yangtze R.)

Chongqing

Three Gorges Dam

Dongting Hu

Poyang Hu

Nanjing

Wuhan   Hangzhou

Shanghai

East China Sea

Ryukyu Islands

Wenzhou

Korea Strait

Kunming

Mekong R.

Salween R.

Xi Jiang (West R.)

Guangzhou

Hong Kong

Macao

VIETNAM

MYANMAR

THAILAND   LAOS

Gulf of Tonkin

Hainan

South China Sea

Taiwan Strait

Taipei

TAIWAN

Luzon Strait

Philippine Sea

PHILIPPINES

☆ National capital

• Other city

···· Great Wall

0   250   500 miles

0   250   500 kilometers

Two-Point Equidistant Projection

EAST ASIA

613

Regional Patterns

These two pages contain a graph and three thematic maps. The graph shows the religions of East Asia. The maps show other important features of East Asia: its vegetation, languages, and population density. After studying these two pages, answer the questions below in your notebook.

Making Comparisons

1. Where is most of the population located in China? Why might people have settled in these areas rather than in other areas?

2. Which is the smallest country in East Asia?

3. What is the vegetation in much of southern China, Taiwan, southern Korea, and southern Japan? How does it differ from the vegetation in Mongolia?

Religions of East Asia

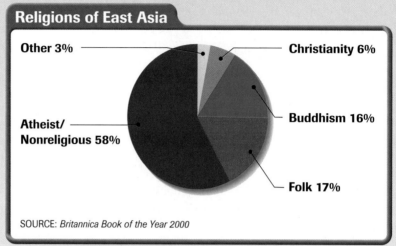

Other 3%

Christianity 6%

Atheist/Nonreligious 58%

Buddhism 16%

Folk 17%

SOURCE: *Britannica Book of the Year 2000*

Vegetation of East Asia

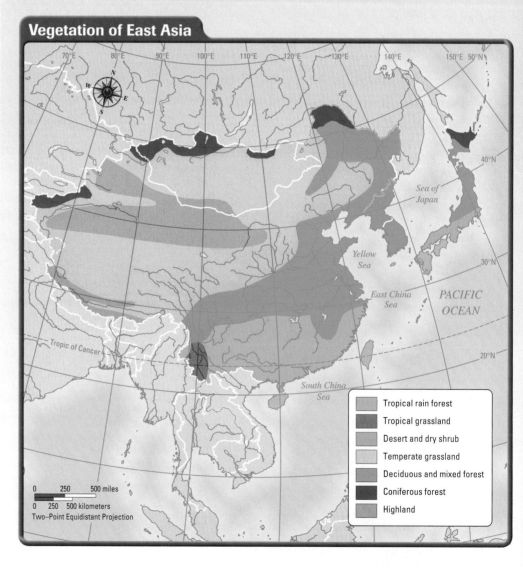

Sea of Japan

Yellow Sea

East China Sea

PACIFIC OCEAN

Tropic of Cancer

South China Sea

	Tropical rain forest
	Tropical grassland
	Desert and dry shrub
	Temperate grassland
	Deciduous and mixed forest
	Coniferous forest
	Highland

0   250   500 miles
0   250   500 kilometers
Two-Point Equidistant Projection

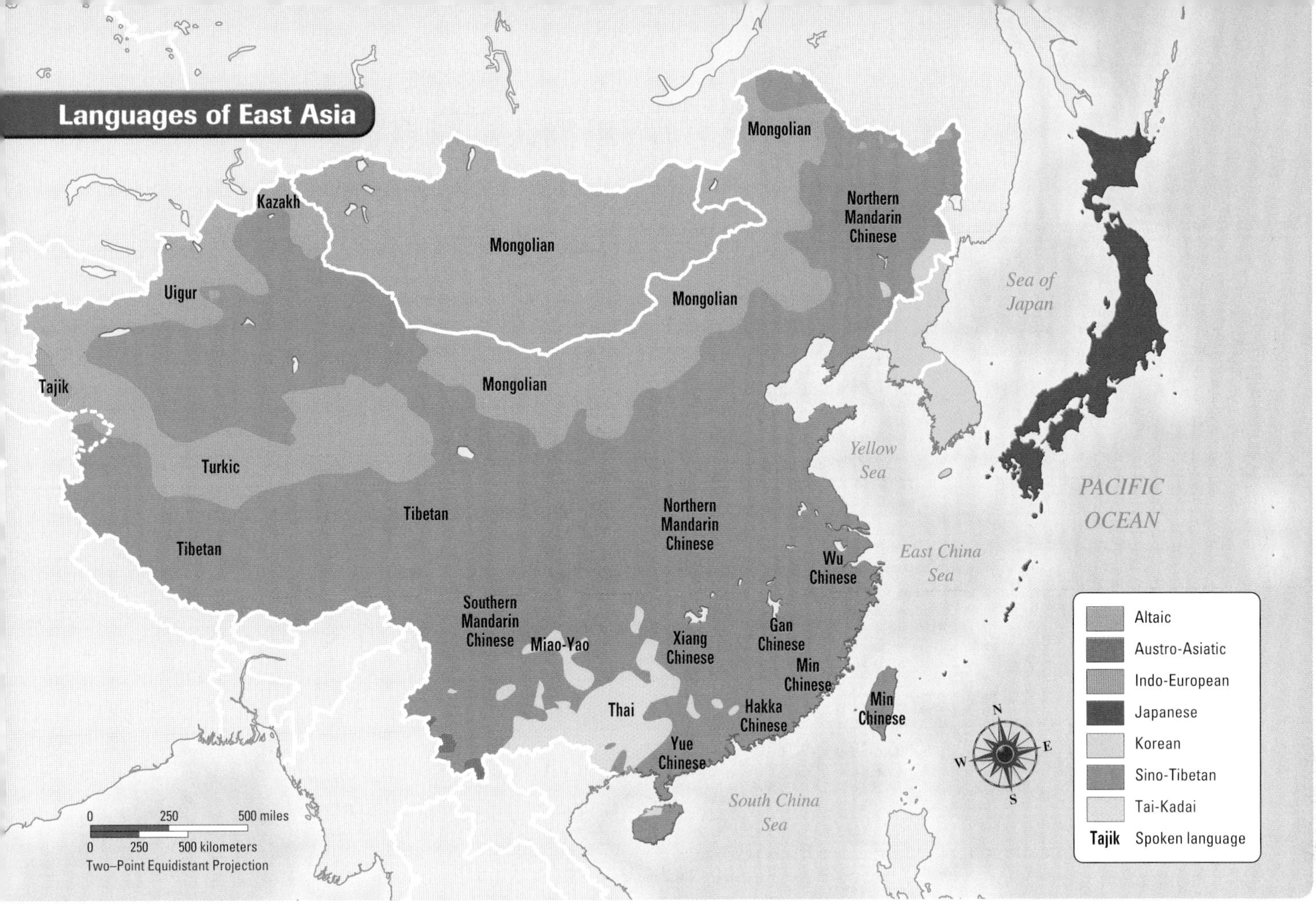

Languages of East Asia

Mongolian

Kazakh

Uigur

Mongolian

Mongolian

Northern Mandarin Chinese

Tajik

Turkic

Tibetan

Tibetan

Sea of Japan

Yellow Sea

PACIFIC OCEAN

East China Sea

Southern Mandarin Chinese

Miao-Yao

Xiang Chinese

Northern Mandarin Chinese

Wu Chinese

Gan Chinese

Min Chinese

Thai

Hakka Chinese

Min Chinese

Yue Chinese

South China Sea

	Altaic
	Austro-Asiatic
	Indo-European
	Japanese
	Korean
	Sino-Tibetan
	Tai-Kadai
Tajik	Spoken language

0   250   500 miles

0   250   500 kilometers

Two–Point Equidistant Projection

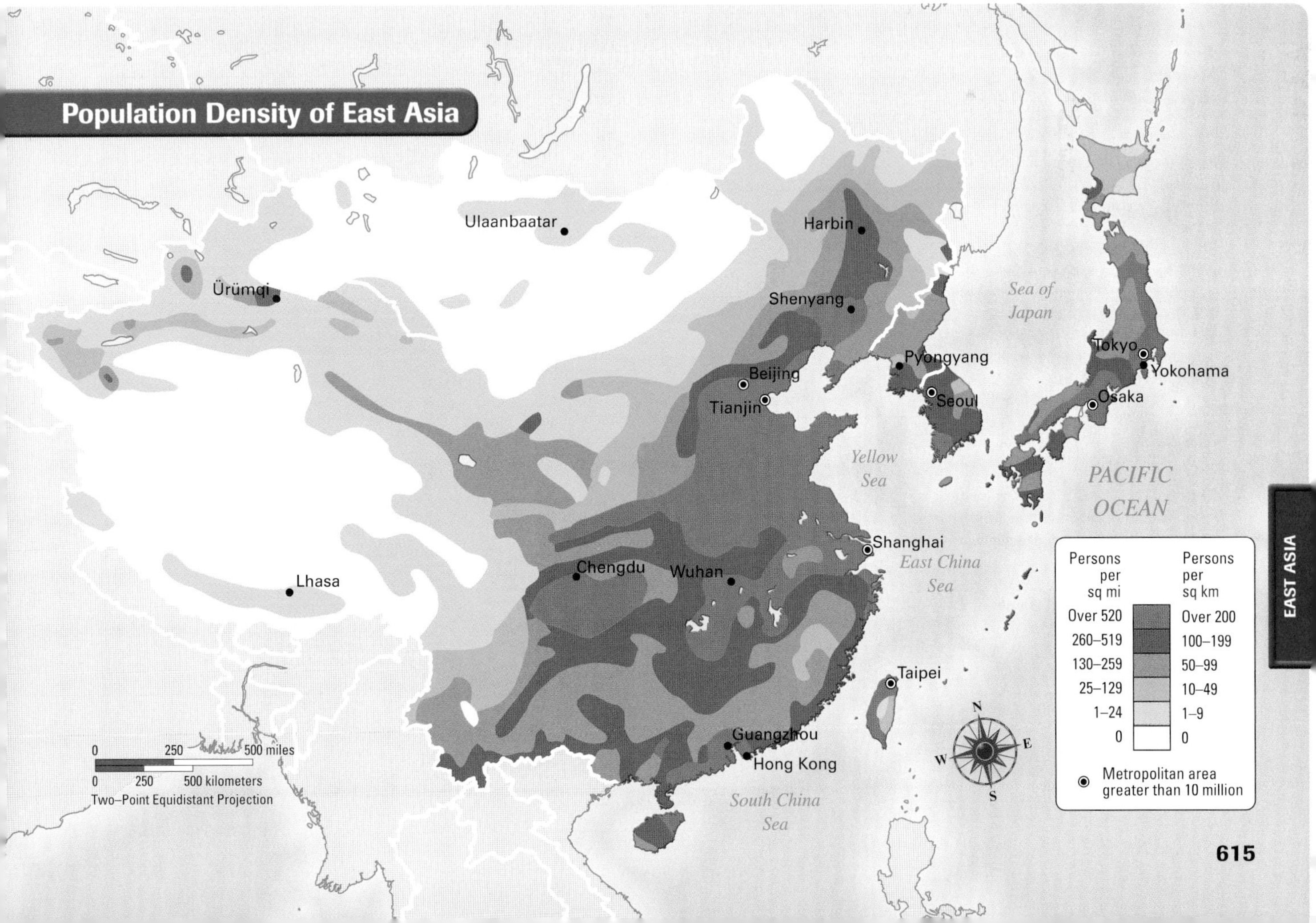

Population Density of East Asia

Ulaanbaatar

Harbin

Ürümqi

Shenyang

Sea of Japan

Pyongyang

Tokyo

Yokohama

Beijing

Seoul

Osaka

Tianjin

Yellow Sea

PACIFIC OCEAN

Lhasa

Chengdu

Wuhan

Shanghai

East China Sea

Taipei

Guangzhou

Hong Kong

South China Sea

Persons per sq mi	Persons per sq km
Over 520	Over 200
260–519	100–199
130–259	50–99
25–129	10–49
1–24	1–9
0	0

⊙ Metropolitan area greater than 10 million

0   250   500 miles

0   250   500 kilometers

Two–Point Equidistant Projection

EAST ASIA

615

Unit ATLAS

Regional Data File

Study the charts on the countries of East Asia. In your notebook, answer these questions.

Making Comparisons

1. Which countries have the most people? Locate them on the map. Are they also the largest countries in terms of total area?

2. In which part of the region are the highest elevations located? What might this suggest about settlement patterns in the region?

Sources:
Europa World Year Book 2000
Human Development Report 2000,
 United Nations
International Data Base, 2000, U.S.
 Census Bureau online
*Merriam-Webster's Geographical
 Dictionary,* 1997
Statesman's Yearbook 2001
2000 World Population Data Sheet,
 Population Reference Bureau online
U.S. Census Bureau, 2000 Census
WHO Estimates of Health Personnel,
 World Health Organization online
World Almanac and Book of Facts 2001
World Education Report 2000, UNESCO
 online
World Factbook 2000, CIA online
N/A = not available

Notes:
* Figures do not include Hong Kong or Macao, both Special Administrative Regions.
ª A comparison of the prices of the same items in different countries is used to figure these data.
ᵇ Includes land and water, when figures are available.

For updated statistics on East Asia . . .

DATA UPDATE
CLASSZONE.COM

Country Flag	Country/ Capital	Population (2000)	Life Expectancy (years) (2000)	Birthrate (per 1,000 pop.) (2000)	Infant Mortality (per 1,000 live births) (2000)
	China* Beijing	1,264,536,000	71	15	31.4
	Japan Tokyo	126,876,000	80	9	3.5
	Mongolia Ulaanbaatar	2,472,000	63	20	34.1
	North Korea Pyongyang	21,688,000	70	21	26.0
	South Korea Seoul	47,275,000	74	14	11.0
	Taiwan Taipei	22,256,000	75	13	6.6
	United States Washington, D.C.	281,422,000	77	15	7.0

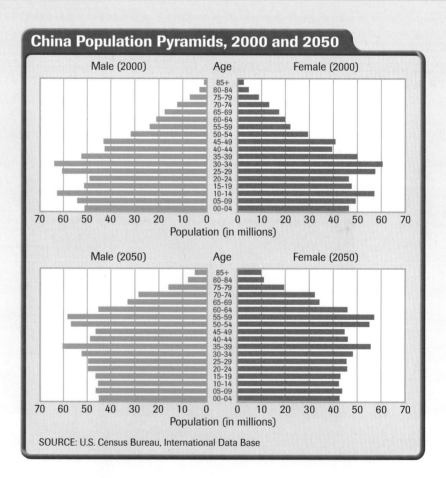

China Population Pyramids, 2000 and 2050

SOURCE: U.S. Census Bureau, International Data Base

Doctors (per 100,000 pop.) (1995–1998)	GDP[a] (billions $US) (1998–1999)	Import/Export[a] (billions $US) (1998–1999)	Literacy Rate (percentage) (1998)	Televisions (per 1,000 pop.) (1998)	Passenger Cars (per 1,000 pop.) (1996–1997)	Total Area[b] (square miles)	
162	4,800.0	165.8 / 194.9	82	205	4	3,704,427	
193	2,950.0	275.4 / 413.0	99	684	367	143,619	
243	6.1	0.472 / 0.317	83	45	8	604,247	
N/A	22.6	0.859 / 0.680	99	48	N/A	46,609	
127	625.7	104.4 / 144.0	98	334	165	38,022	
N/A	357.0	91.5 / 121.6	94	395	198	13,887	
251	9,255.0	820.8 / 663.0	97	847	489	3,787,319	

Profile of East Asia

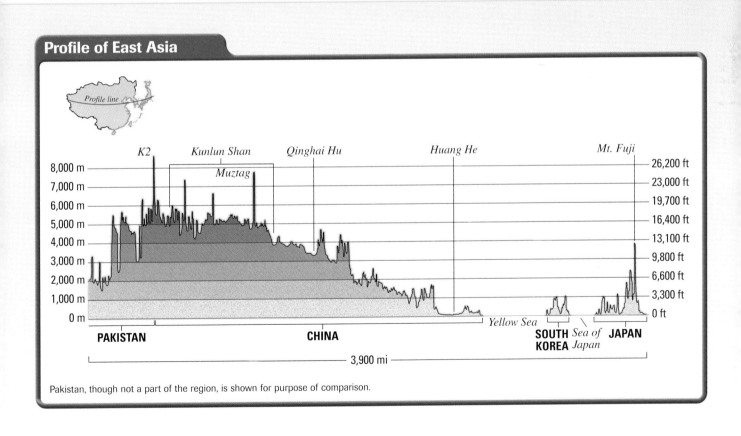

Pakistan, though not a part of the region, is shown for purpose of comparison.

Chapter 27

PHYSICAL GEOGRAPHY OF EAST ASIA
A Rugged Terrain

SECTION 1
Landforms and Resources

SECTION 2
Climate and Vegetation

SECTION 3
Human–Environment Interaction

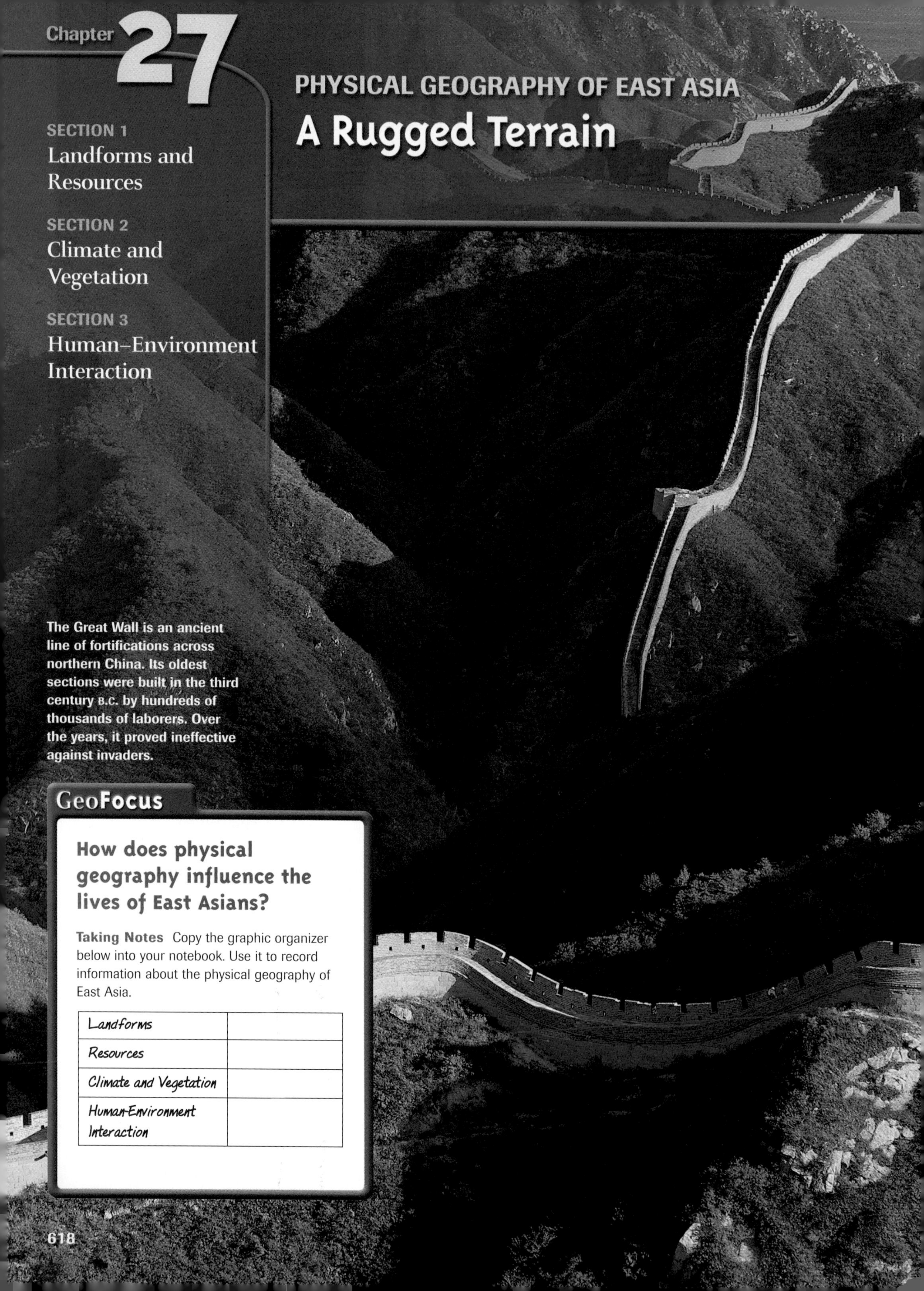

The Great Wall is an ancient line of fortifications across northern China. Its oldest sections were built in the third century B.C. by hundreds of thousands of laborers. Over the years, it proved ineffective against invaders.

GeoFocus

How does physical geography influence the lives of East Asians?

Taking Notes Copy the graphic organizer below into your notebook. Use it to record information about the physical geography of East Asia.

Landforms	
Resources	
Climate and Vegetation	
Human-Environment Interaction	

Forests are also abundant in the region. China, Japan, Taiwan, and both North Korea and South Korea all have forest resources. Japan has been able to keep most of its forests in reserve by buying timber and other forest products from other regions of the world.

MINERAL AND ENERGY RESOURCES China has large energy reserves of petroleum, coal, and natural gas, and Korea has coal reserves. Japan also has deposits of coal. China's resources have enabled it to be self-sufficient for much of its history. In contrast, Japan's shortage of resources has forced it to trade for what it needs.

China's mineral resources include iron ore, tungsten, manganese, molybdenum, magnesite, lead, zinc, and copper. North and South Korea possess important tungsten, gold, and silver reserves. Japan has reserves of lead, silver, and coal.

WATER RESOURCES China's long river systems are important to the country's economy. They provide crop irrigation, hydroelectric power, and transportation. To control flooding on the Chang Jiang and produce more electricity, China is building the Three Gorges Dam. (See pages 628–630.) The Huang He and Xi Jiang also provide hydroelectric power and a means of transportation.

Geographic Thinking

Seeing Patterns
In what ways might river systems be important to an economy?

People in East Asia look to the sea for food. In fact, Japan has developed one of the largest fishing industries in the world. Japanese factory ships process huge amounts of seafood for human consumption throughout the world, as well as in Japan.

You will read about East Asia's climate zones in the next section. You will also read about its vegetation.

Geography TODAY

The Japanese Fishing Industry

There is great competition among the world's nations to harvest the resources of the sea. Sophisticated and mechanized factory ships process the catch while still at sea.

Japan's fishing industry is larger than that of the United States or any country in Western Europe. Fleets of Japanese fishing vessels, such as the sea bass fishing boat shown below, trawl the oceans far from Japan to bring fish back to the home islands. Tuna, mackerel, salmon, and cod are eaten by the Japanese.

SECTION Assessment

1 Places & Terms

Identify each of the following places and terms.

- Kunlun Mountains
- Qinling Shandi Mountains
- Huang He
- Chang Jiang
- Xi Jiang

2 Taking Notes

PLACE Review the notes you took for this section.

Landforms	
Resources	

- What types of landforms are found in East Asia?
- What are their relative locations?

3 Main Ideas

a. How might the river basins of China have affected settlement patterns?

b. How are the landforms of East Asia an advantage to life in the region?

c. What effect might natural resources have had on the development of East Asia?

4 Geographic Thinking

Drawing Conclusions How might China's three large river systems have affected the development of agriculture and trade in the area? **Think about:**

- the obstacles that mountains and deserts present to agriculture
- the network of travel and communication offered by a river system

GeoActivity

SEEING PATTERNS Pair with a partner and draw a **map** of East Asia's rivers and mountains. Use arrows to indicate the directions the rivers flow. Why do the three main rivers of China flow all the way east across the continent even though their headwaters begin in the mountains of the west?

EAST ASIA

Map and Graph Skills

Interpreting a Contour Map

Suppose that you are vacationing on the Japanese island of Hokkaido. As part of your trip, you will be climbing Mount Asahi, the highest point on the island. The members of your group decide to study a contour map to understand the challenge that faces you. You can use a contour map to get a better idea of elevation and the steepness of the mountain.

THE LANGUAGE OF MAPS A **contour map** shows elevations and surface configuration by means of contour lines. Contour lines are lines on a map that show points of equal elevation. These lines are also called isolines. Numbers on the contour lines show the elevation in meters.

Elevation on Hokkaido

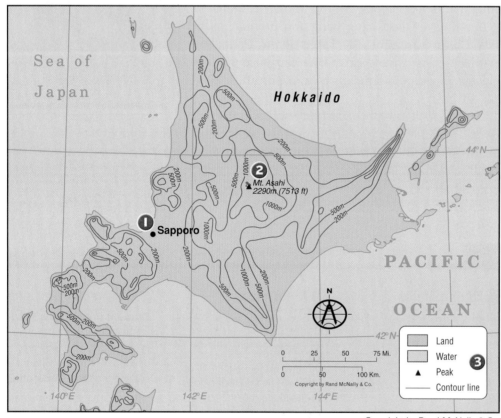

Copyright by Rand McNally & Co.

① Sapporo, the largest city on the island, is situated at a low elevation.

② Mount Asahi is the highest point on the island.

③ The key shows that Mount Asahi is a peak. The key shows that the red lines are contour lines. If you were to follow one contour line around its entire perimeter, you would remain at the same elevation throughout your walk.

Map and Graph Skills Assessment

1. Seeing Patterns
How high, in meters, is Mount Asahi? What is the elevation of the last contour line on the map before the peak?

2. Making Decisions
From what direction of the compass would you approach Mount Asahi if you wanted to make the steepest climb?

3. Drawing Conclusions
Where on the island do the isolines converge most densely to show a very dramatic increase in elevation?

Climate and Vegetation

Main Ideas
• East Asia has a dry highland climate in the west.
• The region has a humid climate in the east.

Places & Terms
typhoon

Taklimakan Desert

Gobi Desert

CONNECT TO THE ISSUES
POPULATION To feed its population, East Asian countries have had to farm in highly productive ways.

A HUMAN PERSPECTIVE Kublai Khan was the ruler of the Mongol Empire (which included China) in the 13th century. In 1281, the Great Khan sent a huge fleet against Japan. A **typhoon**—a tropical storm that occurs in the western Pacific—swept across the Sea of Japan and sank the Mongol ships or dashed them against the rocky Japanese shore. The typhoon had changed the course of history. Typhoons occur in parts of East Asia, but in other ways the weather is similar to that of the United States. Both are at the same latitude, and both have similar climate zones.

High Latitude Climate Zones

The climates in the highest latitudes present a serious challenge to all but the most hardy nomads and herders. These zones generally have severely cold climates. In addition, they tend to be very dry.

SUBARCTIC Subarctic climate zones occur in a small sliver along Mongolia's and China's northern borders with Russia. The summers in these areas range from cool to cold. The winters are brutally cold, testing the survival skills of the inhabitants. The climate is generally dry.

The typical vegetation of this region is the northern evergreen forest. Varieties of mosses and lichens also grow on rocks and tree trunks throughout subarctic zones.

HIGHLAND Highland climates are found mostly in western China. The temperature in highland zones varies with latitude and elevation. In general, the farther north the latitude and the higher the elevation, the colder the climate. The severe climate and topography of the western highlands are two of the reasons that the area is sparsely populated.

The vegetation in the highlands also varies with elevation. Forests and alpine tundra are the typical vegetation. Vast tundras reach as far as the eye can see. Tundras have no trees, and the soil a few feet below the surface is permanently frozen. In this environment, only mosses, lichens, and shrubs can grow. Because of the cold and the difficulty of growing crops, few people scratch out a living here.

HUMAN-ENVIRONMENT INTERACTION A 78-year-old woman tends sheep from the back of a camel in a semiarid zone typical of Mongolia.
What does the occupation of sheepherding and livestock grazing suggest about the vegetation in Mongolia?

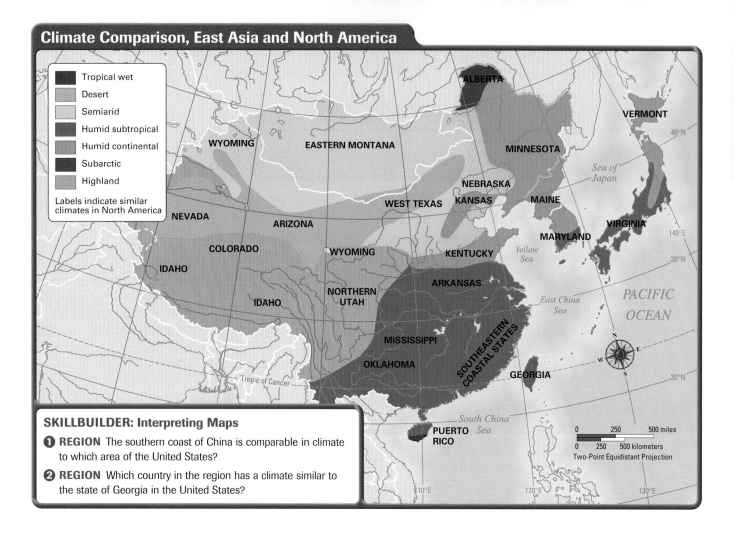

Climate Comparison, East Asia and North America

Legend:
- Tropical wet
- Desert
- Semiarid
- Humid subtropical
- Humid continental
- Subarctic
- Highland

Labels indicate similar climates in North America

SKILLBUILDER: Interpreting Maps

❶ **REGION** The southern coast of China is comparable in climate to which area of the United States?

❷ **REGION** Which country in the region has a climate similar to the state of Georgia in the United States?

Mid-Latitude Zones

Mid-latitude zones are much more comfortable to live in because of their moderate climates. The land is productive, and the rainfall is sufficient for agriculture. An important resource of these zones is their forests.

HUMID CONTINENTAL Northeastern China, North Korea, northern South Korea, and northern Japan all have humid continental climates. The forests of the region are mainly coniferous in the humid continental zone. Temperate grasslands ideal for grazing are also found in these areas. However, over the years agriculture has transformed the landscape and replaced many of the forests.

HUMID SUBTROPICAL Southeastern China, southern South Korea, southern Japan, and northern Taiwan are in a humid subtropical zone. The forests in such zones are both deciduous and coniferous. The broad-leafed, deciduous trees are usually found in the north. The coniferous forests are especially typical of areas with sandy soils in the south. However, loggers and farmers have greatly reduced the forests in the southeast. ▶

CONNECT TO THE ISSUES
POPULATION
◀ Why might most of East Asia's population be centered in the mid-latitude zones?

Dry Zones

Dry zones of the region include both steppes and deserts. There is relatively little vegetation. These zones are not well suited to agriculture and so have not been much settled by people. Instead, nomads have used the semiarid areas to graze livestock.

SEMIARID Parts of the Mongolian Plateau make up the semiarid zones of the region. The vegetation of semiarid zones consists mainly of short grasses, which provide food for grazing animals and livestock.

DESERT Most of the deserts in the region are found in the west central area of the mainland. The **Taklimakan Desert** is located in western China between the Tian Shan and Kunlun Mountains. The **Gobi Desert** is located in northern China and southeast Mongolia. The Gobi is a prime area for finding dinosaur fossils, since thousands of these animals roamed through the region millions of years ago. ◀B

Geographic Thinking ◀

Making Comparisons
B▷ Why might the dry zones of the region be less densely populated?

Tropical Zones

The tropical zones of East Asia contain mainly wet climates. The most common vegetation is the rain forest.

TROPICAL WET The tropical climate zone in East Asia is fairly small. It includes a small strip of land along China's southeastern coast, the island of Hainan, and the southern tip of Taiwan. These areas have high temperatures, heavy rainfall, and high humidity every month of the year. The tropical rain forest in these places is made up of tall dense forests of broadleaf trees.

In the next section, you will read how human-environment interactions affect the quality of life in rural China and urban Japan.

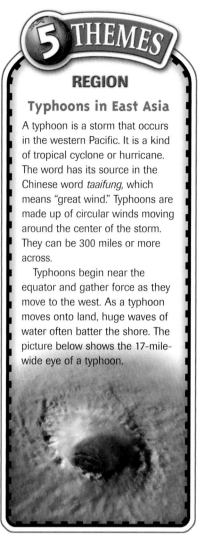

5 THEMES

REGION

Typhoons in East Asia

A typhoon is a storm that occurs in the western Pacific. It is a kind of tropical cyclone or hurricane. The word has its source in the Chinese word *taaifung*, which means "great wind." Typhoons are made up of circular winds moving around the center of the storm. They can be 300 miles or more across.

Typhoons begin near the equator and gather force as they move to the west. As a typhoon moves onto land, huge waves of water often batter the shore. The picture below shows the 17-mile-wide eye of a typhoon.

SECTION 2 Assessment

1 Places & Terms

Identify each of the following places and terms.

• typhoon
• Taklimakan Desert
• Gobi Desert

2 Taking Notes

PLACE Review the notes you took for this section.

Climate	
Vegetation	

• What types of climate are found in East Asia?
• What vegetation characterizes the western reaches of China?

3 Main Ideas

a. In what ways are the climates of the United States and China similar?

b. What effect might severe weather (such as typhoons) have on crops?

c. What has been the human impact on mid-latitude climate zones in the region?

4 Geographic Thinking

Making Inferences How might the climate and vegetation of East Asia have affected patterns of settlement in the region? **Think about:**

• the impact of deserts, steppes, and tundra on patterns of settlement

S See Skillbuilder Handbook, page R4.

EAST ASIA

GeoActivity

EXPLORING LOCAL GEOGRAPHY East Asia has many kinds of climate. Pair with a partner and make a **poster** that shows the climate of East Asia in which you would most want to live. Include photographs, postcards, maps, and charts. Is there any location in the United States that is similar to your preferred climate?

Human–Environment Interaction

A HUMAN PERSPECTIVE Hundreds of thousands of Chinese died in floods in the 20th century. Most of these deaths were caused by the flooding of the Chang Jiang and the Huang He rivers. These vast river floodplains are home to, and help feed, hundreds of millions of people, and this makes people vulnerable to the rivers' wrath. In addition to the many deaths, the flooding has also forced millions of people to abandon their homes. You will read more about one such flood in Chapter 28 (pages 640–641). But since the early 1990s, the Chinese have been building an enormous new dam on the Chang Jiang that will help to control flooding. This is one example of how East Asians have shaped their environment.

The Three Gorges Dam

The **Three Gorges Dam** is being built on the Chang Jiang in China. The dam will, in part, help to control flooding along the great river, the third longest in the world after the Nile and the Amazon. But the dam is also expected to generate power and to allow ships to sail farther into China.

Main Ideas
- The Chinese are building the Three Gorges Dam to control flooding.
- The Japanese have developed creative ways to use their limited amounts of land.

Places & Terms

Three Gorges Dam

PCBs

landfill

CONNECT TO THE ISSUES
PHYSICAL FORCES One reason why the Three Gorges Dam is being built is to control flooding of the Chang Jiang.

Building the Three Gorges Dam

The picture at right, below, shows the city of Fengdu and its suspension bridge. The picture at far right, enhanced by computer imaging, shows how the rising water created by the Three Gorges Dam will cover all of Fengdu except for a few hilltops. The city's population of 65,000 will be moved to a new location.

AN ENGINEERING FEAT The Three Gorges Dam is China's largest construction project and will be the world's biggest dam. When completed, the dam will tower more than 600 feet high and will span a valley more than one mile wide. This dam will create a reservoir nearly 400 miles long. At least 1,000 towns and villages will disappear under the waters of the reservoir when the dam is completed.

POSITIVE EFFECTS The building of the Three Gorges Dam is a complicated issue because it will have both positive and negative effects. Experts disagree about whether the dam should be built. But the Chinese government, which began construction of the dam in 1993, argues that the dam will have three positive effects.

First, the dam will help control the frequent flooding of the Chang Jiang, which causes great damage and loss of life. This is critical because the Chang Jiang irrigates about half of China's crops. Also, the river drains about one-fifth of China's total land area.

Second, the dam will generate huge amounts of electrical power. Giant turbines will produce electricity that will be hooked up to electrical grids in central and eastern China. This will improve the reliability of electricity throughout China. By some estimates, the dam's turbines will produce about 10 percent of China's electrical power. (See the bar chart below for a comparison of the projected generating capacity of the Three Gorges Dam with other large dams.)

Finally, the dam will make it easier for ships to reach China's interior. A series of locks along the river will raise ocean-going ships up from the river to the reservoir. The Chang Jiang carries more than half of the goods moving on China's interior waterways. The dam and the locks will increase shipping capacity and decrease shipping costs. ◀**A**

Geographic Thinking ◀

Seeing Patterns
A ▷ What are three benefits of building the dam?

Facts and Figures

- Length of river: 3,964 miles
- Length of reservoir: 370 miles
- Height of dam: 610 feet
- Width of dam: 1.3 miles
- Number of turbines: 26, generating 18,200 megawatts of electricity

- Lives lost to flooding: about one million deaths in 20th century
- Location of dam: about 1,500 miles from the ocean
- Many hundreds of miles from headwaters in western mountains of China

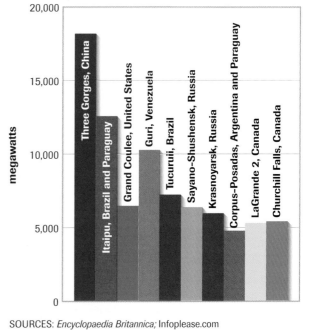

Electric Generating Capacity of World's Largest Dams

megawatts

- Three Gorges, China
- Itaipu, Brazil and Paraguay
- Grand Coulee, United States
- Guri, Venezuela
- Tucuruii, Brazil
- Sayano-Shushensk, Russia
- Krasnoyarsk, Russia
- Corpus-Posadas, Argentina and Paraguay
- LaGrande 2, Canada
- Churchill Falls, Canada

SOURCES: *Encyclopaedia Britannica;* Infoplease.com

Human–Environment Interaction **629**

HUMAN–ENVIRONMENT INTERACTION The river dolphin, the white crane, and the alligator are just three of the species endangered by the construction of the Three Gorges Dam. **Why might the dam be a threat to various species?**

NEGATIVE EFFECTS Most observers agree that the Three Gorges Dam will also have negative effects. The central issue is whether the negative impact on the environment will be greater than the positive benefits.

First, the human costs of the dam will be enormous. Huge numbers of people will have to be moved— somewhere between one million and two million people. Also, hundreds of historical sites and scenic spots will be submerged.

Second, the dam is likely to cost more money than originally anticipated. The Chinese government first estimated the cost at approximately $11 billion dollars. However, other estimates now place the cost closer to $75 billion. A number of banks and other financial institutions have chosen not to participate in the financing of the dam because of their concerns about the cost.

Third, environmental concerns about the dam trouble many observers. The giant reservoir created by the dam will put hundreds of square miles of land under water. This will reduce the habitat of many animals. It is feared that abandoned factories submerged under the reservoir may leak contaminating chemicals into the water. The huge reservoir will affect the climate and temperature of the region as well as the plant and animal life. Such species as the alligator, leopard, sturgeon, white crane, and river dolphin may not survive.

The Three Gorges Dam is scheduled to be completed in 2009. However, the Chinese government has not been careful in protecting the environment from the consequences of building the dam. Some international groups are reluctant to invest in the project because of environmental concerns, and this might delay its completion.

Use of Space in Urban Japan

Throughout history, the geographic challenges facing Japan have been different from those facing China. One of the most important challenges is that Japan is made up of a series of mountainous islands. Most of the cities are on the coasts of these islands. But because of nearby mountains, many of the cities cannot expand to absorb any more of the Japanese population, which is about 127 million people. Tokyo is a good example. One of the world's largest cities, it holds more than 25 million people. There is, however, no more land for the city to grow.

CROWDED LIVING AND WORKING SPACES More than 60 percent of the Japanese people live on only about three percent of the land. The population is clustered along the narrow flat coastal plains.

These plains are among the most densely populated areas in the world. The largest cities in Japan are Tokyo, Yokohama, Osaka, Nagoya, and Sapporo. Close to 80 percent of the people in Japan live in cities.

Partly because of their large populations, some Japanese cities have become very polluted. For example, in the 1950s and 1960s, a number

Geographic Thinking

Seeing Patterns
B What might be some negative effects of the dam?

Geographic Thinking

Using the Atlas
C Use the map on page 615. Why might the Japanese people live on such a small percentage of coastal land?

of Japanese cities experienced poisoning from mercury and **PCBs**—industrial pollutants that build up in animal tissue and can cause disease and birth defects. PCBs were banned in 1977. However, cars and factories still cause massive levels of air and noise pollution.

ADAPTING TO LIMITED SPACE

The Japanese have shown great ingenuity in adapting to limited space. Because of the cost of land, houses are small by American standards. The rooms are separated by sliding screens and are sparsely furnished. People sleep on thin mattresses called futons that can be rolled up and stored during the day.

HUMAN–ENVIRONMENT INTERACTION
Capsule hotels in Japan provide tiny rooms for overnight guests.

Many people, especially in the biggest cities, live in apartments. It is not uncommon for a family of four to live in a one-bedroom apartment. Some Japanese attempt to escape the overcrowding by moving away from the city to distant suburbs, but they must commute for two or even three hours a day to and from work.

One of the solutions to the shortage of space is landfill. **Landfill** is a method of solid waste disposal in which refuse is buried between layers of dirt to fill in or reclaim low-lying ground. The Japanese have used landfill to reclaim land for most of the major cities along the coast. Tokyo, for example, has built factories and refineries on landfill sites. One result of the use of landfill sites has been to enlarge some of Japan's ports. These reclaimed areas are designed to handle the great number of ships that sail in and out of the port.

You will explore more about how East Asians live in the next chapter, on human geography.

Assessment

① Places & Terms

Identify and explain the significance of each in the region.

- Three Gorges Dam
- PCBs
- landfill

② Taking Notes

HUMAN–ENVIRONMENT INTERACTION Review the notes you took for this section.

Human–Environment Interaction

- Which of the examples in this chapter illustrate human adaptation to the environment?
- Which examples illustrate an environment changed by humans?

③ Main Ideas

a. What might be a positive effect of the Three Gorges Dam?

b. What might be a negative effect of the Three Gorges Dam?

c. Why are most of Japan's large cities located along its coast?

④ Geographic Thinking

Determining Cause and Effect What were some of the reasons that led to the building of the Three Gorges Dam? **Think about:**

- the effects of living near an unpredictable river

RESEARCH LINKS
CLASSZONE.COM

GeoActivity

ASKING GEOGRAPHIC QUESTIONS Pair with a partner and research a dam in the United States to compare with the Three Gorges Dam. Devise three geographic questions about the dams, such as "How much concrete was used in the construction of the dams?" Then make a **chart** or **graph** in which you provide data to answer the questions. Be sure to identify your sources.

EAST ASIA

VISUAL SUMMARY
PHYSICAL GEOGRAPHY OF EAST ASIA

Landforms

Major Mountain Ranges: Himalayas, Kunlun, Altun, Altay, Qinling Shandi

Major Rivers: Huang He, Chang Jiang, Xi Jiang

Major Deserts: Taklimakan, Gobi

Major Plateaus and Plains: Plateau of Tibet, Tarim Pendi Basin, Mongolian Plateau, Manchurian Plain, North China Plain

Resources

- China, Mongolia, and North Korea have significant natural resources.
- Japan, South Korea, and Taiwan have limited natural resources.

Climate and Vegetation

- East Asia has a dry continental climate in the west and a humid climate in the east.
- Its mid-latitude zones, both humid continental and humid subtropical, are the most densely populated areas.

Human-Environment Interaction

- The Three Gorges Dam is being built along the Chang Jiang to control flooding.
- Urban Japan is very crowded, and people must adapt to space limitations.

Reviewing Places & Terms

A. Briefly explain the importance of each of the following.

1. Kunlun Mountains
2. Huang He
3. Chang Jiang
4. Xi Jiang
5. typhoon
6. Taklimakan Desert
7. Gobi Desert
8. Three Gorges Dam
9. PCBs
10. landfill

B. Answer the questions about vocabulary in complete sentences.

11. On which river will the Three Gorges Dam attempt to control flooding?
12. What is another name for a tropical cyclone or hurricane?
13. What is the source of two of China's great rivers?
14. Which river joins with others to form an estuary between Hong Kong and Macao?
15. How have landfill sites been used in Tokyo?
16. Where in the region is there a rich supply of dinosaur fossils?
17. What has contributed to the poisoning and pollution of the environment in Japanese cities?
18. Which desert is located in western China near the Kunlun Mountains?
19. Which river is known as "China's Sorrow"?
20. What project is supposed to contain flooding?

Main Ideas

Landforms and Resources (pp. 619-624)

1. Why are the Kunlun Mountains especially important to China?
2. What is the approximate size of the Gobi Desert?
3. What are some of the important islands off the coast of China?
4. Why are China's three river systems so important to the country?

Climate and Vegetation (pp. 625-627)

5. In which latitude and climate zones is most of China's productive agricultural land located?
6. What landforms make up the dry zones of the region?
7. What two factors affect vegetation and temperature in the highland climate?

Human-Environment Interaction (pp. 628-631)

8. What will be some benefits of the Three Gorges Dam?
9. What will be some drawbacks of the dam?
10. What are some of the ways in which the Japanese have adapted to living in a crowded space?

Critical Thinking

1. Using Your Notes

Use your completed chart to answer these questions.

Landforms	
Resources	

a. Where are the highest mountains in China located?

b. What are some energy resources found in abundance in China and Korea?

2. Geographic Themes

a. **LOCATION** Where is the largest desert found in East Asia?

b. **REGION** Write a sentence or two describing the settlement patterns of East Asia in terms of its mountains and coasts.

3. Identifying Themes

Based on landforms and climate, which areas of East Asia would be the least agriculturally productive? Which of the five themes are reflected in your answer?

4. Making Decisions

What factors must people in China consider when they are trying to decide what to do about flooding along one of their great rivers?

5. Drawing Conclusions

How does a typhoon create so much damage?

Additional Test Practice, pp. S1–S37

TEST PRACTICE
CLASSZONE.COM

Geographic Skills: Interpreting Maps

Precipitation in East Asia

Use the map at right to answer the following questions.

1. **REGION** Which parts of the region have the least precipitation?

2. **REGION** Which parts of the region have the most precipitation?

3. **MOVEMENT** How might precipitation patterns have affected settlement in the region?

GeoActivity

Create a way to display the map information in graph form. Be sure to list the six countries of the region by name in your graph.

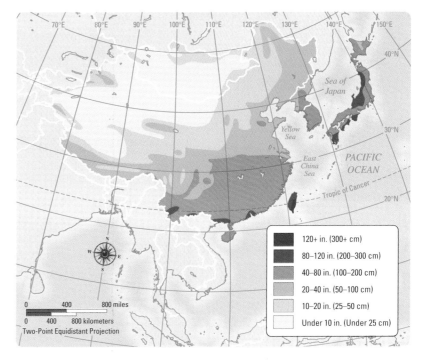

INTERNET ACTIVITY

Use the links at **classzone.com** to do research on the most productive agricultural regions of East Asia. You might focus on the impact that precipitation has had on settlement patterns and crop growth.

Creating Multimedia Presentations Combine charts, maps, or other visual images in an electronic presentation that shows the most productive farming areas and the most common crops in the region.

HUMAN GEOGRAPHY OF EAST ASIA
Shared Cultural Traditions

Four Subregions of East Asia

China
Mongolia and Taiwan
North Korea and South Korea
Japan

MONGOLIA

NORTH KOREA

Sea of Japan

JAPAN

Korea Bay

Bo Hai

SOUTH KOREA

Yellow Sea

CHINA

East China Sea

PACIFIC OCEAN

TAIWAN

Luzon Strait

Tropic of Cancer

Bay of Bengal

Gulf of Tonkin

South China Sea

Equator

0   250   500 miles
0   250   500 kilometers
Two-Point Equidistant Projection

GeoFocus

How has China influenced the cultures of East Asia?

Taking Notes In your notebook, copy a cluster diagram like the one below. For each subregion of East Asia, take notes about its history, economics, culture, and modern life.

China

Mongolia and Taiwan

East Asia

The Koreas: North and South

Japan

China

Main Ideas
- China is the world's most populous country.
- China has been the dominant culture of East Asia since ancient times.

Places & Terms

dynasty

spheres of influence

Boxer Rebellion

Mao Zedong

Confucianism

Taoism

Buddhism

CONNECT TO THE ISSUES
POPULATION China's huge population puts a great strain on the environment.

A HUMAN PERSPECTIVE In ancient times, China had been open to attack from nomadic horsemen who roamed the plains of northern China and Mongolia. Around 220 B.C., the emperor Shi Huangdi decided to build the Great Wall of China by closing the gaps between smaller walls built by earlier rulers. Hundreds of thousands of peasants were used as forced labor to build the Great Wall. The workers hauled and dumped millions of tons of rubble to fill the core of the wall. From the Yellow Sea in the east to the Gobi desert in the west, the Great Wall twisted and turned for thousands of miles, protecting and isolating China from the barbarian warriors beyond its borders.

China's Early History

China is the world's oldest continuous civilization. The beginnings of that civilization extend back into the mists of prehistory. Because of China's geography—the long distances that separated it from Europe and other continents—it followed its own direction.

EARLY CIVILIZATION AND THE DYNASTIES China has been a settled society for more than 4,000 years. In its earliest days, China was made up of a number of Stone Age cultures. Then it was ruled by dynasties. A **dynasty** is a series of rulers from the same family. The first Chinese dynasty was the Shang. This dynasty arose during the 1700s B.C. It ruled a central area in China for about 600 years until it was overthrown by the Zhou Dynasty, which ruled part of northern China.

The next important dynasty, the Qin (chihn), gave its name to China. In 221 B.C., the Qin Dynasty united a number of smaller states under a strong central government and established an empire. The first Qin emperor was Shi Huangdi, the builder of the Great Wall. The Chinese empire, ruled by different dynasties, lasted for more than 2,000 years.

Another important Chinese dynasty was that of the Han. These rulers pushed the empire into central Asia, home to many nomadic tribes. Many other dynasties followed over the centuries.

In 1644, the Manchu people of Manchuria invaded China and established the Qing (chihng) Dynasty. In 1911, the Manchus were overthrown by revolutionaries, and this ended the dynasties and the Chinese empire.

PLACE Thousands of life-sized terra cotta (clay) soldiers have been unearthed by archaeologists near the tomb of the emperor Shi Huangdi near Xian, China.

China Opens Up to the World

Even though China remained isolated from other regions for centuries, that started to change in the 13th century. At that time, European travelers began to visit China. Marco Polo, for example, traveled from Venice, Italy, to China in the 13th century and wrote a book about his adventures, *The Travels of Marco Polo.*

China and Europe had few contacts until the 19th century, when European powers sought access to Chinese markets. At that point, China had a weak military and an ineffective government. Europeans took advantage of China and forced it to sign a series of treaties that granted special privileges to the Europeans. Consequently, China was carved up into **spheres of influence** controlled by Britain, France, Germany, Russia, and Japan. This outside control angered China, which burst forth in the **Boxer Rebellion** of 1900. Chinese militants attacked and killed Europeans and Chinese Christians in China. A multinational force of about 20,000 soldiers finally defeated the Boxers.

BACKGROUND
The Boxers were a secret society whose Chinese name meant "fists of righteous unity."

REVOLUTION AND CHANGE After the Boxer Rebellion, the Qing Dynasty, founded by the Manchus, attempted to reform the Chinese government, but it was too late. Many individuals and groups wanted to form a republic, which would give the people a voice in their government. In 1912, Sun Yat-sen and others founded the *Kuomintang,* or Nationalist Party. However, the republic, led by Sun Yat-sen, was undermined by civil war throughout China.

When Sun Yat-sen died in 1925, a general named Chiang Kai-shek took over the Nationalist Party. Chiang's troops fought against the warlords of China and united most of the country in the 1920s. However, throughout the 1920s and 1930s, the Chinese Communist Party became an increasingly powerful force in China.

The Nationalists and the Communists fought for control of China. In 1949, the Communists, under the leadership of **Mao Zedong,** finally defeated the Nationalists. Mao and the Communists ruled mainland China (now called The People's Republic of China) from Beijing. Chiang Kai-shek and the Nationalists fled to the island of Taiwan.

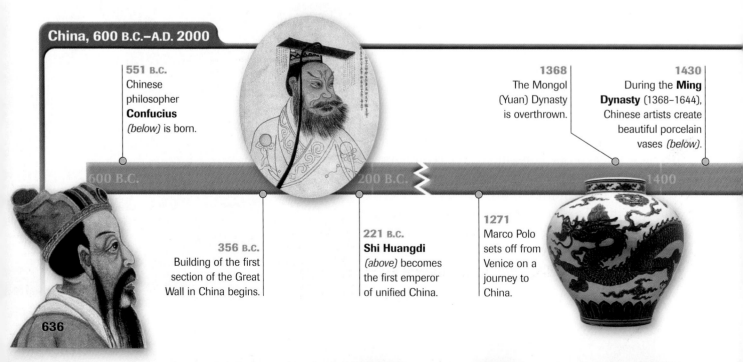

China, 600 B.C.–A.D. 2000

551 B.C.
Chinese philosopher **Confucius** *(below)* is born.

356 B.C.
Building of the first section of the Great Wall in China begins.

221 B.C.
Shi Huangdi *(above)* becomes the first emperor of unified China.

1271
Marco Polo sets off from Venice on a journey to China.

1368
The Mongol (Yuan) Dynasty is overthrown.

1430
During the **Ming Dynasty** (1368–1644), Chinese artists create beautiful porcelain vases *(below).*

600 B.C.   200 B.C.   1400

After Mao died in 1976, Deng Xiaoping, a moderate, became China's most powerful leader through the 1980s. In 1993, Jiang Zemin became president and Zhu Rongji became premier in 1998. Both focused their attention on developing China's economy.

Rural and Industrial Economies

When the Communist Party came to power in China in 1949, its leaders promised to modernize China by encouraging the growth of industry. From the 1950s through the 1970s, the central government tried to do this by planning all economic activities. That approach led to more failures than successes. Since the 1980s, though, China has allowed the marketplace and the consumer to play a role in the economy. As a result, China now has one of the fastest growing economies in the world.

THE RURAL ECONOMY In spite of this economic growth, China remains a largely rural society, self-sufficient in agriculture. Its great river valleys provide rich soil for crops such as rice to feed the vast population. Most of China's workers—about 60 percent—work on farms.

Farming is possible only on about 13 percent of China's land because so much of western China is made up of mountains and deserts. Even so, China manages to grow enough food to feed its people. Much of the population is concentrated in the areas where food can be grown.

CONNECT TO THE ISSUES
POPULATION
A▶ Why is so much of China's population in the east and so little in the west?

The eastern river basins of China produce crops such as rice, maize, wheat, and sweet potatoes. This productivity is aided by the long growing season in southern China. Farmers there can grow two or more crops on the same land during each year. ◀A

THE INDUSTRIAL ECONOMY The industrial heartland of China is in the northeast. Here are abundant resources important to manufacturing, such as coal, iron ore, and oil. (See map, page 622.) In addition, the northeast has better transportation systems than the rest of the country.

Shanghai leads China as a center of manufacturing and is one of the great industrial centers in the world. Other Chinese cities with many factories and industries include Beijing and Tianjin. Southeastern China

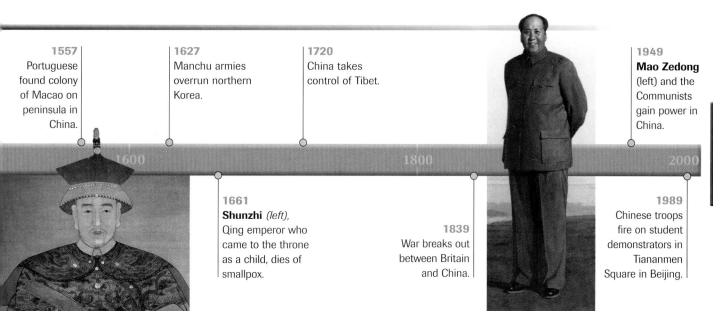

1557 Portuguese found colony of Macao on peninsula in China.

1627 Manchu armies overrun northern Korea.

1720 China takes control of Tibet.

1949 **Mao Zedong** (left) and the Communists gain power in China.

EAST ASIA

1600

1800

2000

1661 **Shunzhi** *(left),* Qing emperor who came to the throne as a child, dies of smallpox.

1839 War breaks out between Britain and China.

1989 Chinese troops fire on student demonstrators in Tiananmen Square in Beijing.

has industrial centers in Guangzhou, Hangzhou, Suzhou, Wuhan, and Wuxi.

China has developed heavy industries, such as steel and machinery. It also produces consumer goods. For example, the country has a huge textile (cloth) industry that produces goods for the home market and export. Many textiles are exported to the United States. ▶

A Rich and Complex Culture

As the world's oldest civilization, China has one of the world's richest cultures. The country has highly developed art, architecture, literature, painting, sculpture, pottery, printing, music, and theater. In all these areas, the Chinese have made influential contributions to the cultures of Korea, Japan, and other countries in the region.

FROM POTTERY TO PAINTING Some of the earliest Chinese works of art have been found in burial sites. Pottery, bronze vessels, and jade disks have been discovered in the excavation of old tombs. In addition, paintings have been found on tiles decorating the walls of tombs. Chinese artists created beautiful works using different materials, such as clay, bronze, jade, ivory, and lacquer.

CHINESE INVENTIONS The Chinese introduced many inventions to the world, such as paper, printing, and gunpowder. Other Chinese inventions include the compass, porcelain, and silk cloth.

RELIGIOUS AND ETHICAL TRADITIONS China has three major religions or ethical traditions. The beliefs of most people include elements of all three. Those traditions have influenced beliefs throughout the region.

Confucius was a Chinese philosopher who lived from 551 to 479 B.C. He believed in respect for the past and for one's ancestors. He thought that in an orderly society, children should obey their parents and parents should obey the government and emperor. He stressed the importance of education in a well-run society. His thinking about the importance of order, education, and hierarchy in a well-ordered society is called **Confucianism.**

Taoism gets its name from a book called the *Tao-te Ching,* based on the teaching of Lao-tzu, who lived in the sixth century B.C. He believed in the importance of preserving and restoring harmony in the individual and in the universe. He also thought the government should leave the people alone and do as little as possible. Another of his major beliefs was that the individual should seek harmony with nature.

Buddhism came to China from India and grew into an important religion in China by the 300s A.D. Confucianism and Taoism influenced Buddhism as it developed in China. Among ideas important in Buddhism are rebirth and the end of the rebirth cycle.

CONNECT TO THE ISSUES
TRADE
B Why might trade between the United States and China be important to both countries?

BACKGROUND
Other important Chinese art forms include calligraphy and brush painting.

Chinese Artifacts

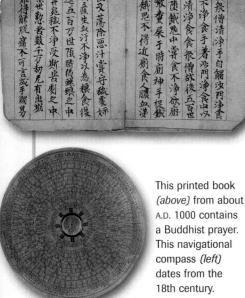

An ancient Chinese coin *(above left)* is from about 450 B.C. The jade pendant *(above right)* is from about 250 B.C.

This printed book *(above)* from about A.D. 1000 contains a Buddhist prayer. This navigational compass *(left)* dates from the 18th century.

The Most Populous Country

One out of every five people in the world lives in China. This makes it the most populous country in the world.

POPULATION PATTERNS China's estimated population in the year 2000 was about 1.3 billion. Somewhere between 30 and 40 Chinese cities have populations of more than one million people. Many of China's 22 provinces have more people than entire countries. In the year 2000, Henan province was estimated to have a population of about 93 million people—more than the population of Great Britain.

Seventy percent of the people live in 12 provinces located in the east. (See map, page 615.) About 6 percent of the people live in the west on 55 percent of the land.

HEALTH CARE One of the great achievements of China since 1950 has been to provide health care for its enormous and far-flung population. The country has pursued a dual strategy in developing its health-care system.

On the one hand, people make use of traditional Chinese medicines, including herbal remedies. Acupuncture is another important part of Chinese medicine.

On the other hand, China's doctors also use Western medicine to treat disease. Western drugs and surgery have their place in the treatment of illness. Most Chinese cities have hospitals, and the villages have clinics staffed by trained medical workers called "barefoot doctors."

In the next section, you will read about two of China's neighbors, Mongolia and Taiwan. China has greatly influenced both places.

Geographic Thinking

Seeing Patterns
What does the immense size of China suggest about its future?

Connect to the Issues

POPULATION
One-Child Policy

Because of concerns about a rapidly expanding population, China in 1979 adopted a policy of one child per family. In addition, the country has age restrictions for marriage. A man must be 22 and a woman 20 before they can marry. Those policies have reduced China's birthrate dramatically.

However, the government policy of one child per family has run into opposition. Rural families, in particular, feel the need for more than one child to help work on their farms. Because of these problems, the government has relaxed the one-child policy.

SECTION 1 Assessment

1 Places & Terms

Identify each of the following places and terms.

- dynasty
- spheres of influence
- Boxer Rebellion
- Mao Zedong
- Confucianism
- Taoism
- Buddhism

2 Taking Notes

REGION Use your notes to answer the questions below.

- What are aspects of China's cultural legacy?
- What are some Chinese dynasties?

3 Main Ideas

a. Why is China's rural economy still so important?

b. What are some of China's most important religious ideas?

c. Why is population such an important issue in China?

4 Geographic Thinking

Making Generalizations
How has China's rugged terrain affected its relations with other countries and civilizations? **Think about:**

- the mountains and deserts to the west
- the ocean to the east

S See Skillbuilder Handbook, page R6.

SEEING PATTERNS Pair with a partner and investigate an invention of the Chinese, such as printing or the compass. Then present your findings to the class in a brief **oral report** accompanied by an illustration of the invention.

Disasters!

Chang Jiang (Yangtze River) Flood of 1931

Throughout Chinese history, the flooding of the Chang Jiang has cost millions of lives. On average, the Chang Jiang has caused a major flood about every 50 years, although in the past century or so the floods have been more frequent. The floods of 1931 and 1954 were particularly devastating. The 1931 flood resulted from monsoon rains. In May and June of that year, six enormous waves poured down the river, demolishing dams and dikes. More than 35,000 square miles of land were flooded and many thousands of people died. Floods along the Chang Jiang continue to the present day. Bad floods occurred in both 1996 and 1998.

Nanjing was one of the cities in China that remained underwater for weeks because of the 1931 flood.

CHINA

Nanjing

Wuhan

Chang Jiang (Yangtze R.)

Wuchang, **Han**yang, and **Han**kou are three cities that make up one huge urban complex called Wuhan. Much of **Wuhan** remained underwater for more than four months in 1931. The water ranged from 6 feet to 20 feet in depth.

The Three Gorges Dam is currently under construction to control the flooding of the Chang Jiang.

In the city of Hankou during the flood, wealthy people traveled in boats while poor tradespeople waded up to their necks through the water.

This panoramic aerial view of one of the Chinese cities flooded in 1931 was taken by Charles Lindbergh. He was the American aviator who had made the first solo flight across the Atlantic Ocean in 1927.

Along the Chang Jiang, human labor is still essential for flood control. These laborers work with shovels and other tools to fortify the banks of the river with dirt to prevent flooding.

GeoActivity

UNDERSTANDING FLOODS

Working with a partner, use the Internet to research one of the floods listed below. Then create a **presentation** about it.

- Create a diagram showing the extent of the flood, the damage caused by it, and the number of lives lost.
- Add a map of the affected region.
- Write a paragraph explaining how the flood affected the people and life of the region.

RESEARCH LINKS
CLASSZONE.COM

GeoData

OTHER DEADLY RIVER FLOODS

1850

1887
Huang He in northeastern China; possibly more than 1,000,000 people killed

1889
Johnstown, Pennsylvania, on May 31; about 2,200 deaths (more than any other river flood in U.S. history)

1911
Chang Jiang in China; 100,000 killed

1937
Mississippi and Ohio rivers; about 250 killed

1988
Three major rivers in Bangladesh; about 1,600 deaths

1993
Mississippi River; millions of acres flooded; about 50 dead

1998
Chang Jiang in China during July and August; about 4,000 dead

2000

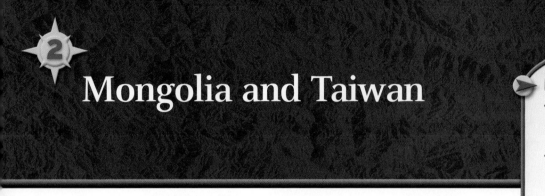

2 Mongolia and Taiwan

Main Ideas

- Taiwan and Mongolia have developed in the shadow of their giant neighbor—China.
- The countries of the region include both capitalist and socialist economies.

Places & Terms

economic tiger

Pacific Rim

CONNECT TO THE ISSUES

TRADE Trade has helped Taiwan achieve prosperity, while Mongolia has not been as economically successful.

A HUMAN PERSPECTIVE The Mongols of the Asian steppe lived their lives on horseback. In 1206, a great leader named Temujin (later called Genghis Khan) united the Mongol clans and led them in conquering much of Asia. He is reported to have said, "Man's greatest good fortune is to chase and defeat his enemy, seize his total possessions, leave his married women weeping and wailing, and ride his horse." The Mongols eventually created the largest unified land empire in history, extending from the Pacific coast of China westward into Europe.

A History of Nomads and Traders

The histories of Mongolia and Taiwan have been closely connected to that of China.

THE MONGOLIAN EMPIRE The Mongols were nomadic herders for thousands of years. Mongol history was changed forever by Genghis Khan, a title that means "supreme conqueror." Genghis Khan died in 1227, having conquered all of Central Asia and begun the conquest of

A Mongol Army on the Move

A Mongol army was like a moving city. The cavalry of 10,000 was accompanied by an even greater number of family members and by tens of thousands of horses and livestock.

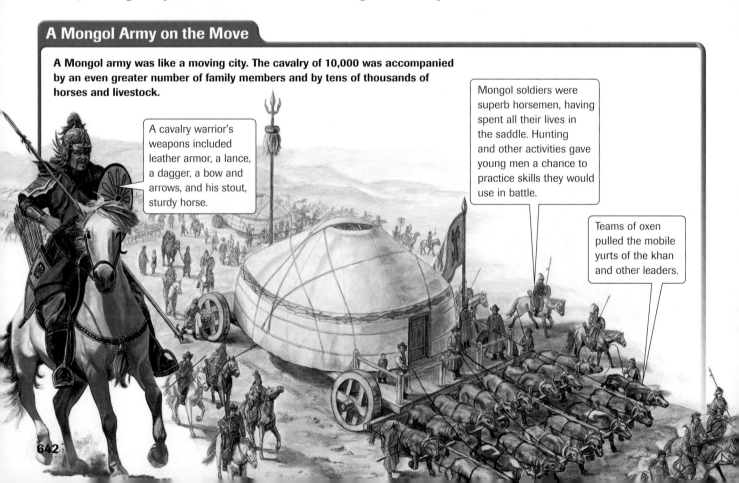

A cavalry warrior's weapons included leather armor, a lance, a dagger, a bow and arrows, and his stout, sturdy horse.

Mongol soldiers were superb horsemen, having spent all their lives in the saddle. Hunting and other activities gave young men a chance to practice skills they would use in battle.

Teams of oxen pulled the mobile yurts of the khan and other leaders.

China. He was succeeded by his son Ogadai, who continued his policies of conquest and expansion. Mongol armies commanded by other sons and grandsons of Genghis Khan moved east, west, and south out of Mongolia.

The Mongol empire broke up in the 1300s. Eventually the Chinese gained control of Mongolia in the 17th century. The Chinese ruled Mongolia for hundreds of years. Only in 1911 were the Mongolians finally able to push the Chinese out and achieve their independence.

Under the influence of its powerful neighbor Russia, Mongolia became the Mongolian People's Republic in 1924. For about 72 years, the Communists ruled Mongolia. However, after the fall of the Soviet Union in 1989, the Communist Party in Mongolia lost its power. The country began moving toward political democracy and a free-enterprise economy. ◁A

⊕ **Geographic Thinking**

Seeing Patterns
A▷ What are some of the countries that have controlled or been controlled by Mongolia over the centuries?

The Mongol Empire, 1294

SKILLBUILDER: Interpreting Maps

❶ **REGION** Which khanate controlled Mongolia and China?

❷ **MOVEMENT** What object may have restricted movement between the Gobi desert and the heartland of China?

TAIWAN'S LINK TO CHINA The island of Taiwan experienced many prehistoric migrations from southern China and southeast Asia. Malay and Polynesian peoples also settled there. Over the centuries, other settlers and groups of people from China settled on the island. In the sixth century, for example, some Han Chinese arrived. Later, when famine struck Fujian province in the 17th century, a large number of Chinese migrated from the mainland. That contributed to the large Chinese settlements on the island. The Manchu Dynasty conquered Taiwan in 1683. (See Unit Atlas, page 613.)

The Japanese seized Taiwan (then called Formosa) after winning a war with China in 1895. Japan kept the island until its defeat in World War II. Then Chinese Nationalists took control of the island as part of their fight with the Communists for control of mainland China. When the Nationalists lost to the Communists in 1949, they moved their government to Taiwan. There they established the Republic of China. However, the People's Republic of China has never recognized Taiwan as a separate country and considers it a province.

Cultures of Mongolia and Taiwan

China is a cultural hearth that has influenced its neighbors. It has been the source for many of the important ideas and inventions that have shaped Mongolia and Taiwan and the rest of the region.

MONGOLIA Mongolia has both ruled and been ruled by China. Kublai Khan was the Mongol emperor of China when Marco Polo visited in the 13th century. In the mid-14th century, the Chinese rose up against their

Mongol rulers and drove them out of China. In the 17th century, the Chinese under the Manchus conquered Mongolia, which they ruled for hundreds of years. This interaction produced a profound cultural influence as the Mongols adopted many aspects of Chinese culture.

The most important festival in Mongolia is the annual Naadam festival of the Three Games of Men. The festival, which dates back 2,300 years, begins each year on July 11. The three games are wrestling, archery, and horse racing. The competitors are highly skilled, and winners receive titles proclaiming their abilities. All of these contests have their roots in the ancient way of life of the Mongolian people.

TAIWAN Unlike Mongolia, Taiwan has a population that is almost exclusively Chinese. Thus, the culture of the island is Chinese. The capital city of Taipei includes Buddhist temples as well as museums of Chinese art. The island has many universities and about 30 daily newspapers. The population is well-educated, and most of the people speak the official language of Northern Chinese (also called Mandarin).

The people of Taiwan combine a number of religious and ethical beliefs. More than 90 percent practice a blend of Buddhism, Confucianism, and Taoism. A small number are Christian and an even smaller percentage practice other religions.

BACKGROUND The population of Taiwan is one of the best educated in Asia, second only to that of Japan.

Two Very Different Economies

The economies of Mongolia and Taiwan have roots in the past. Raising livestock, a part of the nomadic life, is at the core of the Mongolian economy. Because Taiwan is an island, trade is key to its economy.

ECONOMIC PROSPECTS FOR MONGOLIA A large part of the population of Mongolia still engages in herding and managing livestock. For centuries, the economy was based on the nomadic herding of sheep, goats, camels, horses, and cattle. More goats are being raised to meet the demands of the cashmere industry, which uses soft wool from goats of the region. Of the millions of animals kept in herds in the country, nearly a third are sheep. Animals and animal products are used for domestic consumption as well as for export.

Although livestock remains the basis of the economy, Mongolia is now committed to the development of other industries. Under the Communist government, the state owned and operated most of the factories in the country. The Soviets guided Mongolia's economy for about 70 years. When the Soviet Union fell

HUMAN-ENVIRONMENT INTERACTION A Mongolian mother and daughter use red paint to mark the horns of their goats.
What purpose might marking goats serve?

apart, Mongolia was one of the first Communist countries to attempt to shift to a market economy. The transition has been difficult as the country has turned increasingly from a Soviet-style managed economy to a free-market economy.

Mongolia has large deposits of fuels such as coal and petroleum. It also has rich deposits of metals such as copper, gold, and iron. Those resources are used in both manufacturing and construction, industries which are of growing importance to the economy.

TAIWAN'S ECONOMIC SUCCESS Taiwan has one of the world's most successful economies. It has succeeded despite the fact that it has few natural resources. However, it has a highly trained and motivated work force.

Taiwan's prosperity is based on its strong manufacturing industries and its trade with other nations. Among the most successful products of its factories are radios, televisions, calculators, and computers. Taiwanese companies sell their products around the world.

Taiwan is considered one of the economic tigers of Asia, along with Singapore and South Korea. An **economic tiger** is a nation that has rapid economic growth due to cheap labor, high technology, and aggressive exports. It is one of the very prosperous economies of the western Pacific. These economies are highly industrialized and trade with nations around the world. They are part of the **Pacific Rim**—the countries surrounding the Pacific Ocean. The Pacific Rim is an economic and social region. It includes the countries of East Asia, Southeast Asia, Australia, New Zealand, Chile, and the west coast of the United States. ◀ B

Geographic Thinking

Making Comparisons

B▶ What are some differences between the economies of Mongolia and Taiwan?

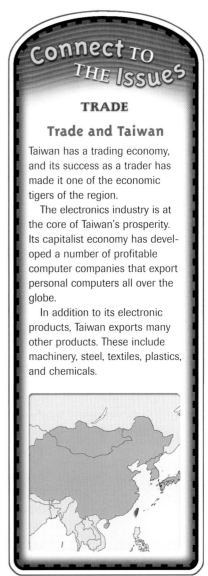

Connect TO THE Issues

TRADE

Trade and Taiwan

Taiwan has a trading economy, and its success as a trader has made it one of the economic tigers of the region.

The electronics industry is at the core of Taiwan's prosperity. Its capitalist economy has developed a number of profitable computer companies that export personal computers all over the globe.

In addition to its electronic products, Taiwan exports many other products. These include machinery, steel, textiles, plastics, and chemicals.

Daily Life in Mongolia and Taiwan

The daily life of people in Mongolia and Taiwan shows traditional influences as well as modern influences. This blending of old and new can be seen in both work and play.

HERDING IN MONGOLIA As you learned earlier in this section, the people of Mongolia were nomads who guided their animals from grassland to grassland. The land through which they traveled has an unpredictable, hostile environment. The climate is extreme. Long, cold winters lasting six months alternate with short, hot summers of only two months. Severe winter weather makes it difficult for livestock to survive. Bad weather can kill animals from intense cold and starvation.

Nomads live in tents called yurts that are made of felt covered with leather. This is the traditional form of shelter in Mongolia. Yurts can even be found in the capital of Ulaanbaatar.

Today, many of the people of Mongolia still spend their days raising sheep, cattle, and goats. Some still follow the nomadic way of life, but most people care for livestock on farms and ranches. Often these farms have small villages in the center, with shops, offices, and houses.

EAST ASIA

MOVEMENT
Taiwan's team celebrates winning the Little League World Series in Williamsport, Pennsylvania, in 1996.

WESTERN INFLUENCES IN TAIWAN Although Mongolia remains relatively isolated from the West, Taiwan has opened itself to many Western influences.

For example, baseball has become popular in Taiwan and in other parts of Asia, particularly Japan. As a part of this general interest in the sport, Little League baseball has also become popular in parts of Asia.

Little League became popular after World War II. In 1974, the United States banned teams from foreign countries from the Little League World Series. In part, that was a response to the success of Taiwan's teams which, throughout the 1970s, dominated the World Series. However, they were restored to competition in 1976. By the 1980s, there were leagues in the United States and 30 other countries.

In the next section, you will read about two countries that share one peninsula: North Korea and South Korea.

Assessment

1 Places & Terms

Identify each of the following places and terms.

- economic tiger
- Pacific Rim

2 Taking Notes

REGION Use your notes to answer the questions below.

Mongolia & Taiwan — East Asia

- How are the economies of Mongolia and Taiwan different from one another?
- What effect did Genghis Khan have on the history of the region?

3 Main Ideas

a. In which ways has China influenced its neighbors?

b. What are some of the characteristics of an economic tiger?

c. In what ways does the modern life of Mongolia and Taiwan show a blending of ancient and modern traditions?

4 Geographic Thinking

Drawing Conclusions
How might the locations of Mongolia and Taiwan have made them open to the influence of China? **Think about:**

- the relative locations of Taiwan and Mongolia

GeoActivity

SEEING PATTERNS Pair with a partner and do Internet research on Little League baseball in Taiwan or some other country in East Asia. Create a **poster** showing various teams in the region. You might include photographs and charts in your poster, listing the names of teams, their win-loss records, and any other information your research turns up.

The Koreas: North and South

A HUMAN PERSPECTIVE Korea is surrounded by water on three sides and by mountains on its northern border. In the 17th and 18th centuries, Korea chose self-protected isolation and became known as "the hermit kingdom." This isolation has continued in North Korea, which has little contact with other nations even today. However, that may be changing.

A Divided Peninsula

Korea is a peninsula. To the east lies the Sea of Japan. To the west lies the Yellow Sea. To the south lies the Korea Strait. To the north lie China and Russian Siberia. Korea's location has shaped its history.

ANCIENT KOREA AND FOREIGN INFLUENCES The ancestors of today's Koreans probably migrated into the peninsula from Manchuria and North China many thousands of years ago. Over the course of the centuries, different clans or groups controlled different parts of the country. About 2000 B.C., the first state, called Chosen, arose in Korea.

Around 100 B.C., China conquered the northern half of the peninsula. This began the history of invasions by China and Japan. Because of its location, Korea has been a buffer between the two countries.

After being partially conquered by China, the Koreans gradually won back their territory. By the late 300s, the **Three Kingdoms** had formed in the peninsula. These were Koguryo in the northeast, Paekche in the southwest, and Silla in the southeast. In the 660s, Silla conquered the other two kingdoms and controlled the peninsula for hundreds of years.

Main Ideas
• The Korean peninsula is divided into two separate countries.
• North Korea is a Communist country, and South Korea is a republic.

Places & Terms
Three Kingdoms

Seoul

Pyongyang

CONNECT TO THE ISSUES
TRADE South Korea is one of the economic tigers of the region, and much of its prosperity depends upon industry and trade.

PLACE Kyungbok Palace is located in Seoul, South Korea. **What does the setting of the palace amidst the bustle of Seoul suggest about the culture?**

EAST ASIA

647

① North Korea Invasion, 1950 — Area occupied by Communist forces
② UN Offensive, 1950 — Area occupied by UN forces
③ Chinese Offensive, 1950 — Movement of Communist forces
④ Stalemate and Armistice, 1953 — Movement of UN forces

0   100   200 miles
0   100   200 kilometers   Lambert Conformal Conic Projection

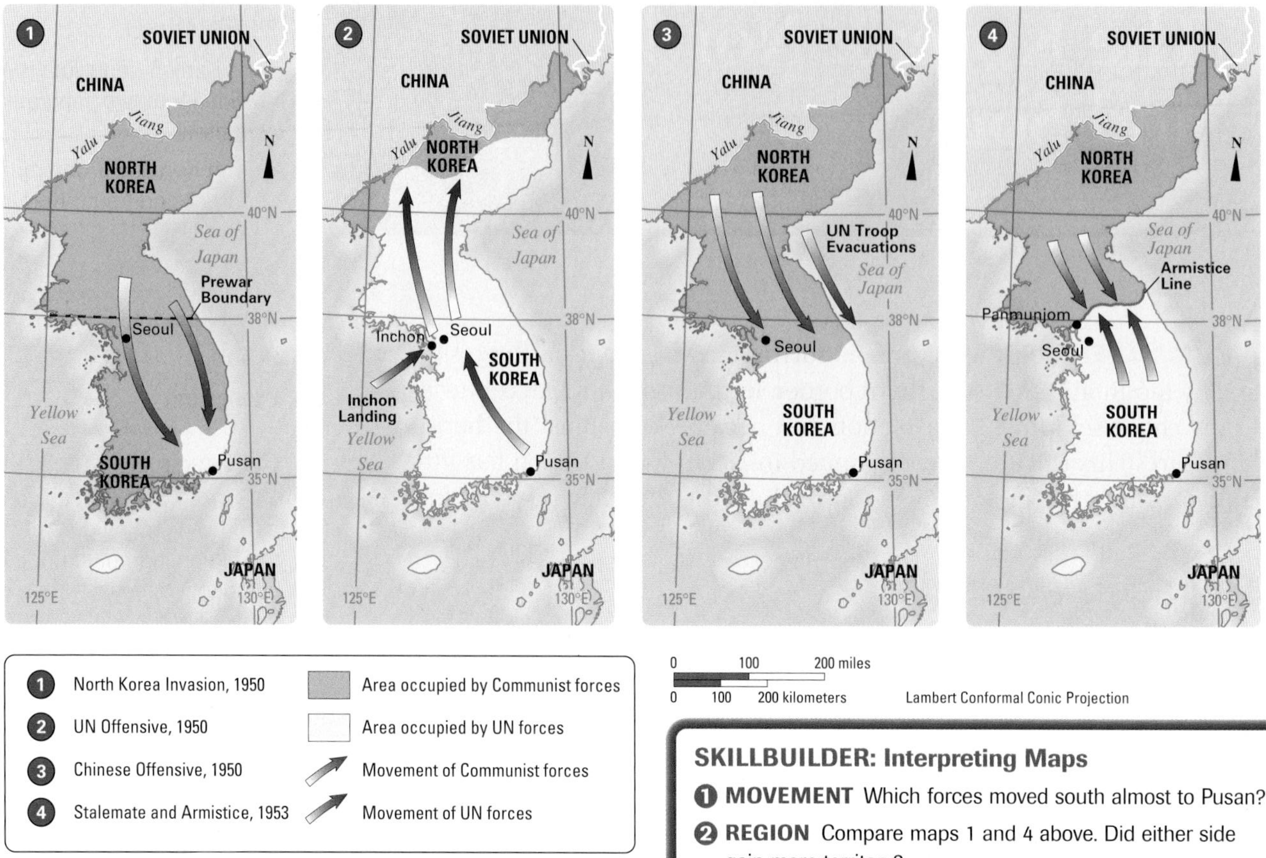

SKILLBUILDER: Interpreting Maps

❶ **MOVEMENT** Which forces moved south almost to Pusan?

❷ **REGION** Compare maps 1 and 4 above. Did either side gain more territory?

In 1392, a general named Yi Songgye became ruler of Korea. He founded a dynasty that lasted for hundreds of years. But the dynasty ended in 1910, when Japan took control of the entire peninsula. The Japanese ruled Korea until they were defeated in World War II in 1945.

TWO KOREAS: NORTH AND SOUTH After Japan's defeat in the war, the northern part of Korea was controlled by the Soviet Union, and the southern half was supported by the United States. In 1950, Korean troops from the North invaded South Korea, starting the Korean War. The war ended in 1953 with a treaty that divided the peninsula between the Communist state of North Korea and the democratic country of South Korea. The two nations remained hostile toward each other, but in the year 2000, they began discussions on reuniting.

Influences on Korean Culture

The shadow cast by China has fallen across the Korean peninsula. Korean culture, including language, art, and religion, shows this influence. More recently, western economic influences have been very important.

THE CHINESE INFLUENCE In philosophy and religion, Korea has adapted many ideas from China. Confucianism (see Section 1) is a system of teachings based on the beliefs of the Chinese scholar Confucius. His ideas stressing social order have influenced many Koreans. Buddhism, which came to Korea by way of China, has also influenced many Koreans.

OTHER CULTURAL INFLUENCES Since World War II, two major influences have had a profound effect on Korea. First, Communism has molded the culture of North Korea. Non-Communist South Korea, on the other hand, has been greatly influenced by Western culture.

In North Korea, the government only allows art that glorifies Communism or the folk tradition. In South Korea, artists have more freedom of expression. They work with themes drawn from their own history and culture, as well as themes drawn from Western art.

Moving Toward Unity

The most important recent development in North Korea and South Korea is the movement toward unification. However, the communist North and democratic South must overcome years of mutual hostility.

AN ARMED SOCIETY After World War II, both North Korea and South Korea built up huge armies. The armed forces of South Korea number more than 600,000 soldiers and sailors. The armed forces of North Korea are even larger, numbering well over one million.

Both countries have existed with large armies and the threat of another war for many years. Only recently has there been an attempt to defuse the situation to prevent an outbreak of war. War has been a real possibility along the border between North Korea and South Korea, which is guarded by nearly 2 million troops on both sides. ◀**A**

A SINGLE FLAG There are signs of hope, however. In June 2000, the leaders of both Koreas held a summit meeting at which they declared their intention to reunite the two countries. Shortly after, the defense

Geographic Thinking ◀

Seeing Patterns
A▷ What have been the main differences dividing North Korea and South Korea?

---

growing up in...South Korea

Young people, like most other South Koreans, follow at least some of the teachings of Confucius. For example, education is highly valued. The state requires by law that students obtain a primary education, and this schooling is free. The majority of children attend secondary schools. More than one million students attend college-level schools in South Korea.

However, in addition to traditional ideas and ways of life, there is a strong western influence in South Korea. This can be seen in the Western clothes worn by these students as they enjoy an outing in the Nampodong shopping district in Pusan, South Korea.

If you lived in South Korea, you would pass these milestones:

- You would be required by law to attend school through 6th grade.
- You would next attend middle school—grades 7 through 9.
- You would then probably attend high school—grades 10 through 12.
- You would be able to vote at age 20.
- The average age for a first marriage is 29 for men and 26 for women.
- The average age of women at the birth of their first child is 27.

chiefs of the two Koreas met and agreed to reduce tensions along their border. They agreed to discuss clearing land mines so they could rebuild a rail link between the two countries. Perhaps most importantly, families in North Korea and South Korea were allowed to visit each other.

At the summer Olympics held in Sydney, Australia, in 2000, there was another sign of a thaw. The two Koreas marched into the Olympic Stadium under a new flag designed for a single, unified Korea.

Economic and Human Resources

Before the Korean War, the economies of North Korea and South Korea were agricultural. After the war, industry gained in importance in both countries. In many ways, the resources of each country balance one another.

ECONOMIC PATTERNS If North Korea and South Korea reunite, they will form an economic powerhouse. North Korea will be able to provide natural resources and raw materials for South Korea's industries.

South Korea, like Taiwan, is one of the economic tigers of Asia. It is a highly successful and competitive economy. It has the world's largest shipbuilding industry, as well as large automobile, steel, and chemical industries. South Korea is today one of the world's top trading nations.

POPULATION PATTERNS Most of the people in Korea live on plains along the coast or in river valleys among the mountains of the peninsula. South Korea has 45 percent of the Korean peninsula's land area but about 66 percent of its people. **Seoul** is by far the largest city in South Korea, with a population of more than 10 million. The largest city in North Korea is **Pyongyang,** with more than 2.5 million people.

In the next section, you will read about the history, culture, economics, and daily life in Japan.

Assessment

1 Places & Terms

Identify and explain the significance of each of the following in the region.

• Three Kingdoms

• Seoul

• Pyongyang

2 Taking Notes

REGION Use your notes to answer the questions below.

East Asia
The Koreas

• In which ways has China influenced the culture of Korea?

• Which countries in the region have invaded Korea?

3 Main Ideas

a. What impact has the border between North Korea and South Korea had upon life in both countries?

b. How is the economy of South Korea different from that of North Korea?

c. Which two major influences have shaped North Korea and South Korea since World War II?

4 Geographic Thinking

Drawing Conclusions How has Korea's physical location affected its history? **Think about:**

• the definition of a peninsula

• the location of Korea's neighbors

S **See Skillbuilder Handbook, page R5.**

SEEING PATTERNS Both Taiwan and South Korea are considered economic tigers of East Asia. What are some characteristics that they share? Make a **Venn diagram** showing the similarities and differences between the two.

Japan

Main Ideas
- Japan has an ancient culture and traditions.
- Japan is the economic giant of East Asia.

Places & Terms

samurai

shogun

CONNECT TO THE ISSUES
PHYSICAL FORCES Japan is vulnerable to devastating earthquakes and huge ocean waves because of its location.

A HUMAN PERSPECTIVE The Japanese flag shows a red sun against a white background. The red sun symbolizes Amaterasu, the sun goddess. According to myth, the Japanese emperor and his family are descended from the goddess. The Japanese call their country *Nippon,* which means "source of the sun." The name *Japan* may have come from a Chinese phrase meaning "origin of the sun," or it may have come from *Chipangu,* a name for the country recorded by Marco Polo.

Samurai and Shogun

Japan lies east of China—toward the rising sun. In their earliest history, the Japanese were close enough to China to feel its civilizing effects, but they were far enough away to be protected from invasion.

ANCIENT JAPAN The original inhabitants of Japan may have come to the islands from the mainland of Asia and from the South Pacific. There is some evidence to suggest that the ancestors of today's Japanese came eastward through Siberia and Korea and entered Japan. By about 1,500 years ago, most of Japan was actively growing food, such as rice. Weapons and tools made of bronze and iron were introduced, along with textiles.

Until the A.D. 300s, Japan was not a unified country. It was made up of hundreds of clans ruling separate territories. Then, by the fifth century, the Yamato clan had become the ruling clan. It claimed descent from the sun goddess, and by the seventh century, its leaders called themselves emperors of Japan.

In 794, the rulers moved the capital to the city of Heian (modern Kyoto). The era from 794 to 1185 is called the Heian period. During this time, Japan's central government was strong, but eventually the great landowners and clan chiefs began to act as independent rulers.

Professional soldiers called **samurai** served the interests of the landowners and clan chiefs. The samurai (the word means "one who guards") served as a bodyguard of warriors loyal to the leader of a clan.

THE SHOGUNS In 1192, after a struggle between two powerful clans, the Japanese emperor created the position of shogun. The **shogun** was the general of the emperor's army with the powers of a military dictator.

PLACE Takeda Shingen was one of the greatest samurai leaders in 16th century Japan.
What does this painting seem to suggest are some of the qualities of the samurai warrior?

EAST ASIA

All officials, judges, and armies were under his authority. The shoguns appointed governors, called *daimyo,* to each province. They were responsible for maintaining order.

Rule by the shoguns lasted for about 700 years. During those years, the Japanese fought off Mongol invasions and saw the arrival of Portuguese traders, who brought Christianity and firearms to Japan in the 1500s. In 1853, Commodore Matthew Perry's arrival to Japan from the United States ended Japan's isolation. In 1868, the last shogun resigned, and the emperor became head of the government.

EMERGING WORLD POWER During the late 19th century, Japan's government began bringing Japan into the modern age. By the early 20th century, Japan had become a major power.

During the early years of the 20th century, Japan expanded its empire. (See map on next page.) Its interests and those of the United States came into conflict in the Pacific region. On December 7, 1941, the Japanese launched a surprise attack on the U.S. naval base at Pearl Harbor in Hawaii. The attack brought the United States into World War II, which ended with Japan's defeat and surrender in 1945. ◢

After World War II, the United States headed the occupation of Japan and introduced political and economic reforms. Eventually Japan became a democracy—a constitutional monarchy with an emperor and an elected parliament.

Geographic Thinking◀

Seeing Patterns
◢ How might Japan's 20th century empire have reflected its history?

An Economic Powerhouse

After its defeat in World War II, Japan transformed itself into one of the world's most powerful economies. It experienced an economic boom, even though it has few natural resources. Japan is second only to the United States in the size of its economy.

PEOPLE AND PRODUCTS The population of Japan is more than 126 million. About 75 percent of Japan's people live in cities. Sixty percent of the people live on 2.7 percent of the land. Japan has few minorities, and those few are often discriminated against.

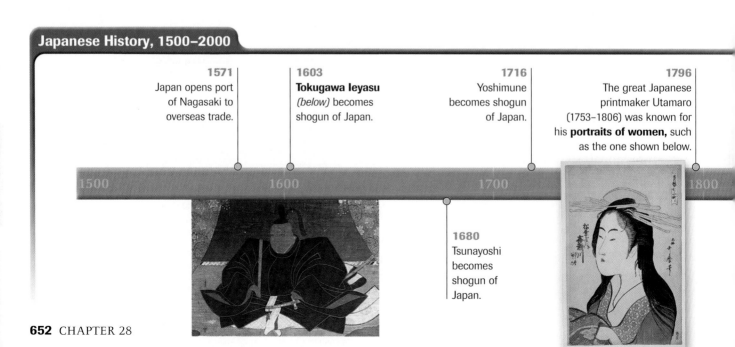

Japanese History, 1500–2000

1571
Japan opens port of Nagasaki to overseas trade.

1603
Tokugawa Ieyasu *(below)* becomes shogun of Japan.

1716
Yoshimune becomes shogun of Japan.

1796
The great Japanese printmaker Utamaro (1753–1806) was known for his **portraits of women,** such as the one shown below.

1500   1600   1700   1800

1680
Tsunayoshi becomes shogun of Japan.

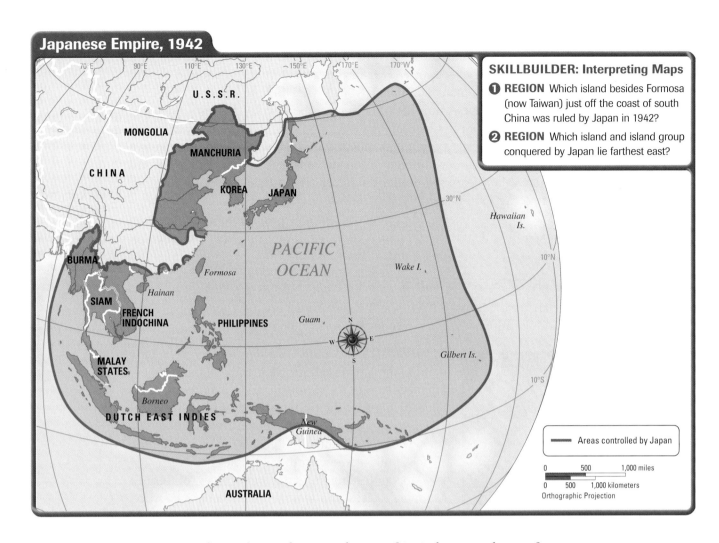

Japanese Empire, 1942

U.S.S.R.

MONGOLIA

MANCHURIA

CHINA

KOREA

JAPAN

BURMA

Formosa

Hainan

SIAM

FRENCH
INDOCHINA

PHILIPPINES

Guam

PACIFIC
OCEAN

Hawaiian
Is.

30°N

10°N

Wake I.

MALAY
STATES

Gilbert Is.

10°S

Borneo

DUTCH EAST INDIES

New
Guinea

AUSTRALIA

70°E 90°E 110°E 130°E 150°E 170°E 170°W

Areas controlled by Japan

0   500   1,000 miles

0   500   1,000 kilometers

Orthographic Projection

SKILLBUILDER: Interpreting Maps

❶ **REGION** Which island besides Formosa (now Taiwan) just off the coast of south China was ruled by Japan in 1942?

❷ **REGION** Which island and island group conquered by Japan lie farthest east?

Most of Japan's population and most of its industry and manufacturing are located in a corridor hundreds of miles long along the east coast of the main island of Honshu, with Tokyo as its anchor. The people who live in this corridor form the work force that produces goods sold around the world.

Manufacturing and trade are at the heart of Japan's economy. Japan imports most of the natural resources for its industrial needs. Among the resources it imports are coal and petroleum. Then it uses those resources and others to manufacture products for export to the global market. Among the most important of those products are cars, trucks, and electronic equipment such as televisions and computers.

A strong alliance between business and government has been one of the reasons for Japan's economic success during the second half of the 20th century. After the war, the United States gave economic assistance to Japan. Financial support from the government helped Japanese businesses develop products to market abroad.

1945
The **mushroom cloud** *(below)* is from an atomic bomb dropped on Nagasaki on August 9, 1945.

1900

2000

1853
A Japanese woodcut shows **Commodore Perry** *(above)* upon his arrival in Japan.

ECONOMIC SLOWDOWN After four decades of rapid growth, Japan's economy began to slow down in the 1990s. As the economic growth rate declined, many companies scaled back their operations, and some went bankrupt. A number of reasons accounted for this slowdown.

Other economies in East Asia, such as those of Taiwan, South Korea, and Hong Kong, provided competition. Then, when the economies of Southeast Asia encountered problems, Japanese investments there lost value. Many banks proved vulnerable. The Japanese stock market suffered big losses. Also, the Japanese people tended to save rather than spend. As a result, the economy became even more dependent on exports, which declined because of competition from other countries. ▶

CONNECT TO THE ISSUES
TRADE
◀B How are the economies of the region connected?

Japanese Culture

Japanese culture reflects the influences of both East and West. From these influences, Japan has developed its own unique culture.

PLACE Toyozo Arakawa, one of Japan's leading potters, was named a "Living National Treasure" in 1955.
What does the naming of a person as a national treasure say about a culture?

A TRADITIONAL PEOPLE In developing their early culture, the Japanese borrowed from China. Japanese language, religion, art, music, and government were all influenced by the Chinese.

The city of Kyoto is a monument to Japanese culture. The city contains Buddhist temples and Shinto shrines built of wood in the old style. The entire city is a living testament to Japanese ideas of beauty. Gardens, palaces, and temples all reflect a very spare, elegant, and refined style. In Kyoto and throughout Japan, great emphasis is placed on achieving harmony between a building and its natural surroundings.

Traditional drama is still performed in Japan. Noh plays developed during the 14th century. They deal with subjects drawn from history and legend and are performed by actors wearing masks. In the 17th century, Kabuki plays developed. They have colorful scenery, an exaggerated acting style, and vivid costumes.

Japanese painting was influenced by Chinese techniques and themes. Many early Japanese paintings show Buddhist themes that often came to Japan by way of China. Some examples of Japanese artistic works include long picture scrolls, ink paintings, and wood-block prints.

WESTERN INFLUENCES Since the day in 1853 when Commodore Perry sailed his fleet into Tokyo Bay, Japan has been open to Western influences. Those influences are visible in modern-day Japan.

Sports like baseball, golf, sumo wrestling, soccer, and tennis are popular in Japan. The clothes worn by most people are Western in style, although traditional clothing is worn on special occasions.

Western music is also popular in Japan. Rock music is popular among younger Japanese. They listen to Western groups and form rock bands of their own. Many cities in Japan have symphony orchestras that play Western classical music. Jazz is also popular.

Japan has been successful at balancing its traditional styles in art, theater, music, and architecture with influences from the West.

BACKGROUND
A tradition of print-making native to Japan is called *ukiyo-e,* which means "pictures of the floating world," the Japanese term for scenes from everyday life.

Life in Today's Japan

The people of Japan are educated and disciplined. This work force has enabled Japan to achieve prosperity.

EDUCATION Japan's educational system is highly structured. Students often attend school six days a week. They have a shorter summer vacation than American students—just six weeks in late July and August. Students attend six years of elementary school and three years of junior high school. Education is free during those years. Then they spend three years in high school. At the same time, many students attend classes at private schools called *juku* to help get them into good colleges.

Competition among students is high to gain admission to the best universities. Japan has more than 1,000 universities and technical colleges. Universities that rank at the top of the educational system include the University of Tokyo, Kyoto University, Keio University, and Waseda University.

CHANGES IN SOCIETY The Japanese are making some changes in the way their society is run. People are now increasingly demanding an end to pollution and overcrowding. Furthermore, workers at all skill levels are asking for shorter workdays and more vacation time.

In the next chapter, you will read about three important issues in East Asia. These include trade, the pressures of a large population, and the dangers posed by volcanoes around the Pacific Ocean.

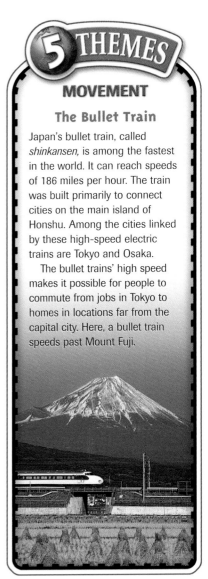

5 THEMES

MOVEMENT

The Bullet Train

Japan's bullet train, called *shinkansen*, is among the fastest in the world. It can reach speeds of 186 miles per hour. The train was built primarily to connect cities on the main island of Honshu. Among the cities linked by these high-speed electric trains are Tokyo and Osaka.

The bullet trains' high speed makes it possible for people to commute from jobs in Tokyo to homes in locations far from the capital city. Here, a bullet train speeds past Mount Fuji.

SECTION 4 Assessment

❶ Places & Terms

Identify and explain the significance of each of the following in the region.

- samurai
- shogun

❷ Taking Notes

REGION Use your notes to answer the questions below.

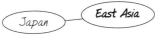

Japan — East Asia

- What happened to Japan in World War II?
- What is the importance of education in Japan today?

❸ Main Ideas

a. What is the basis of Japan's economic prosperity?

b. What are some examples of traditional Japanese culture?

c. How did the Western world influence Japan beginning in the 19th century?

❹ Geographic Thinking

Making Inferences How might Japan's isolation and its uniform population have both helped and hindered it in its attempts to achieve prosperity? **Think about:**

- the advantages of uniformity
- the importance of creativity

GeoActivity

SEEING PATTERNS Japan has some very distinctive cultural forms, such as Kabuki theater and sumo wrestling. Present a brief **report** to the class on some aspect of Japanese culture, illustrated by visuals that you have found in your research.

EAST ASIA

Comparing Cultures

Masks

Masks are coverings that disguise the face. Most cultures use masks for a variety of purposes. Followers sometimes wear ceremonial masks during religious celebrations. Actors wear theatrical masks during performances such as those in the classical drama of ancient Greece, China, and Japan. Mourners sometimes placed burial masks over the faces of the dead before they were buried. In ancient Egypt, they placed the mask directly on the mummy or else on the mummy case. Participants sometimes wear festival masks during celebrations such as Mardi Gras in New Orleans or Carnival in Rio de Janeiro.

United States

Japan

Angola

Indonesia

A masked dancer in Bali, Indonesia, performs a ritual dance. Balinese dancers move to the music of gongs and flutes. In their dances, each movement and gesture helps to tell the story.

This mask from Angola represents a female ancestor with an elaborate headdress. A member of the Chokwe culture in Africa created this mask out of wood and fibers in the 20th century.

Native American ceremonial masks were used to calm angry spirits. This mask is a product of the Iroquois culture of the northeast woodlands and was used in healing ceremonies.

Japanese masks and costumes are worn by a performer in a Noh drama, the classical drama of Japan. Masked performers create music and dance in a highly stylized manner.

GeoActivity

MAKING MASKS

Use the Internet to research how to make different kinds of masks. Choose materials that are easy to obtain. Then make a **mask** that you will show to the class.

- Use a technique about which you have found information.
- Write a description of the procedure you followed to make the mask.
- Display your mask in an area set aside in the classroom.

RESEARCH LINKS
CLASSZONE.COM

GeoData

ODD FACTS ABOUT MASKS

- In Europe, masks have been discovered that date back as early as 30,000 years ago to Paleolithic times.
- The solid gold death mask of the pharaoh Tutankhamen, which covered the head of his mummy, weighs 22.5 pounds.
- Masks were worn by the performers of tragedies and comedies in ancient Greece.
- The Senesi people of New Guinea use masks that include skirts that cover much of the body.
- The Aleuts of Alaska cover the faces of their dead with wooden masks.
- Death masks made of plaster are sometimes put on the face of the dead to preserve their features for posterity. Death masks exist for Napoleon Bonaparte and Ludwig van Beethoven.
- The mask worn by actor Clayton Moore in the television show *The Lone Ranger* was sold at auction for $33,000.

VISUAL SUMMARY
HUMAN GEOGRAPHY OF EAST ASIA

Subregions of East Asia

○ China
- China has more people than any other country in the world.
- It is about the same size as the United States in area.
- It has been the dominant culture in the region since ancient times.

○ Mongolia and Taiwan
- The histories of Mongolia and Taiwan have been closely linked with that of China.
- They have pursued separate paths of development—Mongolia has had a managed economy, while Taiwan has a capitalist economy based on manufacturing and trade.

● The Koreas: North and South
- The Korean peninsula is divided into two separate countries: Communist North Korea and capitalist South Korea.
- Recently, the two countries have begun discussing the possibility of becoming one country.

● Japan
- Japan is a great industrial power.
- It has managed to achieve economic prosperity despite its small land area and limited resources.

Reviewing Places & Terms

A. Briefly explain the importance of each of the following.

1. dynasty	**6.** Three Kingdoms
2. Boxer Rebellion	**7.** Seoul
3. Mao Zedong	**8.** Pyongyang
4. Confucianism	**9.** samurai
5. Pacific Rim	**10.** shogun

B. Answer the questions about vocabulary in complete sentences.

11. Which area extends from New Zealand in the western Pacific to Chile in the eastern Pacific?

12. What term means "one who guards"?

13. What is the largest city in North Korea?

14. Which city in the Koreas has about 10 million residents?

15. What event did it take a multinational force of 20,000 soldiers to end?

16. Which term describes a leader with the powers of a military dictator?

17. In which system of thought was there respect for the past and one's ancestors?

18. Who ruled the People's Republic of China from 1949 to 1976?

19. The Shang and the Han are examples of what?

20. Koguryo, Paekche, and Silla made up what?

Main Ideas

China (pp. 635–641)

1. In what ways has China influenced other cultures in the region?

2. How is China able to feed its enormous population?

3. What are some of the basic beliefs of Confucianism?

Mongolia and Taiwan (pp. 642–646)

4. What kind of economy does Mongolia have, and what activity is at its core?

5. What kind of economy does Taiwan have?

The Koreas: North and South (pp. 647–650)

6. Why did North Korea become a communist state and South Korea a democracy?

7. Why is South Korea considered an economic tiger?

Japan (pp. 651–657)

8. Why did Japan emerge onto the world scene in the 19th century?

9. Why is the city of Kyoto in Japan important?

10. Where does Japan get its resources, and how does it use them in its industries?

Critical Thinking

1. Using Your Notes

Use your completed chart to answer these questions.

a. What are some of the ways in which China has influenced the culture of East Asia?

b. What seems to be the general direction of economic development in the region?

2. Geographic Themes

a. **HUMAN-ENVIRONMENT INTERACTION** How have the river basins of eastern China supported a high population density?

b. **REGION** What are some of the natural barriers that have provided isolation or security to the different countries of the region?

3. Identifying Themes

Interaction between cultures occurred throughout the region. What are some of the consequences of this interaction? Which of the five themes are reflected in your answer?

4. Making Inferences

What might be the effect of innovations of modern life, such as computers and the Internet, on the development of democracy and free-market economies in the region?

5. Making Comparisons

How would you compare the economic prosperity and success of managed and capitalist economies in the region?

Additional Test Practice, pp. S1–S37

TEST PRACTICE
CLASSZONE.COM

Geographic Skills: Interpreting Graphs

Stock Market in South Korea

Use the graph at right to answer the following questions.

1. **ANALYZING DATA** When did the stock market in South Korea reach its lowest level?

2. **MAKING COMPARISONS** What was its highest level before its plunge?

3. **DRAWING CONCLUSIONS** What level did it reach by the year 2000? What does this suggest about the economy of South Korea?

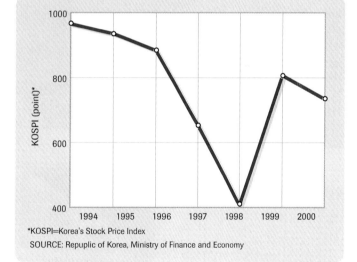

*KOSPI=Korea's Stock Price Index

SOURCE: Repulic of Korea, Ministry of Finance and Economy

The stock market in South Korea has seen dramatic ups and downs corresponding to the economic crises in the region in recent years.

GeoActivity

Research stock market activity in one or more of the other countries in the region. Show your findings in a graph tracking stock market activity for the late 1990s.

INTERNET ACTIVITY

Use the links at **classzone.com** to do research on the Mongol conquests. Focus on the reasons for the success of their conquests and whether the results of their conquests were mainly negative or positive.

Analyzing Data Present the results of your research in a chart that shows the positive and negative effects of the Mongol conquests.

Chapter 29

TODAY'S ISSUES
East Asia

SECTION 1
The Ring of Fire

SECTION 2
Trade and Prosperity

CASE STUDY
POPULATION AND THE QUALITY OF LIFE

For more on these issues in East Asia . . .

CURRENT EVENTS
CLASSZONE.COM

A bus teeters on the edge of a highway torn apart by an earthquake in Kobe, Japan, in 1995.

GeoFocus

How do people in East Asia deal with issues of a rapidly changing society?

Taking Notes In your notebook, copy a cause-and-effect chart like the one below. Then take notes on causes and effects of some aspect of each issue.

	Causes	Effects
Issue 1: Ring of Fire		
Issue 2: Trade		
Case Study: Population		

The Ring of Fire

How might people in East Asia prepare for earthquakes and volcanoes?

Main Ideas

- The islands of Japan form part of a geologically active area called the Ring of Fire.
- Because of its location, Japan has faced disastrous earthquakes, volcanic eruptions, and tsunamis.

Places & Terms

Ring of Fire

Great Kanto earthquake

tsunami

A HUMAN PERSPECTIVE On January 17, 1995, at 5:46 A.M., a severe earthquake rocked Kobe, Japan's sixth largest city. When the dust settled and the last of the fires burned out, about 6,000 people lay dead, and more than 40,000 suffered injuries. The government quickly began rebuilding the port city, but psychologists warned that reviving the spirit of Kobe's people would take time. Many lost family members. Entire neighborhoods vanished. A year after the quake, nearly 50,000 people were still living in temporary shelters, and anger grew against the government. Clearly, much more than glass, steel, bricks, and mortar would be needed to bring Kobe fully back to life.

Physical Forces in the Ring of Fire

Like Kobe, many Japanese cities are threatened by earthquakes. This is because Japan is part of the **Ring of Fire**—a chain of volcanoes that line the Pacific Rim. (See the map on the next page.)

SHIFTING PLATES As you learned in Unit 1, the outer crust of the earth is made up of a number of shifting tectonic plates that continually bump and slide into each other. When a dense oceanic plate meets a less dense continental plate, the oceanic plate slides under the continental plate in a process called subduction. The area where the oceanic crust is subducted is called a trench.

In East Asia, the Pacific oceanic plate encounters the Eurasian continental plate. When the oceanic plate moves under the continental plate, it crumples the continental crust, building mountains and volcanoes such as those that form the Ring of Fire.

At the same time, tremendous stress builds up along the edges of the plates. The stress keeps building until eventually the plates move suddenly and violently. The result is an earthquake.

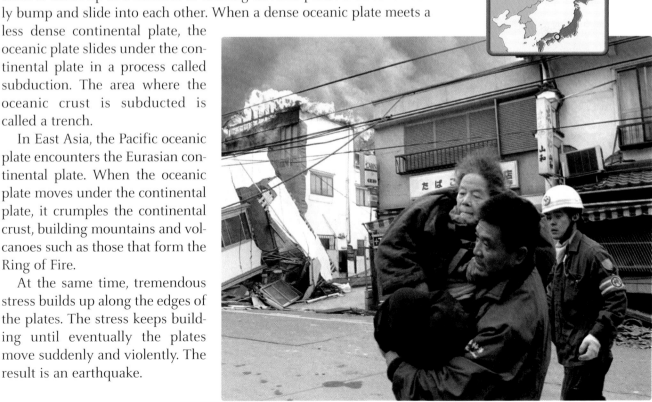

HUMAN-ENVIRONMENT INTERACTION An elderly woman is carried from a collapsing building during the earthquake in Kobe, Japan, in 1995.
What damage is apparent in the photograph?

EAST ASIA

The Geology of Japan

The Japanese islands exist because of subduction. The islands were formed by volcanoes created as the Pacific plate slid under the Eurasian plate. But the same forces that build islands can also destroy them.

VOLCANOES Living along the Ring of Fire means living with volcanic activity. From the time historical records were first kept, at least 60 volcanoes have been active on the islands of Japan. In fact, the best-known landform in Japan, Mt. Fuji, is a volcano.

EARTHQUAKES AND TSUNAMIS Earthquakes like the one that destroyed Kobe are common in Japan. An average of 1,000 quakes occur there each year. Most are too mild to affect people's lives. Some, however, cause many deaths and massive destruction. In 1923, the **Great Kanto earthquake** and the fires it caused killed an estimated 140,000 people and left the city of Tokyo in ruins. The quake partially or completely destroyed nearly 700,000 homes.

Another geological threat to Japan comes from the sea. When an earthquake occurs under the ocean floor, part of the floor moves. If the quake is strong enough, this shift may produce a **tsunami,** a huge wave of great destructive power. Underwater volcanic eruptions and coastal landslides can also cause tsunamis. Some waves have reached heights of over 100 feet.

Geographic Thinking

Making Comparisons
Ⓐ How many lives were lost in the Great Kanto earthquake compared to the Kobe earthquake?

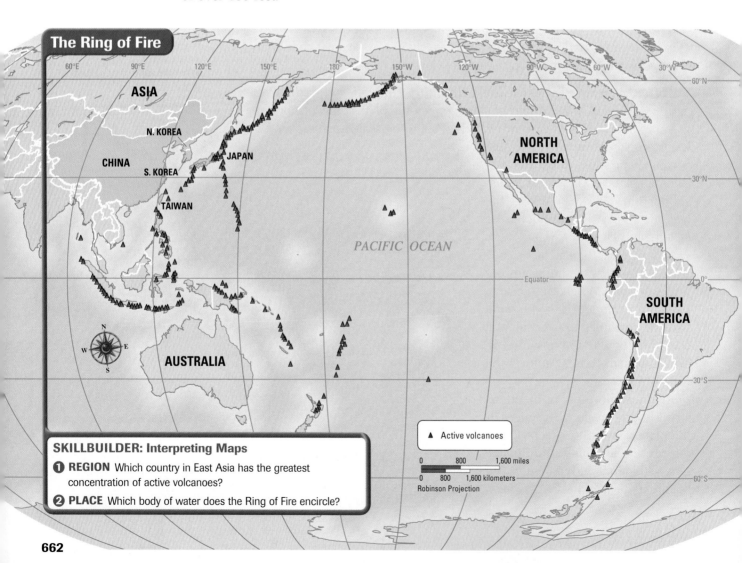

The Ring of Fire

ASIA

N. KOREA

CHINA

JAPAN

S. KOREA

TAIWAN

NORTH AMERICA

PACIFIC OCEAN

Equator

SOUTH AMERICA

AUSTRALIA

▲ Active volcanoes

0   800   1,600 miles
0   800   1,600 kilometers
Robinson Projection

SKILLBUILDER: Interpreting Maps

❶ **REGION** Which country in East Asia has the greatest concentration of active volcanoes?

❷ **PLACE** Which body of water does the Ring of Fire encircle?

Preparing for Disasters

For thousands of years, people have tried to predict when natural disasters will occur. At the dawn of the 21st century, they are still trying. Vulnerable nations like Japan are working to improve their defenses against the destructive power of geological forces.

PROBLEMS Many older buildings in Japan are not as likely to withstand earthquakes as newer buildings. In addition, some buildings have been constructed on ground or landfill that is not very stable. Underground gas lines are likely to rupture in the event of an earthquake, and leaking gas can catch fire. Crowded blocks and narrow streets spread the fires and hinder rescue operations.

SOLUTIONS Japan has established a strict building code. Whenever a quake rocks some area of the nation, engineers are quick to study how different types of buildings withstood the heaving ground beneath them. The results of their studies affect building codes governing construction materials and techniques. This has made newer buildings safer than older ones.

Because of the dangers, the Japanese people understand the importance of being prepared for disasters. Schoolchildren participate in yearly disaster drills with local fire-fighters. Organizations like the Japanese Red Cross Society and the Asia Pacific Disaster Management Center offer courses on disaster preparedness and management.

Japan and the other countries along the Ring of Fire cannot change the geology that shapes their land. They can, however, learn more about it and prepare to deal with disaster when it strikes next.

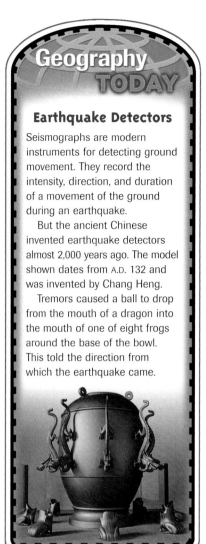

Geography TODAY

Earthquake Detectors

Seismographs are modern instruments for detecting ground movement. They record the intensity, direction, and duration of a movement of the ground during an earthquake.

But the ancient Chinese invented earthquake detectors almost 2,000 years ago. The model shown dates from A.D. 132 and was invented by Chang Heng.

Tremors caused a ball to drop from the mouth of a dragon into the mouth of one of eight frogs around the base of the bowl. This told the direction from which the earthquake came.

SECTION Assessment

1 Places & Terms

Identify and explain the following places and terms.

- Ring of Fire
- Great Kanto earthquake
- tsunami

2 Taking Notes

HUMAN-ENVIRONMENT INTERACTION Review the notes you took for this section.

	Causes	Effects
Issue 1: Ring of Fire		

- What was the effect of subduction on Japan?
- What causes tsunamis?

3 Main Ideas

a. What are some of the natural disasters that can strike around the Ring of Fire?

b. What role do shifting plates play in earthquakes?

c. What organizations help the Japanese prepare for natural disasters?

4 Geographic Thinking

Making Inferences How will Japan respond in the future to natural disasters such as earthquakes? **Think about:**

- how it has responded so far
- its location and the frequency of earthquakes there

 See Skillbuilder Handbook, page R4.

EXPLORING LOCAL GEOGRAPHY Pair with a partner and research the natural disasters that might possibly occur where you live—flood, tornado, hurricane, earthquake, and so forth. Then develop an **Emergency Procedures brochure** that lists the steps you would take to deal with such an emergency.

EAST ASIA

Interpreting a Proportional Circle Map

The earthquake that devastated Kobe, Japan, in 1995 measured 6.8 on the Richter scale, which is a scale for measuring the magnitude of earthquakes. About 6,000 people died and many thousands more were injured. Although the Kobe quake was the most destructive in recent years, there have been many others in Japan in the 1990s. Some of these were more powerful than the Kobe quake but they did not do as much damage.

THE LANGUAGE OF MAPS A **proportional circle map** shows the relative sizes of objects or events, such as earthquakes. This map shows major earthquakes in Japan during a ten-year period beginning in 1991. The larger the circle on the map, the greater the magnitude of the earthquake as measured by the scale.

Major Earthquakes in Japan, 1991–2000

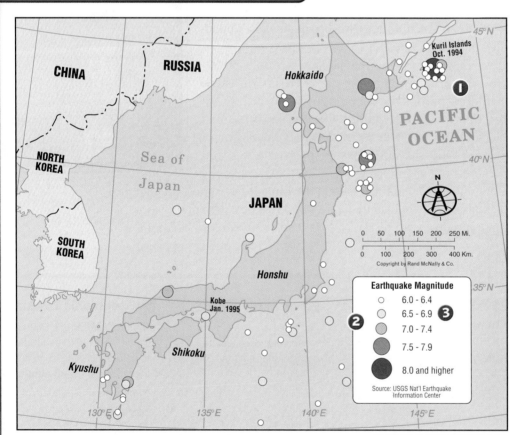

Copyright by Rand McNally & Co.

❶ A cluster of circles indicates that an area is prone to frequent quakes.

❷ The key explains that the bigger and darker a circle is on the map, the greater the size and intensity of the quake.

❸ Values on an earthquake magnitude scale are typically between 1 and 9. This map shows earthquakes with a magnitude of 6 and higher. Each increase of .5 represents an increase in released energy. Scales for measuring earthquakes include the Richter, the moment magnitude, and others.

Map and Graph Skills Assessment

1. Analyzing Data
What was the intensity of the earthquake that struck Kobe?

2. Making Comparisons
On which islands did the most powerful quake occur in this period? In what range did it fall, as measured by the scale?

3. Making Inferences
Why do you think the quake you identified in question 2 was not as destructive as the Kobe quake?

Trade and Prosperity

What are some benefits of global trade?

Main Ideas

• East Asian economies became global powerhouses in the 1970s and 1980s.

• The decline of Asian economies in the 1990s created a crisis that spread around the globe.

Places & Terms

UNICEF

global economy

Jakota Triangle

recession

sweatshop

A HUMAN PERSPECTIVE At the beginning of the 1990s, the economies of East Asia were growing very rapidly. Unfortunately, there was a dark side to this prosperity. In 1995, <u>**UNICEF (the United Nations Children's Fund)**</u> reported that more than half a million children in East Asia were working in factories or begging on the streets. UNICEF regional director Daniel Brooks noted that, due to fast-paced economic growth, "We are seeing the erosion of family values and that includes the exploitation of children." This is one of the important issues facing the region.

Opening Doors

The process by which East Asia became an economic powerhouse took centuries. Until the 1500s, the nations of East Asia had been isolated from the rest of the world. As Western demand for Asian products grew, European traders used a variety of means—including force—to end East Asia's isolation.

Eventually, the economies of the region were to emerge as major players in the global economy. However, foreign intervention and world war lay ahead before East Asian nations achieved widespread prosperity.

OPENING TO THE WEST By the 1800s, the nations of Europe had signed treaties that gave them distinct spheres of influence in the East. These were areas where they could control trade without interference from other Western nations. In 1853, Commodore Matthew Perry set sail from the United States to Japan to persuade the Japanese to establish trade and diplomatic relations with the United States. The naval warships that accompanied Perry intimidated Japan into opening its doors to the United States and the West.

MOVEMENT Japan exports about 4½ million vehicles each year. Here, cars are about to be loaded onto a boat in the port of Nagoya.

EAST ASIA

INDUSTRIALIZATION AND GLOBALIZATION After World War II, the nations of East Asia began industrializing, using cheap labor to produce goods for trade. Trade between East and West steadily increased. The labels "Made in China" and "Made in Japan" on goods became very common in the United States and Europe.

At the same time, regional economies, which had evolved from national economies, began to merge. Eventually, a **global economy** developed, in which nations became dependent on each other for goods and services. For example, Japan imported many natural resources from around the world and then transformed those resources into manufactured goods that it sold around the globe. The nations of East Asia used their supplies of cheap labor to become manufacturing powerhouses. The World Bank described this boom as an "economic miracle."

Powerful Economies of East Asia

During the 1980s and early 1990s, many Asian economies did very well. The most powerful of the Pacific Rim nations of East Asia—Japan, Taiwan, and South Korea—enjoyed record prosperity. These three countries formed a part of a zone of prosperity referred to by some as the **Jakota Triangle**—**Ja**pan, **Ko**rea (South), and **Ta**iwan. By the mid-1990s, however, these economies were experiencing problems.

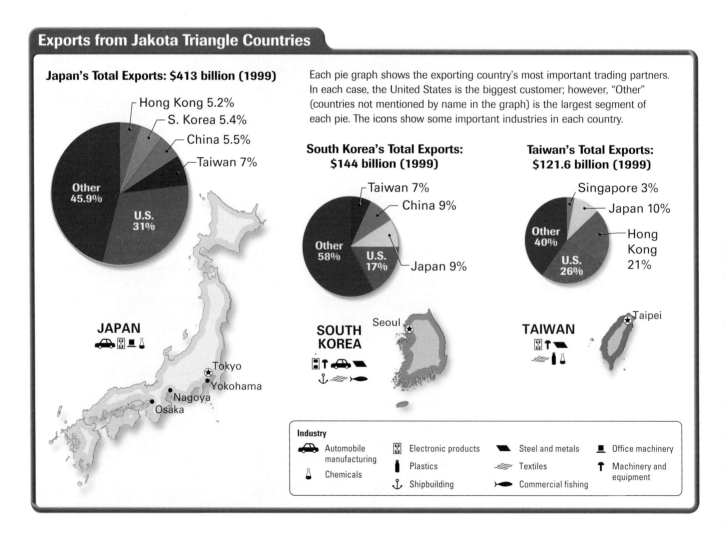

Exports from Jakota Triangle Countries

Japan's Total Exports: $413 billion (1999)

- Hong Kong 5.2%
- S. Korea 5.4%
- China 5.5%
- Taiwan 7%
- Other 45.9%
- U.S. 31%

Each pie graph shows the exporting country's most important trading partners. In each case, the United States is the biggest customer; however, "Other" (countries not mentioned by name in the graph) is the largest segment of each pie. The icons show some important industries in each country.

South Korea's Total Exports: $144 billion (1999)

- Taiwan 7%
- China 9%
- Other 58%
- U.S. 17%
- Japan 9%

Taiwan's Total Exports: $121.6 billion (1999)

- Singapore 3%
- Japan 10%
- Hong Kong 21%
- Other 40%
- U.S. 26%

JAPAN
- Tokyo
- Yokohama
- Nagoya
- Osaka

SOUTH KOREA
- Seoul

TAIWAN
- Taipei

Industry
- Automobile manufacturing
- Chemicals
- Electronic products
- Plastics
- Shipbuilding
- Steel and metals
- Textiles
- Commercial fishing
- Office machinery
- Machinery and equipment

ECONOMIC PROBLEMS ARISE Although some East Asian economies appeared healthy, they were burdened by debt and mismanagement. The Asian economic miracle had been based in part on efficiency and innovation. It also had been built partly on the sacrifices of very poor and very young workers, who were paid low wages.

In the mid-1990s, a series of banks and other companies went bankrupt (could not pay their debts). This sparked panic among foreign investors, who began selling their Asian stocks and currency. In some countries, riots broke out. In Japan and South Korea, ruling politicians had to resign. Japan's economy entered a **recession**—an extended decline in general business activity. The Asian economic miracle had come to an end. South Korea and Taiwan also experienced recessions.

Geographic Thinking

Seeing Patterns
A▷ What were some of the factors that led to recession in the region?

A GLOBAL RIPPLE EFFECT Because the economies of many nations are interconnected, the crisis in Asia spread throughout the world. Uncertainty led to concern at the New York Stock Exchange and other national exchanges. To prevent a global economic downturn, the World Bank and the International Monetary Fund stepped in, lending money to East Asian countries that promised reform. This began to reverse the downslide, but the world had learned an important lesson—a global economy could threaten prosperity as well as improve it.

THE PROMISE OF REFORM The economic crisis led to an awareness in East Asia that serious reform was necessary. Reform would have to include increased wages for adult workers, as well as a ban on child-labor and forced-labor practices. It would also mean an end to **sweatshops.** These are workplaces where people work long hours for pennies under poor conditions. At the dawn of the 21st century, reforms had begun, and Asian economies were showing new signs of life.

In the next section, you will read about the expanding population of East Asia. The growth in population has had an impact on the quality of life in the region.

SECTION 2 ▷ Assessment

① Places & Terms

Identify and explain the following places and terms.

- UNICEF
- global economy
- Jakota Triangle
- recession
- sweatshop

② Taking Notes

HUMAN-ENVIRONMENT INTERACTION Review the notes you took for this section.

	Causes	Effects
Issue 2: Trade		

- Why is trade important to the economies of the region?
- How did the people of East Asia make possible the "economic miracle"?

③ Main Ideas

a. How was the prosperity of East Asia linked to the wider world?

b. What were some of the consequences of economic development in the region?

c. What were some of the causes of economic decline in the region?

④ Geographic Thinking

Making Inferences Why might changes in the global economy have a greater effect on South Korea and Taiwan than on China and Mongolia? **Think about:**

- the global economy
- agriculture and industry

RESEARCH LINKS
CLASSZONE.COM

EAST ASIA

GeoActivity

SEEING PATTERNS Pair with a partner and choose one country in the region that is heavily dependent on trade—for example, Japan, South Korea, or Taiwan. Then use the Internet to find out how that country's economy did in the year 2000. Give a **class report** on whether the economy is improving.

POPULATION AND
THE QUALITY OF LIFE

What pressures does population put on the environment?

Trams, buses, and
people crowd the
streets of Hong Kong.

**The Voyageur Experience
in World Geography**
China: Food for a Billion Plus

Because East Asia has changed so much, it's hard to imagine how different the region looked 50 years ago. Today, some of the countries and cities of the region are among the most prosperous in the world. In Japan, South Korea, and Taiwan, the statistics on per capita income, length of life, and literacy are all high. Despite recent problems, the economies are generally prosperous, as can be seen in the glittering shopping districts and luxurious residential neighborhoods of Tokyo, Seoul, and Taipei. But it wasn't always that way. If the big problem of the past was industrializing, today it is managing population.

Patterns of Population

Many of the countries of East Asia have been so successful in dealing with the basic problems of feeding their people and industrializing that they now face other problems. Several of these problems are caused by the expanding populations in the region.

THE SITUATION AT MID-CENTURY At the middle of the 20th century, the nations of East Asia ranked among the least developed in the world. In fact, statistics on health, literacy, fertility, and economics in East Asia mirrored those of the poorest region of the world—sub-Saharan Africa. Widespread poverty was the norm. Life expectancy was short. Fertility rates were high, as were infant and maternal death rates. In 1950, East Asian women often married young and gave birth to six children on average during their lifetimes. Most economies remained rural.

Addressing Population Problems

Policy makers in the region understood that population control was key to solving a wide range of social and economic woes. Among the successful programs were those that stressed education and family planning.

ENVIRONMENTAL STRESS Unrestricted population growth put tremendous strain on the quality of life in the region and on the environment. Food production on existing farmland was barely adequate. The absence of basic sanitation fouled the region's water supplies. In some countries, such as China, the water tables were drained to dangerously low levels. Fortunately, the governments of East Asia recognized this catastrophe-in-the-making. They moved quickly to reverse course.

PROBLEMS AND POLICIES Aggressive family planning programs were begun in the region. Birth rates began leveling off and then dropping. By the year 2000, women were marrying much later and giving birth to an average of 2.5 children. In China alone, the birth rate dropped from 6.22 children per woman in 1950–1955 to just 1.82 in the year 2000.

IMPRESSIVE RESULTS This drop in birth rates, combined with industrialization, led to fast economic growth. By the 1990s, the economies of East Asia were booming, transforming social and economic conditions. In just over a generation, the region's quality of life has improved to the point where life expectancy and literacy rates are among the highest in the world.

The Quality of Life

Although these changes in East Asia have been dramatic, they have not solved all of the region's problems. Some countries in the region, such as China and Japan, are among the most populous in the world. Furthermore, life expectancy in East Asia has increased from 41 years in the period 1950–1955 to 69 years in the year 2000.

SOME ONGOING PROBLEMS The huge populations of the region continue to put pressure on the environment. Even if China were to maintain a modest growth rate of one percent a year, it would still add 13 million people to its population annually.

The growing populations are concentrated in the cities of the region, where they must be provided with housing, sanitation, and transportation. Pollution, overcrowding, and flooding are all problems that are made worse by an expanding population.

However, not all family planning programs were well received. Some citizens criticized China's one-child-per-family policy as harsh and an assault on their rights. In the face of such criticism, the region's family planning efforts were expanded.

Despite these difficulties, East Asia has shown the world that rapid social and economic progress are possible. This requires that people and their leaders join hands with the world community to make difficult decisions and put in place sound policies.

A case study project on population follows on the next two pages.

SEE PRIMARY SOURCE A

SEE PRIMARY SOURCE D

Population

Some Major Cities of East Asia, 1995–1999

City	Population (in millions)
Shanghai, China	13.58
Beijing, China	11.30
Seoul, South Korea	10.29
Tianjin, China	9.42
Tokyo, Japan	7.85
Hong Kong, China	6.84
Shenyang, China	5.12
Guangzhou, China	4.49
Wuhan, China	4.25
Pusan, South Korea	3.87
Chongqing, China	3.47
Xian, China	2.97
Nanjing, China	2.96
Taipei, Taiwan	2.60
Osaka, Japan	2.48

SOURCE: The Statesman's Yearbook (2001)

SKILLBUILDER: Interpreting Charts

❶ **HUMAN-ENVIRONMENT INTERACTION** What are the two largest cities in South Korea?

❷ **REGION** Which country on the chart has most of the largest cities?

EAST ASIA

PROJECT

A Visual Presentation

Primary sources A, B, C, D, and E offer assessments of East Asia's population challenges. Use these resources along with your own research to prepare maps, graphs, and charts that tell a story about population and quality of life in one nation of East Asia.

RESEARCH LINKS
CLASSZONE.COM

Suggested Steps

1. Choose one East Asian nation to study. Search for information that can be presented visually in charts and graphs. The visuals you create should explain some aspect of the nation's population and quality of life.

2. Use online and print resources to research your topic.

3. Look for information that shows relationships between population and quality of life. For example, one chart might illustrate declining birth rates while another shows rising literacy rates.

4. Include several different types of visuals: pie graphs, line and bar graphs, pictograms, population distribution maps, and so on.

5. Try to make your visuals as colorful as possible. Use color to make the information easier to understand.

6. Prepare a brief oral explanation of your visuals and the story they tell.

Materials and Supplies
- posterboard
- color markers
- computer with Internet access
- books, newspapers, and magazines
- printer

PRIMARY SOURCE A

Bar Graph *This bar graph, prepared from U.S. Census Bureau statistics, shows where and by how much population is expected to grow from 2000 to 2050.*

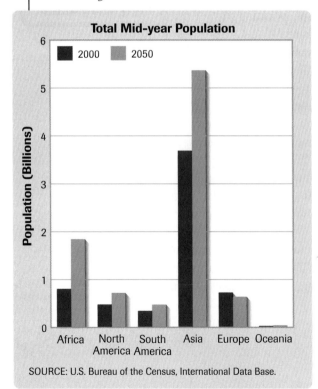

SOURCE: U.S. Bureau of the Census, International Data Base.

PRIMARY SOURCE B

Policy Statement *On a trip to Hong Kong in 1998, U.S. President Bill Clinton discussed the issue of pollution in China. He noted that overcrowding and industrialization had led to serious environmental problems that would only get worse if not addressed. The following CNN news story quotes some of Clinton's remarks.*

Clinton addressed a contentious [controversial] issue separating the two countries—global warming. He also announced a series of clean air and water measures to help China, which has five of the most polluted cities in the world, according to environmentalists. . . .

"You know better than I that polluted air and water are threatening your remarkable progress," Clinton said. "Smog has caused entire Chinese cities to disappear from satellite photographs, and respiratory illness is China's number one health problem."

News Analysis *In this article from Asiaweek.com, the author addresses an interesting problem posed by population growth in Asia.*

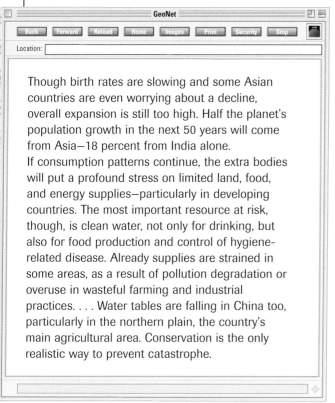

GeoNet

Back Forward Reload Home Images Print Security Stop

Location:

Though birth rates are slowing and some Asian countries are even worrying about a decline, overall expansion is still too high. Half the planet's population growth in the next 50 years will come from Asia—18 percent from India alone.

If consumption patterns continue, the extra bodies will put a profound stress on limited land, food, and energy supplies—particularly in developing countries. The most important resource at risk, though, is clean water, not only for drinking, but also for food production and control of hygiene-related disease. Already supplies are strained in some areas, as a result of pollution degradation or overuse in wasteful farming and industrial practices. . . . Water tables are falling in China too, particularly in the northern plain, the country's main agricultural area. Conservation is the only realistic way to prevent catastrophe.

Fact Sheet *In 1997, Population Action International produced a fact sheet that helped explain the relationship between population control and development in East Asia.*

- **A shift to smaller families produced three important demographic changes: slower growth in the number of school-age children, a lower ratio of dependents to working-age adults, and a reduced rate of labor-force growth.** These alone were not enough to create the educated work force, high wages and savings rates, and the capital-intensive industries that now characterize the [region]. But linked to an enterprising business sector, wise public investment, and an equitable education system, demographic change soon became economic opportunity. . . .

- **With fewer children, households placed more of their earnings in savings, and governments reduced public expenditures.** In 1960, there were only 1.3 working-age adults for each child in . . . South Korea, Taiwan, Singapore, and Hong Kong. Because families chose to have fewer children, by 1995 there were 3.1 working-age adults for each child, dramatically reducing the dependency burden and allowing families to save more of their incomes.

Political Cartoon *This cartoon was created by Nick Anderson in 1999. It shows one cartoonist's viewpoint of the effect of a rapidly expanding population on the natural environment.*

WORLD POPULATION TOPS 6 BILLION

THERE GOES THE NEIGHBORHOOD.

PROJECT CheckList

Have I . . .

✔ fully researched my topic?

✔ created informative, colorful visuals that make my report clear and interesting?

✔ used charts and graphs to tell a story about population issues in East Asia?

✔ practiced explaining my report?

✔ anticipated questions others might ask and prepared answers?

VISUAL SUMMARY
TODAY'S ISSUES IN EAST ASIA

Environment

The Ring of Fire
- Parts of East Asia are located along the northwestern edge of the Pacific Ocean's Ring of Fire.
- The heavily populated areas of East Asia (especially Japan) are endangered by the earthquakes, volcanic eruptions, and tsunamis along the Ring of Fire.

Economics

Trade and Prosperity
- Most of the nations of East Asia have prospered from trade with each other and with other parts of the world.
- In the second half of the 20th century, many countries in East Asia developed powerful economies.
- In the 1990s, there was a decline in the economies of the region but they have begun to recover.

Population

Case Study: Population and the Quality of Life
- East Asia has a huge population.
- Despite a reduced birth rate, the population in the region will continue to grow well into the 21st century.
- A growing population affects the quality of life in a nation.

Reviewing Places & Terms

A. Briefly explain the importance of each of the following.

1. Ring of Fire
2. Great Kanto earthquake
3. tsunami
4. UNICEF
5. global economy
6. Jakota Triangle
7. recession
8. sweatshop

B. Answer the questions about vocabulary in complete sentences.

9. How many people were killed and how many homes destroyed in the Great Kanto earthquake?
10. What is the basic cause of the physical events that characterize the Ring of Fire?
11. Upon what is the prosperity of the Jakota Triangle primarily based?
12. Why are sweatshops profitable?
13. What sorts of natural disasters occur around the Ring of Fire?
14. How does Japan participate in the global economy?
15. How might economic reform in East Asia affect sweatshops?
16. What besides earthquake damage made the Great Kanto earthquake so destructive?
17. What are three causes of tsunamis?
18. Which countries in the region experienced a recession?
19. What sorts of economies make up the Jakota Triangle?
20. With what issues does UNICEF concern itself?

Main Ideas

The Ring of Fire (pp. 661-664)

1. What causes an earthquake?
2. Why are the Japanese islands so unstable?
3. What are some Japanese organizations that help prepare for disasters?

Trade and Prosperity (pp. 665-667)

4. What effect did Western nations have on economic development in East Asia?
5. What is the connection between industrialization and globalization?
6. What are some of the things that went wrong in the economies of the region?

Case Study: Population and the Quality of Life (pp. 668-671)

7. What are some examples of the stress that population growth puts on the environment?
8. What are some effective ways to manage population growth?
9. How developed was East Asia in the middle of the 20th century?
10. How had East Asia changed by the beginning of the 21st century?

Critical Thinking

1. Using Your Notes

Use your completed chart to answer these questions.

	Causes	Effects
Issue 1: Ring of Fire		
Issue 2: Trade		

a. What are some of the effects of the Ring of Fire?

b. What role did labor play in the booming economies of East Asia after World War II?

2. Geographic Themes

a. **REGION** What are some of the ways that people respond to the dangers of living in the Ring of Fire?

b. **HUMAN-ENVIRONMENT INTERACTION** How does a rising population put a strain on the environment?

3. Identifying Themes

What might be some of the advantages of reducing population growth in the region? Which of the five themes apply to this situation?

4. Determining Cause and Effect

What might be the connection between population and trade in some of the economies of the region?

5. Making Inferences

Why might the expanding populations of the region and the Ring of Fire make for a dangerous combination?

Additional Test Practice, pp. S1–S37

TEST PRACTICE
CLASSZONE.COM

Geographic Skills: Interpreting Graphs

World Population and Growth

Use the graph to answer the questions.

1. **ANALYZING DATA** What was the population of the world in the year 1?

2. **MAKING COMPARISONS** How long did it take for the world's population to double from the year 1?

3. **MAKING COMPARISONS** How many years might it take for the world's population to double after 1974? What is the total expected to be in 2028?

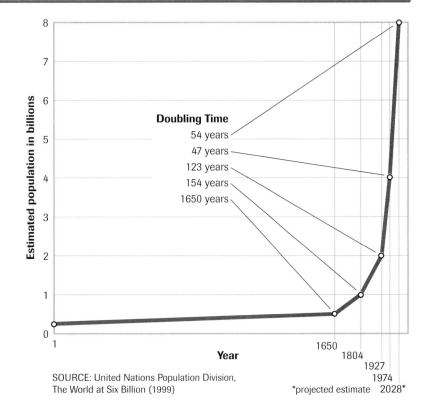

SOURCE: United Nations Population Division, The World at Six Billion (1999)

GeoActivity

Do research to create a bar graph showing population growth and doubling time in one country in the region. Compare it with a bar graph showing the same information for the United States. Display the two bar graphs side by side.

INTERNET ACTIVITY

Use the links at **classzone.com** to do research about the Ring of Fire. Focus on major eruptions, earthquakes, and tsunamis in the region.

Creating Multimedia Presentations Combine charts, maps, or other visual images in a presentation showing strategies to prepare for natural disasters along the Ring of Fire.

Southeast Asia, Oceania, and Antarctica

Ranging from flat plateaus to volcanic peaks, this region has diverse landforms. The vast Pacific Ocean links the scattered parts of this region together.

REGION Towering cliffs covered with snow and ice are a distinctive characteristic of the landscape of Antarctica.

MOVEMENT Traders travel the rivers of Thailand to sell produce and other goods in that country's famous floating markets.

GeoData

REGION Oceania includes the Pacific Islands not considered to be part of Southeast Asia. Some people include New Zealand and Australia, even though Australia is a continent, not an island.

LOCATION Australia is known as the "Land Down Under." It is the only inhabited continent to lie completely in the Southern Hemisphere.

HUMAN–ENVIRONMENT INTERACTION Farmers have adapted to the region's varied environments. They use terraced fields on steep Southeast Asian slopes and irrigate arid parts of Australia.

For more information on Southeast Asia, Oceania, and Antarctica . . .

RESEARCH LINKS
CLASSZONE.COM

Unit PREVIEW 10

Today's Issues in Southeast Asia, Oceania, and Antarctica

Today, Southeast Asia, Oceania, and Antarctica face the issues previewed here. As you read Chapters 30 and 31, you will learn helpful background information. You will study the issues themselves in Chapter 32.

In a small group, answer the questions below. Then participate in a class discussion of your answers.

Exploring the Issues

1. **LAND CLAIMS** Search the Internet for information about Aboriginal land claims in Australia. What are the different sides in the conflict?

2. **INDUSTRIALISM** Make a list of the possible results of industrial growth, both positive and negative. How might a country reduce the negative effects?

3. **ENVIRONMENTAL CHANGE** Consider news stories that you've heard about global warming and the ozone hole. What are some of the predicted effects? Make a list of all the effects you can remember.

For more on these issues in Southeast Asia, Oceania, and Antarctica . . .

CURRENT EVENTS
CLASSZONE.COM

LAND CLAIMS

Should native people be given back their ancestors' land?

These two Aboriginal men are elders of the Wuthathi people. They have come to bury the skull of an ancestor in their homeland. Aboriginal people feel a strong spiritual connection to their land and do not want to be separated from it even in death.

INDUSTRIALISM

How does industrialization affect cities?

This slum in Jakarta, Indonesia, shows how difficult it is to provide adequate housing for the thousands of people who move to cities seeking factory jobs.

CASESTUDY

How have people changed the atmosphere?

The green and blue areas in these satellite images show where the ozone layer over Antarctica is thinnest. Ozone in the stratosphere, a layer of the atmosphere, protects the living things of earth from harmful ultraviolet radiation.

ENVIRONMENTAL CHANGE

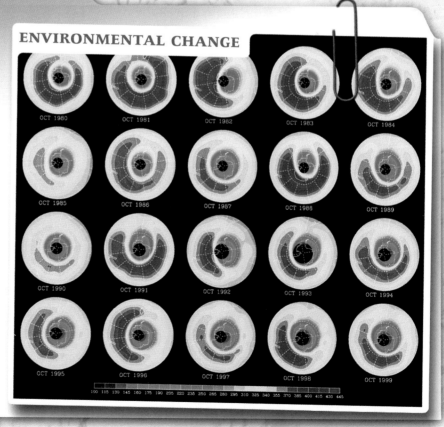

OCT 1980 · OCT 1981 · OCT 1982 · OCT 1983 · OCT 1984

OCT 1985 · OCT 1986 · OCT 1987 · OCT 1988 · OCT 1989

OCT 1990 · OCT 1991 · OCT 1992 · OCT 1993 · OCT 1994

OCT 1995 · OCT 1996 · OCT 1997 · OCT 1998 · OCT 1999

100 115 130 145 160 175 190 205 220 235 250 265 280 295 310 325 340 355 370 385 400 415 430 445

SE ASIA & OCEANIA

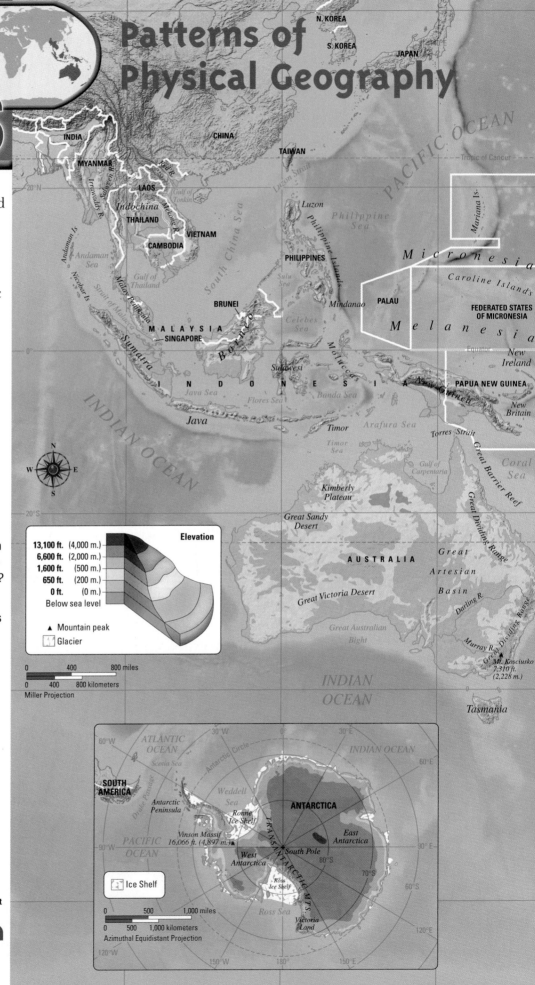

Patterns of Physical Geography

Use the Unit Atlas to add to your knowledge of Southeast Asia, Oceania, and Antarctica. As you look at the maps and charts, notice geographic patterns and specific details about the region. For example, the chart gives details about large islands in the region.

After studying the pictures, graphs, and physical map on these two pages, jot down in your notebook answers to the questions below.

Making Comparisons

1. How does the population of the region compare to that of the United States?

2. What is the world's largest island? How does its area compare to the combined area of New Guinea, Borneo, and Sumatra?

3. Which countries of this region would you consider flat? Which would you consider mountainous?

For updated statistics on Southeast Asia, Oceania, and Antarctica . . .

DATA UPDATE
CLASSZONE.COM

Elevation

13,100 ft.	(4,000 m.)
6,600 ft.	(2,000 m.)
1,600 ft.	(500 m.)
650 ft.	(200 m.)
0 ft.	(0 m.)

Below sea level

▲ Mountain peak

Glacier

0   400   800 miles
0   400   800 kilometers
Miller Projection

Ice Shelf

0   500   1,000 miles
0   500   1,000 kilometers
Azimuthal Equidistant Projection

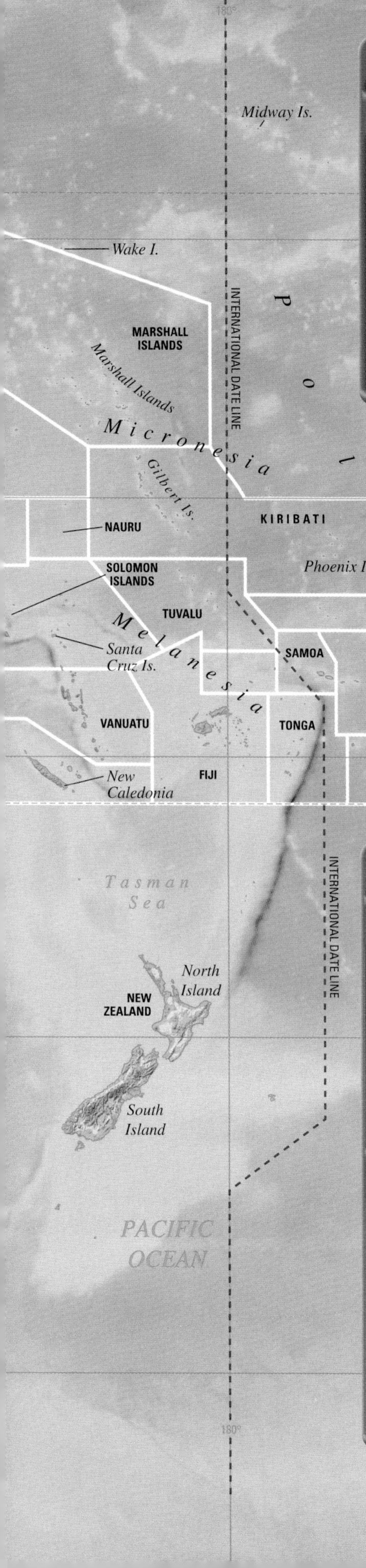

Comparing Data

Landmass

Southeast Asia and Oceania
5,065,224.45 sq mi

Continental United States
3,165,630 sq mi

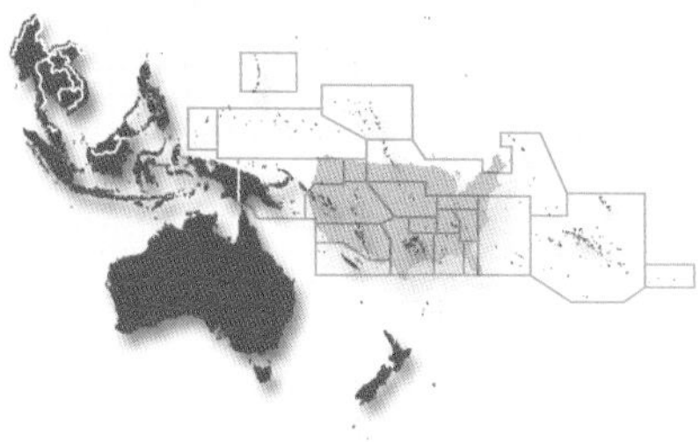

Midway Is.

Wake I.

MARSHALL ISLANDS

Marshall Islands

INTERNATIONAL DATE LINE

P o l y

M i c r o n e s i a

Gilbert Is.

NAURU

KIRIBATI

Line Islands

Equator 0°

SOLOMON ISLANDS

Phoenix Is.

n

TUVALU

M e l a n e s i a

Santa Cruz Is.

SAMOA

e

Cook Islands

Society Is.

Tuamotu Archipelago

VANUATU

TONGA

s

Tahiti

New Caledonia

FIJI

i

20°S

a

Tasman Sea

INTERNATIONAL DATE LINE

North Island

NEW ZEALAND

South Island

PACIFIC OCEAN

Tropic of Capricorn

Comparing Data

Population

*Antarctica is not included because it has no permanent population.

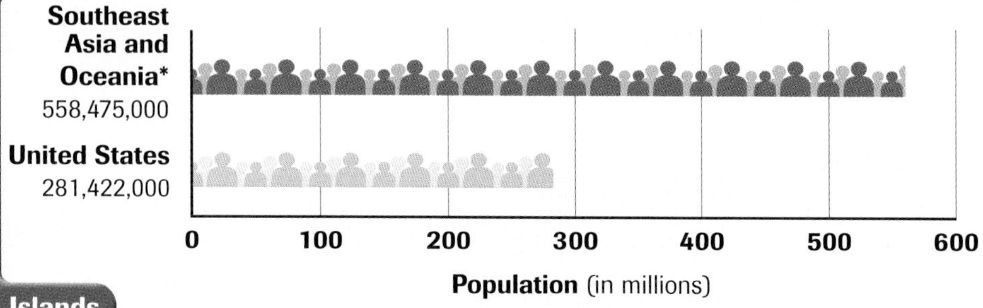

Southeast Asia and Oceania*
558,475,000

United States
281,422,000

| 0 | 100 | 200 | 300 | 400 | 500 | 600 |

Population (in millions)

Islands

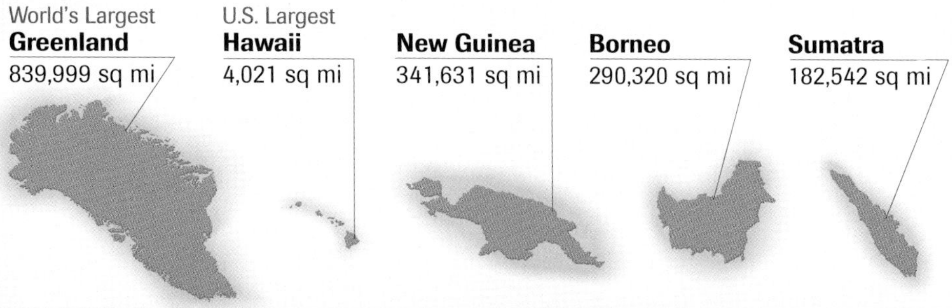

| World's Largest **Greenland** 839,999 sq mi | U.S. Largest **Hawaii** 4,021 sq mi | **New Guinea** 341,631 sq mi | **Borneo** 290,320 sq mi | **Sumatra** 182,542 sq mi |

SE ASIA & OCEANIA

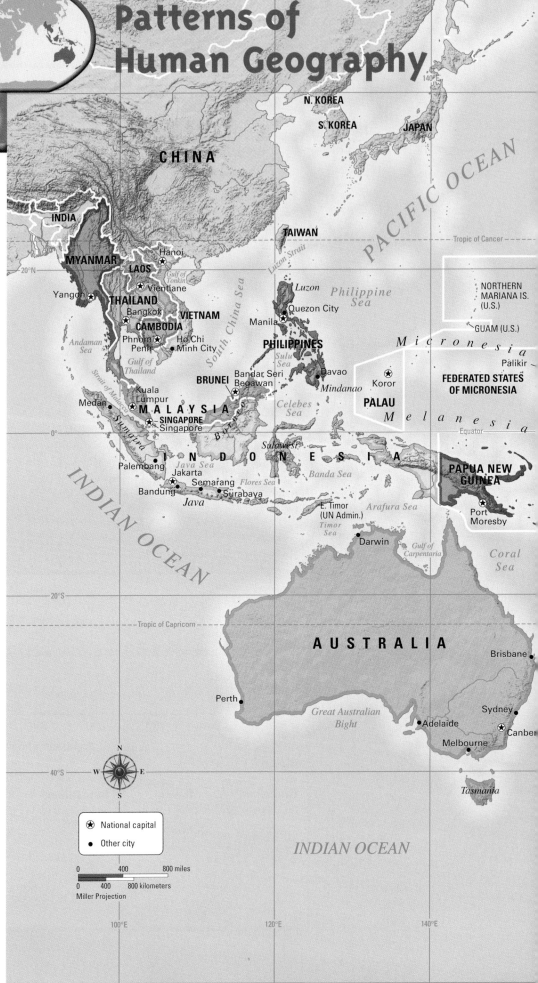

Patterns of Human Geography

Study the map on page 681 to learn about ancient kingdoms and empires of Southeast Asia and the map on both pages to learn about the present-day nations of the region. Then write in your notebook the answers to these questions.

Making Comparisons

1. Which ancient kingdoms or empires have names similar to present-day countries in Southeast Asia? How do their locations compare?

2. Which are the largest countries in the region?

3. Which country includes part of the Asian mainland and part of a large island?

Map labels:

CHINA
N. KOREA
S. KOREA
JAPAN
TAIWAN
INDIA
MYANMAR
Hanoi
LAOS
Gulf of Tonkin
Yangon
Vientiane
THAILAND
Bangkok
VIETNAM
CAMBODIA
Phnom Penh
Ho Chi Minh City
Andaman Sea
Gulf of Thailand
PHILIPPINES
Manila
Quezon City
Luzon
Luzon Strait
Philippine Sea
South China Sea
Sulu Sea
PACIFIC OCEAN
Tropic of Cancer
NORTHERN MARIANA IS. (U.S.)
GUAM (U.S.)
Micronesia
Palikir
FEDERATED STATES OF MICRONESIA
Bandar Seri Begawan
BRUNEI
Davao
Mindanao
Koror
PALAU
Melanesia
Kuala Lumpur
Medan
MALAYSIA
SINGAPORE
Singapore
Borneo
Celebes Sea
Strait of Malacca
Sumatra
Palembang
INDONESIA
Java Sea
Jakarta
Sulawesi
Banda Sea
Equator
PAPUA NEW GUINEA
Bandung
Semarang
Surabaya
Flores Sea
Java
E. Timor (UN Admin.)
Timor Sea
Arafura Sea
Port Moresby
INDIAN OCEAN
Darwin
Gulf of Carpentaria
Coral Sea
Tropic of Capricorn
AUSTRALIA
Brisbane
Perth
Great Australian Bight
Adelaide
Sydney
Canberra
Melbourne
Tasmania
INDIAN OCEAN

★ National capital
• Other city

0 400 800 miles
0 400 800 kilometers
Miller Projection

20°N
0°
20°S
40°S
100°E
120°E
140°E

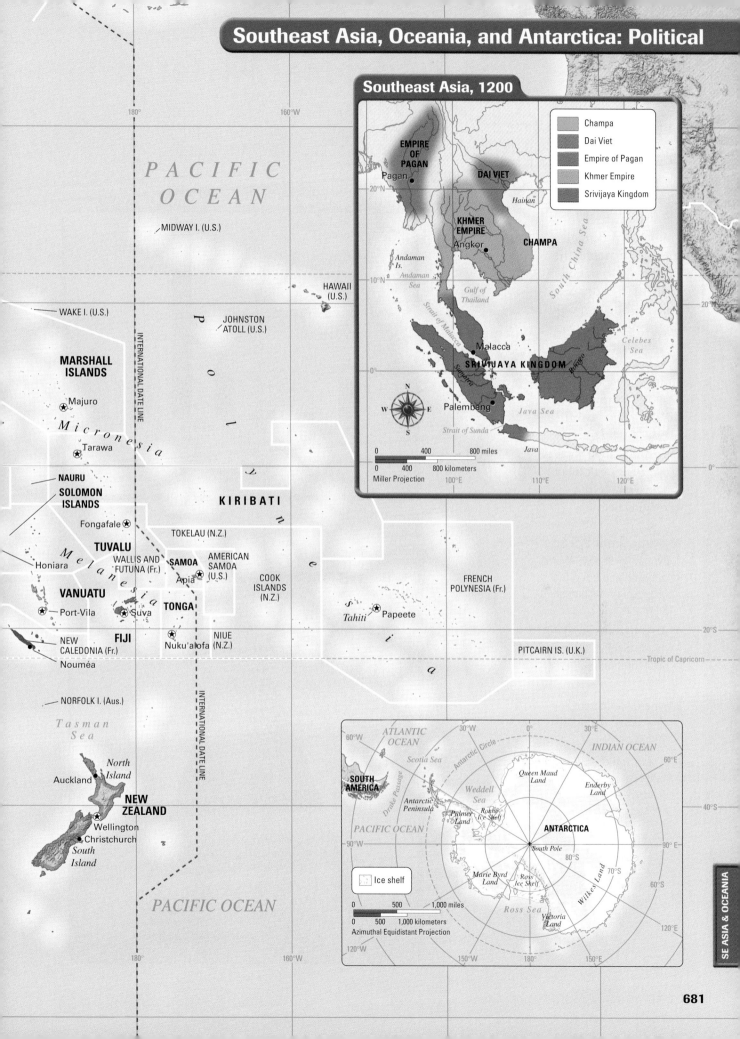

Southeast Asia, 1200

	Champa
	Dai Viet
	Empire of Pagan
	Khmer Empire
	Srivijaya Kingdom

EMPIRE OF PAGAN

Pagan

DAI VIET

KHMER EMPIRE

Angkor

CHAMPA

Hainan

Andaman Is.

Andaman Sea

South China Sea

Gulf of Thailand

Strait of Malacca

Malacca

SRIVIJAYA KINGDOM

Sumatra

Borneo

Celebes Sea

Palembang

Java Sea

Strait of Sunda

Java

N
W E
S

0   400   800 miles
0   400   800 kilometers
Miller Projection

20°N
10°N
0°
20°
100°E   110°E   120°E

PACIFIC OCEAN

MIDWAY I. (U.S.)

HAWAII (U.S.)

WAKE I. (U.S.)

JOHNSTON ATOLL (U.S.)

MARSHALL ISLANDS

Majuro

Micronesia

Tarawa

NAURU

SOLOMON ISLANDS

KIRIBATI

Fongafale

TOKELAU (N.Z.)

Melanesia

Honiara

TUVALU

WALLIS AND FUTUNA (Fr.)

SAMOA

Apia

AMERICAN SAMOA (U.S.)

COOK ISLANDS (N.Z.)

FRENCH POLYNESIA (Fr.)

VANUATU

Port-Vila

Suva

TONGA

Tahiti

Papeete

FIJI

NEW CALEDONIA (Fr.)

Nouméa

Nuku'alofa

NIUE (N.Z.)

PITCAIRN IS. (U.K.)

Polynesia

INTERNATIONAL DATE LINE

NORFOLK I. (Aus.)

Tasman Sea

North Island

Auckland

NEW ZEALAND

Wellington

Christchurch

South Island

PACIFIC OCEAN

180°   160°W   180°

20°S

Tropic of Capricorn

Ice shelf

ATLANTIC OCEAN

Scotia Sea

SOUTH AMERICA

Antarctic Circle

INDIAN OCEAN

Queen Maud Land

Enderby Land

Weddell Sea

Antarctic Peninsula

Palmer Land

Ronne Ice Shelf

ANTARCTICA

South Pole

PACIFIC OCEAN

Marie Byrd Land

Ross Ice Shelf

Wilkes Land

Ross Sea

Victoria Land

0   500   1,000 miles
0   500   1,000 kilometers
Azimuthal Equidistant Projection

60°W   30°W   0°   30°E   60°E
90°W   90°E
80°S   70°S   60°S
120°W   150°W   180°   150°E   120°E

Unit ATLAS

Regional Patterns

These two pages contain graphs and thematic maps. The graphs show the percentage of ethnic Chinese in Southeast Asian populations and the number of active volcanoes in the region. One map shows the climates of the region. The other shows the major religions of the region. After studying the graphs and maps, jot down in your notebook the answers to the questions below.

Making Comparisons

1. Which Southeast Asian nation has the highest proportion of Chinese in its population?

2. What percentage of the region's active volcanoes are found in Southeast Asia?

3. Where are the coldest climates to be found in the region?

4. Would you describe this as a region of religious diversity? Why or why not?

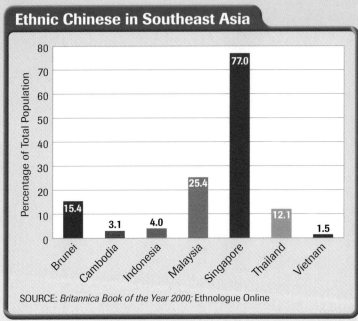

Ethnic Chinese in Southeast Asia

Percentage of Total Population

- Brunei 15.4
- Cambodia 3.1
- Indonesia 4.0
- Malaysia 25.4
- Singapore 77.0
- Thailand 12.1
- Vietnam 1.5

SOURCE: *Britannica Book of the Year 2000;* Ethnologue Online

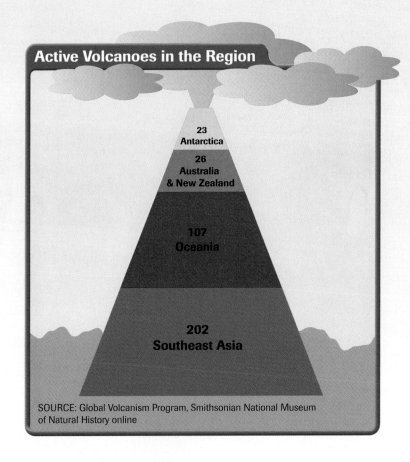

Active Volcanoes in the Region

- 23 Antarctica
- 26 Australia & New Zealand
- 107 Oceania
- 202 Southeast Asia

SOURCE: Global Volcanism Program, Smithsonian National Museum of Natural History online

Climates of the Region

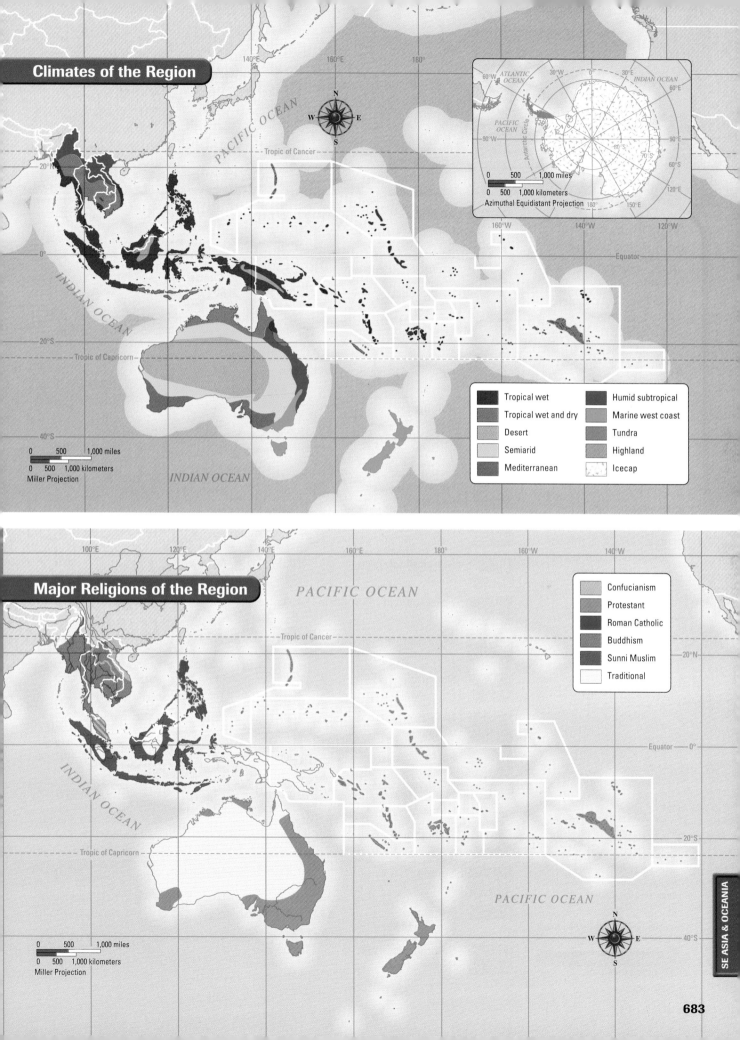

Legend (Climates):
- Tropical wet
- Tropical wet and dry
- Desert
- Semiarid
- Mediterranean
- Humid subtropical
- Marine west coast
- Tundra
- Highland
- Icecap

0   500   1,000 miles
0   500   1,000 kilometers
Miller Projection

PACIFIC OCEAN

INDIAN OCEAN

Tropic of Cancer

Equator

Tropic of Capricorn

20°N
0°
20°S
40°S
140°E   160°E   180°

ATLANTIC OCEAN
PACIFIC OCEAN
INDIAN OCEAN
Antarctic Circle
0   500   1,000 miles
0   500   1,000 kilometers
Azimuthal Equidistant Projection

Major Religions of the Region

Legend (Religions):
- Confucianism
- Protestant
- Roman Catholic
- Buddhism
- Sunni Muslim
- Traditional

PACIFIC OCEAN

INDIAN OCEAN

Tropic of Cancer

Equator

Tropic of Capricorn

PACIFIC OCEAN

100°E   120°E   140°E   160°E   180°   160°W   140°W

20°N
0°
20°S
40°S

0   500   1,000 miles
0   500   1,000 kilometers
Miller Projection

SE ASIA & OCEANIA

683

Unit ATLAS

Study the charts on the countries of this region.

Making Comparisons

1. Compare the population and total area of Australia to that of the United States. What conclusions can you draw?

2. Make a list of the top three countries in population. What is the difference in population between the top two countries?

3. Make a list of the top three countries in total area. How does this list compare to your list of the most populous countries?

(continued on page 686)

Notes:
ᵃ A comparison of the prices of the same items is used to figure these data.
ᵇ Includes land and water, when figures are available.
* East Timor became an independent country on May 20, 2002.

For updated statistics on Southeast Asia, Oceania, and Antarctica . . .

DATA UPDATE
CLASSZONE.COM

Country Flag	Country/ Capital	Population (2000)	Life Expectancy (years) (2000)	Birthrate (per 1,000, pop.) (2000 estimate)	Infant Mortality (per 1,000 live births) (2000)
	Australia Canberra	19,200,000	79	13	5.3
	Brunei Bandar Seri Begawan	331,000	71	25	24.0
	Cambodia Phnom Penh	12,127,000	56	38	80.8
	East Timor* Dili	737,000	50	25	120.9
	Fiji Suva	811,000	67	22	12.9
	Indonesia Jakarta	212,207,000	64	24	45.7
	Kiribati Tarawa	92,000	62	33	62.0
	Laos Vientiane	5,218,000	51	41	104.0
	Malaysia Kuala Lumpur	23,253,000	73	25	7.9
	Marshall Islands Majuro	68,000	65	26	30.5
	Fed. States of Micronesia Palikir	119,000	66	33	46.0
	Myanmar Yangon	48,852,000	54	30	82.5
	Nauru (no capital)	12,000	61	19	25.0
	New Zealand Wellington	3,836,000	77	15	5.5
	Palau Koror	19,000	67	18	19.2
	Papua New Guinea Port Moresby	4,810,000	56	34	77.0
	Philippines Manila	80,298,000	67	29	35.3
	Samoa Apia	176,000	68	31	25.0
	Singapore Singapore City	4,001,000	78	13	3.2
	Solomon Islands Honiara	434,000	71	37	25.3

Doctors (per 100,000 pop.) (1994–1999)	GDP[a] (billions $US) (1998–1999)	Import/Export[a] (billions $US) (1998–1999)	Literacy Rate (percentage) (1996–1998)	Televisions (per 1,000 pop.) (1998)	Passenger Cars (per 1,000 pop.) (1996–1997)	Total Area[b] (square miles)
240	416.2	67.0 / 58.0	100	495	474	2,967,909
85	5.6	1.24 / 2.04	88	239	477	2,226
30	8.2	1.2 / 0.821	65 (1993)	9	1.2	69,898
N.A.	0.415 (2001 est.)	0.237 / .008 (2001 est.)	48	79	23	5,641
48	5.9	0.612 / 0.393	92	18	38	7,055
16	610.0	21.6 / 48.0	84	66	12	779,675
30	0.074	0.033 / 0.006	90	N/A	N/A	277
24	7.0	0.497 / 0.271	57	9	1.7	91,428
66	229.1	61.5 / 83.5	84	164	143	128,727
42	0.105 (1997)	0.058 / 0.028 (1997)	93 (1994)	N/A	N/A	70
57	0.240 (1997)	0.151 / 0.073 (1996)	90 (1991)	N/A	N/A	1,055
30	121.0 (1996)	1.829 / 0.886 (1996)	83	5	0.7	261,789
157	0.100 (1993)	0.019 / 0.025 (1991)	99	N/A	N/A	8.2
217	63.8	11.2 / 12.2	100	514	391	103,736
110	0.160 (1997)	0.072 / 0.014 (1996)	98 (1990)	N/A	N/A	191
7	11.6	1.0 / 1.9	72	4	5	178,260
123	282.0	30.7 / 34.8	95	49	9	115,651
34	0.485	0.097 / 0.021	98	41	7	1,209
163	98.0	111.0 / 114.0	91	361	95	225
14	1.21	0.144 / 0.142	54	6	N/A	11,500

Regional Data File

Country Flag	Country/ Capital	Population (2000)	Life Expectancy (years) (2000)	Birthrate (per 1,000, pop.) (2000 estimate)	Infant Mortality (per 1,000 live births) (2000)
	Thailand Bangkok	62,043,000	73	16	22.4
	Tonga Nuku'alofa	108,000	71	27	19.0
	Tuvalu Fongafale	10,838	64	22	24.8
	Vanuatu Port-Vila	195,000	65	35	39.0
	Vietnam Hanoi	78,697,000	66	20	36.7
	United States Washington, D.C.	281,422,000	77	15	7.0

Making Comparisons

(continued)

4. Which countries have a literacy rate below 60 percent?

5. For the countries you identified in question 4, look at their ratio of doctors to population. Is it high or low compared to other countries? What might be the relationship between literacy rate and number of doctors?

Sources:
ASEAN statistics online
Europa World Year Book 2000
Human Development Report 2000, United Nations
International Data Base, 2000, U.S. Census Bureau online
Merriam-Webster's Geographical Dictionary, 1997
Statesman's Yearbook 2001
2000 World Population Data Sheet, Population Reference Bureau online
WHO Estimates of Health Personnel, World Health Organization online
World Almanac and Book of Facts 2001
World Factbook 2000, CIA online
N/A = not available

Notes:
[a]A comparison of the prices of the same items is used to figure these data.
[b]Includes land and water, when figures are available.

Territories and Possessions in Oceania

Name	Status
American Samoa	U.S. territory*
Cook Islands	Self-governing area in free association with New Zealand
French Polynesia	French overseas territory
Guam	U.S. territory*
Irian Jaya	Indonesian province
Midway Islands	U.S. possession*
New Caledonia	French overseas territory
Niue	Self-governing area in free association with New Zealand
Norfolk Island	Australian territory
Northern Mariana Islands	U.S. commonwealth*
Pitcairn Islands	British overseas territory
Tokelau	New Zealand territory
Wake Island	U.S. possession*
Wallis and Futuna	French overseas territory

* A commonwealth is a self-governing political unit in voluntary association with the United States; a U.S. territory is not a state but has a governor and a legislature; the U.S. possessions in the Pacific are administered by the Navy.

SOURCE: *World Book Encyclopedia 2000*

Doctors (per 100,000 pop.) (1994–1999)	GDP[a] (billions $US) (1998–1999)	Import/Export[a] (billions $US) (1998–1999)	Literacy Rate (percentage) (1996–1998)	Televisions (per 1,000 pop.) (1998)	Passenger Cars (per 1,000 pop.) (1996–1997)	Total Area[b] (square miles)	
24	388.7	45.0 / 58.5	94	189	25	198,455	
44	0.238	0.069 / 0.008	93 (1992)	16	31	270	
30	0.008 (1995)	0.004 / 0.0002 (1989)	95	N/A	N/A	9	
12	0.245	0.076 / 0.034	36	13	21	5,700	
48	143.1	11.6 / 11.5	94	43	1	130,468	
251	9,255.0	820.8 / 663.0	97	847	489	3,787,319	

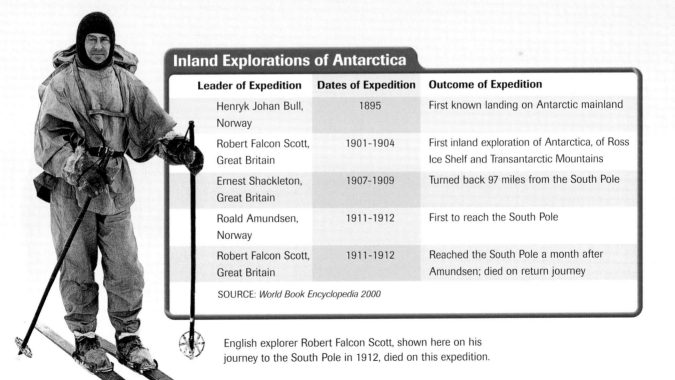

Inland Explorations of Antarctica

Leader of Expedition	Dates of Expedition	Outcome of Expedition
Henryk Johan Bull, Norway	1895	First known landing on Antarctic mainland
Robert Falcon Scott, Great Britain	1901–1904	First inland exploration of Antarctica, of Ross Ice Shelf and Transantarctic Mountains
Ernest Shackleton, Great Britain	1907–1909	Turned back 97 miles from the South Pole
Roald Amundsen, Norway	1911–1912	First to reach the South Pole
Robert Falcon Scott, Great Britain	1911–1912	Reached the South Pole a month after Amundsen; died on return journey

SOURCE: *World Book Encyclopedia 2000*

English explorer Robert Falcon Scott, shown here on his journey to the South Pole in 1912, died on this expedition.

Chapter 30

SECTION 1
Landforms and Resources

SECTION 2
Climate and Vegetation

SECTION 3
Human–Environment Interaction

PHYSICAL GEOGRAPHY OF SOUTHEAST ASIA, OCEANIA, AND ANTARCTICA

A Region of Extremes

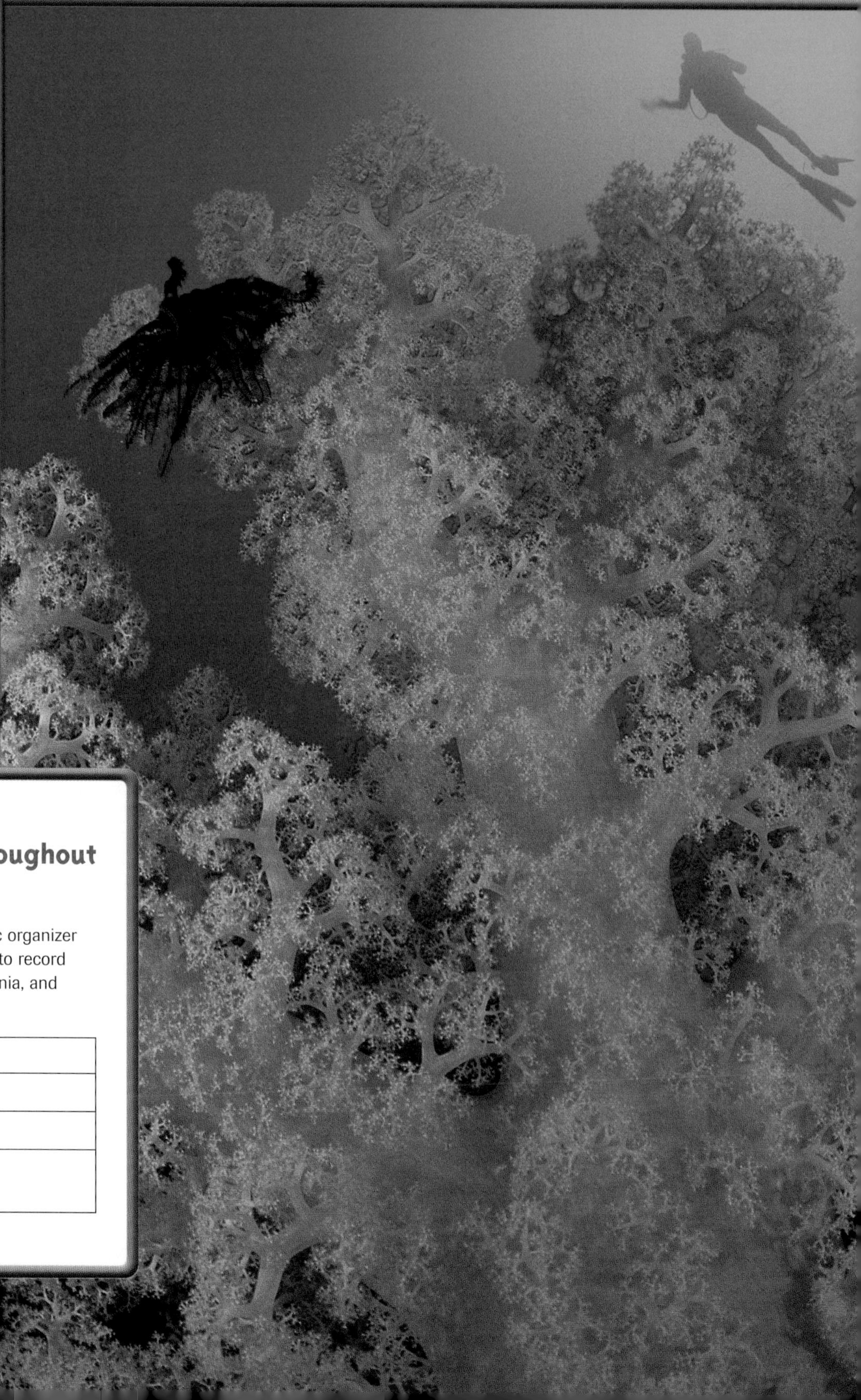

Scuba divers in Australia's Great Barrier Reef can observe some of its more than 1,500 species of fish and approximately 400 species of coral.

GeoFocus

How does physical geography vary throughout this vast region?

Taking Notes Copy the graphic organizer below into your notebook. Use it to record facts about Southeast Asia, Oceania, and Antarctica.

Landforms	
Resources	
Climate and Vegetation	
Human-Environment Interaction	

Landforms and Resources

Main Ideas

• This region includes two peninsulas of Asia, two continents, and more than 20,000 islands.

• Its landforms include mountains, plateaus, and major river systems.

Places & Terms

archipelago   low island

Oceania   Great Barrier

high island   Reef

CONNECT TO THE ISSUES

INDUSTRIALIZATION Some countries of this region have used their resources to develop industry, with mixed results.

A HUMAN PERSPECTIVE The Aeta (EE•duh) people of the Philippines lived on the volcano Mount Pinatubo for generations. They knew this volcano so well that they timed the planting and harvesting of their crops by the amount of steam rising from a vent on its slope. In 1991, the Aeta noticed changes in the mountain and concluded that it was about to erupt. Tens of thousands of Aeta fled their homes as did countless other Filipinos. Pinatubo did erupt for the first time in 600 years, spewing ash for miles. Since then, many of the Aeta have formed new communities, but they still miss their homeland. As their story shows, the geologic processes that destroy landforms also disrupt human lives.

Southeast Asia: Mainland and Islands

Southeast Asia has two distinct subregions: the southeastern corner of the Asian mainland and a great number of islands. Both the mainland and the islands have many high mountains.

PENINSULAS AND ISLANDS The most noticeable feature of mainland Southeast Asia is that it lies on two peninsulas. The Indochinese Peninsula, located south of China, has a rectangular shape. In contrast, the Malay Peninsula is a narrow strip of land about 700 miles long, stretching south from the mainland and then curving southeast. It serves as a bridge between the mainland and islands.

Most of the islands of Southeast Asia are found in archipelagoes. An **archipelago** is a set of closely grouped islands, which sometimes form a curved arc. The Philippines and the islands of Indonesia are part of the Malay Archipelago. (See the map on page 680.) A few Southeast Asian islands, such as Borneo, are actually the high points of a submerged section of the Eurasian plate.

MOUNTAINS AND VOLCANOES On the map at right, you can see that the mainland has several mountain ranges, such as the Annamese Cordillera, running roughly north and south. These ranges fan out from a mountainous area to the north.

Southeast Asian Mountains and Rivers

Map labels: Arakan Yoma, Irrawaddy R., Salween R., Red R., Annamese Cordillera, Mekong R., Gulf of Tonkin, Andaman Sea, Bilauktaung Range, Chao Phraya, Indochinese Peninsula, 15°N, Malay Peninsula, Gulf of Thailand, 10°N, South China Sea, 100°E, 105°E

SKILLBUILDER: Interpreting Maps

❶ **PLACE** Which mountain chain lies east of the Mekong River?

❷ **LOCATION** How would you describe the relative location of the Chao Phraya?

SE ASIA & OCEANIA

Island Formation in the Pacific

High (volcanic) Islands

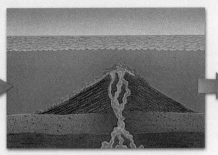

1. Magma sometimes erupts through cracks in the ocean floor.

2. Over time, layers of lava can build up to form a volcanic cone.

3. Some volcanic cones rise above sea level and become islands.

Low (coral) Islands

1. Some corals form reefs on the sides of volcanic islands.

2. As the island erodes, the reef continues to grow upward.

3. In time, only the low islands of the reef remain.

On the islands, most of the mountains are of volcanic origin. Southeast Asia is part of the Pacific Ring of Fire that you read about in Chapter 29. Volcanic eruptions and earthquakes are natural disasters that frequently occur in this region. (See pages 710–711.)

RIVERS AND COASTLINES The mainland has several large rivers that run from the north through the valleys between the mountain ranges. Near the coast these rivers spread out into fertile deltas. For example, the Mekong (MAY·KAWNG) River begins in China and crosses several Southeast Asian nations before becoming a wide delta on Vietnam's coast. Millions of people rely on the Mekong for farming and fishing.

Southeast Asia's peninsulas and islands give it a long, irregular coastline with many ports. As you can imagine, this has encouraged a great deal of seagoing travel and trade.

RESOURCES Fertile soil is a valuable resource in Southeast Asia. Volcanic activity and flooding rivers both add nutrients back to the soil and keep it rich. Southeast Asians also have access to large numbers of fish in the rivers and nearby seas. Parts of the region have mineral resources, such as petroleum, tin, and gems, which industry can use.

Lands of the Pacific and Antarctica

No one knows how many islands exist in the Pacific Ocean, but some geographers estimate that there are more than 20,000. As a group, the Pacific Islands are called **Oceania.** (The Philippines, Indonesia, and other islands near the mainland are not considered part of Oceania because their people have cultural ties to Asia.) In the southwestern

The high island in the background is Bora Bora, a volcanic island in French Polynesia. (See the map on page 681.) Many small coral islands surround it.

Pacific lie New Zealand and Australia, which are often considered part of Oceania, even though Australia is a continent, not an island.

OCEANIA'S MANY ISLANDS One reason geographers don't know the number of islands in Oceania is that it changes. Erosion causes some islands to vanish, while other forces create new islands. Most Pacific islands fall into two categories: **high islands** are created by volcanoes, and **low islands** are made of coral reefs. Although a few of Oceania's islands are large, most are small. If you added the land area of all the islands together, the total would be smaller than the area of Alaska.

Oceania is not rich in resources. The low islands have poor soil, and most of the islands lack minerals. But New Caledonia has nickel, chromium, and iron; New Guinea has copper, gold, and oil; Nauru has phosphate; and both Fiji and the Solomon Islands have gold. The general scarcity of resources has made it difficult to develop industry.

MAJESTIC NEW ZEALAND New Zealand has two main islands, North Island and South Island. Running down the center of South Island is a 300-mile-long mountain range, the Southern Alps. This range has 16 peaks over 10,000 feet high and more than 360 glaciers. Several rivers flow down the eastern slopes to the ocean.

North Island has hilly ranges and a volcanic plateau, but it is much less mountainous than South Island. North Island has fertile farmland and forest that support the lumber industry. In addition, its coastline has natural harbors that are used for seaports. Like South Island, North Island has many rivers running from the mountains to the sea. ◀A

New Zealand has few mineral resources. However, its swift-flowing rivers have allowed its people to build dams that generate electricity.

Geographic Thinking

Seeing Patterns
Ⓐ Judging from the information in this paragraph, what products do you think New Zealand exports?

Also, North Island has a volcanic area with underground steam. Engineers have found ways to use this steam to power generators.

FLAT AUSTRALIA The land mass known as Australia is the smallest continent on earth. It is also the flattest. Near the eastern coast, running roughly parallel to it, is a chain of highlands called the Great Dividing Range. Unlike New Zealand's mountains, few of these peaks rise higher than 5,000 feet. To the west of this range stretches a vast expanse of plains and plateaus, broken by only a few mountains.

Many other differences exist between Australia and New Zealand. For example, Australia has very few rivers. The largest is the Murray River, which flows into the Southern Ocean. Forestry is not a major industry in Australia, but the country is rich in minerals. It is the world's leading supplier of bauxite, diamonds, opals, lead, and coal.

Along Australia's northeast coast lies one of the wonders of nature. The **Great Barrier Reef** is often called the world's largest coral reef, although it is really a 1,250-mile chain of more than 2,500 reefs and islands. Some 400 species of coral are found there.

BACKGROUND
People in Australia and New Zealand call the waters around Antarctica the Southern Ocean.

ICY ANTARCTICA Antarctica is the fifth largest continent. Generally circular in shape, it is centered on the South Pole. Its topography is hidden by a thick ice sheet, but under the ice lies a varied landscape. The Transantarctic Mountains divide the continent in two. East Antarctica is a plateau surrounded by mountains and valleys. West Antarctica is a group of separate islands linked only by the ice that covers them. **B**

Geographic Thinking

Making Comparisons
B What are similarities and differences between the physical geographies of Australia and Antarctica?

Antarctica's ice sheet is the largest supply of fresh water in the world. Geologists believe that resources such as coal, minerals, and perhaps even petroleum may lie beneath the ice. But in 1991, 26 nations agreed not to mine Antarctica for 50 years. In the next section, you will read about Antarctica's harsh climate as well as the climates of Southeast Asia, Oceania, Australia, and New Zealand.

SECTION **Assessment**

1 Places & Terms
Identify these terms and explain their importance in the region's physical geography.
- archipelago
- Oceania
- high island
- low island
- Great Barrier Reef

2 Taking Notes
PLACE Review the notes you took for this section.

Landforms	
Resources	

- What river begins in China and flows to the Vietnamese coast?
- What are the two main islands of New Zealand?

3 Main Ideas
a. What are the main resources of Southeast Asia?
b. What different resources do Australia and New Zealand have?
c. What landform divides the continent of Antarctica?

4 Geographic Thinking
Seeing Patterns By what processes do low islands replace high islands? **Think about:**
- the process that causes some islands to disappear
- the diagrams on page 690

RESEARCH LINKS
CLASSZONE.COM

GeoActivity

SEEING PATTERNS Do research to learn about the ways that humans have damaged the Great Barrier Reef. Write the script for a **public service announcement,** telling visitors to Australia what behaviors to avoid. You might also include visuals of the Great Barrier Reef. Use standard grammar, spelling, punctuation, and sentence structure in your script.

Interpreting a Relief Map

Two activities that are popular in New Zealand are mountain climbing and skiing. The relief map below shows mountainous areas, which are suitable to those activities. The mountains also provide some regions of New Zealand with spectacular scenery—especially in the Southern Alps of South Island.

THE LANGUAGE OF MAPS A **relief map** illustrates the differences in elevation that are found in a region. It does this with a combination of colors and shading. The lowest elevations are shown in green, and various shades of brown represent progressively higher elevations. The gray shading shows the locations of mountainous landforms.

New Zealand: Physical

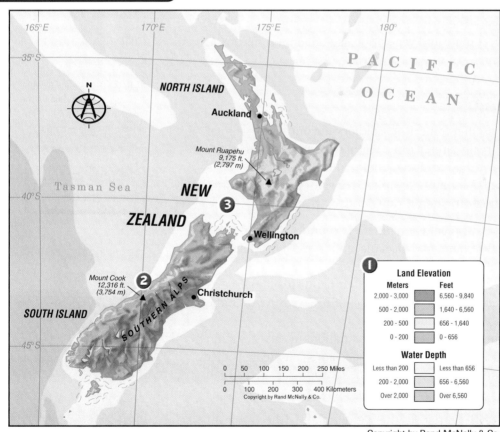

Copyright by Rand McNally & Co.

① The key illustrates the colors used on the map and the range of elevation that each color represents.

② The symbol for peak is ▲. The map also shows the peaks' names and elevations.

③ This map clearly shows the difference between the physical geographies of North Island and South Island.

Map and Graph Skills Assessment

1. Seeing Patterns
Which of New Zealand's two large islands is more mountainous?

2. Drawing Conclusions
How high is Mount Cook?

3. Making Inferences
Which island is better suited to farming? Why?

Climate and Vegetation

Main Ideas

• This region's climates range from tropical to desert to polar icecap.

• There is a great diversity of plant and animal life, including some species found nowhere else in the world.

Places & Terms

outback

CONNECT TO THE ISSUES
ENVIRONMENTAL CHANGE The hole in the ozone layer, located over Antarctica, has affected the climate of this region.

A HUMAN PERSPECTIVE During the Vietnam War, American troops were sent to fight in unfamiliar Southeast Asia. Among the hardships they endured was the tropical climate. Few had ever lived in a place that had a monsoon season with constant rain. One soldier wrote to his wife, "We live in mud and rain. I'm so sick of rain that it is sometimes unbearable. At night the mosquitoes plague me. . . . The rain drips on me until I go to sleep from exhaustion."

Another soldier wrote to a friend about the vegetation: "Try to imagine grass 8 to 15 feet high so thick as to cut visibility to one yard, possessing razor-sharp edges. Then try to imagine walking through it." As these letters make clear, climate and vegetation can create serious obstacles to military operations—or other activities.

Widespread Tropics

Although the conditions that American soldiers encountered seemed unusual to them, they really aren't rare. Vietnam is just one of many countries in this region with a tropical climate. In fact, tropical climates cover most of Southeast Asia and Oceania. Tropical climates fall into two categories, depending on when it rains during the year.

YEAR-ROUND RAINS A tropical wet climate is found in coastal parts of Myanmar, Thailand, Vietnam, and Oceania, and in most of Malaysia, Indonesia, and the Philippines. Temperatures are high. For example, most of Southeast Asia has an average annual temperature of 80°F. Parts of Southeast Asia receive over 100 inches of rain a year, with some places receiving more than 200 inches.

Although the climate is fairly consistent, variations do exist within the region. Elevation, ocean breezes, and other factors can create cooler temperatures. For example, Indonesia has some locations at such a high elevation that they have glaciers. (See the infographic on page 56.)

WET AND DRY SEASONS Bordering the wet climate is the tropical wet and dry climate, in which monsoons shape the weather. As you read in Unit 8, monsoons are winds that cause wet and dry seasons. This climate is found in parts of Myanmar, Thailand, Laos, Cambodia, and Vietnam—generally to the north or inland of the wet climate. Parts of Oceania and northern Australia also have this climate.

Although temperatures are consistently hot, rainfall varies greatly within the climate zone. Local conditions and

PLACE The Rafflesia, which is native to Indonesia, is the world's largest flower. It is almost three feet across.

landforms can affect precipitation amounts. For example, mountains create rain shadows.

Areas with monsoons often experience disastrous weather. During the wet season, typhoons can occur in Southeast Asia and Oceania.

TROPICAL PLANTS Southeast Asia has one of the greatest diversities of vegetation of any region. For example, it has a remarkable number of tree species. Near the equator are tropical evergreen forests, while deciduous forests are more common in the wet and dry climate zone. Teak, a valuable tree that Asians harvest commercially, comes from these deciduous forests. Southeast Asia also has many types of plants.

In general, Oceania does not have diverse vegetation. The low islands have poor soil and small amounts of rain, so plants don't grow well. Some high islands have rich, volcanic soil and plentiful rain. These islands have abundant flowers and trees, such as the coconut palm.

Bands of Moderate Climate

Australia is the only inhabited continent that lies completely in the Southern Hemisphere. New Zealand is even farther south. Australia and New Zealand have generally moderate climates.

HOT SUMMERS, MILD WINTERS As Section 1 explained, a mountain chain runs parallel to the east coast of Australia. The strip between the mountains and ocean is divided mostly into two climate zones. The northern part of this strip has a humid subtropical climate, with hot summers, mild winters, and heavy rainfall. It is one of Australia's wettest regions, receiving an average of 126 inches of rain a year. This climate also exists in northern Vietnam, Laos, Thailand, and Myanmar.

BACKGROUND
The tropical rain forests of the Philippines alone have more than 3,000 species of trees and 8,000 species of wild plants.

Geographic Thinking

Using the Atlas
A▷ Find this mountain range on the map on page 678. What is it called?

SE ASIA & OCEANIA

Climate and Vegetation **695**

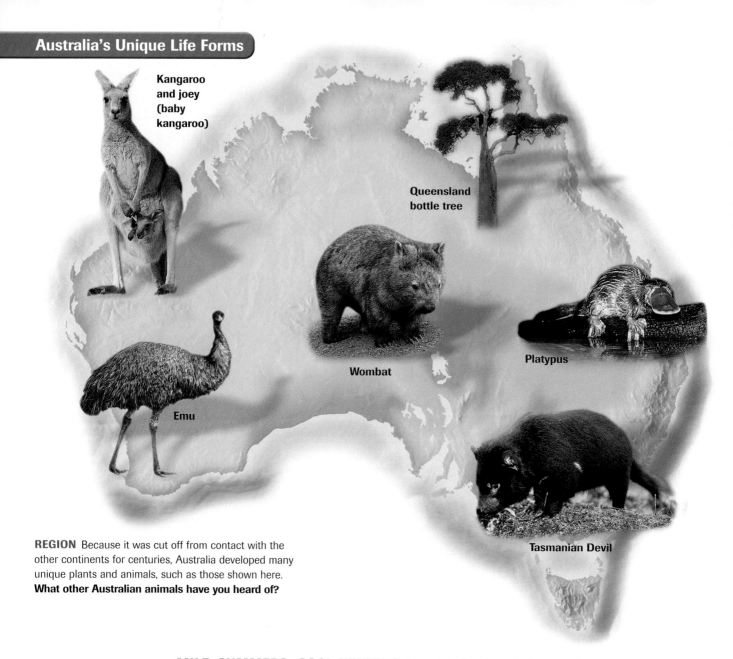

Kangaroo and joey (baby kangaroo)

Queensland bottle tree

Wombat

Platypus

Emu

Tasmanian Devil

REGION Because it was cut off from contact with the other continents for centuries, Australia developed many unique plants and animals, such as those shown here. **What other Australian animals have you heard of?**

MILD SUMMERS, COOL WINTERS New Zealand and the southern part of Australia's east coast share a marine west coast climate. The seasons have mild temperatures because ocean breezes warm the land in the winter and cool it in the summer. New Zealand's forests consist primarily of evergreens and tree ferns, which thrive in such a climate.

New Zealand receives rainfall year-round, although the amount varies dramatically from one part of the country to another. For example, the mountains of South Island cause rain to fall on their western slopes, so the eastern part of the island is dryer. Mountains change the climate in another way. The mountainous inland areas of New Zealand are cooler than the coastal areas. Temperatures drop about three-and-a-half degrees for every 1,000-foot rise in elevation.

Mountains influence Australia's climates, too. The Great Dividing Range forces moisture-bearing winds to rise and shed their rain before moving inland. For that reason, the marine west coast and humid subtropical climates exist only on the east coast. That coast is Australia's most heavily populated region. The moist coasts are also the only parts of Australia with enough rain for trees that grow taller than 300 feet.

BACKGROUND
This rate of temperature reduction is true in almost all mountain ranges.

MOVEMENT U.S. sailors and Bikini Islanders load supplies before the evacuation of Bikini Atoll in 1946. **Why do you think they used U.S. Navy landing craft instead of privately owned boats?**

BACKGROUND
The two-piece bikini bathing suit was named after the Bikini test because designers claimed the suit was "explosive."

of the Bikini Atoll and contaminated the entire area with high levels of radiation. Many islanders were injured or became ill.

LONG-TERM EFFECTS In the meantime, the Bikini Islanders remained exiled from their homeland. The first atoll to which they were moved proved to be unable to support inhabitants, so in 1948, they were moved to the island of Kili. But they soon grew unhappy because conditions there made it impossible to grow enough food or to engage in fishing.

In the late 1960s, the United States government declared Bikini Atoll safe for humans, and some islanders returned home. Then, in 1978, doctors discovered dangerous levels of radiation in the islanders' bodies. The affected islanders had to leave again. A cleanup began in 1988, but no one knows when Bikini Atoll will again be suitable for human life.

In Chapter 31, you will read more about the history and culture of Oceania, Southeast Asia, Australia, and New Zealand.

Assessment

1 Places & Terms

Identify these terms and explain their importance in the region's physical geography.

- voyaging canoe
- outrigger canoe
- atoll
- Bikini Atoll

2 Taking Notes

HUMAN-ENVIRONMENT INTERACTION Review the notes you took for this section.

Human-Environment Interaction

- What is an example of humans adapting to the environment?
- What are examples of humans altering the environment?

3 Main Ideas

a. How did Pacific Islanders navigate the ocean in ancient times?

b. How have Australians tried to control the rabbit problem?

c. Why have the Bikini Islanders been unable to return home?

4 Geographic Thinking

Determining Cause and Effect What do the atomic tests on Bikini reveal about the long-term effects of using atomic weapons?

Think about:

- how the blasts affect people and the environment

RESEARCH LINKS
CLASSZONE.COM

MAKING COMPARISONS Do research to learn about French atomic tests in the Pacific. Create a **chart** comparing the French tests to the U.S. tests. You might use such categories as location, impact on people, and current policy about the tests.

SE ASIA & OCEANIA

Chapter 30 Assessment

VISUAL SUMMARY
PHYSICAL GEOGRAPHY OF SOUTHEAST ASIA, OCEANIA, AND ANTARCTICA

Landforms

Southeast Asia: Indochinese Peninsula; Malay Peninsula; Malay Archipelago; mountain ranges and rivers

Oceania: high islands; low islands; New Zealand—South Island and North Island; Australia—Great Dividing Range, Murray River

Antarctica: Transantarctic Mountains; East Antarctica and West Antarctica

Resources

- Southeast Asia has fish, fertile soil, and mineral resources.
- Oceania is generally poor in resources. Some islands have minerals.
- New Zealand has fertile farmland, forests, and rivers. Australia is rich in minerals.

Climate and Vegetation

- Southeast Asia and Oceania have tropical or subtropical climates. Southeast Asia has a great diversity of vegetation.
- Australia has moderate climates on its coasts and arid climates inland. New Zealand has a marine west coast climate.
- Antarctica is a polar desert.

Human-Environment Interaction

- The people who settled the Pacific Islands navigated using traditional methods and doubled-hulled canoes.
- Imported rabbits severely damaged the vegetation of Australia.
- U.S. atomic tests contaminated the Bikini Atoll with radiation.

Reviewing Places & Terms

A. Briefly explain the importance of each of the following.

1. archipelago
2. Oceania
3. high island
4. low island
5. Great Barrier Reef
6. outback
7. voyaging canoe
8. outrigger canoe
9. atoll
10. Bikini Atoll

B. Answer the questions about vocabulary in complete sentences.

11. Which of the terms above are related to Australia?
12. Are atolls high islands or low islands? Explain.
13. Is the Great Barrier Reef most closely related to high islands or low islands? Explain.
14. Which would a tourist be more likely to visit, the outback or the Great Barrier Reef? Why?
15. Which of the subregions contain archipelagoes?
16. Where in Oceania are outrigger canoes used?
17. What are the important features of voyaging canoes?
18. Which of the terms above is associated with human damage to the environment?
19. Are high islands or low islands more likely to have prosperous economies? Why?
20. Which term or terms name a place in Oceania?

Main Ideas

Landforms and Resources (pp. 689-693)

1. What are the two distinct subregions of Southeast Asia?
2. What is the physical pattern formed by the mountain ranges and rivers of mainland Southeast Asia?
3. For what purpose do engineers use the underground steam found in the volcanic area of New Zealand?
4. What is one of the many differences between the physical geographies of Australia and New Zealand?

Climate and Vegetation (pp. 694-697)

5. Where is the tropical wet and dry climate found?
6. How does the Great Dividing Range influence Australia's climate?
7. What are the main plants and animals of Antarctica?

Human-Environment Interaction (pp. 698-701)

8. On the navigation charts of Pacific Islanders, what did the shells represent?
9. Why did the rabbit population grow so quickly in Australia?
10. Why have the Bikini Islanders been unhappy with the places where the U.S. government resettled them?

Critical Thinking

1. Using Your Notes

Use your completed chart to answer these questions.

Landforms	
Resources	

   a. What subregion has a large diversity of both landforms and vegetation?

   b. How did the type of vegetation found in Australia make it an unsuitable place for the introduction of rabbits?

2. Geographic Themes

   a. **LOCATION** Where are the tropical climates of this region located relative to the equator?

   b. **MOVEMENT** How does the physical geography of Southeast Asia encourage movement?

3. Identifying Themes

Consider the way that Pacific Islanders used shell maps (see page 698). How does the use of such maps demonstrate all five themes of geography?

4. Determining Cause and Effect

What are some of the negative and positive effects of volcanic activity in Southeast Asia?

5. Identifying and Solving Problems

In general, Oceania has few resources. What problem does this create for Pacific Islanders, and how might they solve it?

Additional Test Practice, pp. S1–S37

TEST PRACTICE
CLASSZONE.COM

Geographic Skills: Interpreting Maps

Bikini Atoll

Use the map to answer the following questions.

1. **PLACE** Which channel leads to the lagoon inside the Bikini Atoll?

2. **LOCATION** What is the absolute location of the Bravo test site?

3. **MOVEMENT** How far did radiation travel from the Bravo test site in order to contaminate Bikini Island?

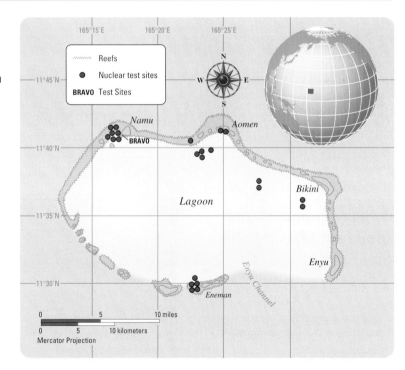

GeoActivity

In addition to Bikini Atoll, other atolls and islands were contaminated with radiation from the U.S. atomic-weapons tests. Do research to learn the names and locations of these islands and atolls. Then create a map showing the full area of radiation contamination.

INTERNET ACTIVITY

Use the links at **classzone.com** to do research about active volcanoes in Southeast Asia. Look for such information as location and recent eruptions.

Writing About Geography Write a report of your findings. Include a map of the volcanoes and a chart listing recent eruptions. List the Web sites that were your sources.

Chapter 31

SECTION 1
Southeast Asia

SECTION 2
Oceania

SECTION 3
Australia, New Zealand, and Antarctica

HUMAN GEOGRAPHY OF SOUTHEAST ASIA, OCEANIA, AND ANTARCTICA
Migration and Conquest

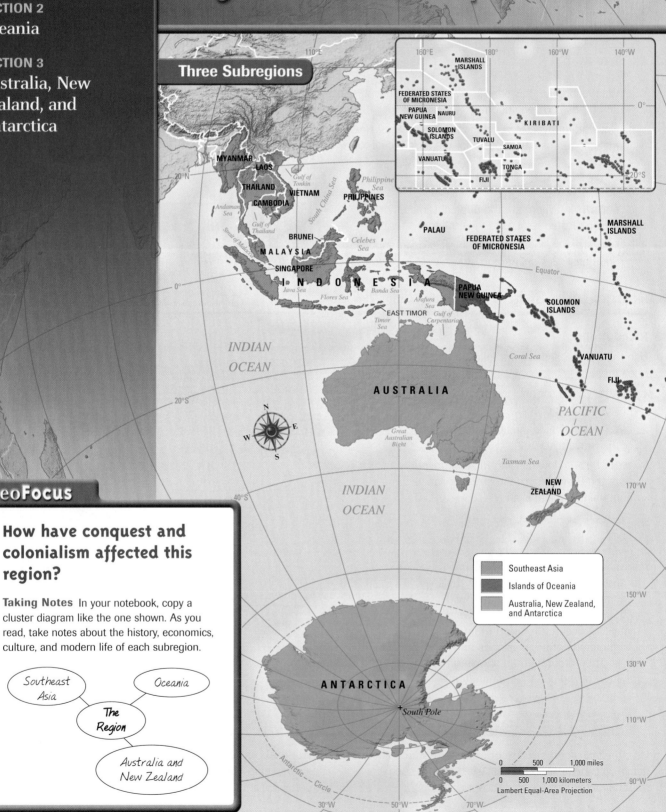

Three Subregions

Southeast Asia
Islands of Oceania
Australia, New Zealand, and Antarctica

Lambert Equal-Area Projection

GeoFocus

How have conquest and colonialism affected this region?

Taking Notes In your notebook, copy a cluster diagram like the one shown. As you read, take notes about the history, economics, culture, and modern life of each subregion.

Southeast Asia — The Region — Oceania — Australia and New Zealand

Southeast Asia

Main Ideas

• Influenced by China and India, Southeast Asia developed many vibrant, complex cultures.

• European colonialism left a legacy that continues to affect the region's politics and economics.

Places & Terms

mandala

Khmer Empire

Indochina

Vietnam War

ASEAN

CONNECT TO THE ISSUES
INDUSTRIALIZATION
Since 1960, many Southeast Asian nations industrialized, while others lagged behind.

A HUMAN PERSPECTIVE Much of Southeast Asia is haunted by its colonial past. One example is the divided island of Timor. The Netherlands ruled Western Timor, later part of Indonesia. Portugal ruled East Timor. In 1975, East Timor declared itself an independent state (even though some people living there wanted to join Indonesia). In response, Indonesia invaded the new nation and ruled it for 24 years.

In 1999, Indonesia let East Timor vote on the choice of limited self-rule within Indonesia or independence. When most voters chose independence, pro-Indonesia militias reacted with violence. The United Nations stepped in and helped East Timor prepare for nationhood. In May 2002, the country gained its independence.

The new nation is also one of the poorest. However, the development of a natural gas field in the Timor Sea should help solve East Timor's economic challenges. In fact, the revenue from the sale of the gas is expected to guarantee the new nation a steady income until 2020.

A Long History of Diversity

Since ancient times, many cultures have influenced Southeast Asia, yet it has retained its own character. Today the region includes the nations of Brunei, Cambodia, Indonesia, Laos, Malaysia, Myanmar, the Philippines, Singapore, Thailand, and Vietnam.

EARLY HISTORY China and India influenced ancient Southeast Asia. China ruled northern Vietnam from 111 B.C. to A.D. 939. Chinese art, technology, political ideas, and ethical beliefs shaped Vietnam's culture. Hinduism and Buddhism spread from India and influenced religion and art in much of Southeast Asia. Yet, Southeast Asia kept some of its own traditions, such as more equal roles for women.

Early Southeast Asian states didn't have set borders. Instead, they were ***mandalas,*** states organized as rings of power around a central court. Those regions of power changed in size over time. A *mandala's* region might overlap that of a neighbor, so rulers had to make alliances for a state to survive. The **Khmer Empire** was a powerful *mandala* that lasted roughly from the 9th to the 15th centuries in what is now Cambodia.

MOVEMENT The temple complex of Angkor Wat in Cambodia was built in the 1100s and dedicated to the Hindu god Vishnu.
How does this temple illustrate the movement of ideas?

POWERFUL STATES The years 1300 through 1800 were important to Southeast Asia's development. Five powerful states existed where Myanmar, Vietnam, Thailand, Java, and the Malay Peninsula are now. Those states were similar to *mandalas* but were larger and more complex. Trade within the region was important to their economies.

During that period, the Burmese, the Vietnamese, the Thai, and the Javanese each began to define their national identities. Urbanization, or the growth of large cities, also took place. For instance, Malacca, on the Malay Peninsula, grew to have about 100,000 people in the early 1500s.

BACKGROUND
The Burmese are the people of Myanmar, which used to be called Burma; the Javanese live in Indonesia.

Colonialism and Its Aftermath

Southeast Asian states not only traded with each other but also with merchants from Arabia and India, who brought Islam to Southeast Asia. Islam attracted many followers, especially in the islands.

EUROPEAN CONTROL Large numbers of Europeans began to arrive in Southeast Asia in 1509. At that time, Europeans had little interest in setting up colonies there, except for the Spanish, who took over the Philippines. Instead, the goal of most Europeans was to obtain wealth.

Europeans used various business methods to take over much of Southeast Asia's trade. As the region's wealth flowed to Europe, local control in Southeast Asian states declined. By the 20th century, Europeans had made all of Southeast Asia except Siam (now Thailand) into colonies. **A**

Colonialism changed Southeast Asia. First, colonial rulers set up centralized, bureaucratic governments with set routines and regulations. Second, Europeans forced the colonies to produce commodities that would help Europe's economy. They included rubber, sugar, rice, tea, and coffee. Third, colonialism had the unintended effect of sparking nationalism. Groups that never had been allies united against European rule. And Southeast Asians who gained Western education learned about political ideas such as self-rule.

Geographic Thinking

Seeing Patterns
A Why would a loss of wealth cause local control to weaken?

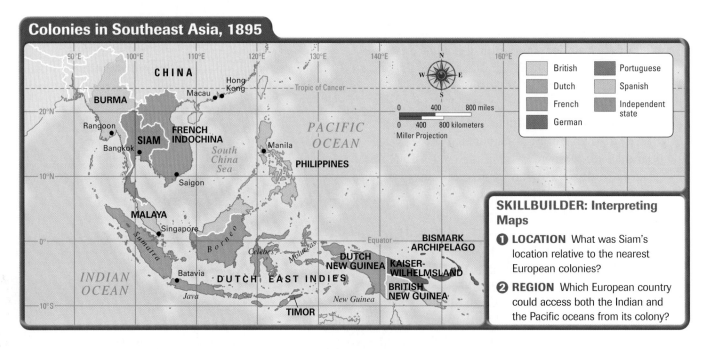

Colonies in Southeast Asia, 1895

British
Dutch
French
German
Portuguese
Spanish
Independent state

SKILLBUILDER: Interpreting Maps

❶ **LOCATION** What was Siam's location relative to the nearest European colonies?

❷ **REGION** Which European country could access both the Indian and the Pacific oceans from its colony?

INDEPENDENCE Claiming to take back "Asia for Asians," Japan occupied Southeast Asia during World War II. Southeast Asians soon realized that Japan was exploiting the region for its own benefit just as Europe had. But unlike the Europeans, the Japanese put Southeast Asians in leadership roles, which gave them valuable experience.

After the war ended, Southeast Asian leaders sought independence. Several Southeast Asian nations gained their freedom peacefully. Indonesia had to fight from 1945 to 1949 to gain independence from the Dutch.

BACKGROUND
The name *Indochina* refers to the Indian and Chinese influences on the region. The colony took up only part of the Indochinese Peninsula.

Indochina, a French colony made up of Cambodia, Laos, and Vietnam, suffered decades of turmoil. The Vietnamese defeated the French in 1954, winning independence for Cambodia, Laos, North Vietnam, and South Vietnam. The United States became involved in South Vietnam to prevent its takeover by Communist North Vietnam. The resulting conflict was the **Vietnam War** (1957-1975). In 1973, the United States withdrew. In 1975, South Vietnam surrendered, and Vietnam became one country, ruled by Communists. Also in that year, Communists took over both Cambodia and Laos.

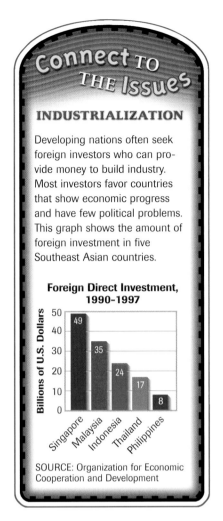

Connect to the Issues

INDUSTRIALIZATION

Developing nations often seek foreign investors who can provide money to build industry. Most investors favor countries that show economic progress and have few political problems. This graph shows the amount of foreign investment in five Southeast Asian countries.

Foreign Direct Investment, 1990-1997

Billions of U.S. Dollars

Singapore	Malaysia	Indonesia	Thailand	Philippines
49	35	24	17	8

SOURCE: Organization for Economic Cooperation and Development

An Uneven Economy

Agriculture is the main source of livelihood in Southeast Asia. Several nations began to industrialize in the 1960s, but industry is unevenly distributed across the region.

TRADITIONAL ECONOMIES The people of Cambodia, Myanmar, Laos, and Vietnam depend mostly on agriculture for income. Rice is the chief food crop in those countries, as it is in almost every Southeast Asian nation. Myanmar is heavily forested and produces much of the world's teak, a yellowish-brown wood valued for its durability.

The lack of industry has many causes. The Vietnam War destroyed factories and roads. Thousands of refugees fled Vietnam, Laos, and Cambodia after the war, reducing the work force. Political turmoil, especially in Cambodia and Myanmar, has continued to block growth.

But some economic growth has occurred. For example, Vietnam has built industry and sought foreign investment and trade.

INDUSTRY AND FINANCE In general, Brunei, Indonesia, Malaysia, the Philippines, Singapore, and Thailand have more highly developed economies than others in the region. Those countries have long been members of **ASEAN,** the Association of Southeast Asian Nations, an alliance that promotes economic growth and peace in the region. (The other four Southeast Asian countries did not join ASEAN until 1995 or later.)

CONNECT TO THE ISSUES
INDUSTRIALIZ-ATION
B▷ What further effects do you think industrialization will have on Southeast Asia?

Although these countries didn't begin to industrialize extensively until the 1960s, manufacturing has grown quickly. The processing of agricultural products is the chief industry. Other industries include the production of textiles, clothing, and electronic products. Service industries are also important. For example, Singapore is a center of finance. ◀B

Energy sources and mining are significant. Brunei receives most of its wealth from petroleum and natural gas reserves, but they are expected to run out in the early 2000s. Southeast Asia's mineral resources include tin, which is found mostly in Indonesia, Malaysia, and Thailand.

A Rich Mosaic of Culture

Although Southeast Asia has absorbed many influences from other regions, it has used them to create a culture that is distinctly its own.

RELIGIOUS DIVERSITY Southeast Asia has much religious diversity. Buddhism is widespread in the region, while the Philippines is mostly Catholic (as a result of Spanish rule), and Indonesia and Brunei are mostly Muslim. In addition, some Southeast Asians practice Hinduism, and others follow traditional local beliefs.

RICH ARTISTIC LEGACY Buddhism and Hinduism have influenced the region's sculpture and architecture. Perhaps the most famous example is the ancient temple complex of Angkor Wat in what is now Cambodia. (See page 705.) Thailand's Buddhist temples are modern examples of religious architecture.

Southeast Asia is also famous for its performing arts and literature. For example, Thailand and Indonesia have traditional forms of dance, in which richly costumed dancers act out stories. In Vietnam, poetry is highly respected. Nearly all Vietnamese know at least part of the 3,253-line poem "Kim van Kieu," which is about love and sacrifice.

Geographic Thinking

Using the Atlas
C Use the map on page 683 to learn about the major religions in Southeast Asia. What do you notice about the places where Catholicism and Islam are practiced?

growing up in... Thailand

About 95 percent of the people who live in Thailand are Buddhists and follow an ancient tradition of Buddhism that stresses the importance of being a monk. This has led to a unique custom. During their late teens or early twenties, many Thai men become monks for a short time.

The new monks go to live in a monastery where they meditate and study Buddhist teachings. They also shave their heads, wear saffron (orange-yellow) robes, and give up their worldly possessions. Some Thai men remain monks their whole lives, but most leave the monastery after a short period, usually a few weeks or months. After his time as a monk, a young man is considered ready for adult life.

If you grew up in Thailand, you would pass the following milestones:

- At your birth, your parents might ask a Buddhist monk to help them choose your name.
- You would have to attend school for 6 years, between ages 7 and 14. Although higher education is available, very few people can afford it.
- You could vote at age 18.
- If you were a man 18 years of age, you might be drafted to serve in the army.

Changing Lifestyles

Most Southeast Asians live in rural villages and follow traditional ways. However, a growing number of people are moving to cities and leading more modern lives—a trend taking place all around the world.

THE VILLAGES In many Southeast Asian villages, people live in wood houses built on stilts for protection against floods. Roofs are usually made of thatch, although wealthy families may have a tin roof. In Laos, Myanmar, and Thailand, most villages have a Buddhist temple that serves as the center of social life. In Indonesia, most villages have a group of leaders who govern by a system that stresses cooperation.

Some Southeast Asian villagers still wear traditional clothing, such as the *longyi*—a long, tightly wrapped skirt—of Myanmar. Yet modern conveniences are slowly beginning to change village life. For instance, listening to the radio is common in Indonesia and Thailand.

PLACE People waiting at a bus stop in Kuala Lumpur, Malaysia, wear Western clothes and traditional Muslim attire.
What does this scene show about diversity in Malaysia?

Geographic Thinking

Using the Atlas
Use the map on page 680 to locate Kuala Lumpur and Singapore. How far apart are these two major cities?

THE CITIES Kuala Lumpur, Malaysia, and Singapore are examples of bustling cities with towering skyscrapers and modern business districts. In Southeast Asian cities, most people live in apartments.

But there is a shortage of housing for the large numbers of people migrating to cities for jobs. Many of them live in makeshift shacks in slums. The dangers of doing that were shown by a disaster in Manila, Philippines. Hundreds of people had built shanties at a city dump. In July 2000, after a typhoon weakened a tower of garbage, it crashed onto those shacks and burst into flames. More than 100 people died.

Another region facing the changes caused by rural-to-urban migration is Oceania. You will read about that region in Section 2.

SECTION 1 Assessment

1 Places & Terms

Identify these terms and explain their importance in the region's history or culture.

- *mandala*
- Khmer Empire
- Indochina
- Vietnam War
- ASEAN

2 Taking Notes

PLACE Review the notes you took for this section.

- Where did powerful states exist during the period 1300 to 1800?
- What is the only country in the region that wasn't a colony?

3 Main Ideas

a. How did China and India influence Southeast Asia?

b. How did the Vietnam War affect the economy?

c. What is village life like in Southeast Asia?

4 Geographic Thinking

Drawing Conclusions How has ASEAN helped to create a region within a region?
Think about:

- the goals of ASEAN
- differences between longtime and more recent ASEAN members

See Skillbuilder Handbook, page R5.

MAKING COMPARISONS Choose two Southeast Asian nations and research their similarities and differences. Create a **chart** comparing the two countries by using such categories as languages, religions, main economic activities, and types of government.

Disasters!

Krakatoa

Imagine an explosion so destructive that it sends volcanic ash 50 miles into the air and so loud that people hear it about 3,000 miles away. In 1883, the Indonesian volcano Krakatoa (also spelled *Krakatau*) erupted in an explosion that created those effects. But the blast was only the beginning of the disaster. The eruption caused the volcano to collapse into the sea and triggered a series of deadly tsunamis, or giant waves. The greatest of those towered 120 feet high. The tsunami swept the coasts of Java and Sumatra, killing more than 36,000 people.

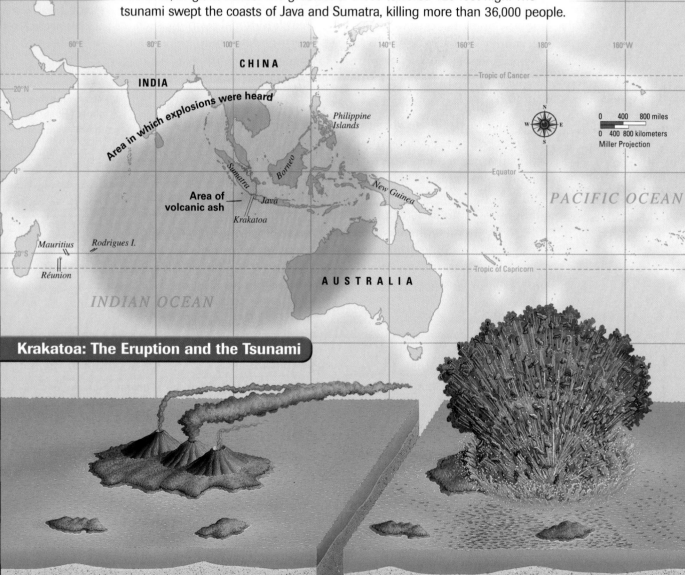

Krakatoa: The Eruption and the Tsunami

① Krakatoa was a main island consisting of three overlapping volcano cones, plus two small islands.

② After a few months of minor volcanic activity, Krakatoa blew up violently on August 27, 1883.

In the Sunda Strait, in 1927, lava began to flow through a crack in the sea floor beneath the site of the old island. By 1928, a new island was born and named Anak Krakatoa, which means "child of Krakatoa." The island is still volcanically active, but it is not considered dangerous.

GeoActivity

PREPARING A NEWSCAST
Working with a partner, use the Internet to research one of the other volcanoes listed below. Create a **television newscast** about the disaster.

* Sketch a map showing the volcano and the region affected by lava, ash, mud slides, or tsunamis.
* Create drawings, diagrams, or graphs about the disaster.
* Write a script for the newscast.

RESEARCH LINKS
CLASSZONE.COM

GeoData

DUST IN THE WIND
* Krakatoa threw so much ash and dust into the air that temperatures dropped by about 0.9°F around the world.
* The dust filtered light and caused spectacular sunsets around the world for about a year.
* The dust in the atmosphere also made the moon look shades of green and blue.

FIVE DEADLIEST VOLCANOES SINCE 1500

1792
Unzen, Japan:
15,000 dead

1815
Tambora, Indonesia:
92,000 dead

1883
Krakatoa, Indonesia:
36,000 dead

1902
Mount Pelée, Martinique:
30,000 dead

1985
Nevado del Ruiz, Colombia:
25,000 dead

③ All three cones disappeared, leaving only a small island. Massive amounts of sea water were displaced. The disturbance of the ocean created giant tsunamis. The tsunamis destroyed about 163 villages.

Oceania

Main Ideas

- Settled in ancient times by migrating Southeast Asians, Oceania developed three cultural regions.
- Contact with Europeans and Americans disrupted the islanders' traditional ways of life.

Places & Terms

Micronesia

Melanesia

Polynesia

subsistence activities

copra

taro

CONNECT TO THE ISSUES

ENVIRONMENTAL CHANGE A possible rise in sea level from global warming threatens some islands.

A HUMAN PERSPECTIVE Noah Idechong has fought to protect the sea life of Palau, an archipelago east of the Philippines. Palauans have always earned their living by fishing, but in the 1980s, many species of fish were in danger of extinction because they were such popular menu items in Asian restaurants. Idechong began to study the problem in 1988.

His efforts paid off. In 1994, the year Palau became independent, it banned the export of certain species, and fish populations grew again. However, in 2000, the government planned building projects that would help the economy but strain the environment. Idechong kept working to save wildlife. He said, "Palau right now needs . . . people who can say what they want Palau to look like 50 years from now." In other words, Palauans need to decide what to preserve in the face of change.

A History of the Islands

Like Palau, all the nations of Oceania except Nauru are island groups. They are Fiji, Kiribati, Marshall Islands, Federated States of Micronesia, Palau, Papua New Guinea, Samoa, Solomon Islands, Tonga, Tuvalu, and Vanuatu. (Some geographers consider Australia and New Zealand part of Oceania, but those nations are covered in Section 3.)

FIRST ISLANDERS Prehistoric people journeyed from mainland Southeast Asia to nearby Pacific islands using small rafts or canoes and land bridges that have since disappeared. In time, they developed large

PLACE These stone heads are on an altar on Vao, a small island of Vanuatu. They were used in rituals for controlling the weather. **How has time affected the stone heads?**

Cultural Regions of Oceania

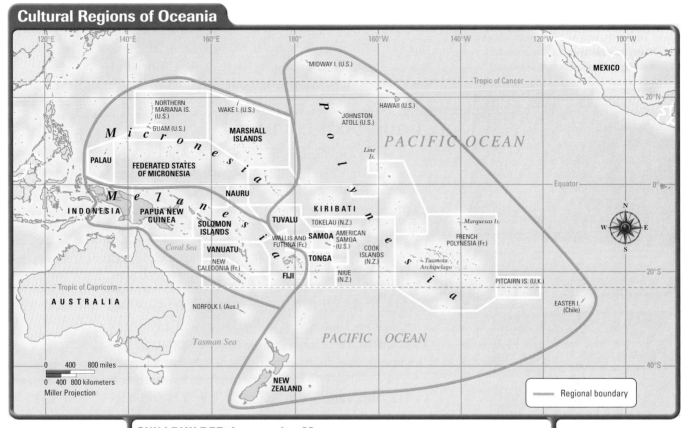

SKILLBUILDER: Interpreting Maps

❶ **REGION** Which of the cultural regions contains islands held by the United States?

❷ **MOVEMENT** Consider what you have learned about ancient migrations of people in the Pacific Ocean. Which cultural region was the last to be settled?

voyaging canoes (see page 699) that enabled them to sail longer distances. For thousands of years, their descendants continued to migrate as far east as Hawaii, as far south as New Zealand, and as far west as Madagascar.

For centuries, the people of Oceania had little contact with the rest of the world, so they developed their own ways of life. Geographers divide Oceania into three regions, defined both by physical geography and culture. The regions are **Micronesia,** meaning "tiny islands," **Melanesia,** meaning "black islands," and **Polynesia,** meaning "many islands."

CONTACT WITH THE WEST Beginning in the 1500s, many Europeans explored the Pacific. Perhaps the most famous was the British captain James Cook, the first European to visit many of the islands.

In the 1800s, European missionaries arrived and tried to convert the islanders to Christianity. Traders came for products such as coconut oil, and sailors hunted whales. Settlers started plantations on which they could grow coconuts, coffee, pineapples, or sugar.

As a result, island societies began to decline. Many islanders died of diseases brought by the Europeans. Western ways often replaced traditional customs. And Europe and the United States took control of the islands and turned them into territories and possessions.

RECENT HISTORY Oceania experienced turmoil in the 20th century. During World War II, the Allies and the Japanese fought fierce battles there to gain control of the Pacific. Afterward, some islands were used as nuclear test sites, not only by the United States (see Chapter 30) but

BACKGROUND
James Cook was also one of the first Europeans to explore Australia and New Zealand. See page 718 for his portrait.

SE ASIA & OCEANIA

Oceania **713**

Many residents of Oceania make a living from traditional activities.

This resident of Fiji is husking coconuts to make copra, or the dried meat of coconuts.

Traditional dances are often performed for tourists. These dancers are from French Polynesia.

Many people of Oceania, such as these Cook Islanders, earn their living from fishing.

also by other countries. Gradually, inhabitants of many of the islands moved toward self-rule. Since 1962, 12 different nations have gained independence. Foreigners still rule the other islands.

A Traditional Economy

Most of Oceania has an economy in which people work not for wages but at **subsistence activities.** These are activities in which a family produces only the food, clothing, and shelter they themselves need. The tiny island of Nauru is an exception. It has a prosperous economy based on the mining of phosphates, used in fertilizer. But Nauru's phosphate deposits are expected to give out early in the 21st century.

AGRICULTURE As Chapter 30 explained, most low islands do not have plentiful or fertile soil. In spite of this, agriculture is the region's main economic activity because many high islands do have soil that supports agriculture. The chief crops are bananas, sugar, cocoa, coffee, and **copra,** which is the dried meat of coconuts. Fishing also provides a significant source of income.

OTHER ECONOMIC ACTIVITIES Since the invention of jet travel, tourism has become very important to the economy of Oceania. This has been a mixed blessing. Although tourists spend money in the islands, they also require hotels, stores, roads, and vehicles. These threaten the islands' environment and traditional ways of life.

A few islands besides Nauru have mining industries. For example, Papua New Guinea is developing a large copper mine with the help of foreign investment. Some industry also exists. Some of the larger towns have factories that produce goods such as coconut oil and soap. As in Southeast Asia, an increasing number of people in the Pacific Islands are moving to cities to find jobs.

Culture of the Islands

Oceania has a culture that blends traditional ways with the cultures of Europe and the United States.

LANGUAGE AND RELIGION Oceania is one of the most linguistically diverse regions in the world. Some 1,100 of the world's languages are spoken there. The people of Papua New Guinea alone speak 823 languages. In addition, many Pacific Islanders speak European languages. English is the most common.

Because of missionaries' work and colonialism, Christianity is the most widely spread religion. Even so, some Pacific Islanders still practice their traditional religions.

Geographic Thinking

Seeing Patterns
Which characteristics of Oceania might account for its high levels of migration to cities?

THE ARTS Many Pacific Islanders produce arts and crafts, such as baskets and mats woven from the leaves of palm trees or carved wooden masks. Some islanders make a living selling such items to tourists.

Island Life

As in Southeast Asia, two distinct ways of life exist on the islands: traditional village life and more modern city life.

TRADITIONAL LIFE Ways of life varied throughout the islands. In Polynesia, most people lived in villages, ranging from small clusters of houses to large walled settlements. The houses were usually wooden with thatched roofs. Generally, a chief led each village. The villages' economies centered on fishing and farming. One major crop was **taro,** a plant with a starchy root. Taro can be eaten boiled, or it can be made into breads, puddings, or a paste called poi.

Many Polynesian societies were warlike and had frequent conflicts. In contrast, Micronesians tended to exist peacefully with their neighbors. Most Micronesians lived in extended family groups. As in Polynesia, they made a living by fishing and farming, with taro being a main crop.

In Melanesia, villages usually existed by the coast where people could fish. Inland, many people practiced shifting cultivation, moving often to let fields regain fertility. Other Melanesians were hunter-gatherers.

RECENT CHANGE Oceania has few cities, but they have been growing as many people move to them for education or jobs. Rapid urban growth has led to sprawling shantytowns and inadequate sanitation facilities. In addition, city dwellers are giving up their traditional ways of life.

But change is also helping Oceania. Modern communications systems can unify countries consisting of scattered island groups and also can link Oceania to the rest of the world. Section 3 will describe the two most westernized nations in the region: Australia and New Zealand.

Geographic Thinking

Making Comparisons
B What other regions of the world that you have studied are experiencing these same problems in their growing cities?

Section 2 Assessment

1 Places & Terms

Identify these terms and explain their importance in the region.

- Micronesia
- Melanesia
- Polynesia
- subsistence activities
- copra
- taro

2 Taking Notes

MOVEMENT Review the notes you took for this section.

- How were the Pacific Islands first settled?
- What type of migration is happening within Oceania today?

3 Main Ideas

a. How did contact with Europeans and Americans affect the societies of the Pacific Islands?

b. What are the chief crops of Oceania?

c. What is distinctive about Oceania in terms of its languages?

4 Geographic Thinking

Determining Cause and Effect How has modern technology both helped and harmed Oceania? **Think about:**

- jet travel
- modern communications

RESEARCH LINKS
CLASSZONE.COM

GeoActivities

SEEING PATTERNS Use the Internet to research several nations and territories in Oceania. Then choose the one that you think would make the best vacation spot. Create a **tourist brochure** that will persuade travelers to visit that place. Check your brochure for correct grammar, spelling, sentence structure, and punctuation.

SE ASIA & OCEANIA

Comparing Cultures

Regional Costumes

Blue jeans have spread around the globe and become a popular item of clothing symbolizing U.S. culture. But traditional items of clothing remain important in many regions of the world. Regional costumes are unique not only because of their styles but also because of the materials from which they are made.

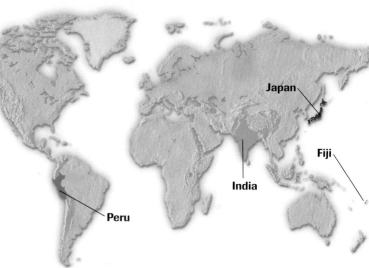

Japan

Fiji

India

Peru

In India, the traditional garment of women for centuries has been the sari— five to seven yards of unstitched cloth wrapped around the body. The most valuable saris are made of silk, but saris of cotton and synthetic fabrics are also common.

In Fiji, traditional outfits are made from tapa cloth, a nonwoven fabric made from the inner bark of trees. This woman is wearing a skirt of tapa cloth. Fijians often decorate the cloth with geometric designs painted in brown, black, or reddish bark dyes.

Although colorful silk kimonos symbolize Japan, neither the fabric nor the robe itself originated there. Silk was first developed in China, and kimonos are patterned after a wide-sleeved Chinese robe, the *p'ao*.

These Indians of Peru wear traditional wool clothing woven from llama hair; llamas are native to South America. Each village has it own set of traditional patterns that are woven into its cloth. Some of the designs indicate local landscapes; others depict animals or historical events.

GeoData

Tapa Cloth
- Tapa cloth is also made in other Melanesian islands, New Guinea, and northern Australia.
- The most popular material for tapa is the inner bark of the mulberry.

Silk
- The Chinese began to produce silk in about 2700 B.C. and kept their methods a secret until about 140 B.C.
- The wide silk or satin sash worn with a kimono is called an obi. It is about 12 feet long.

Cotton
- South Asia was one of the first regions of the world where cotton was cultivated—starting in about 3000 B.C.
- Indian men also wear a type of wrapped garment called a dhoti. Mohandas Gandhi wore a dhoti to show his allegiance to the common Indian man.

Wool
- Wool is the fiber forming the coats of such hairy mammals as sheep, goats, camels, and llamas.
- Intricate textiles have been produced in Peru since about 1000 B.C.

Australia, New Zealand, and Antarctica

Main Ideas

• Both Australia and New Zealand were colonized by Europeans and still have a strong European heritage.

• Because of its harsh climate, Antarctica has no permanent settlements.

Places & Terms

penal colony

Aboriginal people

Maori

Treaty of Waitangi

pakeha

CONNECT TO THE ISSUES
LAND CLAIMS The Aboriginal people of Australia are trying to reclaim ancestral lands.

A HUMAN PERSPECTIVE In 1788, Great Britain founded Sydney, Australia, as a **penal colony**—that is, a place to send prisoners. By the end of the 20th century, Sydney had overcome its origins and earned a reputation as a fun and fascinating international city. That has been due, in part, to a unique combination of physical and cultural geographic assets.

Sydney is located on a deep, beautiful harbor that not only allows the city to function as a port but also provides an arena for sailing and swimming. The mild climate there encourages such outdoor activities. In addition, Sydney has an increasingly diverse population. People who visit the city can view art and dine on food from many cultures.

In 2000, Sydney hosted the Olympic Games. With a physical environment that favors sports and a culture shaped by immigrants, the city seemed a perfect site for an international athletic event.

History: Distant European Outposts

Australia, New Zealand, and Antarctica made up the last region to be explored by Europeans. Australia and New Zealand became British colonies, even though they were already inhabited by people with ancient cultures of their own.

THE ORIGINAL INHABITANTS The **Aboriginal people** migrated to Australia from Asia at least 40,000 years ago. When Europeans arrived in Australia, there were an estimated 500 Aboriginal groups, speaking perhaps 200 different languages. The Aboriginal people had complex

Australia and New Zealand, Prehistory to Today

40,000 B.C.
Australia is gradually settled by Aboriginal people. Their art includes rock paintings like this one in **Kakadu National Park.**

1788
Great Britain starts a penal colony in Australia.

| 40,000 B.C. | 1750 A.D. | 1800 | 1850 |

1769-1770
Captain James Cook
(right) explores New Zealand and Australia.

1851
Gold is discovered in **New South Wales,** Australia.

religious beliefs and social structures but a simple economy; they lived by hunting and gathering.

New Zealand was settled first by the **Maori,** who had migrated there from Polynesia more than 1,000 years ago. The Maori lived by fishing, hunting, and farming.

EARLY EXPLORERS During the 1600s and 1700s, several European explorers sailed in the coastal waters of New Zealand and Australia. Captain James Cook of Britain was the first to explore those two lands—New Zealand in 1769 and Australia's east coast in 1770. Antarctica was first discovered in 1820.

BACKGROUND
The name *Australia* comes from the Latin phrase *Terra Australis Incognita,* which means unknown southern land.

EUROPEAN SETTLEMENT In 1788 Britain began to colonize Australia (called New South Wales until 1820) as a place to send prisoners. Having a colony in Australia also gave Britain more Pacific naval bases. New Zealand was colonized by hunters and whalers from Europe, America, and Australia. No permanent settlements were established in Antarctica because of its cold climate.

In Australia, the British colonists had violent conflicts with the Aboriginal people, many of whom were killed. Even greater numbers of native people died from diseases brought by Europeans.

In New Zealand in 1840, the British and several Maori tribes signed the **Treaty of Waitangi,** giving Britain control over New Zealand. But the English and the Maori translations of the treaty differed. The English version gave Great Britain complete control; the Maori version gave Britain "governorship." Disagreement over who owned the land helped cause the Land Wars that lasted from 1845 to 1847 and from 1860 to 1872. In addition, tens of thousands of Maoris died from diseases.

Gold was discovered in Australia in 1851 and in New Zealand in 1861. Hundreds of thousands of people who dreamed of wealth flocked to the two countries, but few miners grew rich. Most, however, stayed there.

Connect TO THE Issues

MAORI LAND CLAIMS

Like many other cultures, the Maori believed that land could not be sold without tribal consent because the tribe, not individuals, owned it. When the English seized land or tried to buy it without tribal consent, conflicts broke out. Differing views of the Treaty of Waitangi, shown below, also fueled conflicts.

The Land Wars in New Zealand lasted from 1845 to 1847 and from 1860 to 1872. A law passed in 1862 let people buy native lands, and the Maori lost most of their territory. In recent years, the Maori made land claims, and in the 1990s, they won some awards of cash and land.

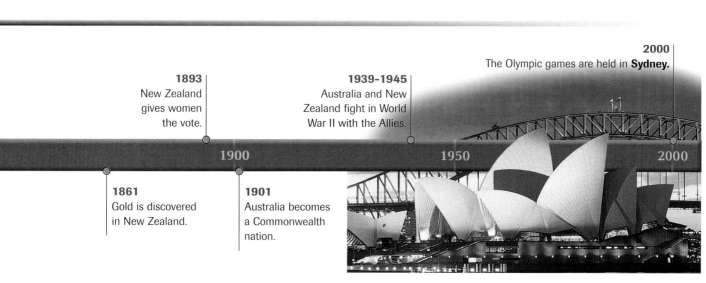

1893
New Zealand gives women the vote.

1939-1945
Australia and New Zealand fight in World War II with the Allies.

2000
The Olympic games are held in **Sydney.**

1900   1950   2000

1861
Gold is discovered in New Zealand.

1901
Australia becomes a Commonwealth nation.

Australia, New Zealand, and Antarctica **719**

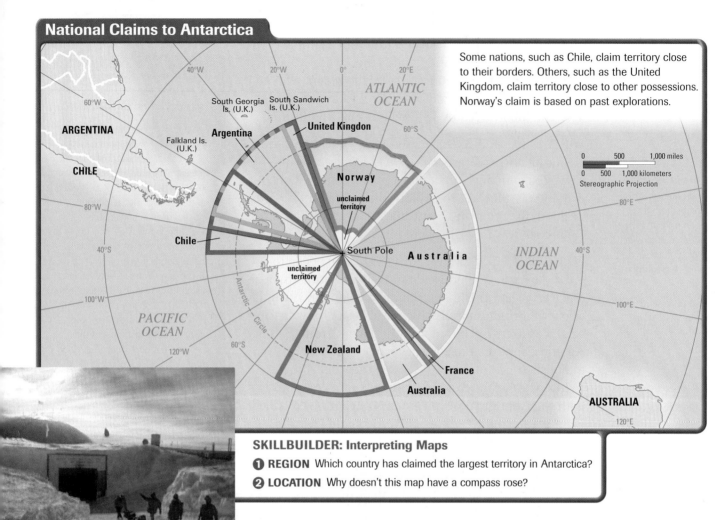

National Claims to Antarctica

Some nations, such as Chile, claim territory close to their borders. Others, such as the United Kingdom, claim territory close to other possessions. Norway's claim is based on past explorations.

SKILLBUILDER: Interpreting Maps

❶ **REGION** Which country has claimed the largest territory in Antarctica?

❷ **LOCATION** Why doesn't this map have a compass rose?

REGION Eighteen nations have scientific research stations in Antarctica. This one is run by U.S. scientists.

Modern Nations

Originally, several colonies existed in Australia, but in 1901, they joined into a single, independent nation. New Zealand became self-governing in 1907. Both Australia and New Zealand remained in the British Commonwealth, which is a free association of Great Britain and several of its former colonies.

RIGHTS AND LAND CLAIMS New Zealanders have a long tradition of concern for equal rights and the welfare of its citizens. In 1893, New Zealand became the first country to grant women the vote. It was also one of the first nations to provide pensions for its senior citizens.

In both Australia and New Zealand, native people generally have less education and higher rates of poverty than other citizens. Attempting to improve their lives, the Aboriginal people and the Maori have made claims for the return of their former lands. (See Chapter 32.) ◁A

ISSUES A recent issue in Australia was a movement to withdraw from the Commonwealth. In 1999, Australia held a referendum on becoming an independent republic, but voters defeated the proposal, because Australians could not agree on how to choose a head of state.

Antarctica remains unsettled. In 1959, 12 countries drafted a treaty preserving the continent for research. By 2000, 18 countries had scientific research stations there. Seven countries have claimed territory in Antarctica, but many other countries do not recognize those claims.

CONNECT TO THE ISSUES

LAND CLAIMS
◁A How might land ownership improve Aboriginal and Maori lives?

Economy: Meat, Wool, and Butter

As Commonwealth members, Australia and New Zealand prospered by exporting food products and wool to the United Kingdom. So neither country developed much industry. But, since 1950, their exports to the United Kingdom have declined. To continue to prosper, Australia and New Zealand must either develop industry or find other trading partners, such as the nations of nearby Asia.

AGRICULTURE Australia and New Zealand are major exporters of farm products. New Zealand earns much of its income by selling butter, cheese, meat, and wool to other countries. Ranching is so widespread in New Zealand that in 1998 the number of farm animals (including 47.6 million sheep and 8.8 million cattle) was 15 times greater than the number of people! Crops include vegetables and fruits. For example, New Zealand is the world's largest producer of kiwi fruit.

Sheep ranching is also important in Australia, which is the largest exporter of wool in the world. Because so much of Australia is arid, less than ten percent of the land is used to grow crops.

MINING Australia earns a large part of its income from mining. It is the world's top producer of diamonds, lead, zinc, and opals. In addition, it is a major producer of bauxite, coal, copper, gold, and iron ore.

The mining industry faces one difficulty. Many deposits lie in the outback, far from cities. As a result, it is expensive to build the roads and buildings necessary for the mines to operate. Because of the high costs of mining and because Australia has historically lacked capital (money or property invested in business), Australian companies have had to rely on foreign investment. Foreign investors control about half the mining industry, so not all the profits stay within Australia. **B**

Geographic Thinking

Seeing Patterns
B What are the pros and cons of foreign investment in industry?

MANUFACTURING AND SERVICE Unlike most developed countries, Australia does not rely heavily on manufacturing. One of the major industries in both Australia and New Zealand is the processing of food products. Because of its forests, New Zealand also produces wood and paper products.

PLACE Sheep ranches dot the New Zealand landscape—here by Mount Egmont.
Why are more ranches than farms found in mountainous areas?

As in all developed countries, service industries have been growing. For example, nearly 65 percent of Australia's jobs are in service industries such as government, communications, and tourism.

THE ECONOMIC FUTURE Both Australia and New Zealand want to develop a more diversified economy that is not so dependent on agriculture. But it will be difficult to develop manufacturing plants that can compete with those in nearby Asia, where the cost of labor is generally lower. Finding a way to maintain prosperity in the face of global economic change is a major issue for these two nations.

Distinctive Cultures

The British colonial past has shaped the cultures of Australia and New Zealand, but they also have developed in distinctive ways.

AUSTRALIA'S CULTURE Most Australians are of British descent, but that proportion is changing because of high rates of immigration from places like Greece, Italy, and Southeast Asia. More than 20 percent of Australians are foreign born. Only about one percent are of Aboriginal descent.

Like the British, Australians drive on the left side of the road, and many enjoy drinking tea. Christianity is the major religion. Australians speak English but also have many colorful terms that are all their own. For example, they call ranches "stations" and wild horses "brumbies."

Australia's environment and history have influenced the arts, too. The Aboriginal people have an ancient tradition of painting human and animal figures. Some of those works can be seen on rock walls around the country. Many Australian painters of European descent have portrayed the landscape. For example, Russell Drysdale is known for his pictures of the outback. Several Australian novelists have written adventure stories about life in the bush country.

NEW ZEALAND'S CULTURE The majority of New Zealanders are of European, mostly British, descent. They are called **pakehas,** a Maori term for white people. The Maori of New Zealand fared somewhat better than the Aboriginal people of Australia; about 15 percent of New Zealand's people are descended from the Maori.

New Zealand's culture blends British and Maori ways. For example, both English and Maori are official languages. Christianity is the main religion, but some churches combine biblical and Maori teachings.

Both cultures have shaped New Zealand's art. Maori art, including intricate woodcarvings and poetic legends, still survives. Western art also thrives. Well-known New Zealand authors have included the novelist Janet Frame and the mystery writer Ngaio Marsh. New Zealand filmmakers Jane Campion and Peter Jackson have made movies that were popular in many countries. And the opera singer Kiri Te Kanawa is admired internationally.

REGION The traditional facial markings of the Maori, shown here, are called *moko.*

🌐 **Geographic Thinking** ←

Making Comparisons
Ⓒ How are the experiences of the Aboriginal people and the Maori similar and different?

Modern Life

Australians and New Zealanders have similar lifestyles. For example, about 70 percent of Australians and 70 percent of New Zealanders own their own homes—usually single-family homes with enough land to grow a small garden.

CITY AND COUNTRY Australia and New Zealand are two of the most urbanized countries in the world; about 85 percent of their people live in cities and towns. Australia's large cities have the usual problems of pollution and traffic jams. In contrast, New Zealand's cities are relatively quiet, uncrowded, and pollution-free because of its small population and lack of industry.

In both Australia and New Zealand, many ranchers live far away from settlements. New Zealand has a good system of roads, even in rural areas, which aids travel. In Australia, many wealthy ranchers own private airplanes to help them cross the vast distances in the country. Some of the largest ranches in Australia can have a total land area of thousands of square miles.

RECREATION Both countries have climates that allow people to spend a great deal of time outdoors. As a result, aquatic sports, tennis, and team sports, such as rugby, cricket, and soccer, are very popular. Australia has developed its own form of football, called Australian rules football. Because New Zealand is mountainous, skiing and mountain climbing are common there.

In Chapter 32, you will read about Aboriginal land claims in Australia, industrialization in Southeast Asia, and global environmental change.

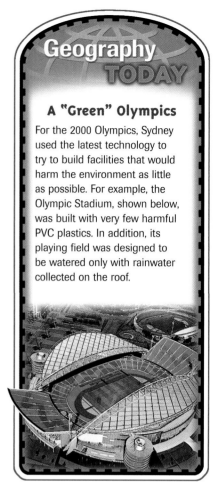

Geography TODAY

A "Green" Olympics

For the 2000 Olympics, Sydney used the latest technology to try to build facilities that would harm the environment as little as possible. For example, the Olympic Stadium, shown below, was built with very few harmful PVC plastics. In addition, its playing field was designed to be watered only with rainwater collected on the roof.

SECTION 3 Assessment

① Places & Terms

Identify these terms and explain their importance in the region's history or culture.

- penal colony
- Aboriginal people
- Maori
- Treaty of Waitangi
- pakeha

② Taking Notes

MOVEMENT Review the notes you took for this section.

The Region → Australia and New Zealand

- What mineral lured colonists to Australia and New Zealand?
- To which country did Australia and New Zealand export wool?

③ Main Ideas

a. How did the Treaty of Waitangi cause a misunderstanding over land ownership in New Zealand?

b. Who owns Antarctica?

c. What are some of the British cultural influences in Australia?

④ Geographic Thinking

Identifying and Solving Problems How do you think Australia and New Zealand can solve their economic problems? **Think about:**

- their need for new markets
- Asia's large population and its relative closeness
- the Asian nations that have thriving industries

EXPLORING LOCAL GEOGRAPHY Do research to learn which farm animals and minerals (if any) are major products of your state. Create a **Venn Diagram** showing those farm animals and minerals that your state has in common with Australia and those that are unique to each.

SE ASIA & OCEANIA

Chapter 30 Assessment

VISUAL SUMMARY
HUMAN GEOGRAPHY OF SOUTHEAST ASIA, OCEANIA, AND ANTARCTICA

Subregions of Southeast Asia

○ **Southeast Asia**

- Southeast Asia was influenced by ancient China and India and later by European colonists.
- After World War II, the nations of Southeast Asia became independent.
- Industrialization and urbanization are taking place in many countries.

● **Oceania**

- After being isolated for centuries, the islands of Oceania came under the control of European countries and the United States.
- Since 1962, 12 nations have gained independence.
- Their economies are generally based on agriculture, tourism, and a small amount of industry.

○ **Australia, New Zealand, and Antarctica**

- The Aboriginal people of Australia and the Maori of New Zealand lost land when European colonists arrived.
- Both Australia and New Zealand are former British colonies that are now Commonwealth nations.
- They both want to diversify their economies.

Reviewing Places & Terms

A. Briefly explain the importance of each of the following.

1. Indochina
2. Vietnam War
3. ASEAN
4. Micronesia
5. Melanesia
6. Polynesia
7. subsistence activities
8. penal colony
9. Aboriginal people
10. Maori

B. Answer the questions about vocabulary in complete sentences.

11. Which of the above terms was a French colony in Southeast Asia?
12. What are the goals of ASEAN?
13. During the Vietnam War, the United States tried to protect South Vietnam from takeover by what group?
14. What are the three cultural regions of Oceania?
15. Which European nation used Australia as a penal colony?
16. Which of the above terms is the region from which the Maori migrated?
17. What is the name of the place to which the Maori migrated?
18. Which of the above terms name regions where you are likely to find subsistence activities?
19. What are some of the subsistence activities found there?
20. How long have Aboriginal people been living in Australia?

Main Ideas

Southeast Asia (pp. 705–711)

1. What were the distinctive characteristics of the states known as *mandalas*?
2. What effect did colonialism have on Southeast Asia?
3. What are some of the major changes that Southeast Asia has undergone since 1960?
4. What are some of the arts for which Southeast Asia is known?

Oceania (pp. 712–717)

5. How far east, south, and west did Pacific Islanders migrate?
6. What caused many island societies to decline starting in the 1800s?
7. What are the major economic activities in Oceania?

Australia, New Zealand, and Antarctica (pp. 718–723)

8. What prevents Australia from benefiting completely from its mining industry?
9. What historic actions demonstrated New Zealanders' concern for equal rights and social welfare?
10. What is the major activity conducted in Antarctica?

Critical Thinking

1. Using Your Notes

Use your completed chart to answer these questions.

a. How does agriculture differ in the three subregions?

b. When and how did various nations of the region gain independence from European control?

2. Geographic Themes

a. **HUMAN-ENVIRONMENT INTERACTION** In what ways has the Pacific Ocean helped to shape the various cultures in this region?

b. **MOVEMENT** What role did migration play in the settling of this region?

3. Identifying Themes

Drawing on what you know about this region, what are general differences between village life and city life? What geographic themes are included in your answer?

4. Seeing Patterns

How did the arrival of Europeans affect Southeast Asia, Oceania, Australia, and New Zealand?

5. Analyzing Data

Use the Regional Data File (pages 684-687) to calculate per capita GDP (total GDP divided by population) for Indonesia, Malaysia, Philippines, Singapore, and Thailand. Rank the countries from highest to lowest. Compare your list to the graph on page 707. What pattern do you notice?

Additional Test Practice, pp. S1–S37

TEST PRACTICE CLASSZONE.COM

Geographic Skills: Interpreting Maps

Population Distribution of Australia

Use the map at the right to answer the following questions.

1. **REGION** How would you describe the population distribution of Australia?

2. **PLACE** Identify the two most heavily populated cities of Australia. What do you notice about their surrounding regions?

3. **PLACE** Judging from this map, would you characterize Australia as a heavily populated or lightly populated country? Explain.

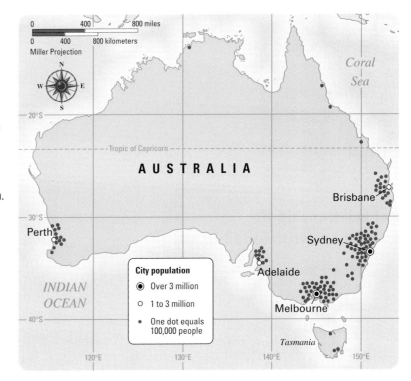

GeoActivity

Copy this map on your own paper. Use the map on page 683 to make a climate map of Australia. Display the maps side by side with a caption explaining the link between climate and population distribution.

 INTERNET ACTIVITY

Use the links at **classzone.com** to do research about two countries from different subregions in this unit. Look for information about government, economic activities, culture, and modern life.

Writing About Geography Write a report comparing the two countries. Include maps, charts, and graphs to help present the information. List the Web sites that were your sources.

32

For more on these issues in Southeast
Asia, Oceania, and Antarctica . . .

 CURRENT EVENTS
CLASSZONE.COM

TODAY'S ISSUES
Southeast Asia, Oceania,
and Antarctica

Gordon and Alick Pablo,
elders of the Wuthathi
Aboriginal people, bring
a 200-year-old skull of
an ancestor to be buried
in their homeland.

GeoFocus

What relationship do
humans have with land?

Taking Notes In your notebook, copy a
cause-and-effect chart like the one shown
below. Then take notes on the causes and
effects of each issue.

	Causes	Effects
Issue 1:
Land Claims		
Issue 2:
Industrialization		
Case Study:
Environmental
Change		

Interpreting a Cartogram

Even though Southeast Asia has been experiencing industrial growth as a region, not all Southeast Asian nations have prospered equally. A table listing the value of industrial output for the ten countries would give this information in numerical form. A cartogram shows the information visually.

THE LANGUAGE OF MAPS A **cartogram** is a special type of map that conveys a set of data, such as population or GDP. The sizes of the nations on the map are adjusted to reflect the amounts of data each one has. The cartogram below shows the value of industrial output for the nations of Southeast Asia.

Industrial Output of Southeast Asia

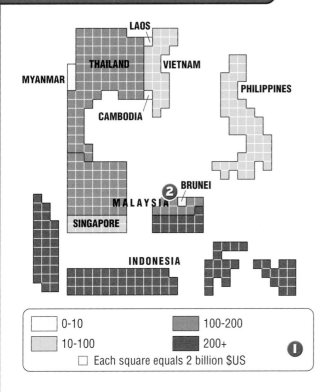

Legend:
- ☐ 0-10
- ☐ 10-100
- ☐ 100-200
- ☐ 200+
- ☐ Each square equals 2 billion $US

❶ The key of this cartogram helps you to intrepret the value of industrial output in two ways. It tells you that each small square equals 2 billion U.S. dollars. It also identifies the colors that the cartogram uses to identify ranges of output.

❷ Cartograms adjust the sizes of countries to convey relative quantities. The countries' shapes are altered because a cartogram uses squares or straight lines.

❸ Comparing a cartogram to a conventional map can show which countries have more or less of the data under study than you would expect from looking at their size alone.

Map and Graph Skills Assessment

1. Analyzing Data
According to the cartogram, how much industrial output does Thailand have?

2. Drawing Conclusions
Which country or countries seem to have a small industrial output compared to their actual size?

3. Drawing Conclusions
Which country or countries seem to have a large industrial output compared to their actual size?

GLOBAL ENVIRONMENTAL CHANGE

How have people changed the atmosphere?

Some people fear that global warming might cause an increase in violent weather.

As you have read in other units, many human activities harm the environment. Among these are the burning of fossil fuels and the use of chemicals such as chlorofluorocarbons (CFCs) in aerosol cans. Many scientists fear these activities are changing the environment in ways that affect the whole world.

Damage to the Environment

Scientists believe that the use of fossil fuels has begun to heat the climate, and the use of chemicals has damaged the ozone layer.

GLOBAL WARMING The burning of fossil fuels releases carbon dioxide (CO_2) into the atmosphere. Carbon dioxide is one of the greenhouse gases—gases that trap the sun's heat. Greenhouse gases serve the useful function of preventing the escape of all the sun's energy into space. Without them, the earth would be cold and lifeless.

Some scientists fear that the atmosphere now has too many greenhouse gases. CO_2 emissions have increased 50 percent since the 1970s. Scientists believe that the increase in CO_2 levels causes the atmosphere to trap too much heat, so temperatures have been gradually rising.

Many people disagree with the theory of global warming. Some say the temperature rise may be due to natural processes. Others say that temperatures have not gone up, that they fall within a normal range.

OZONE HOLE Another change is damage to the ozone layer, which exists high in the atmosphere. It absorbs most of the sun's damaging ultraviolet rays. In the 1970s, scientists discovered a thinning of the ozone layer over Antarctica, often called a hole in the ozone layer. Chemicals such as the chlorine found in CFCs react with ozone and destroy it. Many governments have restricted the use of such chemicals, but others have delayed passing such laws because they are costly for industry.

Looking Toward the Future

Scientists fear that many problems may result from these changes to the environment. Because of that, many people and nations around the world are trying to halt the damage before it is too late.

LONG-TERM EFFECTS One fear about global warming is that even small temperature increases could melt the world's ice caps. This would cause a rise in sea levels that might swamp coastal cities and islands. For example, the low islands of Oceania might disappear.

Some people predict that global warming might change patterns of evaporation and precipitation. This could make violent storms such as typhoons and droughts more common. The location of climate zones and agricultural regions might shift, upsetting the world's economy.

People worry about the ozone layer hole because more ultraviolet rays will reach earth. Ultraviolet rays are linked to such problems as skin cancer, eye damage, and crop damage. Because it lies close to Antarctica, New Zealand may be at higher risk than other regions.

TAKING ACTION In 1992, the UN held the Earth Summit, a conference to discuss ways to pursue economic development while protecting the environment. Representatives of 178 nations attended.

In 1997, the UN held a convention in Kyoto, Japan, to discuss climate change. The conference wrote the Kyoto Protocol, guidelines for developed countries to reduce greenhouse gas emissions. In time, 165 nations signed the treaty. The United States signed the treaty, but the Senate didn't ratify it—fearing that the guidelines might harm U.S. businesses.

On the next two pages are primary sources expressing different views about environmental problems. Use them to form your own opinion.

SEE
PRIMARY SOURCE **D**

SEE
PRIMARY SOURCE **A**

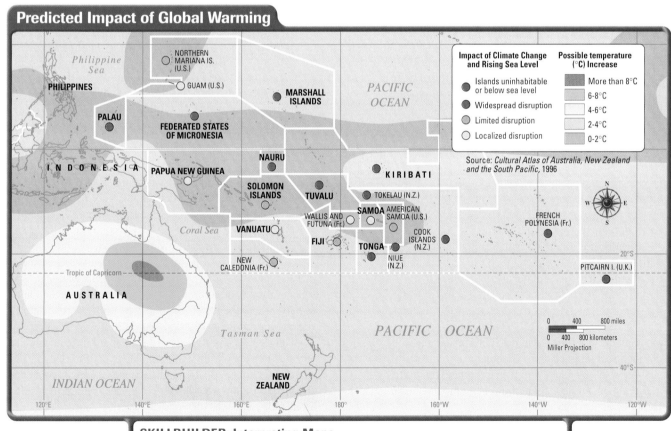

Predicted Impact of Global Warming

Impact of Climate Change and Rising Sea Level
- ● Islands uninhabitable or below sea level
- ◑ Widespread disruption
- ◔ Limited disruption
- ○ Localized disruption

Possible temperature (°C) Increase
- More than 8°C
- 6-8°C
- 4-6°C
- 2-4°C
- 0-2°C

Source: *Cultural Atlas of Australia, New Zealand and the South Pacific,* 1996

Miller Projection

SKILLBUILDER: Interpreting Maps

❶ **PLACE** Where are the greatest temperature increases expected to occur?

❷ **PLACE** What places in Oceania are expected to experience the least disruption?

SE ASIA & OCEANIA

CASESTUDY

PROJECT

Political Cartoon

Primary sources A to E on these two pages present differing opinions on global environmental change. Use these sources and your own research to create a political cartoon expressing your opinion. You might use the Internet and the library for research.

RESEARCH LINKS
CLASSZONE.COM

Suggested Steps

1. Use the sources here and your own research to decide if you believe that global warming and the ozone hole are problems.

2. Draw a pencil sketch of a cartoon expressing your opinion about global environmental change. As you decide what to draw, consider the following questions.

   • Do you think that the theories about environmental change are wrong? If so, why are people so concerned about the issue?

   • Do you think environmental change poses a threat to the world's climate? If so, what should be done?

3. Show the sketch to a friend to see if you have conveyed your point. Use your friend's feedback to make your cartoon more effective.

4. Create your final cartoon. You may wish to draw it lightly in pencil and then ink over the pencil marks. Post the cartoon in class.

Materials and Supplies

   • Samples of political cartoons
   • Drawing paper
   • Pencils and erasers
   • Felt-tip markers
   • Computer
   • Internet access

PRIMARY SOURCE Ⓐ

Educational Pamphlet *In 1994, the United Nations Environment Programme and the World Meteorological Organization published a pamphlet called* Beginners Guide to the Convention *to help people understand the Kyoto Protocol and the reasons for it.*

Human beings seem to be changing the global climate. The results are uncertain, but if current predictions prove correct, the climatic changes over the coming century will be larger than any since the dawn of human civilization.

The principal change to date is in the earth's atmosphere. . . . We have changed, and are continuing to change, the balance of gases that form the atmosphere. This is especially true of such key "greenhouse gases" as carbon dioxide (CO_2), methane (CH_4), and nitrous oxide (N_2O). (Water vapour is the most important greenhouse gas, but human activities do not affect it directly.) . . . Greenhouse gases are vital because they act like a blanket around the earth. Without this natural blanket the earth's surface would be some 30°C colder than it is today.

The problem is that human activity is making the blanket "thicker." . . . The most direct result, says the scientific consensus, is likely to be a "global warming" of 1.5 to 4.5°C over the next 100 years.

PRIMARY SOURCE Ⓑ

Political Commentary *The American Policy Center is a conservative group that wants to promote free enterprise and reduce government regulations. It opposed the Kyoto Protocol and published the article "There is No Global Warming."*

There is no global warming. Period.
   You can't find a real scientist anywhere in the world who can look you in the eye and, without hesitation, . . . say "yes, global warming is with us."
   There is no evidence whatsoever to support such claims. Anyone who tells you that scientific research shows warming trends . . . is wrong. There is no global warming.
   Scientific research through U.S. Government satellite and balloon measurements shows that the temperature is actually cooling—very slightly—.037 degrees Celsius.
   A little research into modern-day temperature trends bears this out. For example, in 1936 the Midwest of the United States experienced 49 consecutive days of temperatures over 90 degrees. There were another 49 consecutive days in 1955. But in 1992 there was only one day over 90 degrees and in 1997 only 5 days.

Data *The National Climatic Data Center collects data on temperature and precipitation. In the graph below, the line at zero represents the average annual world temperature for the period* 1880 to 2000. *The bars show how much the average temperatures for individual years were higher or lower than the average. Scientists use this graph to spot climate trends.*

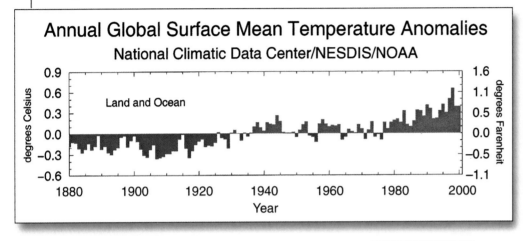

Annual Global Surface Mean Temperature Anomalies
National Climatic Data Center/NESDIS/NOAA

PRIMARY SOURCE D

News Article *On October 10, 2000, the* New York Times *published the article "Record Ozone Hole Refuels Debate on Climate" by Andrew C. Revkin. The article appeared in the science section of the paper.*

The hole that opens in the ozone layer over Antarctica each southern spring formed earlier and grew bigger this year than at any time since satellites have been monitoring the polar atmosphere, scientists have reported.

The finding renewed suspicions among atmospheric scientists that global warming could be indirectly abetting the chemical reactions that destroy ozone, but many still say the growth of the hole could also be the result of natural . . . variations in Antarctic weather and other conditions. . . .

The hole is closely watched because the stratosphere's . . . layer of ozone . . . absorbs ultraviolet rays, which could contribute to skin cancers and cataracts and threaten agriculture and ecosystems if they reached the surface.

PRIMARY SOURCE E

Satellite Images *Satellites took these images of ozone over Antarctica. The color blue represents areas with an extremely low concentration of ozone, while red shows a high concentration.*

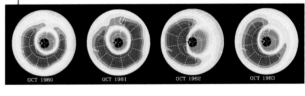

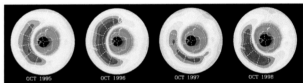

P R O J E C T *CheckList*

Have I . . .

✓ researched opinions on global environmental change?

✓ formed my own opinion based on evidence about the issue?

✓ created an interesting cartoon that clearly expresses that opinion?

✓ created a cartoon that is neat enough to print in a newspaper?

VISUAL SUMMARY
TODAY'S ISSUES IN SOUTHEAST ASIA, OCEANIA, AND ANTARCTICA

Government

Aboriginal Land Claims

- When the British first arrived in Australia, British authorities declared the continent to be empty. They decided they had the right to take the land without making treaties.

- Aboriginal people lost much of their land and had to live on reserves.

- Recent court cases have provided the grounds for Aboriginal people to make land claims. However, the Australian government took steps to limit those land claims.

Economics

Industrialization Sparks Change

- The growth of industry often leads to rapid urban growth. People move to cities because of push-pull factors.

- Industrialization creates higher incomes for many, but in Southeast Asia the income gap remains high. This has the potential to cause social unrest.

- Because of the use of fossil fuels and careless waste disposal, industrialization often causes pollution.

Environment

Global Environmental Change

- Many scientists believe that increases in carbon dioxide emission have caused global temperatures to rise; it is feared that global warming might lead to flooding and an increase in droughts and violent weather.

- The use of chemicals such as CFCs has been linked to a thinning of the protective ozone layer. The hole in the ozone layer may let more ultraviolet rays reach the earth and cause cancer, eye damage, and crop damage.

Reviewing Places & Terms

A. Briefly explain the importance of each of the following.

1. assimilation
2. Stolen Generation
3. Land Rights Act of 1976
4. *Mabo* Case
5. pastoral leases
6. *Wik* Case
7. industrialization
8. push-pull factors

B. Answer the questions about vocabulary in complete sentences.

9. What is the relationship between the terms *assimilation* and *Stolen Generation*?
10. Who owned the Australian lands that were held by pastoral leases?
11. Which of the above terms is a pull factor leading to urban growth?
12. To which Australian territory did the Land Rights Act of 1976 apply?
13. What was the main decision in the *Mabo* Case?
14. What was the main decision in the *Wik* Case?
15. How would you apply the term *push factors* to the experience of the Aboriginal people in Australia?

Main Ideas

Aboriginal Land Claims (pp. 727-729)

1. What does the Aboriginal Tent Embassy symbolize?
2. When the Aboriginal people fought European settlement, what enabled the Europeans to win?
3. How did Eddie Mabo prove his family's land ownership?
4. Why did white Australians fear the *Wik* decision?

Industrialization Sparks Change (pp. 730-733)

5. Why do many people in Southeast Asia move temporarily to cities?
6. How has industrialization affected cities?
7. What effect has industrial growth had on trade and exports?

Global Environmental Change (pp. 734-737)

8. What are greenhouse gases?
9. What are the arguments against the theory of global warming?
10. What health problems may increase because of the hole in the ozone layer?

Critical Thinking

1. Using Your Notes

Use your completed chart to answer these questions.

	Causes	Effects
Issue 1: Land Claims		
Issue 2: Industrialization		

a. What caused the hole in the ozone layer?

b. In what way are some of these issues linked? Explain.

2. Geographic Themes

a. **HUMAN-ENVIRONMENT INTERACTION** How has industrialization affected Southeast Asia's water supplies?

b. **MOVEMENT** What impact might global warming have upon the movement of people?

3. Identifying Themes

Consider what you have learned about Aboriginal land claims, industrialization, and global environmental change. Which of the five geographic themes relate to all three issues? Explain.

4. Determining Cause and Effect

How did the Australian government's policy of taking mixed-race children from their families affect the desire of the Aboriginal people to reclaim lands? Explain.

5. Drawing Conclusions

Overall, do you think industrialization is a positive or negative development for Southeast Asia? Explain.

Additional Test Practice, pp. S1–S37

TEST PRACTICE
CLASSZONE.COM

Geographic Skills: Interpreting Graphs

Annual Industrial Production Growth Rate

Use the graph to answer the following questions.

1. **PLACE** For which country were statistics for the year 2000 not available?

2. **PLACE** How would you describe the pattern of industrial growth in Thailand?

3. **REGION** In which year did Southeast Asia as a whole experience economic problems? How can you tell?

GeoActivity

Research the industrial production growth rate for another Southeast Asian country. Copy this graph on your own paper and add the information for the country you researched.

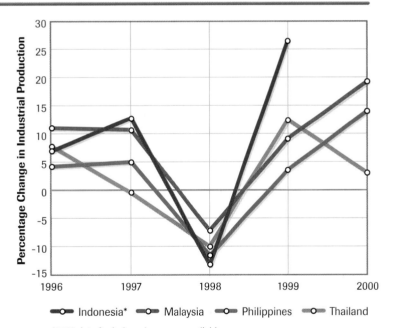

*2000 data for Indonesia were unavailable.
SOURCE: Asia Recovery Information Center online, 2001

INTERNET ACTIVITY

Use the links at **classzone.com** to do research about global warming. Look for additional evidence that either supports or refutes the theory.

Creating a Database Compile statistics that either support or refute the theory of global warming. Present these statistics in tables, charts, or graphs.

World Geography

Reference Section

Contents

1.1 Analyzing Data

Defining the Skill

Analyzing data means studying quantitative information—numbers, proportions, and similar statistics. Data are often presented graphically, in graphs, charts, and maps. When you analyze data, you find patterns, make generalizations and comparisons, and locate facts.

Applying the Skill

The following line graph is titled "World Population Growth." Use the listed strategies to analyze the data presented.

How to Analyze Data

Strategy ① Rephrase the title given for the graphic as a question that can lead you to its main idea. For example: "How has world population growth changed over time?"

Strategy ② To understand how data are displayed, choose one point on the graph. Identify what piece of data is shown at that point. For example, in the line graph, the point on the line that is right above the horizontal number 1000 represents how many billions of people lived in the world in the year 1000—just under one-half billion.

Strategy ③ Make a comparison between two points or other parts on the graph. For example, compare the rate of world population growth between 1000 and 1500 with the rate over the following 500 years. You can see that the population barely grew at all between 1000 and 1500, but increased significantly between 1500 and 2000.

Strategy ④ Answer the question you posed in Strategy 1 in order to summarize data and note a general pattern.

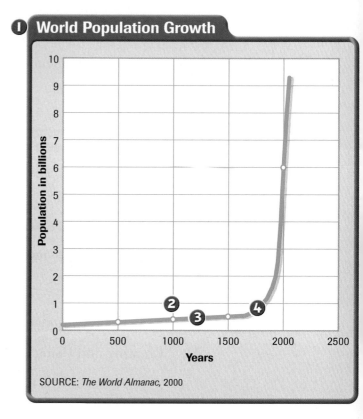

① World Population Growth

SOURCE: *The World Almanac*, 2000

Write a Summary

Summarize the most important idea in your analysis of the data shown. This summary statement might, for example, answer the question suggested by the graph title.

④ *The world's population did not even reach the 1-billion mark until the 1800s, but skyrocketed after that and is on its way to 10 billion.*

Practicing the Skill

Turn to Chapter 31, Section 1, "Southeast Asia." Find the feature on page 707 titled "Industrialization." Analyze the data in the bar graph shown. Write a summary of your analysis.

1.2 Making Comparisons

Defining the Skill

Making comparisons means thinking about similarities and differences. Two or more concepts are grouped together because of shared features, but they are distinguished from one another by other features.

Applying the Skill

The following passage tells about economic development. Use the listed strategies to compare two categories of nations.

How to Make Comparisons

Strategy ❶ Note the concepts being compared. In this passage, categories of economic development are described.

Strategy ❷ Look for words that signal similarities such as *both, same, similar,* and *like.* Look for words that signal differences or contrasts such as *different, in contrast, however,* and *on the other hand.*

Strategy ❸ Sum up what you have learned by telling yourself (a) what concepts are being compared; (b) why they are grouped together; and (c) what their main differences are.

LEVELS OF ECONOMIC DEVELOPMENT

❶ Countries of the world have two different levels of economic development. Developing nations have a low GDP per capita. (GDP is Gross Domestic Product, the value of goods and services produced within a country over a year or other period of time.) Developing nations also have limited development on all levels of economic activities. These countries lack an industrial base and struggle to provide for their citizens' basic needs. Many young countries and former colonies are found in this category.

Developed nations, ❷ on the other hand, are countries with a high per capita income and varied economy. Western European nations, Canada, and the United States are highly developed economies.

Make a Chart

One way to sum up the main points of comparison is with a chart that lists features. The chart below is based on the example passage.

❸	Developing nations	Developed nations
GDP per capita	low	high
Variety of economic activities	limited development; lack of industrial base	varied economy
Examples	young countries, former colonies	Western European nations, Canada, U.S.

Practicing the Skill

Turn to Chapter 5, Section 3, "Human-Environment Interaction." Read "Building Cities" on page 128. Identify the main similarities and differences described, and show them in a chart.

1.3 Making Inferences

Defining the Skill

Making inferences involves using information that is directly stated in the text in order to think of, or infer, ideas that are not directly stated. You use logic and your own experience and knowledge to make inferences.

Applying the Skill

The passage below tells about a feature of the climate of South Asia. Use the listed strategies to make inferences about monsoons.

How to Make Inferences

Strategy ❶ Find statements of fact and other stated ideas, such as opinions and generalizations.

Strategy ❷ Ask yourself questions about the stated facts and ideas. Think of likely answers that are not directly stated. For example, the passage states that dry winds blow between October and May, and moist winds blow between June and September. Ask, "What else can I understand from that information?"

Strategy ❸ Make inferences from the facts and ideas. For example, you might infer that the region has two main seasons—a long dry one and a shorter wet one.

MONSOONS

❶ Although climate varies throughout South Asia, the region as a whole is greatly affected by monsoons, or seasonal winds. ❷ Between October and May, dry winds blow across South Asia from the northeast. ❷ Between June and September, the winds reverse and blow in from the southwest, bringing moist air from the ocean. ❶ Heavy rains fall, especially in the southern and eastern portions of South Asia.

❶ Rainfall is crucial to life on the subcontinent. Yet the monsoons can cause severe hardship for millions of South Asians, especially those living in the lowlands of India and Bangladesh. The monsoons are also highly unpredictable. Some areas may get too little rain, while others get too much. The monsoons are an essential but difficult feature of life in South Asia.

Make a Chart

A chart can show the inferences made from stated facts and ideas. The chart below is based on the passage you just read.

❶ Stated Facts and Ideas	❷ Questions	❸ Inferences
The direction of the winds shifts seasonally, from the northeast to the southwest.	What causes the wind patterns to change?	Wind patterns change as Earth changes its position relative to the sun.
Heavy rains follow from winds coming from the ocean.	How do ocean winds carry water?	Water evaporates from the ocean, is carried by the air, and condenses over land.
The monsoons can cause severe hardship, especially in the lowlands.	What problems do the monsoons cause in the lowlands?	Damaging floods can result from monsoon rains.

Practicing the Skill

Turn to Chapter 25, Section 2, "India's Neighbors: Pakistan and Bangladesh." Read the subsection "New Countries, Ancient Lands," on pages 573–574. Use the facts and ideas to infer other ideas. Show your inferences in a chart.

1.4 Drawing Conclusions

Defining the Skill

Drawing conclusions means combining factual information with your own reasoning to formulate a statement that is likely to be true. To draw conclusions, look at the facts and think about what they mean.

Applying the Skill

The following passage offers facts about two of the world's largest lakes. Use the listed strategies to draw conclusions about the information.

How to Draw Conclusions

Strategy 1 Read carefully to identify and understand the statements of fact, the items of information that can be proved true.

Strategy 2 Think about which facts fit together and how they fit. List the facts in a diagram and use your own experiences to understand how the facts relate to each other.

Strategy 3 Come up with a statement, different from one given in the text, that draws a conclusion about the factual information.

TWO LARGE LAKES OF CENTRAL ASIA

1 The Caspian Sea, which is actually a saltwater lake, stretches for nearly 750 miles from north to south, making it the largest inland sea in the world. **1** Recently, the Caspian's water levels have been rising, and have flooded many surrounding villages and towns. **1** The sea now stands over two yards higher than it did in 1978. Nobody is certain what is causing the change. But scientists say possible causes might include climate change or more water flowing off deforested land.

1 The Aral Sea, another of the world's largest lakes, lies east of the Caspian. **1** Unlike the Caspian, the Aral Sea is shrinking. **1** Extensive irrigation projects have diverted water away from the lake. **1** Since 1960, the Aral has lost about 80 percent of its water volume.

Make a Diagram

A diagram can highlight the facts that fit together to point to a conclusion. The diagram below shows a conclusion that can be drawn from the passage above.

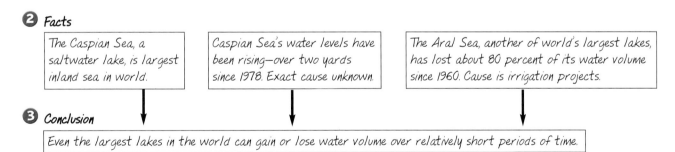

2 Facts

| The Caspian Sea, a saltwater lake, is largest inland sea in world. | Caspian Sea's water levels have been rising—over two yards since 1978. Exact cause unknown. | The Aral Sea, another of world's largest lakes, has lost about 80 percent of its water volume since 1960. Cause is irrigation projects. |

3 Conclusion

Even the largest lakes in the world can gain or lose water volume over relatively short periods of time.

Practicing the Skill

Turn to Chapter 10, Section 1. Read the subsection "Native Americans and the Spanish Conquest" on page 216. Make a diagram to show selected facts and the conclusion you drew from them.

1.5 Making Generalizations

Defining the Skill

Making a generalization means making a broad statement that applies to a number of examples. Generalizations can be made from examples given in one passage, in several sources, or from graphic aids.

Applying the Skill

The following two passages present examples on the same topic. Use the listed strategies to make a generalization based on the examples.

How to Make Generalizations

Strategy ❶ Note the examples given on the same topic.

Strategy ❷ Use a term such as *generally* or *usually* as you decide what the examples have in common.

Strategy ❸ Formulate a logical, general statement that applies to all examples.

OCEANS AND MOUNTAINS

❶ The Canadian coastal ranges prevent the warming of Canada's interior by blocking warm Pacific air. ❶ In the United States, the western mountains trap Pacific moisture. This makes the climate in the lands to the west of the mountains rainy and those to the east very dry.

The North Atlantic Drift, a current of warm water from the tropics, flows near Europe's west coast. The prevailing westerlies, which blow west to east, pick up warmth from this current and carry it over Europe. ❶ No large mountain ranges block the winds, so the influence of the westerlies extends far inland.

Make a Diagram

A diagram can show how examples add up to a generalization. The diagram below is based on the passages you just read.

❶ Example: Canadian coastal ranges block warm Pacific air from reaching Canada's interior.

+

❶ Example: Western mountains of the United States trap Pacific moisture, making the lands on the Pacific side moist and the eastern side dry.

+

❶ Example: Atlantic Ocean warmth is carried over Europe because no large mountain ranges block the winds.

=

❷❸ Generalization: Mountains generally prevent ocean air from traveling farther inland.

Practicing the Skill

Find passages about the humid continental climate of the United States and Canada (page 124), of Europe (page 279), and of East Asia (page 626). Make a diagram to show examples and a generalization.

1.6 Making Decisions

Defining the Skill

Making decisions means choosing between two or more courses of action. When you analyze the decisions people have made, you think about the needs they were trying to meet and the consequences of each choice.

Applying the Skill

The following passage describes the problem of rapid population growth facing the Chinese government. Use the listed strategies to analyze the decisions made.

How to Make Decisions

Strategy ❶ Look for a statement of the difficulty. Think about the choices facing the group.

Strategy ❷ Consider possible consequences of each choice.

Strategy ❸ Identify the decisions that were made.

Strategy ❹ Identify actual consequences.

CONTROLLING CHINA'S POPULATION

One out of every five people in the world lives in China. China's estimated population in the year 2000 was about 1.3 billion. ❶ Because of concerns about a rapidly expanding population, ❸ China in 1979 adopted a policy of one child per family. In addition, the country has age restrictions for marriage—a man must be 22 and a woman 20 before they can marry. ❹ These policies have reduced China's birthrate dramatically.

❹ However, the government policy of one child per family has run into opposition. Rural families, in particular, feel the need for more than one child to help work on their farms. ❹ As a result, the government has relaxed the one-child policy.

Make a Flow Chart

The process of decision-making can be shown in a flow chart. The flow chart below summarizes the decisions described in the passage you just read.

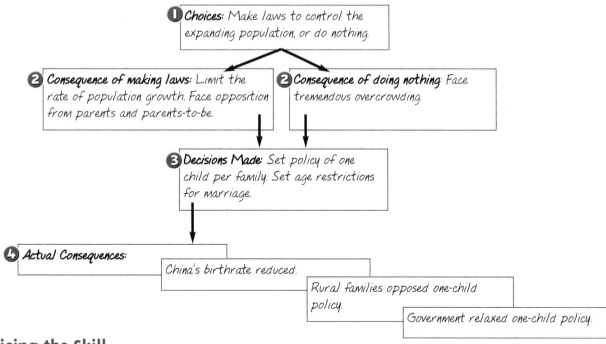

❶ **Choices:** Make laws to control the expanding population, or do nothing.

❷ **Consequence of making laws:** Limit the rate of population growth. Face opposition from parents and parents-to-be.

❷ **Consequence of doing nothing:** Face tremendous overcrowding.

❸ **Decisions Made:** Set policy of one child per family. Set age restrictions for marriage.

❹ **Actual Consequences:**

China's birthrate reduced.

Rural families opposed one-child policy.

Government relaxed one-child policy.

Practicing the Skill

Turn to Chapter 23, Section 1. Read "New Industry Requires More Workers," on pages 525–526. Make a flow chart to show the choices faced by the nations' governments and the consequences of the decisions made.

1.7 Seeing Patterns

Defining the Skill

Seeing patterns involves seeing the overall shape, organization, or trend of geographic characteristics. It often means noting variations or contrasts, and thinking about the "rules" that describe them and could apply to similar situations. Seasonal weather cycles are one example of a pattern; economic changes are another. Graphs, maps, charts, and text passages are all sources of information that help you see patterns.

Applying the Skill

The passage below tells about the economics of oil in North Africa. Use the listed strategies to think about the pattern described.

How to See Patterns

Strategy ❶ Note any directly stated main ideas about details of geographic characteristics, or changes and contrasts. (If none is directly stated, try to make your own statement of comparison, based on the details in the passage.)

Strategy ❷ Notice examples that support the ideas.

Strategy ❸ Use the word *pattern* in a question about the information. For this passage, you could ask, "What economic pattern is seen in the oil-producing nations of North Africa?" Your answer will sum up the pattern you see. (The chart below has a possible answer.)

Make a Chart

Make a chart to sum up the pattern. The chart below organizes information from the passage you have just read.

AN OIL-BASED ECONOMY

❶ Oil has transformed the economies of some North African countries, including Libya, Algeria, and Tunisia. ❷ In Algeria, oil has surpassed farm products as the major export and source of revenue. Furthermore, oil makes up about 99 percent of Libya's exports.

❶ Although oil has helped the economies of these countries, it has also caused some problems. ❷ Libya, Algeria, and Tunisia face shortages of skilled labor to carry out this work. For example, Libya's labor force cannot meet the demands of the oil industry because of a lack of training and education. Oil companies are forced to give many high-paying jobs to foreign workers. Even within the oil industry, overall unemployment is still a problem. As a result, large numbers of North Africans have migrated to Europe in search of jobs.

❶ Main Ideas About Contrasts and Changes	❷ Examples	❸ Summary Statement of Pattern
The oil industry has transformed the economies of some North African countries.	Algeria—oil major export and revenue source. Libya—oil about 99 percent of exports.	A single industry can power the economy of a nation, but an unskilled labor force may not benefit.
Oil helps the economy but also causes problems.	Libya, Algeria, Tunisia face shortages of skilled workers. Libya—labor force lacks training and education. Foreign skilled workers get high-paying jobs. Unemployment, emigration.	

Practicing the Skill

Turn to Chapter 19, Section 5, "Southern Africa." Read the subsection "Success at a Cost" on pages 455–456. Use the information in it to sum up the pattern you see. Use standard grammar, sentence structure, and punctuation in your summary.

1.8 Determining Cause and Effect

Defining the Skill

A **cause** is why something happens. An **effect** is what happens. A single cause can lead to one effect or multiple effects. One effect can have multiple causes. Cause-effect chains are also common, in which a cause leads to an effect that becomes the cause of another effect, and so on.

Applying the Skill

The following paragraphs sum up major events in the recent European past. Use the listed strategies to analyze the cause-effect relationships.

How to Determine Cause and Effect

Strategy ① Use the word *why* to formulate questions about the topic of the passage. Example: *Why was there conflict in Europe?* The answers you find will be the causes.

Strategy ② Look for words such as *because, cause, in order to,* and *reason,* which signal causes. Look for words such as *so, consequence,* and *result,* which signal effects.

Strategy ③ Restate the cause-effect connections in your own words or in a diagram.

① CONFLICT IN EUROPE

Western Europe experienced industrial growth in the 1800s. ② Industrialism caused European nations to set up colonies in other lands in order to gain raw materials and markets. Many European nations saw each other as rivals in the race to gain colonies. ② The nationalistic rivalry and competition for colonies among European nations helped cause World War I. The Allied Powers (including France) fought the Central Powers (Germany, Austria-Hungary, and their allies). The Allies won and imposed harsh terms on Germany. ② German resentment over those terms helped cause World War II, in which Germany, led by Adolf Hitler and the Nazis, tried to conquer Europe.

Make a Diagram

A diagram can show how causes and effects are connected. Because the example passage tells how one event led to another, a cause-effect chain is a useful way to diagram its major ideas.

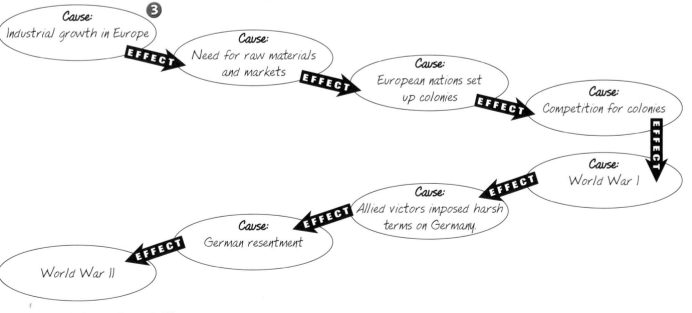

Practicing the Skill

Turn to Chapter 10, Section 4. Read the subsection "Native Peoples and Portuguese Conquest" on page 236. Make a diagram to show major cause-effect connections.

1.9 Identifying and Solving Problems

Defining the Skill

Identifying and solving problems means analyzing the difficulties that are faced by individuals and groups. You determine why the difficulties exist, how people try to overcome them, and what solutions, if any, are achieved.

Applying the Skill

The following paragraph describes a general problem related to the issue of national boundaries, and offers a particular African nation as an example. Use the listed strategies to understand the problem-solution connection.

How to Identify Problems and Solutions

Strategy ❶ Look for a statement of the problem. Note words such as *problem, conflict, difficulty,* or *controversy.* Use the details to ask yourself why the problem exists, and why people wish to overcome it.

Strategy ❷ Identify attempts to solve the problem.

Strategy ❸ Think about the outcome. Ask yourself whether the problem is solved, or whether the outcome is likely to lead to more difficulties.

> **ARTIFICIAL NATIONAL BOUNDARIES**
> Africa is a good example of how ❶ boundary lines can divide groups of people or put groups that have long been enemies together in one state. When parts of Africa were divided by European colonial powers, ❶ the boundary lines for Nigeria enclosed the traditional lands of the Hausa-Fulani people, the Yoruba people, and the Ibo people. Under British control, the three groups were forced to follow British rules. When Britain left, there was controversy over the control of the lands. ❷ One group, the Ibo, attempted to withdraw from Nigeria and form its own nation-state, Biafra. ❸ A civil war resulted, and the attempt to split away failed.

Make a Chart

A chart can help you take notes and sum up important ideas about problems and solutions. The chart below shows problems and solutions in the passage you just read.

❶ Problem	Solution Attempts ❷	Outcome ❸
Nigerian boundary lines artificially enclose the traditional lands of three groups of people.	One group, the Ibo, attempted to form a separate nation-state.	Civil war. Attempt to split away failed.

Practicing the Skill

Turn to Chapter 8, Section 2, "Urban Sprawl." Read "Urban Sprawl's Negative Impact" and "Solutions to Sprawl." Make a chart to sum up the problem and possible solutions. Write a summary of the information presented in your chart using standard grammar, sentence structure, and punctuation.

1.10 Distinguishing Fact from Opinion

Defining the Skill

Facts are dates, numbers, names, and statements that can be proved true. **Opinions** are statements that express beliefs, values, and feelings. Although opinions cannot be proved true or false, they can be supported with facts and logical reasons. In order to decide whether to agree with stated opinions, readers must first separate opinion from fact.

Applying the Skill

The following paragraph tells how human-environment interaction affects climate and vegetation. Use the strategies listed below to distinguish fact from opinion.

How to Distinguish Fact from Opinion

Strategy ① Notice words that reveal the author's beliefs or feelings. In the sample paragraph, *unfortunately* and *careless* show that opinions are being expressed.

Strategy ② Look for statements about future events. These statements are opinions because they cannot be proved.

Strategy ③ Look for facts that are given as supporting reasons for the statements of opinion.

Strategy ④ Identify ways in which you can check the facts.

> **HUMAN IMPACT ON THE ENVIRONMENT**
>
> ① Unfortunately, the damage that humans cause to soil and vegetation is a by-product of human-environment interaction. ③ Fragile biomes such as the tundra are easily damaged. Oil pipelines crisscross tundra regions and ② bring the threat of leakage and spills. . . .
>
> In the United States, millions of people choose to live in the desert southwest, part of a region known as the Sunbelt. ③ The desert land is easily eroded, and housing sub-divisions destroy vegetation. In other regions of the world, ① careless use of the land often leaves it in a condition that ② will not support life, even with sophisticated technological intervention.

Make a Chart

The chart below analyzes the facts and opinions from the passage above.

Opinion ① ②	Supporting Facts ③	④ How to Check Facts
Human-environment interaction results in unfortunate damage to soil and vegetation.	Fragile biomes such as the tundra are easily damaged.	Research current articles about human-caused damage to tundra.
	The desert land of the Sunbelt is easily eroded.	Research current articles about desert erosion in Sunbelt region.
	Housing sub-divisions destroy vegetation.	Research current articles about effects of development on vegetation in desert southwest.
The tundra is threatened with oil leakage and spills.	Oil pipelines crisscross tundra regions.	Research oil-industry and news sources.
Careless use of the land often leaves it in a condition that will not support life, even with sophisticated technological intervention.	None given	

Practicing the Skill

Turn to Chapter 3, Section 2, and read the passage "Global Warming." Show opinions and supporting facts in a chart.

1.11 Creating a Sketch Map

Defining the Skill

When you are reading about routes, regions, landforms, political boundaries, or any other geographical information, try to visualize what is described. One way to clarify the information is by **creating a sketch map.** To sketch your own map, use one or more published maps as guides.

Applying the Skill

After reading the passage below, a student sketched the map shown. Read the listed strategies to see how the map was created.

> **WESTWARD MOVEMENT**
>
> From departure points such as Independence, Missouri, hundreds of thousands of pioneers left in covered wagons bound for the West. They blazed trails that crossed prairie, plains, desert, and mountains, moving toward the Pacific. A wagon train on the Oregon Trail might take up to six months to reach its destination 2,000 miles away.

How to Create a Sketch Map

Strategy ① Choose a title that sums up what you will show in the map.

Strategy ② Consider the purpose of the map as you decide which standard features need to be included. Because the main purpose of this sketch map is to show journeys, it includes a scale of distance. Other maps may require lines of latitude and longitude, for example, and a compass rose.

Strategy ③ Find one or more maps that you can use to guide the placement of elements and labels. For this sketch, the student consulted a historical map and a physical map.

Strategy ④ Create a legend to explain any symbols or colors used.

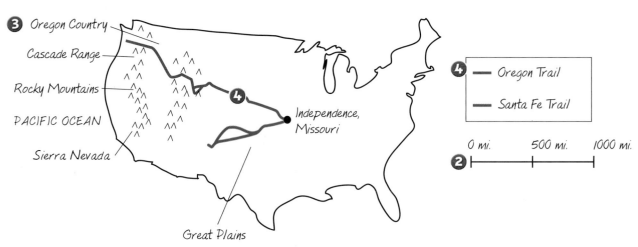

Practicing the Skill

Turn to Chapter 12, Section 1. Read the introductory paragraph "A Human Perspective" on page 273. Create a sketch map of the route of Hannibal's troops. Include the map elements needed to show why Hannibal's achievement was so remarkable.

1.12 Creating Graphs and Charts

Defining the Skill

Whenever your research provides you with information involving numbers and quantities, you can **create graphs and charts** to show patterns in your data. Software programs tend to use the terms *graphs* and *charts* interchangeably. Kinds of graphs and charts include bar graphs, line graphs, pictographs, and pie graphs, which are also called pie charts. The kind you choose depends on your data.

Applying the Skill

The three visuals below are a pie chart, a bar graph, and a line graph. Use the listed strategies to think about their purposes and parts.

How to Create Graphs and Charts

Strategy ❶ Organize your numerical data. Make a table with rows and columns, or use the grid layout of a spreadsheet. The headings in your table or spreadsheet will correspond to labels in your graph.

Strategy ❷ Choose the type of graph to create. Are you showing changes over time? A line graph might be best. Are you making a series of comparisons? Consider a bar graph. Do you want to show how parts make the whole? A pie chart shows percentages.

Strategy ❸ In line and bar graphs, plot the data along the axes. The X-axis is horizontal; the Y-axis is vertical. Make sure that both axes are labeled with words or numbers.

Strategy ❹ Include a legend to indicate what each bar, line, or section represents.

Strategy ❺ Add a title.

Practicing the Skill

Turn to Chapter 6, Section 3. Look at the data listed on page 147, accompanying the subsection "The Midwest." Show the data in two clearly labeled pie charts. Use graphing software if possible. Write a generalization about the information in each chart using standard grammar, sentence structure, and punctuation.

❷ Pie Chart

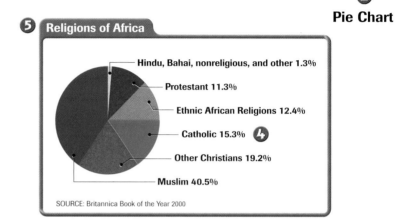

❺ Religions of Africa

- Hindu, Bahai, nonreligious, and other 1.3%
- Protestant 11.3%
- Ethnic African Religions 12.4%
- Catholic 15.3% **❹**
- Other Christians 19.2%
- Muslim 40.5%

SOURCE: Britannica Book of the Year 2000

❷ Bar Graph

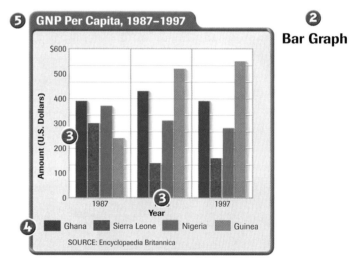

❺ GNP Per Capita, 1987–1997

Amount (U.S. Dollars): $600, 500, 400, 300, 200, 100, 0

Year: 1987, 1992, 1997 **❸**

❹ Ghana ■ Sierra Leone ■ Nigeria ■ Guinea

SOURCE: Encyclopaedia Britannica

❷ Line Graph

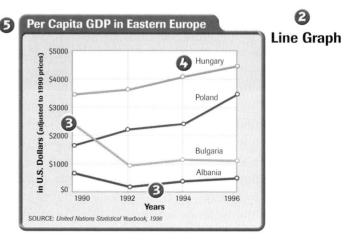

❺ Per Capita GDP in Eastern Europe

in U.S. Dollars (adjusted to 1990 prices): $5000, $4000, $3000, $2000, $1000, $0

❹ Hungary
Poland
Bulgaria
Albania

Years: 1990, 1992, 1994, 1996 **❸**

SOURCE: United Nations Statistical Yearbook, 1996

2.1 Creating a Multimedia Presentation

Defining the Skill

Print is a medium of communication. Video and audio recordings, Web pages, and photographic slides are other examples of media. To **create a multimedia presentation,** you collect and display information so that your audience watches, listens, and learns.

Applying the Skill

A multimedia presentation can incorporate high-tech electronics, but it does not have to. A photo essay with audio background, for example, is also an effective multimedia presentation. Use the listed strategies to create your own multimedia presentation.

How to Create a Multimedia Presentation

Strategy ① Choose a topic that lends itself to multimedia. Consider using still or moving images, a script for one or more speakers, sound effects, and music. You might create a travelogue, for example, in which you show your audience a place, and develop a narrative to go with the visual images.

Strategy ② Research the topic to get a general overview. Then narrow the topic to one of manageable size. Make an outline to show the steps you will take to develop your presentation.

Strategy ③ Collect information. Then select the text, images, and audio you plan to use. Show your plan graphically, using a storyboard format, for example.

Strategy ④ Put your presentation together.

Practicing the Skill

Turn to Chapter 4 and read Section 4, "Urban Geography." Choose a topic that you think will work well for a multimedia presentation. Do research, narrow the topic, and make an outline for a future presentation.

2.2 Creating and Using a Database

Defining the Skill

A **database** is any listing system in which related information is organized so that particular items can be retrieved. An electronic library catalog is an example of a database; new information can be added based on the categories, and users input search terms in order to pull out specific listings. Specialized software programs are used to create large, complex databases. Spreadsheet programs are frequently used to create less complex databases.

Applying the Skill

The table below is part of a database for statistics about the countries of Latin America. Use the listed strategies to understand the organization of a database.

How to Create or Use a Database

Strategy ❶ Identify or name the topic of the database table.

Strategy ❷ Define or identify the categories of data. In a computer database, these categories are called fields, and correspond to column headings. A field can specify names, dates or other numbers, or text.

Strategy ❸ The data in each row of a database table form a record. The records are sorted by a particular field—usually alphabetically, or numerically in ascending or descending order. In the table shown, the records are sorted alphabetically by country name.

Strategy ❹ To find a particular piece of data in an existing database, choose a search criterion. The table shown could lead to a list of all countries in which life expectancy is 70 or lower, for example.

❶ Regional Statistics: Latin America (year 2000 estimates)				
Country/Capital ❷	Population ❷	❷ Life Expectancy in years (1995–2000)	❷Birthrate per 1,000 pop.	Infant Mortality ❷ per 1,000 live births
❸ Antigua and Barbuda/St. John's	68,000	71	22	17.1
❸ Argentina/Buenos Aires	37,048,000	73	19	19.2
❸ Bahamas/Nassau	310,000	74	21	18.4
Barbados/Bridgetown	259,000	76	14	14.2
Belize/Belmopan	254,000	75	32	33.9
Bolivia/La Paz, Sucre	8,281,000	❹ 61	30	67.0
Brazil/Brasilia	170,115,000	❹ 67	21	40.0
Chile/Santiago	15,211,000	75	18	10.5
Colombia/Bogota	40,037,000	❹ 70	26	28.0
Costa Rica/San José	3,589,000	76	22	12.6

Practicing the Skill

Use spreadsheet or database software to input the following fields from the "Regional Data File" for the 50 U.S. states, shown on pages 110–112: Name of State, Population, Population Density, Total Area (square miles). Sort the data (a) alphabetically by name of state and (b) by population, in descending order.

A

Aboriginal people *n.* people who migrated to Australia from Asia at least 40,000 years ago; the original settlers of the land. (p. 718)

absolute location *n.* the exact place on earth where a geographic feature is found. (p. 6)

acculturation *n.* the cultural change that occurs when individuals in a society accept or adopt an innovation. (p. 72)

acquired immune deficiency syndrome (AIDS) *n.* a disease caused by the human immunodeficiency virus, or HIV. (p. 465)

Aksum *n.* an important trading capital from the first to the eighth centuries A.D. in what is now Ethiopia; it flourished due to its location near the Red Sea and the Indian Ocean. (p. 431)

alluvial plain *n.* land that is rich farmland, composed of clay, silt, sand, or gravel deposited by running water. (p. 553)

Amazon River *n.* the second longest river in the world, and one of South America's three major river systems, running about 4,000 miles from west to east, and emptying into the Atlantic Ocean. (p. 203)

Andes Mountains *n.* a large system of mountain ranges located along the Pacific coast of Central and South America. (p. 201)

anti-Semitism *n.* discrimination against Jewish people. (p. 315)

apartheid (uh•PAHRT•HYT) *n.* a policy of complete separation of the races, instituted by the white minority government of South Africa in 1948. (p. 454)

Appalachian Mountains *n.* one of two major mountain chains in the eastern United States and Canada, extending 1,600 miles from Newfoundland south to Alabama. (p. 119)

aqueduct *n.* a structure that carries water over long distances. (p. 292)

aquifer *n.* an underground layer of rock that stores water. (p. 421)

archipelago *n.* a set of closely grouped islands. (pp. 553, 689)

ASEAN *n.* the Association of Southeast Asian Nations, an alliance that promotes economic growth and peace in the region. (p. 707)

Ashanti *n.* a people who live in what is now Ghana, in West Africa, and who are known for their artful weaving of colorful *asasia,* or *kente* cloth. (p. 444)

assimilation *n.* a process whereby a minority group gradually gives up its own culture and adopts the culture of a majority group. (p. 728)

Aswan High Dam *n.* a dam on the Nile River in Egypt, completed in 1970, which increased Egypt's farmable land by 50 percent and protected it from droughts and floods. (p. 426)

Atlantic Provinces *n.* the provinces in Eastern Canada—Prince Edward Island, New Brunswick, Nova Scotia, and Newfoundland. (p. 166)

atmosphere *n.* the layers of gases immediately surrounding the earth. (p. 28)

atoll *n.* a ringlike coral island or string of small islands surrounding a lagoon. (pp. 553, 700)

B

balkanization *n.* the process of breaking up a region into small, mutually hostile units. (p. 311)

Baltic Republics *n.* the countries of Latvia, Lithuania, and Estonia, located on the eastern coast of the Baltic Sea. (p. 361)

Bantu migration *n.* the movement of the Bantu peoples southward throughout Africa, spreading their language and culture, from around 500 B.C. to around A.D. 1000. (p. 448)

basic necessity *n.* food, clothing, and shelter. (p. 593)

Benelux *n.* the economic union of Belgium, the Netherlands, and Luxembourg. (p. 296)

Beringia *n.* a land bridge thought to have connected what are now Siberia and Alaska. (p. 127)

Berlin Conference *n.* a conference of 14 European nations held in 1884–1885 in Berlin, Germany, to establish rules for political control of Africa. (p. 432)

Berlin Wall *n.* a wall erected by East Germany in 1961 to cut the capital of Berlin in two, and later dismantled in 1989. (p. 298)

Bikini Atoll *n.* the isolated reef, located in the Marshall Islands of the central Pacific, that was the site of U.S. nuclear bomb tests, consequently contaminating the atoll with high levels of radiation and driving its inhabitants away. (p. 700)

biodiversity *n.* the variety of organisms within an ecosystem. (p. 245)

biological weapon *n.* a bacterium or virus that can be used to harm or kill people, animals, or plants. (p. 175)

biome *n.* a regional ecosystem. (p. 65)

biosphere *n.* all the parts of the earth where plants and animals live, including the atmosphere, the lithosphere, and the hydrosphere. (p. 28)

birthrate *n.* the number of live births per total population, often expressed per thousand population. (p. 78)

blizzard *n.* a heavy snowstorm with winds of more than 35 miles per hour and reduced visibility of less than one-quarter mile. (p. 52)

Boxer Rebellion *n.* an uprising in China in 1900, spurred by angry Chinese militants, or Boxers, over foreign control; several hundred Europeans, Christians, and Chinese died. (p. 636)

British Columbia *n.* Canada's westernmost province, located within the Rocky Mountain range. (p. 169)

Buddhism *n.* a religion that originated in India about 500 B.C. and spread to China, where it grew into a major religion by A.D. 400. (p. 638)

C

calypso *n.* a style of music that began in Trinidad and combines musical elements from Africa, Spain, and the Caribbean. (p. 227)

Canadian Shield *n.* a northern part of the interior lowlands that is a rocky, flat region covering nearly two million square miles and encircling Hudson Bay. (p. 119)

canopy *n.* the area encompassing the tops of the trees in a rain forest, about 150 feet above ground. (p. 422)

capoeira *n.* a martial art and dance that developed in Brazil from Angolans who were taken there by the Portuguese from Africa. (p. 239)

Carnival *n.* the most colorful feast day in Brazil. (p. 239)

carrying capacity *n.* the number of organisms a piece of land can support without negative effects. (p. 82)

Carthage *n.* one of the great empires of ancient Africa, situated on a triangular peninsula on the Gulf of Tunis on the coast of the Mediterranean Sea. (p. 438)

cartographer *n.* a mapmaker. (p. 10)

cash crop *n.* a crop grown for direct sale, and not for use in a region, such as coffee, tea, and sugar in Africa. (p. 433)

caste system *n.* the Aryan system of social classes in India and one of the cornerstones of Hinduism in which each person is born into a caste and can only move into a different caste through reincarnation. (p. 571)

Caucasus *n.* a region that straddles the Caucasus Mountains and stretches between the Black and Caspian seas. (p. 385)

caudillo (kow•DEE•yoh) *n.* a military dictator or political boss. (p. 249)

Central Asia *n.* a region that includes the republics of Kazakhstan, Kyrgyzstan, Tajikistan, Turkmenistan, and Uzbekistan. (p. 346)

central business district (CBD) *n.* the core of a city, which is almost always based on commercial activity. (p. 89)

cerrado (seh•RAH•doh) *n.* a savanna that has flat terrain and moderate rainfall, which make it suitable for farming. (p. 202)

Chang Jiang *n.* (or Yangtze River) the longest river in Asia, flowing about 3,900 miles from Xizang (Tibet) to the East China Sea. (p. 621)

chaparral *n.* the term, in some locations, for a biome of drought-resistant trees. (p. 66)

Chechnya *n.* one of the republics that remains a part of Russia after the collapse of the Soviet Union despite independence movements and violent upheaval. (p. 386)

chemical weathering *n.* a process that changes rock into a new substance through interactions among elements in the air or water and the minerals in the rock. (p. 43)

chernozem *n.* black topsoil, one of the world's most fertile soils. (p. 345)

cholera *n.* a treatable infectious disease that can be fatal and is caused by a lack of adequate sanitation and a clean water supply. (p. 465)

city *n.* an area that is the center of business and culture and has a large population. (p. 87)

city-state *n.* an autonomous political unit made up of a city and its surrounding lands. (p. 289)

climate *n.* the typical weather conditions at a particular location as observed over time. (p. 50)

coalition *n.* an alliance. (p. 174)

Cold War *n.* the conflict between the United States and the Soviet Union after World War II, called "cold" because it never escalated into open warfare. (p. 363)

collective farm *n.* an enormous farm in the Soviet Union on which a large team of laborers were gathered to work together during Joseph Stalin's reign. (p. 364)

Columbian Exchange *n.* the movement of plants, animals, and diseases between the Eastern and Western hemispheres during the age of exploration. (p. 136)

command economy *n.* a type of economic system in which production of goods and services is determined by a central government, which usually owns the means of production. Also called a planned economy. (pp. 91, 364)

commodity *n.* an agricultural or mining product that can be sold. (p. 462)

communism *n.* a system in which the government holds nearly all political power and the means of production. (p. 83)

confederation *n.* a political union. (p. 156)

Confucianism *n.* a movement based on the teachings of Confucius, a Chinese philosopher who lived about 500 B.C.; Confucius stressed the importance of education in an ordered society in which one respects one's elders and obeys the government. (p. 638)

coniferous *adj.* another word for needleleaf trees. (p. 66)

constitutional monarchy *n.* a government in which the ruler's powers are limited by a constitution and the laws of a nation. (p. 580)

continent *n.* a landmass above water on the earth. (p. 27)

Continental Divide *n.* the line of the highest points in North America that marks the separation between rivers flowing eastward and westward. (p. 120)

continental drift *n.* the hypothesis that all continents were once joined into a supercontinent that split apart over millions of years. (p. 29)

continentality *n.* a region's distance from the moderating influence of the sea. (p. 350)

continental shelf *n.* the earth's surface from the edge of a continent to the deep part of the ocean. (p. 36)

convection *n.* the transfer of heat in the atmosphere by upward motion of the air. (p. 54)

copra *n.* the dried meat of coconuts. (p. 714)

core *n.* the earth's center, made up of iron and nickel; the inner core is solid, and the outer core is liquid. (p. 28)

crude oil *n.* petroleum that has not been processed. (p. 497)

Crusades *n.* a series of wars launched by European Christians in 1096 to capture the Holy Land (Palestine) from Muslims. (p. 291)

crust *n.* the thin rock layer making up the earth's surface. (p. 28)

cultural crossroad *n.* a place where various cultures cross paths. (p. 310)

cultural hearth *n.* the heartland or place of origin of a major culture; a site of innovation from which basic ideas, materials, and technology diffuse to other cultures. (pp. 72, 222)

culture *n.* the total of knowledge, attitudes, and behaviors shared by and passed on by members of a group. (p. 71)

cyclone *n.* a violent storm with fierce winds and heavy rain; the most extreme weather pattern of South Asia. (p. 558)

czar *n.* the emperor of Russia prior to the Russian Revolution of 1917 and the subsequent creation of the Soviet Union in 1922. (p. 362)

D

Dead Sea *n.* a landlocked salt lake between Israel and Jordan that is so salty that almost nothing can live in its waters; it is 1,349 feet below sea level, making it the lowest place on the exposed crust of the earth. (p. 489)

debt-for-nature swap *n.* a debt-reducing deal wherein an organization agrees to pay off a certain amount of government debt in return for government protection of a certain portion of rain forest. (p. 247)

deciduous *adj.* a named characteristic of broadleaf trees, such as maple, oak, birch, and cottonwood. (p. 66)

deforestation *n.* the cutting down and clearing away of trees and forests. (p. 246)

delta *n.* a fan-like landform made of deposited sediment, left by a river that slows as it enters the ocean. (p. 43)

democracy *n.* a type of government in which citizens hold political power either directly or through elected representatives. (p. 83)

desalinization *n.* the removal of salt from ocean water. (p. 496)

desertification *n.* an expansion of dry conditions to moist areas that are next to deserts. (p. 424)

dialect *n.* a version of a language that reflects changes in speech patterns due to class, region, or cultural changes. (p. 73)

dictatorship *n.* a type of government in which an individual or a group holds complete political power. (p. 83)

diffusion *n.* the spread of ideas, inventions, or patterns of behavior to different societies. (p. 72)

dike *n.* an earthen bank used to direct or prevent the passage of water. (p. 282)

distance decay *n.* a term referring to the concept that increasing distances between places tend to reduce interactions among them. (p. 389)

diversify *v.* to increase the variety of products in a country's economy; to promote manufacturing and other industries in order to achieve growth and stability. (p. 462)

Dome of the Rock *n.* a shrine in Jerusalem, located on the Temple Mount, which houses the spot where Muslims believe Muhammad rose into heaven and where Jews believe Abraham prepared to sacrifice his son Isaac to God. (p. 511)

Dominion of Canada *n.* the loose confederation of Ontario (Upper Canada), Quebec (Lower Canada), Nova Scotia, and New Brunswick, created by the British North America Act in 1867. (p. 156)

drainage basin *n.* an area drained by a major river and its tributaries. (p. 33)

drip irrigation *n.* the practice of using small pipes that slowly drip water just above ground to conserve water to use for crops. (p. 496)

drought *n.* a long period without rain or with very minimal rainfall. (p. 53)

dynasty *n.* a series of rulers from the same family. (p. 635)

E

earthquake *n.* a sometimes violent movement of the earth, produced when tectonic plates grind or slip past each other at a fault. (p. 39)

economic system *n.* the way people produce and exchange goods. (p. 91)

economic tiger *n.* a country with rapid economic growth due to cheap labor, high technology, and aggressive exports. (p. 645)

economy *n.* the production and exchange of goods and services among a group of people. (p. 91)

ecosystem *n.* an interdependent community of plants and animals. (p. 65)

El Niño (el NEEN•YOH) *n.* a weather pattern created by the warming of the waters off the west coast of South America, which pushes warm water and heavy rains toward the Americas and produces drought conditions in Australia and Asia. (p. 57)

entrepreneur *n.* a person who starts and builds a business. (p. 575)

epicenter *n.* the point on the earth's surface that corresponds to the location in the earth where an earthquake begins. (p. 39)

equator *n.* the imaginary line that encircles the globe, dividing the earth into northern and southern halves. (p. 6)

equinox *n.* each of the two days in a year on which day and night are equal in length; marks the beginning of spring and autumn. (p. 49)

erosion *n.* the result of weathering on matter, created by the action of wind, water, ice, or gravity. (p. 43)

escarpment *n.* a steep slope with a nearly flat plateau on top. (p. 417)

estuary *n.* a broadened seaward end of a river, where the river's currents meet the ocean's tides. (p. 563)

ethnic cleansing *n.* the policy of trying to eliminate an ethnic group. (p. 320)

ethnic group *n.* a group of people who share language, customs, and a common heritage. (p. 71)

Euphrates River *n.* a river of Southwest Asia, which supported several ancient civilizations and flows through parts of Turkey, Syria, and Iraq and empties into the Persian Gulf. (p. 489)

Eurasia *n.* the combined continent of Europe and Asia. (p. 346)

euro *n.* a common currency proposed by the European Union for its member nations. (p. 305)

European Environmental Agency *n.* an agency that provides the European Union with reliable information about the environment. (p. 324)

Everglades *n.* a large subtropical swampland in Florida of about 4,000 square miles. (p. 126)

export *n.* a product or good that is sold from one economy to another. (p. 140)

F

Fang sculpture *n.* carved boxes containing the skulls and bones of deceased ancestors, created by the Fang, who live in Gabon, southern Cameroon, and Equatorial Guinea. (p. 451)

fault *n.* a fracture in the earth's crust. (p. 39)

federal republic *n.* a nation whose powers are divided among the federal, or national, government and various state and local governments. (p. 139)

feudalism *n.* a political system prevailing in Europe from about the 9th to about the 15th centuries in which a king allowed nobles the use of his land in exchange for their military service and their protection of the land. (p. 297)

fertility rate *n.* the average number of children a woman of childbearing years would have in her lifetime, if she had children at the current rate for her country. (p. 78)

First Nations *n.* a group of Canada's Native American people. (p. 159)

fjord (fyawrd) *n.* a long, narrow, deep inlet of the sea between steep slopes. (p. 273)

folk art *n.* handmade items, such as pottery, woodcarving, and traditional costumes, produced by rural people with traditional lifestyles, instead of by professional artists. (p. 314)

fossil water *n.* water pumped from underground aquifers. (p. 496)

free enterprise *n.* an economic system in which private individuals own most of the resources, technology, and businesses, and can operate them for profit with little control from the government. (p. 140)

frontier *n.* the free, open land in the American West that was available for settlement. (p. 137)

G

Ganges River *n.* river in South Asia; an important water resource flowing more than 1,500 miles from its source in a Himalayan glacier to the Bay of Bengal. (p. 560)

Gaza Strip *n.* a territory along the Mediterranean Sea just northeast of the Sinai Peninsula; part of the land set aside for Palestinians, which was occupied by Israel in 1967. (p. 527)

Geographic Information System (GIS) *n.* technology that uses digital map information to create a databank; different "data layers" can be combined to produce specialized maps. GIS allows geographers to analyze different aspects of a specific place to solve problems. (p. 13)

geography *n.* the study of the distribution and interaction of physical and human features on the earth. (p. 5)

glaciation *n.* the changing of landforms by slowly moving glaciers. (p. 44)

glacier *n.* a large, long-lasting mass of ice that moves because of gravity. (p. 44)

global economy *n.* the merging of regional economies in which nations become dependent on each other for goods and services. (p. 666)

global network *n.* a worldwide interconnected group. (p. 173)

global warming *n.* the buildup of carbon dioxide in the atmosphere, preventing heat from escaping into space and causing rising temperatures and shifting weather patterns. (p. 246)

globe *n.* a three-dimensional representation of the earth. (p. 10)

Gobi Desert *n.* a desert located in northern China and southeast Mongolia, and a prime area for finding dinosaur fossils. (p. 627)

Golan Heights *n.* a hilly plateau overlooking the Jordan River and the Sea of Galilee; a strategic location that has been the site of conflict in Southwest Asia for decades. (p. 487)

Gorée Island *n.* an island off the coast of Senegal that served as a major departure point for slaves during the slave trade. (p. 442)

Great Barrier Reef *n.* a 1,250-mile chain of more than 2,500 reefs and islands along Australia's northeast coast, containing some 400 species of coral. (p. 692)

Great Game *n.* a struggle between the British Empire and the Russian Empire for control of Central Asia in the 19th century. (p. 376)

Great Kanto Earthquake *n.* an earthquake in 1923 in Japan that killed an estimated 140,000 people and left the city of Tokyo in ruins. (p. 662)

Great Lakes *n.* a group of five freshwater lakes of central North America between the United States and Canada; the lakes are Huron, Ontario, Michigan, Erie, and Superior. (p. 121)

Great Plains *n.* a vast grassland of central North America that is largely treeless and ascends to 4,000 feet above sea level. (p. 119)

Great Zimbabwe *n.* a city established in what is now Zimbabwe by the Shona around 1000; it became the capital of a thriving gold-trading area. (p. 453)

greenhouse effect *n.* the layer of gases released by the burning of coal and petroleum that traps solar energy, causing global temperature to increase. (p. 58)

Green Revolution *n.* an agricultural program launched by scientists in the 1960s to develop higher-yielding grain varieties and improve food production by incorporating new farming techniques. (p. 569)

Gross Domestic Product (GDP) *n.* the value of only goods and services produced within a country in a period of time. (p. 95)

Gross National Product (GNP) *n.* the total value of all goods and services produced by a country in a period of time. (p. 94)

ground water *n.* the water held under the earth's surface, often in and around the pores of rock. (p. 33)

guest worker *n.* a largely unskilled laborer, often an immigrant from South and East Asia, brought in to the oil-booming countries to fill job openings that the region's native peoples find culturally or economically unacceptable. (p. 525)

H

hemisphere *n.* each half of the globe. (p. 6)

high islands *n.* Pacific islands created by volcanoes. (p. 691)

Himalaya Mountains *n.* a mountain range in South Asia that includes Mount Everest, the world's tallest mountain peak. (p. 551)

Hinduism *n.* the dominant religion of India. (p. 560)

Holocaust *n.* the Nazi program of mass murder of European Jews during World War II. (p. 298)

Huang He (hwahng huh) *n.* a river in northern China, also called the Yellow River, that starts in the Kunlun Mountains and winds east for about 3,000 miles, emptying into the Yellow Sea. (p. 621)

human resources *n.* the skills and talents of employed people. (p. 531)

humus *n.* organic material in soil. (p. 45)

hurricane *n.* a storm that forms over warm, tropical ocean waters. (p. 51)

hydrologic cycle *n.* the continuous circulation of water among the atmosphere, the oceans, and the earth. (p. 32)

hydrosphere *n.* the waters comprising the earth's surface, including oceans, seas, rivers, lakes, and vapor in the atmosphere. (p. 28)

I

Ijsselmeer (EYE•suhl•MAIR) *n.* a freshwater lake separated from the North Sea by a dike and bordered by polders. (p. 283)

illiteracy *n.* the inability to read or write. (p. 593)

Inca *n.* a member of the Quechen peoples of South America who built a civilization in the Andes Mountains in the 15th and 16th centuries. (p. 230)

Indochina *n.* a French colony comprised of Cambodia, Laos, and Vietnam; it won independence from France in 1954. (p. 707)

industrialization *n.* the growth of industry in a country or a society. (p. 730)

Indus Valley civilization *n.* the largest of the world's first civilizations in what is now Pakistan; this was a highly developed urban civilization, lasting from 2500 B.C. to about 1500 B.C. (p. 573)

infant mortality rate *n.* the number of deaths among infants under age one as measured per thousand live births. (p. 79)

infrastructure *n.* the basic support systems needed to keep an economy going, including power, communications, transportation, water, sanitation, and education systems. (pp. 94, 177, 212)

innovation *n.* taking existing elements of society and creating something new to meet a need. (p. 72)

Institutional Revolutionary Party (PRI) *n.* the political party introduced in 1929 in Mexico that helped to introduce democracy and maintain political stability for much of the 20th century. (p. 218)

Islam *n.* a monotheistic religion based on the teachings of the prophet Muhammad, and the biggest cultural and religious influence in North Africa. (pp. 439, 503)

J

Jakota Triangle *n.* a zone of prosperity during the 1980s and early 1990s—Japan, South Korea, and Taiwan. (p. 666)

Jordan River *n.* a river that serves as a natural boundary between Israel and Jordan, flowing from the mountains of Lebanon with no outlet to the Mediterranean Sea. (p. 489)

junta (HOON•tah) *n.* a government run by generals after a military takeover. (p. 249)

K

Kashmir *n.* a region of northern India and Pakistan over which several destructive wars have been fought. (p. 574)

Khmer Empire *n.* a powerful empire that lasted roughly from the 9th to the 15th centuries in what is now Cambodia. (p. 706)

King Leopold II *n.* the Belgian king who opened up the African interior to European trade along the Congo River and by 1884 controlled the area known as the Congo Free State. (p. 449)

KLA (Kosovo Liberation Army) *n.* a group that fought against Serbian attempts to control the region of Kosovo in the 1990s. (p. 321)

Kunlun Mountains *n.* mountains located in the west of China that are the source of two of China's great rivers, the Huang He (Yellow) and the Chang Jiang (Yangtze). (p. 619)

Kurds *n.* an ethnic group in Southwestern Asia that has occupied Kurdistan, located in Turkey, Iraq, and Iran, for about a thousand years, and who have been involved in clashes with these three countries over land claims for most of the 20th century. (p. 516)

L

landfill *n.* a method of solid waste disposal in which refuse is buried between layers of dirt in order to fill in or reclaim low-lying ground. (p. 631)

landform *n.* a naturally formed feature on the surface of the earth. (p. 33)

landlocked *adj.* having no outlet to the sea. (p. 84)

land reform *n.* the process of breaking up large landholdings to attain a more balanced land distribution among farmers. (pp. 250, 569)

Land Rights Act of 1976 *n.* a special law passed for Aboriginal rights in Australia giving Aboriginal people the right to claim land in the Northern Territory. (p. 728)

Landsat *n.* a series of satellites that orbit more than 100 miles above the earth. Each satellite picks up data in an area 115 miles wide. (p. 12)

latitude (lines) *n.* a set of imaginary lines that run parallel to the equator, and that are used in locating places north or south. The equator is labeled the zero-degree line for latitude. (p. 6)

lava *n.* magma that has reached the earth's surface. (p. 40)

lithosphere *n.* the solid rock portion of the earth's surface. (p. 28)

llanos (LAH•nohs) *n.* a large, grassy, treeless area in South America, used for grazing and farming. (p. 202)

lock *n.* a section of a waterway with closed gates where water levels are raised or lowered, through which ships pass. (p. 129)

loess (LOH•uhs) *n.* wind-blown silt and clay sediment that produces very fertile soil. (p. 44)

longitude (lines) *n.* a set of imaginary lines that go around the earth over the poles, dividing it east and west. The prime meridian is labeled the zero-degree line for longitude. (p. 6)

Louisiana Purchase *n.* the territory, including the region between the Mississippi River and the Rocky Mountains, that the United States purchased from France in 1803. (p. 136)

low islands *n.* Pacific islands made of coral reefs. (p. 691)

M

Mabo Case *n.* in Australia, the law case that upheld Aboriginal Eddie Mabo's land claim by which the Court recognized that Aboriginal people had owned land before the British arrived. (p. 728)

Mackenzie River *n.* Canada's longest river, which is part of a river system that flows across the Northwest Territories to the Arctic Ocean. (p. 121)

magma *n.* the molten rock material formed when solid rock in the earth's mantle or crust melts. (p. 28)

malaria *n.* an infectious disease of the red blood cells, carried by mosquitoes, that is characterized by chills, fever, and sweating. (p. 466)

mandala *n.* in Tibetan Buddhism, a geometric design that symbolizes the universe and aids in meditation. (p. 583)

mandala *n.* a state organized as a ring of power around a central court, which often changed in size over time, and which was used instead of borders in early Southeast Asian states. (p. 705)

mantle *n.* a rock layer about 1,800 miles thick that is between the earth's crust and the earth's core. (p. 28)

Maori *n.* the first settlers of New Zealand, who had migrated from Polynesia more than 1,000 years ago. (p. 719)

Mao Zedong *n.* the leader of the Communists in China who defeated the Nationalists in 1949; he died in 1976. (p. 636)

map projection *n.* a way of mapping the earth's surface that reduces distortion caused by converting three dimensions into two dimensions. (p. 10)

map *n.* a two-dimensional graphic representation of selected parts of the earth's surface. (p. 10)

maquiladora *n.* a factory in Mexico that assembles imported materials into finished goods for export. (p. 220)

market economy *n.* a type of economic system in which production of goods and services is determined by the demand from consumers. Also called a demand economy or capitalism. (pp. 91, 313)

Massif Central (ma•SEEF sahn•TRAHL) *n.* the uplands of France, which account for about one-sixth of French lands. (p. 275)

Mecca *n.* the holiest city of Islam, located in Saudi Arabia, where people make pilgrimages to fulfill Islamic religious duty. (p. 503)

mechanical weathering *n.* natural processes that break rock into smaller pieces. (p. 42)

megalopolis *n.* a region in which several large cities and surrounding areas grow together. (p. 146)

Melanesia *n.* a region in Oceania meaning "black islands." (p. 713)

Meseta (meh•SEH•tah) *n.* the central plateau of Spain. (p. 275)

Mesopotamia *n.* a region in Southwest Asia between the Tigris and the Euphrates rivers, which was the location of some of the earliest civilizations in the world; part of the cultural hearth known as the Fertile Crescent. (p. 516)

métis (may•TEES) *n.* a person of mixed French-Canadian and Native American ancestry. (p. 161)

metropolitan area *n.* a functional area including a city and its surrounding suburbs and exurbs, linked economically. (pp. 87, 148)

microcredit *n.* a small loan available to poor entrepreneurs, to help small businesses grow and raise living standards. (p. 575)

Micronesia *n.* one of three regions in Oceania, meaning "tiny islands." (p. 713)

Midwest *n.* the region that contains the 12 states of the north-central United States. (p. 147)

migration *n.* the movement of peoples within a country or region. (p. 135)

Mississippi River *n.* a major river that runs north-south almost the length of the United States, from Minnesota to the Gulf of Mexico, and is part of the longest river system on the continent. (p. 121)

mistral (MIHS•truhl) *n.* a cold, dry wind from the north. (p. 279)

Mobutu Sese Seko *n.* the leader of Zaire, which is now the Democratic Republic of the Congo, from its independence in the 1960s until 1997. He brought the country's businesses under national control, profited from the reorganization, and used the army to hold power. (p. 450)

monarchy *n.* a type of government in which a ruling family headed by a king or queen holds political power and may or may not share the power with citizen bodies. (p. 83)

monsoon *n.* a seasonal wind, especially in South Asia. (p. 558)

moraine *n.* a ridge or hill of rock carried and finally deposited by a glacier. (p. 44)

mortality rate *n.* the number of deaths per thousand. (p. 79)

mosque *n.* an Islamic place of worship, where Muslims pray facing toward the holy city of Mecca. (p. 504)

Mount Kilimanjaro *n.* a volcano in Tanzania in Africa, also Africa's highest peak. (p. 417)

Mughal Empire *n.* the Muslim empire established by the early 1500s over much of India, which brought with it new customs that sometimes conflicted with those of native Hindus. (p. 568)

Muhammad *n.* the founder and a prophet of Islam, who lived part of his life in the city of Mecca. (p. 503)

multinational *n.* a corporation that engages in business worldwide. (p. 142)

Mutapa Empire *n.* a state founded in the 15th century by a man named Mutota and that extended throughout all of present-day Zimbabwe except the eastern part. (p. 453)

N

Nagorno-Karabakh *n.* the mountainous area of Azerbaijan, fought over by Armenia and Azerbaijan. (p. 386)

nation *n.* a group of people with a common culture living in a territory and having a strong sense of unity. (p. 83)

nationalism *n.* the belief that people should be loyal to their nation, the people with whom they share land, culture, and history. (p. 297)

nation-state *n.* the name of a territory when a nation and a state occupy the same territory. (p. 83)

natural resource *n.* a material on or in the earth, such as a tree, fish, or coal, that has economic value. (p. 93)

needleleaf *adj.* characteristic of trees like pine, fir, and cedar, found in northern regions of North America. (p. 66)

Nelson Mandela *n.* one of the leaders of the African National Congress who led a struggle to end apartheid and was elected president in 1994 in the first all-race election in South Africa. (p. 454)

New England *n.* the six northern states in the Northeast United States—Maine, Vermont, New Hampshire, Massachusetts, Rhode Island, and Connecticut. (p. 145)

Niger delta *n.* delta of the Niger River and an area of Nigeria with rich oil deposits. (p. 424)

Nile River *n.* the world's longest river, flowing over 4,000 miles through the Sudan Basin into Uganda, Sudan, and Egypt. (p. 416)

nomad *n.* a person with no permanent home who moves according to the seasons from place to place in search of food, water, and grazing land. (pp. 127, 378)

nonviolent resistance *n.* a movement that uses all means of protest except violence. (p. 568)

Nordic countries *n.* countries of northern Europe, including Denmark, Finland, Iceland, Norway, and Sweden. (p. 302)

NAFTA (North American Free Trade Agreement) *n.* an important trade agreement creating a huge zone of cooperation on trade and economic issues in North America. (p. 220)

North Atlantic Drift *n.* a current of warm water from the Tropics. (p. 278)

Nunavut *n.* one of Canada's territories and home to many of Canada's Inuit; it was carved out of the eastern half of the Northwest Territories in 1999. (p. 169)

O

oasis *n.* a place where water from an aquifer has reached the surface; it supports vegetation and wildlife. (pp. 421, 492)

Oceania *n.* the group of islands in the Pacific, including Melanesia, Micronesia, and Polynesia. (p. 690)

Olduvai Gorge *n.* a site of fossil beds in northern Tanzania, containing the most continuous known record of humanity over the past 2 million years, including fossils from 65 hominids. (p. 431)

oligarchy (AHL•ih•GAHR•kee) *n.* a government run by a few persons or a small group. (p. 249)

"one-commodity" country *n.* a country that relies on one principal export for much of its earnings. (p. 462)

Ontario *n.* one of Canada's Core Provinces. (p. 167)

OPEC *n.* the Organization of Petroleum Exporting Countries, a group established in 1960 by some oil-producing nations to coordinate policies on selling petroleum products. (p. 505)

Orinoco River *n.* a river mainly in Venezuela and part of South America's northernmost river system. (p. 202)

outback *n.* the dry, unpopulated inland region of Australia. (p. 697)

outrigger canoe *n.* a small ship used in the lagoons of islands where Pacific Islanders settled. (p. 699)

ozone *n.* a chemical created when burning fossil fuels react with sunlight; a form of oxygen. (p. 325)

P

Pacific Rim *n.* an economic and social region including the countries surrounding the Pacific Ocean, extending clockwise from New Zealand in the western Pacific to Chile in the eastern Pacific and including the west coast of the United States. (p. 645)

pakehas n. a Maori term for white people, for the New Zealanders of European descent. (p. 722)

Palestine Liberation Organization (PLO) *n.* a group formed in the 1960s to regain the Arab land in Israel for Palestinian Arabs. (p. 513)

Palestinians *n.* a displaced group of Arabs who lived or still live in the area formerly called Palestine and now called Israel. (p. 527)

pampas (PAHM·puhs) *n.* a vast area of grassland and rich soil in south-central South America. (p. 202)

Panama Canal *n.* a ship canal cut through Panama connecting the Caribbean Sea with the Pacific Ocean. (p. 226)

pandemic *n.* a disease affecting a large population over a wide geographic area. (p. 435)

Paraná River *n.* a river in central South America and one of its three major river systems, originating in the highlands of southern Brazil, travelling about 3,000 miles south and west. (p. 203)

parliament *n.* a representative lawmaking body whose members are elected or appointed and in which legislative and executive functions are combined. (pp. 158, 303)

parliamentary government *n.* a system where legislative and executive functions are combined in a legislature called a parliament. (p. 158)

particulate *n.* a very small particle of liquid or solid matter. (p. 324)

partition *n.* separation; division into two or more territorial units having separate political status. (p. 574)

pastoral lease *n.* in Australia, a huge chunk of land still owned by the government; ranchers take out leases, renting the land from the government. (p. 729)

PCB *n.* an industrial compound that accumulates in animal tissue and can cause harmful effects and birth defects; PCBs were banned in the United States in 1977. (p. 631)

peat *n.* partially decayed plant matter found in bogs. (p. 277)

penal colony *n.* a place to send prisoners. (p. 718)

per capita income *n.* the average amount of money earned by each person in a political unit. (p. 94)

permafrost *n.* permanently frozen ground. (pp. 63, 123)

polder *n.* land that is reclaimed from the sea or other body of water by diking and drainage. (p. 282)

Polynesia *n.* one of three regions in Oceania, meaning "many islands." (p. 713)

population density *n.* the average number of people who live in a measurable area, reached by dividing the number of inhabitants in an area by the amount of land they occupy. (p. 81)

population pyramid *n.* a graphic device that shows gender and age distribution of a population. (p. 79)

postindustrial economy *n.* an economic phase in which manufacturing no longer plays a dominant role. (p. 142)

Prairie Provinces *n.* in Canada, the provinces west of Ontario and Quebec—Manitoba, Saskatchewan, and Alberta. (p. 168)

precipitation *n.* falling water droplets in the form of rain, sleet, snow, or hail. (p. 50)

prevailing westerlies *n.* winds that blow from west to east. (p. 124)

prime meridian *n.* the imaginary line at zero meridian used to measure longitude east to west, and dividing the earth's east and west halves; also called the Greenwich Meridian because it passes through Greenwich, England. (p. 6)

prime minister *n.* the head of a government; the majority party's leader in parliament. (p. 158)

privatization *n.* the selling of government-owned business to private citizens. (p. 388)

province *n.* a political unit. (p. 156)

pull factor *n.* a factor that draws or attracts people to another location. (pp. 81, 211)

push factor *n.* a factor that causes people to leave their homelands and migrate to another region. (pp. 81, 211, 730)

Pyongyang *n.* the largest city in North Korea, with more than 2.5 million people. (p. 650)

Q

Qin Ling Mountains *n.* mountains in southeastern and east-central China; they divide the northern part of China from the southern part. (p. 619)

Quebec *n.* one of Canada's Core Provinces. (p. 167)

Quechua (KEHCH·wuh) *n.* the language of the Inca Empire, now spoken in the Andes highlands. (p. 231)

R

rai n. a kind of popular Algerian music developed in the 1920s by poor urban children that is fast-paced with danceable rhythms; was sometimes used as a form of rebellion to expose political unhappiness. (p. 440)

rain forest *n.* a forest region located in the Tropical Zone with a heavy concentration of different species of broadleaf trees. (pp. 66, 207)

rain shadow *n.* the land on the leeward side of hills or mountains that gets little rain from the descending dry air. (p. 51)

raj n. the period of British rule in India, which lasted for nearly 90 years, from 1857 to 1947. (p. 568)

Ramadan *n.* an Islamic practice of month-long fasting from sunup to sundown. (p. 576)

rate of natural increase *n.* also called population growth rate—the rate at which population is growing, found by subtracting the mortality rate from the birthrate. (p. 79)

recession *n.* an extended period of decline in general business activity. (p. 667)

Red Army *n.* the name of the Soviet Union's military. (p. 371)

refinery *n.* a place where crude oil is converted into useful products. (p. 497)

Reformation *n.* a movement in Western Europe beginning in 1517, when many Christians broke away from the Catholic Church and started Protestant churches; this led to mutual hostility and religious wars that tore apart Europe. (p. 297)

reggae *n.* a style of music that developed in Jamaica in the 1960s and is rooted in African, Caribbean, and American music, often dealing with social problems and religion. (p. 227)

relative location *n.* describes a place in relation to other places around it. (p. 6)

relief *n.* the difference in elevation of a landform from the lowest point to the highest point. (p. 36)

religion *n.* the belief in a supernatural power or powers that are regarded as the creators and maintainers of the universe, as well as the system of beliefs itself. (p. 75)

Renaissance *n.* a time of renewed interest in learning and the arts that lasted from the 14th through 16th centuries; it began in the Italian city-states and spread north to all of Europe. (p. 291)

representative democracy *n.* a government in which the people rule through elected representatives. (p. 139)

republic *n.* a government in which citizens elect representatives to rule on their behalf. (p. 290)

reserve *n.* public land set aside for native peoples by the government. (p. 162)

Richter scale *n.* a way to measure information collected by seismographs to determine the relative strength of an earthquake. (p. 40)

rift valley *n.* a long, thin valley created by the moving apart of the continental plates, present in East Africa, stretching over 4,000 miles from Jordan in Southwest Asia to Mozambique in Southern Africa. (p. 416)

Ring of Fire *n.* the chain of volcanoes that lines the Pacific Rim. (pp. 41, 661)

Rocky Mountains *n.* a major mountain system of the United States and Canada, extending 3,000 miles from Alaska south to New Mexico. (p. 119)

Rub al Khali *n.* also known as the Empty Quarter; one of the largest sandy deserts in the world, covering about 250,000 square miles; located on the Arabian Peninsula. (p. 491)

Russian Revolution *n.* the revolt of 1917, in which the Russian Communist Party, led by V. I. Lenin, took control of the government from the czars. (p. 363)

runoff *n.* rainfall not absorbed by soil, which can carry pesticides and fertilizers from fields into rivers, endangering the food chain. (p. 353)

S

Sahara *n.* the largest desert in the world, stretching 3,000 miles across the African continent, from the Atlantic Ocean to the Red Sea, and measuring 1,200 miles from north to south. (p. 420)

Sahel *n.* a narrow band of dry grassland, running east to west on the southern edge of the Sahara, that is used for farming and herding. (p. 424)

St. Lawrence Seaway *n.* North America's most important deepwater ship route, connecting the Great Lakes to the Atlantic Ocean by way of the St. Lawrence River. (p. 129)

St. Petersburg *n.* the old capital of Russia, established by Peter the Great, who moved it there from Moscow because St. Petersburg provided direct access by sea to Western Europe. (p. 362)

salt flat *n.* flat land made of chemical salts that remain after winds evaporate the moisture in the soil. (p. 492)

samba *n.* a Brazilian dance with African influences. (p. 239)

samurai *n.* a professional soldier in Japan who served the interests of landowners and clan chiefs. (p. 651)

satellite nation *n.* a nation dominated by another country. (p. 312)

savanna *n.* the term for the flat, grassy, mostly treeless plains in the tropical grassland region. (p. 66)

seawork *n.* a structure used to control the sea's destructive impact on human life. (p. 283)

sectionalism *n.* when people place their loyalty to their region, or section, above loyalty to the nation. (p. 136)

sediment *n.* small pieces of rock produced by weathering processes. (p. 42)

seismograph (SYZ•muh•GRAF) *n.* a device that measures the size of the waves created by an earthquake. (p. 39)

Seoul *n.* the largest city in South Korea, with a population of more than ten million people. (p. 650)

Serengeti *n.* an area of East Africa, containing some of the best grasslands in the world and many grazing animals. (p. 422)

service industry *n.* any kind of economic activity that produces a service rather than a product. (p. 142)

Sherpa *n.* a person of Tibetan ancestry in Nepal, who serves as the traditional mountain guide of the Mount Everest region. (p. 582)

Shi'ite *n.* one of the two main branches of Islam including most Iranians and some populations of Iraq and Afghanistan. (p. 516)

shogun *n.* the general of the emperor's army with the powers of a military dictator, a position created by the Japanese emperor in 1192 after a struggle between two powerful clans. (p. 651)

Siberia *n.* a region of central and eastern Russia, stretching from the Ural Mountains to the Pacific Ocean, known for its mineral resources and for being a place of political exile. (p. 349)

Siddhartha Gautama *n.* the founder of Buddhism and known as the Buddha, born in southern Nepal in the sixth century B.C. (p. 582)

Silicon Glen *n.* the section of Scotland between Glasgow and Edinburgh, named for its high concentration of high-tech companies. (p. 305)

Silk Road *n.* the 4,000-mile route between China and the Mediterranean Sea, named for the costly silk acquired in China. (p. 375)

silt *n.* loose sedimentary material containing very small rock particles, formed by river deposits and very fertile. (p. 426)

Sinhalese *n.* an Indo-Aryan people who crossed the strait separating India and Sri Lanka in the sixth century B.C. and who created an advanced civilization there, adopting Buddhism. (p. 584)

sirocco (suh•RAHK•oh) *n.* a hot, steady south wind that blows from North Africa across the Mediterranean Sea into southern Europe, mostly in spring. (p. 279)

slash-and-burn *adj.* a way of clearing fields for planting by cutting trees, brush, and grasses and burning them. (p. 210)

smart growth *n.* the efficient use and conservation of land and other resources. (p. 178)

smog *n.* a brown haze that occurs when gases released by burning fossil fuels react with sunlight. (p. 324)

society *n.* a group that shares a geographic region, a common language, and a sense of identity and culture. (p. 71)

soil *n.* the loose mixture of weathered rock, organic matter, air, and water that supports plant growth. (p. 45)

solar system *n.* consists of the sun and nine known planets, as well as other celestial bodies that orbit the sun. (p. 27)

solstice *n.* either of two times of year when the sun's rays shine directly overhead at noon at the furthest points north or south, and that mark the beginning of summer and winter; in the Northern Hemisphere, the summer solstice is the longest day and the winter solstice the shortest. (p. 49)

South, the *n.* a region that covers about one-fourth of the land area of the United States and contains more than one-third of its population. (p. 148)

South Slav *n.* a person who migrated from Poland or Russia and settled in the Balkan Peninsula around 500. (p. 319)

Spanish conquest *n.* the conquering of the Native Americans by the Spanish. (p. 217)

sphere of influence *n.* a method of dividing foreign control in China, after the country was forced to sign a series of treaties granting special privileges to the Europeans. China was partitioned for control by Britain, France, Germany, and Russia, among others. (p. 636)

state *n.* a political term describing an independent unit that occupies a specific territory and has full control of its internal and external affairs. (p. 83)

stateless nation *n.* a nation of people that does not have a territory to legally occupy, like the Palestinians, Kurds, and Basques. (p. 526)

stateless society *n.* one in which people use lineages, or families whose members are descended from a common ancestor, to govern themselves. (p. 443)

steppe *n.* the term used for the temperate grassland region in the Northern Hemisphere. (p. 66)

Stolen Generation *n.* in Australia, what Aboriginal people today call the 100,000 mixed-raced children who were taken by the government and given to white families to promote assimilation. (p. 728)

storm surge *n.* high water level brought by a cyclone that swamps low-lying areas. (p. 562)

strategic commodity *n.* a resource so important that nations will go to war to ensure its steady supply. (p. 529)

subcontinent *n.* a landmass that is like a continent, only smaller, such as South Asia, which is called the Indian subcontinent. (p. 551)

subsistence activity *n.* an activity in which a family produces only the food, clothing, and shelter they themselves need. (p. 714)

suburb *n.* a political unit or community touching the borders of the central city or touching other suburbs that touch the city. (pp. 87, 138)

sultan *n.* a ruler of a Muslim country. (p. 585)

summer monsoon *n.* the season when winds blow from the southwest across the Indian Ocean toward South Asia, from June through September, with winds stirring up powerful storms and causing severe flooding. (p. 597)

Sunni *n.* one of the two main branches of Islam, comprising about 83 percent of all Muslims, including those in Turkey, Iraq, and Afghanistan. (p. 516)

supra *n.* Georgian (Russian) term for dinner party, with many dishes and courses, toasts, and short speeches. (p. 374)

sustainable community *n.* a community where residents can live and work in harmony with the environment. (p. 178)

sweatshop *n.* a workplace where people work long hours for low pay under poor conditions to enrich manufacturers. (p. 667)

T

taiga *n.* a nearly continuous belt of evergreen coniferous forests across the Northern Hemisphere, in North America and Eurasia. (p. 351)

Taklimakan Desert *n.* a desert located in western China between the Tian Shan and Kunlun mountains. (p. 627)

Taliban *n.* a strict Muslim group in Afghanistan that has imposed rigid rules on society, including prescribed clothing styles for both men and women, restrictions on the appearance of women in public places, and regulations on television, music, and videos. (p. 517)

Tamil *n.* a Dravidian Hindu, who arrived in Sri Lanka in the fourth century, settling in the north while the Sinhalese moved further south. (p. 584)

Taoism *n.* a philosophy based on the book *Tao Te Ching* and the teachings of Lao-Tzu, who lived in China in the sixth century B.C. and believed in preserving and restoring harmony in the individual, with nature, and in the universe, with little interference from the government. (p. 638)

taro *n.* a tropical Asian plant with a starchy root, which can be eaten as a boiled vegetable or made into breads, puddings, or a paste called poi. (p. 715)

tectonic plate *n.* an enormous moving shelf that forms the earth's crust. (p. 37)

Tenochtitlan (teh•NOH•tee•TLAHN) *n.* the ancient Aztec capital, site of Mexico City today. (p. 217)

terpen *n.* high earthen platforms used in seaworks. (p. 283)

terraced farming *n.* an ancient technique for growing crops on hillsides or mountain slopes, using step-like horizontal fields cut into the slopes. (p. 211)

terrorism *n.* the use of, or threatened use of, force or violence against individuals or property for the purpose of intimidating or causing fear for political or social ends. (p. 173)

theocratic *adj.* a form of government in which religious leaders control the government, relying on religious law and consultation with religious scholars. (p. 504)

Three Gorges Dam *n.* a dam begun in the late 20th century on the Chang Jiang in China, to help control flooding, generate power, and allow ships to sail farther into China. (p. 628)

Three Kingdoms *n.* the kingdoms formed in the peninsula of Korea by A.D. 300—Koguryo in the northeast, Paekche in the southwest, and Silla in the southeast. (p. 647)

Tigris River *n.* one of the most important rivers of Southwest Asia; it supported several ancient river valley civilizations, and flows through parts of Turkey, Syria, and Iraq. (p. 489)

tornado *n.* a powerful funnel-shaped column of spiraling air. (p. 51)

topographic map *n.* a general reference map; a representation of natural and man-made features on the earth. (p. 11)

topography *n.* the combined characteristics of landforms and their distribution in a region. (p. 36)

Transcaucasia *n.* a region that consists of the republics of Armenia, Azerbaijan, and Georgia; located between the Caucasus Mountains and the borders of Turkey and Iran. (p. 346)

Trans-Siberian Railroad *n.* a railroad that would eventually link Moscow to the Pacific port of Vladivostok; built between 1891 and 1903. (p. 355)

Treaty of Tordesillas *n.* a treaty between Spain and Portugal in 1494 that gave Portugal control over the land that is present-day Brazil. (p. 236)

Treaty of Waitangi *n.* the treaty signed by the British and Maori in 1840 giving Britain control over New Zealand. (p. 719)

tsunami (TSU•NAH•mee) *n.* a giant ocean wave, caused by an underwater earthquake or volcanic eruption, with great destructive power. (pp. 40, 662)

tuberculosis *n.* a respiratory infection spread by human contact, which often accompanies AIDS. (p. 466)

tundra *n.* the flat treeless lands forming a ring around the Arctic Ocean; the climate region of the Arctic Ocean. (p. 63)

typhoon *n.* a tropical storm, like a hurricane, that occurs in the western Pacific. (pp. 51, 625)

U

USSR *n.* the Union of Soviet Socialist Republics, or Soviet Union, formed in 1922 by the Communists and officially dissolved in 1991. (p. 363)

UNICEF (United Nations Children's Fund) *n.* an international watchdog and relief organization for children. (p. 665)

United Provinces of Central America *n.* the name of Central America after the region declared independence from Mexico in 1823. (p. 223)

upland *n.* a hill or very low mountain that may also contain mesas and high plateaus. (p. 275)

Ural Mountains *n.* the mountain ranges that separate the Northern European and West Siberian plains and used as the dividing line between Europe and Asia. (p. 346)

urban geography *n.* the study of how people use space in cities. (p. 87)

urbanization *n.* the dramatic rise in the number of cities and the changes in lifestyle that result. (p. 88)

urban sprawl *n.* poorly planned development that spreads a city's population over a wider and wider geographic area. (p. 176)

V

Vietnam War *n.* (1954–1975) the military conflict resulting from American involvement in South Vietnam to prevent its takeover by Communist North Vietnam. (p. 707)

volcano *n.* a natural event, formed when magma, gases, and water from the lower part of the crust or mantle collect in underground chambers and eventually erupt and pour out of cracks in the earth's surface. (p. 40)

voyaging canoe *n.* a large ship developed by Pacific Islanders to sail the ocean. (p. 699)

W

wadi *n.* a riverbed that remains dry except during the rainy seasons. (p. 488)

water table *n.* the level at which rock is saturated. (p. 33)

weather *n.* the condition of the atmosphere at a particular location and time. (p. 50)

weathering *n.* physical and chemical processes that change the characteristics of rock on or near the earth's surface, occurring slowly over many years. (p. 42)

West *n.* North American region, consisting of 13 states, that stretches from the Great Plains to the Pacific Ocean and includes Alaska to the north and Hawaii in the Pacific. (p. 148)

West Bank *n.* in Israel, a strip of land on the west side of the Jordan River, originally controlled by Jordan, which is part of the land set aside for Arab Palestinians. (p. 527)

Western Wall *n.* for Jews, the holiest site in Jerusalem; the only remaining portion of the Second Temple, built in 538 B.C. and destroyed in A.D. 70 by the Romans. (p. 510)

Wik Case *n.* in Australia, the court ruled in this case that Aboriginal people could claim land held under a pastoral lease. (p. 729)

winter monsoon *n.* the season when dry winds blow from the northeast across the Himalaya Mountains toward the sea from October through February, sometimes causing drought. (p. 597)

X

Xi Jiang (shee JYAHNG) *n.* also called the West River; the river that flows eastward through southeast China and joins the Pearl River (Zhu Jiang) to flow into the South China Sea, forming an estuary between Hong Kong and Macao. (p. 621)

Y

yurt *n.* a tent of Central Asia's nomads. (p. 379)

Z

Zionism *n.* a movement that began in the 19th century to create and support a Jewish homeland in Palestine. (p. 511)

Zuider Zee (ZEYE•duhr ZAY) *n.* former inlet of the North Sea in the Netherlands. (p. 283)

A

Aboriginal people [Aborígenes] *s.* gente que emigró a Australia desde Asia, hace al menos 40.000 años; los pobladores originales de la tierra. (p. 718)

absolute location [ubicación absoluta] *s.* el lugar exacto en la Tierra donde se encuentra un accidente geográfico. (p. 6)

acculturation [aculturación] *s.* el cambio cultural que ocurre cuando las personas en una sociedad aceptan o adoptan una innovación. (p. 72)

acquired immune deficiency syndrome (AIDS) [síndrome de inmunodeficiencia adquirida (SIDA)] *s.* enfermedad producida por el virus de la inmunodeficiencia humana o VIH. (p. 465)

Aksum [Aksum] *s.* una importante capital comercial desde el s. I al s. VIII de nuestra era, situada en lo que hoy es Etiopía; floreció debido a su ubicación junto al Mar Rojo y el Océano Índico. (p. 431)

alluvial plain [llanura aluvial] *s.* tierra fértil para la labranza, formada por depósitos de arcilla, limo, arena o grava producidos por las aguas corrientes. (p. 553)

Amazon River [Río Amazonas] *s.* el segundo río más largo del mundo y uno de los tres principales sistemas fluviales de América del Sur. Se extiende unas 4.000 millas (6.436 km) de oeste a este y desemboca en el Océano Atlántico. (p. 203)

Andes Mountains [Cordillera de los Andes] *s.* una larga cordillera que se extiende a lo largo de la costa del Pacífico de Centroamérica y América del Sur. (p. 201)

anti-Semitism [antisemitismo] *s.* discriminación contra los judíos. (p. 315)

apartheid [apartheid] (a-par-zeid) *s.* política de separación completa de las razas, implementada por el gobierno de la minoría blanca de Sudáfrica en 1948. (p. 454)

Appalachian Mountains [Montes Apalaches] *s.* una de las dos cordilleras más importantes en la región Este de los Estados Unidos y Canadá, que se extiende 1.600 millas (2.575 km) desde Terranova (Newfoundland) hacia el sur hasta Alabama. (p. 119)

aqueduct [acueducto] *s.* estructura para transportar agua por largas distancias. (p. 292)

aquifer [acuífero] *s.* capa subterránea de roca donde se almacena agua. (p. 421)

archipelago [archipiélago] *s.* grupo de islas cercanas. (pp. 553, 689)

ASEAN [ANSA] *s.* Asociación de Naciones del Sudeste Asiático, una alianza que promueve el desarrollo económico y la paz en la región. (p. 707)

Ashanti [Ashanti] *s.* gente que vive en lo que es ahora Ghana, en África Occidental, renombrada por sus diseños artísticos de ropa asasia o kente que usa la realeza. (p. 444)

assimilation [asimilación] *s.* proceso por el cual un grupo minoritario gradualmente se desprende de su propia cultura y adopta la cultura del grupo mayoritario. (p. 728)

Aswan High Dam [La gran presa de Asuán] *s.* presa en el río Nilo de Egipto, construida en 1970, la cual aumentó las tierras arables de Egipto en un 50 por ciento y las protegió contra las sequías y las inundaciones. (p. 426)

Atlantic Provinces [Las provincias atlánticas] *s.* las provincias en la región este del Canadá: Isla Príncipe Eduardo, Nueva Brunswick, Nueva Escocia y Terranova o Newfoundland. (p. 166)

atmosphere [atmósfera] *s.* las capas gaseosas que envuelven inmediatamente la Tierra. (p. 28)

atoll [atolón] *s.* isla coralina en forma anular o un conjunto de pequeñas islas que rodean una laguna central. (pp. 553, 700)

B

balkanization [balcanización] *s.* proceso por el cual una región se fragmenta en unidades pequeñas, mutuamente hostiles. (p. 311)

Baltic Republics [Países Bálticos] *s.* los países de Latvia, Lituania y Estonia, ubicados en la costa este del mar Báltico. (p. 361)

Bantu migration [migración bantú] *s.* desplazamiento de los pueblos bantú hacia el sur a través de África, que propagaron su lengua y su cultura desde alrededor del año 500 antes de nuestra era hasta alrededor del año 1000 de nuestra era. (p. 448)

basic necessity [necesidades básicas] *s.* alimentos, ropa y vivienda. (p. 593)

Benelux [Benelux] *s.* la unión económica de Bélgica, Países Bajos (Nederland) y Luxemburgo. (p. 296)

Beringia [Behring] *s.* puente de tierra que se cree conectaba lo que son ahora Siberia y Alaska. (p. 127)

Berlin Conference [Conferencia de Berlín] *s.* una conferencia de 14 países europeos realizada en 1884-1885 en Berlín, Alemania, para establecer normas de control político de África. (p. 432)

Berlin Wall [Muro de Berlín] *s.* muro construido por Alemania Oriental en 1961 para dividir la capital de Berlín en dos, derruido en 1989. (p. 298)

Bikini Atoll [Atolón Bikini] *s.* arrecife aislado en las Islas Marshall del Pacífico central, donde se efectuaron experimentos de bombas nucleares estadounidenses, lo que contaminó el atolón con altos niveles de radiación, y ahuyentó a sus habitantes. (p. 700)

biodiversity [biodiversidad] *s.* la variedad de organismos en un ecosistema. (p. 245)

biological weapon (arma biológica) *s.* bacteria o virus que se puede utilizar para dañar o matar personas, animales o plantas. (p. 175)

biome [bioma] *s.* un ecosistema regional. (p. 65)

biosphere [biósfera] *s.* todas las partes de la Tierra donde viven plantas y animales, incluyendo la atmósfera, la litosfera y la hidrosfera. (p. 28)

birthrate [índice de natalidad] *s.* el número de nacimientos vivos por total de la población, con frecuencia expresado por miles de habitantes. (p. 78)

blizzard [ventisca] *s.* tormenta de nieve fuerte con vientos de más de 35 millas (55 km) por hora y visibilidad reducida de menos de un cuarto de milla (0.40 km). (p. 52)

Boxer Rebellion [Guerra de los bóxers] *s.* rebelión en China en 1900, producida por militantes chinos enfurecidos, o bóxers, por el control extranjero; cientos de europeos, cristianos y chinos murieron. (p. 636)

British Columbia [Columbia Británica] *s.* la provincia más occidental de Canadá en las Montañas Rocosas. (p. 169)

Buddhism [Budismo] *s.* religión originda en la India por el año 500 antes de nuestra era, que se extendió hacia China, donde se convirtió en una religión importante alrededor del año 400 de nuestra era. (p. 638)

C

calypso [calypso] *s.* estilo de música que comenzó en Trinidad y combina elementos musicales de África, España y el Caribe. (p. 227)

Canadian Shield [escudo canadiense] *s.* parte norteña de las tierras bajas interiores que es una región rocosa y plana que cubre casi dos millones de millas cuadradas (cinco millones doscientos mil kilómetros cuadrados) y encierra la Bahía de Hudson. (p. 119)

canopy [bóveda] *s.* área que comprende la parte superior de los árboles en una selva tropical, a unos 150 pies (45 metros) sobre el suelo. (p. 422)

capoeira [capoeira] *s.* arte marcial y danza que desarrollaron en Brasil los angolanos que fueron llevados allí desde el África por los portugueses. (p. 239)

Carnival [Carnaval] *s.* el día de fiesta más llamativo de Brasil. (p. 239)

carrying capacity [capacidad de soporte] *s.* número de organismos que un pedazo de terreno puede soportar sin efectos negativos. (p. 82)

Carthage [Cartago] *s.* uno de los grandes imperios de África en la antigüedad, situado en una península triangular en el Golfo de Túnez en la costa del Mar Mediterráneo. (p. 438)

cartographer [cartógrafo] *s.* persona que levanta mapas. (p. 10)

cash crop [cultivo industrial o comercial] *s.* producto cultivado para la venta directa y no para uso en una región, como café, té y azúcar en África. (p. 433)

caste system [sistema de castas] *s.* el sistema ario de clases sociales en la India y uno de los pilares del hinduismo en el cual cada persona nace dentro de una casta y sólo puede pasar a otra casta mediante la reencarnación. (p. 571)

Caucasus [Cáucaso] *s.* región que comprende el sistema montañoso del mismo nombre y se extiende entre el mar Negro y el Caspio. (p. 385)

caudillo [caudillo] *s.* dictador militar o líder político. (p. 249)

Central Asia [Asia Central] *s.* región que incluye las repúblicas de Kazajstán, Kirguistán, Tayikistán, Turkmenistán y Uzbekistán. (p. 346)

central business district (CBD) [distrito comercial central (DCC)] *s.* el centro de una ciudad, en el cual casi siempre se desarrollan actividades comerciales. (p. 89)

cerrado [cerrado] *s.* una sabana que tiene terreno plano y lluvias moderadas, lo que la hace apta para la agricultura. (p. 202)

Chang Jiang [Chang Jiang] *s.* (o Río Yang-tsé) el río más largo del Asia, que fluye unas 3.900 millas (6.275 km) desde Xizang (Tibet) hasta el mar de la China oriental. (p. 621)

Chaparral [chaparral] *s.* término, en algunos lugares, para una bioma de árboles resistentes a la sequía. (p. 66)

Chechnya [Chechenia] *s.* una de las repúblicas que continúa siendo parte de Rusia después del colapso de la Unión Soviética a pesar de los movimientos independentistas y levantamientos violentos. (p. 386)

chemical weathering [meteorización química] *s.* proceso por el cual una roca se convierte en una nueva substancia a través de la interacción entre los elementos en el aire o el agua y los minerales en la roca. (p. 43)

chernozem [quimiozen] *s.* capa superior negra del suelo, una de las tierras más fértiles del mundo. (p. 345)

cholera [cólera] *s.* enfermedad infecciosa tratable que puede ser mortal y es producida por la falta de medidas higiénicas adecuadas y de suministro de agua limpia. (p. 465)

city [ciudad] *s.* zona que es el centro de los negocios y la cultura y tiene una población numerosa. (p. 87)

city-state [ciudad-estado] *s.* una unidad política autónoma compuesta por una ciudad y los terrenos circundantes. (p. 289)

climate [clima] *s.* las condiciones atmosféricas típicas de un lugar específico que se observan con el tiempo. (p. 50)

coalition (coaliciún) *s.* alianza. (p. 174)

Cold War [Guerra Fría] *s.* el conflicto entre los Estados Unidos y la Unión Soviética después de la II Guerra Mundial, llamada "fría" porque nunca se intensificó hasta el grado de convertirse en una guerra abierta. (p. 363)

collective farm [granja colectiva] *s.* un gran equipo de peones reunidos para trabajar juntos en enormes granjas en la Unión Soviética, durante el gobierno de Jósiv Stalin. (p. 364)

Columbian Exchange [Intercambio Colombino] *s.* el intercambio de plantas, animales y enfermedades entre el hemisferio oriental y el hemisferio occidental durante la era de las exploraciones. (p. 136)

command economy [economía dirigida] *s.* tipo de sistema económico en el cual la producción de bienes y servicios es determinada por un gobierno central, el cual usualmente es dueño de los medios de producción. Llamado también "economía planificada". (pp. 91, 364)

commodity [bien de consumo] *s.* un producto agrícola o de minería que se puede vender. (p. 462)

communism [comunismo] *s.* sistema en el cual el gobierno retiene casi todo el poder político y los medios de producción. (p. 83)

confederation [confederación] *s.* una unión política. (p. 156)

Confucianism [Confucianismo] *s.* movimiento basado en las enseñanzas de Confucio, filósofo chino que vivió alrededor del año 500 antes de nuestra era; Confucio enfatizaba la importancia de la educación en una sociedad ordenada en la cual las personas respetan a sus mayores y obedecen al gobierno. (p. 638)

coniferous [conífero] *adj.* otro término para los árboles de hojas perennes y aciculares. (p. 66)

constitutional monarchy [monarquía constitucional] *s.* sistema de gobierno en el cual los poderes del gobernante están limitados por una constitución y las leyes de la nación. (p. 580)

continent [continente] *s.* una masa de tierra firme sobre el agua en la Tierra. (p. 27)

Continental Divide [La Divisoria Continental] *s.* la línea de los picos más altos en América del Norte que marca la separación entre los ríos que fluyen hacia el este y hacia el oeste. (p. 120)

continental drift [deriva de los continentes] *s.* la hipótesis de que los continentes fueron una vez un supercontinente que se dividió lentamente a través de millones de años. (p. 29)

continentality [continentalidad] *s.* la distancia de una región de la influencia moderadora del mar. (p. 350)

continental shelf [plataforma continental] *s.* la superficie de la Tierra desde el borde de un continente hasta la parte profunda del océano. (p. 36)

convection [convección] *s.* la transferencia de calor en la atmósfera por el movimiento ascendente del aire. (p. 54)

copra [copra] *s.* la pulpa seca del coco. (p. 714)

core [centro] *s.* el núcleo de la Tierra, compuesto de hierro y niquel; el centro interior es sólido, el centro exterior es líquido. (p. 28)

crude oil [petróleo crudo] *s.* petróleo que no ha sido procesado. (p. 497)

Crusades [Cruzadas] *s.* una serie de guerras impulsadas por los cristianos europeos en 1096 para recuperar la Tierra Santa (Palestina) de los musulmanes. (p. 291)

crust [corteza] *s.* la capa delgada de rocas que compone la superficie de la Tierra. (p. 28)

cultural crossroad [cruce cultural] *s.* un lugar donde convergen varias culturas. (p. 310)

cultural hearth [centro cultural] *s.* el centro o lugar de origen de una cultura importante; un lugar de innovaciones desde el cual se difunden ideas, materiales y tecnologías fundamentales a otras culturas. (pp. 72, 222)

culture [cultura] *s.* el total de conocimientos, actitudes y comportamientos compartidos y transmitidos por los miembros de un grupo. (p. 71)

cyclone [ciclón] *s.* una tormenta violenta con vientos fuertes y mucha lluvia; el patrón climatológico más extremo del Asia Meridional. (p. 558)

czar [zar] *s.* el emperador de Rusia antes de la Revolución de 1917 y de la subsiguiente creación de la Unión Soviética en 1922. (p. 362)

D

Dead Sea [Mar Muerto] *s.* lago salado, sin salida al mar, entre Israel y Jordania, con un nivel de salinidad tan alto que casi nada puede vivir en sus aguas; se encuentra a 1.349 pies (411 m) por debajo del nivel del mar, lo que lo convierte en el lugar más bajo en la corteza expuesta de la Tierra. (p. 489)

debt-for-nature swap [Intercambio de deuda por naturaleza] *s.* acuerdo para reducir una deuda por el cual una organización acepta pagar cierta cantidad de una deuda gubernamental a cambio de protección gubernamental de cierta parte de una selva tropical. (p. 247)

deciduous [caducifolio] adj. característica de los árboles de hojas anchas, como el arce, el roble, el abedul y el Alamo de Virginia. (p. 66)

deforestation [deforestación] *s.* el corte y la eliminación de árboles y bosques. (p. 246)

delta [delta] *s.* zona de forma de abanico formada por sedimentos depositados dejados por un río que disminuye su velocidad al desembocar en el océano. (p. 43)

democracy [democracia] *s.* tipo de gobierno en el cual los ciudadanos ejercen el poder político sea directamente o mediante representantes elegidos. (p. 83)

desalinization [desalinización] *s.* la eliminación de sal del agua del océano. (p. 496)

desertification [desertización] *s.* ampliación de condiciones secas a zonas húmedas que se encuentran próximas a desiertos. (p. 424)

dialect [dialecto] *s.* una versión de un idioma que refleja cambios en patrones de habla por factores relacionados con cambios de clase, regionales o culturales. (p. 73)

dictatorship [dictadura] *s.* tipo de gobierno en el cual un individuo o grupo de individuos tienen el poder político completo. (p. 83)

diffusion [difusión] *s.* la diseminación de ideas, invenciones o patrones de comportamiento hacia otras sociedades. (p. 72)

dike [dique] *s.* muro de tierra usado para contener o desviar el curso de las aguas. (p. 282)

distance decay [deterioro de la distancia] *s.* término que se refiere al concepto de que a mayor distancia entre dos puntos, menor interacción entre los mismos. (p. 389)

diversify [diversificar] *v.* aumentar la variedad de productos en la economía de un país; promover la industria fabril y otras industrias con el propósito de lograr el desarrollo y la estabilidad. (p. 462)

Dome of the Rock [Cúpula de la Roca] *s.* un santuario en Jerusalén, ubicado en el monte del Templo, que contiene el lugar donde los musulmanes creen que Mahoma se elevó a los cielos y donde los judíos creen que Abraham preparó el sacrificio de su hijo Isaac a Dios. (p. 511)

Dominion of Canada [Dominio de Canadá] *s.* la amplia confederación de Ontario (Alto Canadá), Quebec (Bajo Canadá), Nueva Escocia y Nuevo Brunswick, creada por el Acta de la América del Norte Británica en 1867. (p. 156)

drainage basin [cuenca de drenaje] *s.* una zona drenada por un río importante y sus afluentes. (p. 33)

drip irrigation [irrigación por goteo] *s.* la práctica de usar tubos pequeños que lentamente gotean agua justo sobre el suelo para conservar agua para usarse en los cultivos. (p. 496)

drought [sequía] *s.* un largo período sin lluvia o con precipitación mínima. (p. 53)

dynasty [dinastía] *s.* una serie de gobernantes de la misma familia. (p. 635)

E

earthquake [terremoto] *s.* un movimiento a veces violento de la tierra, producido cuando placas tectónicas se tocan o deslizan una sobre otra en una falla. (p. 39)

economic system [sistema económico] *s.* la forma como la gente produce e intercambia bienes. (p. 91)

economic tiger [tigre económico] *s.* un país con rápido crecimiento económico debido al bajo coste de la mano de obra, la alta tecnología y las exportaciones agresivas. (p. 645)

economy [economía] *s.* la producción y el intercambio de bienes y servicios entre un grupo de personas. (p. 91)

ecosystem [ecosistema] *s.* una comunidad interdependiente de plantas y animales. (p. 65)

El Niño [El Niño] *s.* un patrón meteorológico creado por el calentamiento de las aguas de las costas occidentales de América del Sur, que empuja aguas cálidas y fuertes lluvias hacia el continente americano y produce condiciones de sequía en Australia y Asia. (p. 57)

entrepreneur [empresario] *s.* persona que inicia y desarrolla un negocio. (p. 575)

epicenter [epicentro] *s.* el punto en la superficie terrestre que corresponde a la ubicación en la Tierra donde comienza un terremoto. (p. 39)

equator [ecuador] *s.* la línea imaginaria que rodea la esfera terrestre, dividiendo la Tierra en las mitades norte y sur. (p. 6)

equinox [equinoccio] *s.* cada uno de los dos días del año en los cuales el día y la noche tienen la misma duración; marca el comienzo de la primavera y el otoño. (p. 49)

erosion [erosión] *s.* el resultado del desgaste de la materia producido por la acción del viento, el agua, el hielo o la gravedad. (p. 43)

escarpment [escarpa] *s.* declive empinado de un terreno con una meseta casi plana en la cima. (p. 417)

estuary [estuario] *s.* desembocadura de un río con una amplia apertura por donde las corrientes del río chocan con las mareas del océano. (p. 563)

ethnic cleansing [limpieza étnica] *s.* la política de tratar de eliminar a un grupo étnico. (p. 320)

ethnic group [grupo étnico] *s.* un grupo de personas que comparten un idioma, costumbres y una herencia común. (p. 71)

Euphrates River [Río Éufrates] *s.* un río en el Sudoeste asiático que sirvió de apoyo a varias civilizaciones antiguas, fluye a través de regiones de Turquía, Siria e Irak y desemboca en el Golfo Pérsico. (p. 489)

Eurasia [Eurasia] *s.* los continentes combinados de Europa y Asia. (p. 346)

euro [euro] *s.* moneda común propuesta por la Unión Europea para sus naciones miembros. (p. 305)

European Environmental Agency [Agencia Europea del Medio Ambiente] *s.* esta agencia proporciona a la Unión Europea información confiable sobre el medio ambiente. (p. 324)

Everglades [Everglades] *s.* una amplia zona de terrenos pantanosos subtropicales en la Florida, de cerca de 4.000 millas cuadradas (10.400 kilómetros cuadrados). (p. 126)

export [exportación] *s.* un producto o bien que se vende desde una economía a otra. (p. 140)

F

Fang sculpture [esculturas de los fangs] *s.* cajas talladas que contienen las calaveras y los huesos de los antepasados muertos, creadas por los fangs, que vivieron en Gabón, la región sur de Camerún y Guinea Ecuatorial. (p. 451)

fault [falla] *s.* una fractura en la corteza terrestre. (p. 39)

folk art [arte folclórico] *s.* artículos hechos a mano, como cerámica, objetos tallados en madera y trajes tradicionales, elaborados por habitantes de zonas rurales que llevan estilos de vida tradicionales, no por artistas profesionales. (p. 314)

federal republic [república federal] *s.* una nación cuyos poderes están divididos entre el gobierno federal o nacional y varios gobiernos estatales o locales. (p. 139)

feudalism [feudalismo] *s.* un sistema político imperante en Europa entre el *s.* IX y el *s.* XV, en el cual el rey permitía a los nobles el uso de sus tierras a cambio de servicios militares y la protección de la tierra. (p. 297)

fertility rate [índice de fertilidad] *s.* el número promedio de hijos que una mujer en edad fértil tendría durante su vida si tuviese hijos de acuerdo con el índice vigente para su país. (p. 78)

First Nations [Primeras Naciones] *s.* un grupo de indígenas del Canadá. (p. 159)

fjord [fiordo] *s.* una entrada larga, estrecha y profunda del mar en la tierra entre pendientes empinadas. (p. 273)

fossil water [agua fósil] *s.* agua bombeada desde acuíferos subterráneos. (p. 496)

free enterprise [libre empresa] *s.* sistema económico en el cual individuos privados son dueños de la mayor parte de los recursos, la tecnología y las empresas, y pueden explotarlos para obtener ganancias con poco control del gobierno. (p. 140)

frontier [frontera] *s.* la tierra libre y abierta en el Oeste Norteamericano que estaba disponible para colonización. (p. 137)

G

Ganges River [Río Ganges] *s.* río en el Sur de Asia, un importante recurso acuático que fluye más de 1.500 millas (2.415 km) desde su fuente en un glaciar del Himalaya hasta la Bahía de Bengala. (p. 560)

Gaza Strip [Franja de Gaza] *s.* territorio a lo largo del Mar Mediterráneo, justo al noreste de la Península del Sinaí; parte del territorio asignado a los palestinos y que fue ocupado por Israel en 1967. (p. 527)

Geographic Information System (GIS) [Sistema de Información Geográfica (GIS por sus siglas en inglés)] *s.* tecnología que usa información de mapas digitalizados para crear un banco de datos; diferentes "capas de datos" pueden combinarse para producir mapas especializados. El GIS permite a los geógrafos analizar diferentes aspectos de un lugar específico para resolver problemas. (p. 13)

geography [geografía] *s.* estudio de la distribución y la interacción de las características físicas y humanas de la Tierra. (p. 5)

glaciation [glaciación] *s.* cambios en los accidentes geográficos debidos al lento movimiento de los glaciares. (p. 44)

glacier [glaciar] *s.* una masa de hielo grande y duradera que se mueve debido al efecto de la gravedad. (p. 44)

global economy [economía global] *s.* la fusión de economías regionales por la cual las naciones se vuelven dependientes unas de otras para la producción de bienes y servicios. (p. 666)

global network (red mundial) *s.* grupo que se mantiene conectado alrededor del mundo. (p. 173)

global warming [calentamiento global] *s.* la acumulación de dióxido de carbono (anhídrido carbónico) en la atmósfera, lo que evita que el calor escape al espacio, aumentando las temperaturas y ocasionando cambios en las condiciones meteorológicas. (p. 246)

globe [globo] *s.* una representación tridimensional de la Tierra. (p. 10)

Gobi Desert [Desierto de Gobi] *s.* desierto ubicado en el norte de China y en el sudeste de Mongolia, zona importante para la localización de fósiles de dinosaurios. (p. 627)

Golan Heights [Altos del Golán] *s.* meseta montañosa que se eleva sobre el Río Jordán y el Mar de Galilea; un punto estratégico que ha sido sitio de conflictos en el Sudoeste asiático durante décadas. (p. 487)

Gorée Island [Isla de Gorée] *s.* isla en las costas de Senegal que sirvió como importante punto de partida de esclavos durante el tráfico de esclavos. (p. 442)

Great Barrier Reef [La Gran Barrera de Coral] *s.* una cadena de 1.250 millas (2.000 km) de más de 2.500 arrecifes e islas a lo largo de la costa noreste de Australia, que contiene unas 400 especies de coral. (p. 692)

Great Game [El Gran Juego] *s.* un conflicto entre el Imperio Británico y el Imperio Ruso por el control del Asia Central en el *s.* XIX. (p. 376)

Great Kanto Earthquake [El Gran Terremoto de Kanto] *s.* terremoto ocurrido en 1923 en Japón que causó la muerte de aproximadamente 140.000 personas y dejó la ciudad de Tokio en ruinas. (p. 662)

Great Lakes [Grandes Lagos] *s.* grupo de cinco lagos de agua dulce en la región central de América del Norte entre los Estados Unidos y Canadá; los lagos son el Hurón, el Ontario, el Michigan, el Erie y el Superior. (p. 121)

Great Plains [Grandes Llanuras] *s.* una amplia zona de praderas en la región central de América del Norte, carente de árboles en su mayor parte, que se eleva hasta 4.000 pies (1.200 metros) sobre el nivel del mar. (p. 119)

Great Zimbabwe [El Gran Zimbabwe] *s.* un emplazamiento urbano en lo que es hoy Zimbabwe fundado por los shonas alrededor del año 1000; se convirtió en la capital de una próspera zona de comercio de oro. (p. 453)

greenhouse effect [efecto invernadero] *s.* la capa de gases emitidos por la quema de carbón y petróleo que atrapa la energía solar, elevando la temperatura de la Tierra. (p. 58)

Green Revolution [La Revolución Verde] *s.* programa agrícola lanzado por científicos en la década de 1960 para producir variedades de granos de mayor rendimiento y mejorar la producción de alimentos incorporando nuevas técnicas de labranza. (p. 569)

Gross Domestic Product (GDP) [Producto Interior Bruto (PIB)] *s.* el valor de sólo bienes y servicios producidos en un país durante un período determinado. (p. 95)

Gross National Product (GNP) [Producto Nacional Bruto (PNB)] *s.* el valor total de todos los bienes y servicios producidos por un país durante un período determinado. (p. 94)

ground water [agua subterránea] *s.* agua retenida debajo de la superficie terrestre, con frecuencia en y alrededor de los poros de las rocas. (p. 33)

guest worker [trabajador invitado] *s.* trabajadores poco calificados, a menudo inmigrantes del Sur y el Este de Asia, trasladados a los países productores de petróleo para ocupar puestos de trabajo que las personas nacidas en la región consideran cultural y económicamente inaceptables. (p. 525)

H

hemisphere [hemisferio] *s.* cada una de las dos mitades de la esfera terrestre. (p. 6)

high islands [Islas altas] *s.* islas del Pacífico creadas por volcanes. (p. 691)

Himalaya Mountains [Himalaya] *s.* cordillera del Sur de Asia que incluye el Monte Everest, el pico más alto del mundo. (p. 551)

Hinduism [Hinduismo] *s.* la religión dominante en la India. (p. 560)

Holocaust [Holocausto] *s.* programa de los nazis de asesinatos masivos de judíos europeos durante la II Guerra Mundial. (p. 298)

Huang He [Huang He] *s.* río del Norte de China, llamado también Río Amarillo, que nace en las Montañas Kunlun y se extiende unas 3.000 millas (4,800 km) hacia el Este, desembocando en el mar Amarillo. (p. 621)

human resources [recursos humanos] *s.* las aptitudes y los talentos de la gente que trabaja. (p. 531)

humus [humus] *s.* material orgánico en el suelo. (p. 45)

hurricane [huracán] *s.* una tormenta que se forma sobre las aguas cálidas de los océanos tropicales. (p. 51)

hydrologic cycle [ciclo hidrológico] *s.* la continua circulación de agua entre la atmósfera, los océanos y la Tierra. (p. 32)

hydrosphere [hidrosfera] *s.* las aguas que comprenden la superficie de la Tierra, incluyendo océanos, mares, ríos, lagos y el vapor en la atmósfera. (p. 28)

I

Ijsselmeer [Ijsselmeer] *s.* lago de agua dulce separado del Mar del Norte por un dique y bordeado por pólders. (p. 283)

illiteracy [analfabetismo] *s.* la incapacidad de leer o escribir. (p. 593)

Inca [Inca] *s.* miembro del pueblo quechua en América del Sur, que desarrolló una civilización en los Andes en los siglos XV y XVI. (p. 230)

Indochina [Indochina] *s.* colonia francesa compuesta por Camboya, Laos y Vietnam; obtuvo la independencia de Francia en 1954. (p. 707)

industrialization [industrialización] *s.* el desarrollo de la industria en un país o en una sociedad. (p. 730)

Indus Valley civilization [Civilización del Valle del Indo] *s.* la más grande de las primeras civilizaciones del mundo en lo que hoy es Pakistán; fue una civilización urbana altamente desarrollada, que duró desde el 2500 hasta cerca del 1500 antes de nuestra era. (p. 573)

infant mortality rate [índice de mortalidad infantil] *s.* el número de muertes de niños menores de un año, calculado por cada mil nacimientos vivos. (p. 79)

infrastructure [infraestructura] *s.* los sistemas básicos de apoyo necesarios para mantener una economía en desarrollo, que incluyen sistemas de suministro de energía, comunicaciones, transporte, aguas, servicios sanitarios y educación. (pp. 94, 177, 212)

innovation [innovación] *s.* tomar los elementos existentes en una sociedad para crear algo nuevo con el propósito de satisfacer una necesidad. (p. 72)

Institutional Revolutionary Party (PRI) [Partido Revolucionario Institucional (PRI)] *s.* partido político creado en México, en 1929, que ayudó a introducir la democracia y mantener la estabilidad política durante la mayor parte del siglo XX. (p. 218)

Islam [Islam] *s.* religión monoteísta basada en las enseñanzas del profeta Mahoma y la mayor influencia cultural y religiosa en el Norte de África. (pp. 439, 503)

J

Jakota Triangle [Triángulo de Jakota] *s.* zona de prosperidad en la década de 1980 y comienzos de la década de 1990, que comprende Japón, Corea del Sur y Taiwán. (p. 666)

Jordan River [Río Jordán] *s.* río que sirve como frontera natural entre Israel y Jordania, y fluye desde los montes de Líbano sin desembocar en el Mar Mediterráneo. (p. 489)

junta [junta] *s.* gobierno dirigido por generales después de un golpe militar. (p. 249)

K

Kashmir [Cachemira (Kashmir)] *s.* región del Norte de la India y Pakistán sobre la que se han librado varias guerras destructivas. (p. 574)

Khmer Empire [Imperio Khmer] *s.* poderoso imperio que duró aproximadamente del siglo IX al siglo XV, en lo que hoy es Camboya. (p. 706)

King Leopold II [Rey Leopoldo II] *s.* rey de Bélgica que abrió el interior del África al comercio europeo a lo largo del río Congo y para 1884 controlaba la zona conocida como el Estado Libre del Congo. (p. 449)

KLA (Kosovo Liberation Army) [ELK (Ejército de Liberación de Kosovo)] *s.* grupo que combatió contra los intentos de los serbios de controlar la región de Kosovo en la década de 1990. (p. 321)

Kunlun Mountains [Cordillera Kunlun] *s.* cordillera ubicada en el Oeste de China que es la fuente de dos de los principales ríos de China, el Huang He (río Amarillo) y el Chang Jiang (Yangtzé) (p. 619)

Kurds [Kurdos] *s.* grupo étnico en el sudoeste de Asia, que ha ocupado la región de Kurdistán, ubicada en Turquía, Irac e Irán, por cerca de mil años, y que ha estado involucrado en enfrentamientos con estos tres países por recobrar tierras durante la mayor parte del siglo XX. (p. 516)

L

landfill [vertedero] *s.* método de eliminación de residuos sólidos por el cual los residuos son enterrados entre capas de tierra con el propósito de rellenar o recuperar terrenos bajos. (p. 631)

landform [accidente geográfico] *s.* una característica de la superficie terrestre formada naturalmente. (p. 33)

landlocked [sin litoral] *adj.* que no tiene salida al mar. (p. 84)

land reform [reforma agraria] *s.* proceso por el cual se dividen grandes latifundios con el propósito de lograr una distribución más equitativa de la tierra entre los agricultores. (pp. 250, 569)

Land Rights Act of 1976 [Ley de Derechos de Tierra de 1976] *s.* una ley especial promulgada en beneficio de los derechos de los aborígenes en Australia, dándoles el derecho de reclamar tierras en el Territorio Norte. (p. 728)

Landsat [Landsat] *s.* una serie de satélites que orbitan a más de 100 millas (160 km) sobre la Tierra. Cada satélite recoge información en una zona de 115 millas (185 km) de ancho. (p. 12)

latitude (lines) [latitudes (líneas)] *s.* un conjunto de líneas imaginarias que corren paralelas al ecuador, las cuales son usadas para localizar lugares al Norte y al Sur. El ecuador es denominado la línea de cero grados de latitud. (p. 6)

lava [lava] *s.* magma que ha llegado hasta la superficie terrestre. (p. 40)

lithosphere [litosfera] *s.* la capa de roca sólida de la superficie terrestre. (p. 28)

llanos [llanos] *s.* una extensa zona de praderas sin árboles de América del Sur, utilizada para pastoreo y labranza. (p. 202)

lock [esclusa] *s.* una sección de una vía acuática con puertas de entrada y salida donde se llenan o vacían de agua los espacios entre las mismas, a través de las cuales pasan los barcos. (p. 129)

loess [loess] *s.* sedimentos de limo o arcilla depositados por el viento que producen tierras muy fértiles. (p. 44)

longitude (lines) [longitud (líneas)] *s.* un conjunto de líneas imaginarias que circundan la Tierra por los polos, dividiéndola en las zonas Este y Oeste. El primer meridiano (meridiano de Greenwich) ha sido designado como la línea de cero grados para longitud. (p. 6)

Louisiana Purchase [La Compra de Louisiana] *s.* el territorio, incluyendo la región entre el río Mississippi y las Montañas Rocosas, que los Estados Unidos compró a Francia en 1803. (p. 136)

low islands [Islas bajas] *s.* islas del Pacífico formadas por arrecifes de coral. (p. 691)

M

Mabo Case [el caso *Mabo*] *s.* en Australia, el proceso jurídico por el cual se declaró con lugar la reclamación de tierra del aborigen Eddie Mabo, por medio del cual el tribunal reconoció que los aborígenes eran dueños de tierras antes de la llegada de los británicos. (p. 728)

Mackenzie River [Río Mackenzie] *s.* el río más largo del Canadá, el cual es parte de un sistema fluvial que fluye a lo largo de los Territorios del Noroeste hasta el Océano Ártico. (p. 121)

magma [magma] *s.* material de roca fundida creada cuando roca sólida en el manto o corteza funde. (p. 28)

malaria [malaria] *s.* enfermedad infecciosa de los glóbulos rojos propagada por mosquitos, que se caracteriza por escalofríos, fiebre y sudor. (p. 466)

mandala [mandala] *s.* diseño geométrico usado en el budismo tibetano como símbolo del universo y que ayuda en la meditación. (p. 583)

mandala [mandala] *s.* un estado organizado como un anillo de poder alrededor de una corte central, que con frecuencia cambiaba de tamaño con el tiempo y que era usado en lugar de fronteras en los antiguos estados del sudeste asiático. (p. 705)

mantle [manto] *s.* una capa de roca de unas 1.800 millas (2.896 km) que està entre la corteza y el centro de la Tierra. (p. 28)

Maori [Maori] *s.* los primeros pobladores de Nueva Zelanda, que emigraron de Polinesia hace más de 1.000 años. (p. 719)

Mao Zedong [Mao Zedong] *s.* líder de China comunista que derrotó a los Nacionalistas en 1949; falleció en 1976. (p. 636)

map projection [proyección cartográfica] *s.* una forma de trazar el mapa de la superficie de la Tierra que reduce la distorsión causada convirtiendo tres dimensiones en dos dimensiones. (p. 10)

map [mapa] *s.* representación gráfica bidimensional de partes selectas de la superficie de la Tierra. (p. 10)

maquiladora [maquiladora] *s.* fábrica en México que ensambla materiales importados para convertirlos en productos acabados de exportación. (p. 220)

market economy [economía de mercado] *s.* tipo de sistema económico en el cual la producción de bienes y servicios se determina por la demanda de los consumidores. Llamado también economía de demanda o capitalismo. (pp. 91, 313)

Massif Central [Massif Central] *s.* las tierras altas de Francia, que abarcan un sexto del territorio francés. (p. 275)

Mecca [Meca] *s.* la ciudad más sagrada del Islam, situada en Arabia Saudita, a la que la gente hace peregrinaciones para cumplir con deberes religiosos islámicos. (p. 503)

mechanical weathering [meteorización mecánica] *s.* proceso natural por el cual las rocas se descomponen en pedazos más pequeños. (p. 42)

megalopolis [megalópolis] *s.* una región en la cual varias ciudades grandes y las áreas circundantes se unen. (p. 146)

Melanesia [Melanesia] *s.* región en Oceanía que significa "islas negras." (p. 713)

Meseta [Meseta] *s.* la planicie central de España. (p. 275)

Mesopotamia [Mesopotamia] *s.* una región en el sudoeste asiático entre los ríos Tigris y Eufrates, donde se desarrollaron algunas de las civilizaciones más antiguas del mundo; parte del corazón cultural denominado la Media Luna de las tierras fértiles. (p. 516)

métis [métis] *s.* una persona con antepasados franco-canadienses e indígenas americanos. (p. 161)

metropolitan area [área metropolitana] *s.* área funcional que incluye una ciudad y los suburbios y exurbios que la rodean, económicamente ligados entre sí. (pp. 87, 148)

microcredit [microcrédito] *s.* un pequeño préstamo disponible a los empresarios de escasos recursos para ayudar a las empresas pequeñas a desarrollarse y elevar los niveles de vida. (p. 575)

Micronesia [Micronesia] *s.* una de las tres regiones de Oceanía, el nombre significa "islas pequeñas". (p. 713)

Midwest [El Medio-Oeste] *s.* la región que contiene los 12 estados de la zona Norte-Central de los Estados Unidos. (p. 147)

migration [migración] *s.* el desplazamiento de gente dentro de un mismo país o región. (p. 135)

Mississippi River [Río Mississippi] *s.* un importante río que fluye de norte a sur por casi todo el largo de los Estados Unidos, desde Minnesota hasta el Golfo de México y forma parte del sistema fluvial más largo del continente. (p. 121)

mistral [mistral] *s.* viento frío y seco del norte. (p. 279)

Mobutu Sese Seko [Mobutu Sese Seko] *s.* líder de Zaire, la actual República Democrática del Congo, desde su independencia en la década de 1960 hasta 1997. Puso los negocios del país bajo el control nacional, se benefició de la reorganización y utilizó el ejército para conservar el poder. (p. 450)

monarchy [monarquía] *s.* tipo de gobierno en el cual una familia gobernante dirigida por un rey o una reina, tiene el poder y puede o no compartirlo con organismos ciudadanos. (p. 83)

monsoon [monzón] *s.* viento estacional, especialmente en el Asia meridional. (p. 558)

moraine [morrena] *s.* una cadena o colina de rocas transportada y finalmente depositada por un glaciar. (p. 44)

mortality rate [índice de mortalidad] *s.* el número de muertes por cada mil. (p. 79)

mosque [mezquita] *s.* un lugar de culto islámico, donde los mahometanos rezan con el rostro orientado hacia la ciudad sagrada de la Meca. (p. 504)

Mount Kilimanjaro [Monte Kilimanjaro] *s.* un volcán en Tanzania en el Africa, es el pico más alto del Africa. (p. 417)

Mughal Empire [Imperio Mughal] *s.* el imperio musulmán establecido a comienzos del siglo XVI y que se extendió por gran parte de la India, importando nuevas costumbres que algunas veces entraban en conflicto con las de los hindúes nativos. (p. 568)

Muhammad [Mahoma] *s.* fundador y profeta del Islam, que vivió parte de su vida en la ciudad de la Meca. (p. 503)

multinational [multinacional] *s.* una compañía que realiza operaciones comerciales en todo el mundo. (p. 142)

Mutapa Empire [Imperio de Monomotapa] *s.* un estado fundado en el siglo XV por un hombre llamado Mutota y que se extendió por todo lo que hoy es Zimbabwe excepto su parte oriental. (p. 453)

N

Nagorno-Karabakh [Nagorno-Karabakh] *s.* la zona montañosa de Azerbaiján, por la cual combatieron Armenia y Azerbaiján. (p. 386)

nation [nación] *s.* un grupo de personas con una cultura común que viven en un territorio y tienen un fuerte sentimiento de unidad. (p. 83)

nationalism [nacionalismo] *s.* la creencia de que la gente tiene que ser leal con su nación y con las demás personas con la que comparte la tierra, la cultura y la historia. (p. 297)

nation-state [nación-estado] *s.* nombre de un territorio cuando una nación y un estado ocupan el mismo territorio. (p. 83)

natural resource [recurso natural] *s.* un material sobre o dentro de la Tierra, como un árbol, un pez o el carbón, que tiene valor económico. (p. 93)

needleleaf [acicular] *adj.* característica de las hojas de ciertos árboles como el pino, el abeto y el cedro, que se encuentran en las regiones del norte de América del Norte. (p. 66)

Nelson Mandela [Nelson Mandela] *s.* uno de los líderes del Congreso Nacional Africano que dirigió la lucha contra el apartheid y fue elegido presidente en 1994, en las primeras elecciones multirraciales de Sudáfrica. (p. 454)

New England [Nueva Inglaterra] *s.* los seis estados del norte en la región noreste de los Estados Unidos: Maine, Vermont, New Hampshire, Massachusetts, Rhode Island y Connecticut. (p. 145)

Niger delta [Delta del Níger] *s.* delta del río Níger y zona de Nigeria rica en depósitos de petróleo. (p. 424)

Nile River [Río Nilo] *s.* el río más largo del mundo, que recorre más de 4.000 millas (6.436 km) a través de la cuenca del Sudán, hasta Uganda, el Sudán y Egipto. (p. 416)

nomad [nómada] *s.* persona que no tiene residencia permanente y se traslada según las estaciones de un lugar a otro en busca de alimentos, agua y tierra para pastoreo. (p. 127, 378)

nonviolent resistance [resistencia pacífica] *s.* un movimiento que usa todos los medios de protesta excepto la violencia. (p. 568)

Nordic countries [países nórdicos] *s.* países del norte de Europa, entre ellos Dinamarca, Finlandia, Islandia, Noruega y Suecia. (p. 302)

NAFTA (North American Free Trade Agreement) [NAFTA (Tratado de Libre Comercio de América del Norte)] *s.* un acuerdo comercial importante que creó una amplia zona de cooperación sobre asuntos comerciales y económicos en América del Norte. (p. 220)

North Atlantic Drift [Corriente del Atlántico Norte] *s.* una corriente de agua cálida proveniente de los Trópicos. (p. 278)

Nunavut [Nunavut] *s.* uno de los territories del Canadá, donde viven muchos de los esquimales del Canadá; fue forjado de la mitad este de los Territorios Noroestes en 1999. (p. 169)

O

oasis [oasis] *s.* un lugar donde agua de un acuífero ha llegado hasta la superficie; permite el desarrollo de vegetación y fauna. (pp. 421, 492)

Oceania [Oceanía] *s.* grupo de islas del Pacífico, que incluye Melanesia, Micronesia y Polinesia. (p. 690)

Olduvai Gorge [Garganta Olduvai] *s.* un lugar de capas fosilíferas en el norte de Tanzania, que contiene el historial más continuo que se conoce de vida humana en los últimos 2 millones de años, incluyendo fósiles de 65 homínidos. (p. 431)

oligarchy [oligarquía] *s.* un gobierno dirigido por unas cuantas personas o un pequeño grupo. (p. 249)

"one-commodity" country [país de "un solo producto"] *s.* país que depende de un producto de exportación principal para muchos de sus ingresos. (p. 462)

Ontario [Ontario] *s.* una de las Provincias más importantes del Canadá. (p. 167)

OPEC [OPEP] *s.* Organización de Países Exportadores de Petróleo, grupo establecido en 1960 por algunos países productores de petróleo para coordinar políticas sobre venta de productos de petróleo. (p. 505)

Orinoco River [Río Orinoco] *s.* río que corre principalmente por Venezuela y forma parte del sistema fluvial más hacia el norte de América del Sur. (p. 202)

outback ["outback"] *s.* zona seca y despoblada en el interior de Australia. (p. 697)

outrigger canoe [canoa con balancines] *s.* una embarcación pequeña usada en las lagunas de islas en las que se establecieron isleños del Pacífico. (p. 699)

ozone [ozono] *s.* una substancia química que se produce cuando combustibles fósiles en combustión reaccionan con la luz del sol; una forma de oxígeno. (p. 325)

P

Pacific Rim [Cuenca del Pacífico] *s.* una región económica y social que incluye los países que rodean el Océano Pacífico, la cual se extiende en el sentido de las manecillas del reloj desde Nueva Zelanda en la región occidental del Pacífico hasta Chile en la región oriental del Pacífico e incluye la costa oeste de los Estados Unidos. (p. 645)

pakehas [pakehas] *s.* término maorí para designar a las personas blancas, a los neozelandeses de ascendencia europea. (p. 722)

Palestine Liberation Organization (PLO) [Organización para la Liberación de Palestina (OLP)] *s.* grupo formado en la década de 1960 para recuperar las tierras árabes en Israel para los árabes palestinos. (p. 513)

Palestinians [Palestinos] *s.* grupo de árabes desplazados que vivían o viven todavía en la zona llamada anteriormente Palestina y ahora denominada Israel. (p. 527)

pampas [pampas] *s.* amplia zona de praderas y tierras fértiles en la región sur-central de América del Sur. (p. 202)

Panama Canal [Canal de Panamá] *s.* canal para embarcaciones a través de Panamá que conecta el Mar Caribe con el océano Pacífico. (p. 226)

pandemic [pandemia] *s.* una enfermedad que afecta a un gran número de habitantes de una amplia zona geográfica. (p. 435)

Paraná River [Río Paraná] *s.* río en la región central de América del Sur y uno de sus tres sistemas fluviales más importantes, que nace en las tierras altas del Sur del Brasil y fluye unas 3.000 millas (4.827 km) hacia el sur y el oeste. (p. 203)

parliament [parlamento] *s.* cuerpo legislativo representativo cuyos miembros son elegidos o designados y cuyas funciones legislativas y ejecutivas están combinadas. (pp. 158, 303)

parliamentary government [gobierno parlamentario] *s.* sistema en el cual las funciones legislativas y ejecutivas están combinadas en una legislatura llamada parlamento. (p. 158)

particulate [macropartícula] *s.* partícula muy pequeña de materia líquida o sólida. (p. 324)

partition [partición] *s.* separación; división en dos o más unidades territoriales con estatus político separado. (p. 574)

pastoral lease [arrendamiento pastoral] *s.* en Australia, gran extensión de terreno que todavía es propiedad del gobierno; los rancheros arriendan la tierra del gobierno. (p. 729)

PCB [PCB (policlorobifenilo)] *s.* un compuesto industrial que se acumula en el tejido animal y puede ocasionar efectos perjudiciales y defectos congénitos; el PCB fue prohibido en los Estados Unidos en 1977. (p. 631)

peat [turba] *s.* materia vegetal parcialmente descompuesta que se encuentra en las turberas. (p. 277)

penal colony [colonia penal] *s.* lugar donde son enviados los prisioneros. (p. 718)

per capita income [ingreso per cápita] *s.* la cantidad de dinero promedio que gana una persona en una unidad política. (p. 94)

permafrost [permafrost (pergelisol)] *s.* terreno permanentemente congelado. (pp. 63, 123)

polder [pólder] *s.* terreno protegido contra el mar u otra masa de agua mediante diques o drenaje. (p. 282)

Polynesia [Polinesia] *s.* una de las tres regiones de Oceanía, cuyo nombre significa, "muchas islas". (p. 713)

population density [densidad poblacional] *s.* el número promedio de habitantes de una zona mensurable, el cual se obtiene dividiendo el número de habitantes en la zona por la cantidad de tierra que ocupan. (p. 81)

population pyramid [pirámide poblacional] *s.* un dispositivo gráfico que muestra la distribución de una población por sexo y edad. (p. 79)

postindustrial economy [economía postindustrial] *s.* fase económica en la cual la manufactura no desempeña un papel dominante. (p. 142)

Prairie Provinces [Las Provincias de las Praderas] *s.* en Canadá, las provincias que se encuentran al oeste de Ontario y Quebec: Manitoba, Saskatchewan y Alberta. (p. 168)

precipitation [precipitación] *s.* gotas de agua que caen en forma de lluvia, aguanieve, nieve o granizo. (p. 50)

prevailing westerlies [vientos del oeste predominantes] *s.* vientos que soplan de oeste a este. (p. 124)

prime meridian [primer meridiano] *s.* la línea imaginaria a cero meridiano usada para medir longitud de este a oeste, y que divide la Tierra en dos mitades, este y oeste; llamado meridiano de Greenwich porque pasa por Greenwich, Inglaterra. (p. 6)

prime minister [primer ministro] *s.* la cabeza del gobierno; el líder del partido de la mayoría en el parlamento. (p. 158)

privatization [privatización] *s.* la venta de empresas propiedad del Estado a ciudadanos privados. (p. 388)

province [provincia] *s.* una unidad política. (p. 156)

pull factor [factor de atracción] *s.* un factor que atrae o arrastra a personas a otro lugar. (pp. 81, 211)

push factor [factor de empuje] *s.* un factor que hace que la gente abandone sus hogares y emigre a otra región. (pp. 81, 211, 730)

Pyongyang [Pyongyang] *s.* la ciudad más grande de Corea del Norte, con más de 2.500 millones de habitantes. (p. 650)

Q

Qin Ling Mountains [Montañas de Qin Ling] *s.* montañas de la región sudeste y este central de China; dividen la parte norte de China de la parte sur. (p. 619)

Quebec [Quebec] *s.* una de las provincias más importantes del Canadá. (p. 167)

Quechua [Quechua] *s.* idioma del Imperio Inca, hablado actualmente en las tierras de la zona andina. (p. 231)

R

rai [rai] *s.* tipo de música popular argelina compuesta en la década de 1920 por niños pobres de las zonas urbanas, con ritmos rápidos bailables; algunas veces se usó como forma de rebeldía para expresar el descontento político. (p. 440)

rain forest [selva tropical] *s.* una región selvática ubicada en la Zona Tropical con una gran concentración de diferentes especies de árboles de hojas anchas. (pp. 66, 207)

rain shadow [sombra de lluvia] *s.* la tierra del lado de sotavento de colinas o montañas que recibe muy poca lluvia del aire seco descendiente. (p. 51)

raj [raj] *s.* el período de gobierno británico en la India que duró cerca de 200 años, de 1857 a 1947. (p. 568)

Ramadan [Ramadán] *s.* práctica islámica de ayunar un mes desde que sale el sol hasta que se pone. (p. 576)

rate of natural increase [tasa de aumento natural] *s.* llamada también tasa de crecimiento demográfico; la tasa de crecimiento poblacional, que se encuentra restando la tasa de mortalidad de la tasa de natalidad. (p. 79)

recession [recesión] *s.* un período prolongado de descenso en la actividad comercial general. (p. 667)

Red Army [Ejército Rojo] *s.* nombre del ejército de la Unión Soviética. (p. 371)

refinery [refinería] *s.* lugar donde el petróleo crudo es convertido en productos útiles. (p. 497)

Reformation [Reforma] *s.* movimiento en Europa Occidental iniciado en 1517, cuando muchos cristianos se desligaron de la Iglesia Católica para fundar iglesias protestantes; esto produjo hostilidades mutuas y guerras religiosas que desgarraron a Europa. (p. 297)

reggae [reggae] *s.* un estilo de música creado en Jamaica en la década de 1960, que tiene sus raíces en la música africana, caribeña y americana, con frecuencia trata sobre problemas sociales y religión. (p. 227)

relative location [ubicación relativa] *s.* describe un lugar en relación con otros lugares que lo rodean. (p. 6)

relief [relieve] *s.* la diferencia en elevación de una forma fisiográfica, desde el punto más bajo hasta el punto más alto. (p. 36)

religion [religión] *s.* la creencia en un poder o poderes sobrenaturales que se consideran como los creadores y conservadores del universo, así como el propio sistema de creencias. (p. 75)

Renaissance [Renacimiento] *s.* época de renovado interés por la educación y las artes que duró del *s.* XIV al *s.* XVI; comenzó en los estados-ciudades italianos y se extendió hacia el norte por toda Europa. (p. 291)

representative democracy [democracia representativa] *s.* un gobierno en el cual el pueblo gobierna mediante sus representantes elegidos. (p. 139)

republic [república] *s.* gobierno en el cual los ciudadanos eligen a sus representantes para que gobiernen en su nombre. (p. 290)

reserve [reserva] *s.* terrenos públicos destinados por el gobierno para los pueblos indígenas. (p. 162)

Richter scale [escala de Richter] *s.* una forma de medir información registrada por los sismógrafos para determinar la fuerza relativa de un terremoto. (p. 40)

rift valley [Valle del Rift] *s.* un valle largo y delgado creado por la separación de las placas continentales, presente en África Oriental, el cual se prolonga por 4.000 millas (6.436 km) desde Jordania en el Sudoeste asiático hasta Mozambique en el Sur de África. (p. 416)

Ring of Fire [El Anillo de Fuego] *s.* la cadena de volcanes que bordea la cuenca del Pacífico. (pp. 41, 661)

Rocky Mountains [Las Montañas Rocosas] *s.* un importante sistema montañoso de los Estados Unidos y el Canadá que se extiende por 3.000 millas (4.827 km) desde Alaska hacia el Sur hasta Nuevo México. (p. 119)

Rub al Khali [Rub' Al Jali] *s.* conocido también como el Cuarto Vacío, uno de los desiertos arenosos más grandes del mundo, abarca cerca de 250.000 millas cuadradas (650 mil kilómetros cuadrados); ubicado en la Península Arábiga. (p. 491)

Russian Revolution [Revolución Rusa] *s.* la revuelta de 1917 por la cual el Partido Comunista Ruso dirigido por V. I. Lenin, tomó el control del gobierno de los zares. (p. 363)

runoff [escorrentía] *s.* agua de lluvia no absorbida por el suelo y que puede transportar pesticidas y fertilizantes de los campos a los ríos, poniendo en peligro la cadena alimentaria. (p. 353)

S

Sahara [Sahara] *s.* el desierto más grande del mundo, que se extiende 3.000 millas (4.827 km) por el continente africano, desde el Océano Atlántico hasta el Mar Rojo, y mide 1.200 millas (1.930 km) de norte a sur. (p. 420)

Sahel [Sahel] *s.* una banda estrecha de pradera seca, que se extiende de este a oeste en el borde sur del Sahara, usada para agricultura y pastoreo. (p. 424)

St. Lawrence Seaway [La Ruta Marítima del San Lorenzo] *s.* la ruta de barcos de aguas profundas más importante de América del Norte, la cual conecta los Grandes Lagos con el Océano Atlántico a través del Río San Lorenzo. (p. 129)

St. Petersburg [San Petersburgo] *s.* la vieja capital de Rusia, fundada por Pedro el Grande, que trasladó la capital allí desde Moscú, debido a que San Petersburgo tenía acceso directo por mar hacia Europa Occidental. (p. 362)

salt flat [salinas] *s.* terrenos planos formados por sales químicas que permanecen después de que los vientos evaporan la humedad del suelo. (p. 492)

samba [samba] *s.* danza brasileña con influencia africana. (p. 239)

samurai [samurai] *s.* soldado profesional japonés al servicio de los intereses de terratenientes y jefes de clanes. (p. 651)

satellite nation [país satélite] *s.* un país dominado por otro. (p. 312)

savanna [sabana] *s.* término para describir las llanuras herbáceas que carecen de árboles en su mayor parte, en la región de las praderas tropicales. (p. 66)

seawork [espigón] *s.* una estructura utilizada para controlar el impacto destructivo del mar en la vida humana. (p. 283)

sectionalism [faccionalismo] *s.* cuando la gente pone su lealtad a su región o sección por encima de la lealtad al país. (p. 136)

sediment [sedimento] *s.* pequeños trozos de roca producidos por la acción de los elementos. (p. 42)

seismograph [sismógrafo] *s.* un dispositivo para medir el tamaño de las ondas creadas por un terremoto. (p. 39)

Seoul [Seúl] *s.* la ciudad más grande de Corea del Sur, con una población de más de diez millones de habitantes. (p. 650)

Serengeti [Serengeti] *s.* zona de África Oriental, que tiene muchas de las mejores praderas del mundo y muchos animales de pastoreo. (p. 422)

service industry [industria de servicios] *s.* cualquier tipo de actividad económica que produce servicios en vez de productos. (p. 142)

Sherpa [Sherpa] *s.* una persona de ascendencia tibetiana en Nepal, que trabaja como guía tradicional en la región del Monte Everest. (p. 582)

Shi'ite [Shiita] *s.* una de las dos principales ramas del Islam, que incluye a la mayoría de los iraníes y parte de las poblaciones de Irak y Afganistán. (p. 516)

shogun [Shogun] *s.* el general del ejército del emperador con poderes de dictador militar, una posición creada por el emperador del Japón en 1192 después de una lucha entre dos clanes poderosos. (p. 651)

Siberia [Siberia] *s.* región del centro y la zona este de Rusia que se extiende desde los Montes Urales hasta el Océano Pacífico, conocida por sus recursos minerales y por ser un lugar de exilio político. (p. 349)

Siddhartha Gautama [Siddhartha Gautama] *s.* fundador del budismo y conocido como Buda, nacido en el Sur de Nepal en el siglo sexto antes de nuestra era. (p. 582)

Silicon Glen [Silicon Glen] *s.* sección de Escocia entre Glasgow y Edimburgo, así denominada por su alta concentración de compañías de alta tecnología. (p. 305)

Silk Road [La Ruta de la Seda] *s.* la ruta de 4.000 millas (6.436 km) entre China y el Mar Mediterráneo, así llamada por la costosa seda adquirida en China. (p. 375)

silt [limo] *s.* material sedimentario suelto que contiene partículas de roca muy pequeñas, formado por depósitos de ríos y muy fértil. (p. 426)

Sinhalese [cingalés] *s.* pueblo indo-ario que cruzó el estrecho que separa la India y Sri Lanka en el siglo sexto antes de nuestra era y creó una civilización avanzada, adoptando el budismo. (p. 584)

sirocco [siroco] *s.* viento cálido y constante del Sur que sopla desde África del Norte a través del Mar Mediterráneo hasta el Sur de Europa, usualmente en la primavera. (p. 279)

slash-and-burn [cortar y quemar] *s.* método para despejar los campos para plantar, que consiste en cortar y quemar árboles, arbustos y hierbas. (p. 210)

smart growth [crecimiento inteligente] *s.* el uso eficiente y la conservación de la tierra y otros recursos. (p. 178)

smog [smog] *s.* una niebla marrón que se produce cuando los gases liberados por combustibles fósiles en combustión reaccionan con la luz solar. (p. 324)

society [sociedad] *s.* un grupo que comparte una región geográfica, un idioma común y un sentido de identidad y cultura. (p. 71)

soil [suelo] *s.* la mezcla suelta de roca meteorizada, material orgánico, aire y agua que permiten el crecimiento de las plantas. (p. 45)

solar system [sistema solar] *s.* se compone del sol y nueve planetas conocidos, así como otros cuerpos celestes que gravitan alrededor del sol. (p. 27)

solstice [solsticio] *s.* cualquiera de dos épocas en el año cuando los rayos solares brillan directamente arriba al mediodía en los puntos más alejados al norte o al sur, y que marcan el comienzo del verano o el invierno; en el Hemisferio Norte, el solsticio de verano es el día más largo y el solsticio de invierno, el más corto. (p. 49)

South, the [sur, el] *s.* una región que cubre aproximadamente un cuarto de la superficie terrestre de los Estado Unidos y contiene más de un tercio de su población. (p. 148)

South Slav [eslavo del sur] *s.* una persona que emigró de Polonia y Rusia y se estableció en la Península Balcánica alrededor del año 500. (p. 319)

Spanish conquest [conquista española] *s.* la conquista de los pueblos indígenas americanos por los españoles. (p. 217)

sphere of influence [esfera de influencia] *s.* un método de dividir el control extranjero en China, después de que el país fuera obligado a firmar una serie de tratados otorgando privilegios especiales a los europeos. China fue dividida para el control por Gran Bretaña, Francia, Alemania y Rusia, entre otras potencias. (p. 636)

state [estado] *s.* término político para describir una unidad independiente que ocupa un territorio específico y tiene pleno control de sus asuntos internos y externos. (p. 83)

stateless nation [nación sin estado] *s.* un pueblo que no tiene un territorio que pueda ocupar legalmente, como los palestinos, los kurdos y los vascos. (p. 526)

stateless society [sociedad sin estado] *s.* una sociedad en la cual la gente usa linajes o familias cuyos miembros descienden de un antepasado común para gobernarse. (p. 443)

steppe [estepa] *s.* término usado para la región de praderas templadas en el Hemisferio Norte. (p. 66)

Stolen Generation [La Generación Robada] *s.* en Australia, término que utilizan los aborígenes actualmente para denominar a los 100.000 niños de raza mixta que fueron tomados por el gobierno y entregados a familias blancas para promover la asimilación. (p. 728)

storm surge [olas ciclónicas] *s.* altos niveles de agua producidos por un ciclón que inunda zonas de bajo nivel. (p. 562)

strategic commodity [recurso estratégico] *s.* un recurso tan importante que las naciones están dispuestas a ir a la guerra para asegurar el suministro continuo del mismo. (p. 529)

subcontinent [subcontinente] *s.* una masa de tierra similar a un continente, aunque de menor extensión, como Asia del Sur, denominada el subcontinente Indio. (p. 551)

subsistence activity [actividad de subsistencia] *s.* una actividad en la cual una familia produce únicamente los alimentos, la ropa y la vivienda que necesita. (p. 714)

suburb [suburbio] *s.* una unidad o comunidad política que linda con la ciudad central o con otros suburbios que lindan con la ciudad. (pp. 87, 138)

sultan [sultán] *s.* el gobernante de un país musulmán. (p. 585)

summer monsoon [monzón húmedo (verano)] *s.* la estación cuando los vientos soplan desde el sudoeste a través del Océano Índico hacia Asia del Sur, desde junio hasta septiembre, cuando los vientos producen fuertes tormentas y graves inundaciones. (p. 597)

Sunni [Suni] *s.* una de las dos principales ramas del Islam, la cual abarca cerca del 83 por ciento de todos los musulmanes, incluyendo los que viven en Turquía, Irak y Afganistán. (p. 516)

supra [supra] *s.* término georgiano (ruso) para designar una cena con muchos platos, brindis y discursos cortos. (p. 374)

sustainable community [comunidad sostenible] *s.* una comunidad cuyos residentes pueden vivir y trabajar en armonía con el medio ambiente. (p. 178)

sweatshop [fábrica explotadora] *s.* un lugar de trabajo donde se trabajan largas horas por salario bajo y en malas condiciones para enriquecer a los fabricantes. (p. 667)

T

taiga [taiga] *s.* una faja casi continua de bosques coníferos de hojas perennes, a través del Hemisferio Norte en América del Norte y Eurasia. (p. 351)

Taklimakan Desert [Takla-Makan] *s.* desierto ubicado en la región occidental de China entre las montañas de Tian Shan y Kunlún. (p. 627)

Taliban [Talibán] *s.* un grupo musulmán estricto en Afganistán que ha impuesto reglas muy rígidas en la sociedad, incluyendo estilos de vestuario para hombres y mujeres, restricciones en la apariencia de las mujeres en lugares públicos y reglamentos para la televisión, la música y los videos. (p. 517)

Tamil [Tamil] *s.* hindú dravídico que llegó a Sri Lanka en el *s.* IV y se estableció en el norte, mientras los sinhaleses se trasladaron más al sur. (p. 584)

Taoism [Taoísmo] *s.* filosofía basada en el libro Tao Te Ching y las enseñanzas de Lao-Tsé, que vivió en China en el siglo VI antes de nuestra era, quien creía en conservar y restaurar la armonía dentro del individuo, con la naturaleza y con el universo, con poca intervención del gobierno. (p. 638)

taro [taro] *s.* planta tropical de Asia con raíz a base de féculas, la cual se puede comer como un vegetal hervido o preparada como pan, budín o una pasta llamada "poi". (p. 715)

tectonic plate [placa tectónica] *s.* una enorme plataforma móvil que forma la corteza de la Tierra. (p. 37)

Tenochtitlan [Tenochtitlán] *s.* la antigua capital de los aztecas, donde se encuentra la Ciudad de México en la actualidad. (p. 217)

terpen [terpén] *s.* plataformas altas de tierra de barro usadas en trabajos de mar. (p. 283)

terraced farming [cultivo en andenes] *s.* una técnica antigua para cultivar la tierra en laderas o pendientes de montañas, utilizando campos horizontales a manera de peldaños, cortados en las pendientes. (p. 211)

terrorism (terrorismo) *s.* uso ilegal o amenazante de la fuerza, o violencia, contra individuos o propiedades, con el propÙsito de intimidar o causar temor con fines polÌticos o sociales. (p. 173)

theocratic [teocrático] *adj.* una forma de gobierno en la cual líderes religiosos controlan el gobierno con leyes religiosas y consultando con eruditos religiosos. (p. 504)

Three Gorges Dam [Presa de las Tres Gargantas] *s.* una presa que se comenzó a construir a finales del siglo. XX en Chang Jiang, China, para ayudar a controlar las inundaciones, generar energía y permitir que los barcos naveguen más hacia el interior de China. (p. 628)

Three Kingdoms [Los Tres Reinos] *s.* los reinos formados en la península de Corea por el año 300 de nuestra era: Koguryo en el norte, Paikche en el sudoeste y Silia en el sudeste. (p. 647)

Tigris River [Río Tigris] *s.* uno de los ríos más importantes del Sudoeste Asiático; sirvió de base a varias civilizaciones antiguas en el valle del río y fluye por partes de Turquía, Siria e Irak. (p. 489)

tornado [tornado] *s.* una poderosa columna de aire en espiral en forma de túnel. (p. 51)

topographic map [mapa topográfico] *s.* un mapa para referencia general; representación de características terrestres, naturales y hechas por el hombre. (p. 11)

topography [topografía] *s.* las características combinadas de formas fisiográficas y su distribución en una región. (p. 36)

Transcaucasia [Transcaucasia] *s.* una región compuesta por las repúblicas de Armenia, Azerbaiján y Georgia; situada entre el Cáucaso y las fronteras de Turquía e Irán. (p. 346)

Trans-Siberian Railroad [Ferrocarril Transiberiano] *s.* un ferrocarril que uniría Moscú con el Puerto de Vladivostok en el Pacífico; construido entre 1891 y 1903. (p. 355)

Treaty of Tordesillas [Tratado de Tordesillas] *s.* un tratado entre España y Portugal firmado en 1494, por el cual Portugal obtuvo el control de la tierra que hoy constituye el Brasil. (p. 236)

Treaty of Waitangi [Tratado de Waitangi] *s.* tratado firmado por los británicos y los maorís en 1840, por el cual Gran Bretaña obtuvo el control de Nueva Zelanda. (p. 719)

tsunami [tsunami] *s.* una ola oceánica gigantesca, producida por un terremoto o erupción volcánica subacuático, con gran poder de destrucción. (p. 40, 662)

tuberculosis [tuberculosis] *s.* una infección respiratoria propagada a través del contacto humano, que con frecuencia acompaña al SIDA. (p. 466)

tundra [tundra] *s.* las tierras planas sin árboles que forman un aro alrededor del Océano Ártico; la región climática del Océano Ártico. (p. 63)

typhoon [tifón] *s.* una tormenta tropical, como un huracán, que se da en la región occidental del Pacífico. (pp. 51, 625)

U

USSR [URSS] *s.* la Unión de Repúblicas Socialistas Soviéticas o Unión Soviética, formada en 1922 por los comunistas y disuelta oficialmente en 1991. (p. 363)

UNICEF (United Nations Children's Fund) [UNICEF (Fondo de las Naciones Unidas para la Infancia)] *s.* organización internacional de vigilancia y ayuda para los niños. (p. 665)

United Provinces of Central America [Provincias Unidas de Centroamérica] *s.* nombre de Centroamérica después de que la región declaró su independencia de México en 1823. (p. 223)

upland [tierras altas] *s.* una colina o una montaña muy baja que también puede contener mesas y planicies altas. (p. 275)

Ural Mountains [Montes Urales] *s.* la cordillera que separa las planicies del norte de Europa y Siberia Occidental y utilizada como la línea divisoria entre Europa y Asia. (p. 346)

urban geography [geografía urbana] *s.* el estudio de cómo las personas utilizan el espacio en las ciudades. (p. 87)

urbanization [urbanización] *s.* el dramático aumento en el número de ciudades y los cambios en estilos de vida que resultan del mismo. (p. 88)

urban sprawl [expansión urbana descontrolada] *s.* desarrollo mal planificado que extiende la población de una ciudad por una zona geográfica cada vez más amplia. (p. 176)

V

Vietnam War [Guerra de Vietnam] *s.* (1954-1975) el conflicto militar producido por la intervención de Estados Unidos en Vietnam del Sur para evitar su apoderamiento por los comunistas de Vietnam del Norte. (p. 707)

volcano [volcán] *s.* un evento natural, formado cuando magma, gases y agua de la parte inferior de la corteza o capa se acumulan en cámaras subterráneas y posteriormente hacen erupción y surgen por grietas en la superficie terrestre. (p. 40)

voyaging canoe [canoas viajeras] *s.* una embarcación grande construida por habitantes de las islas del Pacífico para navegar por el océano. (p. 699)

W

wadi [wadi] *s.* lecho de un río que permanece seco excepto durante la estación lluviosa. (p. 488)

water table [nivel hidrostático] *s.* el nivel en el cual las rocas se saturan. (p. 33)

weather [clima] *s.* las condiciones atmosféricas en un lugar y tiempo específicos. (p. 50)

weathering [meteorización] *s.* proceso químico y físico que cambia las características de las rocas en o cerca de la superficie terrestre, lo cual ocurre lentamente durante el lapso de muchos años. (p. 42)

West [Oeste] *s.* región de América del Norte compuesta por 13 estados, que se extiende desde las Grandes Llanuras hasta el Océano Pacífico e incluye Alaska por el norte y Hawaii en el Pacífico. (p. 148)

West Bank [Cisjordania] *s.* en Israel, una franja de tierra en el lado oeste del Río Jordán, originalmente controlada por Jordania, que forma parte de la tierra destinada para los árabes palestinos. (p. 527)

Western Wall [El Muro de los Lamentos] *s.* para los judíos, el sitio más sagrado de Jerusalén; lo único que queda del Segundo Templo, construido en 538 antes de nuestra era y destruido en el 70 de nuestra era por los romanos. (p. 510)

***Wik* Case** [el caso *Wik*] *s.* en Australia, los tribunales dispusieron en este caso que los aborígenes pueden reclamar tierras retenidas bajo arrendamiento pastoral. (p. 729)

winter monsoon [monzón seco (invierno)] *s.* la estación cuando los vientos secos soplan desde el noreste a través de los montes Himalaya hacia el mar desde octubre hasta febrero, algunas veces causando sequías. (p. 597)

X

Xi Jiang [Xi Jiang] *s.* llamado también el Río del Oeste; río que fluye hacia el este a través del sudeste de China y se une con el Río Perla (Zhu Jiang) para desembocar en el Mar del Sur de la China, formando un estuario entre Hong Kong y Macao. (p. 621)

Y

yurt [yurt] *s.* una tienda de nómadas del Asia Central. (p. 379)

Z

Zionism [sionismo] *s.* movimiento iniciado en el siglo. XIX para crear y apoyar una patria judía en Palestina. (p. 51)

Zuider Zee [Zuiderzee] *s.* antiguo lago interior de los Países Bajos en el Mar del Norte. (p. 283)

The U.S. Geological Survey (USGS) is the primary source for latitudes and longitudes. USGS reports these figures for any geographic feature (or political division) as averages of all the values within that feature. These averages help to locate on a map such large or extended features as continents, seas, rivers, or mountain ranges. However, one must look at a map to learn the overall shape, size, and extent of any geographic feature.

Abkhazia (43°00′N/41°00′E) A republic in northwestern Georgia in Transcaucasia, 386, *m385*

Abu Dhabi (24°28′N/54°22′E) The capital of United Arab Emirates, 484

Abuja (9°05′N/7°32′E) The capital of Nigeria, 412

Accra (5°33′N/0°13′W) The capital of Ghana, 410

Addis Ababa (9°02′N/38°42′E) The capital of Ethiopia, 433

Adriatic Sea (43°30′N/14°27′E) An arm of the Mediterranean Sea, bounded by Italy, Croatia, Yugoslavia, and Albania, 281

Afghanistan (33°00′N/65°00′E) A country in the northeast region of Southwest Asia, 516

Africa (10°00′N/22°00′E) The second largest continent; bounded by the Mediterranean Sea, the Indian Ocean, the Red Sea, and the Atlantic Ocean, 28, *m29*

Ahmadabad (23°04′N/72°38′E) A city in western India, 570, *m569*

Al Jawlan (also called the Golan Heights) (33°00′N/35°45′E) A hilly plateau in Syria overlooking the Jordan River and the Sea of Galilee, 487

Alabama (32°45′N/86°45′W) A state in the southern United States, 108

Alaska (64°00′N/150°00′W) A U.S. state that is northwest of Canada, 148

Albania (41°00′N/20°00′E) A country in Eastern Europe, 308

Albany (42°40′N/73°48′W) The capital of New York, 110

Alberta (55°00′N/115°00′W) A Prairie Province of Canada, 168

Aleutian Islands (52°06′N/173°30′W) Rugged, treeless islands that extend in an arc off the coast of Alaska, 121

Algeria (28°00′N/3°00′E) A country in North Africa, 438

Algiers (36°46′N/3°03′E) The capital of Algeria, 408

Alps (46°25′N/10°00′E) A European mountain range that arcs across France, Italy, Germany, Switzerland, Austria, and into the Balkan Peninsula, 272, *m271*

Altai Mountains (48°00′N/90°00′E) A mountain system in Central Asia, 346

Amazon rain forest A large, tropical forest located in north-central South America, 208

Amazon River (0°10′S/49°00′W) The world's second longest river; flows from northern Peru across northern Brazil to the Atlantic Ocean, 203, *m203*

American Samoa (14°21′S/170°31′W) A U.S. territory in the Pacific made up of the eastern islands of the Samoan archipelago, 112

Amman (31°57′N/35°56′E) The capital of Jordan, 484

Amsterdam (52°21′N/4°55′E) The capital of the Netherlands, 268

Amu Darya (43°40′N/59°01′E) A river that flows from the Pamir Mountains across south Central Asia to the southern Aral Sea, 347

An-Nafud (28°30′N/40°30′E) A desert in the northern part of the Arabian Peninsula, 492

Anatolia (39°00′N/35°00′E) A peninsula in northwestern Southwest Asia occupied by Turkey, 487, *m488*

Andes (20°00′S/67°00′W) A mountain range that runs down the Pacific coast of South America, 201, *m203*

Andorra (42°30′N/1°30′E) A tiny country between France and Spain in the Pyrenees, 266

Andorra la Vella (42°33′N/1°26′E) The capital of Andorra, 266

Angola (12°30′S/18°30′E) A country in Southern Africa, 453, *m454*

Ankara (39°56′N/32°52′E) The capital of Turkey, 484

Annamese Cordillera (17°00′N/106°00′E) A mountain range in Southeast Asia, 689, *m689*

Annapolis (38°59′N/76°30′W) The capital of Maryland, 110

Antananarivo (18°55′S/47°31′E) The capital of Madagascar, 410

Antarctica (90°00′S) A continent located mostly south of the Antarctic Circle; bounded by the Atlantic, Pacific, and Indian oceans, 28, *m29*

Antigua and Barbuda (17°03′N/61°48′W) A country that consists of islands in the eastern Caribbean Sea, 196

Apennines (43°00′N/13°00′E) A European mountain range that runs down the center of Italy, dividing the Italian Peninsula from east to west, 272

Apia (13°50′S/171°44′W) The capital of Samoa, 684

Appalachian Mountains (40°00′N/78°00′W) A North American mountain chain that runs north to south about 1,600 miles from Newfoundland to Alabama, *m118*, 119

Arabian Peninsula (25°00′N/45°00′E) A peninsula separated from the continent of Africa by the Red Sea on the southwest and separated from Iran by the Persian Gulf on the east, 487

Arabian Sea (20°00′N/65°00′E) The northwest area of the Indian Ocean; lies between the Arabian Peninsula and western India, 488

Aral Sea (45°00′N/60°00′E) An inland sea in Central Asia, 348

Ararat, Mount (39°40′N/44°24′E) A mountain in Turkey, 478

Arctic Ocean (66°40′N/167°55′W) The world's smallest ocean; surrounds the North Pole between North America and Eurasia, 32

Argentina (34°00′S/64°00′W) A country in southern South America, 230, *m234*

Arizona (34°30′N/111°30′W) A state in the western United States, 108

Arkansas (34°45′N/92°30′W) A state in the southern United States, 148

Armenia (40°00′N/45°00′E) A country in Transcaucasia, 370, *m370*

Ashgabat (37°57′N/58°23′E) The capital of Turkmenistan, 342

Asia The largest continent; bounded by the Pacific Ocean, Europe, the Arctic Ocean, and the Indian Ocean, 28, *m29*

Asmara (15°20′N/38°56′E) The capital of Eritrea, 408

Astana (51°11′N/71°26′E) The capital of Kazakhstan, 342

Asunción (25°16′S/57°40′W) The capital of Paraguay, 198

Aswan High Dam (23°57′N/32°52′E) A dam on the Nile River in southern Egypt, 426, *m427*

Atacama Desert (24°30′S/69°15′W) An arid region in northern Chile, 209

Athens (37°45′N/23°30′E) The capital of Greece, 266

Atlanta (33°45′N/84°23′W) The capital of Georgia, 148

Atlantic Coastal Plain (35°00′W/79°00′W) A flat plain that begins as narrow lowland in the northeastern United States and widens as it extends southward, along the Atlantic coast, into Florida, *m118,* 119

Atlantic Ocean (10°00′N/25°00′W) The world's second largest ocean; extends from the Arctic to Antarctica and from the eastern Americas to western Europe and Africa, 32

Atlantic Provinces An area of eastern Canada that includes Prince Edward Island, New Brunswick, Nova Scotia, and Newfoundland, 166

Atlas Mountains (32°00′N/2°00′W) A mountain range in North Africa, 423

Augusta (44°20′N/69°44′W) The capital of Maine, 110

Austin (30°16′N/97°45′W) The capital of Texas, 112

Australia (25°00′S/135°00′E) The smallest continent; southeast of Asia, bounded by the Pacific and Indian oceans, and occupied by the country of Australia 28, *m29*

Austria (47°20′N/13°20′E) A country in Western Europe, 294

Azerbaijan (40°30′N/47°30′E) A country in Transcaucasia, 370, *m370*

Baffin (68°30′N/70°00′W) A large island in northern Canada, 121

Baghdad (33°20′N/44°24′E) The capital of Iraq, 484

Bahamas (24°00′N/76°00′W) A country that consists of a group of islands in the Atlantic Ocean near southeastern Florida, 203

Bahrain (26°00′N/50°30′E) A country in the Arabian Peninsula of Southwest Asia, 503

Baikal, Lake (54°00′N/109°00′E) The deepest lake in the world; located in south-central Russia, 348

Bairiki (1°19′N/172°58′E) The capital of Kiribati, 684

Baku (40°22′N/49°54′E) The capital of Azerbaijan, 371

Balkan Mountains (43°15′N/25°00′E) A European mountain range that blocks the Balkan Peninsula from the rest of Europe, 272

Balkan Peninsula (44°00′N/23°00′E) A southeastern peninsula of Europe; bounded by the Adriatic and Aegean Seas and occupied by numerous countries, *m271,* 272

Baltic States An area between Russia and the Baltic Sea that consists of Latvia, Lithuania, and Estonia, 361

Baltic Sea (56°00′N/18°00′E) An arm of the Atlantic Ocean bounded by Finland, Sweden, Denmark, Germany, Poland, Russia, Lithuanian, Latvia, and Estonia, 271, *m271*

Bamako (12°39′N/8°00′W) The capital of Mali, 410

Bandar Seri Begawan The capital of Brunei, 684

Bandung (6°56′S/107°35′E) A city in Indonesia, 732

Bangkok (13°45′N/100°31′E) The capital of Thailand, 686

Bangladesh (24°00′N/90°00′E) A country in South Asia, 573

Bangui (4°22′N/18°35′E) The capital of Central African Republic, 408

Banjul (13°27′N/16°35′W) The capital of Gambia, 410

Barbados (13°10′N/59°32′W) An island country in the eastern Lesser Antilles in the Atlantic Ocean, 196

Basseterre (17°18′N/62°43′W) The capital of Saint Kitts and Nevis, 198

Baton Rouge (30°27′N/91°09′W) The capital of Louisiana, 108

Bay of Bengal (15°00′N/90°00′E) An arm of the Indian Ocean that lies between eastern India and Southeast Asia, 552, *m554*

Beijing (39°56′N/116°23′E) The capital of China, 637

Beirut (33°52′N/35°31′E) The capital of Lebanon, 484

Belarus (53°00′N/28°00′E) A country that is west of Russia, 361

Belgium (50°50′N/4°00′E) A country in Western Europe, 294

Belgrade (44°49′N/20°28′E) The capital of Yugoslavia, 268

Belize (17°15′N/88°45′W) A country in Central America, 223

Belmopan (17°15′N/88°46′W) The capital of Belize, 196

Benelux The Western European countries of Belgium, the Netherlands, and Luxembourg, 294

Benin (9°30′N/2°15′E) A country in West Africa, 442

Berlin (52°31′N/13°24′E) The capital of Germany, 266

Bern (46°55′N/7°28′E) The capital of Switzerland, 268

Bhutan (27°30′N/90°30′E) A country in South Asia, 580

Bikini (11°35′N/165°23′E) A coral island of the Marshall Islands, 700

Bishkek (42°54′N/74°36′E) The capital of Kyrgyzstan, 342

Bismarck (46°48′N/100°46′W) The capital of North Dakota, 110

Bissau (11°51′N/15°35′W) The capital of Guinea-Bissau, 410

Black Sea (43°00′N/35°00′E) A sea situated between northern Turkey and southwestern Russia, 488, *m488*

Blanc, Mont (45°55′N/6°55′E) A mountain on the border of France and Italy, 260

Blue Ridge Mountains (35°30′N/82°50′W) A North American mountain range located in the southern part of the Appalachian system, 119

Bogotá (4°36′N/74°05′W) The capital of Columbia, 211

Boise (43°38′N/116°11′W) The capital of Idaho, 108

Bolivia (17°00′S/65°00′W) A country in central South America, 230, *m234*

Bombay (also called Mumbai) (18°59′N/72°50′W) A city in western India, *m569,* 570

Bosnia and Herzegovina (44°15′N/17°50′E) A country in Eastern Europe, 308

Bosporus Strait (41°00′N/29°00′E) A narrow waterway in northwest Turkey that connects the Black Sea and the Sea of Marmara, 488

Boston (42°18′N/71°05′W) A seaport city and capital of Massachusetts, 137, *m145*

Botswana (22°00′S/24°00′E) A country in Southern Africa, 453, *m454*

Brahmaputra River (24°02′N/90°59′E) A river starting in China that flows east, then west and south through Bangladesh where it joins the Ganges River; together they form a huge delta before entering the Bay of Bengal, 553

Brasília (15°47′S/47°55′W) The capital of Brazil, 239

Bratislava (48°09′N/17°07′E) The capital of Slovakia, 268

Brazil (10°00′S/55°00′W) A country in central South America, 236

Brazilian Highlands (18°00′S/47°00′W) A mountainous area in southeastern Brazil, 202, *m203*

Brazzaville (4°16′S/15°17′E) The capital of Republic of Congo, 408

Bridgetown (13°06′N/59°37′W) The capital of Barbados, 196

British Columbia (55°00′N/125°00′W) The Pacific Province of western Canada, 169

Brunei (4°30′N/114°40′E) A Southeast Asian country on the island of Borneo, 705

Brussels (50°50′N/4°20′E) The capital of Belgium, 266

Bucharest (44°26'N/26°06'E) The capital of Romania, 268

Budapest (47°30'N/19°05'E) The capital of Hungary, 266

Buenos Aires (34°35'S/58°40'W) The capital of Argentina, 211

Bujumbura (3°23'S/29°22'E) The capital of Burundi, 408

Bulgaria (43°00'N/25°00'E) A country in Eastern Europe, 308

Burkina Faso (13°00'N/2°00'W) A country in West Africa, 442

Burundi (3°00'S/29°30'E) A country in East Africa, 431

Cairo (30°03'N/31°15'E) The capital of Egypt, 408

California (37°15'N/119°45'W) A state in the western United States, 149

Cambodia (13°00'N/105°00'E) A country in Southeast Asia, 705

Cameroon (6°00'N/12°00'E) A country in Central Africa, 448, *m450*

Cameroon, Mount (4°12'N/9°11'E) A mountain in Cameroon, 417

Canada (60°00'N/96°00'W) A country in northern North America that consists of ten provinces and three territories, 117

Canadian Shield (55°00'N/90°00'W) A rocky, flat region that encircles Hudson Bay, *m118, 119*

Canberra (35°17'S/149°13'E) The capital of Australia, 684

Cape Verde (16°00'N/24°00'W) A country formed by a group of islands in West Africa, 442

Caracas (10°30'N/66°55'W) The capital of Venezuela, 198

Caribbean Islands Three major groups of islands: the Bahamas (in the Atlantic Ocean near southeastern Florida) and the Greater Antilles and the Lesser Antilles (in the Caribbean Sea), 203

Caribbean Sea (15°00'N/75°00'W) A body of water bounded by South America, Central America, the Gulf of Mexico, and the Greater Antilles, 191

Carson City (39°10'N/119°43'W) The capital of Nevada, 110

Carthage (36°51'N/10°20'E) A city located in northeastern Tunisia; in ancient times, the center of the Carthaginian Empire, 438

Cascade Range (44°43'N/122°03'W) A North American mountain range that runs parallel to the Pacific coastline from California to British Columbia, Canada, 120

Caspian Sea (42°00'N/50°00'E) A lake that lies between southeast Europe and western Asia, 348

Castries (14°00'N/61°00'W) The capital of Saint Lucia, 198

Catskill Mountains (42°15'N/74°15'W) A North American mountain range located in the northern part of the Appalachian system, 119

Caucasus (42°00'N/45°00'E) A region that includes the Caucasus Mountains, which stretch between the Black and Caspian seas, 385, *m385*

Caucasus Mountains (42°30'N/45°00'E) A mountain range that stretches across the isthmus that separates the Black and Caspian seas, 346

Central Africa A region of Africa that includes Cameroon, Central African Republic, Democratic Republic of the Congo, Republic of the Congo, Equatorial Guinea, Gabon, and São Tomé & Príncipe, 448, *m450*

Central African Republic (7°00'N/21°00'E) A country in Central Africa, 448, *m450*

Central America A Latin American subregion bounded by Mexico, the Caribbean Sea, the Pacific Ocean, and South America, 223

Central Asia A region that includes Kazakhstan, Kyrgyzstan, Tajikistan, Turkmenistan, and Uzbekistan, 375

Central Siberian Plateau (66°00'N/106°00'E) A plateau in central Russia, 345

Chad (15°00'N/19°00'E) A country in West Africa, 442

Chad, Lake (13°20'N/14°00'E) A lake in western Chad; on the borders of Nigeria, Niger, and Chad, 425

Chang Jiang (also called Yangtze River) (31°47'N/121°08'E) The longest river in Asia; flows from Xizang (Tibet) across China to the East China Sea, *m620, 621*

Charleston (38°21'N/81°38'W) The capital of West Virginia, 112

Charlotte Amalie The capital of the U.S. Virgin Islands, 112

Charlottetown (46°14'N/63°08'W) The capital of Prince Edward Island, Canada, 114

Chechnya (43°18'N/45°42'E) A Russian republic in the Caucasus, 385, *m385*

Chennai (also called Madras) (13°05'N/80°17'E) A city in southern India, *m569, 570*

Chernobyl (51°16'N/30°14'E) A city in north-central Ukraine, 392

Cheyenne (41°08'N/104°49'W) The capital of Wyoming, 112

Chicago (41°50'N/87°41'W) Located in northeastern Illinois; the largest city in the Midwest, 137

Chile (30°00'S/71°00'W) A country that runs along the southern Pacific coast of South America, 230, *m234*

China (35°00'N/105°00'E) A country in East Asia, 635

Chisinau (47°00'N/28°51'E) The capital of Moldova, 342

Chittagong A city in Bangladesh, 573

Chittagong Hills (23°00'N/92°15'E) A hilly region in southeastern Bangladesh, 576

Chota Nagpur Plateau (23°00'N/85°00'E) A plateau in India that is northeast of the Deccan Plateau, 552

Cincinnati (39°09'N/84°31'W) A city in southwestern Ohio, 147

Cleveland (41°29'N/81°40'W) A city in northern Ohio, 137

Colombo (6°56'N/79°51'E) The capital of Sri Lanka, 585

Colorado (39°00'N/105°30'W) A state in the western United States, 108

Colorado River (31°54'N/114°57'W) A river that rises in the Rocky Mountains, flow through the southwestern United States, and empties into the Gulf of California in northwest Mexico, 149, *m149*

Colombia (4°00'N/72°00'W) A country in northern South America, 230, *m234*

Columbia (34°00'N/81°02'W) The capital of South Carolina, 112

Columbus (39°59'N/82°59'W) The capital of Ohio, 110

Comoros (12°10'S/44°15'E) A country formed by a group of islands in Southern Africa, 453, *m454*

Conakry (9°31'N/13°43'W) The capital of Guinea, 410

Concord (43°14'N/71°34'W) The capital of New Hampshire, 110

Congo River (6°04'S/12°24'E) A river in Central Africa that flows through the Democratic Republic of the Congo and empties into the Atlantic Ocean, 416

Congo, Democratic Republic of the (0°00'N/25°00'E) A country in Central Africa, 448, *m450*

Congo, Republic of the (1°00'S/15°00'E) A country in Central Africa, 408

Connecticut (41°40'N/72°40'W) A state in New England in the northeastern United States, 145

Continental Divide The line of highest points in the Rockies that marks the separation between rivers flowing eastward and westward, 120

Copenhagen (55°40'N/12°35'E) The capital of Denmark, 266

Core Provinces An area located in east central Canada that includes Quebec and Ontario, 167

Corsica (42°00′N/9°00′E) An French-owned island in the western Mediterranean Sea, 272

Costa Rica (10°00′N/84°00′W) A country in Central America, 223

Côte d'Ivoire (8°00′N/5°00′W) A country in West Africa, 442

Crete (35°15′N/24°45′E) A Greek-owned island in the eastern Mediterranean Sea, 272

Croatia (45°10′N/15°30′E) A country in Eastern Europe, 308

Cuba (21°30′N/80°00′W) An island country in the Caribbean Sea, 226

Cyprus (35°00′N/33°00′E) A Southwest Asian island country in the Mediterranean Sea, south of Turkey, 513

Czech Republic (49°45′N/15°00′E) A country in Eastern Europe, 308

Dagestan (43°00′N/47°00′E) A Russian republic in the Caucasus, 385, *m385*

Dakar (14°40′N/17°26′W) The capital of Senegal, 412

Dallas–Fort Worth A metropolitan area in east central Texas formed by the rapid growth of Dallas (32°47′N/96°48′W) and Ft. Worth (32°44′N/97°19′W), 148

Damascus (33°30′N/36°18′E) The capital of Syria, 484

Damavand, Mount (35°57′N/52°07′E) A mountain in Iran, 478

Danube River (45°20′N/29°40′E) A European river that flows from southwest Germany, across southeast Europe, and into the Black Sea, 273, *m273*

Dardenelles Strait A narrow waterway in northwest Turkey that joins the Sea of Marmara and the Aegean Sea, 488

Dasht-e Kavir Desert A salt flat desert in central Iran, 492

Dasht-e Lut Desert A salt flat desert in eastern Iran, 492

Dead Sea (31°30′N/35°30′E) A landlocked salt lake that lies between Israel and Jordan, 489

Deccan Plateau (14°00′N/77°00′E) A large plateau in central India, 552

Delaware (39°00′N/75°30′W) A state in the southern United States (sometimes included with the Middle Atlantic states), 108

Denali (also called Mount McKinley) (63°04′N/151°00′W) North America's highest mountain; located in Alaska, 120

Denmark (56°00′N/10°00′E) A Northern European country located on the Jutland Peninsula, 300

Denver (39°44′N/104°59′W) The capital of Colorado, 108

Des Moines (41°36′N/93°37′W) The capital of Iowa, 108

Detroit (42°20′N/83°03′W) A city in southeastern Michigan, 137

Dhaka (23°43′N/90°25′′E) The capital of Bangladesh, 548

District of Columbia (38°54′N/77°02′W) A federal district in the eastern United States; occupied by the city of Washington, 108

Djibouti (11°30′N/42°30′E) A country in East Africa, 463

Djibouti, the city of (11°36′N/43°09′E) The capital of Djibouti, 408

Dnieper River (46°30′N/32°18′E) A river that flows from west-central Russia through Belarus and Ukraine to the Black Sea, 361

Dodoma (6°10′S/35°45′E) The capital of Tanzania, 412

Doha (25°17′N/51°32′E) The capital of Qatar, 484

Dom (46°06′N/7°51′E) A mountain in Switzerland, 260

Dominica (15°30′N/61°20′W) An island country in the eastern Caribbean Sea, 196

Dominican Republic (19°00′N/70°40′W) A country that occupies the eastern two-thirds of the island of Hispaniola in the Caribbean Sea, 196

Dover (39°09′N/75°31′W) The capital of Delaware, 108

Dublin (53°20′N/6°15′W) The capital of Ireland, 266

Durham (36°00′N/78°54′W) A city in North Carolina, 178

Dushanbe (38°34′N/68°46′E) The capital of Tajikistan, 342

East Africa A region in Africa that includes Burundi, Djibouti, Eritrea, Ethiopia, Kenya, Rwanda, Seychelles, Somalia, Tanzania, and Uganda, 431

East Antarctica (80°00′S/80°00′E) A major region in Antarctica lying on the Indian Ocean side of the Transantarctic Mountains, 692

East Asia A region that includes China, Japan, Mongolia, Taiwan, North Korea, and South Korea, 619

Eastern Europe A region that includes Albania, Bosnia and Herzegovina, Bulgaria, Croatia, the Czech Republic, Hungary, Macedonia, Poland, Romania, Slovakia, and Yugoslavia, 308

Eastern Ghats (14°00′N/78°50′E) A mountain range that runs along the east coast of India, 552

Eastern Mediterranean Region A region that includes Lebanon, Syria, Jordan, and Israel, 510

Ecuador (2°00′S/77°30′W) A country in northwestern South America, 230, *m234*

Edinburgh (55°57′N/3°12′W) A city in Scotland, 303

Edmonton (53°33′N/113°30′W) The capital of Alberta, Canada, 114

Egypt (27°00′N/30°00′E) A country in North Africa, 438

El Salvador (13°50′N/88°55′W) A country in Central America, 223

Elbe River (53°50′N/9°00′E) A European river that runs north from the Czech Republic, across Germany, and into the North Sea, 260

Elburz Mountains (36°00′N/53°00′E) A mountain range in northern Iran, 488

Ellesmere (79°00′N/82°00′W) A large island in northern Canada, 121

Equatorial Guinea (2°00′N/10°00′E) A country in Central Africa, 448, *m450*

Erie Canal Opened in 1825; crosses upstate New York and is used as a water link between the Atlantic Ocean and the Great Lakes; now part of the New York State Barge Canal System, 129

Erie, Lake (42°15′N/81°25′W) One of the Great Lakes of North America, *m118*, 121

Eritrea (15°00′N/39°00′E) A country in East Africa, 431

Estonia (59°00′N/26°00′E) A country west of Russia; one of the Baltic Republics, 361

Ethiopia (8°00′N/38°00′E) A country in East Africa, 431

Ethiopian Highlands A mountainous area in Ethiopia, 417

Euphrates River (31°00′N/47°25′E) A river that rises in central Turkey, flows southeast through Syria and Iraq, and joins the Tigris River; together they form the Shatt al Arab, which flows into the Persian Gulf, *m488*, 489

Eurasia The combination of Europe and Asia; some consider it to be a single continent, 346

Europe A peninsula of the Eurasian land mass; a continent bounded by Asia, the Mediterranean Sea, the Atlantic Ocean, and the Arctic Ocean, 28, *m29*

Everest, Mount (27°59′N/86°56′E) The world's tallest mountain; located on the border of Nepal and China, 551

Everglades (26°05′N/80°46′W) A huge swampland in southern Florida that covers about 4,000 square miles, 126

Farakka dam (24°49′N/87°56′E) A dam that crosses the Ganges River at a point just before it enters Bangladesh, 599

Feni River (22°46′N/91°26′E) A river in Bangladesh, 562

Fiji (18°00′S/178°00′E) A country that consists of an island group of Oceania in the Pacific Ocean, 712, *m713*

Finland (64°00′N/26°00′E) A country in Northern Europe, 300

Florence (43°46′N/11°15′E) A city in north-central Italy, 291

Florida (28°45′N/82°30′W) A state in the southern United States, 108

Fongafale (8°31′S/179°13′E) The capital of Tuvalu, 686

Foraker, Mount (62°58′N/151°24′E) A mountain in Alaska in the United States, 102

France (46°00′N/2°00′E) A country in Western Europe, 294

Frankfort (38°12′N/84°52′W) The capital of Kentucky, 108

Fredericton (45°57′N/66°38′W) The capital of New Brunswick, Canada, 114

Freetown (8°29′N/13°14′W) The capital of Sierra Leone, 412

French Guiana (4°20′N/53°00′W) A French overseas department in northern South America, 230, *m234*

Fuji, Mount (35°22′N/138°43′E) A volcanic mountain in Japan, 662

Gaborone (24°39′S/25°55′E) The capital of Botswana, 408

Gabon (1°00′S/11°45′E) A country in Central Africa, 448, *m450*

Galilee, Sea of (also called Lake Kinneret) (32°48′N/35°35′E) A freshwater lake in northeastern Israel, 495

Gambia (13°30′N/15°30′W) A country in West Africa, 442

Ganges River (23°22′N/90°32′E) A river that rises in the central Himalayas, flows eastward across northern India, and joins the Brahmaputra river; together they form a huge delta before entering the Bay of Bengal, 553

Ganges-Brahmaputra River delta (23°00′N/89°00′E) A triangular area of land formed by the mouth of the combined Ganges and Brahmaputra rivers, 553

Gaza Strip (31°25′N/34°20′E) A territory along the Mediterranean Sea just northeast of the Sinai Peninsula, 527

Georgetown (6°48′N/58°10′W) The capital of Guyana, 198

Georgia (42°00′N/43°30′E) A country in Transcaucasia, 370, *m370*

Georgia, the U.S. state of (32°45′N/83°30′W) A state in the southern United States, 108

Germany (51°30′N/10°30′E) A country in Western Europe, 294

Ghana (8°00′N/2°00′W) A country in West Africa, 442

Glasgow (55°50′N/4°15′W) A city in Scotland, 303

Gobi Desert (44°00′N/105°00′E) A large desert that stretches from northern China into Mongolia, 627

Godwin Austen, Mount (also called K2) (35°52′N/76°34′E) The world's second tallest mountain; located in northern Pakistan, 552

Golan Heights (also called Al Jawlan) (33°00′N/35°45′E) Located in Syria; a hilly plateau overlooking the Jordan River and the Sea of Galilee, 487

Gorèe Island (14°40′N/17°24′W) An island off the coast of Senegal; served as one of the busiest departure points for slaves during the slave trade from the mid-1500s to the mid-1800s, 442

Granada (37°11′N/3°36′W) A city in southern Spain, 291

Great Barrier Reef (18°00′S/146°50′E) A chain of more than 2,500 reefs and islands; located off the northeast coast of Australia, 692

Great Britain (54°00′N/2°00′W) An island consisting of England, Scotland, and Wales and located north of France, 272

Great Dividing Range (25°00′S/147°00′E) A mountain range near the eastern coast of Australia, 692

Great Drakensberg An escarpment in Southern Africa, 417

Great Lakes (45°41′N/84°26′W) A chain of five large lakes—Huron, Ontario, Michigan, Erie, and Superior—located in central North America, *m118*, 121

Great Plains (39°25′N/101°18′W) A largely treeless area that extends from the Interior Plains to the Rocky Mountains, *m118*, 119

Great Smoky Mountains (35°35′N/83°31′W) A North American mountain range located in the southern part of the Appalachian system, 119

Greater Antilles (20°00′N/74°00′W) A group of islands in the Caribbean Sea; includes Cuba, Jamaica, Hispaniola, and Puerto Rico, 203

Greece (39°00′N/22°00′E) A Mediterranean country of Europe, 287

Green Mountains (42°34′N/72°36′W) A North American mountain range located in the northern part of the Appalachian system, 119

Greenland (72°00′N/40°00′W) The largest island in the world; bounded by the northern Atlantic Ocean and the Arctic Ocean and owned by Denmark, 272

Greenwich (51°28′N/0°00′E) A town in England and site of the Royal Observatory, through which the prime meridian, or longitude 0°, passes, 6

Grenada (12°07′N/61°40′W) A country consisting of the island of Grenada and the southern Grenadines in the Caribbean Sea, 196

Grozny (43°18′N/45°42′E) The capital of Chechnya, *m385*, 386

Guam (13°27′N/144°44′E) A U.S. territory and island in the Pacific, 112

Guangzhou (23°07′N/113°15′E) A city in China, 637

Guatemala (15°30′N/90°15′W) A country in Central America, 223

Guatemala City (14°38′N/90°31′W) The capital of Guatemala, 196

Guiana Highlands (4°00′N/60°00′W) A mountain range in northeast South America, 202, *m203*

Guinea (11°00′N/10°00′W) A country in West Africa, 442

Guinea-Bissau (12°00′N/15°00′W) A country in West Africa, 442

Gulf Coastal Plain A broad plain that stretches along the Gulf of Mexico from Florida into Texas, 119

Gulf of Mexico (26°00′N/91°00′W) An arm of the Atlantic Ocean bordering on eastern Mexico, the southeastern United States, and Cuba, *m118*, 121

Guyana (5°00′N/59°00′W) A country in northern South America, 230, *m234*

Hagatna (13°28′N/144°45′E) The capital of Guam, 112

Hainan (19°00′N/109°30′E) An island off the coast of southern China, *m620*, 621

Haiti (19°00′N/72°25′W) A country that occupies the western one-third of the island of Hispaniola in the Caribbean Sea, 198

Halabja (35°21′N/45°54′E) A city in Iraq, 516

Halifax (44°39′N/63°36′W) The capital of Nova Scotia, Canada, 166

Hangzhou (30°15′N/120°10′E) A city in China, 637

Hanoi (21°02′N/105°51′E) The capital of Vietnam, 686

Harare (17°50′S/31°03′E) The capital of Zimbabwe, 412

Harrisburg (40°16′N/76°53′W) The capital of Pennsylvania, 112

Hartford (41°46′N/72°41′W) The capital of Connecticut, 108

Havana (23°08′N/82°22′W) The capital of Cuba, 196

Hawaii (20°45′N/156°30′W) A state in the western United States that consists of several islands in the central Pacific, 108

Hawaiian Islands (20°45′N/156°30′W) A chain of islands in the central Pacific Ocean that make up the state of Hawaii, 121

Hejaz Mountains A mountain range on the southwest corner of the Arabian Peninsula; part of the western region of Saudi Arabia, 488

Helena (46°36′N/112°02′W) The capital of Montana, 110

Helsinki (60°11′N/24°56′E) The capital of Finland, 266

Hermon, Mount (33°25′N/35°51′E) A mountain on the border between Lebanon and Syria, 478

Himalaya Mountains (28°00′N/84°00′E) The world's highest mountain range; located in Nepal, Bhutan, northern India, and southwestern China, 551

Hindu Kush (35°00′N/71°00′E) A mountain range in eastern Afghanistan and northern Pakistan, 488

Holland (52°30′N/5°45′E) Another name for the Netherlands, a country in Northern Europe, 280

Hollywood (34°06′N/118°20′W) A district of Los Angeles, California; the center of the motion picture industry in the United States, 143

Honduras (15°00′N/86°30′W) A country in Central America, 223

Hong Kong (22°15′N/114°10′E) A region on the coast of southeastern China; includes Hong Kong Island and nearby areas, 621

Honiara (9°26′S/159°57′E) The capital of the Solomon Islands, 684

Honolulu (21°18′N/157°52′W) The capital of Hawaii, 108

Honshu (36°00′N/138°00′E) The largest island of Japan, 653

Houston (31°20′N/95°25′W) A city in southeastern Texas, 148

Huang He (also called Yellow River) (37°45′N/119°05′E) A Chinese river that rises in the Kunlun Mountains, flows east for about 3,000 miles, and empties into the Yellow Sea, *m620,* 621

Hudson Bay (52°52′N/102°25′W) An extended bay located in Canada, 119

Hungary (47°00′N/20°00′E) A country in Eastern Europe, 308

Huron, Lake (44°30′N/82°00′W) The second largest of the Great Lakes of North America, *m118,* 121

Iberian Peninsula (40°00′N/5°00′W) A southwestern peninsula of Europe bounded by France, the Mediterranean Sea, and the Atlantic Ocean and occupied by Spain and Portugal, *m271,* 272

Iceland (65°00′N/18°00′W) An island country in the North Atlantic, northwest of Great Britain, 300

Idaho (44°30′N/114°15′W) A state in the western United States, 108

Ijsselmeer (52°49′N/5°15′E) A freshwater lake in the Netherlands, 281

Illinois (40°00′N/89°15′W) A state in the Midwest of the United States, 108

India (20°00′N/77°00′E) A country in South Asia, 567

Indian Ocean (10°00′S/70°00′E) The world's third largest ocean; extends from southern Asia to Antarctica and from western Australia to eastern Africa, 32

Indiana (40°00′N/86°15′W) A state in the Midwest of the United States, 108

Indo-Gangetic Plain (also called the Northern Indian Plain) (27°00′N/80°00′E) A plain in northern India and Bangladesh that lies between the Deccan Plateau and the northern mountains, 552

Indochina A peninsula located south of China; includes Myanmar, Thailand, Laos, Cambodia, Vietnam, and West Malaysia, 689

Indonesia (5°00′S/120°00′E) A Southeast Asian country that consists of several islands between the Asian mainland and Australia, 705

Indus River (24°20′N/67°47′E) A river that flows west and then south through Pakistan to the Arabian Sea, 553

Indus Valley (29°00′N/71°00′E) A valley formed by the Indus River in Pakistan, 567

Ingushetia (43°13′N/44°47′E) A Russian republic in the Caucasus, 385, *m385*

Interior Plains A lowland area that extends from the Appalachians to about 300 miles west of the Mississippi River, *m118,* 119

Iowa (42°00′N/93°30′W) A state in the Midwest of the United States, 108

Iqaluit (63°44′N/68°30′W) The capital of Nunavut, Canada, 114

Iran (32°00′N/53°00′E) A country in the northeast region of Southwest Asia, 516, *m516*

Iraq (33°00′N/44°00′E) A country in the northeast region of Southwest Asia, 516, *m516*

Ireland (53°00′N/8°00′W) A country occupying most of the island of Ireland, which is west of Great Britain, 272

Islamabad (33°42′N/73°10′E) The capital of Pakistan, 548

Israel (31°30′N/34°45′E) A country in the Eastern Mediterranean in Southwest Asia, 512, *m512*

Issyk-Kul, Lake (42°25′N/77°15′E) A lake in Kyrgyzstan, 336

Italian Peninsula A southern peninsula of Europe bounded by the Mediterranean Sea, the Adriatic Sea, France, Switzerland, Austria, and Slovenia and occupied by Italy, *m271,* 272

Italy (42°50′N/12°50′E) A Mediterranean country of Europe, 266

Jackson (32°18′N/90°11′W) The capital of Mississippi, 110

Jaffna Peninsula (9°45′N/80°10′E) A peninsula on the northern tip of the island of Sri Lanka, 585

Jakarta (6°10′S/106°49′E) The capital of Indonesia, 684

Jamaica (18°15′N/77°30′W) An island country in the Caribbean Sea, 198

Jamestown (37°19′N/78°18′W) Founded in 1607; the first permanent English settlement in the United States, 136

Japan (36°00′N/138°00′E) An East Asian country consisting of several islands in the western Pacific Ocean, 651, *m652*

Jefferson City (38°35′N/92°10′W) The capital of Missouri, 110

Jerusalem (31°47′N/35°13′E) A holy city for Jews, Muslims, and Christians; also the capital of Israel, 510, *m512*

Johannesburg (26°12′S/28°05′E) The largest city in South Africa, 457

Jordan (31°00′N/36°00′E) A country in the Eastern Mediterranean in Southwest Asia, 511, *m512*

Jordan River A river that flows south from Syria through the Sea of Galilee to the Dead Sea, 489

Juneau (58°18′N/134°25′W) The capital of Alaska, 108

Jutland Peninsula A Northern European peninsula that consists of Denmark and northern Germany, 271, *m271*

K2 (also called Mount Godwin Austen) (35°52′N/76°34E) The world's second tallest mountain; located in northern Pakistan, 552

Kabul (34°31′N/69°11′E) The capital of Afghanistan, 484

Kalahari (23°00′S/22°00′E) A desert in Southern Africa, 420

Kaliningrad (54°43′N/20°30′E) A city in western Russia, 345

Kamchatka (56°00′N/160°00′E) A peninsula of northeastern Russia bounded by Sea of Okhotsk and the Bering Sea, 346

Kampala (0°19′N/32°35′E) The capital of Uganda, 412

Kanchenjunga (27°42′N/88°08′E) A mountain on the border of India and Nepal, 542

Kansas (38°30′N/98°30′W) A state in the Midwest of the United States, 108

Kansas City (Kansas) (39°07′N/94°44′W) A city in northeastern Kansas, 147

Kansas City (Missouri) (39°05′N/94°35′W) A city in western Missouri, 147

Kara Kum (39°00′N/60°00′E) A large, black sand desert in Central Asia, 352

Karakoram Mountain Range (34°00′N/78°00′E) A mountain range in northern Pakistan, northern India, and southwestern China, 552

Karnataka Plateau A plateau in Karnataka, a state in southwestern India, 552

Kashmir A territory located at the foot of the Himalayas in northern Pakistan, northern India, and southwestern China, 600, *m601*

Kathmandu (27°43′N/85°19′E) The capital of Nepal, 581

Kathmandu Valley (27°40′N/85°21′E) A valley in Nepal, 582

Kazakhstan (48°00′N/68°00′E) A country in Central Asia, 375

Kentucky (37°30′N/85°15′W) A state in the southern United States, 108

Kenya (1°00′N/38°00′E) A country in East Africa, 431

Kenya, Mount (0°10′S/37°20′E) A volcanic mountain in East Africa, 417

Khartoum (15°45′N/32°30′E) The capital of Sudan, 425

Khyber Pass (34°04′N/71°13′E) A major land route through the Safed Koh Mountains, 552

Kiev (50°26′N/30°31′E) The capital of the Ukraine, 345

Kigali (1°57′S/30°04′E) The capital of Rwanda, 412

Kilimanjaro, Mount (3°04′S/37°22′E) Africa's highest peak; a volcanic mountain in Tanzania in East Africa, 417

Kingston (18°00′N/76°48′W) The capital of Jamaica, 198

Kingstown (13°08′N/61°13′W) The capital of Saint Vincent and the Grenadines, 198

Kinneret, Lake (also called the Sea of Galilee) (32°48′N/35°35′E) A freshwater lake in northeastern Israel, 495

Kinshasha (4°20′S/15°19′E) The capital of the Democratic Republic of the Congo, 408

Kiribati (5°00′S/170°00′W) A country that consists of an island group of Oceania in the Pacific Ocean, 712, *m713*

Kjolen Mountains (65°00′N/14°00′E) An upland area of Scandinavia, 273

Kobe (34°41′N/135°10′E) A city in Japan, 661

Kolkata (Calcutta) (22°34′N/88°22′E) A city in eastern India, *m569*, 570, 599

Kongur, Mount (38°40′N/75°21′E) A mountain in China, 611

Korea Strait (34°00′N/129°00′E) A waterway between southern South Korea and southwestern Japan that connects the Sea of Japan with the East China Sea, 647

Korean Peninsula A peninsula bounded by the Yellow Sea and the Sea of Japan and occupied by North Korea and South Korea, 620

Koror (7°20′N/134°28′E) The capital of Palau, 684

Kosovo (42°35′N/21°00′E) A province of southern Yugoslavia, 319

Kuala Lumpur (3°10′N/101°42′E) The capital of Malaysia, 709

Kunlun Mountains (36°00′N/84°00′E) A mountain range in western China, 619, *m620*

Kuril Islands (46°10′N/152°00′E) A chain of Russian-owned islands that extend off the southern tip of Kamchatka Peninsula; Japan claims ownership of some of the islands, 346

Kuwait (29°30′N/47°45′E) A country in the Arabian Peninsula of Southwest Asia, 503

Kuwait City (29°22′N/47°59′E) The capital of Kuwait, 484

Kyoto (35°00′N/135°45′E) A city in Japan, 654

Kyrgyzstan (41°00′N/75°00′E) A country in Central Asia, 375

Kyzyl Kum (42°30′N/64°30′E) A large red sand desert in Central Asia, 352

La Paz (16°30′S/68°09′W) The administrative capital of Bolivia, 196

Labrador A section of Newfoundland, Canada, 167

Lansing (42°44′N/84°33′W) The capital of Michigan, 110

Laos (18°00′N/105°00′E) A country in Southeast Asia, 705

Las Vegas (36°11′N/115°08′W) A city in southern Nevada, 149, *m149*

Latin America A region that includes Mexico, Central America, the Caribbean, and South America, 201, *m203*

Latvia (57°00′N/25°00′E) A country west of Russia; one of the Baltic Republics, 361

Lebanon (33°50′N/35°50′E) A country in the Eastern Mediterranean in Southwest Asia, 511, *m512*

Leizhou Peninsula (20°40′N/110°05′E) A peninsula in southern China between the South China Sea and the Gulf of Tonkin, 620

Lena (72°20′N/126°37′E) A river that flows through east-central Russia and empties into the Laptev Sea, 347

Lesotho (29°30′S/28°15′E) A country in Southern Africa, 453, *m454*

Lesser Antilles (15°00′N/61°00′W) A group of islands southeast of Puerto Rico; divided into the Leeward Islands and Windward Islands, 203

Liberia (6°30′N/9°30′W) A country in West Africa, 442

Libreville (0°23′N/9°27′E) The capital of Gabon, 410

Libya (25°00′W/17°00′E) A country in North Africa, 438

Libyan Desert (24°00′N/25°00′E) A desert in northeast Africa; the northeast section of the Sahara, 420

Liechtenstein (47°10′N/9°32′E) A country in Western Europe, 294

Lilongwe (13°59′S/33°47′E) The capital of Malawi, 410

Lima (12°03′S/77°03′W) The capital of Peru, 211

Lincoln (40°48′N/96°40′W) The capital of Nebraska, 110

Lisbon (38°43′N/9°08′W) The capital of Portugal, 268

Lithuania (56°00′N/24°00′E) One of the Baltic Republics, 361

Little Rock (35°45′N/92°17′W) The capital of Arkansas, 108

Ljubljana (46°03′N/14°31′E) The capital of Slovenia, 268

Llanos (5°00′N/70°00′W) Vast plains located in Colombia and Venezuela, 202, *m203*

Logan, Mount (60°34′N/140°25′W) A mountain in Canada, 102

Lombardy (45°40′N/9°30′E) A region in northern Italy, 273

Lomè (6°08N/1°13′E) The capital of Togo, 412

London (51°30′N/0°08′W) The capital of the United Kingdom, 304

Los Angeles (34°03′N/118°15′W) A major seaport city in southwestern California, 128, *m149*

Louisiana (31°00′N/92°00′W) A state in the southern United States, 148

Lowell (42°39′N/71°19′W) A city in Massachusetts that became a textile center by the 1840s, 136

Luanda (8°50′S/13°14′E) The capital of Angola, 408

Lusaka (15°25′S/28°17′E) The capital of Zambia, 412

Luxembourg (49°45′N/6°10′E) A country in Western Europe, 294

Luxembourg City (49°37′N/6°08′E) The capital of Luxembourg, 266

Macao Peninsula A peninsula in southeast China just west of Hong Kong, 620

Macedonia (41°50′N/22°00′E) A country in Eastern Europe, 308

MacKenzie River (69°20′N/134°00′W) The largest river in Canada; flows across the Northwest Territories to the Arctic Ocean, 121

Madagascar A country in Southern Africa, 453, m454

Madison (43°05′N/89°23′W) The capital of Wisconsin, 112

Madras (also called Chennai) (13°05′N/80°17′E) A city in southern India, m569, 570

Madrid (40°24′N/3°41′W) The capital of Spain, 268

Maine (45°30′N/69°15′W) A state in New England in the northeastern United States, 145

Majuro (7°06′S/171°23′E) The capital of the Marshall Islands, 684

Makalu (27°55′N/87°08′E) A mountain on the border of China and Nepal, 542

Malabo The capital of Equatorial Guinea, 408

Malawi (3°21′N/8°40′E) A country in Southern Africa, 453, m454

Malay Archipelago (0°00′N/120°00′E) An island group of Southeast Asia; separates the Pacific and Indian oceans and includes the Philippines, Malaysia, and the islands of Indonesia, 689

Malay Peninsula (6°00′N/102°00′E) A narrow strip of land about 700 miles long that stretches south from the Indochinese Peninsula and then curves southeast, 689

Malaysia (2°30′N/112°30′E) A Southeast Asian country that occupies part of the island of Borneo and the southern end of the Malay Peninsula, 705

Maldives (3°12′N/73°00′E) A South Asian country that occupies a chain of islands in the Indian Ocean off the southwest coast of India, 584

Male (4°10′N/73°30′E) The capital of the Maldives, 548

Mali (17°00′N/4°00′W) A country in West Africa, 442

Malta (35°55′N/14°26′E) An island country in the Mediterranean Sea, off the southern coast of Sicily, 268

Managua (12°09′N/86°25′W) The capital of Nicaragua, 198

Manama (26°14′N/50°35′E) The capital of Bahrain, 484

Manchurian Plain (44°00′N/124°00′E) A plain in northeastern China, 620, m620

Manila (14°35′N/121°00′E) The capital of the Philippines, 684

Manitoba (55°00′N/97°00′W) A Prairie Province of Canada, 168

Maputo (25°58′S/32°35′E) The capital of Mozambique, 410

Marrakesh (31°38′N/8°00′W) A city in Morocco, m439, 440

Marshall Islands (10°00′N/167°00′E) A country that consists of an island group of Oceania in the Pacific Ocean, 712, m713

Maryland (39°00′N/76°45′W) A state in the southern United States (sometimes included with the Middle Atlantic states), 110

Maseru (29°19′S/27°29′E) The capital of Lesotho, 410

Massachusetts (42°15′N/71°30′W) A state in New England in the northeastern United States, 145

Massif Central Uplands located in central France, 273

Mauritania (20°00′N/12°00′W) A country in West Africa, 442

Mauritius (20°18′S/57°35′E) An island country in Southern Africa, 453, m454

Mbabane (26°19′S/31°08′E) The capital of Swaziland, 412

McKinley, Mount (also called Denali) (63°04′N/151°00′W) North America's highest mountain; located in Alaska, 120

Mecca (21°26′N/39°50′E) The holiest city of Islam; located in western Saudi Arabia, 503

Mediterranean Europe A region that includes Spain, Italy, and Greece, 287

Mediterranean Sea (35°00′N/20°00′E) An inland sea bounded by southern Europe, northern Africa, and southwestern Asia, 277

Mekong River (10°15′N/105°55′E) A river that begins in China and crosses several Southeast Asian nations before becoming a wide delta on Vietnam's coast, m689, 690

Melanesia (12°00′S/160°00′E) A region in southwestern Oceania that consists of numerous Pacific islands, 713, m713

Meseta (41°00′N/4°00′W) A central plateau of Spain, 273

Mexico (23°00′N/102°00′W) A Latin American country bounded by the United States, the Gulf of Mexico, the Pacific Ocean, and Central America, 217

Mexico City (19°26′N/99°08′W) The largest city in Latin America and the capital of Mexico, 211

Miami (25°46′N/80°12′W) A city in southern Florida, 148

Michigan (44°15′N/85°30′W) A state in the Midwest of the United States, 110

Michigan, Lake (43°20′N/87°10′W) The third largest of the Great Lakes of North America, m118, 121

Micronesia (9°00′N/155°00′E) A region in northwestern Oceania that consists of numerous Pacific islands, 713, m713

Micronesia, Federated States of (5°00′N/152°00′E) A country that consists of an island group of Oceania in the Pacific Ocean, 712, m713

Mid–Atlantic Ridge (0°00′N/20°00′W) The mountain range on the ocean floor; extends for thousands of miles north to south through the middle of the Atlantic Ocean, 36

Middle Atlantic States The states of Pennsylvania, New York, and New Jersey (sometimes Maryland and Delaware are included in this group), 145

Midwest An area of the north-central United States that includes Michigan, Ohio, Indiana, Illinois, Wisconsin, Minnesota, Iowa, Missouri, Kansas, Nebraska, North Dakota, and South Dakota, 147

Milwaukee (43°04′N/87°58′W) A city in southeastern Wisconsin, 147

Minneapolis (44°58′N/93°16′W) A city in eastern Minnesota, 147

Minnesota (46°15′N/94°15′W) A state in the Midwest of the United States, 110

Minsk (53°54′N/27°34′E) The capital of Belarus, 342

Mississippi (32°45′N/89°45′W) A state in the southern United States, 110

Mississippi River (29°09′N/89°15′W) A North American river that runs from Minnesota to the Gulf of Mexico, m118, 121

Missouri (38°15′N/92°30′W) A state in the Midwest of the United States, 110

Missouri River (38°49′N/90°07′W) A North American river that runs from the Rocky Mountains in southwest Montana into the Mississippi River, m118, 121

Mogadishu (2°04′N/45°22′E) The capital of Somalia, 412

Mojave Desert (35°25′N/115°35′W) A desert in southern California, 124

Moldova (47°00′N/29°00′E) A country that lies between the Ukraine and Romania, 361

Monaco (43°44′N/7°24′E) A principality bounded by France and the Mediterranean Sea, 268

Monaco, the village of (43°44′N/7°25′E) The capital of Monaco, 268

Mongolia (46°00′N/105°00′E) A country in East Asia, 642

Mongolian Plateau A plateau in Mongolia and northeastern China, 620, *m620*

Monrovia (6°19′N/10°48′W) The capital of Liberia, 421

Montana (47°00′N/109°45′W) A state in the western United States, 110

Montevideo (34°51′S/56°10′W) The capital of Uruguay, 198

Montgomery (32°22′N/86°18′W) The capital of Alabama, 108

Montpelier (44°16′N/72°34′W) The capital of Vermont, 112

Montreal (45°30′N/73°35′W) Located in Quebec, Canada; the second largest metropolitan area in the country, 168

Morocco (32°00′N/5°00′W) A country in North Africa, 438

Moroni (11°42′S/43°14′E) The capital of Comoros, 408

Moscow (55°45′N/37°37′E) The capital of Russia, 366

Mozambique (18°15′S/35°00′E) A country in Southern Africa, 453, *m454*

Mumbai (also called Bombay) (18°59′N/72°50′W) A city in western India, *m569*, 570

Murray River (35°22′S/139°22′E) The largest river of Australia; flows into an arm of the Indian Ocean, 692

Muscat (23°37′N/58°36′E) The capital of Oman, 484

Myanmar (22°00′N/98°00′E) A country in Southeast Asia, 705

N'Djamena (12°07′N/15°03′E) The capital of Chad, 408

Nagorno-Karabakh (40°00′N/46°35′E) A mountainous republic of Azerbaijan, 386

Nagoya (35°10′N/136°55′E) A city in Japan, 630

Nairobi (1°17′S/36°49′E) The capital of Kenya, 410

Namib (23°00′S/15°00′E) A desert in southwest Africa, 420

Namibia (22°00′S/17°00′E) A country in Southern Africa, 453, *m454*

Nashville (36°10′N/86°47′W) The capital of Tennessee, 112

Nassau (25°05′N/77°21′W) The capital of the Bahamas, 196

Nasser, Lake (22°50′N/32°30′E) A lake created by the Aswan High Dam; lies in southern Egypt and northern Sudan, 426, *m427*

Nauru (0°32′S/166°55′E) An island country of Oceania in the Pacific Ocean, 712, *m713*

Nebraska (41°30′N/99°45′W) A state in the Midwest of the United States, 110

Negev (30°34′N/34°43′E) A desert area that occupies southern parts of Israel, 492

Nepal (28°00′N/84°00′E) A country in South Asia, 580

Netherlands (52°30′N/5°45′E) A country in Western Europe, 294

Nevada (39°15′N/116°45′W) A state in the western United States, 110

New Brunswick (46°30′N/66°45′W) An Atlantic Province of Canada, 166

New Caledonia (21°30′S/165°30′E) A French overseas territory that consists of an island group of Oceania in the Pacific Ocean, 691

New Delhi (28°36′N/77°12′E) The capital of India, 548

New England An area of the northeastern United States that includes Maine, Vermont, New Hampshire, Massachusetts, Rhode Island, and Connecticut, 145

New Guinea (5°00′S/140°00′E) An island north of Australia; occupied by the countries of Indonesia (west half) and Papua New Guinea (east half), 679

New Hampshire (43°40′N/71°30′W) A state in New England in the northeastern United States, 145

New Jersey (40°10′N/74°30′W) A Middle Atlantic state of the eastern United States, 145

New Mexico (34°30′N/106°00′W) A state in the western United States, 110

New Orleans (29°58′N/90°04′W) A city in southeastern Louisiana, 148

New York (43°00′N/75°30′W) A Middle Atlantic state in the eastern United States, 145

New York City (40°41′N/73°59′W) A major seaport city located in southeastern New York State, *m145*, 147

New Zealand (42°00′S/174°00′E) A country that consists of several islands (including North Island and South Island) in the Pacific Ocean off the southeast coast of Australia; part of Oceania, 691

Newfoundland (52°00′N/56°00′W) An Atlantic Province of Canada, 166

Newfoundland Island (49°00′N/56°00′W) An island that makes up part of Newfoundland, Canada, 167

Niamey (13°31′N/2°07′E) The capital of Niger, 412

Nicaragua (13°00′N/85°00′W) A country in Central America, 223

Nicosia (35°10′N/33°22′E) The capital of Cyprus, 484

Niger (16°00′N/8°00′E) A country in West Africa, 442

Niger River (5°33′N/6°33′E) A river that begins in Guinea and flows toward the Sahara (northeast); it then cuts through Nigeria and empties into the Gulf of Guinea, 416

Niger River Delta (4°50′N/6°00′E) A triangular area of land formed by the mouth of the Niger River, 425

Nigeria (10°00′N/8°00′E) A country in West Africa, 442

Nile River Delta (31°00′N/31°00′E) A triangular area of land formed by the mouth of the Nile River, 416

Nile River (l30°10′N/31°06′E) The world's longest river; flows through Uganda, Sudan, and Egypt and empties into the Mediterranean Sea, 416

Nordic countries The Northern European countries of Denmark, Finland, Iceland, Norway, and Sweden, 300

Norilsk (69°20′N/88°06′E) A Siberian mining center in Russia, 354

North Africa A region of Africa that includes Algeria, Egypt, Libya, Morocco, Sudan, and Tunisia, 438

North America The northern continent of the Western Hemisphere; bounded by the Arctic Ocean, the Atlantic Ocean, the Pacific Ocean, and the Caribbean Sea, 28, *m29*

North Carolina (35°30′N/80°00′W) A state in the southern United States, 110

North China Plain (34°00′N/116°00′E) A plain in northern China, 620, *m620*

North Dakota (47°30′N/100°00′W) A state in the Midwest of the United States, 110

North Island (37°20′S/173°30′E) The northern of the two main islands that make up New Zealand, 691

North Korea (40°00′N/127°00′E) A country in East Asia, 648, *m648*

North Ossetia (43°00′N/44°15′E) A Russian republic in the Caucasus, 385, *m385*

North Pole (90°00′N) The northern end of the earth's axis of rotation; a point located in the Arctic Ocean, 56

North Sea (55°20′N/3°00′E) An arm of the Atlantic Ocean bounded by Norway, Denmark, Germany, the Netherlands, England, and Scotland, 271, *m271*

Northeast Region of Southwest Asia A region that includes Turkey, Iran, Iraq, and Afghanistan, 516

Northern Europe A region that includes the United Kingdom, Ireland, Denmark, Finland, Iceland, Norway, and Sweden, 300

Northern European Plain A fertile plain that stretches across parts of France, Belgium, the Netherlands, Denmark, Germany, and Poland, 273

Northern Indian Plain (also called the Indo-Gangetic Plain) (27°00′N/80°00′E) A plain in northern India that lies between the Deccan Plateau and the northern mountains, 552

Northern Plains of Afghanistan A plain in northern Afghanistan, 488

Northwest Territories (65°00′N/118°00′W) A territory in north-central Canada, 169

Norway (62°00′N/10°00′E) A Northern European country that occupies the western part of the Scandinavian Peninsula, 300

Norwegian Sea (70°00′N/5°00′E) An extension of the Atlantic Ocean off the northwest coast of Norway, 271, *m271*

Nouakchott (18°06′N/15°57′W) The capital of Mauritania, 410

Nova Scotia (45°00′N/63°00′W) An Atlantic Province of Canada, 166

Nuku'alofa (21°08′S/125°12′W) The capital of Tonga, 686

Nunavut (70°00′N/90°00′W) A territory in north-central Canada; home to many of Canada's Inuit, 169

Ob River (66°45′N/69°30′E) A river in central Russia that flows into the Gulf of Ob, 347

Oceania A region that includes Australia, New Zealand, and the Pacific Islands (but not the Philippines, Indonesia, and other islands near Asia), 690

Ogallala Aquifer The largest aquifer in the United States; stretches from South Dakota to Texas, 33

Ohio (40°15′N/83°00′W) A state in the Midwest of the United States, 110

Ohio River (36°59′N/89°08′W) A North American river that runs from the Allegheny Mountains into the Mississippi River, *m118,* 121

Oklahoma (35°30′N/97°30′W) A state in the south-central United States, 110

Oklahoma City (35°28′N/97°31′W) The capital of Oklahoma, 110

Olduvai Gorge (2°58′S/35°22′E) A ravine in northern Tanzania that contains archeological sites rich in fossils, 431

Olympia (47°02′N/122°54′W) The capital of Washington, 112

Omaha (41°16′N/95°56′W) A city in eastern Nebraska, 147

Oman (21°00′N/57°00′E) A country in the Arabian Peninsula of Southwest Asia, 503

Ontario (50°00′N/86°00′W) A Core Province of Canada, 167

Ontario, Lake (43°40′N/78°00′W) The smallest of the Great Lakes of North America, *m118,* 121

Oregon (44°00′N/120°03′W) A state in the western United States, 148

Orinoco River (41°12′S/68°15′W) A river that flows through the northern part of South America, mainly in Venezuela, 202

Osaka (35°57′N/137°16′E) A city in Japan, 630

Oslo (59°55′N/10°45′E) The capital of Norway, 268

Ouagadougou (12°22′N/1°31′W) The capital of Burkina Faso, 408

outback The dry, interior region of Australia, 697

Pacific Ocean (0°00′N/180°00′E) The world's largest ocean; extends from the Arctic Circle and northeastern Asia to Antarctica and from the western Americas to eastern Asia and Australia, 32

Paektu, Mount (41°59′N/128°05′E) A mountain in Korea, 611

Pago Pago (14°17′S/170°42′W) The capital of American Samoa, 112

Pakistan (30°00′N/70°00′E) A country in South Asia, 573

Palau (7°30′N/134°27′E) A country that consists of an island group of Oceania in the Pacific Ocean, 712, *m713*

Palestine (18°30′N/73°45′W) A historical region of Southwest Asia located between the Mediterranean Sea and the Jordan River, 511

Palikir (6°55′N/158°09′E) The capital of the Federated States of Micronesia, 684

pampas (35°00′S/63°00′W) Vast plains located in Uruguay and south-central Argentina, 202 *m203*

Panama (9°00′N/80°00′W) A country in Central America, 223

Panama Canal (9°20′N/79°55′W) A canal that cuts through Panama and connects the Caribbean Sea and Pacific Ocean, 226, *m226*

Panama City (8°58′N/79°32′W) The capital of Panama, 198

Papua New Guinea (6°00′S/147°00′E) A country that consists of an island group of Oceania in the Pacific Ocean, 712, *m713*

Paraguay (22°60′S/57°60′W) A country in south-central South America, 230, *m234*

Paramaribo (5°52′N/55°10′W) The capital of Suriname, 198

Paraná River (33°43′S/59°15′W) A river that begins in the highlands of Brazil and runs south and west through Paraguay and Argentina. It then turns eastward and empties into the Atlantic, 203, *m203*

Paris (48°52′N/2°20′E) The capital of France, 266

Patagonia (44°00′S/68°00′W) A region of South America, mostly in Argentina, 209

Pearl River (also called Zhu Jiang) (22°46′N/113°38′E) A river in southeastern China that joins the Xi Jiang (or West River) and empties into the South China Sea, 622

Pennsylvania (40°45′N/77°45′W) A Middle Atlantic state of the eastern United States, 145

Persian Gulf (27°00′N/51°00′E) An extension of the Arabian Sea situated between the Arabian Peninsula and Iran, 487, *m488*

Peru (10°00′S/76°00′W) A country in western South America, 230, *m234*

Philadelphia (39°57′N/75°10′W) A city in southeastern Pennsylvania, 146, *m145*

Philippines, the (13°00′N/122°00′E) A Southeast Asian country that occupies about 7,100 islands east of the Asian mainland and northeast of Borneo, 712, *m713*

Phnom Penh (11°33′N/104°55′E) The capital of Cambodia, 684

Phoenix (33°27′N/112°4′W) The capital of Arizona, 149, *m149*

Pierre (44°22′N/100°21′W) The capital of South Dakota, 112

Pinatubo, Mount (15°08′N/120°21′E) A volcanic mountain in the Philippines, 689

Pittsburgh (40°26′N/79°60′W) A city in southwestern Pennsylvania, 137

Plateau of Tibet (33°00′N/92°00′E) A vast tableland in south-central Asia; mostly in Tibet, but extends into China, 619, *m620*

Poland (52°00′N/20°00′E) A country in Eastern Europe, 308

Polynesia (10°0′S/150°00′W) A region in central Oceania that consists of numerous Pacific islands, 713, *m713*

Port Louis (20°10′S/57°30′E) The capital of Mauritius, 410

Port Moresby (9°28′S/147°12′E) The capital of Papua New Guinea, 684

Port-of-Spain (10°39′N/61°31′W) The capital of Trinidad and Tobago, 198

Port-au-Prince (18°32′N/72°20′W) The capital of Haiti, 198

Port-Vila (17°44′S/168°19′E) The capital of Vanuatu, 686

Porto-Novo (6°29′N/2°37′E) The capital of Benin, 408

Portugal (39°04′N/8°14′W) A country located on the Iberian Peninsula; bounded by the Atlantic Ocean and Spain, 268

Prague (50°5′N/14°28′E) The capital of the Czech Republic, 312

Praia (14°55′N/23°31′W) The capital of Cape Verde Island, 408

Prairie Provinces An area located in west-central Canada that includes Manitoba, Saskatchewan, and Alberta, 168

Pretoria (25°42′S/28°14′E), **Cape Town** (33°55′S/18°25′E), **Bloemfontein** (20°08′S/26°12′E) The capital towns of South Africa, 412

Prince Edward Island (46°20′N/63°30′W) An Atlantic Province of Canada, 166

Providence (41°49′N/71°25′W) The capital of Rhode Island, 112

Puerto Rico (18°15′N/66°30′W) An island in the Caribbean Sea that is a self-governing commonwealth in union with the United States, 112

Puerto Vallarta (20°37′N/105°15′W) A city in Mexico, 213

Pyongyang (39°1′N/125°45′E) The capital of North Korea, 616

Pyrenees (42°40′N/1°00′E) A European mountain range that forms a border between France and Spain, *m271, 272*

Qatar (25°30′/51°15′E) A country in the Arabian Peninsula of Southwest Asia, 503

Qinling Shandi Mountains (33°30′N/108°36′E) A mountain range that divides the northern part of China from the southern part, 619

Quebec (54°00′N/72°00′W) A Core Province of Canada, 167

Quebec City (46°49′N/71°15′W) The capital of Quebec, Canada, 167

Quito (0°13′S/78°30′W) The capital of Ecuador, 196

Rabat (34°02′N/6°50′W) The capital of Morocco, 410

Raleigh (35°46′N/78°38′W) The capital of North Carolina, 110

Red Sea (19°00′N/39°30′E) A long, narrow sea situated between northeast Africa and the Arabian Peninsula, 487, *m488*

Regina (50°27′N/104°37′W) The capital of Saskatchewan, Canada, 114

Reykjavik (64°09′N/21°57′W) The capital of Iceland, 266

Rhine River (51°58′N/4°05′E) A European river that flows from the interior of Europe north to the North Sea, 273, *m273*

Rhode Island (41°45′N/71°30′W) A state in New England in the northeastern United States, 145

Richmond (37°33′N/77°28′W) The capital of Virginia, 112

Riga (56°57′N/24°06′E) The capital of Latvia, 342

Ring of Fire A chain of volcanoes that line the Pacific Rim, *m37,* 41, 661, *m662*

Rio de Janeiro (22°54′S/43°14′W) A city in Brazil, 239

Rio de la Plata The name for the last stretch of the Paraná River before it empties into the Atlantic Ocean between Argentina and Uruguay, 203

Rio Grande (23°51′N/102°59′W) A river that forms part of the Mexican-U.S. border, 102

Riyadh (24°38′N/46°46′E) The capital of Saudi Arabia, 484

Rocky Mountains (43°22′N/110°55′W) A mountain system in the western United States and Canada that extends about 3,000 miles from the Arctic to the Mexican frontier, 119

Romania (46°00′N/25°00′E) A country in Eastern Europe, 308

Rome (41°54′N/12°29′E) The capital of Italy, 266

Rosa, Monte (45°55′N/7°53′E) A mountain on the border of Switzerland and Italy, 260

Roseau (15°18′N/61°24′W) The capital of Dominica, 196

Rub al-Khali (21°00′N/51°00′E) A large desert in the southern part of the Arabian Peninsula, 491

Russia An empire that extended from eastern Europe across north-central Asia to the Pacific Ocean, 361, *m362*

Russia and the Republics A region that stretches across much of eastern Europe and north-central Asia to the Pacific; consists of Russia and 14 other countries (republics), 345

Russian Far East The area of eastern Russia, 345

Rwanda, Republic of (2°00′S/30°00′E) A country in East Africa, 431

Sacramento (38°35′N/121°30′W) The capital of California, 108

Sahara (26°00′N/13°00′E) The largest desert in the world; stretches across northern Africa, from the Atlantic Ocean to the Red Sea, 420

Saint Augustine (29°54′N/81°19′W) Founded in 1565; the oldest permanent European settlement in the United States, 135

Saint Elias, Mount (60°18′N/140°56′W) A mountain located on the U.S.-Canada border, 102

Saint George's (12°03′N/61°45′W) The capital of Grenada, 196

Saint John (45°16′N/66°04′W) A city in New Brunswick, Canada, 166

Saint John's (17°07′N/61°51′W) The capital of Antigua and Barbuda, 196

Saint John's, Canada (47°28′N/52°18′W) The capital of Newfoundland, Canada, 114

Saint Kitts and Nevis, Federation of (17°20′N/62°45′W) A country consisting of the islands of Saint Kitts, Nevis, and Sombrero in the Caribbean Sea, 198

Saint Lawrence River (49°30′N/65°00′W) A Canadian river that flows from Lake Ontario to the Gulf of Saint Lawrence, *m118,* 129

Saint Lawrence Seaway (45°20′N/74°50′W) A waterway that connects the Great Lakes to the Atlantic Ocean by way of the Saint Lawrence River, 129

Saint Louis (38°38′N/90°12′W) A city in eastern Missouri, 147

Saint Lucia (13°53′N/60°58′W) An island country in the Caribbean Sea, 198

Saint Paul (44°57′N/93°06′W) The capital of Minnesota, 147

Saint Petersburg (59°54′N/30°16′E) A city in western Russia, 362

Saint Vincent and the Grenadines (13°05′N/61°12′W) A country consisting of Saint Vincent Island and the northern islets of the Grenadines in the Caribbean Sea, 198

Sakhalin Island An island, governed by a Russian Federation, off the east coast of Russia, 346

Salem (44°57′N/123°02′W) The capital of Oregon, 110

Salt Lake City (40°46′N/111°53′W) The capital of Utah, 112

Samoa, Independent State of (13°35′S/172°20′W) A country that consists of an island group of Oceania in the Pacific Ocean, 712

San Antonio (29°25′N/98°30′W) A city in southern Texas, 148

San Diego (34°43′N/117°09′W) A city in southwestern California, *m149,* 150

San Francisco (37°47′N/122°25′W) A city in western California, 150

San José (9°56′N/84°05′W) The capital of Costa Rica, 196

San Juan (18°28′N/66°06′W) The capital of Puerto Rico, 112

San Marino, Republic of (43°56′N/12°25′E) A tiny country surrounded by Italy, 268

San Marino, the city of (43°55′N/12°28′E) The capital of San Marino, 268

San Salvador (13°42′N/89°12′W) The capital of El Salvador, 196

Sanaa (15°21′N/44°12′E) The capital of Yemen, 484

Santa Fe (35°41′N/105°56′W) The capital of New Mexico, 110

Santiago (33°27′S/70°40′W) The capital of Chile, 211

Santo Domingo (18°28′N/69°54′W) The capital of the Dominican Republic, 196

São Paulo (23°32′S/46°37′W) A city in Brazil, 239

São Tomé (00°20′N/6°44′E) The capital of São Tomé and Príncipe, 412

São Tomé and Príncipe (1°00′N/7°00′E) An island country off the coast of Gabon in Central Africa, 448, *m450*

Sapporo (43°03′N/141°21′E) A city in Japan, 630

Sarajevo (43°51′N/18°23′E) The capital of Bosnia and Herzegovina, 266

Sardinia (40°00′N/9°00′E) An autonomous region of Italy; an island in the central Mediterranean Sea, 272

Saskatchewan (54°00′N/106°00′W) A Prairie Province of Canada, 168

Saudi Arabia (25°00′N/45°00′E) A country that occupies most of the Arabian Peninsula of Southwest Asia, 503

Scandinavian Peninsula (65°00′N/16°00′E) A European peninsula bounded by the Norwegian Sea, the North Sea, and the Baltic Sea; occupied by Norway, Sweden, and Denmark, 271, *m271*

Sea of Japan (43°30′N/135°45′E) An enclosed arm of the Pacific Ocean; bounded by Japan, North Korea, South Korea, and Russia, 647

Seattle (47°36′N/122°20′W) A city in northwestern Washington, 150

Senegal (14°00′N/14°00′W) A country in West Africa, 442

Seoul (37°34′N/126°60′E) The capital of South Korea, 616

Serengeti Plains (3°25′S/38°00′E) A tropical grassland in East Africa, 422

Seychelles (4°35′S/55°40′E) A country formed by a group of islands off the east coast of Africa; part of East Africa, 431

Shandong Peninsula (37°00′N/121°00′E) A peninsula in northeastern China bounded by the Bo Hai and the Yellow seas, 620

Shanghai (31°14′N/121°28′E) A city in China, 637

Siberia (60°00′N/100°00′E) A region, largely in Russia, that lies on the continent of north-central Asia, 349

Sicily (37°45′N/14°15′E) An autonomous region of Italy; an island located off the coast of southern Italy, 272

Sierra Leone (8°30′N/11°30′W) A country in West Africa, 442

Sierra Madre A mountain range that runs down Mexico, 201, *m203*

Sierra Nevada Mountains (37°42′N/119°19′W) A North American mountain range that runs parallel to the Pacific coastline from California to British Columbia in Canada, 120

Silicon Glen An area in Scotland that has many high-tech companies, 303

Silicon Valley An area in western California known for high-technology industries, 141

Sinai Peninsula (29°30′N/34°00′E) A peninsula at the north end of the Red Sea; situated between the Gulf of Suez on the west and the Gulf of Aqaba on the east, 527

Singapore (1°22′N/103°48′E) A Southeast Asian country that occupies Singapore Island and nearby smaller islands, 705

Singapore City (1°18′N/103°51′E) The capital of Singapore, 709

Skopje (42°00′N/21°26′E) The capital of Macedonia, 268

Slovak Republic (48°40′N/19°30′E) A country in Eastern Europe, 308

Slovenia, Republic of (46°15′N/15°10′E) A country in Eastern Europe, 308

Sofia (42°41′N/23°19′E) The capital of Bulgaria, 266

Solomon Islands (8°00′S/159°00′E) A country that consists of an island group of Oceania in the Pacific Ocean, 712, *m713*

Somalia (6°00′N/48°00′E) A country in East Africa, 431

Sonoran Desert An arid region in west North America, 124

South, the The south-central and southeastern area of the United States that includes the states of Maryland, Delaware, West Virginia, Virginia, North Carolina, South Carolina, Kentucky, Tennessee, Georgia, Florida, Alabama, Mississippi, Louisiana, Arkansas, Texas, and Oklahoma, 148

South Africa, Republic of (30°00′S/26°00′E) A country in Southern Africa, 453, *m454*

South America The southern continent of the Western Hemisphere bounded by the Caribbean Sea, the Atlantic Ocean, and the Pacific Ocean, 28, *m29*

South Asia A region that includes Afghanistan, India, Pakistan, Bangladesh, Bhutan, Nepal, Sri Lanka, and the Maldives, 551

South Carolina (34°00′N/81°00′W) A state in the southern United States, 112

South China Sea (15°00′N/115°00E) An arm of the Pacific Ocean bounded by southeastern China, Taiwan, Borneo, the Philippines, and Indochina, *m620,* 622

South Dakota (44°30′N/100°15′W) A state in the Midwest of the United States, 112

South Island (43°00′S/171°00′E) The southern of the two islands that make up New Zealand, 691

South Korea, Republic of (37°00′N/127°30′E) A country in East Asia, 648, *m648*

South Pole (9°S) The southern end of the earth's axis of rotation; a point located in Antarctica, 56

Southeast Asia A region that includes the countries of Brunei, Cambodia, Indonesia, Laos, Malaysia, Myanmar, the Philippines, Singapore, Thailand, and Vietnam, 705

Southern Africa A region of Africa that includes Angola, Botswana, Comoros, Lesotho, Madagascar, Malawi, Mauritius, Mozambique, Namibia, South Africa, Swaziland, Zambia, and Zimbabwe, 453, *m454*

Southern Alps (43°30′S/170°30′E) A mountain range on South Island, New Zealand, 691

Southwest Asia A region that includes Bahrain, Iran, Iraq, Israel, Jordan Kuwait, Lebanon, Oman, Qatar, Saudi Arabia, Syria, Turkey, United Arab Emirates, and Yemen, 487

Spain (40°00′N/4°00′W) A Mediterranean country in Europe, 268

Springfield (39°48′N/89°39′W) The capital of Illinois, 108

Sri Lanka, Democractic Socialist Republic of (7°00′N/81°00′E) A South Asian island country off the southeast coast of India, 584

steppe A grassland that extends from southern Ukraine through northern Kazakhstan to the Altai Mountains, 352

Stockholm (59°20′N/18°03′E) The capital of Sweden, 268

Straits of Hormuz (26°37′N/56°30′E) A narrow waterway between Oman and southern Iran that connects the Persian Gulf with the Gulf of Oman, 488

Sucre (19°03′S/65°16′W) The constitutional capital of Bolivia, 196

Sudan (15°00′N/30°00′E) A country in North Africa, 438

Suez Canal (29°55′N/32°33′E) A canal connecting the Mediterranean Sea with the Red Sea, 487

Sumatra (0°00′N/102°00′E) An island southwest of the Malay Peninsula; part of Indonesia, 678

Superior, Lake (47°38′N/89°20′W) The largest of the Great Lakes of North America, *m118,* 121

Suriname (4°00′N/56°00′W) A country in northern South America, 230, *m234*

Suva (18°08′S/178°25′E) The capital of Fiji, 684

Suzhou (33°38′N/116°59′E) A city in China, 637

Swaziland (26°30′S/31°30′E) A country in Southern Africa, 453, *m454*

Sweden (62°00′N/15°00′E) A Northern European country that occupies the eastern part of the Scandinavian Peninsula, 300

Switzerland (47°00′N/8°00′E) A country in Western Europe, 294

Syria (35°00′N/38°00′E) A country in the Eastern Mediterranean region of Southwest Asia, 511, *m512*

Syrian Desert (32°00′N/40°00′E) A desert that extends north from the An-Nafud Desert and separates the coastal regions of Lebanon, Israel, and Syria from the Tigris and Euphrates valleys, *m488,* 492

taiga (56°04′N/85°05′E) A large forest holding the world's largest timber reserve; located south of the tundra in Russia, 352

Taipei (25°01′N/121°27′E) The capital of Taiwan, 616

Taiwan (24°00′N/121°00′E) An island country off the coast of southeastern China, 642

Tajikistan (39°00′N/71°00′E) A country in Central Asia, 375

Taklimakan (39°00′N/83°00′E) A desert in western China, 627

Tallahassee (30°26′N/84°17′W) The capital of Florida, 108

Tallinn (59°26′N/24°44′E) The capital of Estonia, 342

Tampa-Saint Petersburg A metropolitan area in western Florida formed by the growth of two cities, Tampa (27°57′N/82°28′W) and Saint Petersburg (27°46′N/82°41′W), 148

Tanganyika, Lake (6°00′S/29°30′E) The longest freshwater lake in the world; forms the border between The Democratic Republic of the Congo and Tanzania, 417

Tanzania (6°00′S/35°00′E) A country in East Africa, 431

Tarim Basin (41°00′N/84°00′E) A lowland area in western China, 619

Tashkent (41°19′N/69°15′E) The capital of Uzbekistan, 342

Taurus Mountains (37°00′N/33°00′E) A mountain range in southern Turkey, 488

Tbilisi (41°43′N/44°47′E) The capital of Georgia, 342

Tegucigalpa (14°06′N/87°13′W) The capital of Honduras, 198

Tehran (35°40′N/51°25′E) The capital of Iran, 484

Tennessee (35°45′N/86°15′W) A state in the southern United States, 112

Texas (31°15′N/99°15′W) A state in the south-central United States, 148

Thailand (15°00′N/100°00′E) A country in Southeast Asia, 705

Thar Desert (27°00′N/71°00′E) A desert in southeastern Pakistan and northwestern India, 558

Thimphu (27°29′N/89°36′E) The capital of Bhutan, 548

Three Gorges Dam A dam under construction in 2001 by China; eventually it will span a valley more than one mile wide, 629

Tian Shan (42°00′N/80°00′E) A mountain range in Central Asia, 346

Tianjin (39°09′N/117°11′E) A city in China, 637

Tibesti Mountains (21°30′N/17°30′E) A mountain range in the Sahara, 417

Tierra del Fuego (54°00′S/70°00′W) The southernmost tip of South America, 201

Tigris River (31°00′N/47°25′E) A river that rises in eastern Turkey, flows southeast through Iraq, and joins the Euphrates River; together they form the Shatt al Arab, which flows into the Persian Gulf, *m488,* 489

Timor (10°08′S/125°00′E) An island of southeast Indonesia, 705

Tiranë (41°20′N/19°49′E) The capital of Albania, 266

Togo (8°00′N/1°10′E) A country in West Africa, 442

Tokyo (35°41′N/139°45′E) One of the largest cities of the world; the capital of Japan, 630

Tonga (20°00′S/175°00′W) A country that consists of an island group of Oceania in the Pacific Ocean, 712, *m713*

Topeka (39°03′N/95°41′W) The capital of Kansas, 108

Toronto (43°40′N/79°25′W) The capital of Ontario, Canada; the most populous city in the country, 168

Transantarctic Mountains (85°00′S/175°00′W) Mountain ranges in Antarctica, 692

Transcaucasia (42°00′N/45°00′E) A region bounded by Russia, the Caspian Sea, the Black Sea, Turkey, and Iran and consisting of Armenia, Azerbaijan, and Georgia, 346

Trenton (40°13′N/74°45′W) The capital of New Jersey, 110

Trinidad and Tobago (11°00′N/61°00′W) A country consisting of the islands of Trinidad and Tobago in the Atlantic Ocean near northeast Venezuela, 198

Tripoli (32°54′N/13°11′E) The capital of Libya, 410

Tucson (32°13′N/110°56′W) A city in southern Arizona, 149, *m149*

Tunis (36°48′N/10°11′E) The capital of Tunisia, 412

Tunis, Gulf of (36°58′N/10°46′E) An inlet of the Mediterranean Sea near the city of Tunis, 438

Tunisia (34°00′N/9°00′E) A country in North Africa, 438

Turkey (39°00′N/35°00′E) A country in the northwest region of Southwest Asia, 516, *m516*

Turkmenistan (40°00′N/60°00′E) A country in Central Asia, 375

Tuvalu (8°00′S/178°00′E) A country that consists of an island group of Oceania in the Pacific Ocean, 712, *m713*

U.S. Virgin Islands A U.S. territory that consists of the southwest group of the Virgin Islands, 112

Uganda (2°00′N/33°00′E) A country in East Africa, 431

Ukraine (49°00′N/32°00′E) A country that is west of Russia, 361

Ulaanbaatar (47°55′N/106°55′E) The capital of Mongolia, 616

United Arab Emirates (24°00′N/54°00′E) A country in the Arabian Peninsula of Southwest Asia, 503

United Kingdom (54°00′N/4°00′W) A Northern European nation consisting of England, Wales, Scotland, and Northern Ireland, 300

United States (38°00′N/110°00′W) A country in North America that consists of 50 states, the District of Columbia, and four territories, 117

Ural Mountains (60°00′N/60°00′E) A mountain range that runs north and south in western Russia; some consider it as the border between Europe and Asia, 345

Uruguay (33°00′S/56°00′W) A country in southern South America, 230, *m234*

Utah (39°15′N/111°45′W) A state in the western United States, 112

Uzbekistan (41°00′N/64°00′E) A country in Central Asia, 375

Vaduz (47°08′N/9°31′E) The capital of Liechtenstein, 266

Valletta (35°54′N/14°31′E) The capital of Malta, 268

Vancouver (49°15′N/123°08′W) A city in British Columbia, 169

Vanuatu (16°00′S/167°00′E) A country that consists of an island group of Oceania in the Pacific Ocean, 712, *m713*

Vatican City (41°54′N/12°27′E) An independent papal state located near Rome, Italy, 268

Venezuela (8°00′N/66°00′W) A country in northern South America, 230, *m234*

Venice (45°26′N/12°20′E) A city in northeastern Italy, 281

Verkhoyansk (67°35′N/133°27′E) A city in Siberia in Russia, 354

Vermont (44°00′N/72°45′W) A state in New England in the northeastern United States, 145

Victoria (48°26′N/123°21′W) The capital of British Columbia, Canada, 169

Victoria (4°37′S/55°27′E) The capital of Seychelles, 412

Victoria Island (71°00′N/110°00′W) A large island in northern Canada, 121

Victoria, Lake (1°00′S/33°00′E) The second largest freshwater lake in the world; lies in East Africa, 417

Vienna (48°12′N/16°22′E) The capital of Austria, 266

Vientiane (17°58′N/102°36′E) The capital of Laos, 684

Vietnam (16°00′N/106°00′E) A country in Southeast Asia, 705

Vilnius (54°41′N/25°19′E) The capital of Lithuania, 342

Vindhya Mountains (24°37′N/82°00′E) A mountain range in central India, 552

Virginia (37°30′N/78°30′W) A state in the southern United States, 148

Volga River (45°51′N/47°58′E) The longest river in Europe; rises near Moscow, flows east and then south, and empties into the Caspian Sea, 347

Warsaw (52°15′N/21°00′E) The capital of Poland, 268

Washington (47°30′N/120°30′W) A state in the northwestern United States, 148

Washington, D.C. (38°54′N/77°02′W) The capitol of the United States, *m145,* 147

Wellington (41°18′S/174°47′E) The capital of New Zealand, 684

West, the An area of the United States that stretches from the Great Plains to the Pacific Ocean and includes Alaska and Hawaii. Other states in this area are Montana, Wyoming, Colorado, New Mexico, Arizona, Utah, Idaho, Washington, Oregon, Nevada, and California, 148

West Africa A region of Africa that includes Benin, Burkina Faso, Cape Verde, Chad, Côte d'Ivoire, Gambia, Ghana, Guinea, Guinea-Bissau, Liberia, Mali, Mauritania, Niger, Nigeria, Senegal, Sierra Leone, and Togo, 442

West Antarctica A region of Antarctica; a group of islands of Antarctica linked by the ice that covers them, 692

West Bank (31°40′N/35°15′E) A strip of land on the west side of the Jordan River, 527

West Siberian Plain (60°00′N/75°00′E) A plain in west central Russia, 345

West Virginia (38°30′N/80°30′W) A state in the southern United States, 112

Western Europe A region that includes France, Germany, Austria, Liechtenstein, Switzerland, Belgium, the Netherlands, and Luxembourg, 294

Western Ghats (14°00′N/75°00′E) A mountain range that runs along the west coast of India, 552

Western Republics Countries located west of Russia; they include Ukraine, Belarus, Moldova, Latvia, Lithuania, and Estonia, 361

Whitehorse (60°43′N/135°03′W) The capital of the Yukon Territory of Canada, 114

Windhoek (22°34′S/17°05′E) The capital of Namibia, 410

Winnipeg (49°53′N/97°10′W) The capital of Manitoba, Canada, 114

Wisconsin (44°30′N/90°00′W) A state in the Midwest of the United States, 112

Wuhan (30°35′N/114°16′E) A city in China, 637

Wuxi (31°36′N/120°17′E) A city in China, 637

Wyoming (43°00′N/107°30′W) A state in the western United States, 112

Xi Jiang (also called West River) (22°48′N/113°03′E) A river that flows through southeast China, joins the Pearl River (Zhu Jiang) and empties into the South China Sea, 621, *m620*

Yalu River (39°56′N/124°19′E) A river that forms the border between North Korea and China, 622, *m620*

Yamoussoukro (6°49′N/5°17′W) The capital of Côte d'Ivoire, 408

Yangon (16°47′N/96°10′E) The capital of Myanmar, 684

Yangtze River (also called Chang Jiang) (31°47′N/121°08′E) The longest river in Asia; flows from Xizang (Tibet) across China to the East China Sea, *m620,* 621

Yaoundé (3°52′N/11°31′E) The capital of Cameroon, 408

Yellow River (also called Huang He) (37°45′N/119°05′E) A Chinese river that rises in the Kunlun Mountains, flows east for about 3,000 miles, and empties into the Yellow Sea, *m620,* 621

Yellow Sea (36°00′N/124°00′E) An arm of the Pacific Ocean between the Korean Peninsula and northeastern China, *m620,* 621

Yellowknife (62°27′N/114°21′W) The capital of the Northwest Territories of Canada, 114

Yemen (15°16′N/42°35′E) A country in the Arabian Peninsula of Southwest Asia, 503

Yenisey River (71°50′N/82°40′E) A river that flows through central Russia and empties into the Kara Sea, 347

Yerevan (40°11′N/44°30′E) The capital of Armenia, *m370,* 371

Yugoslavia (43°45′N/20°45′E) A country in Eastern Europe, 308

Yukon Territory (63°00′N/136°00′W) A territory in northwestern Canada, 169

Zagreb (45°48′N/16°00′E) The capital of Croatia, 266

Zagros Mountains (33°40′N/47°00′E) A mountain range in western Iran, 488, *m488*

Zambia (15°00′S/30°00′E) A country in Southern Africa, 453, *m454*

Zimbabwe (19°00′S/29°00′E) A country in Southern Africa, 453, *m454*

INDEX

W

wadis, 488
Waialeale, Mount, 126
Waitangi, Treaty of, *i719*
war against terrorism
   in Afghanistan, 174, 517, *i517,* 518, 519
   Bush administration and, 174, 517
Washington, D.C., *m16, m179*
water, 33, 149, 490, 495–496, *c537*
waterfalls, *i32, i414,* 416
weapons of mass destruction, 518
weather, 12, 50–53, 126, 350, 354
weather extremes
   blizzards, 52
   droughts, 53
   floods, 53
   hurricanes, 51
   tornadoes, 51
   typhoons, 51
   in United States and Canada, 126
weathering, 42–43
   defined, 42
West, the (United States), *m134, i137,*
      148–149
West Africa, 416
   art in, 444–445
   economic activities in, 443–444
   empires of, *m442,* 443
   gross national product, 1987–1997,
      c443
   history of, 442–443
   music in, 445
West Bank, 513, 527, 532
Western Europe
   art in, 300
   economic activities in, 298–300
   history of, 296–298
   modern life in, 301
   Reformation, the, 297
Western Ghats, 552
Western Hemisphere, *i6*
Western Wall, 510
West Germany, 298, *m317*
West Indies, 203
West Siberian Plain, 346
Wik Case, 729
winds
   climate effects, *i54, i57*
   erosion, 44
   global wind currents, *i54*
   khamsin, 42
   mistral, 279
   sirocco, 279
   tornadoes, 51–52
   over tropical ocean waters, 51

westerlies in Europe, 278
   willy-willies, 51
Windward Islands, 203
women's roles, 595, 720
   in Arabian Peninsula, 506, *i507*
   farming in United States, 135
   in North Africa, *i441*
   in Southwest Asia, *i519,* 529, 531
wool, 721
World Trade and Development Act, 471
World Trade Center, 173. *See also*
      September 11 terrorist attack.
World War I, 298, *c311*
World War II, *c311,* 312, 652, 713

X

Xi Jiang, *c610,* 621, 622
Xizang Plateau, 619

Y

Yalu Jiang, *m84,* 622
Yangtze River. *See* Chang Jiang.
Yellow River. *See* Huang He.
Yemen, *c484–485,* 503, 505, *c537*
Yenisey River, 347–348
Yugoslavia, *c270–271,* 310, 311, 313, 319,
      i321, m322
Yukon Territory, *c114–115,* 123, *m157,*
      168–169
yurts, *i72, i379, i380,* 645

Z

Zagros Mountains, 487, 488
Zaire, 471
Zambezi River, *i414*
Zambia, *c412–413,* 417, 453
Zeugma, *i497*
Zhu Jiang, 622
Zhu Rongji, 637
Zimbabwe, *c412–413,* 451, 453
Zuider Zee, 283
Zulu, 454, *i455,* 457

Acknowledgments

This product contains proprietary property of **MAPQUEST.COM**.
Unauthorized use, including copying, of this product, is expressly prohibited.

TEXT ACKNOWLEDGMENTS

Unit 1, Adapted "Figure 9.51," from *Physical Geography, Second Edition.* Copyright © 1989, 1992 by West Publishing Company. Reprinted by permission of Thomson Learning.

Excerpt from "Flight of a Lifetime" by James Irwin, from *The Greatest Adventure.* Copyright © 1994 by the Association of Space Explorers.

Excerpt from "To Really See It, Leave It" by Sally Ride, from *The Greatest Adventure.* Copyright © 1994 by the Association of Space Explorers.

Unit 3, Excerpt from "Brazilian police go on trial for murder of street children" by Marina Mirabella, from CNN World News, April 26, 1996. Copyright © 1996 by Cable News Network, Inc. Reprinted by permission of Cable News Network, Inc.

Excerpt from "Hope for the no-hopers," from the *Economist,* December 23, 2000. Copyright © 2000 by the *Economist.* Reprinted by permission of the *Economist.*

Excerpt from "Rich-poor gap as wide as ever in Latin America" by Steve Gutkin, from *The Times of India Online,* September 5, 2000. Copyright © 2000 by Times Internet Limited. Reprinted by permission of Times Internet Limited. All rights reserved.

Unit 4, Excerpt from "Poland Opens Door to West, and Chills Blow Both Ways" by Edmund L. Andrews, from the *New York Times,* June 21, 1999. Copyright © 1999 by the *New York Times.* Reprinted by permission of the *New York Times.*

Excerpt from "Britain and the Single Currency," from Global Britain Briefing Note, No. 1, January 25, 1999. Copyright © 1999 by Global Britain. Reprinted by permission of Global Britain.

Unit 5, Excerpt from "Workers Bid Ill-Fated Chernobyl a Bitter Farewell" by Michael Wines, from the *New York Times,* December 15, 2000. Copyright © 2000 by the *New York Times.* Reprinted by permission of the *New York Times.*

Excerpt from "Reducing Russian Dangers," from the *New York Times,* January 21, 1999. Copyright © 1999 by the *New York Times.* Reprinted by permission of the *New York Times.*

Unit 6, Excerpt from "Cash Strapped African Leaders Beg To Be Re-Colonized" by Ron Daniels, from the *Black World Today,* August 1, 1999. Copyright © 1999 by the *Black World Today.* Reprinted by permission of the *Black World Today.*

Excerpt from "An African Success Story," from the *New York Times,* January 8, 2001. Copyright © 2001 by the *New York Times.* Reprinted by permission of the *New York Times.*

Unit 7, Excerpt from "All Sides Resist Plan by Clinton For the Mideast" by John Kifner, from the *New York Times,* December 31, 2000. Copyright © 2000 by the *New York Times.* Reprinted by permission of the *New York Times.*

Excerpt from "The Price of Peace Will Be Paid in Dreams" by John F. Kifner, from the *New York Times,* December 31, 2000. Copyright © 2000 by the *New York Times.* Reprinted by permission of the *New York Times.*

Excerpt from "A City That Echoes Eternity" by Kenneth L. Woodward, from *Newsweek,* July 24, 2000. Copyright © 2000 by Newsweek, Inc. Reprinted by permission of Newsweek, Inc. All rights reserved.

Excerpts from "City of Jerusalem," from *United Nations: General Assembly Resolution 181,* November 29, 1947. Copyright © 1947 by the United Nations. Reprinted by permission of the United Nations.

Unit 8, Excerpt from President K. R. Narayanan's state dinner toast, March 21, 2000. Copyright © 2001 by the Embassy of India, Press and Information. Reprinted by permission of the Embassy of India.

Unit 9, Excerpt from "Clinton in Hong Kong, prods China on environment," from CNN.com, July 2, 1998. Copyright © 1998 by Cable News Network, Inc. Reprinted by permission of Cable News Network, Inc.

Excerpt from "Six Billion People," from *Asiaweek.com,* October 29, 1999, Vol. 25, No. 43. Copyright © 1999 by Asiaweek. Reprinted by permission of Asiaweek.

Unit 10, Excerpt from "Record Ozone Hole Refuels Debate on Climate" by Andrew C. Revkin, from the *New York Times,* October 10, 2000. Copyright © 2000 by the *New York Times.* Reprinted by permission of the *New York Times.*

Excerpts from *Dear America, Letters Home from Vietnam* edited by Bernard Edelman. Copyright © 1985 by The New York Vietnam Veterans Memorial Commission. Reprinted by permission of Simon & Schuster.

Excerpt from "Heroes for the Planet" by Terry McCarthy, from *Time,* Apr–May 2000. Copyright © 2000 by Time, Inc. Reprinted by permission of Time, Inc.

Excerpt from "There is No Global Warming," from the American Policy Center. Copyright © 2001 by the American Policy Center. Reprinted by permission of the American Policy Center.

The editors have made every effort to trace the ownership of all copyrighted material found in this book and to make full acknowledgment for its use. Omissions brought to our attention will be corrected in a subsequent edition.

ART CREDITS

Cover front *background* Copyright © Orbital Imaging Corporation and processing by NASA Goddard Space Flight Center. Image provided by ORBIMAGE; *insets, top to bottom* Copyright © Hugh Sitton/Stone; Copyright © George Hunter/H. Armstrong Roberts; Copyright © James Martin/Stone; Copyright © Michael Dunning/Image Bank.

Front Matter *ii–iii background* Copyright © Orbital Imaging Corporation and processing by NASA Goddard Space Flight Center. Image provided by ORBIMAGE; *ii top, left to right* Copyright © Hugh Sitton/Tony Stone Images; Copyright © George Hunter/ H. Armstrong Roberts; Copyright © James Martin/Tony Stone Images; *iii right* Copyright © Michael Dunning/Image Bank; *v, vi–vii background* Earth from space; *viii top* Copyright © Schafer & Hill/Tony Stone Images; *bottom* Copyright © 1994 Jeffrey Aaronson/Network Aspen. All rights reserved; *ix top* SuperStock; *bottom* Copyright © Kevin Miller/Tony Stone Images; *x top* Copyright © Loren McIntyre; *bottom* Copyright © Ary Diesendruck/Tony Stone Images; *xi top* Copyright © Eye Ubiquitous/Corbis; *bottom* Ric Ergenbright Photography; *xii top* Sovfoto/Eastfoto; *bottom* Copyright © David Sutherland/Tony Stone Images; *xiii top* Copyright © Daryl Balfour/Tony Stone Images; *center* Copyright © Mitch Reardon/Tony Stone Images; *bottom* Copyright © M. Thonig/H. Armstrong Roberts; *xiv top* Copyright © Ali Kazuyoshi Namachi/Pacific Press; *bottom* Copyright © John Egan/Hutchison Picture Library; *xv top* Copyright © Paul Harris/Tony Stone Images; *bottom* Copyright © Martin Puddy/Tony Stone Images; *xvi top* Copyright © David Ball/Tony Stone Images; *center* Copyright © Tony Stone Images; *bottom* Copyright © John Lamb/Tony Stone Images; *xvii top* Copyright © Roger Near/Tony Stone Images; *center* Copyright © Nicholas DeVore/Tony Stone Images; *bottom* Copyright © David Austen/Tony Stone Images. *xxiv* Copyright © Dani/Jeske/Animals Animals; Copyright © Patti Murray/Earth Scenes; Copyright © J. Carnemolla/Australian Picture Library; Copyright © John Cancalosi/Peter Arnold Inc.; Copyright © Norman Owen Tomalin/Bruce Coleman; Copyright © Tim Flach/Tony Stone Images; *xxv top* The Granger Collection, New York; *bottom* Copyright © 2001 The New Yorker Collection from cartoonbank.com. All rights reserved.

Maps pp. A1–A25, 64, 102–115, 131, 179, 190–199, 206, 248, 262–271, 281, 322, 336–343, 357, 391, 402–413, 419, 464, 478–485, 494, 528, 542–549, 559, 596, 610–617, 624, 664, 678–687, © Rand McNally & Company. All rights reserved.

Units **Unit 1, 2–3** *background* Copyright © Earth Imaging/Tony Stone Images; **2** *foreground* Copyright © Schafer & Hill/Tony Stone Images; **3** *foreground* Copyright © Damir Sagolj/Reuters/Archive Photos; **4** *background, left* MapQuest.com; **7** Copyright © Eduardo Garcia/FPG International/PictureQuest; **8** Copyright © Alan Weiner/Liaison; **10** Globe (c. 1492), Martin Benhaim. Bibliotheque Nationale, Paris/Giraudon/Art Resource, New York; **11** *background* Copyright © Paul Morrell/Tony Stone Images; *top left* National Oceanic and Atmospheric Administration/Department of Commerce; **12** *top left* Copyright © Bob Krist/ eStock Photography/PictureQuest; *top center* Copyright © PhotoDisc/Getty Images; **13** Copyright © Erwin and Peggy Bauer/Bruce Coleman Inc.; **14** *background* National Oceanic and Atmospheric Administration/Department of Commerce; *top* AP/Wide World Photos; *bottom left* Science Museum/Science & Society Picture Library, London; *bottom right* Copyright © Owen Franken/Stock Boston/PictureQuest; **26** Copyright © Paul Morrell/Tony Stone Images; **30–31** *background,* **31** Illustrations by Roberta Polfus; **32** Copyright © SuperStock; **33** Illustration by Stephen R. Wagner; **34–35** Illustration by Ken Goldammer; **36** Copyright © Earth Satellite Corporation/Science Photo Library/Photo Researchers Inc.; **38–39** Illustration by Roberta Polfus; **40** Copyright © Chip Hires/Liaison; **41** Copyright © HO/Reuters/Archive Photos; **42** Copyright © David Muench/Corbis; **43** AP/Wide World Photos; **44** Copyright © Jerryl Hout/ Bruce Coleman Inc.; **48** Copyright © SuperStock; **50** Illustration by Stephen R. Wagner; **51** National Oceanic and Atmospheric Administration/Department of Commerce; **52** *inset* Copyright © Chris Johns/Tony Stone Images; **53** Getty News Services; **59** Copyright © Vladpans/eStock Photography/PictureQuest; **60** *top* Copyright © SuperStock; *bottom* Copyright © 1994 Michael Fogden/Bruce Coleman Inc./PictureQuest; **61** *top insert* Copyright © 1999 Picture Finders Ltd./ eStock Photography/PictureQuest; *bottom insert* Copyright © Daniel J. Cox/ Liaison; **62** Copyright © Charlie Waite/Tony Stone Images; **63** Copyright 1993 © Richard Olsenius/Black Star/PictureQuest; **67** *left* Used by permission, Utah State Historical Society. All rights reserved; *right* Copyright © Tom Till Photography; **70** Copyright © 1994 David Hiser/Photographers/Aspen/PictureQuest; **72** Copyright © Adrian Arbib/Corbis; **73** Courtesy Medmaster, Inc. Miami, Florida; **77** Copyright © Werner Forman/Corbis; **81** Copyright © 1994 Jeffrey Aaronson/Network Aspen. All rights reserved; **83** Copyright © 1960 Sergio Larrain/Magnum Photos; **85** Copyright © D. E. Cox/Tony Stone Images; **87** Copyright © SuperStock; **88** Copyright © Earth Imaging/Tony Stone Images; **91** Copyright © SuperStock; **92** Photograph by Sharon Hoogstraten; **94** *top row, pots* Copyright © 1999 Stockbyte/PictureQuest; *bicycles* Copyright © Image Club/EyeWire; *middle row, kettles and sieves* Copyright © 1999 Stockbyte/PictureQuest; *televisions* Copyright © Digital Stock/Corbis; *stereos* Copyright © Jeffrey Coolidge/Image Bank; *bottom row, bicycle* Copyright © Image Club/EyeWire; *radios* Copyright © Digital Stock/Corbis; *VCR* Copyright © 1999–2001 PhotoDisc/Getty Images; **95** *top row, bicycles* Copyright © Image Club/EyeWire; *radios and telephone* Copyright ©

Robert Frerck/Odyssey/Chicago; **445** Copyright © Deborah Feingold/Archive Photos; **446** *clockwise from top right* Copyright © Trip/Dinodia; Copyright © 1996 Christopher Liu/ChinaStock; Copyright © Trip/H. Rogers; **447** *top* SuperStock; *bottom* Copyright © Robert Frerck/Odyssey/Chicago; **449** Cartoon by Linley Sambourne (1906). Mary Evans Picture Library, London; **451** *When There is Work, the Village Expands,* Mode Muntu. Collection of Guy de Plaen; **452** Copyright © 1995 Malcolm Linton/Liaison; **453** Robert Aberman/Art Resource, New York; **455** *from top to bottom* The Granger Collection, New York; Copyright © Hulton/Archive Swimming Pool Database-Big Picture; The Royal Collection copyright © 2001 Her Majesty Queen Elizabeth II; Copyright © Mark Peters/Sipa Press; **456** Copyright © Robert Frerck/Odyssey/Chicago; **460** Copyright © 1968 Rene Burri/Magnum Photos; **461** Photograph by Bjorn Klingwall; **463** Copyright © Piotr Jaxa/The Rolex Awards for Enterprise; **465** Copyright © Malcolm Linton/Liaison; **467** *top right* Copyright © Wendy Stone/Liaison; **468** Corbis-Bettmann; **469** *bottom* AP/Wide World Photos; **471** Copyright © 1996 Alan King. National Archives of Canada, Ottawa, ref. No. R2905-116.

Unit 7, 474–475 *background* Copyright © 1995 WorldSat International and J. Knighton/Photo Researchers, Inc.; **474** *left* Copyright © Sylvain Grandadam/Tony Stone Images; *right* Copyright © Bernard Gerard/Hutchison Picture Library; **475** Copyright © Ali Kazuyoshi Nomachi/Pacific Press Service; **476** Copyright © John Egan/Hutchison Picture Library; **477** *top* Copyright © Robert Azzi/Woodfin Camp; *bottom* Mark Fiore. San Francisco, California. www.markfiore.com; **484–485** MapQuest.com; **486** Copyright © Peter Sanders; **487** Copyright © Richard T. Nowitz/Corbis; **489** Copyright © Roger Antrobus/Corbis; **490** AP/Wide World Photos; **491** Copyright © Alistair Duncan/DK Images; **493** Copyright © Nik Wheeler/Corbis; **495** Copyright © Alan Puzey/Tony Stone Images; **496** *background* Copyright © PhotoDisc/Getty Images; *clockwise from top left* Copyright © Richard T. Nowitz/NGS Image Collection; Copyright © Trip/H. Rogers; Copyright © Peter J. Ochs II; Copyright © Christopher Rennie/Robert Harding Picture Library; **497** Getty News Services, #372147_06; **503** Copyright © Nabeel Turner/Tony Stone Images; **504** Scala/Art Resource, New York; **506** Copyright © 1985 Robert Azzi/Woodfin Camp; **507** Copyright © 2001 Stern Magazine/Black Star; **508–509** *background* Copyright © Hugh Sitton/Tony Stone Images; **508** SEF/Art Resource, New York; **509** *top* SuperStock; *bottom* Copyright © 1995 David Hiser/Photographers/Aspen/PictureQuest; **510** AP/Wide World Photos; **511** Copyright © Ilene Perlman/Stock Boston/PictureQuest; **513, 514** AP/Wide World Photos; **515** Copyright © Trip/H. Rogers; **517** Copyright © Reuters New Media Inc./Corbis; **518** Copyright © Leo Touchet/Woodfin Camp; **519** Copyright © R. Ashworth/ Robert Harding Picture Library; **521** *top* Copyright © ABC Basin Ajansi/Liaison; *bottom* AP/Wide World Photos; **524** Copyright © John Egan/Hutchison Picture Library; **525** Copyright © Thomas Hartwell/TimePix; **527** Copyright © Will Yurman/Liaison; **530** *top* Copyright © Ali Kazuyoshi Nomachi/Pacific Press Service; *bottom* Copyright © Baron Wolman; **531** Copyright © David & Peter Turnley/Corbis; **532** *top* AP/Wide World Photos; *bottom left* Corbis-Bettmann; *bottom right* Copyright © Dirck Halstead/Liaison; **535** Mark Fiore. San Francisco, California. www.markfiore.com.

Unit 8, 538–539 *background* Copyright © Earth Imaging/Tony Stone Images; **538** Copyright © Paul Harris/Tony Stone Images; **539** *left* Copyright © 1997 Neal Beidleman/Network Aspen; *right* Copyright © Dominic Sansoni; **540** Copyright © James Strachan/Robert Harding Picture Library; **541** *top* Copyright © Chip Hires/Liaison; *bottom* Copyright © 1998 Carlson/Milwaukee Journal Sentinel; **550** Copyright © Jon Sparks/Corbis; **551** Illustration by Stephen R. Wagner; **552** Digital image copyright © 1996 Corbis. Original image courtesy of NASA/Corbis; **553** Copyright © E. Valentin/Photo Researchers, Inc.; **554** *inset* Copyright © Dominic Sansoni; **555** Copyright © Takeshi Takahara/Photo Researchers, Inc.; **556** Copyright © Brian Vikander/Corbis; **558** Copyright © E. Hanumantha Rao/Photo Researchers, Inc.; **560–561** Copyright © David Sutherland/Tony Stone Images; **562** Copyright © Pablo Bartholomew/Liaison; *inset right* Copyright © St. Franklin/Magnum Photos; **568** *top* Corbis-Bettmann; *bottom left* The Granger Collection, New York; *bottom right* Copyright © Roderick Johnson/Images of India; **570** Copyright © 1978 Dilip Mehta/Contact Press Images/PictureQuest; **571** Copyright © Anthony Cassidy/Tony Stone Images; **572** The Art Archive/Victoria & Albert Museum, London/Eileen Tweedy. Ref. no. AA328654; **573** Copyright © Photobank Photo Library; **575** *left* Copyright © David Sanger/Network Aspen; *right* Copyright © Brecel/J. Hodalic/Liaison; **576** Copyright © Robert Nickelsberg/TimePix; **577** Copyright © Beatrice Kiener/Liaison; **578–579** *background* National Oceanic and Atmospheric Administration/Department of Commerce; **578** *bottom* Illustration by Stephen R. Wagner; **579** *top* Copyright © Topham Picturepoint; *bottom* Jim Holmes/Eye Ubiquitous; **580** Thrikheb or "throne cover." Wool, silk, and cotton, 152 cm x 70 cm. Private collection, Oltromare, Geneva, Switzerland. Photograph copyright © Erich Lessing/ Art Resource; **581** Copyright © John Callahan/Tony Stone Images; **582** *background* Illustration by Stephen R. Wagner; *top inset* Copyright © Alison Wright/ Stock Boston/PictureQuest; **583** Copyright © Barbara Bussell/Network Aspen; **584** Copyright © Nokelsberg/Liaison; **585** Copyright © Hugh Sitton/Tony Stone Images; **586** Copyright © Martin Puddy/Tony Stone Images; **587** Copyright © Pete Seaward/Tony Stone Images; **588** *left* Copyright © Greig Cranna/Stock Boston/PictureQuest; *right* Copyright © Wolfgang Kaehler/Corbis; **589** *top* Copyright © 1999 Pictor International/Pictor International, Ltd./PictureQuest *bottom* AP/Wide World Photos; **592** Copyright © James Strachan/Robert Harding Picture Library; **593** Copyright © Bruno Barbey/ Magnum Photos; **595** Copyright © Jeroen Snjdera/Images of India; **598** Illustrations by Stephen R. Wagner; **599** *top* Courtesy David Cumming/ Hodder Wayland Picture Library; *bottom* Copyright © Trip/H. Rogers; **600** *top* Copyright © Ric Ergenbright/Corbis *bottom left* Corbis-Bettmann; *bottom right* Reuters/Archive Photos; **603** Copyright © 1998 Carlson/Milwaukee Journal Sentinel.

Unit 9, 606–607 *background* Copyright © Tom Van Sant/Geosphere Project/Planetary Visions/Science Photo Library/Photo Researchers, Inc.; **606** SuperStock; **607** *left* Copyright © David Ball/Tony Stone Images; *right* SuperStock; **608** John Pryke/Reuters/Archive Photos; **609** *top* Copyright © Hugh Sitton/Tony Stone Images; *bottom* SuperStock; **618** SuperStock; **619** Copyright © Tony Stone Images; **621** Copyright © Keren Su/Corbis; **623** Copyright © Michael S. Yamashita/Corbis; **625** Copyright © Cynthia M. Beall/NGS Image Collection; **627** Digital image copyright © 1996 Corbis. Original image courtesy of NASA/Corbis; **628** *right* Copyright © 1997 Bob Sacha; **629** *left* NG Maps/NGS Image Collection; **630** *top to bottom* Copyright © Luo Wenfa/ChinaStock; Copyright © 1995 Wang Xinlin/ChinaStock; Copyright © ChinaStock; **631** Copyright © M. Bertinetti/White Star; **635** Copyright © Julian Calder/Tony Stone Images; **636** *left* Bibliothéque Nationale, Paris/Bridgeman Art Library, London/SuperStock; *center* By permission of The British Library, London; *right* The Granger Collection, New York; **637** *left* Copyright © ChinaStock; *right* The Granger Collection, New York; **638** *top left* Copyright © 1995 Christopher Liu/ChinaStock; *top right* The Metropolitan Museum of Art, New York. Gift of Ernest Erickson Foundation, Inc., 1985 (1985.214.99). Photograph copyright © 1986 the Metropolitan Museum of Art; *center* By permission of The British Library, London; *bottom* The Granger Collection, New York; **639** Copyright © Alain le Caromeu/Panos Pictures; **640–641** *background,* **641** *top* UPI/Corbis-Bettmann; **641** *bottom* Copyright © Tom Nebbia/Corbis; **642** Illustration by Patrick Whalen; **644** Copyright © Cynthia M. Beall & Melvyn C. Goldstein/NGS Image Collection; **646** AP/Wide World Photos; **647** Copyright © Bob Thomas/Tony Stone Images; **649** Copyright © Jean-Léo Dugast/Panos Pictures; **651** Copyright © Japan Archive/TimePix; **652** *left* The Granger Collection, New York; *right* Portrait of Kisegawa of Matsubaya (c. 1796), Kitagaw Utamoro. Fitzwilliam Museum, University of Cambridge, UK/The Bridgeman Art Library, London; **653** *bottom left* The Granger Collection, New York; *bottom right* Corbis-Bettmann; **654** Copyright © 1976 Kenneth Love; **655** SuperStock; **656** *left* Manu Sassoonian/Art Resource, New York; *right* SuperStock; **657** *top* Werner Forman Archive/Private collection, New York/Art Resource, New York; *bottom* SuperStock; **660** John Pryke/Reuters/Archive Photos; **661** AP/Wide World Photos; **663** Copyright © Bzad/ChinaStock; **665** Photograph by Manabu Watabe; **668** Copyright © John Lamb/Tony Stone Images; **671** Copyright © 2000 Nick Anderson, The Washington Post Writers Group. Reprinted with permission.

Unit 10, 674 *left* Copyright © Tony Stone Images: *right* Copyright © Roger Near/Tony Stone Images; **676** Copyright © Thad Samuels/NGS Image Collection; **677** *top* Copyright © Sergio Dorantes/Sygma; *bottom* National Oceanic and Atmospheric Administration/Department of Commerce; **687** The Granger Collection, New York; **688** Copyright © David Doubilet; **690** Illustrations by Stephen R. Wagner; **691** Copyright © David Moore/Black Star; **694** Copyright © Dani/Jeske/Earth Scenes; **695** Copyright © Denis Waugh/ Tony Stone Images; **696** Copyright © Dani/Jeske/Animals Animals; Copyright © Patti Murray/Earth Scenes; Copyright © J. Carnemolla/Australian Picture Library; Copyright © John Cancalosi/Peter Arnold Inc.; Copyright © Norman Owen Tomalin/Bruce Coleman; Copyright © Tim Flach/Tony Stone Images; **698** Copyright © Walter Edwards/NGS Image Collection; **699** Copyright © Greg Taylor; **700** Australian Picture Library; **701** Copyright © U.S. Government Air Force/NGS Image Collection; **705** Copyright © Glen Allison/Tony Stone Images; **708** Copyright © W. Robert Moore/NGS Image Collection; **709** Copyright © 1984 Jonathan Kirn/Liaison; **710–711** *bottom background* Illustration by Stephen R. Wagner; **711** *top* Copyright © Georg Gerster/Photo Researchers Inc.; **712** G. Burenhult Productions, Tjörnarp, Sweden; **714** *top* Copyright © Peter Stone/Pacific Stock; *center* Copyright © Joe Carini/Pacific Stock; *bottom* Copyright © Nicholas DeVore/Tony Stone Images; **716** *left* Copyright © Robert Frerck/Odyssey; *right* Copyright © 1991 Buddy Mays/FPG International; **717** *top* Copyright © Orion Press/Pacific Stock; *bottom* Copyright © 1988 Lee Kuhn/FPG International; **718** *left* Copyright © D. Johanson/Institute of Human Origins; *center* Detail of James Cook (1776), John Webber. Oil on canvas. The Granger Collection, New York; *right* Copyright © Douglass Bagli/Australian Picture Library; **719** *top* Australian Picture Library; *bottom* Copyright © Christopher Arnesen/Tony Stone Images; **720** Copyright © George F. Mobley/NGS Image Collection; **721** SuperStock; **722** Australian Picture Library/NZPL; **723** AP/Wide World Photos; **726** Copyright © Thad Samuels/NGS Image Collection; **727** credit to come; **729** Copyright © David Austen/Tony Stone Images; **730** Darren Whiteside/Reuters/Archive Photos; **731** Copyright © Sergio Dorantes/Sygma; **734** Copyright © Wolfgang Kaehler; **737** *top* National Climatic Data Center/NESDIS/NOAA; *bottom* National Oceanic and Atmosphere Administration/Department of Commerce.

1 2 3 4 5 6 7 8 9 – VEI – 09 08 07 06 05 04 03